Lecture Notes in Artificial Intelligence   8148

Subseries of Lecture Notes in Computer Science

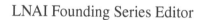

Pedro Cabalar   Tran Cao Son (Eds.)

Logic Programming and Nonmonotonic Reasoning

12th International Conference, LPNMR 2013
Corunna, Spain, September 15-19, 2013
Proceedings

 Springer

Volume Editors

Pedro Cabalar
University of Corunna
Department of Computer Science
Campus de Elviña s/n
15071 Corunna, Spain
E-mail: pedro.cabalar@udc.es

Tran Cao Son
New Mexico State University
Department of Computer Science
1290 Frenger Mall, SH 123
P.O. Box 30001, MSC CS
Las Cruces, NM 88003, USA
E-mail: tson@cs.nmsu.edu

ISSN 0302-9743   e-ISSN 1611-3349
ISBN 978-3-642-40563-1   e-ISBN 978-3-642-40564-8
DOI 10.1007/978-3-642-40564-8
Springer Heidelberg New York Dordrecht London

Library of Congress Control Number: 2013946023

CR Subject Classification (1998): I.2.3, I.2.4, F.1.1, F.4.1, D.1.6, G.2

LNCS Sublibrary: SL 7 – Artificial Intelligence

Preface

This volume contains the papers presented at the 12th International Conference on Logic Programming and Nonmonotonic Reasoning (LPNMR 2013) held during September 15–19, 2013, in Corunna.

LPNMR is a forum for exchanging ideas on declarative logic programming, nonmonotonic reasoning, and knowledge representation. The aim of the conference is to facilitate interaction between researchers interested in the design and implementation of logic-based programming languages and database systems, and researchers who work in the areas of knowledge representation and nonmonotonic reasoning. LPNMR strives to encompass theoretical and experimental studies that have led or will lead to the construction of practical systems for declarative programming and knowledge representation.

LPNMR 2013 was the 12th event in the series of international conferences on Logic Programming and Nonmonotonic Reasoning. Past editions were held in Washington, D.C., USA (1991), Lisbon, Portugal (1993), Lexington, Kentucky, USA (1995), Dagstuhl, Germany (1997), El Paso, Texas, USA (1999), Vienna, Austria (2001), Fort Lauderdale, Florida, USA (2004), Diamante, Italy (2005), Tempe, Arizona, USA (2007), Potsdam, Germany (2009) and Vancouver, Canada (2011).

LPNMR 2013 received 91 submissions in three categories (technical papers, applications, and system descriptions) and two different formats (long and short papers). Each submission was reviewed by at least three Program Committee members. The final list of 53 accepted papers consists of 33 technical papers (22 long, 11 short), 12 applications (nine long, three short), and eight system descriptions (three long, five short).

The conference program featured the presentations of all accepted papers plus three invited talks by Gerhard Brewka, Robert Kowalski, and James Delgrande whose corresponding abstracts are included in these proceedings (as an extended abstract in the first two cases, and as an appendix to this preface in the third case).

The conference also hosted four workshops and the award ceremony of the Fourth ASP Competition, held and organized prior to the conference by Mario Alviano, Francesco Calimeri, Guenther Charwat, Minh Dao-Tran, Carmine Dodaro, Giovambattista Ianni, Thomas Krennwallner, Martin Kronegger, Johannes Oetsch, Andreas Pfandler, Joerg Puehrer, Christoph Redl, Francesco Ricca, Patrik Schneider, Martin Schwengerer, Lara Katharina Spendier, Johannes Peter Wallner, and Guohui Xiao at the University of Calabria, Italy, and the Institutes of Information Systems and of Computer Language at Vienna University of Technology, Austria.

We would like to thank the members of the Program Committee and the additional reviewers for their efforts to produce fair and thorough evaluations of the submitted papers, the Workshop Chair Marcello Balduccini, the local Organizing Committee, especially Felicidad Aguado, Martín Diéguez, Javier Parapar, Gilberto Pérez, Concepción Vidal, and of course the authors of the scientific papers. Furthermore we are grateful to the sponsors for their generous support: *Artificial Intelligence Journal*, the Assocation of Logic Programming (ALP), Red IEMath-Galicia, the CITIC Research Center, the Spanish Association for Artificial Intelligence (AEPIA), the European Regional Development Fund (European Commission), the University of Corunna, and New Mexico State University at Las Cruces. Last, but not least, we thank the people of EasyChair for providing resources and a marvelous conference management system.

September 2013   Pedro Cabalar
                                                   Tran Cao Son

Organization

Program Committee

Jose Julio Alferes	Universidade Nova de Lisboa, Portugal
Chitta Baral	Arizona State University, USA
Leopoldo Bertossi	Carleton University, Canada
Gerhard Brewka	Leipzig University, Germany
Pedro Cabalar	University of Corunna, Spain
Stefania Costantini	Università di L'Aquila, Italy
Marina De Vos	University of Bath, UK
James Delgrande	Simon Fraser University, Canada
Marc Denecker	K.U.Leuven, Belgium
Yannis Dimopoulos	University of Cyprus
Juergen Dix	TU Clausthal, Germany
Agostino Dovier	Università di Udine, Italy
Thomas Eiter	Vienna University of Technology, Austria
Esra Erdem	Sabanci University, Turkey
Wolfgang Faber	University of Calabria, Italy
Michael Fink	Vienna University of Technology, Austria
Andrea Formisano	Università di Perugia, Italy
Martin Gebser	University of Potsdam, Germany
Michael Gelfond	Texas Tech University, USA
Giovambattista Ianni	University of Calabria, Italy
Tomi Janhunen	Aalto University, Finland
Antonis Kakas	University of Cyprus
Joohyung Lee	Arizona State University, USA
Vladimir Lifschitz	University of Texas, USA
Fangzhen Lin	HKUST, SAR China
Jorge Lobo	Universitat Pompeu Fabra, Spain
Robert Mercer	The University of Western Ontario, Canada
Alessandra Mileo	University of Milano-Bicocca, Italy
Mauricio Osorio	UDLA, Mexico
Ramon Otero	University of Corunna, USA
David Pearce	Universidad Politécnica de Madrid, Spain
Axel Polleres	Siemens AG Österreich / DERI, National University of Ireland, Galway

Enrico Pontelli	New Mexico State University, USA
Alessandro Provetti	University of Messina, Italy
Chiaki Sakama	Wakayama University, Japan
Torsten Schaub	University of Potsdam, Germany
John Schlipf	University of Cincinnati, USA
Tran Cao Son	New Mexico State University, USA
Terrance Swift	Universidade Nova de Lisboa, Portugal
Eugenia Ternovska	Simon Fraser University, Canada
Hans Tompits	Vienna University of Technology, Austria
Mirek Truszczynski	University of Kentucky, USA
Agustín Valverde	Universidad de Málaga, Spain
Kewen Wang	Griffith University, Australia
Yisong Wang	Guizhou University, China
Stefan Woltran	Vienna University of Technology, Austria
Jia-Huai You	University of Alberta, Canada
Yan Zhang	University of Western Sydney, Australia
Yi Zhou	University of Western Sydney, Australia

Additional Reviewers

Acosta-Guadarrama, Juan C.	Ji, Jianmin
Alviano, Mario	Jost, Holger
Berger, Gerald	Kaminski, Roland
Bliem, Bernhard	Karimi, Arash
Bogaerts, Bart	Kaufmann, Benjamin
Bozzano, Marco	Krennwallner, Thomas
Bulling, Nils	König, Arne
Calimeri, Francesco	Leite, Joao
Campeotto, Federico	Liu, Guohua
Charwat, Guenther	Michael, Loizos
Chen, Yin	Miculan, Marino
De Cat, Broes	Mu, Kedian
Devriendt, Jo	Navarro Perez, Juan Antonio
Ding, Ning	Nieves, Juan Carlos
Dvorak, Wolfgang	Obermeier, Philipp
Ensan, Alireza	Oetsch, Johannes
Falkner, Andreas	Oikarinen, Emilia
Fichte, Johannes	Ojeda-Aciego, Manuel
Harrison, Amelia	Perri, Simona
Heljanko, Keijo	Pieris, Andreas
Hogan, Aidan	Popovici, Matei
Hutter, Frank	Pührer, Jörg
Inclezan, Daniela	Quintarelli, Elisa
Inoue, Katsumi	Redl, Christoph
Jansen, Joachim	Ricca, Francesco

Romero, Javier
Sabuncu, Orkunt
Scalabrin, Simone
Schiele, Gregor
Schneider, Marius
Schüller, Peter
Stepanova, Daria
Tari, Luis
Tasharrofi, Shahab
Van Hertum, Pieter
Vennekens, Joost

Viegas Damásio, Carlos
Wallner, Johannes P.
Weinzierl, Antonius
Xiao, Guohui
Yang, Bo
Yang, Fangkai
Zhang, Heng
Zhang, Yingqian
Zhang, Yuanlin
Zhang, Zhiqiang
Zhuang, Zhiqiang

Sponsors

UNIÓN EUROPEA
FONDO EUROPEO DE
DESENVOLVEMENTO REXIONAL
"Unha maneira de facer Europa"

Managing Change in Answer Set Programs: A Logical Approach

— Invited Talk —

James Delgrande

Simon Fraser University, Burnaby, B.C., Canada

`jim@cs.sfu.ca`

Abstract

Answer set programming (ASP) is an appealing, declarative approach for representing problems in knowledge representation and reasoning. While answer sets have a conceptually simple syntactic characterization, ASP has been shown to be applicable to a wide range of practical problems, and efficient implementations are now available. However, as is the case with any body of knowledge, a logic program is not a static object in general, but rather it will evolve and be subject to change. Such change may come about as a result of adding to (or removing from) the program, importing the contents of one program into another, merging programs, or in some other fashion modifying the knowledge in the program. In classical logic, the problem of handling such change has been thoroughly investigated. The seminal AGM approach provides a general and widely accepted framework for this purpose. While this approach focusses on revision and contraction, related work, such as merging, has stayed close to the AGM paradigm.

Until recently, the study of change in ASP has been addressed at the level of the program, focussing on the revision (and update) of logic programs. In revision, for example, a typical approach is to begin with a sequence of answer set programs, and determine answer sets based on a priority ordering among the programs or among rules in the programs. An advantage of such approaches is that they are readily implementable. On the other hand, the underlying non-monotonicity of logic programs makes it difficult to study formal properties of an approach.

Recently a more logical view of ASP has emerged. Central to this view is the (monotonic) concept of SE-models, which underlies the answer-set semantics of logic programs. In characterizing an AS program by its set of SE models, one can deal with a program at an abstract, syntax-independent level. I suggest that this is an appropriate level for dealing with change in logic programs, complementing earlier syntax-dependent approaches. To this end, I review such work dealing with change in logic programs, beginning with the much simpler, but nonetheless relevant, case of Horn theories. From this, I touch on work concerning AS revision, with respect to both specific approaches and logical characterisations.

This leads naturally to approaches for merging logic programs. While logic program contraction has not been addressed, I discuss what can be regarded as an extreme case of contraction, that of forgetting. I finish with some thoughts on open problems and issues, and future directions. The overall conclusion is that classical belief change is readily applicable to ASP, via its model theoretic basis.

Table of Contents

Towards Reactive Multi-Context Systems

Gerhard Brewka

Leipzig University, Informatics Institute, Postfach 100920, 04009 Leipzig, Germany
brewka@informatik.uni-leipzig.de

Abstract. Among the challenges faced by the area of knowledge representation (KR) are the following ones: firstly, knowledge represented in different knowledge representation languages needs to be integrated, and secondly, certain applications have specific needs not typically fulfilled by standard KR systems. What we have in mind here are applications where reasoners, rather than being called by the user in order to answer some specific query, run online and have to deal with a continuous stream of information. In this paper we argue that multi-context systems (MCS) are adequate tools for both challenges. The original MCS approach was introduced to handle the integration problem in a principled way. A later extension to so-called managed MCS appears to provide relevant functionality for the second challenge. In this paper we review both MCS and managed MCS and discuss how the latter approach needs to be further developed for online applications.

1 Introduction

Research in knowledge representation (KR) and, more generally, information technology faces at least the following two problems:

1. A large variety of formats and languages for representing knowledge has been produced. A wealth of tools and formalisms is now available, including rather basic ones like databases or the more recent triple-stores, and more expressive ones like ontology languages (e.g., description logics), temporal and modal logics, nonmonotonic logics, or logic programs under answer set semantics, to name just a few.
   This diversity of formalisms poses some important challenges. There are many situations where the integration of the knowledge represented in such diverse formalisms is crucial. But how can this be achieved in a principled way?
2. Most of the tools providing reasoning services for KR languages were developed for offline usage: given a knowledge base (KB) computation is one-shot, triggered by a user, through a specific query or a request to compute, say, an answer set. This is the right thing for specific types of applications, for instance expert systems, configuration or planning problems where a specific answer to a problem instance is needed at a particular point in time. However, there are different kinds of applications where a reasoning system is continuously online and observes a particular system, informing the user in case something unforeseen/uninteded happens. This different usage of KR systems again poses important challenges.

Let's illustrate these problems with two examples. For the first problem, assume your home town's hospital has

P. Cabalar and T.C. Son (Eds.): LPNMR 2013, LNAI 8148, pp. 1–10, 2013.
© Springer-Verlag Berlin Heidelberg 2013

- a patient database (e.g. an Oracle database),
- a disease ontology (written in a particular description logic),
- an ontology of the human body (using OWL),
- an expert system describing the effects of different medications (using a nonmonotonic reasoning formalism, say disjunctive logic programming).

Needless to say that it is in the patients' best interest to integrate all available knowledge. But how? Translating everything into a single all-purpose formalism is certainly not a solution. First of all, no standardized, universal knowledge representation language exists, and there are very good reasons for this (e.g. specific modeling needs or complexity considerations). Secondly, even if there were such a language, most probably remodeling the information would be too cumbersome and costly. What seems to be needed is a principled way of integrating knowledge expressed in different formats/languages/logics.

For the second problem consider an assisted living scenario where people in need of support live in an apartment which is equipped with various sensors, e.g. smoke detectors, cameras, and body sensors measuring relevant body functions (e.g. pulse, blood pressure). A reasoning system continuously receives sensor information. The task is to detect emergencies (health problems, forgotten medication, overheating stove,...) and cause adequate reactions (e.g. turning off the electricity, calling the ambulance, ringing an alarm). Apparently the system is not explicitly called by a user to become active. It rather is continuously online and must be able to process a continuous stream of information rather than a fixed KB.

This shift in view clearly has an impact on KR formalisms. Most importantly, since the system is permanently online, the available information continuously grows. This obviously cannot go on forever as the KB needs to be kept in a manageable size. We thus need principled ways of forgetting/disregarding information. In the literature one often finds sliding window techniques where information is kept for a specific, predefined period of time and forgotten if it falls out of this time window. We believe this approach is far too inflexible. What is needed is a dynamic, situation dependent way of determining whether information needs to be kept or can be given up. Ideally we would like our online KR system to guarantee specific response times; although it may be very difficult to come up with such guarantees, it is certainly necessary to find means to identify and focus on relevant parts of the available information. Moreover, although the definition of the semantics of the underlying KR formalism certainly remains essential, we also need to impose procedural aspects reflecting the necessary modifications of the KB. This leads to a new, additional focus on *runs* of the system, rather than single evaluations.

We believe nonmonotonic multi-context systems (MCS) [1] are promising tools for addressing both problems. The original MCS framework was explicitly developed to handle problem 1, the integration of diverse KR formalisms. In Sect. 2 we recall the basic ideas underlying this approach. In a nutshell, an MCS consists of reasoning units - called contexts for historical reasons [8] - where each unit can be connected with other units via so-called bridge rules. The collection of bridge rules associated with a context specifies additional beliefs the context is willing to accept depending on what is believed by connected contexts. The contexts themselves can be viewed as parts of an agent's knowledge connected to other parts, but they can also be viewed as single agents willing to listen to and incorporate information from other agents.

The original framework was aimed at modelling the flow of information among contexts, consequently the addition of information to a context was the only possible operation. Of course, it is easy to imagine other operations one may want to perform, for instance revisions which keep the underlying KB consistent instead of simple additions. To capture arbitrary operations MCS were later generalized to so called managed MCS (mMCS) [3], a general and flexible framework that we will briefly discuss in Sect. 3. The possibility to have arbitrary operators is what, as we believe, makes mMCS suitable tools for the kind of online applications we discussed earlier. Sect. 4 describes some of the necessary steps that need to be done. In particular, what is required is an instantiation of the general mMCS framework with operations suitable to model focussing and forgetting. The systems we have in mind are *reactive* in the sense that they modify themselves to keep system performance up and in response to potential emergencies.

In cases where knowledge integration is not an issue (that is, where a single context is sufficient) and where moreover the underlying KR formalism is rule based the separation between context knowledge base and bridge rules may become obsolete. We briefly illustrate this in Sect. 5 using ASP as the underlying formalism. Sect. 6 concludes and points to open research questions.

2   Nonmonotonic Multi-Context Systems

The basic idea underlying MCS is to leave the diverse formalisms and knowledge bases untouched, and to equip each context with a collection of so-called bridge rules in order to model the necessary information flow among contexts.

Bridge rules are similar to logic programming rules (including default negation), with an important difference: they allow to access other contexts in their bodies. Using bridge rules has several advantages: the specification of the information flow is fully declarative; moreover, information - rather than simply being passed on as is - can be modified in various ways:

- we may translate a piece of information into the language/format of another context,
- we may pass on an abstraction of the original information, leaving out unnecessary details,
- we may select or hide information,
- we may add conclusions to a context based on the absence of information in another one,
- we may use simple encodings of preferences among parent contexts,
- we can even encode voting rules, say based on majorities etc.

The semantics of MCS is defined in terms of equilibria: a belief state assigns a belief set to each context C_i. Intuitively, a belief state is an equilibrium whenever the belief set selected for each C_i is acceptable for C_i's knowledge base *augmented by the heads of C_i's applicable bridge rules.*

The history of MCS started in Trento. Advancing work in [7,9], the Trento School developed monotonic heterogeneous multi-context systems [8] with the aim to integrate different inference systems. Here reasoning within as well as across contexts is mono-tonic. The first, still somewhat limited attempts to include nonmonotonic reasoning were done in [10] and [4]. To allow for reasoning based on the *absence* of informa-tion from a context, in both papers default negation is allowed in the rules. In this way contextual and default reasoning are combined.

The nonmonotonic MCS of [1] substantially generalized these approaches, by ac-commodating *heterogeneous* and both *monotonic* and *nonmonotonic* contexts. They are thus capable of integrating "typical" monotonic logics like description logics or temporal logics, and nonmonotonic formalisms like Reiter's default logic, answer set programs, circumscription, defeasible logic, or theories in autoepistemic logic. The cur-rently most general MCS variant, the so-called managed MCS (mMCS) [3] allow for arbitrary user-defined operations on the context knowledge bases, not just augmenta-tions. They will be discussed in the next section.

Here is a more formal description of multi-context systems as defined in [1]. MCS build on an abstract notion of a *logic* L as a triple (KB_L, BS_L, ACC_L), where KB_L is the set of admissible knowledge bases (KBs) of L, which are sets of KB-elements ("formulas"); BS_L is the set of possible belief sets, whose elements are beliefs; and $ACC_L : KB_L \to 2^{BS_L}$ is a function describing the semantics of L by assigning to each knowledge-base a set of acceptable belief sets.

A *multi-context system (MCS)* $M = (C_1, \ldots, C_n)$ is a collection of contexts $C_i = (L_i, kb_i, br_i)$ where L_i is a logic, $kb_i \in KB_{L_i}$ is a knowledge base and br_i is a set of bridge rules of the form:

$$s \leftarrow (c_1 : p_1), \ldots, (c_j : p_j), not(c_{j+1} : p_{j+1}), \ldots, not(c_m : p_m). \qquad (1)$$

such that $kb \cup \{s\}$ is an element of KB_{L_i}, $c_\ell \in \{1, \ldots, n\}$, and p_ℓ is element of some belief set of BS_{c_ℓ}, for all $1 \leq \ell \leq m$. For a bridge rule r, we denote by $hd(r)$ the formula s while $body(r)$ denotes the set $\{(c_{\ell_1} : p_{\ell_1}) \mid 1 \leq \ell_1 \leq j\} \cup \{not(c_{\ell_2} : p_{\ell_2}) \mid j < \ell_2 \leq m\}$.

A belief state $S = (S_1, \ldots, S_n)$ for M consists of belief sets $S_i \in BS_i, 1 \leq i \leq n$. A bridge rule r of form (1) is applicable wrt. S, denoted by $S \models body(r)$, iff $p_\ell \in S_{c_\ell}$ for $1 \leq \ell \leq j$ and $p_\ell \notin S_{c_\ell}$ for $j < \ell \leq m$. We use $app_i(S) = \{hd(r) \mid r \in br_i \land S \models body(r)\}$ to denote the heads of all applicable bridge rules of context C_i wrt. S.

The semantics of an MCS M is then defined in terms of equilibria, where an *equi-librium* is a belief state $(S_1, \ldots, S_n)$ such that $S_i \in ACC_i(kb_i \cup app_i(S)), 1 \leq i \leq n$.

3   Managed MCS: Beyond Information Flow

Although nonmonotonic MCS are, as we believe, an excellent starting point to address the problems discussed above, the way they integrate knowledge is still somewhat lim-ited: if a bridge rule for a context is applicable, then the rule head is simply added to the context's knowledge base (KB). Although this covers the flow of information, it does not capture other operations one may want to perform on context KBs. For instance, rather than simply *adding* a formula ϕ, we may want to delete some information, or to *revise* the KB with ϕ to avoid inconsistency in the context's belief set. We are thus interested in generalizations of the MCS approach where specific predefined operations on knowledge bases can be performed.

A first step into this direction are argumentation context systems (ACS) [2]. They specialize MCS in one respect, and are more general in another. First of all, in contrast to nonmonotonic MCS they are homogeneous in the sense that all reasoning components in an ACS are of the same type, namely Dung-style argumentation frameworks [5]. The latter are widely used as abstract models of argumentation. However, ACS go beyond MCS in two important aspects:

1. The influence of an ACS module M_1 on another module M_2 can be much stronger than in an MCS. M_1 may not only provide information for M_2 and thus augment the latter, it may directly affect M_2's KB and reasoning mode: M_1 may invalidate arguments or attack relationships in M_2's argumentation framework, and even determine the semantics to be used by M_2.
2. A major focus in ACS is on *inconsistency handling*. Modules are equipped with additional components called *mediators*. The main role of the mediator is to take care of inconsistencies in the information provided by connected modules. It collects the information coming in from connected modules and turns it into a consistent update specification for its module, using a pre-specified consistency handling method which may be based on preference information about other modules.

Managed MCS (mMCS) push the idea of mediators even further. They allow additional operations on knowledge bases to be freely defined; this is akin to management functionality of database systems. We thus call the additional component *context manager*. In a nutshell (and somewhat simplifying) the features of mMCS are as follows:

- Each logic comes with a set of operations O.
- An operational statement is an operation applied to a formula (e.g. insert(p), delete(p), revise(p), ...).
- Bridge rules are as before, except for the heads which now are operational statements.
- A management function: $mng : 2^{Opst} \times KB \to 2^{KB}$, produces a collection of KBs out of set of operational statements and a KB.
- A managed context consists of a logic, a KB, a set of bridge rules (as before), together with the new part, a management function.
- An mMCS is just a collection of managed contexts.

Regarding the semantics, a belief state $S = (S_1, \ldots S_n)$ contains - as before - a belief set for each context. To be an equilibrium S has to satisfy the following condition: the belief set chosen for each context must be acceptable for one of the KBs obtained by applying the management function to the heads of applicable bridge rules and the context's KB. More formally, for all contexts $C_i = (L_i, kb_i, br_i, mng_i)$: let S_i be the belief set chosen for C_i, and let Op_i be the heads of bridge rules in br_i applicable in S. Then S is an equilibrium iff, for $1 \leq i \leq n$,

$$S_i \in ACC_i(kb') \text{ for some } kb' \in mng_i(Op_i, kb_i).$$

Management functions allow us to model all sorts of modifications of a context's knowledge base and thus make mMCS a powerful tool for describing the influence contexts can have on each other. Of course, the framework is very general and needs to be instantiated adequately for particular problems. As a short illustrative example let us consider an instantiation we call revision-based MCS. The main goal here is to keep each context's KB consistent when information is added, that is, we want to guarantee consistency of belief sets in equilibria.

Assume the KB's logic has a single operation inc (include). For a formula p, $inc(p)$ intuitively says: incorporate p consistently into your KB. Two things can go wrong: the formulas to be included in a particular situation

1. may be inconsistent with each other, or
2. may be inconsistent with the context KB.

For 1 we introduce preferences among bridge rules. More precisely, we represent a total preorder on bridge rules by using indexed operations $inc_1, inc_2, \ldots$ where a lower index represents higher priority. Given a collection of indexed inclusion operations we can now identify preferred sets of formulas as follows: we pick a maxi-consistent subset of inc_1-formulas (i.e. formulas appearing as arguments of inc_1), extend the set maxi-consistently with inc_2-formulas etc.

For 2 we assume a consistency preserving base revision operator

$$rev : KB \times KB \to 2^{KB}$$

that is, $rev(kb_1, kb_2)$ may potentially produce alternative outcomes of the revision, however each outcome is consistent whenever kb_2 is. We can now define the management function as follows: for

$$Op = \{inc_1(p_{1,1}), \ldots, inc_1(p_{1,m}), \ldots, inc_k(p_{k,1}), \ldots, inc_k(p_{k,n})\}$$

let:

$$kb' \in mng(Op, kb) \text{ iff } kb' \in rev(kb, F) \text{ for some preferred set } F \text{ of } Op.$$

Each belief set in each equilibrium now is apparently consistent.

Here is a specific example. Let C be a context based on propositional logic, its KB is

$$\{July \to \neg Rain, Rain \to Umbrella, July\}.$$

C has 2 parent contexts; C_1 believes $Rain$, C_2 believes $\neg Rain$. C_1 more reliable wrt. the weather. C thus has the following bridge rules:[1]

$$\{inc_1([\neg]Rain) \leftarrow 1{:}[\neg]Rain;\ inc_2([\neg]Rain) \leftarrow 2{:}[\neg]Rain\}.$$

As C_1 is preferred to C_2 the single preferred set is $\{Rain\}$.

To fully specify the management function we still need to define the revision operator. We do this as follows: $K' \in rev(K, F)$ iff $K' = M \cup F$ for some maximal $M \subseteq K$ consistent with F. Now we obtain the following two acceptable belief sets for C:

$$Th(\{Rain, July \to \neg Rain, Rain \to Umbrella\})$$
$$Th(\{Rain, July, Rain \to Umbrella\}).$$

---

[1] We use square brackets in the rules to represent optional parts; each rule with $[\neg]$ thus actually represents 2 rules.

4   Reactive MCS: A Sketch

In this section we discuss some of the issues that need to be addressed for applications like the assisted living scenario we described in the introduction. We believe managed MCS are an excellent starting point for the following reasons:

- they offer means to integrate sensor information from different sources, handling inconsistencies if needed,
- the management function provides capabilities to modify KBs which appear essential to keep the sizes of knowledge bases manageable.

Nevertheless, the general managed MCS framework obviously needs to be further modified, respectively instantiated, to become reactive. What we aim for is a specialization of the (potentially generalized) managed MCS framework suitable for online applications. Here we identify some of the relevant changes.

First of all, it is useful to introduce different types of contexts:

- observer contexts which are "connected to the real world via sensors; these contexts keep track of (time-stamped) sensor readings,
- analyzer contexts which reason about the current situation and in particular detect emergencies; they obtain relevant information from sensing contexts via bridge rules and generate alarms if needed,
- control contexts which make sure the system focuses on the right issues; this includes dynamically setting adequate time windows for information, increasing the frequency of sensor readings if relevant/dangerous things happen, making sure outdated/irrelevant information is deleted/disregarded to keep the system performance up.

Next, the management function needs to be instantiated adequately for purposes of focusing and forgetting:

- a language of adequate operations for focusing and forgetting needs to be defined,
- ideally the performed operations may also depend on the actual system performance,
- it would be highly useful if the management function were able to restrict recomputations to specific, relevant contexts.

Finally, we anticipate that preferences will play an essential role:

- for inconsistency handling among different sensor readings,
- to handle more important emergencies with high priority,
- to mediate between what's in the current focus and the goal not to overlook important events.

We thus will need to equip MCS with expressive and flexible preference handling capabilities.

As pointed out earlier, in online applications the major object of interest is the system behaviour over time. For this reason we next define runs of reactive MCS, an essential basic notion:

Definition 1. *Let M be a managed MCS with contexts $C_0, \ldots, C_n$ ($C_0, \ldots, C_k$ are observer contexts). Let $Obs = (Obs^0, Obs^1, \ldots)$ be a sequence of observations, that is, for $j \geq 0$, $Obs^j = (Obs_i^j)_{i \leq k}$, where Obs_i^j is the new (sensor) information for context i at step j. A run R of M induced by Obs is a sequence*

$$R = Kb^0, Eq^0, Kb^1, Eq^1, \ldots$$

where

- *$Kb^0 = (Kb_i^0)_{i \leq n}$ is the collection of initial knowledge bases, Eq^0 an equilibrium of Kb^0,*
- *for $j \geq 1$ and $i \leq n$, Kb_i^j is the knowledge base of context C_i produced be the context's management function for the computation of Eq^{j-1}, and $Kb^j = (Kb_i^j)_{i \leq n}$,*
- *for $j \geq 1$, Eq^j is an equilibrium for the knowledge bases*

$$(Kb_0^j \cup Obs_0^j, \ldots, Kb_k^j \cup Obs_k^j, Kb_{k+1}^j, \ldots, Kb_n^j).$$

5 Reactive ASP: A Bottom Up Approach

In the last section we sketched some of the issues that need to be addressed in order to turn MCS into a reactive formalism suitable for online applications. The basic idea was to handle reactivity by adequate operations in bridge rules. We now consider cases where the integration of information from different sources is not an issue and where we work with a single context. The separation between context and bridge rules is still relevant as the bridge rules (which now should better be called operational rules as they do no longer bridge different contexts) implement the focusing and forgetting strategies of the context.

However, if the single context we work with is itself rule based, then strictly speaking the separation of bridge/operational rules from the rest of the program becomes obsolete. We may as well use operational rules within the formalism itself. For instance, assume we use logic programs under answer set semantics.[2] Some of the rules in the program may have operational statements in their heads which simply are interpreted as operations to be performed on the program itself. Again, this allows us to represent the strategy for maintaining the knowledge base manageable declaratively. When the program is run, the strategy is realized by a self-modification of the program.

This is reflected in the following notion of a run of a reactive answer set program (RASP) P, that is an ASP program which has some rules with operational statements in the heads. Intuitively, the behaviour of RASP P is characterized as follows:

1. P computes an answer set S_0, during the computation the current information is frozen until the computation is finished,
2. the set of operations to be performed is read off the answer set S_0 and P is modified accordingly, at the same time observations made since the last computation started are added,

---

[2] For some important steps towards stream reasoning with answer set programming see also the work of Torsten Schaub's group, e.g [6].

3. the modified program computes a new answer set, and so on.

This is captured in the following definition of a run:

Definition 2. *A run of a reactive answer set program P induced by a sequence of sets of observations $(Obs_0, Obs_1, \ldots)$ is a sequence $(S_0, S_1, \ldots)$ of answer sets satisfying the following conditions:*

1. *S_0 is an answer set of $P_0 = P$.*
2. *For $i \geq 0$, S_{i+1} is an answer set of $P_{i+1} = Mod_i(P_i) \cup Obs_i$, where $Mod_i(P_i)$ is the result of modifying P_i according to the operational statements contained in S_i.*

This definition implies new information obtained while the last answer set was computed is always included in the new. modified program. In certain situations, for instance if parts of the new knowledge are outside the current focus, it may even be useful to disregard pieces of the new information entirely. Formally this can be captured by letting $P_{i+1} = Mod_i(P_i \cup Obs_i)$ in item 2 of the definition above.

6 Discussion and Future Work

In this paper we discussed some of the issues that need to be addressed if KR wants to meet the challenges of certain types of applications where continuous online reasoning is required. We sketched a top down approach, instantiating the managed MCS approach accordingly. We also briefly described a bottom up approach based on a related extension of the ASP framework.

We obviously left many questions open. The major open issue is the specification of a suitable language for the operational statements which are relevant for forgetting and focusing. The operations should allow to set the window size for specific sensor/information types dynamically, keeping relevant information available. They also should make it possible to specify the system's focus depending on events pointing to potential problems/emergencies. Focusing may lead to more regular checks for specific information, whereas other information may be looked at only from time to time.

Of course, focusing bears the danger that new problems may be overlooked. Ideally, we would like to have a guarantee that every potential emergency is checked on a regular basis, even if it is not in the current focus. In addition, it would be very useful to take information about the current system performance into account to determine what information to keep and what to give up. This would lead to a notion of resource-aware computation where part of the sensor information made available in each step of a run reveals how the system currently is performing.

The notions of a run we defined both for managed MCS and for reactive ASP is built on a credulous view: in each step a single equilibrium, respectively answer set, is computed and taken as the starting point for the next step. There may be scenarios where a skeptical approach built on what is believed in all (or in some preferred) equilibria/answer sets is more adequate. It may even be useful to switch between the credulous and skeptical approach dynamically.

Finally, if memory is not an issue (but computation time is) then rather than deleting irrelevant information one could as well keep it but put it aside for a certain time.

The available information would thus be divided in a part to be forgotten, a part to be kept but disregarded for the time being, and a part currently in the focus of attention.

In conclusion, a lot remains to be done in KR to fully solve the challenges of integration and online reasoning. Nevertheless, we believe promising ideas are already around and addressing the open problems will definitely be worth it.

Acknowledgements. Some of the ideas presented here are based on discussions with Torsten Schaub and Stefan Ellmauthaler. The presented work was partly funded by Deutsche Forschungsgemeinschaft, grant number FOR 1513.

References

1. Brewka, G., Eiter, T.: Equilibria in heterogeneous nonmonotonic multi-context systems. In: Proc. AAAI 2007, pp. 385–390. AAAI Press (2007)
2. Brewka, G., Eiter, T.: Argumentation context systems: A framework for abstract group argumentation. In: Erdem, E., Lin, F., Schaub, T. (eds.) LPNMR 2009. LNCS, vol. 5753, pp. 44–57. Springer, Heidelberg (2009)
3. Brewka, G., Eiter, T., Fink, M., Weinzierl, A.: Managed multi-context systems. In: Proc. IJCAI 2011, pp. 786–791 (2011)
4. Brewka, G., Roelofsen, F., Serafini, L.: Contextual default reasoning. In: Proc. IJCAI 2007, pp. 268–273 (2007)
5. Dung, P.M.: On the acceptability of arguments and its fundamental role in nonmonotonic reasoning, logic programming and n-person games. Artif. Intell. 77(2), 321–358 (1995)
6. Gebser, M., Grote, T., Kaminski, R., Obermeier, P., Sabuncu, O., Schaub, T.: Stream reasoning with answer set programming: Preliminary report. In: Proc. KR 2012, pp. 613–617 (2012)
7. Giunchiglia, F.: Contextual reasoning. Epistemologia XVI, 345–364 (1993)
8. Giunchiglia, F., Serafini, L.: Multilanguage hierarchical logics or: How we can do without modal logics. Artif. Intell. 65(1), 29–70 (1994)
9. McCarthy, J.: Generality in artificial intelligence. Commun. ACM 30(12), 1029–1035 (1987)
10. Roelofsen, F., Serafini, L.: Minimal and absent information in contexts. In: Proc. IJCAI 2005 (2005)

Logic Programming in the 1970s

Robert Kowalski

Imperial College London
rak@doc.ic.ac.uk

Abstract. Logic programming emerged in the 1970s from debates concerning procedural versus declarative representations of knowledge in artificial intelligence. In those days, declarative representations were associated mainly with bottom-up proof procedures, such as hyper-resolution. The development of logic programming showed that procedural representations could be obtained by applying top-down proof procedures, such as linear resolution, to declarative representations in logical form.

In recent years, logic programming has become more purely declarative, with the development of answer set programming, tabling and the revival of datalog. These recent developments invite comparison with earlier attempts to reconcile procedural and declarative representations of knowledge, and raise the question whether anything has been lost.

Keywords: logic programming, Prolog.

1   What is Logic Programming?

Logic programming can be viewed as the use of logic to perform computation, building upon and extending one of the simplest logics imaginable, namely the logic of Horn clauses. A *Horn clause* is a sentence that can be written in the logical form:

$$A_0 \leftarrow A_1 \wedge ... \wedge A_n \quad \text{where } n \geq 0.$$

where each A_i is an atomic formula of the form $p(t_1,...t_m)$, where p is a predicate symbol, the t_i are terms, and all variables are universally quantified.

Horn clauses are theoretically sufficient for programming and databases. But for non-monotonic reasoning, they need to be extended to clauses of the form:

$$A_0 \leftarrow A_1 \wedge ... \wedge A_n \wedge \; not \; B_1 \wedge ... \wedge not \; B_m \quad \text{where } n \geq 0 \text{ and } m \geq 0$$

where each A_i and B_i is an atomic formula. Sets of clauses in this form are called *normal logic programs*, or just *logic programs*.

In this paper, which is a shorter version of a longer history to be included in [36], I present a personal view of the early history of logic programming. It focuses on logical, rather than on technological issues, and assumes that the reader is already familiar with the basics of logic programming. Other histories that give other perspectives include [4, 7, 10, 11, 17, 23].

P. Cabalar and T.C. Son (Eds.): LPNMR 2013, LNAI 8148, pp. 11–22, 2013.
© Springer-Verlag Berlin Heidelberg 2013

2   The Prehistory of Logic Programming

Horn clauses are a special case of the clausal form of first-order logic, and the earliest implementations of Horn clause programming used the resolution rule of inference [31] developed by (John) Alan Robinson.

2.1   Resolution

The invention of resolution revolutionized the field of automated theorem-proving, and inspired other applications of logic in artificial intelligence. One of the most successful of these applications was the question-answering system QA3 [14], developed by Cordell Green. Green also showed that resolution could be used to automatically generate a program in LISP, from a specification of its input-output relation written in the clausal form of logic.

Green seems to have anticipated the possibility of dispensing with LISP and using only the logical specification of the desired input-output relation, writing [14]:

> "The theorem prover may be considered an "interpreter" for a high-level assertional or declarative language - logic. As is the case with most high-level programming languages the user may be somewhat distant from the efficiency of "logic" programs unless he knows something about the strategies of the system."

However, he does not seem to have pursued these ideas much further. Moreover, there was an additional problem, namely that the resolution strategies of that time behaved unintuitively and were very redundant and inefficient. For example, given a clause of the form $L_1 \lor ... \lor L_n$, and n clauses of the form $\neg L_i \lor C_i$, resolution would derive the same clause $C_1 \lor ... \lor C_n$ redundantly in $n!$ different ways.

2.2   Procedural Representations of Knowledge

Green's ideas were attacked from MIT, where researchers were advocating procedural representations of knowledge. Terry Winograd's PhD thesis [38] gave the most compelling and most influential voice to this opposition. Winograd argued [38, page 232]:

> "Our heads don't contain neat sets of logical axioms from which we can deduce everything through a "proof procedure". Instead we have a large set of heuristics and procedures for solving problems at different levels of generality."

Winograd's procedural alternative to logic was based on Carl Hewitt's language PLANNER [16]. Winograd describes PLANNER in the following terms [38, page 238]:

> "The language is designed so that if we want, we can write theorems in a form which is almost identical to the predicate calculus, so we have the benefits of a uniform system. On the other hand, we have the capability to add as much subject-dependent knowledge as we want, telling theorems about other theorems and proof procedures. The system has an automatic goal-tree backup system, so that

even when we are specifying a particular order in which to do things, we may not know how the system will go about doing them. It will be able to follow our suggestions and try many different theorems to establish a goal, backing up and trying another automatically if one of them leads to a failure ....."

In contrast [38, page 215]:

"A uniform proof procedure gropes its way through the collection of theorems and assertions, according to some general procedure which does not depend on the subject matter. It tries to combine facts which might be relevant, working from the bottom-up."

Winograd's PhD thesis presented a natural language understanding system that was a great advance at the time, and its advocacy of PLANNER was enormously influential.

2.3   Improved Resolution Proof Procedures

At the time that QA3 and PLANNER were being developed, resolution was not very well understood. Perhaps the best known refinement was Robinson's hyper-resolution [32], which, in the case of ground Horn clauses, derives D_0 from the input clause:

$$D_0 \leftarrow B_1 \wedge \ ... \ \wedge B_m$$

and from the input or derived atoms, $B_1, ..., B_m$. The problem with hyper-resolution, as Winograd observed in the passage quoted above, is that it uniformly derives new clauses from the input clauses without paying attention to the problem to be solved.

Linear resolution, discovered independently by Loveland [26], Luckham [28] and Zamov and Sharonov [39], addresses the problem of relevance by focusing on a top clause C_0, which could represent an initial goal:

Let S be a set of clauses. A *linear derivation* of a clause C_n from a top clause $C_0 \in S$ is a sequence of clauses $C_0, ..., C_n$ such that every clause C_{i+1} is a resolvent of C_i with some input clause in S or with some ancestor clause C_j where $j < i$. (It was later realised that ancestor resolution is unnecessary if S is a set of Horn clauses.)

In retrospect, the relationship with PLANNER is obvious. If the top clause C_0 represents an initial goal, then the tree of all linear derivations is a kind of goal tree, and generating the tree top-down is a form of goal-reduction. The tree can be explored using different search strategies. Depth-first search, in particular, can be informed by PLANNER-like strategies that both specify "a particular order in which to do things", but also "back up" automatically in the case of failure.

But the relationship with PLANNER was still obscure, due to the $n!$ redundant ways of resolving upon n literals in the clauses C_i. This redundancy was recognized and solved independently at about the same time by Loveland [27], Reiter [30], and Kowalski and Kuehner [24]. The obvious solution was simply to select a single order for resolving upon the literals in the clauses C_i.

3   The Procedural Interpretation of Horn Clauses

The procedural interpretation of Horn clauses came about during my collaboration with Alain Colmerauer in the summer of 1972 in Marseille. Colmerauer was developing natural language question-answering systems, and I was developing resolution theorem-provers, and attempting to reconcile them with PLANNER-like procedural representations of knowledge.

3.1   The Relationship with Formal Grammars

Colmerauer knew everything there was to know about formal grammars and their application to programming language compilers. During 1967–1970, he created the Q-System [5] at the University of Montreal, which was later used from 1982 to 2001 to translate English weather forecasts into French for Environment Canada. Since 1970, he had been in Marseille, building up a team working on natural language question-answering, investigating SL-resolution [24] for the question-answering component.

My first visit to Marseille was in the summer of 1971, when we investigated the representation of grammars in logical form. We discovered that forward reasoning with hyper-resolution performed bottom-up parsing, while backward reasoning with SL-resolution performed top-down parsing. However, we did not yet see how to represent more general PLANNER-like procedures in logical form.

3.2   Horn Clauses and SLD-Resolution

It was during my second visit to Marseille in April and May of 1972 that the idea of using SL-resolution to execute Horn clause programs emerged. By the end of the summer, Colmerauer's group had developed the first version of Prolog, and used it to implement a natural language question-answering system [6]. I reported an abstract of my own findings at the MFCS conference in Poland in August 1972 [20].

The first Prolog system was an implementation of SL-resolution for the full clausal form of first-order logic, including ancestor resolution. But the idea that Horn clauses were an interesting case was already in the air. Donald Kuehner, in particular, had already been working on bi-directional strategies for Horn clauses [25]. However, the first explicit presentation of the procedural interpretation of Horn clauses appeared in [21]. The abstract begins:

> "The interpretation of predicate logic as a programming language is based upon the interpretation of implications   B if A₁ and .... and Aₙ as procedure declarations, where B is the procedure name and A₁ and .... and Aₙ is the set of procedure calls constituting the procedure body..."

The theorem-prover described in [21] is a variant of SL-resolution without ancestor resolution, to which Maarten van Emden later attached the name SLD-resolution, standing for "selected linear resolution with definite clauses". In fact, SLD-resolution is more general than SL-resolution restricted to Horn clauses, because in SL-resolution atoms (or subgoals) must be resolved upon last-in-first-out, but in SLD-resolution atoms can be resolved upon in any order.

As in the case of linear resolution more generally, the space of all SLD-derivations with a given top clause has the structure of a goal tree, which can be explored using different search strategies. From a logical point of view, it is desirable that the search strategy be complete, so that the proof procedure is guaranteed to find a solution if there is one in the search space. Complete search strategies include breadth-first search and various kinds of best-first and heuristic search. Depth-first search is incomplete in the general case, but it takes up much less space than the alternatives. Moreover, it is complete if the search space is finite, or if there is only one infinite branch that is explored after all of the others.

The different options for selecting atoms to resolve upon and for searching the space of SLD-derivations were left open in [21], but were pinned down in the Marseille Prolog interpreter. In Prolog, subgoals are selected last-in-first-out in the order in which the subgoals are written, and branches of the search space are explored depth-first in the order in which the clauses are written. By choosing the order in which subgoals and clauses are written, a Prolog programmer can exercise a significant amount of control over the efficiency of a program.

3.3   Logic + Control

In those days, there was a wide-spread belief that logic alone is inadequate for problem-solving, and that some way of controlling the problem solver is needed for efficiency. PLANNER combined logic and control in a procedural representation, but in a way that made it difficult to identify the logical component. Logic programs executed with SLD-resolution also combine logic and control, but in a way that makes it possible to read the same program both logically and procedurally. I later expressed this as an equation $A = L + C$ (Algorithm = Logic + Control) [22].

The most straight-forward implication of the equation is that, given a fixed logical representation L, different algorithms can be obtained by applying different control strategies, i.e. $A_1 = L + C_1$ and $A_2 = L + C_2$. Pat Hayes [15], in particular, argued that the logic and control components should be expressed in separate languages, with the logic component L providing a pure, declarative specification of the problem, and the control component C supplying the problem solving strategies needed for an efficient algorithm A. Moreover, he argued against the idea, expressed by the equations $A_1 = L_1 + C$ and $A_2 = L_2 + C$, of using a fixed control strategy C, as in Prolog, and formulating the logic L_i of the problem to obtain a desired algorithm A_i.

The idea of expressing logic and control in different languages has been a recurrent theme in automated theorem-proving, but has had less influence in the field of logic programming. However, Hayes may have anticipated some of the problems that arise when Prolog does not provide sufficiently powerful control to allow a high level representation of the problem to be solved. Here is a simple example, written in Prolog notation:

> *likes(bob, X) :- likes(X, logic).*
> *likes(bob, logic).*
> *:- likes(bob, X).*

Prolog fails to find the solution $X = bob$, because it explores the infinite branch generated by repeatedly using the first clause, without getting a chance to explore the branch generated by the second clause.

These days, SLD-resolution extended with tabling [8, 33, 34] avoids many infinite loops, like the one in this example.

4   The Semantics of Horn Clause Programs

The earliest influences on the development of logic programming had come primarily from automated theorem-proving and artificial intelligence. But researchers in Edinburgh, where I was working at the time, also had strong interests in the theory of computation, and there was a lot of excitement about Dana Scott's recent fixed point semantics for programming languages [35]. Maarten van Emden suggested that we investigate the application of Scott's ideas to Horn clause programs and that we compare the fixed point semantics with the logical semantics.

4.1   What Is the Meaning of a Program?

But first we needed to establish a common ground for the comparison. If we regard computer programs as computing input-output relations, then we can identify the "meaning" (or denotation) of a logic program P with the set of all ground atoms A that can be derived from P, which is expressed by:

$$P \vdash A$$

Here $\vdash$ can represent any derivability relation. Viewed in programming terms, this is analogous to the operational semantics of a programming language.

But viewed in logical terms, this is a proof-theoretic definition, which is not a semantics at all. In logical terms, it is more natural to understand the semantics of P as given by the set of all ground atoms A that are logically implied by P, written:

$$P \models A$$

The operational and model-theoretic semantics are equivalent for any sound and complete notion of derivation.

SL-resolution is sound and complete for arbitrary clauses. So it is sound and complete for Horn clauses in particular. Moreover, ancestor resolution is impossible for Horn clauses. So SL-resolution without ancestor resolution is sound and complete for Horn clause programs. This includes the proof procedure with fixed ordering of subgoals used in Prolog. The completeness of SLD-resolution, with its more dynamic and more liberal selection function, was proved by Robert Hill [18].

Hyper-resolution is also sound and complete for arbitrary clauses, and therefore for Horn clauses as well. Moreover, as we soon discovered, it is equivalent to the construction of the fixed point semantics.

4.2   Fixed Point Semantics

In Dana Scott's fixed point semantics, the denotation of a recursive function is given by its input-output relation. The denotation is constructed by approximation, starting with the empty relation, repeatedly plugging the current approximation of the denotation into the definition of the function, transforming the approximation into a better one, until the complete denotation is obtained in the limit, as the least fixed point.

Applying the same approach to a Horn clause program P, the fixed point semantics uses a similar transformation T_P, called the *immediate consequence operator*, to map a set I of ground atoms representing an approximation of the input-output relations of P into a better and more complete approximation $T_P(I)$. The resulting set $T_P(I)$ is equivalent to the set of all ground atoms that can be derived by applying one step of hyper-resolution to the clauses in $ground(P) \cup I$.

Not only does every Horn clause program P have a fixed point I such that $T_P(I) = I$, but it has a *least fixed point*, $lfp(T_P)$, which is the denotation of P according to the *fixed point semantics*. The least fixed point is also the smallest set of ground atoms I closed under T_P, i.e. the smallest set I such that $T_P(I) \subseteq I$. This alternative characterisation provides a link with the minimal model semantics, as we will see below.

The least fixed point can be constructed, as in Scott's semantics, by starting with the empty set $\{\}$ and repeatedly applying T_P:

$$\text{If } T_P^{\,0} = \{\} \text{ and } T_P^{\,i+1} = T_P(T_P^{\,i}), \text{ then } lfp(T_P) = \cup_{0 \leq i} T_P^{\,i}.$$

The result of the construction is equivalent to the set of all ground atoms that can be derived by applying any number of steps of hyper-resolution to the clauses in $ground(P)$.

The equality $lfp(T_P) = \cup_{0 \leq i} T_P^{\,i}$ is usually proved in fixed point theory by appealing to the Tarski-Knaster theorem. However, in [12], we showed that it follows from the completeness of hyper-resolution and the relationship between least fixed points and minimal models. Here is a sketch of the argument:

$A \in lfp(T_P)$ iff $A \in min(P)$ i.e. least fixed points and minimal models coincide.
$A \in min(P)$ iff $P \models A$ i.e. truth in the minimal model and in all models coincide.
$P \models A$ iff $A \in \cup_{0 \leq i} T_P^{\,i}$ i.e. hyper-resolution is complete.

4.3   Minimal Model Semantics

The minimal model semantics was inspired by the fixed point semantics, but was based on the notion of Herbrand interpretation, which plays a central role in resolution theory.

The key idea of Herbrand interpretations is to identify an interpretation of a set of sentences with the set of all ground atomic sentences that are true in the interpretation. The most important property of Herbrand interpretations is that, in first-order logic, a set of sentences has a model if and only if it has a Herbrand model. This property is a form of the Skolem-Löwenheim-Herbrand theorem.

Thus the model-theoretic denotation of a Horn clause program:

$$M(P) = \{A \mid A \text{ is a ground atom and } P \models A\}$$

is actually a Herbrand interpretation of P in its own right. Moreover, it is easy to show that it is the smallest Herbrand model of P i.e. $M(P) = min(P)$. Therefore:

$$A \in min(P) \text{ iff } P \models A.$$

It is also easy to show that the Herbrand models of P coincide with the Herbrand interpretations that are closed under the immediate consequent operator, i.e.:

$$I \text{ is a Herbrand model of } P \text{ iff } T_P(I) \subseteq I.$$

This is because the immediate consequence operator T_P mimics, not only the definition of hyper-resolution, but also the definition of truth: A set of Horn clauses P is true in a Herbrand interpretation I if and only if, for every clause $A_0 \leftarrow A_1 \wedge ... \wedge A_n$ in $ground(P)$, whenever $A_1 , ... , A_n$ are true in I then A_0 is true in I.

It follows that the least fixed point and the minimal model are the same:

$$lfp(T_P) = min(P).$$

5   Negation as Failure

The practical value of extending Horn clause programs to normal logic programs with negative conditions was recognized from the earliest days of logic programming, as was the obvious way to reason with them - by *negation as failure* (abbreviated as NAF): to prove *not B*, show that all attempts to prove *B* fail. Intuitively, NAF is justified by the assumption that the program contains a complete definition of its predicates. Keith Clark's 1978 paper [2] was the first formal investigation of the semantics of negation as failure.

5.1   The Clark Completion

Clark's solution was to interpret logic programs as a short hand for definitions in if-and-only-if form. In the non-ground case, the logic program needs to be augmented with an equality theory, which mimics the unification algorithm, and which essentially specifies that ground terms are equal if and only if they are syntactically identical. Together with the equality theory, the if-and-only-if form of a normal logic program P is called the *completion* of P, written *comp(P)*.

Although NAF is sound with respect to the completion semantics, it is not complete. For example, if P is the program:

$$p \leftarrow q$$
$$p \leftarrow \neg q$$
$$q \leftarrow q$$

then *comp(P)* implies p, but NAF goes into an infinite loop, failing to show q. With a different semantics, the infinite failure could be interpreted as meaning that *not q* is true, and therefore p is true. The completion semantics does not recognise such infinite failure, because proofs in classical logic are finite. For this reason, the completion semantics is sometimes referred to as the semantics of negation as *finite* failure.

Clark did not investigate the relationship between the completion semantics and the various alternative semantics of Horn clauses. Probably the first such investigation was by Apt and van Emden [1], who showed that if P is a Horn clause program then:

$$I \text{ is a Herbrand model of } comp(P) \text{ iff } T_P(I) = I.$$

5.2   The Analogy with Arithmetic

Clark's 1978 paper was not the first to propose the completion semantics for logic programs. His 1977 paper with Tarnlund [3] proposed using the completion augmented with induction to prove program properties, by analogy with the use of Peano axioms to prove theorems of arithmetic.

Consider, for example, the Horn clause definition of *append(X, Y, Z)*, which holds when the list Z is the concatenation of the list X followed by the list Y:

> *append(nil, X, X)*
> *append(cons(U, X), Y, cons(U, Z))* ← *append(X, Y, Z)*

This is analogous to the definition of *plus(X, Y, Z)*, which holds when $X + Y = Z$:

> *plus(0, X, X)*
> *plus(s(X), Y, s(Z))* ← *plus(X, Y, Z)*

These definitions alone are adequate for computing their denotations. More generally, they are adequate for solving any goal clause, which is an existentially quantified conjunction of atoms. However, they need to be augmented with their completions and induction axioms to prove such properties as the fact that *append* is functional:

$$U = V \leftarrow append(X, Y, U) \wedge append(X, Y, V)$$

Because many such program properties can be expressed in logic programming form, it can be hard to decide whether a clause should be treated as an operational part of a logic program, or as an emergent property of the program. As Nicolas and Gallaire observed [29], a similar problem arises with deductive databases: It can be hard to decide whether a rule should be treated as part of the database, or as an integrity constraint that should be true of the database.

This distinction between clauses that are needed operationally, to define and compute relations, and clauses that are emergent properties is essential for practical applications. Without this distinction, it is easy to write programs that unnecessarily and redundantly mix operational rules and emergent properties. For example, adding to the logic program that defines *append* additional clauses that state the property that *append* is associative would make the program impossibly inefficient.

Arguably, the analogy with arithmetic helps to clarify the relationship between the different semantics of logic programs: It suggests that the completion of a logic program augmented with induction schemas is like the Peano axioms for arithmetic, and the minimal model is like the standard model of arithmetic. The fact that both notions of arithmetic have a place in mathematics suggests that both kinds of "semantics" have a place in logic programming.

Interestingly, the analogy also works in the other direction. The fact that minimal models are the denotations of logic programs shows that the standard model of arithmetic has a syntactic core, which consists of the Horn clauses that define addition and multiplication. Martin Davis [9] makes a similar point, but his core is essentially the completion of the Horn clauses without induction axioms. Arguably, the syntactic core of the standard model arithmetic explains how we can understand what it means for a sentence to be true, even if it cannot be proved.

6   Where Did We Go from Here?

This brief history covers some of the highlights of the development of logic programming from around 1968 to 1978, and it is biased by my own personal reflections. Nonetheless, it suggests a number of questions that may be relevant today:

- In the 1980s, the minimal model semantics of Horn clauses was extended significantly to deal with negation in normal logic programs. As a consequence, the original view in logic programming of computation as deduction has been challenged by an alternative view of computation as model generation. But the model generation view blurs the distinction between clauses that are needed operationally to generate models and clauses that are emergent properties that are true in those models. Would it be useful to pay greater attention to this distinction?

- With the development of answer set programming, tabling and datalog, logic programming has become more declarative. But imperative programming languages continue to dominate the world of practical computing. Can logic programming do more to reconcile and combine declarative and procedural representations in the future?

- In the late 1970s, as evidenced by the logic and databases workshop organized by Gallaire, Minker and Nicolas [13] in Toulouse, logic programming began to show promise as a general-purpose formalism for combining programming and databases. But in the 1980s, logic programming split into a variety of dialects specialized for different classes of applications.

   The recent revival of datalog [19] suggests that the old promise that logic programming might be able to unify different areas of computing may have new prospects. However, the query evaluation strategies associated datalog are mainly bottom-up with magic set transformations used to simulate top-down execution. Are the bottom-up execution methods of datalog really necessary? Or might top-down execution with tabling [37] provide an alternative and perhaps more general approach?

Acknowledgements. Many thanks to Maarten van Emden for helpful comments on the longer version of this paper.

References

1. Apt, K.R., van Emden, M.: Contributions to the Theory of Declarative Knowledge. JACM 29(3), 841–862 (1982)
2. Clark, K.L.: Negation by Failure. In: Gallaire, H., Minker, J. (eds.) Logic and Databases, pp. 293–322. Plenum Press (1978)
3. Clark, K.L., Tärnlund, S.-A.: A First-order Theory of Data and Programs. In: Proceedings of the IFIP Congress 1977, pp. 939–944 (1977)
4. Cohen, J.: A View of the Origins and Development of Prolog. CACM 31, 26–36 (1988)
5. Colmerauer, A.: Les Systèmes Q ou un Formalisme pour Analyser et Synthétiser des Phrases sur Ordinateur. Mimeo, Montréal (1969)
6. Colmerauer, A., Kanoui, H., Pasero, R., Roussel, P.: Un Systeme de Communication Homme-Machine en Francais. Research report, Groupe d'Intelligence Artificielle, Universite d'Aix-Marseille II, Luminy (1973)
7. Colmerauer, A., Roussel, P.: The Birth of Prolog. In: Bergin, T.J., Gibson, R.G. (eds.) History of Programming Languages, pp. 331–367. ACM Press and Addison Wesley (1996)
8. Chen, W., Warren, D.: Tabled Evaluation with Delaying for General Logic Programs. JACM 43, 20–74 (1996)
9. Davis, M.: The Mathematics of Non-monotonic Reasoning. Artificial Intelligence 13(1), 73–80 (1980)
10. Elcock, E.W.: Absys: The First Logic Programming Language—a Retrospective and a Commentary. Journal of Logic Programming 9(1), 1–17 (1990)
11. van Emden, M.: The Early Days of Logic Programming: A Personal Perspective. The Association of Logic Programming Newsletter 19(3) (2006), http://www.cs.kuleuven.ac.be/~dtai/projects/ALP/newsletter/aug06/
12. van Emden, M., Kowalski, R.A.: The Semantics of Predicate Logic as a Programming Language. JACM 23(4), 733–742 (1976)
13. Gallaire, H., Minker, J.: Logic and Data Bases. Plenum Press, New York (1978)
14. Green, C.: Application of Theorem Proving to Problem Solving. In: Walker, D.E., Norton, L.M. (eds.) Proceedings of the 1st International Joint Conference on Artificial Intelligence, pp. 219–239. Morgan Kaufmann (1969)
15. Hayes, P.J.: Computation and Deduction. In: Proceedings of the Second MFCS Symposium, Czechoslovak Academy of Sciences, pp. 105–118 (1973)
16. Hewitt, C.: Procedural Embedding of Knowledge In Planner. In: Proceedings of the 2nd International Joint Conference on Artificial Intelligence. Morgan Kaufmann (1971)
17. Hewitt, C.: Middle History of Logic Programming: Resolution, Planner, Edinburgh LCF, Prolog, Simula, and the Japanese Fifth Generation Project. arXiv preprint arXiv:0904.3036 (2009)
18. Hill, R.: LUSH Resolution and its Completeness. DCL Memo 78. School of Artificial Intelligence, University of Edinburgh (1974)
19. Huang, S.S., Green, T.J., Loo, B.T.: Datalog and Emerging Applications: an Interactive Tutorial. In: Proceedings of the 2011 ACM SIGMOD International Conference on Management of Data, pp. 1213–1216 (2011)
20. Kowalski, R.A.: The Predicate Calculus as a Programming Language (abstract). In: Procedings of the First MFCS Symposium, Jablonna, Poland (1972)
21. Kowalski, R.A.: Predicate Logic as a Programming Language. DCL Memo 70, School of Artificial Intelligence, Univ. of Edinburgh (1973); Proceedings of IFIP, pp. 569–574. North-Holland, Amsterdam (1974)
22. Kowalski, R.A.: Algorithm = Logic+ Control. CACM 22(7), 424–436 (1979)

23. Kowalski, R.A.: The Early History of Logic Programming. CACM 31, 38–43 (1988)
24. Kowalski, R.A., Kuehner, D.: Linear Resolution with Selection Function. Artificial Intelligence Journal 2, 227–260 (1971)
25. Kuehner, D.: Bi-directional Search with Horn Clauses. Edinburgh University (1969)
26. Loveland, D.W.: A Linear Format for Resolution. In: Symposium on Automatic Demonstration, pp. 147–162. Springer, Heidelberg (1970)
27. Loveland, D.W.: A Unifying View of Some Linear Herbrand Procedures. JACM 19(2), 366–384 (1972)
28. Luckham, D.: Refinement Theorems in Resolution Theory. In: Symposium on Automatic Demonstration, pp. 163–190. Springer, Heidelberg (1970)
29. Nicolas, J.M., Gallaire, H.: Database: Theory vs. Interpretation. In: Gallaire, H., Minker, J. (eds.) Logic and Databases. Plenum, New York (1978)
30. Reiter, R.: Two Results on Ordering for Resolution with Merging and Linear Format. JACM 18(4), 630–646 (1971)
31. Robinson, J.A.: Machine-Oriented Logic Based on the Resolution Principle. JACM 12(1), 23–41 (1965)
32. Robinson, J.: Automatic Deduction with Hyper-resolution. International J. Computer Math. 1(3), 227–234 (1965)
33. Sagonas, K., Swift, T., Warren, D.S.: XSB as an Efficient Deductive Database Engine. In: Proceedings of the ACM SIGMOD International Conference on the Management of Data, pp. 442–453 (1994)
34. Tamaki, H., Sato, T.: OLD Resolution with Tabulation. In: Shapiro, E. (ed.) ICLP 1986. LNCS, vol. 225, pp. 84–98. Springer, Heidelberg (1986)
35. Scott, D.: Outline of a Mathematical Theory of Computation. In: Proc. of the Fourth Annual Princeton Conference on Information Sciences and Systems, pp. 169–176 (1970)
36. Siekmann, J., Woods, J.: History of Computational Logic in the Twentieth Century. Elsevier (to appear)
37. Tekle, K.T., Liu, Y.A.: More Efficient Datalog Queries: Subsumptive Tabling beats Magic Sets. In: Proceedings of the 2011 ACM SIGMOD International Conference on Management of Data, pp. 661–672 (2011)
38. Winograd, T.: Procedures as a Representation for Data in a Computer Program for Understanding Natural Language. MIT AI TR-235 (1971) Also: Understanding Natural Language. Academic Press, New York (1972)
39. Zamov, N.K., Sharonov, V.I.: On a Class of Strategies for the Resolution Method. Zapiski Nauchnykh Seminarov POMI 16, 54–64 (1969)

Integrating Temporal Extensions
of Answer Set Programming*

Felicidad Aguado, Gilberto Pérez, and Concepción Vidal

Department of Computer Science
University of Corunna, Spain
{aguado,gilberto.pvega,concepcion.vidalm}@udc.es

Abstract. In this paper we study the relation between the two main extensions of Answer Set Programming with temporal modal operators: *Temporal Equilibrium Logic* (TEL) and *Temporal Answer Sets* (TAS). On the one hand, TEL is a complete non-monotonic logic that results from the combination of Linear-time Temporal Logic (LTL) with Equilibrium Logic. On the other hand, TAS is based on a richer modal approach, Dynamic LTL (DLTL), whereas its non-monotonic part relies on a reduct-based definition for a particular limited syntax. To integrate both approaches, we propose a Dynamic Linear-time extension of Equilibrium Logic (DTEL) that allows accommodating both TEL and TAS as particular cases. With respect to TEL, DTEL incorporates more expressiveness thanks to the addition of dynamic logic operators, whereas with respect to TAS, DTEL provides a complete non-monotonic semantics applicable to arbitrary theories. In the paper, we identify cases in which both formalisms coincide and explain how this relation can be exploited for adapting existing TEL and TAS computation methods to the general case of DTEL.

1 Introduction

Among the frequent applications of Answer Set Programming (ASP) [1], it is quite usual to find temporal scenarios and reasoning problems dealing with discrete time. Although approaches that combine modal or temporal logic with logic programming are not new [2,3,4,5,6] and even a definition of "temporal answer set" dates back to [7], in the recent years there has been a renewed interest in the topic with a more specific focus on the combination of logic programs under the stable models semantics [8] with some kind of modal temporal logic. Two main approaches have been recently adopted: *Temporal Equilibrium Logic* (TEL) [9,10] and *Temporal Answer Sets* (TAS) [11,12]. The two formalisms share some similarities: in both cases, logic programs are extended with modal temporal operators for expressing properties about *linear time*. In the case of TEL, this is done by combining *Equilibrium Logic* [13] (the best-known logical characterisation of ASP) with *Linear-time Temporal Logic* [14,15] (LTL) one of the simplest and most widely studied modal temporal logics. As a result, TEL constitutes a

---

* The authors would like to thank Pedro Cabalar and Martín Diéguez for their suggestions and comments about the contents of this paper. This research was partially supported by Spanish MEC project TIN2009-14562-C05-04.

P. Cabalar and T.C. Son (Eds.): LPNMR 2013, LNAI 8148, pp. 23–35, 2013.

full non-monotonic temporal logic defined for arbitrary temporal theories in the syntax of LTL. On the other hand, TAS relies on a richer modal approach, *Dynamic Linear-time Temporal Logic* [16] (DLTL), which allows modalities on programs formed with regular expressions, including LTL operators as a smaller fragment[1]. However, the definition of TAS uses a syntactic transformation (analogous to Gelfond & Lifschitz's program reduct [8]) that is only defined for theories with a rather restricted syntax.

A first natural question is whether it is possible to get the advantages from both approaches: the richer temporal semantics provided by DLTL together with a complete logical characterisation applicable to arbitrary DLTL theories. A second important question is whether TEL and TAS can be formally related, especially if we consider the syntactic fragment in which both are defined: TAS-like logic programs limited to LTL operators. In this paper we provide a positive answer to both questions proposing a Dynamic Linear-time extension of Equilibrium Logic (DTEL) that allows accommodating both TEL and TAS as particular cases. In the paper, we identify cases in which both formalisms coincide and explain how this relation can be exploited for adapting existing TEL and TAS computation methods to the general case of DTEL.

The rest of the paper is organised as follows. Section 2 defines the new extension and introduces some basic properties. In the next two sections we explain how to respectively embed TEL and TAS in our new proposal. Section 5 presents a variation of the automata-based method from [17] that allows computing DTEL models of an arbitrary DLTL theory. Finally, Section 6 concludes the paper.

2   Dynamic Temporal Equilibrium Logic

In this section we will define the proposed extension we will call *Dynamic Linear-Time Temporal Equilibrium Logic* (DTEL for short). As happens with Equilibrium Logic and with TEL, DTEL is a non-monotonic formalism whose definition consists of two parts: a monotonic basis and a models selection criterion. The monotonic basis is a temporal extension of the intermediate logic of *Here-and-There* [18] (HT). We call this monotonic logic DLTL_{HT}. As a running example, we will use the well-known Yale Shooting scenario from [19] where, in order to kill a turkey, we must shoot a gun that must be previously loaded.

Let the *alphabet* Σ be a non-empty finite set of *actions*. We denote as Σ^* and Σ^ω to respectively stand for the finite and the non-finite words that can be formed with Σ. We also define $\Sigma^\infty \overset{\text{def}}{=} \Sigma^* \cup \Sigma^\omega$. For any $\sigma \in \Sigma^\omega$, we denote by $pref(\sigma)$ the set of its finite prefixes (including the empty word ϵ).

The set of *programs* (regular expressions) generated by Σ is denoted by $Prg(\Sigma)$ and its syntax is given by the grammar:

$$\pi ::= a \mid \pi_0 + \pi_1 \mid \pi_0; \pi_1 \mid \pi^*$$

with $a \in \Sigma$ and $\pi, \pi_0, \pi_1 \in Prg(\Sigma)$. By abuse of notation, we will sometimes identify a finite prefix $\tau = \sigma_1 \ldots \sigma_n$ as the program $\sigma_1; \ldots; \sigma_n$. For example, in the case of the

---

[1] According to [16], DLTL is strictly more expressive than LTL, as it covers full Monadic Second Order Logic for a linear order, something in which LTL is well-known to fail.

Yale Shooting scenario for the set of actions $\Sigma = \{load, shoot, wait\}$ we could write a program like $\pi = (load; shoot)^*$ representing repetitions of the sequence $load; shoot$.

The mapping $|| \cdot || : Prg(\Sigma) \to 2^{\Sigma^*}$ associates to each program a set of finite words (regular set) as follows:

$$
\begin{aligned}
||a|| &\stackrel{\text{def}}{=} \{a\} \\
||\pi_0 + \pi_1|| &\stackrel{\text{def}}{=} ||\pi_0|| \cup ||\pi_1|| \\
||\pi_0; \pi_1|| &\stackrel{\text{def}}{=} \{\tau_0 \tau_1 \mid \tau_0 \in ||\pi_0|| \text{ and } \tau_1 \in ||\pi_1||\} \\
||\pi^*|| &\stackrel{\text{def}}{=} \bigcup_{i \in \omega} ||\pi^i||
\end{aligned}
$$

where

$$
\begin{aligned}
||\pi^0|| &\stackrel{\text{def}}{=} \{\epsilon\} \\
||\pi^{i+1}|| &\stackrel{\text{def}}{=} \{\tau_0 \tau_1 \mid \tau_0 \in ||\pi|| \text{ and } \tau_1 \in ||\pi^i||\} \text{ for every } i \in \omega.
\end{aligned}
$$

Let $\mathcal{P} = \{p_1, p_2, \ldots\}$ be a countable set of atomic propositions. We denote the set of *simple literals* as $Lit_S \stackrel{\text{def}}{=} \{p, \sim p; p \in \mathcal{P}\}$. The syntax of DLTL$_{\text{HT}}$ coincides with DLTL plus the addition of the strong negation operator '$\sim$.' A *well-formed formula F* is defined as follows:

$$
F ::= \bot \mid p \mid \sim F \mid F_1 \vee F_2 \mid F_1 \wedge F_2 \mid F_1 \to F_2 \mid F_1 \, \mathcal{U}^\pi \, F_2 \mid F_1 \, \mathcal{R}^\pi \, F_2
$$

where p is an atom and F, F_1, F_2 are well-formed formulas. The expression $\neg F$ stands for $F \to \bot$, constant $\top$ corresponds to $\neg\bot$ whereas $F_1 \leftrightarrow F_2$ is an abbreviation for $(F_1 \to F_2) \wedge (F_2 \to F_1)$ as usual. We include the following axiom schemata for strong negation:

- $\sim(\alpha \to \beta) \leftrightarrow \alpha \wedge \sim\beta$
- $\sim(\alpha \vee \beta) \leftrightarrow \sim\alpha \wedge \sim\beta$
- $\sim(\alpha \wedge \beta) \leftrightarrow \sim\alpha \vee \sim\beta$
- $\sim\sim\alpha \leftrightarrow \alpha$

- $\sim \neg\alpha \leftrightarrow \alpha$
- (for atomic α) $\sim\alpha \to \neg\alpha$
- $\sim(\alpha \, \mathcal{U}^\pi \, \beta) \leftrightarrow \sim\alpha \, \mathcal{R}^\pi \, \sim\beta$
- $\sim(\alpha \, \mathcal{R}^\pi \, \beta) \leftrightarrow \sim\alpha \, \mathcal{U}^\pi \, \sim\beta$

These axiom schemata are a direct adaptation of Vorob'ev axiomatisation of strong negation [20,21], with the only addition of the interaction between '$\sim$' and the temporal operators. It is not difficult to see that, by exhaustively applying these equivalences from left to right, we can always rewrite a formula into an equivalent one in *strong negation normal form* (SNNF), that is, guaranteeing that the operator '$\sim$' only affects to atoms in $\mathcal{P}$.

Given an infinite word $\sigma \in \Sigma^\omega$, we define a *valuation function* V as a mapping $V : pref(\sigma) \to 2^{Lits}$ assigning a set of literals to each finite prefix of σ. A valuation function V is *consistent* if, for any $\tau \in pref(\sigma)$, $V(\tau)$ does not contain a pair of opposite literals of the form p and $\sim p$ simultaneously. Given two valuation functions V_1, V_2 (wrt σ), we write $V_1 \leq V_2$ when $V_1(\tau) \subseteq V_2(\tau)$ for every $\tau \in pref(\sigma)$. As usual, if $V_1 \leq V_2$ but $V_1 \neq V_2$, we just write $V_1 < V_2$.

Definition 1 (Temporal Interpretation). *A* (temporal) interpretation *of* $DLTL_{HT}$ *is a tuple* $M = (\sigma, V_h, V_t)$ *where* $\sigma \in \Sigma^\omega$ *and* V_h, V_t *are two valuation functions for* σ *such that* V_t *is consistent and* $V_h \leq V_t$. *We say that the interpretation* M *is* total *when* $V_h = V_t$. $\square$

Given a formula α, a prefix $\tau \in pref(\sigma)$ and a temporal interpretation $M = (\sigma, V_h, V_t)$, we define the satisfaction relation $M, \tau \models \alpha$ inductively as follows:

- $M, \tau \models p$ iff $p \in V_h(\tau)$
- $M, \tau \models \sim p$ iff $\sim p \in V_h(\tau)$
- $M, \tau \models \alpha \vee \beta$ iff $M, \tau \models \alpha$ or $M, \tau \models \beta$
- $M, \tau \models \alpha \wedge \beta$ iff $M, \tau \models \alpha$ and $M, \tau \models \beta$
- $M, \tau \models \alpha \rightarrow \beta$ iff for every $w \in \{h, t\}$,
  $$(\sigma, V_w, V_t), \tau \not\models \alpha \text{ or } (\sigma, V_w, V_t), \tau \models \beta$$

- $M, \tau \models \alpha \, \mathcal{U}^\pi \, \beta$ iff there exists $\tau' \in ||\pi||$ such that $\tau\tau' \in pref(\sigma)$ and $M, \tau\tau' \models \beta$, and for every τ'' such that $\epsilon \leq \tau'' < \tau'$, we have $M, \tau\tau'' \models \alpha$.
- $M, \tau \models \alpha \, \mathcal{R}^\pi \, \beta$ iff for every $\tau' \in ||\pi||$ such that $\tau\tau' \in pref(\sigma)$, it is the case that $M, \tau\tau' \models \beta$ or there exists τ'' such that $\epsilon \leq \tau'' < \tau'$ and $M, \tau\tau'' \models \alpha$.

The meaning of $\alpha \, \mathcal{U}^\pi \, \beta$ is similar to the behaviour of "until" in LTL: in principle, we maintain α until a stopping condition β. The difference here is that this behaviour must be satisfied on some trajectory τ' according to π, $\tau' \in ||\pi||$.

Other usual temporal modalities can be defined as derived operators:

$$\langle \pi \rangle \, \alpha \stackrel{\text{def}}{=} \top \, \mathcal{U}^\pi \, \alpha \qquad\qquad \alpha \, \mathcal{U} \, \beta \stackrel{\text{def}}{=} \alpha \, \mathcal{U}^{\Sigma^*} \, \beta$$

$$[\pi] \, \alpha \stackrel{\text{def}}{=} \bot \, \mathcal{R}^\pi \, \alpha \qquad\qquad \alpha \, \mathcal{R} \, \beta \stackrel{\text{def}}{=} \alpha \, \mathcal{R}^{\Sigma^*} \, \beta$$

$$\bigcirc \alpha \stackrel{\text{def}}{=} \bigvee_{a \in \Sigma} \langle a \rangle \, \alpha \qquad\qquad \square \alpha \stackrel{\text{def}}{=} [\Sigma^*] \, \alpha \quad (\equiv \bot \, \mathcal{R} \, \alpha)$$

$$\diamondsuit \alpha \stackrel{\text{def}}{=} \langle \Sigma^* \rangle \, \alpha \quad (\equiv \top \, \mathcal{U} \, \alpha)$$

For instance, if $M = (\sigma, V_h, V_t)$ is an interpretation and $\tau \in pref(\sigma)$, then $M, \tau \models \langle \pi \rangle \, \alpha$ iff there exists $\tau' \in ||\pi||$ such that $\tau\tau' \in pref(\sigma)$ and $M, \tau\tau' \models \alpha$. Analogously, $M, \tau \models [\pi] \, \alpha$ iff for every $\tau' \in ||\pi||$ such that $\tau\tau' \in pref(\sigma)$, then $M, \tau\tau' \models \alpha$.

Back to our running example, and assuming that we have fluents $\{loaded, alive\}$, the following formulas could be used to capture the whole system behaviour

$$F \vee \sim F \tag{1}$$

$$\square \, (\bigcirc F \leftarrow F \wedge \neg \bigcirc \sim F) \tag{2}$$

$$\square \, (\bigcirc \sim F \leftarrow \sim F \wedge \neg \bigcirc F) \tag{3}$$

$$\square \, [load] \, loaded \tag{4}$$

$$\square \, ([shoot] \sim loaded \leftarrow loaded) \tag{5}$$

$$\square \, ([shoot](\sim alive \vee \neg \sim alive) \leftarrow loaded) \tag{6}$$

where (2) and (3) represent the inertia for all fluent F in $\{loaded, alive\}$, and (6) as assuming that the shoot can fail in killing the turkey.

To illustrate the behaviour of modalities with compound regular expressions consider the formulas:

$$[(load; shoot)^*] \sim loaded \tag{7}$$

$$\langle(load; shoot)^*\rangle \sim alive \tag{8}$$

The first formula, (7) would intuitively mean that the gun will be unloaded after *any* repetition of the sequence of actions *load; shoot* a finite number $n \geq 0$ of times, whereas (8) would mean instead that the turkey will be dead after *some* repetition of the same sequence of actions (remember that we may fail the shoot). It is perhaps important to note that, as modalities deal with a single *linear* future, (7) becomes a kind of conditional formula: if the sequence $(load; shoot)^*$ is not in our current trajectory, (7) is trivially true. Contrarily, (8) will *force* a trajectory $(load; shoot)^*$ from the current situation ending with $\sim alive$.

Given a formula α and an interpretation $M = (\sigma, V_h, V_t)$, we say that M is a *model* of α denoted as $M \models \alpha$, iff $M, \epsilon \models \alpha$.

Lemma 1. *For any formula α, any interpretation $M = (\sigma, V_h, V_t)$ and any $\tau \in pref(\sigma)$, the following two conditions hold:*

1. *If $(\sigma, V_h, V_t), \tau \models \alpha$, then $(\sigma, V_t, V_t), \tau \models \alpha$*
2. *$(\sigma, V_h, V_t), \tau \models \neg\alpha$ iff $(\sigma, V_t, V_t), \tau \not\models \alpha$*   □

The following are DLTL$_{HT}$ valid equivalences:

- $\neg(\alpha \, \mathcal{U}^\pi \, \beta) \leftrightarrow \neg\alpha \, \mathcal{R}^\pi \, \neg\beta$
- $\neg \langle\pi\rangle \alpha \leftrightarrow [\pi]\neg\alpha$

- $\neg(\alpha \, \mathcal{R}^\pi \, \beta) \leftrightarrow \neg\alpha \, \mathcal{U}^\pi \, \neg\beta$
- $\neg[\pi]\alpha \leftrightarrow \langle\pi\rangle \neg\alpha$

Definition 2 (DTEL Model). *A total temporal interpretation $M = (\sigma, V_t, V_t)$ is said to be a temporal equilibrium model (DTEL-model for short) of a formula α in DLTL$_{HT}$ if $M \models \alpha$ and there is no $H < T$ such that $(\sigma, V_h, V_t) \models \alpha$.*

Example 1. Suppose we have a computer that, from time to time, sends a request signal to a server. After a request r the computer stays pending p (during several possible waiting actions w) until it receives a server answer a.   □

This behaviour could be captured by the following DTEL theory:

$$\Box \, [r]p \tag{9}$$

$$\Box \, [a] \sim p \tag{10}$$

$$\sim p \tag{11}$$

$$\Box \, (\bigcirc p \leftarrow p \wedge \neg \bigcirc \sim p) \tag{12}$$

$$\Box \, (\bigcirc \sim p \leftarrow \sim p \wedge \neg \bigcirc p) \tag{13}$$

$$\Box \, ([a]\bot \leftarrow \sim p) \tag{14}$$

$$\Box \, ([r]\bot \leftarrow p) \tag{15}$$

Effect axioms (9) and (10) respectively represent that a request sets p to true whereas an answer sets it back to false. (11) guarantees that p is initially false. Formulas (12), (13) represent the inertia law for fluent p. (14) says that an answer a cannot occur when not pending $\sim p$ and a request r cannot occur when pending p.

Suppose that we get several possible runs from this theory but we want to distinguish between those where the last request is eventually answered or not. We use atom *lost* to flag this situation and include a formula as follows:

$$lost \leftarrow \Diamond(p \wedge \Box[a]\bot) \tag{16}$$

That is, there is a future point in which the computer is pending and from that point on is no answer. If we denote by Γ the theory formed by (9)-(16), the temporal equilibrium models of Γ are captured by the Büchi automaton[2] shown in Figure 1. As we can see, from the initial state `init` we can move to the left "sub-automaton" where infinite words are accepted when each request is followed by an answer, or move to the right "sub-automaton" where *lost* becomes true and a request can be unanswered forever.

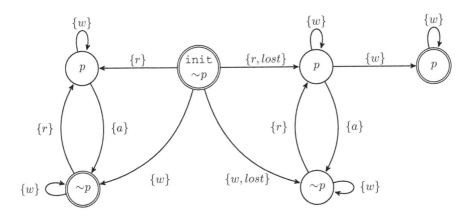

Fig. 1. Büchi automaton for example theory Γ

3   Embedding TEL

Syntactically, TEL is identical to LTL which, in turn, coincides with the DLTL fragment where program modalities do not contain a program π superscript (formally, this means fixing $\pi = \Sigma^*$). In other words, a well-formed formula in TEL follows the syntax:

$$F ::= \bot \mid p \mid \sim F \mid F_1 \vee F_2 \mid F_1 \wedge F_2 \mid F_1 \rightarrow F_2 \mid F_1 \,\mathcal{U}\, F_2 \mid F_1 \,\mathcal{R}\, F_2$$

---

[2] A Büchi automaton is a special kind of automaton that accepts an infinite word when its path visits an accepting state an infinite number of times. In the figure, for readability sake, when a literal is included in *all* outgoing arcs from a state, we show it inside the state.

We keep the same derived operators so that, for instance $\Diamond F \equiv \top\, \mathcal{U} F$ and so on. We also assume the inclusion of Vorob'ev axioms, so that we may assume that $\sim$ is only applied on atoms.

An LTL-interpretation is a sequence of consistent sets of literals $\mathbf{T} = \{T_i\}_{i \geq 0}$. Given two LTL-interpretations $\mathbf{H} = \{H_i\}_{i \geq 0}$ and $\mathbf{T} = \{T_i\}_{i \geq 0}$, we write $\mathbf{H} \subseteq \mathbf{T}$ when $H_i \subseteq T_i$ for all $i \geq 0$. An LTL$_{HT}$-interpretation is a pair $\langle \mathbf{H}, \mathbf{T} \rangle$ of LTL-interpretations such that $\mathbf{H} \subseteq \mathbf{T}$. The satisfaction relation in LTL$_{HT}$ is defined as follows. Given $\mathbf{M} = \langle \mathbf{H}, \mathbf{T} \rangle$

- $\mathbf{M}, i \models L$ iff $L \in H_i$ for any literal L
- $\wedge, \vee, \bot$ as always
- $\mathbf{M}, i \models \alpha \rightarrow \beta$ iff for every $X \in \{\mathbf{H}, \mathbf{T}\}$,

$$X \in \{\mathbf{H}, \mathbf{T}\}, \langle X, \mathbf{T} \rangle, i \not\models \alpha \text{ or } \langle X, \mathbf{T} \rangle, i \models \beta$$

- $\mathbf{M}, i \models \alpha \,\mathcal{U}\, \beta$ iff there exists some $j \geq i$ such that $\mathbf{M}, j \models \beta$, and for every k such that $i \leq k < j$, we have $\mathbf{M}, k \models \alpha$.
- $\mathbf{M}, i \models \alpha \,\mathcal{R}\, \beta$ iff for every $j \geq i$ it is the case that $\mathbf{M}, j \models \beta$ or there exists k such that $i \leq k < j$ and $\mathbf{M}, k \models \alpha$.

Example 2. The unique temporal equilibrium model $\{\mathbf{T}, \mathbf{T}\}$ of $\Box(\neg p \rightarrow \bigcirc p)$ is $T_{2i} = \emptyset$ and $T_{2i+1} = \{p\}$ for any $i \geq 0$. On the other hand, the theory Γ consiting of the formulas:

$$\Box\, (\neg \bigcirc p \rightarrow p) \tag{17}$$

$$\Box\, (\bigcirc p \rightarrow p) \tag{18}$$

has no temporal equilibrium models. The only total model $\langle \mathbf{T}, \mathbf{T} \rangle$ of Γ has the form $T_i = \{p\}$ for any $i \geq 0$. However, it is easy to see that the interpretation $\langle \mathbf{H}, \mathbf{T} \rangle$ with $H_i = \emptyset$ for all $i \geq 0$ is also a THT model, whereas $\mathbf{H} < \mathbf{T}$.

Theorem 1. *Let α be a formula in LTL syntax. Then the LTL$_{HT}$ models of α are in a one-to-one correspondence to (an equivalence class of) the DLTL$_{HT}$ models of α.* $\quad\Box$

Informally speaking, when restricted to LTL syntax, any DLTL$_{HT}$ model (σ, V_h, V_t) can be seen as an LTL$_{HT}$ model where we disregard each prefix $\tau \in pref(\sigma)$ in favour of a simple integer index $i = length(\tau)$. Since for defining temporal equilibrium models in both logics, we impose an analogous models minimisation, we conclude:

Corollary 1. *Let α be a formula in LTL syntax. Then the TEL models of α are in a one-to-one correspondence to (an equivalence class of) the DTEL models of α.* $\quad\Box$

4   Embedding TAS

We recall here some of the contents of [11]. If $a \in \Sigma$ and $l \in Lit_S$ is any simple literal, then $\bigcirc l$ and $[a]l$ are called *temporal fluent literals* (Lit_T). We further define the set $Lit \overset{\text{def}}{=} Lit_S \cup Lit_T \cup \{\bot\}$. An *extended* fluent literal is defined as either t or *not* t with $t \in Lit$.

A *domain description* D over Σ is a tuple $D = (\Pi, \mathcal{C})$ where $\mathcal{C}$ is a set of DLTL formulas called *constraints*, and Π is a set of rules of the forms:

$$\text{Init } t_0 \leftarrow t_1, \ldots, t_m, not\ t_{m+1}, \ldots, not\ t_n \tag{19}$$

$$t_0 \leftarrow t_1, \ldots, t_m, not\ t_{m+1}, \ldots, not\ t_n \tag{20}$$

with the following restrictions:

1. If $t_0 \in Lit_S$, then all $t_i \in Lit_S$ for $i = 1, \ldots, n$.
2. If $t_0 = \bigcirc l$, all the temporal literals in the rule are of the form $\bigcirc l'$
3. If $t_0 = [a]l$, all the temporal literals in the rule are of the form $[a]l'$

As said before, given $\tau = \sigma_1 \ldots \sigma_k$, the expression $[\tau]\alpha$ stands for $[\sigma_1; \ldots; \sigma_k]\alpha$. When $k = 0$, $[\tau]\alpha$ just amounts to α. A *(partial) temporal interpretation* is a pair (σ, S) with $\sigma \in \Sigma^\omega$ and S a set of temporal expressions of the form $[\tau]l$, with $\tau \in pref(\sigma)$ and $l \in Lit_S$, not containing any pair $[\tau]p$ and $[\tau] \sim p$ for any $p \in \mathcal{P}$. Moreover, when $[\tau]p \in S$ iff $[\tau] \sim p \notin S$, then (σ, S) is a *total temporal interpretation*.

The satisfiability of an extended literal $t \in Lit$ with respect to an interpretation (σ, S) and a prefix $\tau \in pref(\sigma)$ is defined as follows, depending on the case:

- $(\sigma, S), \tau \models \top$ and $(\sigma, S), \tau \not\models \bot$
- $(\sigma, S), \tau \models l$ iff $[\tau]l \in S$
- $(\sigma, S), \tau \models \bigcirc l$ iff $[\tau; a]l \in S$ where a is such that $\tau a \in pref(\sigma)$
- $(\sigma, S), \tau \models [a]l$ iff either $\tau a \notin pref(\sigma)$ or $[\tau; a]l \in S$

for any $a \in \Sigma$ and $l \in Lit_S$. The satisfaction of a rule with respect to (σ, S) is as follows:

- $(\sigma, S) \models (19)$ when: if $(\sigma, S), \epsilon \models t_i$ for all $i = 1, \ldots, m$ and $(\sigma, S), \epsilon \not\models t_i$ for all $i = m + 1, \ldots, n$ then $(\sigma, S), \epsilon \models l_0$.
- $(\sigma, S) \models (20)$ when: if for any $\tau \in pref(\sigma)$ such that $(\sigma, S), \tau \models t_i$ for all $i = 1, \ldots, m$ and $(\sigma, S), \tau \not\models t_i$ for all $i = m + 1, \ldots, n$ then $(\sigma, S), \tau \models l_0$.

In order to define the notion of temporal answer set, [11] began considering *positive* programs, that is, those without default negation, i.e., $m = n$ in (19) and (20).

Definition 3 (Temporal Answer Set). (σ, S) *is a* temporal answer set *of Π if S is minimal (with respect to the set inclusion relation) among the S' such that (σ, S') is a partial interpretation satisfying the rules of Π.* $\qquad\square$

For the general case where default negation is allowed, [11] introduced a new kind of rules:

$$[\tau]\,(t_0 \leftarrow t_1, \ldots, t_m) \text{ where } \tau \in pref(\sigma) \tag{21}$$

so that $(\sigma, S) \models (21)$ if, whenever $(\sigma, S), \tau \models t_i$ for $i = 1, \ldots, m$ we have $(\sigma, S), \tau \models t_0$. The *reduct* $\Pi_\tau^{(\sigma, S)}$ of Π relative to (σ, S) and to the prefix $\tau \in pref(\sigma)$ is the set of all rules like (21) such that there exists a rule in Π like (20) satisfying that:

$$(\sigma, S), \tau \not\models t_j \text{ for } j = m + 1, \ldots, n.$$

The program *reduct* is defined as $\Pi^{(\sigma, S)} \stackrel{\text{def}}{=} \bigcup_{\tau \in pref(\sigma)} \Pi_\tau^{(\sigma, S)}$.

Definition 4 (Temporal answer set). (σ, S) *is a* temporal answer set *of* Π *if* (σ, S) *is minimal among the* S' *such that* (σ, S') *is a partial interpretation satisfying the rules of* $\Pi^{(\sigma,S)}$. $\qquad\Box$

Given any partial temporal interpretation (σ, S), we denote by $V_S : pref(\sigma) \to 2^{Lit_S}$ the valuation function defined by

$$V_S(\tau) \stackrel{\text{def}}{=} \{l \in Lit_S \,;\, [\sigma_1; \dots; \sigma_k]l \in S\}$$

for every $\tau = \sigma_1 \dots \sigma_k \in pref(\sigma)$. Conversely, given any valuation function V, we can also define

$$S_V \stackrel{\text{def}}{=} \bigcup_{\tau \in pref(\sigma)} \{[\tau]l \,; l \in V(\tau)\}.$$

establishing a one-to-one correspondence where $S_{V_S} = S$ and $V_{S_V} = V$.

Definition 5 (Extension). (σ, S) *is an* extension *of a domain description* $D = (\Pi, \mathcal{C})$ *if* (σ, S) *is a temporal answer set such that* (σ, V_S, V_S) *satisfies any formula of* $\mathcal{C}$. $\qquad\Box$

For embedding TAS in DLTL$_{HT}$, we have to take into account that the formulas of $\mathcal{C}$ are also formulas of DLTL$_{HT}$ and translate the rules of Π into formulas in DLTL$_{HT}$. For any rule r, we define the formula $\widetilde{r}$ as:

$$\widetilde{r} \stackrel{\text{def}}{=} t_1 \wedge \dots \wedge t_m \wedge \neg t_{m+1} \wedge \dots \wedge \neg t_n \to t_0 \qquad (22)$$

$$\widetilde{r} \stackrel{\text{def}}{=} \Box \, (t_1 \wedge \dots \wedge t_m \wedge \neg t_{m+1} \wedge \dots \wedge \neg t_n \to t_0) \qquad (23)$$

for r of the forms (19) and (20), respectively. Note that for positive programs, $m = n$ and the empty conjunction of negated t_i amounts to $\top$.

Theorem 2. *Take* $D = (\Pi, \mathcal{C})$ *a domain description and* (σ, S) *a temporal interpretation. If we denote by* $\widetilde{\Pi} = \{\widetilde{r} \mid r \in \Pi\}$, *the following assertions are equivalent:*

1. (σ, S) *is a temporal answer set of* Π
2. (σ, V_S, V_S) *is a temporal equilibrium model of* $\widetilde{\Pi}$ $\qquad\Box$

Corollary 2 (Main result). *Take* $D = (\Pi, \mathcal{C})$ *a domain description and* (σ, S) *a temporal interpretation. The following assertions are equivalent:*

1. (σ, S) *is an extension of* D
2. (σ, V_S, V_S) *is a temporal equilibrium model of* $\widetilde{\Pi} \cup \{\neg\neg\alpha \mid \alpha \in \mathcal{C}\}$ $\qquad\Box$

Proof. It follows from the previous result and the fact that, for any $S' \subseteq S$, $\tau \in pref(\sigma)$ and $\alpha \in \mathcal{C}$, the following can be easily checked:

$$(\sigma, V_{S'}, V_S), \tau \models \neg\neg\alpha \text{ iff } (\sigma, V_S, V_S), \tau \models \alpha. \qquad\Box$$

5 Computation of Temporal Equilibrium Models

While Theorem 1 proves that DTEL is a proper extension of TEL, Corollary 2 guarantees that the semantics assigned to TAS programs also coincides with temporal answer sets. Several interesting results follow from this. First, it is obvious that, when we consider the syntactic intersection between TEL and TAS (i.e., TAS programs that exclusively use LTL operators) we have obtained that both semantics are equivalent. In this way, tools for computing temporal equilibrium models [22,23] can be used as tools for TAS in this syntactic fragment. Second, DTEL can be used as a *common logical framework* that subsumes both TEL and TAS into a more expressive formalism without syntactic limitations. Furthermore, given its close relation to TEL, it is possible to adapt some of the methods already available for the latter. In particular, in [17] it was shown how TEL models of an LTL formula α could be computed by performing the following operations:

1. Build automaton $\mathcal{A}_1$ capturing the total $\mathrm{LTL_{HT}}$-models $\langle \mathbf{T}, \mathbf{T} \rangle$ of α. As any total model $\langle \mathbf{T}, \mathbf{T} \rangle$ corresponds to an LTL model $\mathbf{T}$, this step is simply done by applying a standard algorithm for automata construction in LTL [24] (see [17] for further details).
2. Build an automaton $\mathcal{A}_2$ capturing non-total $\mathrm{LTL_{HT}}$-models of α. This time, an encoding of $\mathrm{LTL_{HT}}$ into LTL is previously done by translating α into a new formula α' that uses additional auxiliary atoms. Then, $\mathcal{A}_2$ is built from α' using again [24].
3. Filter auxiliary atoms in $\mathcal{A}_2$ to get $\mathcal{A}_3$. This automaton captures those $\mathbf{T}$ that are total, but not in equilibrium, that is, there is some $\mathbf{H} < \mathbf{T}$ with $\langle \mathbf{H}, \mathbf{T} \rangle \models \Gamma$.
4. Get an automaton from the intersection of $\mathcal{A}_1$ with the negation of $\mathcal{A}_3$ leading to total models that are in equilibrium.

Steps 3 and 4 in this method are actually independent on the modal extension we consider, as they actually perform transformations on Büchi automata. Thus, in order to apply this method for DTEL, the following changes in steps 1 and 2 are required. First, we replace in both steps the automata construction method for LTL from [24] by the methods described in [16] or [25] for the case of DLTL. Second, it only remains to obtain a translation for step 2 adapted to the $\mathrm{DLTL_{HT}}$ case. In other words, for any formula α, obtain α' with an extended signature such that DLTL models for α' correspond to non-total $\mathrm{DLTL_{HT}}$ models of α. Using this, we can assert that the results of [17] can be generalized to $\mathrm{DLTL_{HT}}$ in a straightforward manner.

We describe next this translation and its correctness in a formal way. Suppose α is a $\mathrm{DLTL_{HT}}$ formula over the finite signature $\mathcal{P}$. The DLTL formula α' will be built over an extended signature $\mathcal{P}' = Lit_S \cup \{l' \mid l \in Lit_S\}$. This means that negative literals of the form $\sim p$ are actually considered[3] as atoms "$\sim p$". To put an example, given $\mathcal{P} = \{p\}$ we would get the four atoms $\mathcal{P}' = \{p, (\sim p), p', (\sim p)'\}$. Intuitively, primed literals will represent truth at V_h whereas unprimed ones correspond to V_t. In this way, we can establish a correspondence between $\mathrm{DLTL_{HT}}$ and DLTL interpretations as follows. Given a temporal interpretation $M = (\sigma, V_h, V_t)$ for signature $\mathcal{P}$ we define the DLTL interpretation $H' = (\sigma, U)$ for signature $\mathcal{P}'$ such that for any $\tau \in pref(\sigma)$:

---

[3] Note that the strong negation operator was not originally allowed in DLTL [16].

$$U(\tau) = V_t(\tau) \cup \{l' \mid l \in V_h(\tau)\}.$$

In that case, we write $H' \approx M$.

Lemma 2. *The following one-to-one correspondence can be established:*

(I) *For every DLTL$_{HT}$ interpretation $M = (\sigma, V_h, V_t)$ over signature $\mathcal{P}$, there is a unique DLTL model H' such that $H' \approx M$.*

(II) *For every DLTL model $H' = (\sigma, U)$ such that $H', \epsilon \models \mathcal{G}$ where $\mathcal{G}$ is the formula defined by:*

$$\mathcal{G} \stackrel{\text{def}}{=} \bigwedge_{p \in \mathcal{P}} \Box(p \wedge \sim p \to \bot) \wedge \bigwedge_{l \in Lit_S} \Box(l' \to l)$$

there is a unique DLTL$_{HT}$ model $M = (\sigma, V_h, V_t)$ such that $H' \approx M$. □

Intuitively, $\mathcal{G}$ guarantees that H' corresponds to a well-formed DLTL$_{HT}$ interpretation: the expressions $\Box(p \wedge \sim p \to \bot)$ in $\mathcal{G}$ require consistency regarding strong negation (that is, we cannot have p and $\sim p$ simultaneously) whereas implications $\Box(l' \to l)$ are used to guarantee that $\mathbf{H} \subseteq \mathbf{T}$. In order to complete this model-theoretical correspondence, let us define t as the following translation between formulae:

- $t(\bot) \stackrel{\text{def}}{=} \bot$, $t(l) \stackrel{\text{def}}{=} l'$ for any $l \in Lit_S$
- $t(\alpha \oplus \beta) \stackrel{\text{def}}{=} t(\alpha) \oplus t(\beta)$ with $\oplus \in \{\wedge, \vee, \mathcal{U}^\pi, \mathcal{R}^\pi\}$
- $t(\alpha \to \beta) \stackrel{\text{def}}{=} (\alpha \to \beta) \wedge (t(\alpha) \to t(\beta))$.

The following result can be easily proved by structural induction.

Lemma 3. *Let α be a DLTL$_{HT}$ formula built over the literals in Lit_S and let $M = (\sigma, V_h, V_t)$ and $H' = (\sigma, U)$ be temporal interpretations such that $H' \approx M$. For any $i \geq 0$, we have $H', i \models t(\beta)$ iff $M, i \models \beta$ for every subformula β of α.* □

If $\mathcal{A}_1$ is the Büchi automaton obtained by following a similar construction to [16], the language $L(\mathcal{A}_1)$ accepted by $\mathcal{A}_1$ can be viewed as the set of total DLTL$_{HT}$ models of α.

Consider now the following formula obtained from α:

$$\alpha' \stackrel{\text{def}}{=} \mathcal{G} \wedge t(\alpha) \wedge \bigvee_{l \in Lit_S} \Diamond(l \wedge \neg l')$$

α' characterises the non-total DLTL$_{HT}$ models of the formula α. While $\mathcal{G} \wedge t(\alpha)$ alone would charaterise all DLTL$_{HT}$ models, the disjunction of expressions $\Diamond(l \wedge \neg l')$ guarantees that, at some time point, H_i is a strict subset of T_i, $H_i \subset T_i$. In other words, for any $H' = (\sigma, U)$ such that $H' \approx M$ with $M = (\sigma, V_h, V_t)$, we have that $H', \epsilon \models \bigvee_{l \in Lit_S} \Diamond(l \wedge \neg l')$ if, and only if, there exists $l \in Lit_S$ with $l \in V_t(\tau) \setminus V_h(\tau)$ for some $\tau \in pref(\sigma)$.

Lemma 4. *The set of DLTL models for the formula α' corresponds to the set of non-total DLTL$_{HT}$ models for the temporal formula α.* □

Let us denote by Σ' the alphabet $\Sigma' = 2^{\mathcal{P}'}$ and let $h : \Sigma' \to \Sigma$ be a map (renaming) between the two finite alphabets such that $h(A) = A \cap Lit_S$. The map h can be naturally extended as an homomorphism between finite words, infinite words and as a map between languages. Similary, given a Büchi automaton $\mathcal{A}_2 = (\Sigma', Q, Q_0, \delta, F)$, we write $h(\mathcal{A}_2)$ to denote the Büchi automaton $(\Sigma, Q, Q_0, \delta', F)$ such that if $q \xrightarrow{A} q' \in \delta' \overset{\text{def}}{\Leftrightarrow}$ there is $B \in \Sigma'$ satisfying that $q \xrightarrow{B} q' \in \delta$ and $h(B) = A$. Obviously, $L(h(\mathcal{A}_2)) = h(L(\mathcal{A}_2))$.

Proposition 1. α *has a DTEL model iff* $L(\mathcal{A}_1) \cap (\Sigma^\omega \setminus L(h(\mathcal{A}_2))) \neq \emptyset.$ $\square$

In consequence, the set of DTEL models for a given α forms an ω-regular language.

Proposition 2. *For each DLTL$_{HT}$ formula* α*, one can effectively build a Büchi automaton that accepts exactly the DTEL models for* α*.* $\square$

6   Conclusions

In this paper we have introduced a new temporal extension of Answer Set Programming, called *Dynamic Linear-time Temporal Equilibrium Logic* (DTEL) that covers the two existing temporal modal extensions as particular cases: Temporal Equilibrium Logic (TEL) and Temporal Answer Sets (TAS). This provides a common, more expressive logical framework that allows for studying the formal relations between TEL and TAS and, at the same time, opens the possibility of adapting existing computation methods for the general case.

For future work we plan to use DTEL as a logical framework to determine when theories in TAS can be reduced to TEL allowing the possible addition of auxiliary atoms. We also plan to implement the automata-based method for DTEL described in this paper as an additional option of the tool ABSTEM [23] that currently covers the case of TEL.

References

1. Brewka, G., Eiter, T., Truszczynski, M.: Answer set programming at a glance. Communications of the ACM 54, 92–103 (2011)
2. del Cerro, L.F.: MOLOG: A system that extends Prolog with modal logic. New Generation Computing 4, 35–50 (1986)
3. Gabbay, D.: Modal and temporal logic programming. In: Galton, A. (ed.) Temporal Logics and their Applications, pp. 197–237. Academic Press (1987)
4. Abadi, M., Manna, Z.: Temporal logic programming. Journal of Symbolic Computation 8, 277–295 (1989)
5. Baudinet, M.: A simple proof of the completeness of temporal logic programming. In: del Cerro, L.F., Penttonen, M. (eds.) Intensional Logics for Programming, pp. 51–83. Clarendon Press, Oxford (1992)
6. Baldoni, M., Giordano, L., Martelli, A.: A framework for a modal logic programming. In: Porc. of the Joint International Conference and Symposium on Logic Programming, pp. 52–66 (1996)

7. Cabalar, P.: Temporal answer sets. In: Proc. of the 1999 Joint Conference on Declarative Programming, APPIA-GULP-PRODE 1999, pp. 351–366 (1999)

8. Gelfond, M., Lifschitz, V.: The stable model semantics for logic programming. In: Kowalski, R.A., Bowen, K.A. (eds.) Logic Programming: Proc. of the Fifth International Conference and Symposium, vol. 2, pp. 1070–1080. MIT Press, Cambridge (1988)

9. Cabalar, P., Pérez Vega, G.: Temporal equilibrium logic: A first approach. In: Moreno Díaz, R., Pichler, F., Quesada Arencibia, A. (eds.) EUROCAST 2007. LNCS, vol. 4739, pp. 241–248. Springer, Heidelberg (2007)

10. Aguado, F., Cabalar, P., Diéguez, M., Pérez, G., Vidal, C.: Temporal equilibrium logic: a survey. Journal of Applied Non-Classical Logics (to appear, 2013)

11. Giordano, L., Martelli, A., Dupré, D.T.: Reasoning about actions with temporal answer sets. TPLP 13, 201–225 (2013)

12. Giordano, L., Martelli, A., Dupré, D.T.: Verification of action theories in ASP: A complete bounded model checking approach. In: Lisi, F.A. (ed.) Proceedings of the 9th Italian Convention on Computational Logic, CILC 2012. CEUR Workshop Proceedings, vol. 857, pp. 176–190. CEUR-WS.org (2012)

13. Pearce, D.: A new logical characterisation of stable models and answer sets. In: Dix, J., Przymusinski, T.C., Moniz Pereira, L. (eds.) NMELP 1996. LNCS (LNAI), vol. 1216, pp. 57–70. Springer, Heidelberg (1997)

14. Kamp, J.A.: Tense Logic and the Theory of Linear Order. PhD thesis, University of California at Los Angeles (1968)

15. Manna, Z., Pnueli, A.: The Temporal Logic of Reactive and Concurrent Systems: Specification. Springer (1991)

16. Henriksen, J.G., Thiagarajan, P.S.: Dynamic linear time temporal logic. Annals of Pure and Applied Logic 96, 187–207 (1999)

17. Cabalar, P., Demri, S.: Automata-based computation of temporal equilibrium models. In: Vidal, G. (ed.) LOPSTR 2011. LNCS, vol. 7225, pp. 57–72. Springer, Heidelberg (2012)

18. Heyting, A.: Die formalen Regeln der intuitionistischen Logik. Sitzungsberichte der Preussischen Akademie der Wissenschaften, Physikalisch-mathematische Klasse, 42–56 (1930)

19. Hanks, S., McDermott, D.: Nonmonotonic logic and temporal projection. Artificial Intelligence 33, 379–412 (1987)

20. Vorob'ev, N.N.: A constructive propositional calculus with strong negation. Doklady Akademii Nauk SSR 85, 465–468 (1952) (in Russian)

21. Vorob'ev, N.N.: The problem of deducibility in constructive propositional calculus with strong negation. Doklady Akademii Nauk SSR 85, 689–692 (1952) (in Russian)

22. Cabalar, P., Diéguez, M.: STeLP – A tool for temporal answer set programming. In: Delgrande, J.P., Faber, W. (eds.) LPNMR 2011. LNCS, vol. 6645, pp. 370–375. Springer, Heidelberg (2011)

23. Cabalar, P., Diéguez, M.: ABSTEM – an automata-based solver for temporal equilibrium models (unpublished draft, 2013)

24. Vardi, M., Wolper, P.: Reasoning about infinite computations. Information and Computation 115, 1–37 (1994)

25. Giordano, L., Martelli, A.: Tableau-based automata construction for dynamic linear time temporal logic. Annals of Mathematics and Artificial Intelligence 46, 289–315 (2006)

Forgetting under the Well-Founded Semantics

José Júlio Alferes[1], Matthias Knorr[1], and Kewen Wang[2]

[1] CENTRIA & Departamento de Informática, Faculdade Ciências e Tecnologia,
Universidade Nova de Lisboa, 2829-516 Caparica, Portugal
[2] School of Information and Communication Technology, Griffith University,
Brisbane QLD 4111, Australia

Abstract. In this paper, we develop a notion of forgetting for normal logic programs under the well-founded semantics. We show that a number of desirable properties are satisfied by our approach. Three different algorithms are presented that maintain the computational complexity of the well-founded semantics, while partly keeping its syntactic structure.

1 Introduction

Forgetting has drawn considerable attention in knowledge representation and reasoning. This is witnessed by the fact that forgetting has been introduced in many monotonic and nonmonotonic logics [1,5,9,10,11,12,16,18,19], and in particular, in logic programming [6,15,17].

A potential drawback, common to all these three approaches, is the computational (data) complexity of the answer set semantics, which is **coNP**, while the other common semantics for logic programs, the well-founded semantics (WFS), is in **P**, which may be preferable in applications with huge amounts of data. However, to the best of our knowledge, forgetting under the well-founded semantics has not been considered so far. Therefore, in this paper, we develop a notion of forgetting for normal logic programs under the well-founded semantics. We show that forgetting under the well-founded semantics satisfies the properties in [6]. In particular, our approach approximates semantic forgetting of [6] for normal logic programs under answer set semantics as well as forgetting in classical logic, in the sense that whatever is derivable from a logic program under the well-founded semantics after applying our notion of forgetting, is also derivable in each answer set and classical model after applying semantic and classical forgetting to the logic program and its classical representation, respectively. We also present three different algorithms that maintain the favorable computational complexity of the well-founded semantics when compared to computing answer sets.

2 Preliminaries

A *normal logic program* P, or simply *logic program*, is a finite set of rules r of the form $h \leftarrow a_1, \dots, a_n, not\, b_1, \dots, not\, b_m$ where h, a_i, and b_j, with $1 \le i \le n$ and $1 \le j \le m$, are all propositional atoms over a given alphabet Σ.

P. Cabalar and T.C. Son (Eds.): LPNMR 2013, LNAI 8148, pp. 36–41, 2013.

Given a rule r, we distinguish the *head* of r as $head(r) = h$, and the *body* of r, $body(r) = body^+(r) \cup not\ body^-(r)$, where $body^+(r) = \{a_1, \ldots, a_n\}$, $body^-(r) = \{b_1, \ldots, b_m\}$ and, for a set S of atoms, $not\ S = \{not\ q \mid q \in S\}$. Rule r is *positive* if $body^-(r) = \emptyset$, *negative* if $body^+(r) = \emptyset$, and a *fact* if $body(r) = \emptyset$.

Given a logic program P, B_P is the set of all atoms appearing in P, and $Lit_P = B_P \cup not\ B_P$. Also, $heads(P)$ denotes the set $\{p \mid p = head(r) \wedge r \in P\}$.

A *three-valued interpretation* $I = I^+ \cup not\ I^-$ with $I^+, I^- \subseteq B_P$ and $I^+ \cap I^- = \emptyset$. Informally, I^+ and I^- contain the atoms that are true and false in I, respectively. Any atom appearing neither in I^+ nor in I^- is undefined.

We recall the definition of the well-founded semantics based on the alternating fixpoint [7]. Given a logic program P and $S \subseteq B_P$, we define $\Gamma_P(S) = least(P^S)$ where $P^S = \{head(r) \leftarrow body^+(r) \mid r \in P, body^-(r) \cap S = \emptyset\}$ and $least(P^S)$ is the least model of the positive logic program P^S. The square of Γ_P, Γ_P^2, is a monotonic operator and thus has both a least fixpoint, $lfp(\Gamma_P^2)$, and a greatest fixpoint $gfp(\Gamma_P^2)$. We obtain the well-founded model $WFM(P)$ of a normal logic program P as $WFM(P) = lfp(\Gamma_P^2) \cup not\ (B_P \setminus gfp(\Gamma_P^2))$.

Two programs P and P' are *equivalent (under WFS)*, denoted by $P \equiv_{wf} P'$, iff $WFM(P) = WFM(P')$. Finally, the inference relation under the WFS is defined for any literal $q \in Lit(P)$ as follows: $P \models_{wf} q$ iff $q \in WFM(P)$.

3   Forgetting under the Well-Founded Semantics

When defining forgetting of an atom p in a given logic program P, we want to obtain a new logic program P' such that it does not contain any occurrence of p or its default negation $not\ p$. Additionally, we want to ensure that only the derivation for p (and $not\ p$) is affected, keeping P' and P equivalent w.r.t. all derivable literals excluding p (and $not\ p$). We want to achieve this based on the semantics rather than the syntax and ground it in forgetting in classical logic.

So, we semantically define the result of forgetting under the WFS by determining the well-founded model, and then providing a logic program that excludes p syntactically, and whose well-founded model excludes (only) p semantically.

Definition 1. *Let P be a logic program and p an atom. The result of* forgetting *about p in P, denoted* forget(P, p), *is a logic program P' such that the following two conditions are satisfied:*

(1) $B_{P'} \subseteq B_P \setminus \{p\}$, i.e., p does not occur in P', and
(2) $WFM(P') = WFM(P) \setminus (\{p\} \cup \{not\ p\})$

This definition obviously does not introduce new symbols (cf. (F2) in [6]). In the rest of this section, we assume P, P' logic programs and p an atom, and show a number of desirable properties. The first one corresponds to (F3) in [6].

Proposition 2. *For any $l \in Lit \setminus (\{p\} \cup \{not\ p\})$,* forget$(P, p) \models_{wf} l$ *iff $P \models_{wf} l$.*

Our definition of forgetting also implies that there are syntactically different logic programs that correspond to forget(P, p). However, as we show next, all

Algorithm $\mathsf{forget}_1(P,p)$

Input: Normal logic program P and an atom p in P.

Output: A normal logic program P' representing $\mathsf{forget}(P,p)$.

Method:

Step 1. Compute the well-founded model $WFM(P)$ of P.
Step 2. Let M be the three-valued interpretation obtained from $WFM(P)$ by removing p and $not\,p$. Construct a new logic program with $B_{P'} = B_P \setminus \{p\}$ whose well-founded model is exactly M:
$P' = \{a \leftarrow .\mid a \in M^+\} \cup \{a \leftarrow not\,a.\mid a \in B_{P'} \setminus (M^+ \cup M^-)\}$.
Step 3. Output P' as $\mathsf{forget}(P,p)$.

Fig. 1. Algorithm $\mathsf{forget}_1(P,p)$

results of forgetting about p in P are equivalent w.r.t. the well-founded semantics. So, we simply use $\mathsf{forget}(P,p)$ as a generic notation representing any syntactic variant of all semantically equivalent results of forgetting about p in P.

Proposition 3. *If P' and P'' are two results of $\mathsf{forget}(P,p)$, then $P' \equiv_{wf} P''$.*

Forgetting also preserves equivalence on $\equiv_{wf}$ (cf. (F4) in [6]).

Proposition 4. *If $P \equiv_{wf} P'$, then $\mathsf{forget}(P,p) \equiv_{wf} \mathsf{forget}(P',p)$.*

However, our definition of forgetting preserves neither strong nor uniform equivalence. Intuitively, the reason is that Def. 1 only specifies the change on the semantics but not the precise syntactic form of the resulting program.

We may also generalize the definition of forgetting to a set of atoms S in the obvious way and show that the elements of the set can be forgotten one-by-one.

Proposition 5. *Let P be a logic program and $S = \{q_1, \ldots, q_n\}$ a set of atoms. Then $\mathsf{forget}(P,S) \equiv_{wf} \mathsf{forget}(\mathsf{forget}(P,q_1), \ldots, q_n)$.*

We show that our notion of forgetting is faithful w.r.t. semantic forgetting in ASP [6] as follows, which also links to classical forgetting (cf. (F1) in [6]).

Theorem 6. *Let P be a logic program and p, q atoms.*

1. *If $q \in WFM(\mathsf{forget}(P,p))$, then $q \in M$ for all $M \in \mathcal{AS}(\mathsf{forget}_{ASP}(P,p))$.*
2. *If $not\,q \in WFM(\mathsf{forget}(P,p))$, then $q \notin M$ for all $M \in \mathcal{AS}(\mathsf{forget}_{ASP}(P,p))$.*

4   Computation of Forgetting

4.1   Naïve Semantics-Based Algorithm

Def. 1 naturally leads to an algorithm for computing the result of forgetting about p in a given logic program P: compute the well-founded model M of P and construct a logic program from scratch corresponding to $WFM(\mathsf{forget}(P,p))$. This idea is captured in Algorithm $\mathsf{forget}_1(P,p)$ shown in Fig. 1.

Algorithm forget$_2(P,p)$

Input: Normal logic program P and an atom p in P.

Output: A normal logic program P' representing forget(P,p).

Method:

Step 1. Query for the truth value of p in $WFM(P)$ of P (e.g., using XSB).

Step 2. Remove all rules whose head is p. Moreover, given the obtained truth value of p in $WFM(P)$, execute one of the three cases:

  t: Remove all rules that contain *not p* in the body, and remove p from all the remaining rule bodies.
  u: Substitute p and *not p* in each body of a rule r in P by *not head*(r).
  f: Remove all rules that contain p in the body, and remove *not p* from all the remaining rule bodies.

Step 3. Output the result P' as forget(P,p).

Fig. 2. Algorithm forget$_2(P,p)$

4.2 Query-Based Algorithm

Algorithm forget$_1(P,p)$ has two shortcomings. First, the syntactical structure of the original logic program is completely lost, which is not desirable if the rules are subject to later update or change: the author would be forced to begin from scratch, since the originally intended connections in the rules were lost in the process. Second, the computation is not particularly efficient, e.g., if we consider a huge number of rules from which we want to forget one atom p only.

In the following, we tackle the shortcomings of forget$_1(P,p)$ based on the fact that the WFS is relevant, in the sense that it allows us to query for one atom in a top-down manner without having to compute the entire model.[1] This means that we only consider a limited number of rules in which the query/goal or one of its subsequent subgoals appear. Once the truth value of p is determined, we only make minimal changes to accommodate the forgetting of p: if p is true (resp. false), then body atoms (resp. entire rules) are removed appropriately; if p is undefined, then all occurrences of p (and *not p*) are substituted by the default negation of the rule head, thus ensuring that the rule head will be undefined, unless it is true because of another rule in P whose body is true in $WFM(P)$. The resulting algorithm forget$_2(P,p)$ is shown in Fig. 2.

4.3 Forgetting as Program Transformations

What if we could actually avoid computing the well-founded-model at all? We investigate how to compute forget(P,p) using syntactic program transformations instead, thereby handling (F5) and completing the match to the criteria in [6].

---

[1] See, e.g., XSB (http://xsb.sourceforge.net) for an implementation.

Algorithm forget$_3(P, p)$

Input: Normal logic program P and an atom p in P.

Output: A normal logic program P' representing forget(P, p).

Method:

Step 1. Compute $\hat{P}$ by exhaustively applying the transformation rules in $\mapsto_X$ to P.

Step 2. If neither $p \leftarrow . \in \hat{P}$ nor $p \notin heads(\hat{P})$, then substitute p and $not\, p$ in each body of a rule r in $\hat{P}$ by $not\, head(r)$. After that, remove all rules whose head is p.

Step 3. Output the result P' as forget(P, p).

Fig. 3. Algorithm forget$_3(P, p)$

The basic idea builds on a set of program transformations $\mapsto_X$ [3], which is a refinement of [2] for the WFS, avoiding the potential exponential size of the resulting program in [2] yielding the program remainder $\hat{P}$. It is shown in [3] that $\mapsto_X$ is always terminating and confluent and that the remainder resulting from applying these syntactic transformations to P relates to the well-founded model $WFM(P)$ in the following way: $p \in WFM(P)$ iff $p \leftarrow . \in \hat{P}$ and $not\, p \in WFM(P)$ iff $p \notin heads(\hat{P})$. We can use this to create the algorithm forget$_3(P, p)$ shown in Fig. 3 which syntactically computes the result of forget(P, p).

Theorem 7. *Given logic program P and atom p, forget$_x(P, p)$, $1 \leq x \leq 3$, computes a correct result of forget(P, p), terminates, and computing P' is in* **P***.*

5   Conclusions

We have developed a notion of semantic forgetting under the well-founded semantics and presented three different algorithms for computing the result of such forgetting, and in each case the computational complexity is in **P**.

In terms of future work, we intend to pursue different lines of investigation. First, we may consider a notion of forgetting that also preserves strong equivalence for different programs, similar to [15] for the answer set semantics, possibly based on HT^2 [4] or adapting work on updates using SE-models [13,14]. An important issue then is whether the result is again expressible as a normal logic program. Second, since forgetting has been considered for description logics (DLs), we may also consider forgetting in formalisms that combine DLs and non-monotonic logic programming rules under WFS, such as [8].

Acknowledgments. J. Alferes and M. Knorr were partially supported by FCT (Fundação para a Ciência e a Tecnologia) under project "ERRO – Efficient Reasoning with Rules and Ontologies" (PTDC/EIA-CCO/121823/2010), and M. Knorr also by FCT Grant SFRH/BPD/86970/2012. K. Wang was partially supported by Australian Research Council under grants DP110101042 and DP1093652.

References

1. Antoniou, G., Eiter, T., Wang, K.: Forgetting for defeasible logic. In: Bjørner, N., Voronkov, A. (eds.) LPAR-18 2012. LNCS, vol. 7180, pp. 77–91. Springer, Heidelberg (2012)
2. Brass, S., Dix, J.: Semantics of disjunctive logic programs based on partial evaluation. J. Log. Program. 38(3), 167–312 (1999)
3. Brass, S., Dix, J., Freitag, B., Zukowski, U.: Transformation-based bottom-up computation of the well-founded model. TPLP 1(5), 497–538 (2001)
4. Cabalar, P., Odintsov, S.P., Pearce, D.: Logical foundations of well-founded semantics. In: Doherty, P., Mylopoulos, J., Welty, C.A. (eds.) KR, pp. 25–35. AAAI Press (2006)
5. van Ditmarsch, H.P., Herzig, A., Lang, J., Marquis, P.: Introspective forgetting. In: Wobcke, W., Zhang, M. (eds.) AI 2008. LNCS (LNAI), vol. 5360, pp. 18–29. Springer, Heidelberg (2008)
6. Eiter, T., Wang, K.: Semantic forgetting in answer set programming. Artif. Intell. 172(14), 1644–1672 (2008)
7. Gelder, A.V.: The alternating fixpoint of logic programs with negation. J. Comput. Syst. Sci. 47(1), 185–221 (1993)
8. Knorr, M., Alferes, J.J., Hitzler, P.: Local closed world reasoning with description logics under the well-founded semantics. Artif. Intell. 175(9-10), 1528–1554 (2011)
9. Kontchakov, R., Wolter, F., Zakharyaschev, M.: Logic-based ontology comparison and module extraction, with an application to DL-Lite. Artif. Intell. 174(15), 1093–1141 (2010)
10. Lang, J., Liberatore, P., Marquis, P.: Propositional independence: Formula-variable independence and forgetting. J. Artif. Intell. Res. (JAIR) 18, 391–443 (2003)
11. Lin, F., Reiter, R.: Forget it! In: Proceedings of the AAAI Fall Symposium on Relevance, pp. 154–159 (1994)
12. Lutz, C., Wolter, F.: Foundations for uniform interpolation and forgetting in expressive description logics. In: Walsh, T. (ed.) IJCAI, pp. 989–995. IJCAI/AAAI (2011)
13. Slota, M., Leite, J.: On semantic update operators for answer-set programs. In: Coelho, H., Studer, R., Wooldridge, M. (eds.) ECAI. Frontiers in Artificial Intelligence and Applications, vol. 215, pp. 957–962. IOS Press (2010)
14. Slota, M., Leite, J.: Robust equivalence models for semantic updates of answer-set programs. In: Brewka, G., Eiter, T., McIlraith, S.A. (eds.) KR, pp. 158–168. AAAI Press (2012)
15. Wang, Y., Zhang, Y., Zhou, Y., Zhang, M.: Forgetting in logic programs under strong equivalence. In: Brewka, G., Eiter, T., McIlraith, S.A. (eds.) KR, pp. 643–647. AAAI Press (2012)
16. Wang, Z., Wang, K., Topor, R.W., Pan, J.Z.: Forgetting for knowledge bases in DL-Lite. Ann. Math. Artif. Intell. 58(1-2), 117–151 (2010)
17. Zhang, Y., Foo, N.Y., Wang, K.: Solving logic program conflict through strong and weak forgettings. In: Kaelbling, L.P., Saffiotti, A. (eds.) IJCAI, pp. 627–634. Professional Book Center (2005)
18. Zhang, Y., Zhou, Y.: Knowledge forgetting: Properties and applications. Artif. Intell. 173(16-17), 1525–1537 (2009)
19. Zhou, Y., Zhang, Y.: Bounded forgetting. In: Burgard, W., Roth, D. (eds.) AAAI, pp. 280–285. AAAI Press (2011)

The Fourth Answer Set Programming Competition: Preliminary Report[*]

Mario Alviano[1], Francesco Calimeri[1], Günther Charwat[2], Minh Dao-Tran[2],
Carmine Dodaro[1], Giovambattista Ianni[1], Thomas Krennwallner[2],
Martin Kronegger[2], Johannes Oetsch[2], Andreas Pfandler[2], Jörg Pührer[2],
Christoph Redl[2], Francesco Ricca[1], Patrik Schneider[2], Martin Schwengerer[2],
Lara Katharina Spendier[3], Johannes Peter Wallner[2], and Guohui Xiao[2]

[1] Dipartimento di Matematica e Informatica, Università della Calabria, Italy
[2] Institute of Information Systems, Vienna University of Technology, Austria
[3] Institute of Computer Languages, Vienna University of Technology, Austria

Abstract. Answer Set Programming is a well-established paradigm of declarative programming in close relationship with other declarative formalisms such as SAT Modulo Theories, Constraint Handling Rules, PDDL and many others. Since its first informal editions, ASP systems are compared in the nowadays customary ASP Competition. The fourth ASP Competition, held in 2012/2013, is the sequel to previous editions and it was jointly organized by University of Calabria (Italy) and the Vienna University of Technology (Austria). Participants competed on a selected collection of benchmark problems, taken from a variety of research areas and real world applications. The Competition featured two tracks: the Model& Solve Track, held on an open problem encoding, on an open language basis, and open to any kind of system based on a declarative specification paradigm; and the System Track, held on the basis of fixed, public problem encodings, written in a standard ASP language.

1 Introduction

Answer Set Programming is a declarative approach to knowledge representation and programming proposed in the area of nonmonotonic reasoning and logic programming [9, 11, 23–25, 35, 36, 43, 46]. Among the advantages of ASP are its declarative nature combined with a comparatively high expressive power [19, 42]. After pioneering work [10, 42, 49, 50], several systems supporting ASP and its variants are born from the initial offspring [2, 3, 16, 18, 31, 33, 37, 39–42, 44, 45, 47, 49, 52].

Since the first informal editions (Dagstuhl 2002 and 2005), ASP systems are compared in the nowadays customary ASP Competition series [20, 16, 34], which reached now its fourth official edition. The Fourth ASP Competition featured two tracks: the Model& Solve Track, held on an open problem encoding, open language basis, and

[*] This research is supported by the Austrian Science Fund (FWF) projects P20841 and P24090. Carmine Dodaro is partly supported by the European Commission, European Social Fund and Regione Calabria.

P. Cabalar and T.C. Son (Eds.): LPNMR 2013, LNAI 8148, pp. 42–53, 2013.

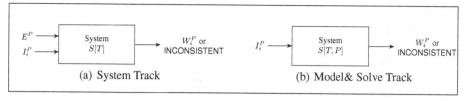

Fig. 1. Competition Setting

open to any system based on a declarative specification paradigm; and the System Track, held on the basis of fixed problem encodings, written in a standard ASP language.

In this paper we illustrate the overall setting of the fourth ASP Competition, its participants and the benchmark suite. A more detailed report, including a complete description of the entire Competition, outcomes of non-participant systems, and comparisons with other state-of-the-art systems is under preparation. The competition had 23 participants which were evaluated on a suite of 27 benchmark domains, for each of which about 30 instances were selected, for a total of about 50'000 separate benchmark runs. Results of the competition were disclosed during the LPNMR 2013 conference.

The remainder of this paper is structured as follows. In Section 2 we illustrate the competition format, especially discussing updates which were introduced with respect to the previous editions. In section 3 we illustrate the new standard language ASP-Core-2. Section 4 illustrates the benchmark problems used in this edition and Section 5 presents the participants to the Competition.

2   Format of the Fourth ASP Competition

We illustrate here the settings of the competition focusing on changes introduced with respect to the Third Competition's edition.

Competition format. The 4th ASP Competition retains the distinction between Model& Solve and System Track. Both tracks run on a selected suite of benchmark domains, which were chosen during an open *Call for Problems* stage.

The System Track was conceived with the aim of *(i)* fostering the standardization of the ASP input language, and *(ii)* let the competitors compare each other in fixed, predefined conditions, excluding e.g., domain-tailored evaluation heuristics and custom problem encodings. The System Track is run as follows (Figure 1-a): for each problem P a corresponding, fixed, declarative specification E^P of P, and a number of instances $I_1^P, \ldots, I_n^P$, are given. Each participant system $S[T]$, for T a participating team, is fed with all the couples $\langle E^P, I_i^P \rangle$, and challenged to produce a *witness* solution to $\langle E^P, I_i^P \rangle$ (denoted by W_i^P) or to report that no witness exist, within a predefined amount of allowed time. A score is awarded to each $S[T]$ per each benchmark, as detailed later in this Section. Importantly, problem encodings were fixed for all participants: specialized solutions on a per-problem basis were not allowed, and problems were specified in the recently-released ASP-Core-2 language. This setting has been introduced in order to give a fair, objective measure of what one can expect when switching from a system to another, while keeping all other conditions fixed, such as the problem encoding and the default solver settings and heuristics.

Differently from the System Track, the Model& Solve Track has been instead left open to any (bundle of) solver systems loosely based on a declarative specification language. Thus no constraints were set on the declarative language used for encoding solutions to be solved by participants' systems. Indeed, the spirit of this Track is to (i) encourage the development of new expressive declarative constructs and/or new modeling paradigms; (ii) to foster the exchange of ideas between communities in close relationships with ASP; (iii) and, to stimulate the development of new ad-hoc solving methods, refined problem specifications and solving heuristics, on a per benchmark domain basis.

In more detail, each participant team T was allowed to present a version $S[T, P]$ of their system(s) possibly customized for each problem domain P in terms of solving heuristics and declarative problem specification. Each system $S[T, P]$, for T a participating team, is challenged to solve some instances of problem P. $S[T, P]$ is expected to produce, within a predefined amount of allowed time, a *witness* solution for each instance in input (or to report that no witness exists). For both tracks, a total score is awarded to each team T summing up the scores obtained by each $S[T, P]$ (or by $S[T]$) on each benchmark, as detailed below.

Scoring system. The competition scoring system was inherited from the third edition of the competition and improved in some specific aspects. In detail, each participant is awarded of a score per each benchmark P proportional to: *a)* the percentage of instances solved within time ($S_{solve}(P)$); *b)* the evaluation time ($S_{time}(P)$); and *c)* the quality of the computed solution in case of optimization problems ($S_{opt}(P)$).[1]

Comparing the scoring system with the one of the former edition, some adjustments were introduced to the logarithmic time scoring quota S_{time}, which has been redefined as follows:

$$S_{time}(P) = \frac{100 - \alpha}{N\gamma} \sum_{i=1}^{N} \left(1 - \left(\frac{\log(\max(1, t_i) + s)}{\log(t_{out} + s)}\right)\right)$$

where P is the problem domain at hand; t_{out} is the maximum allowed time; t_i the time spent by system S while solving instance i (t_i is assumed to be lesser or equal to t_{out}); s is a factor which mildens the logarithmic behavior of S_{time}; γ is a normalization factor (having an effect detailed below); and α is a percentage factor balancing the impact of $S_{time}(P)$ w.r.t. the $S_{solve}(P)$ quota. Indeed, $S_{solve}(P)$ assigns a score that is linearly proportional to the percentage of solved instances for P as follows:

$$S_{solve}(P) = \alpha \frac{N_P}{N}$$

where N_P is the number of instances of problem P solved by S within the timeout.

As in the third edition of the Competition, $S_{time}(P)$ is specified in order to take into account the "perceived" performance of a system according to a logarithmic scoring.

---

[1] Solution quality is intended in terms of normalized percent distance from the optimal solution.

Moreover, the parameters of $S_{time}(P)$ were set in order to obtain a reasonable behavior that is expected to be stable w.r.t. minor fluctuations in measured execution times. In particular, we set for this edition of the competition $s = 10$ (it was previously set to 1) to avoid excessive differences in scoring when solving time was below 10 seconds; also, the correction $\max(1, t_i)$ prevents any score difference at all when t_i is below 1 second. In this way there is basically no difference in assigned score when execution times are very low and close to the order of magnitude of measurement errors. As in the previous competition, α was set to 0.5, so that the time and the instance quota are evenly balanced; finally, γ was chosen in such a way that the time score quota awarded for solving a single instance i within the timeout (i.e., $0 \le t_i \le t_{out}$) is normalized in the range $[0, (100 - \alpha)/N]$, thus we set

$$\gamma = 1 - \frac{\log(1 + s)}{\log(t_{out} + s)}$$

Other improvements were made w.r.t. the scoring system employed in the 3rd edition, e.g., with the introduction of a formal averaging policy for coping with multiple runs of the benchmark suite and other minor refinements. The scoring system of the 4th ASP Competition is extensively described in [5].

Instance selection process. Concerning instance selection, we introduced in this edition an new method for the random selection of instances, by taking into account that (i) the selection process should depend on a unique, not controllable by the organizer, random seed value; (ii) instances should be roughly ordered by some difficulty criterion provided by domain maintainers; (iii) hash values of instance files, and the fixed ordering of instances should be known before the Competition run; (iv) the random sequence used for selection should be unique and applied systematically to each benchmark domain, i.e. it must be impossible in practice, for organizers, to possibly forge the selection of instances in one domain without altering, out of control, the selection of instances in the other domains. (v) the selection method should approximately select a set of instances with a good balance between "hard" and "easy" instances.

The above considerations led us to adopt a variant of the statistical systematic sampling technique for the instance selection process. In detail, let S be the a random seed value chosen from an objective random source, and R be the number of instances per benchmark to be selected. Let D a benchmark domain, L_D its list of available instances, made available from benchmark domain maintainers roughly sorted by difficulty level, with $|L_D| = N_D$. We denote as $L_D[i]$ the i-th instance. over the whole family L_D, as follows: let $Start, Perturb_1, \ldots, Perturb_R$ be values systematically generated from S where $Start$ ranges from 0 to 1 and each $Perturb_i$ ranges from -1.5 to 1.5. Then, we set $Step = \frac{N_D}{R}$ and $Start_D = Step * Start$. Then we select, for all i $(1 \le i \le R)$, all the instances

$$L_D\big[\text{round}(\max(0, \min(N_D, Start_D + i * Step + Perturb_i)))\big]$$

Here round(n) is n rounded to the nearest integer. When $Step > Perturb_{i+1} - Perturb_i$ for some i, we conventionally selected $L_D[h + 1]$ as the $(i + 1)$-th instance, for h the index of the i-th instance.

Software and Hardware settings. The Competition has been run using the purposely de-veloped VCWC environment (Versioning Competition Workflow Compiler) [17]. This tool takes as input the participating solvers and dedicated benchmark sets and generates a workflow description for executing all necessary (sub-)tasks for generating the final solver rankings and statistics. As jobs may fail during the execution, VCWC supports a gradual refinement of the competition workflow and allows to add or update solvers, instances, benchmarks, or further runs after the machinery has been brought up. Gener-ated jobs where scheduled on the Competition hardware using the HTCondor [51] high throughput computing platform.

Concerning hardware, the competition has been run on several Ubuntu Server 12.04 LTS x86-64 machines featuring two AMD Opteron Magny-Cours 6176 SE CPUs (total of 24 cores) running at 2.3 GHz with 128GiB of physical RAM. To accommodate multi-core evaluations, runs were classified into sequential and parallel. Sequential runs have been evaluated in a single-core Linux control group, while parallel runs were limited to a six-core control group; all of the six cores form one NUMA node to prevent memory access overhead to remote NUMA nodes. For both kind of runs only memory with the lowest distance to the corresponding NUMA node has been used. The memory reserved to each control group was constrained to 6 GiB (1 GiB = 1 gibibyte = 2^{30} bytes), while the total CPU time available was 600 seconds. Competitors were instructed about how to reproduce the software environment, in order to properly prepare and test their systems.

Reproducibility of the results. The committee did not disclose any submitted material until the end of the Competition; nonetheless, willingly participants were allowed to share their own work at any moment. In order to guarantee transparency and repro-ducibility of the Competition results, all participants were asked to agree that any kind of submitted material (system binaries, scripts, problems encodings, etc.) was to be made public after the Competition.

3   Competition Language Overview

Since the first Edition of the competition, *standardization* has always been one of the main goals of the ASP Competition Series. The efforts to find a common language basis, started with the LPNMR 2004 language draft [6], and prosecuted with the ASP-Core [15] standard adopted in 3rd edition of the Competition. ASP-Core was published along with the ASP-RfC proposal, which preceded the work of the *ASP Standardiza-tion Working Group*, that produced the ASP-Core-2 standard, adopted for the System Track in the 4th edition of the Competition. The ideas that guided the work are in the trail of the latest version of the standard: to safeguard the original A-Prolog language [36]; to include, as extensions, a number of features both desirable and mature; and, eventually, to have a language with non-ambiguous semantics over which widespread consensus has been reached. The basis of ASP-Core-2 is hence a rule language allowing disjunctive heads and strong and negation-as-failure (NAF) negation, with no need for domain predicates. Arithmetic and Herbrand-interpreted functional terms are explicitly allowed, as well as aggregate literals and queries; choice atoms and weak constraints complete the list of new features.

The semantics of non-ground ASP-Core-2 programs extends the traditional notion of Herbrand interpretation. Concerning the semantics of propositional programs, it is based on [36], extended to aggregates according to [26]; choice atoms [49] are treated in terms of a proper translation. To promote declarative programming as well as practical system implementations, a number of restrictions are imposed. For instance, semantics is restricted to programs containing non-recursive aggregates; reasonable restrictions are applied for ensuring that function symbols, integers and arithmetic built-in predicates are finitely handled.

The ASP-Core-2 specification is rich in new features and is partially backward-compatible with older common input formats. Participants were thus allowed to join the System Track using slightly syntactically different problem encodings. Each statement of alternative problem encodings was kept in strict one-to-one correspondence with the reference ASP-Core-2 encoding.

The work on standardization is beyond the scope of the 4th ASP Competition, and new features (such as *maximize/minimize* statements for optimization, and more) have lately been incorporated into the standard. The detailed ASP-Core-2 language specification used for this Competition can be found at [12], while the ongoing standardization activity can be followed at [13].

4   Benchmark Suite

The benchmark suite has been constructed during a *Call for problems* stage, after which 26 benchmark domains were selected, 13 of which were confirmed from the previous edition. The whole collection was suitable for a proper ASP-Core-2 [12] specification. All 26 problems appeared in the System Track, while the Model& Solve Track featured only 15 domains. The complete list of benchmarks, whose details are available at [4], is reported in Table 1. Concerning legacy benchmark domains, problem maintainers were asked to produce refined specifications and/or better instances sets whenever necessary. The presence of a star (*) in the fourth column means that the corresponding problem was changed in its specifications w.r.t. its third Competition version. The selection criteria for problems aimed to collect a number of domains as balanced as possible in terms of (i) academic vs applicative provenance, (ii) computational complexity, type of domain and type of reasoning task, and (iii) research group provenance.

Problems belonged to a variety of areas, like general artificial intelligence, databases, formal logics, graph theory, planning, natural sciences and scheduling; in addition, the benchmark suite included a synthetic domain and some combinatorial and puzzle problems. Concerning the type of reasoning task to be executed in each domain, we kept the categorization in term of of *Search*, *Query* and *Optimization* problems[2].

Problems were further classified according to their computational complexity in the categories P (polynomially solvable), NP (NP-Hard), Beyond-NP (more than NP-Hard). Apart from this categorization, we classified in the Opt category (optimization problems) all the problems in which the minimization/maximization of a numerical goal function could be identified. The first three categories approximately reflect the

---

[2] The reader is referred to [14] for details concerning the three categories.

Table 1. 4th ASP Competition – Benchmark List

ID	Problem Name	Category	Domain	2011	M&S Track
N01	Permutation Pattern Matching	NP	Combinatorial	NO	YES
N02	Valves Location	Opt	Combinatorial	NO	YES
N04	Connected Maximum-density Still Life	Opt	AI	NO	YES
N05	Graceful Graphs	NP	Graph	NO	YES
N06	Bottle Filling	NP	Combinatorial	NO	YES
N07	Nomystery	NP	Planning	NO	YES
N08	Sokoban	NP	Planning	YES*	YES
N09	Ricochet Robot	NP	Puzzle	NO	YES
O10	Crossing Minimization	Opt	Graph	YES	YES
O11	Reachability	P	Graph	YES*	YES
O12	Strategic Companies	Σ_2^P	AI	YES*	YES
O13	Solitaire	NP	Puzzle	YES	YES
O14	Weighted Sequence	NP	Database	YES	YES
O15	Stable Marriage	P*	Graph	YES	YES
O16	Incremental Scheduling	NP	Scheduling	YES	YES
N17	Qualitative Spatial Temporal Reasoning	NP	Formal logic	NO	NO
N18	Chemical Classification	P*	Natural Sciences	NO	NO
N19	Abstract Dialectical Frameworks Well-founded Model	Opt	Formal logic	NO	NO
N20	Visit-all	NP	Planning	NO	NO
N21	Complex Optimization of Answer Sets	Σ_2^P	Synthetic	NO	NO
N22	Knight Tour with Holes	NP	Puzzle	YES*	NO
O23	Maximal Clique	Opt	Graph	YES	NO
O24	Labyrinth	NP	Puzzle	YES	NO
O25	Minimal Diagnosis	Σ_2^P	Diagnosis	YES	NO
O26	Hanoi Tower	NP	AI	YES	NO
O27	Graph Colouring	NP	Graph	YES	NO

data complexity [48] of the underlying decisional problem, with some exception. In particular, STABLE MARRIAGE [27, 38], for which polynomial algorithms are known, has been re-proposed in the System Track with a natural declarative encoding which makes usage of the Guess & Check paradigm; also, the CHEMICAL CLASSIFICATION benchmark featured sets of Horn rules as input instances, thus, strictly speaking, it is to be considered a *NP* problem under *combined* complexity. It is worth noting that the computational complexity of a problem has also impact on features of solvers which were put under testing. Polynomial problems are mostly, but not exclusively, useful for testing grounding modules, while the role of model generator modules is more prominent when benchmarking is done in domains in the *NP* category.

5   Participants

In this Section we briefly present all participants; we refer the reader to the official Competition website [14] for further details.

System Track Participants.   The System Track of the Competition featured 16 systems; these can be roughly grouped into two main classes: (i) "native" systems, which exploit techniques purposely conceived/adapted for dealing with logic programs under the stable models semantics, and (ii) "translation-based" systems, which (roughly), at some stage of the evaluation process, produce an intermediate specification in a different formalism; such specification is then fed to an external solver. The first category includes clasp – and variants thereof – and DLV+*wasp*, while the second counts IDP3 (which is FO(.)-based), LP2BV-1 and LP2BV-2 (relying on SMT solvers), LP2MIP and LP2MIP-MT (relying on integer programming tools), and LPD2SAT, LP2SAT-MT and LP2SOLRED-MT (relying on SAT solvers). Interestingly, several parallel (multi-threaded) solutions are officially present in this edition of the Competition; such systems are denoted by means of the "-mt" suffix.

- The group from University of Potsdam presented a number of solvers. clasp [33] is an answer set solver for (extended) normal logic programs featuring state-of-the-art techniques from the area of Boolean constraint solving. claspfolio [29] chooses the best suited configuration of clasp to process the given input program, according to machine-learning techniques. claspD-2 [31] is an extension of clasp that allows for solving disjunctive logic programs using a new approach to disjunctive ASP solving that aims at an equitable interplay between "generating" and "testing" solver units, and claspD-2 is a version supporting the ASP-Core-2 standard [12]. Multi-threaded versions clasp-mt [30], claspfolio-mt and claspD-2-mt were also present.
- The research group from Aalto University presented different solvers, all of them working by means of translations. With LP2BV-1 and LP2BV-2 [47], a given ASP program is grounded by Gringo, simplified by Smodels, normalized by getting rid of extended rule types (e.g., choice rules), translated to bit vectors and finally solved by *BOOLECTOR* for LP2BV-1 and *Z3* for LP2BV-1. LP2SAT, LP2SAT-MT [39] and LP2SOLRED-MT [39, 52] work similarly, but rely on translations to SAT rather than bit vectors; *PRECOSAT, PLINGELING* and *GLUCORED* work under the hood for LP2SAT, LP2SAT-MT, and LP2SOLRED-MT, respectively. LP2MIP [45] and LP2MIP-MT, finally, translate to mixed integer programs, which are processed by *CPLEX*.
- The team from KU Leuven presented IDP3, using FO(ID,Agg) + Lua as input language [53]. Model generation/optimization was achieved by lifted unit propagation + grounding with bounds (possibly using XSB for evaluating definitions) and applying MiniSat(ID) as search algorithm.
- *wasp*+DLV. *wasp* [2] is a native ASP solver built upon a number of techniques originally introduced in SAT, which were extended and properly combined with techniques specifically defined for solving disjunctive ASP programs. Among them are restarts, constraints learning and backjumping. Grounding is carried out by an enhanced version of the DLV grounder able to cope with the ASP-Core-2 features. Team members were affiliated to the University of Calabria.

Model& Solve Track Participants. Seven teams participated to the Model& Solve Track, each presenting a custom approach, often explicitly differentiated depending on the domain problem at hand: short descriptions follow.

- B-Prolog [54] provides several tools for tackling combinatorial optimization problems, including tabling for dynamic programming problems, CLP(FD) for CSPs, and a compiler that translates CSPs into SAT.
- Enfragmo [1] is a grounding-based solver. From a given input (expressed in multi-sorted first order logic extended with arithmetic and aggregate operators) a propositional CNF formula is produced, representing the solutions to the instance, which is processed by a SAT solver.
- EZCSP [7, 8] freely combines different ASP (such as Gringo/clasp, Clingo, Clingcon, possibly extended for supporting non-Herbrand functions) and CP (such as B-Prolog) solvers to be selected according to the features of the target domain.
- IDP3 [53] is the same system participating in the System Track, with proper custom options depending on the benchmark problem; IDP2 [53] consists of the grounder GidL and the search algorithm MiniSat(ID).
- *inca* [21, 22] implements Constraint Answer Set Programming (CASP) via Lazy Nogood Generation (LNG) and a selection of translation techniques. It integrates Gringo (for grounding CASP specifications), clasp, and a small collection of constraint propagators enhanced with LNG capacities [22].
- The team of Potassco [32] used Gringo 3, clasp 2, and iClingo 3 (an incremental ASP system [28] implemented on top of Clingo, featuring a combined grounder and solver that keep previous states while increasing an incremental parameter, trying to avoid re-producing already computed knowledge). Search settings were manually chosen (w.r.t. few instances) per problem class.

Acknowledgments. All of us feel honored of the appointment of TU Vienna and University of Calabria as host institutions: we want to thank all the members of the Database and Artificial Intelligence Group, the Knowledge-based Systems Group, and the Theory and Logic Group of Vienna University of Technology (TU Vienna), as well as the Computer Science Group at the Department of Mathematics and Computer Science of University of Calabria (Unical) for their invaluable collaboration, which made this event possible. A special thanks goes to all the members of the ASP Standardization Working Group and to all the members of the scientific community which authored, reviewed and helped in setting up all problem domains; and, of course, to the participating teams, whose feedback, once again, significantly helped at improving competition rules and benchmark specifications. We also want to acknowledge Thomas Eiter as Head of the Institute for Information Systems of the Vienna University of Technology, and Nicola Leone as the Director of the Department of Mathematics and Computer Science of University of Calabria, which provided us with human and technical resources. Eventually, we want to give a special thank to Pedro Cabalar and Tran Cao Son for their support as LPNMR-2013 conference chairs and proceedings editors.

References

1. Aavani, A., Wu, X(N.), Tasharrofi, S., Ternovska, E., Mitchell, D.: Enfragmo: A system for modelling and solving search problems with logic. In: Bjørner, N., Voronkov, A. (eds.) LPAR-18 2012. LNCS, vol. 7180, pp. 15–22. Springer, Heidelberg (2012)
2. Alviano, M., Dodaro, C., Faber, W., Leone, N., Ricca, F.: WASP: A native ASP solver based on constraint learning. In: Cabalar, P., Corunna, Son, T.C. (eds.) LPNMR 2013. LNCS, vol. 8148, pp. 55–67. Springer, Heidelberg (2013)
3. Anger, C., Gebser, M., Linke, T., Neumann, A., Schaub, T.: The nomore++ Approach to Answer Set Solving. In: Sutcliffe, G., Voronkov, A. (eds.) LPAR 2005. LNCS (LNAI), vol. 3835, pp. 95–109. Springer, Heidelberg (2005)
4. 4th ASP Competition Organizing Committee, T.: Official Problem Suite (2013),
https://www.mat.unical.it/aspcomp2013/OfficialProblemSuite
5. 4th ASP Competition Organizing Committee, T.: Rules and Scoring (2013),
https://www.mat.unical.it/aspcomp2013/ParticipationRules
6. Core language for ASP solver competitions, minutes of the steering committee meeting at LPNMR 2004 (2004),
https://www.mat.unical.it/aspcomp2011/files/Corelang2004.pdf
7. Balduccini, M.: Representing Constraint Satisfaction Problems in Answer Set Programming. In: ICLP 2009 Workshop on Answer Set Programming and Other Computing Paradigms (ASPOCP 2009) (July 2009)
8. Balduccini, M.: An Answer Set Solver for non-Herbrand Programs: Progress Report. In: Costa, V.S., Dovier, A. (eds.) Technical Communications of the 28th International Conference on Logic Programming (ICLP 2012). Schloss Dagstuhl-Leibniz-Zentrum für Informatik (2012)
9. Baral, C.: Knowledge Representation, Reasoning and Declarative Problem Solving. Cambridge University Press (2003)
10. Bell, C., Nerode, A., Ng, R.T., Subrahmanian, V.: Mixed Integer Programming Methods for Computing Nonmonotonic Deductive Databases. Journal of the ACM 41, 1178–1215 (1994)
11. Calimeri, F., Cozza, S., Ianni, G., Leone, N.: Enhancing asp by functions: Decidable classes and implementation techniques. In: Fox, M., Poole, D. (eds.) AAAI. AAAI Press (2010)
12. Calimeri, F., Faber, W., Gebser, M., Ianni, G., Kaminski, R., Krennwallner, T., Leone, N., Ricca, F., Schaub, T.: ASP-Core-2: 4th ASP Competition Official Input Language Format (2013),
http://www.mat.unical.it/aspcomp2013/files/ASP-CORE-2.01c.pdf
13. Calimeri, F., Faber, W., Gebser, M., Ianni, G., Kaminski, R., Krennwallner, T., Leone, N., Ricca, F., Schaub, T.: ASP Standardization Activity (2013),
http://www.mat.unical.it/aspcomp2013/ASPStandardization/
14. Calimeri, F., Ianni, G., Krenwallner, T., Ricca, F.: The 4th ASP Competition Organizing Committee: The Fourth Answer Set Programming Competition homepage (2013),
http://www.mat.unical.it/aspcomp2013/
15. Calimeri, F., Ianni, G., Ricca, F.: Third ASP Competition, File and language formats (2011),
http://www.mat.unical.it/aspcomp2011/
files/LanguageSpecifications.pdf
16. Calimeri, F., Ianni, G., Ricca, F.: The third open answer set programming competition. Theory and Practice of Logic Programming FirstView, 1–19 (2012),
http://dx.doi.org/10.1017/S1471068412000105

17. Charwat, G., Ianni, G., Krennwallner, T., Kronegger, M., Pfandler, A., Redl, C., Schwengerer, M., Spendier, L., Wallner, J.P., Xiao, G.: VCWC: A versioning competition workflow compiler. In: Cabalar, P., Son, T.C. (eds.) LPNMR 2013. LNCS (LNAI), vol. 8148, pp. 233–238. Springer, Heidelberg (2013), http://www.kr.tuwien.ac.at/staff/tkren/pub/2013/lpnmr2013-vcwc.pdf

18. Dal Palù, A., Dovier, A., Pontelli, E., Rossi, G.: GASP: Answer set programming with lazy grounding. Fundamenta Informaticae 96(3), 297–322 (2009)

19. Dantsin, E., Eiter, T., Gottlob, G., Voronkov, A.: Complexity and Expressive Power of Logic Programming. ACM Computing Surveys 33(3), 374–425 (2001)

20. Denecker, M., Vennekens, J., Bond, S., Gebser, M., Truszczyński, M.: The Second Answer Set Programming Competition. In: Erdem, E., Lin, F., Schaub, T. (eds.) LPNMR 2009. LNCS, vol. 5753, pp. 637–654. Springer, Heidelberg (2009)

21. Drescher, C., Walsh, T.: A translational approach to constraint answer set solving. Theory and Practice of Logic Programming 10(4-6), 465–480 (2010)

22. Drescher, C., Walsh, T.: Answer set solving with lazy nogood generation. In: Dovier, A., Costa, V.S. (eds.) ICLP (Technical Communications). LIPIcs, vol. 17, pp. 188–200. Schloss Dagstuhl - Leibniz-Zentrum fuer Informatik (2012)

23. Eiter, T., Faber, W., Leone, N., Pfeifer, G.: Declarative Problem-Solving Using the DLV System. In: Minker, J. (ed.) Logic-Based Artificial Intelligence, pp. 79–103. Kluwer Academic Publishers (2000)

24. Eiter, T., Gottlob, G., Mannila, H.: Disjunctive Datalog. ACM Transactions on Database Systems 22(3), 364–418 (1997)

25. Eiter, T., Ianni, G., Krennwallner, T.: Answer Set Programming: A Primer. In: Tessaris, S., Franconi, E., Eiter, T., Gutierrez, C., Handschuh, S., Rousset, M.-C., Schmidt, R.A. (eds.) Reasoning Web 2009. LNCS, vol. 5689, pp. 40–110. Springer, Heidelberg (2009)

26. Faber, W., Leone, N., Pfeifer, G.: Semantics and complexity of recursive aggregates in answer set programming. Artificial Intelligence 175(1), 278–298 (2011)

27. Falkner, A., Haselböck, A., Schenner, G.: Modeling Technical Product Configuration Problems. In: Proceedings of ECAI 2010 Workshop on Configuration, Lisbon, Portugal, pp. 40–46 (2010)

28. Gebser, M., Kaminski, R., Kaufmann, B., Ostrowski, M., Schaub, T., Thiele, S.: Engineering an incremental ASP solver. In: Garcia de la Banda, M., Pontelli, E. (eds.) ICLP 2008. LNCS, vol. 5366, pp. 190–205. Springer, Heidelberg (2008)

29. Gebser, M., Kaminski, R., Kaufmann, B., Schaub, T., Schneider, M.T., Ziller, S.: A portfolio solver for answer set programming: Preliminary report. In: Delgrande, J.P., Faber, W. (eds.) LPNMR 2011. LNCS, vol. 6645, pp. 352–357. Springer, Heidelberg (2011)

30. Gebser, M., Kaufmann, B., Schaub, T.: Multi-threaded ASP solving with clasp. Theory and Practice of Logic Programming 12(4-5), 525–545 (2012)

31. Gebser, M., Kaufmann, B., Schaub, T.: Advanced conflict-driven disjunctive answer set solving. In: Rossi, F. (ed.) Proceedings of the Twenty-Third International Joint Conference on Artificial Intelligence (IJCAI 2013). IJCAI/AAAI (to appear, 2013)

32. Gebser, M., Kaufmann, B., Kaminski, R., Ostrowski, M., Schaub, T., Schneider, M.T.: Potassco: The Potsdam Answer Set Solving Collection. AI Communications 24(2), 107–124 (2011)

33. Gebser, M., Kaufmann, B., Schaub, T.: Conflict-driven answer set solving: From theory to practice. Artificial Intelligence 187-188, 52–89 (2012)

34. Gebser, M., Liu, L., Namasivayam, G., Neumann, A., Schaub, T., Truszczyński, M.: The first answer set programming system competition. In: Baral, C., Brewka, G., Schlipf, J. (eds.) LPNMR 2007. LNCS (LNAI), vol. 4483, pp. 3–17. Springer, Heidelberg (2007)

35. Gelfond, M., Leone, N.: Logic Programming and Knowledge Representation – the A-Prolog perspective. Artificial Intelligence 138(1-2), 3–38 (2002)

36. Gelfond, M., Lifschitz, V.: Classical Negation in Logic Programs and Disjunctive Databases. New Generation Computing 9, 365–385 (1991)
37. Giunchiglia, E., Lierler, Y., Maratea, M.: Answer set programming based on propositional satisfiability. Journal of Automated Reasoning 36(4), 345–377 (2006)
38. Gusfield, D., Irving, R.W.: The stable marriage problem: structure and algorithms. MIT Press, Cambridge (1989)
39. Janhunen, T., Niemelä, I.: Compact translations of non-disjunctive answer set programs to propositional clauses. In: Balduccini, M., Son, T.C. (eds.) Logic Programming, Knowledge Representation, and Nonmonotonic Reasoning. LNCS, vol. 6565, pp. 111–130. Springer, Heidelberg (2011)
40. Janhunen, T., Niemelä, I., Seipel, D., Simons, P., You, J.H.: Unfolding Partiality and Disjunctions in Stable Model Semantics. ACM Transactions on Computational Logic 7(1), 1–37 (2006)
41. Lefèvre, C., Nicolas, P.: The first version of a new ASP solver: ASPeRiX. In: Erdem, E., Lin, F., Schaub, T. (eds.) LPNMR 2009. LNCS, vol. 5753, pp. 522–527. Springer, Heidelberg (2009)
42. Leone, N., Pfeifer, G., Faber, W., Eiter, T., Gottlob, G., Perri, S., Scarcello, F.: The DLV System for Knowledge Representation and Reasoning. ACM Transactions on Computational Logic 7(3), 499–562 (2006)
43. Lifschitz, V.: Answer Set Planning. In: Schreye, D.D. (ed.) Proceedings of the 16th International Conference on Logic Programming (ICLP 1999), pp. 23–37. The MIT Press, Las Cruces (1999)
44. Lin, F., Zhao, Y.: ASSAT: Computing Answer Sets of a Logic Program by SAT Solvers. Artificial Intelligence 157(1-2), 115–137 (2004)
45. Liu, G., Janhunen, T., Niemelä, I.: Answer set programming via mixed integer programming. In: 13th International Conference on Principles of Knowledge Representation and Reasoning (KR 2012), pp. 32–42 (2012)
46. Marek, V.W., Truszczyński, M.: Stable Models and an Alternative Logic Programming Paradigm. In: Apt, K.R., Marek, V.W., Truszczyński, M., Warren, D.S. (eds.) The Logic Programming Paradigm – A 25-Year Perspective, pp. 375–398. Springer (1999)
47. Nguyen, M., Janhunen, T., Niemelä, I.: Translating answer-set programs into bit-vector logic. CoRR abs/1108.5837 (2011)
48. Papadimitriou, C.H.: Computational Complexity. Addison-Wesley (1994)
49. Simons, P., Niemelä, I., Soininen, T.: Extending and Implementing the Stable Model Semantics. Artificial Intelligence 138, 181–234 (2002)
50. Subrahmanian, V., Nau, D., Vago, C.: WFS + Branch and Bound = Stable Models. IEEE Transactions on Knowledge and Data Engineering 7(3), 362–377 (1995)
51. Thain, D., Tannenbaum, T., Livny, M.: Distributed computing in practice: the Condor experience. Concurrency and Computation: Practice and Experience 17(2-4), 323–356 (2005)
52. Wieringa, S., Heljanko, K.: Concurrent clause strengthening. In: Järvisalo, M., Van Gelder, A. (eds.) SAT 2013. LNCS, vol. 7962, pp. 116–132. Springer, Heidelberg (2013)
53. Wittocx, J., Mariën, M., Denecker, M.: The IDP system: a model expansion system for an extension of classical logic. In: Denecker, M. (ed.) International Workshop on Logic and Search (Lash), pp. 153–165 (2008)
54. Zhou, N.F.: The language features and architecture of B-Prolog. Theory and Practice of Logic Programming 12(1-2), 189–218 (2012)

WASP: A Native ASP Solver
Based on Constraint Learning[*]

Mario Alviano, Carmine Dodaro, Wolfgang Faber,
Nicola Leone, and Francesco Ricca

Department of Mathematics and Computer Science,
University of Calabria, 87036 Rende, Italy
{alviano,dodaro,faber,leone,ricca}@mat.unical.it

Abstract. This paper introduces WASP, an ASP solver handling disjunctive logic programs under the stable model semantics. WASP implements techniques originally introduced for SAT solving that have been extended to cope with ASP programs. Among them are restarts, conflict-driven constraint learning and backjumping. Moreover, WASP combines these SAT-based techniques with optimization methods that have been specifically designed for ASP computation, such as source pointers enhancing unfounded-sets computation, forward and backward inference operators based on atom support, and techniques for stable model checking. Concerning the branching heuristics, WASP adopts the BerkMin criterion hybridized with look-ahead techniques. The paper also reports on the results of experiments, in which WASP has been run on the system track of the third ASP Competition.

1 Introduction

Answer Set Programming (ASP) [1] is a declarative programming paradigm which has been proposed in the area of non-monotonic reasoning and logic programming. The idea of ASP is to represent a given computational problem by a logic program whose answer sets correspond to solutions, and then use a solver to find them.

The ASP language considered here allows disjunction in rule heads and nonmonotonic negation in rule bodies. These features make ASP very expressive; all problems in the second level of the polynomial hierarchy are indeed expressible in ASP [2]. Therefore, ASP is strictly more expressive than SAT (unless $P = NP$). Despite the intrinsic complexity of the evaluation of ASP, after twenty years of research many efficient ASP systems have been developed. (e.g. [3–5]).The availability of robust implementations made ASP a powerful tool for developing advanced applications in the areas of Artificial Intelligence, Information Integration, and Knowledge Management; for example, ASP has been used in industrial applications [6], and for team-building [7], semantic-based information extraction [8], and e-tourism [9]. These applications of ASP have confirmed the viability of the use of ASP. Nonetheless, the interest in developing more

---

[*] This research has been partly supported by project PIA KnowRex POR FESR 2007- 2013 BURC n. 49 s.s. n. 1 16/12/2010, by MIUR project FRAME PON01_02477/4, and by the European Commission, European Social Fund and Regione Calabria.

P. Cabalar and T.C. Son (Eds.): LPNMR 2013, LNAI 8148, pp. 54–66, 2013.

effective and faster systems is still a crucial and challenging research topic, as witnessed by the results of the ASP Competition series (see e.g. [10]).

This paper provides a contribution in the aforementioned context. In particular, we present a new ASP solver for propositional programs called WASP. The new system is inspired by several techniques that were originally introduced for SAT solving, like the Davis-Putnam-Logemann-Loveland (DPLL) backtracking search algorithm [11], *clause learning* [12], *backjumping* [13], *restarts* [14], and *conflict-driven heuristics* [15] in the style of BerkMin [16]. The mentioned SAT-solving methods have been adapted and combined with state-of-the-art pruning techniques adopted by modern native disjunctive ASP systems [3–5]. In particular, the role of Boolean Constraint Propagation in SAT-solvers (based only on *unit propagation* inference rule) is taken by a procedure combining several of inference rules. Those rules combine an extension of the well-founded operator for disjunctive programs with a number of techniques based on ASP program properties (see, e.g., [17]). In particular, WASP implements techniques specifically designed for ASP computation, such as source pointers [18] enhanced unfounded-set computation, *native* forward and backward inference operators based on atom support [17]. Moreover, WASP uses a branching heuristics based on a mixed approach between BerkMin-like heuristics and look-ahead which takes into account minimality of answer sets, a requirement not present in SAT solving. Finally, stable model checking, which is a co-NP-complete problem for disjunctive logic programs, is implemented relying on the rewriting method of [19] and by calling MiniSAT [20].

In the following, after briefly introducing ASP, we describe the new system WASP, whose source available at `http://www.mat.unical.it/ricca/wasp`. We start from the solving strategy and present the design choices regarding propagation, constraint learning, restarts, and the heuristics. We also report on an experiment in which we have run WASP on all instances used in the third ASP Competition [10]. In particular, we compare our system with all participants and analyze in detail the impact of our design choices. Finally, we discuss related work and draw the conclusion.

2 Preliminaries

Let $\mathcal{A}$ be a countable set of propositional atoms. A *literal* is either an atom (a positive literal), or an atom preceded by the *negation as failure* symbol not (a negative literal). A *program* is a finite set of rules of the following form:

$$p_1 \vee \cdots \vee p_n :\text{-} q_1, \ldots, q_j, \text{not } q_{j+1}, \ldots, \text{not } q_m \qquad (1)$$

where $p_1, \ldots, p_n, q_1, \ldots, q_m$ are atoms and $n \geq 0$, $m \geq j \geq 0$. The disjunction $p_1 \vee \cdots \vee p_n$ is called head, and the conjunction $q_1, \ldots, q_j, \text{not } q_{j+1}, \ldots, \text{not } q_m$ is referred to as body. For a rule r of the form (1), the following notation is also used: $H(r)$ denotes the set of head atoms; $B(r)$ denotes the set of body literals; $B^+(r)$ and $B^-(r)$ denote the set of atoms appearing in positive and negative body literals, respectively; $C(r) := \overline{H(r)} \cup B(r)$ is the nogood representation of r [4]. In the following a rule r is said to be regular if $|H(r)| \geq 1$, and a constraint if $|H(r)| = 0$. Moreover, the complement of a literal ℓ is denoted $\overline{\ell}$, i.e., $\overline{a} = \text{not } a$ and $\overline{\text{not } a} = a$ for an atom a. This notation extends to sets of literals, i.e., $\overline{L} := \{\overline{\ell} \mid \ell \in L\}$ for a set of literals L.

An *interpretation* I is a set of literals, i.e., $I \subseteq \mathcal{A} \cup \overline{\mathcal{A}}$. Intuitively, literals in I are true, literals whose complements are in I are false, and all other literals are undefined. I is total if there are no undefined literals, and I is inconsistent if there is $a \in \mathcal{A}$ such that $\{a, \text{not } a\} \subseteq I$. An interpretation I satisfies a rule r if $C(r) \cap \overline{I} \neq \emptyset$, while I violates r if $C(r) \subseteq I$. A *model* of a program $\mathcal{P}$ is a total interpretation satisfying all rules of $\mathcal{P}$. The semantics of a program $\mathcal{P}$ is given by the set of its answer sets (or stable models) [1], where a total interpretation M is an answer set (or stable model) for $\mathcal{P}$ if and only if M is a subset-minimal model of the reduct $\mathcal{P}^M$ obtained by deleting from $\mathcal{P}$ each rule r such that $B^-(r) \cap I \neq \emptyset$, and then by removing all the negative literals from the remaining rules.

3   Answer Set Computation

In this section we review the algorithms and the heuristics implemented in WASP. For reasons of presentation, we have considerably simplified the procedures in order to focus on the main principles.

3.1   Main Algorithm

An answer set of a given propositional program $\mathcal{P}$ is computed in WASP by using Algorithm 1, which is similar to the Davis-Putnam procedure in SAT solvers. The process starts with an empty interpretation I in input. Function Propagate extends I with those literals that can be deterministically inferred (line 2) and keeps track of the reasons of each inference by building a representation of the so-called *implication graph* [15]. This function is similar to unit propagation as employed by SAT solvers, but also uses the peculiarities of ASP for making further inferences (e.g., it uses the knowledge that every answer set is a minimal model). Propagate, described in more detail in Section 3.2, returns false if an inconsistency (or conflict) is detected, true otherwise. If Propagate returns true and I is total (line 3), *CheckModel* is invoked (line 4) to verify that I is an answer set by using the techniques described in [19]. In particular, for non head-cycles-free programs the check is co-NP-complete [21] and implemented by a call to the SAT solver MiniSAT [20]. If the stability check succeeds, I is returned; otherwise, I contains some unfounded sets which are analyzed by the procedure *AnalyzeConflictAndLearnConstraints* (described later). Otherwise, if there are undefined literals in I, a heuristic criterion is used to chose one, say ℓ. Then computation proceeds with a recursive call to *ComputeAnswerSet* on $I \cup \{\ell\}$ (lines 6–7). In case the recursive call returns an answer set, the computation ends returning it (lines 8–9). Otherwise, the algorithm unrolls choices until consistency of I is restored (backjumping; lines 10–11), and the computation resumes by propagating the consequences of constraints learned by the conflict analysis. Conflicts detected during propagation are analyzed by procedure *AnalyzeConflictAndLearnConstraints* (line 12; described in Section 3.3).

This general procedure is usually complemented with some heuristic techniques that control the number of learned constraints (which may be exponential in number), and possibly restart the computation to explore different branches of the search tree. Our restart policy is based on the sequence of thresholds introduced in [22], while our learned constraint elimination policy is described in Section 3.4.

Algorithm 1. Compute Answer Set

    Input : An interpretation I for a program P
    Output: An answer set for P or *Incoherent*

1 **begin**
2   **while** Propagate(I) **do**
3   **if** I *is total* **then**
4   **if** CheckModel(I) **then return** I;
5   **break**;   // goto 12
6   ℓ := ChooseUndefinedLiteral();
7   I' := ComputeAnswerSet($I \cup \{\ell\}$);
8   **if** $I' \neq$ *Incoherent* **then**
9   **return** I';
10   **if** *there are violated learned constraints* **then**
11   **return** *Incoherent*;
12   AnalyzeConflictAndLearnConstraints(I);
13   **return** *Incoherent*;

3.2 Propagation

WASP implements a number of deterministic inference rules for pruning the search space during answer set computation. These propagation rules are named *unit, support*, and *well-founded*. During the propagation of deterministic inferences, implication relationships among literals are stored in the implication graph. Each node ℓ in the implication graph is labelled with a *decision level* representing the number of nested calls to AnswerSetComputation at the point in which ℓ has been derived. Note that the implication graph contains at most one node for each atom unless a conflict is derived, in which case for some atom a both a and its negation are in the graph. In the following, we describe the propagation rules and how the implication graph is updated in WASP.

Unit Propagation. An undefined literal ℓ is inferred by unit propagation if there is a rule r that can be satisfied only by ℓ, i.e., r is such that $\overline{\ell} \in C(r)$ and $C(r) \setminus \{\overline{\ell}\} \subseteq I$. In the implication graph we add node ℓ, and arc (ℓ', ℓ) for each literal $\ell' \in C(r) \setminus \{\overline{\ell}\}$.

Support Propagation. Answer sets are supported models, i.e., for each atom a in an answer set there is a (supporting) rule r such that $a \in H(r)$, $B(r) \subseteq I$ and $H(r) \cap I = \{a\}$. Support is on the basis of two propagation rules named *forward* and *backward*.

Forward propagation derives as false all atoms for which there are no candidate supporting rules. More formally, literal not a is derived if for each rule r having a in the head $\overline{I} \cap C(r) \setminus \{\text{not } a\} \neq \emptyset$ holds. In the implication graph, a node not a is introduced, and for each rule r having a in the head an arc $(\overline{\ell}, \text{not } a)$, where $\ell \in C(r) \setminus \{\text{not } a\}$, is added. Within WASP, literal ℓ is the first literal that satisfied r in chronological order of derivation, which is called *first satisfier* of r in the following.

Backward propagation occurs when for a true atom there is only one candidate supporting rule. More in detail, if there are an atom $a \in I$ and a rule r such that $a \in H(r)$ and for each other rule r' having a in the head $\overline{I} \cap C(r') \setminus \{\text{not } a\} \neq \emptyset$ holds, then all literals in $C(r) \setminus I$ are inferred. Concerning the implication graph, we add node ℓ and arc (a, ℓ) for each $\ell \in C(r) \setminus I$. Moreover, for each $\ell \in C(r) \setminus I$ and r' having a in the head and different from r, we add an arc (ℓ', ℓ), where ℓ' is the first satisfier of r'.

Well-Founded Propagation.   Self-supporting truth is not admitted in answer sets. According to this property, a set X of atoms is *unfounded* if for each r such that $H(r) \cap X \neq \emptyset$ at least one of the following conditions is satisfied: (i) $B(r) \cap \bar{I} \neq \emptyset$; (ii) $B^+(r) \cap X \neq \emptyset$; (iii) $I \cap H(r) \setminus X \neq \emptyset$. Intuitively, atoms in X can have support only by themselves, and can thus be derived false.

To compute unfounded sets we adopted *source pointers* [18]. Roughly, for each atom we set a rule r as its candidate supporting rule, referred to as its source pointer. Source pointers are constrained to not introduce self-supporting atoms, and are updated during the computation. Atoms without source pointers form an unfounded set and are thus derived false. Concerning the implication graph, for each atom $a \in X$, node not a is added. Moreover, for each $a \in X$ and for each rule r having a in the head and first satisfier ℓ of r ($\ell \notin X$), arc $(\ell, \text{not } a)$ is added. Note that since $\ell \notin X$ the implication graph is acyclic.

3.3 Constraint Learning

Constraint learning acquires information from conflicts in order to avoid exploring the same search branch several times. In WASP there are two causes of conflicts: failed propagation and stability check failures.

Learning from Propagation.   In this case, our learning schema is based on the concept of the first Unique Implication Point (UIP) [15]. A node n in the implication graph is a UIP for a decision level d if all paths from the literal chosen at the level d to the conflict literals pass through n. We calculate UIPs only for the decision level of conflicts, and more precisely the one closest to the conflict, which is called the first UIP. Our learning schema is as follows: Let u be the first UIP. Let L be the set of literals different form u occurring in a path from u to the conflict literals. The learned constraint comprises u and each literal ℓ such that the decision level of ℓ is lower than the one of u and there is an arc (ℓ, ℓ') in the implication graph for some $\ell' \in L$.

Learning from Model Check Failure.   Answer set candidates are checked for stability by function *CheckModel* in Algorithm 1. If a model M is not stable, an unfounded set $X \subseteq M$ is computed. X represents the reason for the stability check failure. Thus, we learn a constraint c containing all atoms from X and first satisfiers of possible supporting rules for atoms in X. More formally, a literal ℓ is in $B(c)$ if either $\ell \in X$ or ℓ is the first satisfier of some rule r s.t. $H(r) \cap X \neq \emptyset$ and $B^+(r) \cap X = \emptyset$.

3.4 Heuristics

A crucial role is played by the heuristic criteria used for both selecting branching literals and removing learned constraints.

Branching Heuristic.   Concerning the branching heuristics, implemented by function ChooseUndefinedLiteral in Algorithm 1, we adopt a mixed approach between look-back and look-ahead techniques. The idea is to record statistics on atoms involved in

---

Function ChooseUndefinedLiteral

Output: A branching literal

1 **begin**
2   **if** *there is no learned constraint* **then**
3   a := MostOccurrentAtom();
4   **return** MostOccurrentPolarity(a);
5   **if** *there is an undefined learned constraint* **then**
6   c := MostRecentUndefinedLearnedConstraint() ;
7   Candidates := AtomsWithHighestCV(c);
8   **if** $|$Candidates$| = 1$ **then**
9   **return** HighestGCVPolarity(Candidates);
10   a := AtomCancellingMoreRules(Candidates);
11   **return** PolarityCancellingMoreRules(a);
12   a := AtomWithHighestCV();
13   **return** LookAhead(a) ;

---

constraint learning so to prefer those involved in most recent conflicts (look-back), and in some case the branching literal is chosen by estimating the effects of its propagation (look-ahead). More in detail, WASP implements a variant of the criterion used in the BerkMin SAT solver [23]. In this technique each literal ℓ is associated with counters $cv(\ell)$ and $gcv(\ell)$, initially set to zero. When a new constraint is learned, counters for all literals occurring in the constraint are increased by one. Moreover, counters are also updated during the computation of the first UIP: If a literal ℓ is traversed in the implication graph, the associated counters are increased by one, and counters $cv(\cdot)$ are divided by 4 every 100 conflicts. Thus, literals that are often involved in conflicts will have larger values of $cv(\cdot)$ and $gcv(\cdot)$, where counters $cv(\cdot)$ give more importance to literals involved in recent conflicts.

The branching criterion is reported in function ChooseUndefinedLiteral. Initially, there is no learned constraint (line 2), and the algorithm selects the atom, say a, occurring most frequently in rules. Then, the most occurrent literal of a and not a is returned. After learning some constraints, two possible scenarios may happen. If there are undefined learned constraints (line 5), the one that was learned more recently, say c, is considered, and the atoms having the highest value of $cv(\cdot)$ are *candidate* choices. If there is only one candidate, say a, then the literal between a and not a having the maximum value of $gcv(\cdot)$ is returned. (If $gcv(a) = gcv(\text{not } a)$ then not a is returned.) If there are several candidates, an ASP specific criterion is used for estimating the effect of the candidates on the number of potentially supporting rules. In particular, let a be an atom occurring most often in unsatisfied regular rules. The heuristic chooses the literal between a and not a that satisfies the largest number of rules.

The second scenario happens when all learned constraints are satisfied. In this case one atom, say a, having the highest value of $cv(\cdot)$ is selected, and a look-ahead procedure is called to determine the most promising polarity (lines 12–13). Actually, a look-ahead step is performed by propagating both a and not a, and the impact of the two assumptions on answer set computation is estimated by summing up the number of inferred atoms and the number of rules that have been satisfied. The literal having greater impact is chosen, and in case of a tie the negative literal is preferred. It is important to note that if one of the two propagations fails, the conflict is analyzed as described in Section 3.3, and a constraint is learned and propagated. Possibly, after the propagation, a new branching literal is selected applying the above criterion.

Deletion of Constraints. The number of learned constraints can grow exponentially, and this may cause a performance degradation. A heuristic is employed for deleting some of them, typically the ones that are not involved often in the more recent conflicts. To this end, learned constraints are associated with activity counters as implemented in the SAT solver MiniSAT [20]. The activity counters measure the number of times a constraint was involved in the derivation of a conflict. Once the number of learned constraints is greater than one third of the size (in number of rules) of the original program, constraint deletion is performed as follows: First, all constraints having an activity counter smaller than a threshold are removed (as in MiniSAT) if they are *unlocked*. A constraint c is *unlocked* if $C(c) \setminus I \neq \emptyset$ (roughly, c is undefined and not directly involved in propagations). If this cancellation step removes less than half of the constraints, an additional deletion step is performed. In particular, unlocked constraints having activity less than the average are removed possibly until the number of constraints halves. Note that the second cancellation step is done differently in MiniSAT; our policy seems to be effective in practice for ASP.

4   Experiments

In this section we report the results of an experiment assessing the performance of WASP. In particular, we first compare WASP with all participants of the System Track of the 3rd ASP Competition. Then, we analyze in detail the behavior of WASP in specific domains that help to understand strengths and weaknesses of our solver. The experiments were run on the very same benchmarks, hardware and execution platform used in the 3rd ASP Competition [10]. In particular, we used a four core Intel Xeon CPU X3430 2.4 GHz, with 4 GB of physical RAM and PAE enabled, running Linux Debian Lenny (32bit). As in the competition settings, WASP was benchmarked with just one of the four processors enabled, and time and memory limits set to 600 seconds and 3 GiB (1 GiB = 2^{30} bytes), respectively. Execution times and memory consumptions were measured by the same programs and scripts employed in the competition. In particular, we used the Benchmark Tool Run (http://fmv.jku.at/run/).

We have run WASP on the official instances of the System Track of the 3rd ASP Competition [10]. In this paper we consider only problems featuring unstratified or disjunctive encodings, thus avoiding instances that are already solved by the grounders. More in detail, we consider all problems in the NP and Beyond-NP categories, and the polynomial problems Stable Marriage and Partners Unit Polynomial. The competition suite included planning domains, temporal and spatial scheduling problems, combinatorial puzzles, graph problems, and a number of real-world domains in which ASP was applied. (See [10] for an exhaustive description of the benchmarks.)

WASP was coupled with a custom version of the DLV grounder properly adapted to work with our solver. We report the results in Table 1 together with the official results

Table 1. Scores on the 3rd ASP Competition benchmark

System		Total	P	NP	BNP	StableMarriage	PartnerUnitsPolynomial	SokobanDecision	KnightTour	DisjunctiveScheduling	PackingProblem	Labyrinth	MCS Querying	Numberlink	HanoiTower	GraphColouring	Solitaire	Weight-AssignmentTree	MazeGeneration	StrategicCompanies	MinimalDiagnosis
claspd	Score	668	13	552	103	0	13	68	68	30	0	65	75	69	31	19	11	20	96	12	91
	Inst	425	10	355	60	0	10	45	40	25	0	45	50	40	25	10	10	15	50	10	50
	Time	243	3	197	43	0	3	23	28	5	0	20	25	29	6	9	1	5	46	2	41
wasp	Score	663	46	553	64	40	6	32	68	34	0	64	73	59	36	15	37	54	81	0	64
	Inst	465	40	380	45	35	5	25	40	25	0	45	50	35	30	10	25	45	50	0	45
	Time	198	6	173	19	5	1	7	28	9	0	19	23	24	6	5	12	9	31	0	19
claspfolio	Score	627	18	609	-	5	13	66	65	37	0	63	75	64	47	55	21	21	95	-	-
	Inst	400	15	385	-	5	10	45	35	25	0	40	50	35	35	40	15	15	50	-	-
	Time	227	3	224	-	0	3	21	30	12	0	23	25	29	12	15	6	6	45	-	-
clasp	Score	617	20	597	-	6	14	78	63	38	0	78	75	65	39	23	21	21	96	-	-
	Inst	385	15	370	-	5	10	50	35	25	0	50	50	35	30	15	15	15	50	-	-
	Time	232	5	227	-	1	4	28	28	13	0	28	25	30	9	8	6	6	46	-	-
idp	Score	597	0	597	-	0	0	64	74	38	0	52	75	70	65	18	38	8	95	-	-
	Inst	370	0	370	-	0	0	45	45	25	0	30	50	40	45	10	25	5	50	-	-
	Time	227	0	227	-	0	0	19	29	13	0	22	25	30	20	8	13	3	45	-	-
cmodels	Score	582	0	510	72	0	0	67	56	21	0	62	75	30	51	29	18	6	95	0	72
	Inst	380	0	335	45	0	0	45	30	20	0	45	50	20	35	20	15	5	50	0	45
	Time	202	0	175	27	0	0	22	26	1	0	17	25	10	16	9	3	1	45	0	27
lp2diffz3	Score	394	0	394	-	0	0	42	55	0	0	0	70	45	47	27	25	0	83	-	-
	Inst	270	0	270	-	0	0	30	35	0	0	0	50	30	35	20	20	0	50	-	-
	Time	124	0	124	-	0	0	12	20	0	0	0	20	15	12	7	5	0	33	-	-
sup	Score	357	11	346	-	0	11	52	40	37	0	58	72	0	31	16	15	25	0	-	-
	Inst	250	10	240	-	0	10	35	25	25	0	40	50	0	25	10	10	20	0	-	-
	Time	107	1	106	-	0	1	17	15	12	0	18	22	0	6	6	5	5	0	-	-
lp2sat2gmsat	Score	321	11	310	-	0	11	36	10	32	0	46	71	22	47	17	29	0	0	-	-
	Inst	235	10	225	-	0	10	30	5	25	0	35	50	15	35	10	20	0	0	-	-
	Time	86	1	85	-	0	1	6	5	7	0	11	21	7	12	7	9	0	0	-	-
lp2sat2msat	Score	307	5	302	-	0	5	39	0	32	0	52	71	15	47	17	29	0	0	-	-
	Inst	225	5	220	-	0	5	30	0	25	0	40	50	10	35	10	20	0	0	-	-
	Time	82	0	82	-	0	0	9	0	7	0	12	21	5	12	7	9	0	0	-	-
lp2sat2lmsat	Score	301	0	301	-	0	0	35	0	32	0	53	71	17	47	17	29	0	0	-	-
	Inst	220	0	220	-	0	0	30	0	25	0	40	50	10	35	10	20	0	0	-	-
	Time	81	0	81	-	0	0	5	0	7	0	13	21	7	12	7	9	0	0	-	-
smodels	Score	269	0	269	-	0	0	0	55	36	0	9	53	27	0	0	0	0	89	-	-
	Inst	165	0	165	-	0	0	0	30	25	0	5	35	20	0	0	0	0	50	-	-
	Time	104	0	104	-	0	0	0	25	11	0	4	18	7	0	0	0	0	39	-	-

Column groups: **Cumulative** (Total, P, NP, BNP); **P** (StableMarriage, PartnerUnitsPolynomial); **NP** (SokobanDecision, KnightTour, DisjunctiveScheduling, PackingProblem, Labyrinth, MCS Querying, Numberlink, HanoiTower, GraphColouring, Solitaire, Weight-AssignmentTree, MazeGeneration); **Bnd-NP** (StrategicCompanies, MinimalDiagnosis).

of all participants of the competition. The results are expressed in terms of scores computed with the same methods adopted in the competition. Roughly, the instance score can be obtained multiplying by 5 the number of solved instances within the timeout, whereas the time score is computed according to a logarithmic function of the execution times (details can be found in `http://www.mat.unical.it/aspcomp2011/files/scoringdetails.pdf`). The first column contains the total scores, followed by columns containing data aggregated first by benchmark class, and then by problem name. For each solver we report total score, instance score and time score in separate rows. A dash in the table indicates that the corresponding solver cannot handle the corresponding instances of a specific class/problem. We observe that the only solvers able to deal with Beyond-NP problems are claspD, Cmodels and WASP.

As a general comment, by looking at Table 1 we can say that WASP is comparable in performance with the best solvers in the group, and scored just 5 points less than claspD. WASP solved more instances in overall, obtaining 40 points more than claspD for the instance score. However, WASP performs worse than the five best solvers in terms of raw speed, and in particular its time score is 45 points less than claspD. If we analyze the results by problem class, we observe that WASP is the best solver in the P category, and it is comparable to claspD but follows claspfolio, clasp and idp in NP. In Beyond-NP, where claspD is the best solver, WASP solves the same instances as Cmodels but it is slower than this latter. These results already outline some advantages and weaknesses of our implementation. In particular, a weakness of WASP, also affecting claspD when compared with clasp, is that handling unrestricted disjunction can cause a reduction in performance in the NP category, which is however compensated in the total score by the additional points earned in the Beyond-NP category. In this category WASP performs similar to Cmodels, which can be justified by the similar learning strategy in case of model checking failures that the two solvers adopt. Nonetheless, both the efficiency of implementation and the interplay between the main algorithm and model checker has to be significantly improved to fill the gap with claspD. The strength of WASP in P can be explained as a combination of two factors. First, WASP uses the DLV grounder, which in some cases produces a smaller program than Gringo. The second factor is that WASP often requires less memory than the best five alternative solvers. This behavior makes a sensitive difference in this category, where the instances of Stable Marriage are large in size. We will analyze the issue of memory usage in more detail later, but it is worth mentioning that WASP runs out of memory only 23 times in total, which is less than any of the five best systems. In fact, according to the data reported on the Web site of the Competition, claspD, clasp, claspfolio, idp, and Cmodels ran out of memory 63, 61, 56, 63, and 74 times, respectively.

Analyzing the results in more detail, there are some specific benchmark problems where the differences among WASP and the five best participants in the competition are significant. In these cases, the differing behaviors can be explained by different choices made during design and implementation of solvers. Analyzing Table 1 from left to right, the first of these problems is StableMarriage, which belongs to the P category. As previously pointed out, in this category the combination of DLV (grounder) and WASP needs less memory, which explains the result. Then there is SokobanDecision, in which WASP performed worse than several other solvers. To understand the

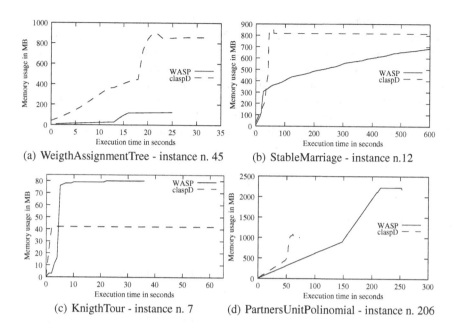

(a) WeigthAssignmentTree - instance n. 45   (b) StableMarriage - instance n.12

(c) KnigthTour - instance n. 7   (d) PartnersUnitPolinomial - instance n. 206

Fig. 1. Memory usage in WASP and claspD

reason, we ran some additional experiments (not report due to space constraints), from which we observed that both (i) the default heuristic of WASP is not suitable for this problem; and (ii) profiling revealed that the implementation of well-founded propagation in WASP causes considerable overhead. Concerning (i), we verified that selecting a different criterion (e.g., standard BerkMin) can sensibly improve performance. Eventually, WASP solves more instances of Weight-Assignment Tree than any alternative. Here WASP, featuring a native implementation of support propagations, is advantaged over other solvers, like clasp(D), Cmodels and idp, which apply Clark's completion. This is a rewriting technique that adds additional symbols and auxiliary rules to encode the support requirement. It is known from the literature that adding these additional symbols can lead to better performance [24], nonetheless, in this case they seem to cause higher memory usage and slower propagation.

Some additional observations can be made by studying in more detail memory usage of WASP and claspD. To this end we report in Figure 1 four plots depicting the memory consumption during the solvers' execution. In particular, Figure 1(a) reports the result for an instance of WeightAssignmentTree, a problem in the NP category whose encoding is unstratified. In this case WASP performs better than claspD both in memory and time. We observed that the output of DLV is five times smaller than the output of Gringo, which can justify the memory required by claspD up to 18 seconds. At that point of execution, claspD doubles its memory consumption, which could be a side effect of Clark's completion. Figure 1(b) shows the result for an instance of Stable-Marriage that neither WASP nor claspD solve in the allotted time. StableMarriage is a problem in the P category and its encoding is unstratified. Also in this case we observe that WASP is less memory demanding than claspD. Figure 1(c) depicts the results for an instance of KnightTour, a problem in the NP category whose encoding is recursive. In this case claspD requires half of the memory consumed by WASP, which is an insight that our current implementation of the well-founded propagation is not optimal in terms of memory consumption. Nonetheless, WASP is faster than claspD for the tested instance. Finally, Figure 1(d) reports the result for PartnersUnitPolinomial, a problem in the P category whose encoding is recursive. In this case claspD performs better than WASP both in memory and size, which is partially due to DLV. In fact, even if DLV and Gringo output programs of the same size for the tested instance, DLV terminated in 150 seconds, while Gringo just requires 50 seconds. Again, we note that WASP requires more than two times the memory used by claspD in this unstratified encoding.

5 Related Work

WASP is inspired by several techniques that were originally introduced for SAT solving, like the DPLL backtracking search algorithm [11], clause learning [12], backjumping [13], restarts [14], and conflict-driven heuristics [15] in the style of BerkMin [16]. Actually, some of the techniques adopted in WASP, including backjumping and lookback heuristics, were first introduced for Constraint Satisfaction, and then successfully applied to SAT and QBF solving. Some of these techniques were already adapted in modern non-disjunctive ASP solvers like Smodels$_{cc}$ [25], clasp [4], Smodels [18], and solvers supporting disjunction like Cmodels3 [5], and DLV [26].

More in detail, WASP differs from non-native solvers like Cmodels3 [5] that are based on a rewriting into a propositional formula and an external SAT solver. Nonetheless, our learning strategy for stability check failures is similar to that of Cmodels3. Concerning native solvers, WASP implements native support propagation rules and model checking techniques similar to DLV [3]. However, we implement look-back techniques borrowed from CP and SAT which are not present in DLV. In fact, DLV implements a systematic backtracking without learning and adopts look-ahead heuristics. We also mention an extension of DLV [26] that implements backjumping and lookback heuristics, which however does not include clause learning, restarts, and does not use an implication graph for determining the reasons of the conflicts. WASP uses an implication graph which is similar to the one implemented in Smodels$_{cc}$ [25]. Nonetheless, there is an important difference between these two implication graphs. In fact, the first one is guaranteed to be acyclic while the latter might be cyclic due to the well founded propagation.

Our solver is more similar to clasp and its extension to disjunctive programs claspD. In fact, source pointers, backjumping, learning, restarts, and look-back heuristics are also used by clasp(D). There are nonetheless several differences with WASP. The first difference is that clasp(D) use Clark's completion for modeling support, while WASP features a native implementation of support propagation (which caused major performance differences in our experiments). Also, minimality is handled by learning nogoods (called loop formulas) in clasp(D). It turns out that clasp(D) almost relies on unit propagation and thus uses an implication graph that is more similar to SAT solvers. Furthermore, there are differences concerning the restart policy, constraint deletion and branching heuristics. WASP adopts as default a policy based on the sequence of thresholds introduced in [22], whereas clasp(D) employs by default a different policy based on geometric series. Concerning deletion of learned constraints, WASP adopts a criterion inspired by MiniSAT. In clasp(D) a more involved criterion is adopted, where constraints are cancelled when the size of the program grows up to a crescent threshold depending on the number of restarts. Nonetheless, the program can grow in clasp(D) up to three times the size of the original input, while WASP limits the growth of the program to one third. Clasp(D) and WASP adopt a branching heuristics based on Berk-Min [16] with differences in the intermediate steps of the selection procedure. WASP extends the original BerkMin heuristics by using a look-ahead technique in place of the *"two"* function calculating the number of binary clauses in the neighborhood of a literal ℓ. Moreover, WASP introduces an additional criterion based on supportedness of answer sets for selecting among heuristically-equivalent candidate literals of the last undefined learned constraint. Clasp(D) instead uses as intermediate step a variant of the MOMS criterion. MOMS estimates the effect of the candidate literals in short clauses and is convenient for clasp(D) because Clark's completion produces many binary constraints.

6 Conclusion

In this paper we presented a new native ASP solver for propositional programs called WASP. WASP builds upon a number of techniques originally introduced in the neighboring fields of CP and SAT, which are extended and properly combined with techniques specifically defined for solving disjunctive ASP programs. The performance of WASP was assessed and compared with alternative implementations by running the System Track of the 3rd ASP Competition. Our analysis shows that WASP is efficient and can compete with the state-of-the-art solvers on this benchmark. The effects of a native implementation of support propagations in a learning-based ASP solver is also discussed, showing that this design choice pays off in terms of memory usage and time performance in specific benchmark domains. The experiments also outline some specific weakness of the implementation (e.g., in Beyond NP domains), which will be subject of future work. It is worth pointing out that the implementation of WASP is still in an initial phase, yet the results obtained up to now are encouraging.

References

1. Gelfond, M., Lifschitz, V.: Classical Negation in Logic Programs and Disjunctive Databases. New Generation Computing 9, 365–385 (1991)
2. Eiter, T., Gottlob, G., Mannila, H.: Disjunctive Datalog. ACM Transactions on Database Systems 22, 364–418 (1997)
3. Alviano, M., Faber, W., Leone, N., Perri, S., Pfeifer, G., Terracina, G.: The disjunctive datalog system DLV. In: de Moor, O., Gottlob, G., Furche, T., Sellers, A. (eds.) Datalog 2010. LNCS, vol. 6702, pp. 282–301. Springer, Heidelberg (2011)
4. Gebser, M., Kaufmann, B., Neumann, A., Schaub, T.: Conflict-driven answer set solving. In: Twentieth International Joint Conference on Artificial Intelligence, IJCAI 2007, pp. 386–392. Morgan Kaufmann Publishers (2007)
5. Lierler, Y., Maratea, M.: Cmodels-2: SAT-based Answer Set Solver Enhanced to Non-tight Programs. In: Lifschitz, V., Niemelä, I. (eds.) LPNMR 2004. LNCS (LNAI), vol. 2923, pp. 346–350. Springer, Heidelberg (2003)

6. Grasso, G., Iiritano, S., Leone, N., Ricca, F.: Some DLV applications for knowledge management. In: Erdem, E., Lin, F., Schaub, T. (eds.) LPNMR 2009. LNCS, vol. 5753, pp. 591–597. Springer, Heidelberg (2009)

7. Ricca, F., Grasso, G., Alviano, M., Manna, M., Lio, V., Iiritano, S., Leone, N.: Team-building with answer set programming in the gioia-tauro seaport. Theory and Practice of Logic Programming 12, 361–381 (2012)

8. Manna, M., Oro, E., Ruffolo, M., Alviano, M., Leone, N.: The HiLeX system for semantic information extraction. In: Hameurlain, A., Küng, J., Wagner, R. (eds.) TLDKS V. LNCS, vol. 7100, pp. 91–125. Springer, Heidelberg (2012)

9. Ricca, F., Alviano, M., Dimasi, A., Grasso, G., Ielpa, S.M., Iiritano, S., Manna, M., Leone, N.: A Logic-Based System for e-Tourism. Fundamenta Informaticae 105, 35–55 (2010)

10. Calimeri, F., et al.: The Third Answer Set Programming Competition: Preliminary Report of the System Competition Track. In: Delgrande, J.P., Faber, W. (eds.) LPNMR 2011. LNCS, vol. 6645, pp. 388–403. Springer, Heidelberg (2011)

11. Davis, M., Logemann, G., Loveland, D.: A Machine Program for Theorem Proving. Communications of the ACM 5, 394–397 (1962)

12. Zhang, L., Madigan, C.F., Moskewicz, M.W., Malik, S.: Efficient Conflict Driven Learning in Boolean Satisfiability Solver. In: Proceedings of ICCAD 2001, pp. 279–285 (2001)

13. Gaschnig, J.: Performance measurement and analysis of certain search algorithms. PhD thesis, Carnegie Mellon University, Pittsburgh, PA, USA (1979) TR CMU-CS-79-124

14. Gomes, C.P., Selman, B., Kautz, H.A.: Boosting Combinatorial Search Through Randomization. In: Proceedings of AAAI/IAAI 1998, pp. 431–437. AAAI Press (1998)

15. Moskewicz, M.W., Madigan, C.F., Zhao, Y., Zhang, L., Malik, S.: Chaff: Engineering an Efficient SAT Solver. In: Proceedings of DAC 2001, pp. 530–535. ACM (2001)

16. Goldberg, E., Novikov, Y.: BerkMin: A Fast and Robust Sat-Solver. In: Design, Automation and Test in Europe Conference and Exposition, DATE 2002, Paris, France, pp. 142–149. IEEE Computer Society (2002)

17. Faber, W., Leone, N., Pfeifer, G.: Pushing Goal Derivation in DLP Computations. In: Gelfond, M., Leone, N., Pfeifer, G. (eds.) LPNMR 1999. LNCS (LNAI), vol. 1730, pp. 177–191. Springer, Heidelberg (1999)

18. Simons, P., Niemelä, I., Soininen, T.: Extending and Implementing the Stable Model Semantics. Artificial Intelligence 138, 181–234 (2002)

19. Koch, C., Leone, N., Pfeifer, G.: Enhancing Disjunctive Logic Programming Systems by SAT Checkers. Artificial Intelligence 15, 177–212 (2003)

20. Eén, N., Sörensson, N.: An Extensible SAT-solver. In: Giunchiglia, E., Tacchella, A. (eds.) SAT 2003. LNCS, vol. 2919, pp. 502–518. Springer, Heidelberg (2004)

21. Ben-Eliyahu, R., Dechter, R.: Propositional Semantics for Disjunctive Logic Programs. Annals of Mathematics and Artificial Intelligence 12, 53–87 (1994)

22. Luby, M., Sinclair, A., Zuckerman, D.: Optimal speedup of las vegas algorithms. Inf. Process. Lett. 47, 173–180 (1993)

23. Goldberg, E., Novikov, Y.: Berkmin: A fast and robust sat-solver. Discrete Appl. Math. 155, 1549–1561 (2007)

24. Gebser, M., Schaub, T.: Tableau calculi for answer set programming. In: Etalle, S., Truszczyński, M. (eds.) ICLP 2006. LNCS, vol. 4079, pp. 11–25. Springer, Heidelberg (2006)

25. Ward, J., Schlipf, J.: Answer Set Programming with Clause Learning. In: Lifschitz, V., Niemelä, I. (eds.) LPNMR 2004. LNCS (LNAI), vol. 2923, pp. 302–313. Springer, Heidelberg (2003)

26. Ricca, F., Faber, W., Leone, N.: A Backjumping Technique for Disjunctive Logic Programming. AI Communications 19, 155–172 (2006)

The Complexity Boundary of Answer Set Programming with Generalized Atoms under the FLP Semantics

Mario Alviano and Wolfgang Faber

Department of Mathematics
University of Calabria
87030 Rende (CS), Italy
{alviano,faber}@mat.unical.it

Abstract. In recent years, Answer Set Programming (ASP), logic programming under the stable model or answer set semantics, has seen several extensions by generalizing the notion of an atom in these programs: be it aggregate atoms, HEX atoms, generalized quantifiers, or abstract constraints, the idea is to have more complicated satisfaction patterns in the lattice of Herbrand interpretations than traditional, simple atoms. In this paper we refer to any of these constructs as generalized atoms. It is known that programs with generalized atoms that have monotonic or antimonotonic satisfaction patterns do not increase complexity with respect to programs with simple atoms (if satisfaction of the generalized atoms can be decided in polynomial time) under most semantics. It is also known that generalized atoms that are nonmonotonic (being neither monotonic nor antimonotonic) can, but need not, increase the complexity by one level in the polynomial hierarchy if non-disjunctive programs under the FLP semantics are considered. In this paper we provide the precise boundary of this complexity gap: programs with convex generalized atom never increase complexity, while allowing a single non-convex generalized atom (under reasonable conditions) always does. We also discuss several implications of this result in practice.

1 Introduction

Various extensions of the basic Answer Set Programming language have been proposed by allowing more general atoms in rule bodies, for example aggregate atoms, HEX atoms, dl-atoms, generalized quantifiers, or abstract constraints. The FLP semantics defined in [5] provides a semantics to all of these extensions, as it treats all body elements in the same way (and it coincides with the traditional ASP semantics when no generalized atoms are present). The complexity analyses reported in [5] show that in programs with single simple atom rule heads, the main complexity tasks do not increase when the generalized atoms present are monotonic or antimonotonic ($coNP$-complete for cautious reasoning), but there is an increase in complexity otherwise (Π_2^P-complete for cautious reasoning). These complexity results hold under the assumptions of dealing with propositional programs and that determining the satisfaction of a generalized atom in an interpretation can be done in polynomial time. Also throughout this paper, we will work under these assumptions.

However, there are several examples of generalized atoms that are nonmonotonic (neither monotonic nor antimonotonic), for which reasoning is still in $coNP$. Examples

P. Cabalar and T.C. Son (Eds.): LPNMR 2013, LNAI 8148, pp. 67–72, 2013.
© Springer-Verlag Berlin Heidelberg 2013

for such easy nonmonotonic generalized atoms are count aggregates with an equality guard, cardinality constraints with upper and lower bounds, or weight constraints with non-negative weights and upper and lower guards. All of these have the property of being convex, which can be thought of as a conjunction of monotonic and antimonotonic. Convex generalized atoms have been studied in [7], and it is implicit in there, and in general not hard to see that there is no increase in complexity in the presence of atoms of this kind.

In this paper, we show that convex generalized atoms are indeed the only ones for which cautious reasoning under the FLP semantics remains in $coNP$. Our main result is that when a language allows any kind of non-convex generalized atom, Π_2^P-hardness of cautious reasoning can be established. We just require two basic properties of generalized atoms: they should be closed under renaming of atoms, and only a subset of all available (simple) atoms should be relevant for the satisfaction of a single generalized atom (this subset is the domain of the generalized atom). All types of generalized atoms that we are aware of meet these assumptions. Essentially, the first requirement means that it is possible to rename the simple atoms in the representation of a generalized atom while retaining its semantic properties, while the second means that modifying truth values of simple atoms that are irrelevant to the general atom does not alter its semantic behavior.

Our result has several implications that are discussed in more detail in section 4. The main ones concern implementation and rewriting issues, but also simpler identification of the complexity of ASP extensions. In the following, we will present a simple language for our study in section 2; essentially, we view a rule body as a single "structure" that takes the role of a generalized atom (sufficiently detailed and expressive, since the FLP semantics treats rule bodies monolithically anyway and because convexity is closed under conjunction). In section 3 we present our main theorem and its proof, and in section 4 we wrap up.

2   Syntax and Semantics

Let $\mathcal{U}$ be a fixed, countable set of propositional atoms. An interpretation I is a subset of $\mathcal{U}$. A structure S on $\mathcal{U}$ is a mapping of interpretations into Boolean truth values. Each structure S has an associated domain $D_S \subset \mathcal{U}$, indicating those atoms that are relevant to the structure. A general rule r is of the following form:

$$H(r) \leftarrow B(r) \tag{1}$$

where $H(r)$ is a propositional atom in $\mathcal{U}$ referred as the head of r, and $B(r)$ is a structure on $\mathcal{U}$ called the body of r. No particular assumption is made on the syntax of $B(r)$, in the case of normal propositional logic programs these structures are conjunctions of literals. We assume that structures are closed under propositional variants, that is, for any structure S, also $S\sigma$ is a structure for any bijection $\sigma : \mathcal{U} \to \mathcal{U}$, the associated domain is $D_{S\sigma} = \{\sigma(a) \mid a \in D_S\}$.

A general program P is a set of general rules. By datalog^S we refer to the class of programs that may contain only the following rule bodies: structures corresponding to conjunctions of atoms, S, or any of its variants $S\sigma$.

Let $I \subseteq \mathcal{U}$ be an interpretation. I is a model for a structure S, denoted $I \models S$, if S maps I to true. Otherwise, if S maps I to false, I is not a model of S, denoted $I \not\models S$. We require that atoms outside the domain of S are irrelevant for modelhood, that is, for any interpretation I and $X \subseteq \mathcal{U} \setminus D_S$ it holds that $I \models S$ if and only if $I \cup X \models S$. Moreover, for any bijection $\sigma : \mathcal{U} \rightarrow \mathcal{U}$, let $I\sigma = \{\sigma(a) \mid a \in I\}$, and we require that $I\sigma \models S\sigma$ if and only if $I \models S$. I is a model of a rule r of the form (1), denoted $I \models r$, if $H(r) \in I$ whenever $I \models B(r)$. I is a model of a program P, denoted $I \models P$, if $I \models r$ for every rule $r \in P$.

The FLP reduct P^I of a program P with respect to I is defined as the set $\{r \mid r \in P \wedge I \models B(r)\}$. I is a stable model of P if $I \models P^I$ and for each $J \subset I$ it holds that $J \not\models P^I$. A propositional atom a is a cautious consequence of a program P, denoted $P \models_c a$, if a belongs to all stable models of P.

Structures can be characterized in terms of *monotonicity* as follows: Let S be a structure. S is monotonic if for all pairs X, Y of interpretations such that $X \subset Y$, $X \models S$ implies $Y \models S$. S is antimonotonic if for all pairs Y, Z of interpretations such that $Y \subset Z$, $Z \models S$ implies $Y \models S$. S is convex if for all triples X, Y, Z of interpretations such that $X \subset Y \subset Z$, $X \models S$ and $Z \models S$ implies $Y \models S$.

3   Main Complexity Result

It is known that cautious reasoning over answer set programs with generalized atoms under FLP semantics is Π_2^P-complete in general. It is also known that the complexity drops to $coNP$ if structures in body rules are constrained to be convex. This appears to be "folklore" knowledge and can be argued to follow from results in [7]. An easy way to see membership in $coNP$ is that all convex structures can be decomposed into a conjunction of a monotonic and an antimonotonic structure, for which membership in $coNP$ has been shown in [5].

We will therefore focus on showing that convex structures define the precise boundary between the first and the second level of the polynomial hierarchy. In fact, we prove that any extension of datalog by at least one non-convex structure and its variants raises the complexity of cautious reasoning to the second level of the polynomial hierarchy.

The hardness proof is similar to the reduction from 2QBF to disjunctive logic programs as presented in [2]. This reduction was adapted to nondisjunctive programs with nonmonotonic aggregates in [5], and a similar adaption to weight constraints was presented independently in [6]. The fundamental tool in these adaptations in terms of structures is the availability of structures S_1, S_2 that allow for encoding "need to have either atom x^T or x^F, or both of them, but the latter only upon forcing the truth of both atoms." S_1, S_2 have domains $D_{S_1} = D_{S_2} = \{x^T, x^F\}$ and the following satisfaction patterns:

$$\emptyset \models S_1 \quad \{x^T\} \models S_1 \quad \{x^F\} \not\models S_1 \quad \{x^T, x^F\} \models S_1$$
$$\emptyset \models S_2 \quad \{x^T\} \not\models S_2 \quad \{x^F\} \models S_2 \quad \{x^T, x^F\} \models S_2$$

A program that meets the specification is $P = \{x^T \leftarrow S_1, x^F \leftarrow S_2\}$. Indeed, $\emptyset$ is not an answer set of P as $P^\emptyset = P$ and $\emptyset \not\models P$ (so also any extension of P can never have an answer set containing neither x^T nor x^F). Both $\{x^T\}$ and $\{x^F\}$ are answer sets of P, because the reducts cancel one appropriate rule. $\{x^T, x^F\}$ is not an answer set of

P because of minimality ($P^{\{x^T, x^F\}} = P$ and $\{x^T, x^F\} \models P$, but also $\{x^T\} \models P$ and $\{x^F\} \models P$), but can become an answer set in an extension of P that forces the truth of both x^T and x^F.

A crucial observation is that S_1 and S_2 are not just nonmonotonic, but also non-convex. The main idea of our new proof is that any non-convex structure S that is closed under propositional variants can take over the role of both S_1 and S_2. For such an S, we will create appropriate variants $S\sigma^T$ and $S\sigma^F$ that use indexed copies of x^T and x^F in order to obtain the required multitudes of elements:

$$
\begin{array}{lll}
\{a_1, \ldots, a_p\} \models S & \{x_1^T, \ldots, x_p^T\} \models S\sigma^T & \{x_1^F, \ldots, x_p^F\} \models S\sigma^F \\
\{a_1, \ldots, a_p, \ldots, a_q\} \not\models S & \{x_1^T, \ldots, x_q^T\} \not\models S\sigma^T & \{x_1^F, \ldots, x_q^F\} \not\models S\sigma^F \\
\{a_1, \ldots, a_p, \ldots, a_q, \ldots a_r\} \models S & \{x_1^T, \ldots, x_r^T\} \models S\sigma^T & \{x_1^F, \ldots, x_r^F\} \models S\sigma^F
\end{array}
$$

We can then create a program P' acting like P by using x_q^T, x_q^F, $S\sigma^F$ and $S\sigma^T$ in place of x^T, x^F, S_1 and S_2, respectively. In addition, we need some auxiliary rules for the following purposes: to force $x_1^T, \ldots, x_p^T, x_1^F, \ldots, x_p^F$ to hold always; to require the same truth value for $x_{p+1}^T, \ldots, x_q^T$ and similar for $x_{p+1}^F, \ldots, x_q^F$; to force truth of $x_{p+1}^T, \ldots, x_r^T$ whenever any of $x_{q+1}^T, \ldots, x_r^T$ is true and to force truth of $x_{p+1}^F, \ldots, x_r^F$ whenever any of $x_{q+1}^F, \ldots, x_r^F$ is true. The resulting program can then give rise to answer sets containing either x_q^T or x_q^F, or both x_q^T, x_q^F when they are forced in an extension of the program. In particular, the answer sets of P' are the following: $\{x_1^T, \ldots, x_q^T, x_1^F, \ldots, x_p^F\}$, corresponding to $\{x^T\}$; and $\{x_1^T, \ldots, x_p^T, x_1^F, \ldots, x_q^F\}$, corresponding to $\{x^F\}$. Model $\{x_1^T, \ldots, x_r^T, x_1^F, \ldots, x_r^F\}$ instead is not an answer set of P' because of minimality, but it can be turned into an answer set by extending the program suitably. In the proof, we need to make the assumption that all symbols x_i^T and x_j^F are outside the domain D_S, which is not problematic if $\mathcal{U}$ is sufficiently large.

Theorem 1. *Let S be any non-convex structure on a set $\{a_1, \ldots, a_s\}$. Cautious reasoning over* datalog^S *is Π_2^P-hard.*

Proof. Deciding validity of a QBF $\Psi = \forall x_1 \cdots \forall x_m \exists y_1 \cdots \exists y_n E$, where E is in 3CNF, is a well-known Π_2^P-hard problem. Formula Ψ is equivalent to $\neg\Psi'$, where $\Psi' = \exists x_1 \cdots \exists x_m \forall y_1 \cdots \forall y_n E'$, and E' is a 3DNF equivalent to $\neg E$ and obtained by applying De Morgan's laws. To prove the claim we construct a datalog^S program P_Ψ such that $P_\Psi \models_c w$ (w a fresh atom) if and only if Ψ is valid, i.e., iff Ψ' is invalid.

Since S is a non-convex structure by assumption, there are interpretations A, B, C such that $A \subset B \subset C$, $A \models S$ and $C \models S$ but $B \not\models S$. Without loss of generality, let $A = \{a_1, \ldots, a_p\}$, $B = \{a_1, \ldots, a_q\}$ and $C = \{a_1, \ldots, a_r\}$, for $0 \le p < q < r \le s$. Let $E' = (l_{1,1} \wedge l_{1,2} \wedge l_{1,3}) \vee \cdots \vee (l_{k,1} \wedge l_{k,2} \wedge l_{k,3})$, for some $k \ge 1$.

Program $P_{\Psi'}$ is reported in Fig. 1, where $\sigma_i^T(a_j) = x_{i,j}^T$ and $\sigma_i^F(a_j) = x_{i,j}^F$ for all $i = 1, \ldots, m$ and $j = 1, \ldots, q$; $\theta_i^T(a_j) = y_{i,j}^T$ and $\theta_i^F(a_j) = y_{i,j}^F$ for all $i = 1, \ldots, n$ and $j = 1, \ldots, q$; $\mu(x_i) = x_{i,r}^T$ and $\mu(\neg x_i) = x_{i,r}^F$ for all $i = 1, \ldots, m$; $\mu(y_i) = y_{i,r}^T$ and $\mu(\neg y_i) = y_{i,r}^F$ for all $i = 1, \ldots, n$.

Rules (2)–(9) represent one copy of the program P' discussed earlier for each of the x_i and y_j ($i = 1, \ldots, m$; $j = 1, \ldots, n$), and so force each answer set of P_Ψ to contain at least one of $x_{i,q}^T$, $x_{i,q}^F$, and $y_{j,q}^T$, $y_{j,q}^F$, respectively, encoding an assignment

$$
\begin{array}{lll}
x_{i,j}^{T} \leftarrow & x_{i,j}^{F} \leftarrow & i \in \{1,\dots,m\},\ j \in \{1,\dots,p\} & (2) \\
x_{i,j}^{T} \leftarrow x_{i,k}^{T} & x_{i,j}^{F} \leftarrow x_{i,k}^{F} & i \in \{1,\dots,m\},\ j,k \in \{p+1,\dots,q\} & (3) \\
x_{i,j}^{T} \leftarrow x_{i,k}^{T} & x_{i,j}^{F} \leftarrow x_{i,k}^{F} & i \in \{1,\dots,m\},\ j \in \{p+1,\dots,r\},\ k \in \{q+1,\dots,r\} & (4) \\
x_{i,q}^{T} \leftarrow S\sigma_{i}^{F} & x_{i,q}^{F} \leftarrow S\sigma_{i}^{T} & i \in \{1,\dots,m\} & (5) \\
y_{i,j}^{T} \leftarrow & y_{i,j}^{F} \leftarrow & i \in \{1,\dots,n\},\ j \in \{1,\dots,p\} & (6) \\
y_{i,j}^{T} \leftarrow y_{i,k}^{T} & y_{i,j}^{F} \leftarrow y_{i,k}^{F} & i \in \{1,\dots,n\},\ j,k \in \{p+1,\dots,q\} & (7) \\
y_{i,j}^{T} \leftarrow y_{i,k}^{T} & y_{i,j}^{F} \leftarrow y_{i,k}^{F} & i \in \{1,\dots,n\},\ j \in \{p+1,\dots,r\},\ k \in \{q+1,\dots,r\} & (8) \\
y_{i,q}^{T} \leftarrow S\theta_{i}^{F} & y_{i,q}^{F} \leftarrow S\theta_{i}^{T} & i \in \{1,\dots,n\} & (9) \\
y_{i,j}^{T} \leftarrow sat & y_{i,j}^{F} \leftarrow sat & i \in \{1,\dots,n\},\ j \in \{p+1,\dots,r\} & (10) \\
sat \leftarrow \mu(l_{i,1}),\mu(l_{i,2}),\ \mu(l_{i,3}) & & i \in \{1,\dots,k\} & (11) \\
a_{j} \leftarrow & & j \in \{1,\dots,p\} & (12) \\
a_{j} \leftarrow a_{k} & & j,k \in \{p+1,\dots,q\} & (13) \\
a_{j} \leftarrow sat & & j \in \{p+1,\dots,q\} & (14) \\
w \leftarrow S & & & (15)
\end{array}
$$

Fig. 1. Program $P_{\Psi'}$

of the propositional variables in Ψ'. Rules (10) are used to simulate universality of the y variables, as described later. Having an assignment, rules (11) derive sat if the assignment satisfies some disjunct of E' (and hence also E' itself). Finally, rules (12)–(15) derive w if sat is false.

We first show that Ψ not valid implies $P_{\Psi} \not\models_{c} w$. If Ψ is not valid, Ψ' is valid. Hence, there is an assignment ν for $x_{1},\dots,x_{m}$ such that no extension to $y_{1},\dots,y_{n}$ satisfies E, i.e., all these extensions satisfy E'. Consider the following model of P_{Ψ}:

$$
\begin{aligned}
M = \ & \{x_{i,j}^{T} \mid \nu(x_{i}) = 1,\ i = 1,\dots,m,\ j = p+1,\dots,q\} \\
\cup\ & \{x_{i,j}^{F} \mid \nu(x_{i}) = 0,\ i = 1,\dots,m,\ j = p+1,\dots,q\} \\
\cup\ & \{x_{i,j}^{T}, x_{i,j}^{F} \mid i = 1,\dots,m,\ j = 1,\dots,p\} \\
\cup\ & \{y_{i,j}^{T}, y_{i,j}^{F} \mid i = 1,\dots,n,\ j = 1,\dots,r\} \\
\cup\ & \{a_{j} \mid j = 1,\dots,q\} \cup \{sat\}
\end{aligned}
$$

We claim that M is a stable model of P_{Ψ}. Consider $I \subseteq M$ such that $I \models P_{\Psi}^{M}$. I contains all x atoms in M due to rules (2)–(5). I also contains an assignment for the y variables because of rules (6)–(9). Since any assignment for the ys satisfies at least a disjunct of E', from rules (11) we derive $sat \in I$. Hence, rules (10) force all y atoms to belong to I, and thus $I = M$ holds, which proves that M is a stable model of P_{Ψ}.

Now we show that $P_{\Psi} \not\models_{c} w$ implies that Ψ is not valid. To this end, let M be a stable model of P_{Ψ} such that $w \notin M$. Hence, by rule (15) we have that $M \not\models S$. Since $A \subseteq M$ because of rules (12), in order to have $M \not\models S$, atoms in B have to belong to M. These atoms can be supported only by rules (13)–(14), from which $sat \in M$ follows. From $sat \in M$ and rules (10), we have $y_{i,q}^{T}, y_{i,q}^{F} \in M$ for all $i = 1,\dots,n$. And M contains either $x_{i,q}^{T}$ or $x_{i,q}^{F}$ for $i = 1,\dots,m$ because of rules (2)–(5). Suppose by contradiction that Ψ is valid. Thus, for all assignments of $x_{1},\dots,x_{m}$, there is an assignment for $y_{1},\dots,y_{n}$ such that E is true, i.e., E' is false. Let ν be an assignment satisfying E and such that $\nu(x_{i}) = 1$ if $x_{i,q}^{T} \in M$ and $\nu(x_{i}) = 0$ if $x_{i,q}^{F} \in M$ for all $i = 1,\dots,m$.

Consider $I = M \setminus \{sat\} \setminus \{y_{i,j}^T, y_{i,j}^F \mid i = 1, \ldots, n, \, j = q+1, \ldots, r\} \setminus \{y_{i,j}^T \mid \nu(y_i) = 0, \, i = 1, \ldots, n, \, j = p+1, \ldots, q\} \setminus \{y_{i,j}^F \mid \nu(y_i) = 1, \, i = 1, \ldots, n, \, j = p+1, \ldots, q\}$.
Since ν satisfies E, ν does not satisfy E', i.e., no disjunct of E' is satisfied by ν. Hence, all rules (11) are satisfied, and thus $I \models P_\Psi^M$, contradicting the assumption that M is a stable model of P_Ψ, and so Ψ is not valid.   □

4 Discussion

Our results have several consequences. First of all, from our proof it is easy to see that convex generalized atoms also form the complexity boundary for deciding whether a program has an answer set (in this case the boundary is between NP and Σ_2^P) and for checking whether an interpretation is an answer set of a program (from P to $coNP$). It also means that for programs containing only convex structures, techniques as those presented in [1] can be used for computing answer sets, while the presence of any non-convex structure requires more complex techniques such as those presented in [4]. There are several examples for convex structures that are easy to identify syntactically: count aggregates with equality guards, sum aggregates with positive summands and equality guards, dl-atoms that do not involve ⊖ and rely on a tractable Description Logic [3]. However many others are in general not convex, for example sum aggregates that involve both positive and negative summands, times aggregates that involve the factor 0, average aggregates, dl-atoms with ⊖, and so on. It is still possible to find special cases of such structures that are convex, but that requires deeper analyses.

The results also immediately imply impossibility results for rewritability: unless the polynomial hierarchy collapses to its first level, it is not possible to rewrite a program with non-convex structures into one containing only convex structures (for example, a program not containing any generalized atoms), unless disjunction or similar constructs are allowed in rule heads.

The results obtained in this work apply only to the FLP semantics. Whether the results carry over in any way to other semantics is unclear and left to future work.

References

1. Alviano, M., Calimeri, F., Faber, W., Leone, N., Perri, S.: Unfounded Sets and Well-Founded Semantics of Answer Set Programs with Aggregates. JAIR 42, 487–527 (2011)
2. Eiter, T., Gottlob, G.: On the Computational Cost of Disjunctive Logic Programming: Propositional Case. AMAI 15(3/4), 289–323 (1995)
3. Eiter, T., Ianni, G., Lukasiewicz, T., Schindlauer, R., Tompits, H.: Combining answer set programming with description logics for the semantic web. Artif. Intell. 172(12-13), 1495–1539 (2008)
4. Faber, W.: Unfounded Sets for Disjunctive Logic Programs with Arbitrary Aggregates. In: Baral, C., Greco, G., Leone, N., Terracina, G. (eds.) LPNMR 2005. LNCS (LNAI), vol. 3662, pp. 40–52. Springer, Heidelberg (2005)
5. Faber, W., Leone, N., Pfeifer, G.: Semantics and complexity of recursive aggregates in answer set programming. AI 175(1), 278–298 (2011), special Issue: John McCarthy's Legacy
6. Ferraris, P.: Answer Sets for Propositional Theories. In: Baral, C., Greco, G., Leone, N., Terracina, G. (eds.) LPNMR 2005. LNCS (LNAI), vol. 3662, pp. 119–131. Springer, Heidelberg (2005)
7. Liu, L., Truszczyński, M.: Properties and applications of programs with monotone and convex constraints. JAIR 27, 299–334 (2006)

ARVis: Visualizing Relations between Answer Sets*

Thomas Ambroz, Günther Charwat, Andreas Jusits,
Johannes Peter Wallner, and Stefan Woltran

Vienna University of Technology, Institute of Information Systems, Austria

Abstract. Answer set programming (ASP) is nowadays one of the most
popular modeling languages in the areas of Knowledge Representation
and Artificial Intelligence. Hereby one represents the problem at hand
in such a way that each model of the ASP program corresponds to one
solution of the original problem. In recent years, several tools which
support the user in developing ASP applications have been introduced.
However, explicit treatment of one of the main aspects of ASP, multiple
solutions, has received less attention within these tools. In this work,
we present a novel system to visualize *relations* between answer sets of
a given program. The core idea of the system is that the user specifies
the concept of a relation by an ASP program itself. This yields a highly
flexible system that suggests potential applications beyond development
environments, e.g., applications in the field of abduction, which we will
discuss in a case study.

Keywords: Answer set programming, Systems, Abduction.

1 Introduction

Answer set programming [1] (ASP, for short) is a declarative problem solving
paradigm, rooted in logic programming and non-monotonic reasoning. The main
idea of ASP is to represent the problem at hand in such a way that the models of
the ASP program characterize the solutions of the original problem. Due to continuous refinements over the last decade state-of-the-art answer set solvers [7,12]
nowadays support a rich language and are capable of solving hard problems efficiently. This made ASP one of today's most popular modeling languages in
Knowledge Representation and Reasoning (KRR).

As a next step to promote ASP, tools for user support are required, in particular systems featuring useful graphical user interfaces (GUIs). Recent development in this direction includes the following systems: ASPViz [3] visualizes
the answer sets of a given program based on a second program that contains the
visualization information specified via predefined predicates. IDPDraw[1] works

---

* Supported by the Austrian Science Fund (FWF) under grant P25607, and by the
  Vienna University of Technology special fund "Innovative Projekte" (9006.09/008).
[1] https://dtai.cs.kuleuven.be/krr/software/visualisation

P. Cabalar and T.C. Son (Eds.): LPNMR 2013, LNAI 8148, pp. 73–78, 2013.

in a similar fashion, but additionally supports explicit time point information in order to represent the result in different states. Kara [11], part of the SeaLion development environment [13] for ASP, supports, in contrast to ASPViz and IDPDraw, modifiable visualizations such that the underlying answer sets can be manipulated. ASPIDE [5] includes many tools for ASP development (e.g. auto-completion, code-templates, debugging, profiling and more) as well as a visual editor that allows to "draw" logic programs. The visualization aspects of these tools focus on the representation of *single* answer sets.

To the best of our knowledge there does not exist a tool yet that is capable of visualizing *relations between answer sets*, thus taking care of one of the main features of ASP, the possibility to deliver multiple solutions. Visualizing the relations is of importance in many aspects. Relations can express a certain *preference* criterion over the solutions. In some domains it is, for example, relevant to compare (parts of) solutions w.r.t. to subset inclusion or cardinality. Furthermore, it might be useful to visualize the results of problems that build upon a *graph structure*. Regarding *debugging* of ASP programs, visualizing relations between answer sets that are "close" to each other can help in finding problems in the developed program, e.g., in case too many answer sets are computed.

In this work we present a new tool, ARVis which we believe appropriately closes this gap. The main purpose of ARVis is to visualize answer sets and their relations by means of a directed graph. Nodes represent answer sets of the program at hand, whereas edges represent relations. The relations themselves are specified in ASP via a second program, thereby giving the user high flexibility. The graph representation allows the user to visually compare answer sets and study their structure. Thereby, bugs, preferences or graph structures in general can be recognized much easier than by analyzing the output of the ASP solver directly. Obviously, ARVis is not primarily tailored to performance since a potential exponential number of answer sets from the first program has to be processed by the second program. But this is not only a drawback. It also allows to implement problems via ASP which are too hard to be solved via a single call. Logic-based abduction is such a problem domain, which we use as a case study to illustrate the functioning of ARVis. The system is available at

http://www.dbai.tuwien.ac.at/research/project/arvis/

2   Answer Set Relationship Visualizer

In a nutshell, the "Answer Set Relationship Visualizer" ARVis is intended for the visualization of answer sets and their relations by means of a directed graph. Each node in the graph represents an answer set and a directed edge between two nodes denotes a relation. In general, the answer sets of a first user-specified ASP encoding are passed to a second user-specified encoding which defines the relations between them. ARVis provides an interface for analysis of the resulting graph(s), thereby easing the study of relations between answer sets. In particular the structured and concise representation allows to quickly receive an impression on the problem at hand.

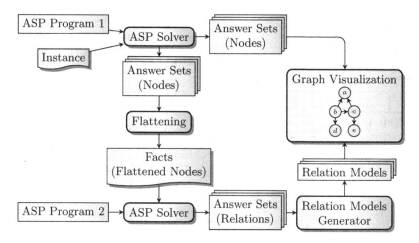

Fig. 1. ARVis: System Architecture

In Fig. 1 an overview of the concept underlying ARVis is given. A problem specified in *ASP Program 1* and an *Instance* are given as input to an ASP solver. Each answer set computed by the solver is represented as a node in the graph visualization. Furthermore, the answer sets are *flattened*, i.e., an id (unique per answer set) is added to the atoms of each answer set. This set of facts, together with *ASP Program 2* (specifying the relations between the answer sets) are again passed to an ASP solver. The answer sets of this call are given to the relations models generator. It is responsible for either generating one graph per answer set or combining all relations into a single graph model. Furthermore, it handles the selection of edge predicates (e.g. for defining normal and highlighted edge predicates of the resulting graph). Finally, the models are handed over to the graph visualization for detailed analysis.

ARVis follows a 5-step approach which is implemented as a GUI wizard that guides the user through the visualization process. Fig. 2 shows the steps from a user perspective. Step (1) and (3) correspond to the input specification as discussed in the previous paragraph. Additionally the user may filter for the desired or necessary predicates contained in the answer sets of the first call (step 2) in order to reduce the amount of data that is flattened. Furthermore, binary predicates that represent (highlighted) edges can be selected in step (4). Finally, the graph is visualized based on the relationship models (step 5). The user may arrange nodes manually or call the built-in algorithms that lay out the graph based on the Kamada-Kawai (KK) layout [9] (minimizes the "tension" between adjacent nodes) or the Fruchterman-Reingold (FR) layout [6] (aims at "nice" graphs with uniform edge lengths). In case the relations models generator is configured to return one graph model per answer set of the second program the user can select the different graphs and (visually) compare them. Any node in the graph can be selected in order to inspect the answer set underlying it. Finally, the graph can be exported as image and all data (answer sets of the two program calls) can be saved for further post-processing with other tools.

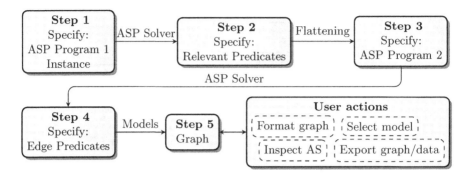

Fig. 2. ARVis: User Perspective

ARVis is implemented in Java and internally uses the graph library Jung[2]. Our tool is capable of handling hundreds of nodes and edges in each graph. ARVis currently supports the ASP solvers DLV [12] and clingo [7], i.e. any user-specified configuration that complies with the solver output format of clasp.

3   Case Study: Abduction

Abduction is a famous non-monotonic reasoning formalism in AI [4] to provide possible explanations for observed behavior. Its propositional variant is very intuitively defined and is composed of manifestations, which one wants to have explained by a subset of a set of hypotheses. Such a set of hypotheses, augmented with a background theory, is a solution to the given abduction problem if (1) it entails the manifestations and (2) is consistent. All of the components of an abduction instance are written as propositional logic formulae. Naturally there may be different solutions to explain a particular manifestation. Such solutions can be ordered according to a preference relation. For instance, we can prefer a solution to another if the set of hypotheses of the first is a subset of the other. Another variant is to relate solutions w.r.t. their cardinality, i.e. one explains the manifestation using fewer hypotheses. One can refine these relations by adding priorities to hypotheses, s.t. it is checked in descending order of priority if one solution is preferred to another w.r.t. the corresponding subsets of the hypotheses. Similarly, one can attach penalties to hypotheses to indicate, e.g., costs. Thus, overall we considered five types of preferences, namely preference w.r.t. subset and cardinality relations with or without priorization and lastly penalization.

Using ARVis we can now investigate preferences between multiple solutions of an abduction instance by creating a graph where the nodes represent the solutions and the directed edges the preference relation. For this we specify the following two programs (available at the system page) for the ARVis workflow:

– *ASP Program 1*: Derives all solutions of the given abduction instance.
– *ASP Program 2*: Relates the solutions with the different preferences.

---

[2] http://jung.sourceforge.net/

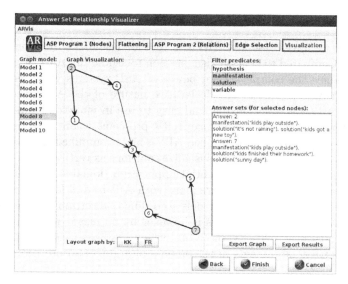

Fig. 3. Example: Visualization of an Abduction Problem

The first program guesses a candidate solution from the set of hypotheses and checks if it entails the manifestations and is consistent w.r.t. the background theory. This can be achieved with the help of the saturation technique. The second program then receives the solutions in form of ASP facts and derives the preference relation. Each answer set of the second program represents a distinct preference relation, where we can additionally derive for each type of preference also the transitive closure. Note that for some of the preference relations considered here it is not possible to compactly encode the problem of finding minimal solutions w.r.t. to the preference relation within one ASP program, due to the corresponding complexity results [4]. Here, our approach of utilizing two ASP programs is not only beneficial for visualization purposes but also allows to handle problems even beyond the second level of the polynomial hierarchy.

Let us outline a part of the second ASP program, which receives the answer sets of the first program (they express solutions via the sol/1 predicate) in flattened form as facts $as(i, sol, s)$, where i stands for an answer set identifier. Preference w.r.t. subset inclusion, for instance, is now achieved by simply comparing answer sets w.r.t. the sol/1 predicate. The edge relation is represented by the prefer/2 predicate and is derived in this case as follows.

$\quad$ notcontained$(X, Y) \leftarrow$ as$(Y, sol, S),$ not as$(X, sol, S),$ as$(X, sol, \_).$
$\quad$ prefer$(X, Y) \leftarrow$ notcontained$(X, Y),$ not notcontained$(Y, X).$

Fig. 3 shows an example visualization for the abduction problem. Here every node represents a solution to the problem, i.e., a set of hypotheses explaining a manifestation. The edges represent preferences between the solutions. Each of the models on the left encode a different preference relation. On the right the answer sets underlying the selected (i.e. highlighted) vertices are shown. The contents of the answer sets are restricted to the filtered predicates. In this case the selected nodes, labeled with "2" and "7", represent subset-minimal solutions.

4 Conclusion

In this paper we have presented ARVis, a tool for visualizing relations between answer sets. Compared to general visualization tools and libraries for graphs our system combines the computational power of ASP systems with visualization aspects. Moreover, thanks to the declarative nature of the ASP encodings we believe that ARVis is quite flexible and easy to use in many domains. Application areas for ARVis are, for instance, KRR problems where relating multiple solutions is a common task. We applied ARVis and exemplified its workflow in a case study for abduction, where we visualize solutions as vertices and preferences over them via edges. Further potential application domains are planning [8], where one may analyze and compare plans with ARVis, and knowledge base revision [10], where the selection of models according to a certain distance measure can be visualized by ARVis. ARVis is however by no means restricted to treat preference or distance relations. In [2] the system has been used for visualization of conflicts between arguments in the domain of argumentation theory.

References

1. Brewka, G., Eiter, T., Truszczyński, M.: Answer set programming at a glance. Commun. ACM 54(12), 92–103 (2011)
2. Charwat, G., Wallner, J.P., Woltran, S.: Utilizing ASP for generating and visualizing argumentation frameworks. In: ASPOCP 2012, pp. 51–65 (2012)
3. Cliffe, O., De Vos, M., Brain, M., Padget, J.: ASPVIZ: Declarative visualisation and animation using answer set programming. In: Garcia de la Banda, M., Pontelli, E. (eds.) ICLP 2008. LNCS, vol. 5366, pp. 724–728. Springer, Heidelberg (2008)
4. Eiter, T., Gottlob, G.: The complexity of logic-based abduction. J. ACM 42(1), 3–42 (1995)
5. Febbraro, O., Reale, K., Ricca, F.: ASPIDE: Integrated development environment for answer set programming. In: Delgrande, J.P., Faber, W. (eds.) LPNMR 2011. LNCS, vol. 6645, pp. 317–330. Springer, Heidelberg (2011)
6. Fruchterman, T.M.J., Reingold, E.M.: Graph drawing by force-directed placement. Softw., Pract. Exper. 21(11), 1129–1164 (1991)
7. Gebser, M., Kaminski, R., Kaufmann, B., Ostrowski, M., Schaub, T., Schneider, M.: Potassco: The Potsdam answer set solving collection. AI Commun. 24(2), 105–124 (2011)
8. Hendler, J.A., Tate, A., Drummond, M.: AI planning: Systems and techniques. AI Magazine 11(2), 61–77 (1990)
9. Kamada, T., Kawai, S.: An algorithm for drawing general undirected graphs. Inf. Process. Lett. 31(1), 7–15 (1989)
10. Katsuno, H., Mendelzon, A.O.: Propositional knowledge base revision and minimal change. Artif. Intell. 52(3), 263–294 (1991)
11. Kloimüllner, C., Oetsch, J., Pührer, J., Tompits, H.: Kara: A system for visualising and visual editing of interpretations for answer-set programs. In: WLP 2011 (2011)
12. Leone, N., Pfeifer, G., Faber, W., Eiter, T., Gottlob, G., Perri, S., Scarcello, F.: The DLV system for knowledge representation and reasoning. ACM Trans. Comput. Log. 7(3), 499–562 (2006)
13. Oetsch, J., Pührer, J., Tompits, H.: The SeaLion has landed: An IDE for answer-set programming – Preliminary report. In: WLP 2011 (2011)

Symbolic System Synthesis Using Answer Set Programming

Benjamin Andres[1], Martin Gebser[1], Torsten Schaub[1],
Christian Haubelt[2], Felix Reimann[3], and Michael Glaß[3]

[1] Institute for Computer Science,
University of Potsdam, Germany
{bandres,gebser,torsten}@cs.uni-potsdam.de
[2] Institute of Applied Microelectronics and Computer Engineering
University of Rostock, Germany
christian.haubelt@uni-rostock.de
[3] Chair for Hardware/Software Co-Design
University of Erlangen-Nuremberg, Germany
{felix.reimann,glass}@cs.fau.de

Abstract. Recently, Boolean Satisfiability (SAT) solving has been proposed to tackle the increasing complexity in high-level system design. Working well for system specifications with a limited amount of routing options, they tend to fail for densely connected computing platforms. This paper proposes an automated system design approach employing Answer Set Programming (ASP). ASP provides a stringent semantics, allowing for an efficient representation of routing options. An automotive case-study illustrates that the proposed ASP-based system design approach is competitive for sparsely connected computing platforms, while it outperforms SAT-based approaches for dense Networks-on-Chip by an order of magnitude.

1 Introduction

Embedded computing systems surround us in our daily life. They are application-specific computing systems embedded into a technical context. Examples of embedded computing systems are automotive, train, and avionic control systems, smart phones, medical devices, home and industrial automation systems, etc. In contrast to general purpose computing systems, embedded computing systems are not only optimized for performance; they additionally have to satisfy power, area, reliability, real-time constraints, to name just a few. As a consequence, the computing platform is adapted to the given application. At the system-level, however, resulting embedded computing platforms are still as complex as heterogeneous multi-processor systems, i.e., several different processing cores are interconnected and the memory subsystem is optimized for the application as well. Finally, the application has to be mapped optimally onto the resulting computing platform. In summary, embedded computing system design includes many interdependent design decisions.

The increasing complexity of interdependent decisions in embedded computing systems design demands for compact design space representations and highly efficient

P. Cabalar and T.C. Son (Eds.): LPNMR 2013, LNAI 8148, pp. 79–91, 2013.

automatic decision engines, resulting in automatic *system synthesis* approaches. Especially, formal methods have shown to be useful in past. (Pseudo-)Boolean Satisfiability (PB/SAT; [2]) solving has been successfully applied in the past to such problems. In particular, explicit modeling of routing decisions in PB formulas has recently enhanced the range of applicability of PB/SAT solvers in synthesizing networked embedded systems [10].

PB/SAT-based approaches to system synthesis work well in the presence of system specifications offering a limited amount of routing options. Such system specifications can be found, e.g., in the automotive or bus-based Multi-Processor System-on-Chip (MPSoC) domain. However, there is a trend towards densely connected networks also in the embedded systems domain. In fact, future MPSoCs are expected to be composed of several hundred processors connected by Networks-on-Chip (NoC) [4]. Hence, system synthesis approaches will face vast design spaces for densely connected networks, resulting in prohibitively long solving times when using PB/SAT-based approaches.

In this paper, we investigate system synthesis scenarios relying on reachability for message routing. We propose a formal approach employing Answer Set Programming (ASP; [1]), a solving paradigm stemming from the area of Knowledge Representation. In contrast to PB/SAT, ASP provides a rich modeling language as well as a more stringent semantics, which allows for succinct design space representations. In particular, ASP supports expressing reachability directly in the modeling language. As a result, much smaller problem descriptions lead to significant reductions in solving time for densely connected networks.

In what follows, we assume some familiarity with ASP, its semantics as well as its basic language constructs. A comprehensive treatment of ASP can be found in [1,7]. Our encodings are written in the input language of *gringo* 3 [6].

After surveying related work, Section 3 introduces the system synthesis setting studied in the sequel. Section 4 provides dedicated ASP formulations of system synthesis. The experiments in Section 5 illustrate the effectiveness of our ASP-based approach. Section 6 concludes the paper.

2   Related Work

Symbolic system synthesis approaches based on Integer Linear Programming (ILP) can be found in the area of hardware/software partitioning (cf. [13]). Such approaches were often limited to the classical bipartitioning problem, i.e., the target platform is composed of a CPU and an FPGA. An extension towards multiple resources and a simple single-hop communication mapping can be found in [3]. In the same work, SAT is reduced to the problem of computing feasible allocations and bindings in platform-based system synthesis approaches, thereby showing that system synthesis is NP-complete. In turn, [8] shows how to reduce the system synthesis problem to SAT in polynomial time; this allows for symbolic SAT-based system synthesis. An analogous approach based on binary decision diagrams is presented in [12]. Since the space requirements of binary decision diagrams may grow exponentially, it could only be applied to small systems. In [11], a first approach to integrate linear constraint checking into SAT-based system synthesis is reported, leading to a PB problem encoding. All aforementioned approaches still use simple single-hop communication as underlying model. However, single-hop communication models are no longer appropriate when designing complex multi-core systems.

In [10], the authors show how to perform symbolic system synthesis including multi-hop communication routing with PB solving techniques. This work is closely related to ours, and we use it as a starting point for the work at hand. We show that the PB-based approach published in [10] does not scale well for system specifications permitting many routing options. By reformulating the PB representation in ASP and exploiting semantic features in expressing reachability, symbolic system synthesis can be applied to more complex system specifications based on densely connected communication networks.

The potential of ASP for system synthesis was already discovered in [9], where it was shown to outperform an ILP-based approach by several orders of magnitude. In contrast to our work, the system synthesis problem considered in [9] does not involve multi-hop communication routing. Moreover, contrary to the genuine ASP encoding(s) developed in Section 4, the one in [9] was derived from an ILP specification without making use of any elaborate ASP features.

3 Symbolic System Synthesis

System synthesis comprises several design phases: the allocation of a computing platform, the binding of tasks onto allocated resources, and the scheduling of tasks for resolving resource conflicts. Each phase, viz., allocation, binding, and scheduling, can be performed either statically or dynamically. We here assume that allocation and binding are accomplished statically, whereas scheduling is realized dynamically. Accordingly, we concentrate on allocation and binding in the sequel.

In order to automate the synthesis of a system implementing an application, the application is modeled by a *task graph* (T, E_T). Its vertices T represent *tasks* and are bipartitioned into *process* tasks P and *communication* tasks C, that is, $T = P \cup C$ and $P \cap C = \emptyset$. The directed edges $E_T \subseteq (P \times C) \cup (C \times P)$ model data and control dependencies between tasks, where every communication task has exactly one predecessor and an arbitrary (positive) number of successors, thus assuming single-source multicast communication.

An exemplary task graph is shown in the upper part of Figure 1. The leftmost task p_s reads data from a sensor and sends it to a master task p_m via communication task c_1. The master task then schedules the workload and passes data via communication task c_2 on to the worker tasks p_1 and p_2. Both workers send their results back to the master via communication tasks c_3 and c_4. Finally, the master uses the combined result to control an actuator task p_a via communication task c_5.

An architecture template, representing all possible instances of a computing platform, is modeled by a *platform graph* (R, E_R). Its vertices R represent resources like processors, buses, memories, etc., and the directed edges $E_R \subseteq R \times R$ model communication connections between them. The lower part of Figure 1 shows a platform graph containing six computational and two communication resources along with 18 connections. Any subgraph of a platform graph constitutes a computing platform instance.

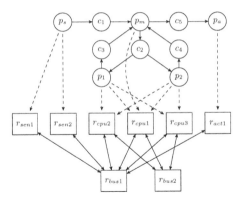

Fig. 1. A system model consisting of a task graph, a platform graph and mapping options

Given a task graph (T, E_T) and a platform graph (R, E_R), mapping options of tasks $t \in T$ are determined by $R_t \subseteq R$, providing resources on which t can be implemented. Assuming that communication tasks can be routed via every resource, the mapping options of process tasks are indicated by dashed arrows in Figure 1, while the (unrestricted) options of communication tasks are not displayed explicitly.

Following [10], the system synthesis problem can be defined as follows. For (T, E_T) and (R, E_R) as above, select an allocation $A \subseteq R$ of resources and a binding $b : T \to 2^R$ such that the following conditions are fulfilled:

- $b(t) \subseteq R_t$ for each task $t \in T$,
- $|b(p)| = 1$ for each process task $p \in P$, and
- for each $(p, c) \in (P \times C) \cap E_T$, there is an arborescence $(b(c), E)$ with root $r \in b(p)$ such that $E \subseteq E_R$ and $\{\hat{r} \mid (c, \hat{p}) \in E_T, \hat{r} \in b(\hat{p})\} \subseteq b(c)$.

These conditions require that each process task is mapped to exactly one resource and that each communication task can be routed (acyclicly) from the sender to resources of its targets.

Figure 2 shows a feasible implementation for the example in Figure 1, consisting of the resource allocation $A = \{r_{sen1}, r_{bus1}, r_{bus2}, r_{cpu1}, r_{cpu2}, r_{cpu3}, r_{act1}\}$ along with the following mapping b:

$$
\begin{aligned}
b(p_s) &= \{r_{sen1}\} & b(c_1) &= \{r_{sen1}, r_{bus1}, r_{cpu1}\} \\
b(p_m) &= \{r_{cpu1}\} & b(c_2) &= \{r_{cpu1}, r_{bus1}, r_{cpu2}, r_{bus2}, r_{cpu3}\} \\
b(p_1) &= \{r_{cpu2}\} & b(c_3) &= \{r_{cpu2}, r_{bus2}, r_{cpu1}\} \\
b(p_2) &= \{r_{cpu3}\} & b(c_4) &= \{r_{cpu3}, r_{bus2}, r_{cpu1}\} \\
b(p_a) &= \{r_{act1}\} & b(c_5) &= \{r_{cpu1}, r_{bus1}, r_{act1}\}
\end{aligned}
$$

For clarity communication task mappings are omitted in Figure 2. Instead, the routing of c_2 is shown. Leading from the resource r_{cpu1} of the master task p_m over r_{bus1}, r_{cpu2}, and r_{bus2} to r_{cpu3}, thus visiting the resources r_{cpu2} and r_{cpu3} of the workers p_1 and p_2.

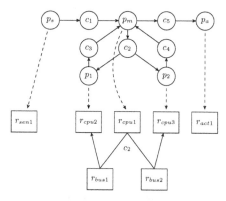

Fig. 2. A feasible implementation of the example system model. Only the routing of c_2 is shown.

PB/SAT-based approaches express system synthesis in terms of (Pseudo-)Boolean formulas. In particular, the PB encoding in [10] relies on the following kinds of Boolean variables:

- a variable $\mathbf{r}$ for each resource $r \in R$, indicating whether r is allocated ($\mathbf{r} = 1$) or not ($\mathbf{r} = 0$),
- a variable $\mathbf{t}_r$ for each task $t \in T$ and each of its mapping options $r \in R_t$, indicating whether t is bound onto r, and
- variables $\mathbf{c}_{r,i}$ for each communication task $c \in C$, its routing options $r \in R_c$, and $i \in \{0, \dots, n\}$ for some integer n, indicating whether c is routed over r at step i.

The following constraints on such variables were used in [10]:

$$\sum_{r \in R_p} \mathbf{p}_r = 1, \qquad \forall p \in P \tag{A}$$

$$\sum_{r \in R_c} \mathbf{c}_{r,0} = 1, \qquad \forall c \in C \tag{B}$$

$$\mathbf{p}_r - \mathbf{c}_{r,0} = 0, \qquad \forall c \in C, p \in \{\hat{p} \mid (\hat{p}, c) \in E_T\}, r \in R_p \cap R_c \tag{C}$$

$$\mathbf{c}_r - \mathbf{p}_r \geq 0, \qquad \forall p \in P, c \in \{\hat{c} \mid (\hat{c}, p) \in E_T\}, r \in R_p \cap R_c \tag{D}$$

$$\sum_{i=0}^{n} \mathbf{c}_{r,i} \leq 1, \qquad \forall c \in C, r \in R_c \tag{E}$$

$$\sum_{i=0}^{n} \mathbf{c}_{r,i} - \mathbf{c}_r \geq 0, \qquad \forall c \in C, r \in R_c \tag{F}$$

$$\mathbf{c}_r - \mathbf{c}_{r,i} \geq 0, \qquad \forall c \in C, r \in R_c, i \in \{0, \dots, n\} \tag{G}$$

$$-\mathbf{c}_{r,i} + \sum_{\hat{r} \in R_c, (\hat{r}, r) \in E_R} \mathbf{c}_{\hat{r}, i-1} \geq 0,$$
$$\forall c \in C, r \in R_c, i \in \{1, \dots, n\} \tag{H}$$

$$\mathbf{r} - \mathbf{p}_r \geq 0, \qquad \forall p \in P, r \in R_p \tag{I}$$

$$\mathbf{r} - \mathbf{c}_r \geq 0, \qquad \forall c \in C, r \in R_c \tag{J}$$

$$-\mathbf{r} + \sum_{p \in P, r \in R_p} \mathbf{p}_r + \sum_{c \in C, r \in R_c} \mathbf{c}_r \geq 0, \qquad \forall r \in R \tag{K}$$

In words, (A) requires each process task to be mapped to exactly one resource. Jointly, (B) and (C) imply that each communication task has exactly one root matching the resource of its sending task. In addition, (D) makes sure that the resources of all targets

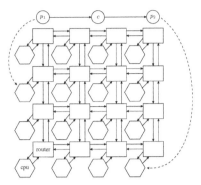

Fig. 3. A possible mapping of two communicating processes to resources connected via a 4x4 mesh network

are among those of a communication task. For excluding cyclic routing, (E) asserts that the step at which a resource is visited upon performing a communication task is unique. By means of (F) and (G), the resources visited at particular steps (i.e., $c_{r,i} = 1$ for some $i \in \{0, \ldots, n\}$) are synchronized with the ones assigned (i.e., $c_r = 1$) to a communication task. The requirement that resources visited at successive steps must be connected in the underlying platform graph is expressed by (H). Finally, (I), (J), and (K) extract allocated resources r, indicated by $\mathbf{r} = 1$, from process and communication tasks such that $\mathbf{p}_r = 1$ or $\mathbf{c}_r = 1$, respectively.

As detailed in [10], a Boolean variable assignment satisfying (A)–(K) provides a feasible implementation via resources r such that $\mathbf{r} = 1$, where each process task p is bound onto the resource r given by $\mathbf{p}_r = 1$ and communication tasks c are routed via resources r according to steps i such that $\mathbf{c}_{r,i} = 1$.

The described approach to system synthesis works well for sparsely connected networks, inducing a limited amount of routing options. However, the representation of routing options, governed by (H), scales proportionally to $|E_R| * |R|$, given that resources may be pairwisely connected and each resource may be visited in the worst case. As a consequence, for densely connected networks, the size required for a step-based representation of routing options can be prohibitively large. For example, let us consider possible routes from (the resource of) a sender p_1 to p_2 available in the 4x4 mesh network shown in Figure 3. The longest of these routes passes all 16 routers and potentially visit any of them at each of the 15 intermediate steps. This yields $16*15 = 240$ instances of (H) per communication task to represent the message exchange between routers. On the other hand, for inductively verifying whether a message reaches its target(s), it is sufficient to consider individual routing hops without relying on an explicit order given by steps. The latter strategy scales linearly to $|E_R|$, thus avoiding a significant blow-up in space. As the semantics of ASP inherently supports efficient representations of inductive concepts like reachability, the potential space savings motivate our desire to switch from the PB-based approach in [10] to using ASP instead.

4   ASP-Based System Synthesis

As common in ASP, we represent the system synthesis problem by facts describing a problem instance along with a generic encoding. To this end, we define the ASP instance for a task graph $(P \cup C, E_T)$ and a platform graph (R, E_R) along with the underlying mapping and routing options, $(R_p)_{p \in P}$ and $(R_c)_{c \in C}$, as follows:

$$
\begin{aligned}
&\{\mathtt{pt(p)}.\mid \mathtt{p} \in P\} \cup \\
&\{\mathtt{send(p,c)}.\mid (\mathtt{p,c}) \in E_T, \mathtt{p} \in P, \mathtt{c} \in C\} \cup \\
&\{\mathtt{read(p,c)}.\mid (\mathtt{c,p}) \in E_T, \mathtt{p} \in P, \mathtt{c} \in C\} \cup \\
&\{\mathtt{pr(p,r)}.\mid \mathtt{p} \in P, \mathtt{r} \in R_p\} \cup \\
&\{\mathtt{cr(c,r)}.\mid \mathtt{c} \in C, \mathtt{r} \in R_c\} \cup \\
&\{\mathtt{edge(r,s)}.\mid (\mathtt{r,s}) \in E_R\} \cup \\
&\{\mathtt{s(i)}.\mid \mathtt{i} \in \{1, \dots, n\}\}
\end{aligned} \tag{1}
$$

While the first six sets capture primary constituents of a problem instance, the introduction of atoms $\mathtt{s(i)}$ for $1 \leq \mathtt{i} \leq n$ is needed to account for the PB formulation in [10] in a faithful way.

Two alternative ASP encodings of system synthesis are shown in Figure 4(a) and 4(b). Essentially, they reformulate the constraints (A)–(K) from Section 3 in the input language of ASP to make sure that every answer set corresponds to a feasible system implementation. To this end, the rule in Line 2 of each encoding specifies that every processing task provided in an instance must be mapped to exactly one of its associated options. Observe that the mapping of processing tasks $\mathtt{p}$ to resources $\mathtt{r}$ is represented by atoms $\mathtt{map(p,r)}$ in an answer set. This provides the basis for further specifying communication routings.

Despite of syntactic differences, the step-oriented encoding ASP(S) in Figure 4(a) stays close to the original PB formulation of constraints, given in (A)–(K). In particular, it uses atoms $\mathtt{reached(c,r,i)}$ to express that some message of communication task $\mathtt{c}$ is routed over resource $\mathtt{r}$ at step $\mathtt{i}$. Note that the omission of lower and upper bounds for the cardinality constraint in the rule form Line 8 means that there is no restriction on the number of atoms constructed by applying the rule. The (trivially satisfied) cardinality constraint is still important because, it allow us to successively construct $\mathtt{reached(c,r,i)}$. Given such atoms, instantiations of the rule in Line 12 (where "_" stands for an unreused anonymous variable) further provide us with projections $\mathtt{reached(c,r)}$. These are used in the integrity constraints in Line 14 and 16, excluding cases where a communication task is routed over the same resource at more than one step or does not reach some of its targets, respectively. Finally, projections via the rules in Line 19 and 20 provide the collection of resources allocated in an admissible system layout, similar to the (redundant) variables $\mathtt{r}$ in (I)–(K).

While the step-oriented encoding ASP(S) aims at being close to the constraints in (A)–(K), the encoding in Figure 4(b), denoted by ASP(R), utilizes ASP's "built-in" support of recursion to implement routing without step counting. To still guarantee an acyclic routing of communication tasks, the idea of ASP(R) is to (recursively) construct non-branching routes from resources of communication targets back to the resource of a sending task, where the construction stops. This recursive approach connects each encountered target resource to exactly one predecessor, where the only exception is due to

```
1   % map each process task to a resource              (A)
2   1 { map(P,R) : pr(P,R) } 1 :- pt(P).

4   % step zero of communication task                  (B,C)
5   reached(C,R,0) :- send(P,C), map(P,R), cr(C,R).
6   % forward steps of communication task              (H)
7   { reached(C,S,I+1) : cr(C,S) : edge(R,S) }
8                       :- reached(C,R,I), s(I+1).

10  % resources of communication task                  (F,G)
11  reached(C,R) :- reached(C,R,_).
12  % reach each resource at most once                 (E)
13   :- reached(C,R), 2 { reached(C,R,_) }.
14  % reach communication target resources             (D)
15   :- read(P,C), map(P,R), not reached(C,R).

17  % allocated resources                              (I,J,K)
18  allocated(R) :- map(_,R).
19  allocated(R) :- reached(_,R).
```

(a) Step-oriented encoding ASP(S).

```
1   % map each process task to a resource              (A)
2   1 { map(P,R) : pr(P,R) } 1 :- pt(P).

4   % root resource of communication task              (B,C)
5   root(C,R) :- send(P,C), map(P,R).
6   % resources of communication task per target
7   sink(C,R,P) :- read(P,C), map(P,R), cr(C,R).
8   sink(C,R,P) :- sink(C,S,P), reached(C,R,S).
9   % reach communication root resource                (D)
10   :- read(P,C), root(C,R), not sink(C,R,P).

12  % resources of communication task                  (F,G)
13  reached(C,R) :- sink(C,R,_).
14  % backward hops of communication task              (E,H)
15  1 { reached(C,R,S) : cr(C,R) : edge(R,S) } 1
16           :- reached(C,S), not root(C,S).

18  % allocated resources                              (I,J,K)
19  allocated(R) :- map(_,R).
20  allocated(R) :- reached(_,R).
```

(b) Recursive encoding ASP(R).

Fig. 4. Two alternative ASP encodings of system synthesis

the sender of a communication task, whose resource, specified by an atom $\mathtt{root}(\mathtt{c},\mathtt{r})$, is not connected back. Finally, the integrity constraint in Line 10 requires that each target of a communication task is located on a route starting at the sender's resource. Note that the target-driven routing approach implemented in ASP(R) intrinsically omits redundant message hops (not leading to communication targets). The same strategy could also be applied in step counting by modifying the constraints in (A)–(K) as well as our previous encoding ASP(S) accordingly. In view of this, the varied encoding idea is not the real achievement of ASP(R), while abolishing one problem dimension by disusing explicit step counters is.

5 Experiments

For evaluating our approach, we conducted systematic experiments contrasting our two ASP encodings, ASP(S) and ASP(R), in terms of design space representation size and solving time. In addition, we compare our methods to the original (sophisticated) PB-based synthesis tool from [10], which like ASP(S) uses steps to express routing. To this end, we consider both a real-world example consisting of a sparsely connected industrial system model as well as series of crafted mesh network system models of varying sizes.

The real-world example models an automotive subsystem including four applications of different criticality and characteristic, amongst others a multimedia/infotainment control and brake-by-wire. Overall, the applications involve 45 process tasks, communicating via 41 messages. The target platform offers 15 Electronic Control Units (ECUs), 9 sensors, and 5 actuators to execute the process tasks. For communication, up to three field buses (CAN or FlexRay), connected by a central gateway, are available. In addition, sensors and actuators are connected to ECUs via LIN buses. The case study, in particular when applying further design constraints, e.g., regarding bus load, can be viewed as a complex specification that tends to max out common synthesis approaches solely based on (greedy) heuristics. However, the PB-based approach solves this problem efficiently, particularly due to the communication topology including a central gateway, resulting in a modest amount of routing options.

We ran the real-world example with the three approaches illustrated in Figure 5, all of which start from a common Java class specifying a system model (like the one shown in Figure 1). With the PB-based approach, the Java specification is directly converted into a PB instance (in OPB format) by the PB generator used also in [10]. Unlike this, with our two ASP-based approaches, the generation of facts describing a problem instance merely requires a syntactic conversion from the Java specification to the format in (1), from where the ASP grounder *gringo* (version 3.0.3) instantiates either of our encodings, ASP(S) or ASP(R), wrt the generated facts. With all three approaches, the generation phase results in standardized text formats, processable by the combined PB and ASP solver *clasp* (version 2.0.3; [5]). Let us note that ASP instance generation, including the conversion to facts and instantiation, runs quickly (only a few seconds on the largest of our benchmarks); on the other hand, PB instance generation can take significant time (up to five hours on the largest benchmarks we tried), which is because the PB generator performs nontrivial simplifications and, in contrast to ASP grounders, is

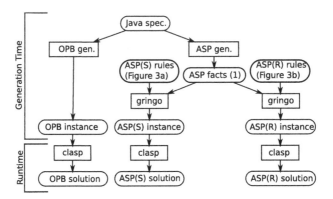

Fig. 5. Workflows of symbolic system synthesis approaches

not optimized towards low-level performance. After instance generation, accomplished offline, we measured (sequential) runtimes of *clasp* on a Linux machine equipped with 3.4GHz Intel Xeon CPUs and 32GB RAM. The search strategies of *clasp* were configured via command-line switches `--heuristic=vsids` and `--save-progress`, which in preliminary experiments turned out to be helpful for solving both PB and ASP instances. Then, the real-world example could be solved by *clasp* in less than a second for all three instance kinds, PB, ASP(S), and ASP(R). As mentioned above, this can be explained by the centralized communication topology in the example, so that routing options are rather limited.

In order to compare the three approaches also on densely connected networks, we generated series of synthesis problems wrt mesh network structures, scaling mesh size and number of process tasks. In these problems, each task can be bound onto a number of processors proportional to mesh size and communicates to one other process task; task mapping options and communication targets were selected randomly. In order to compensate for randomness in problem generation, we report averages over 16 distinct instances per mesh size and task number. Also note that all generated instances are satisfiable. In view of longer runtimes than before, we restricted single runs of *clasp* on a PB, ASP(S), or ASP(R) instance to 300 seconds time. Noise effects are excluded by taking the mean runtime over three (reproducible) runs of *clasp* per instance.

Figure 6 displays average numbers of constraints, as reported by *clasp*, and average runtimes of *clasp*, with timeouts taken as 300 seconds, over mesh networks of quadratic sizes (2x2, 3x3, ...) and increasing task numbers (10, 20, ...), both given along the x-axes; standard deviations are shown as vertical bars through measurements. The average numbers of constraints reported in the left chart provide an indication of problem representation size incurred by PB, ASP(S), and ASP(R). We observe regular scalings here, and ASP(S) is clearly the most space-consuming approach. In fact, the direct PB representation saves about half of the constraints of ASP(S) by virtue of the PB generator's simplifications. However, for larger mesh sizes, the recursive formulation of reachability in ASP(R) yields much more succinct problem representations than ASP(S) and PB, inducing almost one order of magnitude fewer constraints than the latter. Compared to this, the observation that the PB-based approach requires fewer constraints for

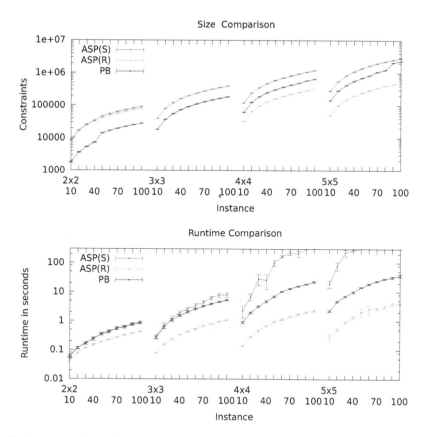

Fig. 6. Average numbers of constraints and runtimes in seconds for mesh networks of varying sizes and task numbers

the smallest instances (with average runtimes in split seconds) is negligible. The corresponding average runtimes in the right chart tightly correlate to representation sizes. While ASP(S) can still cope with small instances, it is drastically worse than PB and ASP(R) from mesh size 4x4 on, and it times out on all instances of size 5x5 with 50 or more tasks. However, as the average runtimes of PB and ASP(R) (the latter again by about one order of magnitude smaller than the former) show, even the larger instances are manageable by means of preprocessing (PB) or avoiding step counting (ASP(R)).

For investigating the further scaling behavior, we applied ASP(R) to larger meshes, using fixed ratios between the number of tasks and available CPUs as shown in Figure 7. (We here omit ASP(S) and PB in view of poor solving performance or long instance generation time, respectively.) While 6x6 mesh networks could easily be solved within seconds, we encountered first timeouts (6 out of 16) on instances of size 7x7 along with 245 process tasks (five per CPU). However, some instances (14 or 3, respectively, out of 16) of size 8x8 could still be solved within the time limit of 300 seconds when given one

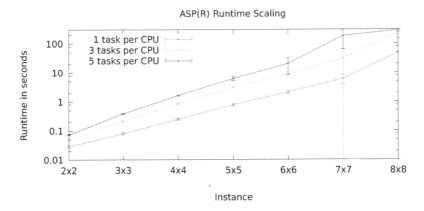

Fig. 7. Average runtimes in seconds for mesh networks of scaled up sizes

or three tasks per CPU, i.e., 64 or 192 tasks in total. Since the problem representation size (cf. numbers of constraints) is linear in the input for ASP(R), the timeouts on large instances are explained by increasing variance of solving performance in view of randomness in problem generation. Regarding the robustness of solving, we conjecture that it can be improved by including domain knowledge in ASP encodings, somewhat similar to simplifications performed by the PB generator, yet specified declaratively by rules rather than implemented by special-purpose procedural components.

6   Conclusion

We proposed a novel approach to system synthesis using ASP. While our naive step-oriented ASP encoding cannot compete with the sophisticated PB/SAT-based approach in [10], the succinct ASP formulation of reachability, as required in multi-hop routing, outperforms previous approaches when applied to densely connected (mesh) networks, providing vast routing options. Such performance gains are made possible by considerably smaller design space representations and accordingly reduced search efforts. Given that ASP solvers like *clasp* also support optimization, the presented ASP approach could be extended to linear and, with some adaptions, even be utilized for non-linear optimization, as previously performed in design space exploration via evolutionary algorithms [11]. At user level, the declarative first-order modeling language of ASP facilitates prototyping as well as adjustment of ASP solutions for new or varied application scenarios, making it a worthwhile alternative to purely propositional formalisms.

Acknowledgments. This work was partially funded by DFG grant SCHA 550/8-3 and SCHA 550/9-1.

References

1. Baral, C.: Knowledge Representation, Reasoning and Declarative Problem Solving. Cambridge University Press (2003)
2. Biere, A., Heule, M., van Maaren, H., Walsh, T. (eds.): Handbook of Satisfiability. Frontiers in Artificial Intelligence and Applications, vol. 185. IOS Press (2009)
3. Blickle, T., Teich, J., Thiele, L.: System-level synthesis using Evolutionary Algorithms. J. Design Automation for Embedded Systems 3(1), 23–58 (1998)
4. Borkar, S.: Thousand core chips: a technology perspective. In: Proc. of DAC 2007, pp. 746–749 (2007)
5. Gebser, M., Kaminski, R., Kaufmann, B., Ostrowski, M., Schaub, T., Schneider, M.: Potassco: The Potsdam answer set solving collection. AI Communications 24(2), 105–124 (2011)
6. Gebser, M., Kaminski, R., Kaufmann, B., Ostrowski, M., Schaub, T., Thiele, S.: A user's guide to `gringo`, `clasp`, `clingo`, and `iclingo`
7. Gebser, M., Kaminski, R., Kaufmann, B., Schaub, T.: Answer Set Solving in Practice. Synthesis Lectures on Artificial Intelligence and Machine Learning. Morgan and Claypool Publishers (2012)
8. Haubelt, C., Teich, J., Feldmann, R., Monien, B.: SAT-Based Techniques in System Design. In: Proc. of DATE 2003, pp. 1168–1169 (2003)
9. Ishebabi, H., Mahr, P., Bobda, C., Gebser, M., Schaub, T.: Answer set vs integer linear programming for automatic synthesis of multiprocessor systems from real-time parallel programs. Journal of Reconfigurable Computing, Article ID 863630 (2009)
10. Lukasiewycz, M., et al.: Combined System Synthesis and Communication Architecture Exploration for MPSoCs. In: Proc. of DATE 2009, pp. 472–477. IEEE Computer Society (2009)
11. Lukasiewycz, M., et al.: Efficient symbolic multi-objective design space exploration. In: Proc. of ASP-DAC 2008, pp. 691–696 (2008)
12. Neema, S.: System Level Synthesis of Adaptive Computing Systems. PhD thesis, Vanderbilt University, Nashville, Tennessee (May 2001)
13. Niemann, R., Marwedel, P.: An Algorithm for Hardware/Software Partitioning Using Mixed Integer Linear Programming. Design Automation for Embedded Systems 2(2), 165–193 (1997)

Accurate Computation of Sensitizable Paths Using Answer Set Programming[*]

Benjamin Andres[1], Matthias Sauer[2], Martin Gebser[1], Tobias Schubert[2],
Bernd Becker[2], and Torsten Schaub[1]

[1] University of Potsdam
August-Bebel-Strasse 89
14482 Potsdam, Germany
{bandres,gebser,torsten}@cs.uni-potsdam.de
[2] Albert-Ludwigs-University Freiburg
Georges-Köhler-Allee 051
79110 Freiburg, Germany
{sauerm,schubert,becker}@informatik.uni-freiburg.de

Abstract. Precise knowledge of the longest sensitizable paths in a circuit is crucial for various tasks in computer-aided design, including timing analysis, performance optimization, delay testing, and speed binning. As delays in today's nanoscale technologies are increasingly affected by statistical parameter variations, there is significant interest in obtaining sets of paths that are within a length range. For instance, such path sets can be used in the emerging areas of *Post-silicon validation and characterization* and *Adaptive Test*. We present an ASP-based method for computing well-defined sets of sensitizable paths within a length range. Unlike previous approaches, the method is accurate and does not rely on a priori relaxations. Experimental results demonstrate the applicability and scalability of our method.

1 Introduction

Precise knowledge of the longest sensitizable paths in a circuit is crucial for various tasks in computer-aided design, including timing analysis, performance optimization, delay testing, and speed binning. However, the delays of individual gates in today's nanoscale technologies are increasingly affected by statistical parameter variations [1]. As a consequence, the longest paths in a circuit depend on the random distribution of circuit features [12] and are thus subject to change in different circuit instances. For this reason, there is significant interest in obtaining sets of paths that are within a length range, in contrast to only the longest nominal path as in classical small delay testing [13]. Among other applications, such path sets can be used in the emerging areas of *Post-silicon validation and characterization* [8] and *Adaptive Test* [14].

Comprehensive test suites are generated and used in the circuit characterization or yield-ramp-up phase. The inputs to be employed in actual volume manufacturing test are chosen based on their observed effectiveness in detecting defects, In general the

---

[*] This work was published as a poster paper in [3].

P. Cabalar and T.C. Son (Eds.): LPNMR 2013, LNAI 8148, pp. 92–101, 2013.

quality of a delay test tends to increase with the delay of the actually tested path. However, a pair t_1 of test inputs may be more effective than a pair t_2, even though t_1 sensitizes a shorter path than t_2. Modeling inaccuracy is one of the reasons leading to such mismatches. For instance, while the sum of gate delays along the path sensitized by t_1 may be smaller than the sum for t_2, the pair t_1 could induce crosstalk or IR-drop, increasing the signal propagation delay along the path. These effects are generally difficult to model during timing analysis, and also affected by process variations. High-quality Automatic Test Pattern Generation (ATPG) methods should be able to control the path length and generate a large number of alternative test pairs that sensitize different paths of predefined length, to be applicable for adaptive test.

While structural paths can be easily extracted from a circuit architecture, many structural paths are not sensitizable and therefore present *false paths* [7]. The usage of such false paths leads to overly pessimistic and inaccurate results. Therefore, determination of path sensitization is required for high-quality results, although it constitutes a challenging task that requires complex path propagation and sensitization rules.

In order to reduce the algorithmic overhead, various methods for the computation of sensitizable paths make use of relaxations [11], making trade-offs between complexity and accuracy. Methods based on the sensitization of structural paths [15,6] restrict the number of paths they consider for accelerating the computation and to limit memory usage. Due to these restrictions, however, they may miss long paths. Recent methods [17,16] based on Boolean Satisfiability (SAT; [5]) have shown good performance results but are limited in the precision of the encoded delay values. As their scaling critically depends on delay resolution, such methods are hardly applicable when high accuracy is required.

We present an exact method for obtaining longest sensitizable paths, using Answer Set Programming (ASP; [4]) to encode the problem. ASP has become a popular approach to declarative problem solving in the field of Knowledge Representation and Reasoning (KRR). Unlike SAT, ASP provides a rich modeling language as well as a stringent semantics, which allows for succinct representations of encodings.

In what follows, we assume some familiarity with ASP, its semantics as well as its basic language constructs. A comprehensive treatment of ASP can be found in [4,10]. Our encodings are written in the input language of *gringo* 3 [9].

The remainder of the paper is structured as follows. Section 2 provides our ASP encoding of sensitizable paths. Experimental results are presented in Section 3. Section 4 concludes the paper.

2 ASP Encoding

The general setting of longest sensitizable path calculation for a Boolean circuit and a *test gate g* is displayed in Figure 1. Observe that gates in the *input cone A_1* influence g, while those in the *output cone A_2* depend on g. Furthermore, the additional gates in A_3 have an impact on gates in A_2. Then, a (longest) sensitizable path including the test gate g is determined by *two* truth assignments (modelling two time frames) to the primary input gates in $A_1 \cup A_3$ such that the truth values of g and one output gate in A_2 get flipped over the two time frames.

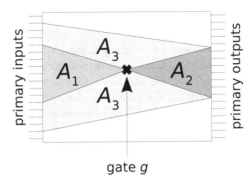

Fig. 1. Input/Output cones for longest sensitizable path calculation

As common in ASP, we represent the problem of calculating a longest sensitizable path by facts describing a problem instance along with a generic encoding. First, a Boolean circuit and gate delays are described by facts of the following form:

`in(g).`	for each primary input gate g.
`nand(g).`	for each non-input gate g.
`out(g).`	for each output gate g.
`test(g).`	for the test gate g.
`wire(g1,g2,r,f).`	for a connection from g1 to g2.

Facts of the first three forms provide constants g standing for input, non-input, and output gates, respectively, of a circuit, where we assume either `in(g)` or `nand(g)` to hold for each gate g. A fact of the fourth form specifies the test gate g in the circuit. Finally, the inputs g1 of a gate g2 are given by facts of the fifth form, where the integers r and f provide the delays of a rising or falling edge at g2, respectively. W.l.o.g., we here limit the attention to NAND non-input gates; further Boolean functions could be handled by extending the encoding in Figure 2. We also assume that output gates do not serve as inputs to other gates.

Our generic encoding of longest sensitizable paths is shown in Figure 2. Given that gate delays are only required for path length maximization, but not for the actual path calculation, the rule in Line 1 projects instances of `wire(G1,G2,R,F)` (given by facts) to connected gates G1 and G2. Then, starting from the test gate in the circuit, the rules in Line 3 to 6 inductively determine all gates of the input and output cone, respectively (cf. A_1 and A_2 in Figure 1). The union of the input and output cone is represented by the instances of `inocone(G)` derived by the rules in Line 8 and 9. In Line 10 and 11, such instances are taken as starting points to inductively determine the set of all relevant gates ($A_1 \cup A_2 \cup A_3$ in Figure 1), given by the derived instances of `allcone(G)`. Note that the rules in Line 1–11 are deterministic, yielding a unique least model relative to facts.

The calculation of a path through the test gate g is implemented by the rules in Line 13 and 14. It starts in Line 13 by choosing exactly one output gate from the output cone, represented by an instance of `path(G)`. In Line 14, the path is continued backwards including exactly one predecessor gate for every non-input gate already on

```
1    wire(G1,G2) :- wire(G1,G2,R,F).

3    inpcone(G2) :- test(G2).
4    inpcone(G1) :- inpcone(G2), wire(G1,G2).
5    outcone(G1) :- test(G1).
6    outcone(G2) :- outcone(G1), wire(G1,G2).

8    inocone(G2) :- inpcone(G2).
9    inocone(G2) :- outcone(G2).
10   allcone(G2) :- inocone(G2).
11   allcone(G1) :- allcone(G2), wire(G1,G2).

13   1 { path(G2) : outcone(G2) : out(G2) } 1.
14   1 { path(G1) : inocone(G1) : wire(G1,G2) } 1 :- path(G2), not in(G2).

16   { one(G1) } :- allcone(G1), in(G1).
17   one(G2) :- allcone(G2), nand(G2), wire(G1,G2), not one(G1).
18   { two(G1) } :- allcone(G1), in(G1).
19   two(G2) :- allcone(G2), nand(G2), wire(G1,G2), not two(G1).

21   flipped(G) :- inocone(G), one(G), not two(G).
22   flipped(G) :- inocone(G), two(G), not one(G).
23   :- path(G), not flipped(G).

25   delay(G2,M) :- path(G1), path(G2), wire(G1,G2),
26   M = #min[ wire(G1,G2,R,F) = R, wire(G1,G2,R,F) = F ].
27   add(G2,R-F) :- path(G1), path(G2), wire(G1,G2,R,F), R > F, two(G2).
28   add(G2,F-R) :- path(G1), path(G2), wire(G1,G2,R,F), R < F, one(G2).

30   #maximize[ delay(G2,M) = M, add(G2,N) = N ].
```

Fig. 2. ASP encoding for calculating longest sensitizable paths in a circuit

the path. Since any path from gates in the input cone to those in the output cone must include g, the restriction of path elements to their union (instances of inocone(G)) makes sure that path(g) holds. Also note that, although path calculation is logically encoded backwards, ASP solving engines are not obliged to proceed in any such order upon searching for answer sets.

The truth assignments needed for checking whether a path at hand is sensitizable are generated by the rules in Line 16 to 19. To this end, for each relevant input gate g1 of the circuit (allcone(g1) and in(g1) hold), choice rules allow for guessing *two* truth values. In fact, the atoms one(g1) and two(g1) express whether g1 is true in the first and the second time frame, respectively. Given the values guessed for input gates, NAND gates g2 are evaluated accordingly, and the outcomes are likewise represented by one(g2) and two(g2). For gates g in the input or output cone, which can possibly belong to a calculated path, the rules in Line 21 and 22 check whether their truth values are sensitizable; if so, it is indicated by deriving flipped(g). Finally, the integrity constraint in Line 23 stipulates that each gate on the calculated path must be flipped, thus denying truth assignments whose transition does not propagate along the whole path.

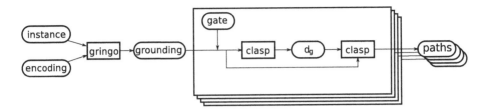

Fig. 3. Workflow of the experiments

In order to calculate the longest sensitizable paths, the rule in Line 25–26 derives a delay incurred whenever two gates g1 and g2 are connected along a path. This delay, given by the minimum of r and f in wire(g1,g2,r,f) (specified by a fact), can be obtained conveniently via *gringo*'s #min aggregate [9]. Furthermore, if r and f diverge, an additional delay r−f is incurred in case that r > f and g2 is flipped to true (Line 27), or f−r when g2 is flipped to false and r < f (Line 28). Note that considering only one(g2) or two(g2), respectively, is sufficient here because the integrity constraint in Line 23 checks that the truth value of g2 is indeed flipped. While (additional) delays derived via the rules in Line 27 and 28 depend on a path and truth assignments, the basic delay in Line 25–26 is obtained as soon as connected gates g1 and g2 are on a path. Since it does not consider truth assignments, the rule in Line 25–26 relies on fewer vagrant prerequisites and is thus "easier to apply" upon searching for answer sets. The main objective of calculating longest sensitizable paths is expressed by the #maximize statement in Line 30, which instructs ASP solving engines to compute answer sets such that the sum of associated gate delays is as large as possible.

3   Experimental Results

We evaluate our method on ISCAS85 and the combinatorial cores of ISCAS89 benchmark circuits, given as gate-level net lists. Path lengths are based on a pin-to-pin delay model with support for different rising-falling delays. The individual delay values have been derived from the Nangate 45nm Open Cell Library [2]. Below, we report sequential runtimes of the ASP solver *clasp* (version 2.0.4) on a Linux machine equipped with 3.07GHz Intel i7 CPUs and 16GB RAM.

Figure 3 shows the workflow for testing a circuit. At the start, the ASP instance describing the circuit and our generic encoding (cf. Figure 2) are grounded by *gringo*. Different from the modeling in Section 2, here, we do not specify a test gate within the ASP instance for *g*, but rather add a corresponding fact after grounding. To obtain a grounding amenable to arbitrary test gates, instead of facts, we used choice rules for a priori leaving a gate to test open. Given that sensitizable paths are computed in a loop over all gates to be analyzed, the reuse of the same grounding saves some overhead by not rerunning *gringo* for each test gate. However, note that such preprocessing "optimization" has no influence on the runtimes of *clasp* and thus does not affect solving time measurements. The grounding augmented with a test gate *g* serves as input for *clasp*, which in its first run performs optimization to identify a longest sensitizable path with maximum delay d_g. With d_g at hand, we further proceed to computing all paths with a delay equal or greater than $r = 0.95 * d_g$. This is accomplished by reinvoking *clasp* with the command-line parameters `--opt-all=r` and `--project` to enumerate all sensitizable paths within the range $[r, d_g]$. While the first parameter informs *clasp* about the quality threshold r for sensitizable paths to enumerate, the second is used to omit repetitions of the same path with different truth assignments. As a consequence, *clasp* enumerates distinct sensitizable paths, whose delay is at least r, without repetitions. An overlaying python program reuses the information of d_g and paths found in previous iterations to decide whether subsequent gates need to be analysed and ensures that *clasp* does not need to calculate the same paths for different gates.

Table 1 displays the runtimes of our method using a length-preserving mapping (avoiding rounding errors) of real-valued gate delays to integers. "Circuit" and "Gates" indicate a particular benchmark circuit along with its number of gates to be tested. The next three columns give statistics for the search for longest sensitizable paths, displaying the average runtime per solver call, the sum of runtimes for all gates in seconds and the number of solver calls needed to calculate d_g for all gates. The three columns below "Path set" provide statistics for the enumeration of distinct sensitizable paths with length at least r. Here, we show the average runtime for enumerating 1000 paths, the sum of runtimes for all gates, and finally the total number of different paths found. The columns below "Total" summarize both computation phases of *clasp*, optimization and enumeration. The first column present the total number *clasp* was called. Finally, the last two columns provide the total solving time of *clasp* for both computation passes and the total runtime needed for the benchmark. Please note, that the smallest resolution for measuring the solving time of *clasp* is $0.01s$. Thus, solving time results for circuits with less than $0.01s$ per gate may be inacurate up to the number of solver calls times $0.01s$.

As can be seen in Table 1, the scaling of our method is primarily dominated by the number of gates in circuits. Over all circuits, the average runtime for processing one test gate is rather low and often within fractions of a second. In addition, our method allows for enumerating the complete set of sensitizable paths within a given range in a single solver call, thus avoiding any expenses due to rerunning our solver. This allows us to enumerate thousands of sensitizable paths and test pattern pairs sensitizing them very efficiently. In fact, the overhead of path set computation compared to optimization in the first phase is relatively small, even for complex circuits. E.g., for the c3540 circuit, 2.26 seconds are on average required for optimization, and 10.42 seconds on average per 1000 enumerated paths. The rather large discrepancy between solving and total runtime for large, computational easy circuits, e.g. cs13207, is explained by the fact that *clasp* currently needs to read the grounded file from the disc for every call. To overcome this bottleneck we hope to utilize *iclingo*, an incremental ASP system implemented on top of *gringo* and *clasp*, in future work as soon as *iclingo* supports `#maximize` statements. This would allow us to analyze all gates of a circuit within a single solver call, thus drastically reducing the disc access. In addition, the *iclingo* could reuse information gained from previously processed gates for solving successive gates, efficiently.

Table 1. Application using exact delay values

Circuit	Gates	Longest path (d_g)			Path set (95%)			Total		
		Time in s per call	Time in s	Calls	Time in s per 1000 paths	Time in s	Paths	Solver calls	Solving Time in s	Total time in s
c0017	6	< 0.01	< 0.01	3	< 0.01	< 0.01	8	7	< 0.01	0.05
c0095	27	< 0.01	< 0.01	7	< 0.01	< 0.01	90	22	< 0.01	0.23
c0432	160	0.05	2.46	53	0.24	4.67	19356	112	7.13	11.67
c0499	202	0.01	0.64	64	0.49	0.94	1928	160	1.58	5.54
c0880	383	< 0.01	0.33	82	0.21	0.77	3617	212	1.10	7.78
c1355	546	0.29	18.87	64	1.36	32.54	23936	160	51.41	63.60
c1908	880	0.25	34.37	137	2.00	64.33	32174	378	98.70	131.22
c2670	1269	0.01	5.30	440	1.41	8.05	5700	1023	13.35	101.79
c3540	1669	2.26	544.32	241	10.42	1125.60	107994	697	1669.92	1799.69
c5315	2307	0.05	25.43	485	2.02	39.65	19603	1206	65.08	266.83
c7552	3513	0.04	24.59	576	1.97	40.77	20745	1622	65.36	444.07
cs00027	10	< 0.01	< 0.01	3	< 0.01	< 0.01	11	7	< 0.01	0.03
cs00208	104	< 0.01	< 0.01	33	< 0.01	< 0.01	97	98	< 0.01	0.62
cs00298	119	< 0.01	< 0.01	48	< 0.01	< 0.01	137	126	< 0.01	1.09
cs00344	160	< 0.01	< 0.01	45	< 0.01	< 0.01	169	125	< 0.01	1.14
cs00349	161	< 0.01	< 0.01	47	< 0.01	< 0.01	170	127	< 0.01	1.93
cs00382	158	< 0.01	< 0.01	48	< 0.01	< 0.01	169	144	< 0.01	2.03
cs00386	159	< 0.01	< 0.01	30	< 0.01	< 0.01	124	142	< 0.01	2.08
cs00400	162	< 0.01	< 0.01	49	< 0.01	< 0.01	184	149	< 0.01	2.15
cs00420	218	< 0.01	0.01	66	< 0.01	< 0.01	305	205	0.01	3.19
cs00444	181	< 0.01	< 0.01	49	< 0.01	< 0.01	210	162	< 0.01	2.45
cs00510	211	< 0.01	< 0.01	58	< 0.01	< 0.01	230	161	< 0.01	2.06
cs00526	194	< 0.01	< 0.01	74	< 0.01	< 0.01	247	190	< 0.01	2.17
cs00641	379	< 0.01	0.01	68	0.06	0.02	326	236	0.03	5.74
cs00713	393	< 0.01	0.01	84	0.06	0.02	309	276	0.03	5.20
cs00820	289	< 0.01	0.02	71	< 0.01	< 0.01	361	273	0.02	5.36
cs00832	287	< 0.01	0.02	73	< 0.01	< 0.01	372	269	0.02	5.02
cs00838	446	< 0.01	0.05	140	0.18	0.15	853	444	0.20	12.31
cs00953	418	< 0.01	0.01	111	0.03	0.01	342	354	0.02	7.48
cs01196	530	< 0.01	0.39	145	0.62	0.34	550	417	0.73	14.90
cs01238	509	< 0.01	0.54	144	0.72	0.42	586	400	0.96	13.39
cs01423	657	< 0.01	1.51	184	0.87	1.94	2236	529	3.45	25.05
cs01488	653	< 0.01	0.02	155	0.02	0.01	517	676	0.03	22.10
cs01494	647	< 0.01	0.02	157	0.02	0.01	521	654	0.03	21.69
cs05378	2779	< 0.01	0.65	506	0.32	1.71	5334	1759	2.36	320.37
cs09234	5597	< 0.01	6.82	795	0.69	13.54	19703	3483	20.36	1091.54
cs13207	8027	0.02	27.15	1332	2.97	53.41	18011	5864	80.56	2833.38
cs15850	9786	0.66	973.07	1480	9.56	3172.45	331964	6322	4145.52	9825.64
cs35932	16353	< 0.01	0.06	5321	0.41	5.9	14321	13463	5.96	16437.94
cs38584	19407	< 0.01	33.23	7266	2.35	65.18	27722	20227	98.41	42700.61

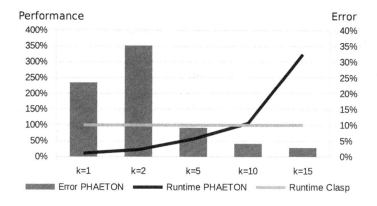

Fig. 4. Comparison with PHAETON [17] using ISCAS85 circuits

In order to demonstrate the scaling of our approach wrt delay accuracy, we also used different mappings of real-valued delays to integers, and corresponding runtime results for the ISCAS85 benchmark set as shown in Table 2. In addition to the exact mode used in the previous experiment, we employed a rounding method to five delay values, shown in the columns labeled with "5". Likewise, we applied rounding to 1000 delay values. As before, we report average runtimes per call in seconds for the two phases of optimizing sensitizable path length and of performing enumeration. Considering the results, we observe that runtimes of *clasp* are almost uninfluenced by the precision of gate delays. This is explained by the fact that weights used in #minimize or #maximize statements do influence the space of answer sets wrt to which optimization and enumeration are applied. In the ISCAS89 benchmark set the solving time per call was almost universally less than $0.01s$.

Table 2. Delay accuracy comparison

Circuit	Time (d_g) per call			Time (95%) per call		
	5	1000	exact	5	1000	exact
c0017	< 0.01	< 0.01	< 0.01	< 0.01	< 0.01	< 0.01
c0095	< 0.01	< 0.01	< 0.01	< 0.01	< 0.01	< 0.01
c0432	0.04	0.05	0.05	0.07	0.08	0.08
c0499	0.01	0.01	0.01	0.01	0.01	0.01
c0880	< 0.01	0.01	< 0.01	0.01	0.01	0.01
c1355	0.13	0.22	0.29	0.15	0.21	0.34
c1908	0.20	0.31	0.25	0.20	0.47	0.27
c2670	0.01	0.01	0.01	0.01	0.02	0.01
c3540	2.24	2.55	2.26	2.50	2.63	2.47
c5315	0.04	0.06	0.05	0.04	0.08	0.05
c7552	0.04	0.05	0.04	0.04	0.07	0.04

We compared our method with an SAT-based approach called "PHAETON" proposed in [17]. The results are shown in Figure 4. The Figure shows the runtime needed by PHAETON to compute 1000 paths for ISCAS85 benchmark circuits with different levels of accuracy indicated by the number of delay steps k. In order to compare the results of the proposed method with PHAETON, the runtime is given as percent on the primary x-axis, with 100% being our method. The secondary x-axis gives the discretization error of PHAETON. As can be seen, for low accuracy levels which result in an average discretization error of around 5%, PHAETON scales better than our optimal approach. However, for increased accuracy levels, the proposed method outperforms PHAETON and is therefore better suited for precise computation of longest sensitizable paths.

4   Conclusions

We presented a method for the accurate computation of sensitizable paths based on a flexible and compact encoding in ASP. Unlike previous methods, our approach does not rely on a priori relaxations and is therefore exact. We demonstrated the applicability and scalability of our method by extensive experiments on ISCAS85 and ISCAS89 benchmark circuits.

Future work includes further efforts to optimize the ASP encoding by incorporating additional rules, with the goal of reducing the search space and helping *clasp* to discard unsatisfactory sensitizable paths faster. Another way to improve runtime is to specialize *clasp*'s search strategy to the problem of calculating (longest) sensitizable paths.

Acknowledgments. Parts of this work are supported by the German Research Foundation under grant GRK 1103, SCHA 550/8-3 and SCHA 550/9-1.

References

1. International Technology Roadmap For Semiconductors, http://www.itrs.net
2. Nangate 45nm open cell library, http://www.nangate.com
3. Andres, B., Sauer, M., Gebser, M., Schubert, T., Becker, B., Schaub, T.: Accurate computation of longest sensitizable paths using answer set programming. In: Drechsler, R., Fey, G. (eds.) Sechste GMM/GI/ITG-Fachtagung für Zuverlässigkeit und Entwurf, ZuE 2012 (2012)
4. Baral, C.: Knowledge Representation, Reasoning and Declarative Problem Solving. Cambridge University Press (2003)
5. Biere, A., Heule, M., van Maaren, H., Walsh, T. (eds.): Handbook of Satisfiability. Frontiers in Artificial Intelligence and Applications, vol. 185. IOS Press (2009)
6. Chung, J., Xiong, J., Zolotov, V., Abraham, J.: Testability driven statistical path selection. In: 2011 48th ACM/EDAC/IEEE Design Automation Conference, DAC, pp. 417–422 (June 2011)
7. Coudert, O.: An efficient algorithm to verify generalized false paths. In: 2010 47th ACM/IEEE Design Automation Conference, DAC, pp. 188–193 (June 2010)
8. Das, P., Gupta, S.K.: On generating vectors for accurate post-silicon delay characterization. In: 2011 20th Asian Test Symposium, ATS, pp. 251–260 (November 2011)

9. Gebser, M., Kaminski, R., Kaufmann, B., Ostrowski, M., Schaub, T., Thiele, S.: A user's guide to `gringo`, `clasp`, `clingo`, and `iclingo`,
   `http://potassco.sourceforge.net`
10. Gebser, M., Kaminski, R., Kaufmann, B., Schaub, T.: Answer Set Solving in Practice. Synthesis Lectures on Artificial Intelligence and Machine Learning. Morgan and Claypool Publishers (2012)
11. Jiang, J., Sauer, M., Czutro, A., Becker, B., Polian, I.: On the optimality of k longest path generation algorithm under memory constraints. In: Design, Automation and Test in Europe, DATE (2012)
12. Killpack, K., Kashyap, C., Chiprout, E.: Silicon speedpath measurement and feedback into eda flows. In: 44th ACM/IEEE Design Automation Conference, DAC 2007, pp. 390–395 (June 2007)
13. Kumar, M.M.V., Tragoudas, S.: High-quality transition fault ATPG for small delay defects. IEEE Transactions on Computer-Aided Design of Integrated Circuits and Systems 26(5), 983–989 (2007)
14. Maxwell, P.: Adaptive test directions. In: 2010 15th IEEE European Test Symposium, ETS, pp. 12–16 (May 2010)
15. Qiu, W., Walker, D.M.H.: An efficient algorithm for finding the k longest testable paths through each gate in a combinational circuit. In: Proceedings of the International Test Conference, ITC 2003, September 30-October 2, vol. 1, pp. 592–601 (2003)
16. Sauer, M., Czutro, A., Schubert, T., Hillebrecht, S., Polian, I., Becker, B.: SAT-based analysis of sensitisable paths. In: 2011 IEEE 14th International Symposium on Design and Diagnostics of Electronic Circuits Systems, DDECS, pp. 93–98 (April 2011)
17. Sauer, M., Jiang, J., Czutro, A., Polian, I., Becker, B.: Efficient SAT-based search for longest sensitisable paths. In: 2011 20th Asian Test Symposium, ATS, pp. 108–113 (November 2011)

HEX Semantics via Approximation Fixpoint Theory*

Christian Antić, Thomas Eiter, and Michael Fink

Institute of Information Systems, Vienna University of Technology
Favoritenstraße 9-11, A-1040 Vienna, Austria
{antic,eiter,fink}@kr.tuwien.ac.at

Abstract. Approximation Fixpoint Theory (AFT) is an algebraic framework for studying fixpoints of possibly nonmonotone lattice operators, and thus extends the fixpoint theory of Tarski and Knaster. In this paper, we uniformly define 2-, and 3-valued (ultimate) answer-set semantics, and well-founded semantics of disjunction-free HEX programs by applying AFT. In the case of disjunctive HEX programs, AFT is not directly applicable. However, we provide a definition of 2-valued (ultimate) answer-set semantics based on non-deterministic approximations and show that answer sets are minimal, supported, and derivable in terms of bottom-up computations. Finally, we extensively compare our semantics to closely related semantics, including constructive dl-program semantics. Since HEX programs are a generic formalism, our results are applicable to a wide range of formalisms.

1 Introduction

HEX programs [10] enrich disjunctive logic programs under the answer-set semantics [12] (ASP programs) by external atoms for software interoperability. As the latter can represent arbitrary computable Boolean functions, HEX programs constitute a powerful extension of ordinary logic programs that has been exploited in a range of applications.[1] Furthermore, they are closely related to other extensions of ASP programs, such as dl-programs (considered below), modular logic programs, or multi-context systems with ASP components (see [8]). The semantics of HEX programs has been defined in terms of FLP-answer sets, which adhere to minimal models or, even more restricting, to models free of unfoundedness. However, FLP-answer sets of HEX programs may permit circular justifications (cf. [18]), and concepts such as well-founded semantics (which is based on unfounded sets) [21] may be cumbersome to define.

Approximation Fixpoint Theory (AFT) [5,7] is an abstract algebraic framework for studying fixpoints of (monotone or nonmonotone) lattice operators in terms of (monotone) *approximations*. In this sense, AFT extends the well-known Tarski-Knaster fixpoint theory to arbitrary lattice operators, with applications in logic programming and non-monotonic reasoning [5,6,7,4,16]; in particular, the major semantics of normal logic programs [5] and of default and autoepistemic logic [6] can be elegantly characterized within the framework of AFT; whole families of 2- and 3-valued semantics

---

* This work was supported by the Austrian Science Fund (FWF) grant P24090.

[1] http://www.kr.tuwien.ac.at/research/systems/dlvhex/
applications.html

P. Cabalar and T.C. Son (Eds.): LPNMR 2013, LNAI 8148, pp. 102–115, 2013.
© Springer-Verlag Berlin Heidelberg 2013

can be obtained, where the underlying fixpoint iteration incorporates a notion of foundedness.

This suggests to use AFT for giving semantics to HEX programs targeted for foundedness, by defining suitable operators in such a way that existing semantics for HEX programs can be reconstructed or refined, in the sense that a subset of the respective answer sets are selected (sound "approximation"). The benefit is twofold: by coinciding semantics, we get operable fixpoint constructions, and by refined semantics, we obtain a sound approximation that is constructive. Furthermore, related semantics emanate in a systematic fashion rather than adhoc, and their relationships are understood at an abstract level and need not be established on an individual basis. Finally, semantics of related formalisms like those mentioned above might be defined in a similar way.

Motivated by this, we consider 2- and 3-valued semantics of HEX programs (the latter has not been considered so far), and provide respective notions of answer sets using AFT. In this way, we reobtain and refine existing semantics of HEX programs and dl-programs. We also consider disjunctive HEX programs (for which AFT is not directly applicable) and define 2-valued answer semantics following a method in [15].

The main contributions of this paper can be summarized as follows:

(1) We define the full class of 3-valued semantics [12,17,21] of *normal* (i.e., *disjunction-free*) HEX programs [10] in a uniform way by applying the AFT framework [5,7] (cf. Section 3). In particular, this class contains 2-, and 3-valued answer-set semantics [12,17], and well-founded semantics [21]. Moreover, we define *ultimate* versions which are the most precise approximation semantics with respect to AFT [7].

(2) We exhaustively compare our semantics with the FLP semantics in [10]. They coincide on *monotone* normal HEX programs, but diverge for arbitrary normal HEX programs: due to constructiveness, each 2-valued AFT answer set is an FLP-answer set but not vice versa (cf. Theorem 2). Also, each 2-valued answer set is *well-supported* [18] which is key to characterize relevant models (cf. Section 5), and thus free of circular justifications. Moreover, our 2-valued and Shen's strongly well-supported answer-set semantics coincide [18] (cf. Theorem 8). However, our AFT approach is more general.

(3) Combining ideas from AFT, logic programs with aggregates [15], and disjunctive logic programming [13], we introduce 2-valued (ultimate) answer sets for disjunctive HEX programs along the lines of [15]. To this end, we translate some of the concepts of AFT to *non-deterministic* operators and use the notion of *computation* [14] to iterate them bottom up; we show that all (ultimate) answer sets are derivable by computations. Furthermore, (ultimate) answer sets are supported models and each 2-valued answer set is an FLP-answer set but not vice vera.

(4) We exploit the results for *description logic (dl-)programs* [9], which can be viewed as special HEX programs whose external atoms amount to so-called *dl-atoms* representing queries to a description logic ontology. Initially, a strong and weak answer set semantics of dl-programs was defined and later a well-founded semantics for *monotone* dl-programs [9]; our results generalize it *a fortiori* to arbitrary dl-programs. It turns out that for *monotone* dl-programs, the semantics coincide for the Fitting approximation Φ_P^{HEX}; however, for general dl-programs, the answer set semantics diverges.

The results of this paper provide further evidence that AFT is a valuable tool to define and study semantics of LP extensions, with well-understood and desired properties.

For space reason, proofs are omitted; they are available in [1], which also provides a more extensive discussion and contains additional results.

2 Preliminaries

2.1 HEX Programs

In this section, we recall HEX programs [10], where we restrict ourselves without loss of generality to the ground (variable-free) case.

Syntax. Let Σ and $\Sigma^\#$ be ranked alphabets of *symbols* and *external symbols*, respectively. Elements from $\Sigma^\#$ are superscripted with # ($f^\#, g^\#, h^\#$ etc.). [2]

In contrast to ordinary logic programs, HEX programs may contain besides *(ordinary) atoms* of the form $(p, c_1, \ldots, c_n) \in \Sigma^{n+1}$, $n \geq 0$, written in familiar form $p(c_1, \ldots, c_n)$, also so called *external atoms*. Formally, an *((m,n)-ary) external atom* has the form $f^\#[\mathbf{i}](\mathbf{o})$ where $f^\# \in \Sigma^\#$, $\mathbf{i} = (i_1 \ldots i_m) \in \Sigma^m$ (=input), $m \geq 0$, and $\mathbf{o} = (o_1 \ldots o_n) \in \Sigma^n$ (=output), $n \geq 0$. We often omit the arguments $\mathbf{i}$ and $\mathbf{o}$ from notation and simply write $f^\#$. A HEX-*atom* is an atom or an external atom. A *rule* has the form

$$a_1 \vee \ldots \vee a_k \leftarrow b_1, \ldots, b_\ell, \sim b_{\ell+1}, \ldots, \sim b_m, \quad k \geq 0, m \geq \ell \geq 0, \quad (1)$$

where $a_1, \ldots, a_k$ are atoms and $b_1, \ldots, b_m$ are HEX-atoms. It will be convenient to define, for a rule r, $H(r) = a_1 \vee \ldots \vee a_k$ *(head)*, $B^+(r) = \{b_1, \ldots, b_\ell\}$, $B^\sim(r) = \{\sim b_{\ell+1}, \ldots, \sim b_m\}$, and $B(r) = B^+(r) \cup B^\sim(r)$ *(body)*. With a slight abuse of notation, we will treat $H(r)$ as the set $\{a_1, \ldots, a_k\}$, i.e., we write, for instance, $a \in H(r)$, $H(r) - \{a\}$ and so on. Finally, a HEX *program* P is a *finite* set of rules of form (1).

FLP Semantics. We denote the set of all atoms (resp., external atoms) occurring in P by At_P (resp., $At_P^\#$). Define the *Herbrand base* of P by $HB_P = At_P \cup At_P^\#$. A *(2-valued) interpretation* I of P is any subset of At_P; for any $p \in \Sigma$, we denote by $p^I = \{\mathbf{c} : p(\mathbf{c}) \in I\}$ its *extension* in I. The set of all interpretations of P is $\mathcal{I}_P = \mathfrak{P}(At_P)$. We associate with every $f^\# \in \Sigma^\#$ a computable interpretation function $f : \mathcal{I}_P \times \Sigma_P^{m+n} \to \{\mathbf{t}, \mathbf{f}\}$. [2]

We define the entailment relation as follows: (i) For an atom a, $I \models a$ if $a \in I$, (ii) for an external atom $f^\#[\mathbf{i}](\mathbf{o})$, $I \models f^\#[\mathbf{i}](\mathbf{o})$ if $f(I, \mathbf{i}, \mathbf{o}) = \mathbf{t}$, (iii) for a rule r, $I \models B(r)$ if $I \models b$ for every $b \in B^+(r)$ and $I \not\models b'$ for every $\sim b' \in B^\sim(r)$, (iv) $I \models r$ if whenever $I \models B(r)$ then $I \models a$ for some $a \in H(r)$, and (v) $I \models P$ if $I \models r$ for each $r \in P$, and in this case we say that I is a *model* of P.

Define the *FLP-reduct of P* relative to I [11] by $fP^I = \{r \in P : I \models B(r)\}$. We say that I is an *FLP-answer set* [10] of P if I is a minimal model of fP^I.

Example 1. Let $P = \{q(a); p(a) \leftarrow p \subseteq^\# q, q(a)\}$, where $p \subseteq^\# q$ is infix notation for $\subseteq^\# [p, q]$, and let $I = \{p(a), q(a)\}$. We interpret $\subseteq^\#$ as set inclusion and define $\subseteq (I, p, q) = \mathbf{t}$ if $p^I \subseteq q^I$ where $p^I = q^I = \{a\}$. Since $fP^I = P$ and I is a minimal model of P, I is an FLP-answer set of P.

---

[2] [10] uses #g and $f_{\#g}$ in place of $g^\#$ and f, respectively, and calls symbols constants.

2.2   Approximation Fixpoint Theory

In this section, we briefly summarize essential notions and results given in [5,7].

In the sequel, we let $(L, \leq)$ denote a complete lattice. We call every $(x_1, x_2) \in L^2$ fulfilling $x_1 \leq x_2$ *consistent* and denote by L^c the set of all such pairs.

Define the *precision ordering* on L^c by $(x_1, x_2) \leq_p (y_1, y_2)$ if $x_1 \leq y_1$ and $y_2 \leq x_2$, i.e. intuitively, (y_1, y_2) is a "tighter" interval inside (x_1, x_2). We identify every $x \in L$ with $(x, x) \in L^c$ and call such pairs *exact*; note that they are the maximal elements w.r.t. $\leq_p$. For $z \in L$ and $(x_1, x_2) \in L^c$, we thus have $(x_1, x_2) \leq_p z$ iff $x_1 \leq z \leq x_2$, and call (x_1, x_2) an *approximation* of z. As distinct exact pairs have no upper bound, $(L^c, \leq_p)$ is not a lattice but a chain-complete poset.

An *operator* on L is any function $O : L \to L$; it is *monotone*, if for every $x, y \in L$ such that $x \leq y$, $O(x) \leq O(y)$. An element $x \in L$ is a *pre-fixpoint of O*, if $O(x) \leq x$, and a *fixpoint of O*, if $O(x) = x$. If existent, we denote the *least fixpoint of O* by $\mathrm{lfp}(O)$.

We now define approximations of O which will play a central role throughout the rest of the paper. We say that an operator $\mathcal{A} : L^c \to L^c$ is an *approximation* of O, if (i) $\mathcal{A}(x, x) = O(x)$ for each $x \in L$, and (ii) $\mathcal{A}$ is monotone with respect to $\leq_p$. Intuitively, $\mathcal{A}$ is a monotone extension of O to L^c. Clearly, $\mathcal{A}$ and O have the same fixpoints in L. Moreover, $\mathcal{A}$ has a least fixpoint $k(\mathcal{A})$, called the *$\mathcal{A}$-Kripke-Kleene fixpoint*.

For every $x, y \in L$, define the *interval* between x and y by $[x, y] = \{z \in L : x \leq z \leq y\}$. Given an element $(x_1, x_2) \in L^c$, we define the *projection* of (x_1, x_2) to the i-th coordinate, $1 \leq i \leq 2$, by $(x_1, x_2)_i = x_i$. We call an element $(x_1, x_2) \in L^c$ *$\mathcal{A}$-reliable*, if $(x_1, x_2) \leq_p \mathcal{A}(x_1, x_2)$, and in this case the restriction of $\mathcal{A}(\,.\,, x_2)_1$ (resp., $\mathcal{A}(x_1, \,.\,)_2$) to $[\bot, x_2]$ (resp., $[x_1, \top]$) is a monotone operator on the complete lattice $([\bot, x_2], \leq)$ (resp., $([x_1, \top], \leq)$). Therefore, $\mathcal{A}(\,.\,, x_2)_1$ (resp., $\mathcal{A}(x_1, \,.\,)_2$) has a least fixpoint in $([\bot, x_2], \leq)$ (resp., $([x_1, \top], \leq)$).

Let (x_1, x_2) be $\mathcal{A}$-reliable. Define the *$\mathcal{A}$-stable revision operator* by $\mathcal{A}^{\downarrow\uparrow}(x_1, x_2) = (\mathcal{A}^{\downarrow}(x_2), \mathcal{A}^{\uparrow}(x_1))$, where $\mathcal{A}^{\downarrow}(x_2) = \mathrm{lfp}(\mathcal{A}(\,.\,, x_2)_1)$ and $\mathcal{A}^{\uparrow}(x_1) = \mathrm{lfp}(\mathcal{A}(x_1, \,.\,)_2)$. Roughly, $\mathcal{A}^{\downarrow}(x_2)$ underestimates every (minimal) fixpoint of O, whereas $\mathcal{A}^{\uparrow}(x_1)$ is an upper bound as tight as possible to the minimal fixpoints of O. The stable revision operator $\mathcal{A}^{\downarrow\uparrow}$ has fixpoints and a least fixpoint on the chain-complete poset $(L_{pr}^{\mathcal{A}}, \leq_p)$ of the so called *$\mathcal{A}$-prudent* pairs $L_{pr}^{\mathcal{A}} = \{(x_1, x_2) \in L^c \mid x_1 \leq \mathcal{A}^{\downarrow}(x_2)\}$. We thus define the *$\mathcal{A}$-well-founded fixpoint* by $w(\mathcal{A}) = \mathrm{lfp}(\mathcal{A}^{\downarrow\uparrow})$. Furthermore, we call every $\mathcal{A}$-reliable fixpoint (x_1, x_2) of $\mathcal{A}^{\downarrow\uparrow}$ an *$\mathcal{A}$-stable fixpoint*, and if in addition $\mathcal{A}$ is an approximation of O and $x_1 = x_2$ an *$\mathcal{A}$-stable fixpoint of O* (note that x_1 is then a fixpoint of O).

Proposition 1 ([7]). *For every $x \in L$, x is an $\mathcal{A}$-stable fixpoint of O iff x is a fixpoint of O and $\mathcal{A}^{\downarrow}(x) = x$.*

In [7] the authors show that there exists a most precise approximation $\mathcal{O}$, called the *ultimate approximation* of O, which can be algebraically characterized in terms of O. Let for $C \in \{\bigwedge, \bigvee\}$ denote $C\,O([x_1, x_2]) = C\,\{O(x) \mid x_1 \leq x \leq x_2\}$.

Theorem 1 ([7]). *The ultimate approximation of O is given, for every $(x_1, x_2) \in L^c$, by $\mathcal{O}(x_1, x_2) = (\bigwedge O([x_1, x_2]), \bigvee O([x_1, x_2]))$.*

We summarize some basic facts about the ultimate approximation and its relationship to every other approximation of O.

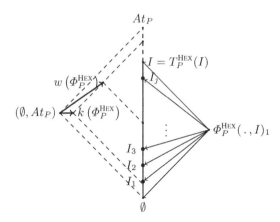

Fig. 1. Illustration of the relations between the Φ_P^{HEX}-Kripke-Kleene-, the Φ_P^{HEX}-well-founded, and the 2-valued Φ_P^{HEX}-answer-set semantics. On the left side: (i) the *Kripe-Kleene fixpoint* $k\left(\Phi_P^{\text{HEX}}\right)$ is the least fixpoint of Φ_P^{HEX}; (ii) the *well-founded fixpoint (least 3-valued stable fixpoint) $w\left(\Phi_P^{\text{HEX}}\right)$* is the least fixpoint of $\Phi_P^{\text{HEX},\downarrow\uparrow}$. On the right side: monotone iteration of the *2-valued Φ_P^{HEX}-stable model I*. If we replace Φ_P^{HEX} by $\mathcal{T}_P^{\text{HEX}}$, we obtain the more precise *ultimate* semantics.

Proposition 2 ([7]). *For every approximation $\mathcal{A}$ of O:*

1. $k(\mathcal{A}) \leq_p w(O)$ and $w(\mathcal{A}) \leq_p w(O)$.
2. *If $k(\mathcal{A})$ (resp., $w(\mathcal{A})$) is exact, then $k(\mathcal{A}) = k(O)$ (resp., $w(\mathcal{A}) = w(O)$) and it is the unique ultimate stable fixpoint of O.*
3. *Every $\mathcal{A}$-stable fixpoint of O is an ultimate stable fixpoint of O and for every ultimate fixpoint x of O, $w(O) \leq_p x$.*

3   Fixpoint Semantics for Normal HEX Programs

In this section, we uniformly extend the 2- and 3-valued answer-set semantics [12,17] and the well-founded semantics [21] of ordinary logic program to the class of normal (i.e., disjunction-free) HEX programs by applying AFT (for a summary, see Figure 1).

In the sequel, let P be a normal HEX program. We can straightforwardly extend the well-known van Emden-Kowalski operator T_P to T_P^{HEX}. A *(consistent) 3-valued interpretation* is a pair (I_1, I_2) such that $I_1 \subseteq I_2$; by $\mathcal{I}_P^c$ we denote the set of all such pairs. The precision ordering $\subseteq_p$ on $\mathcal{I}_P^c$ is given by $(J_1, J_2) \subseteq_p (I_1, I_2)$ if $J_1 \subseteq I_1$ and $I_2 \subseteq J_2$ (cf. Section 2.2). The intuitive meaning is that every $a \in I_1$ (resp., $a \notin I_2$) is *true* (resp., *false*), whereas every $a \in I_2 - I_1$ is *undefined*.

Definition 1. *We identify each $(I_1, I_2) \in \mathcal{I}_P^c$ with the 3-valued evaluation (I_1, I_2) : $HB_P \rightarrow \{\mathbf{t}, \mathbf{f}, \mathbf{u}\}$ defined by:*

1. *For every $a \in At_P$, $(I_1, I_2)(a) = \mathbf{t}$ if $a \in I_1$, $(I_1, I_2)(a) = \mathbf{f}$ if $a \notin I_2$, and $(I_1, I_2)(a) = \mathbf{u}$ if $a \in I_2 - I_1$.*
2. *For every $f^\# \in At_P^\#$, $(I_1, I_2)(f^\#) = \mathbf{t}$ (resp., $(I_1, I_2)(f^\#) = \mathbf{f}$) if $J \models f^\#$ (resp., $J \not\models f^\#$) for every $J \in [I_1, I_2]$, and $(I_1, I_2)(f^\#) = \mathbf{u}$ otherwise.*

We then directly obtain the following two approximations Φ_P^{HEX} and $\mathcal{T}_P^{\text{HEX}}$ of T_P^{HEX}:
(i) The *extended Fitting approximation* Φ_P^{HEX} as an extension of the traditional Fitting operator Φ_P, i.e., $\Phi_P(I_1, I_2) = (I_1', I_2')$ where $I_1' = \{H(r) : r \in P : (I_1, I_2)(B(r)) = \mathbf{t}\}$ and $I_2' = \{H(r) : r \in P : (I_1, I_2)(B(r)) \in \{\mathbf{t}, \mathbf{u}\}\}$ (given the usual extension of 3-valued interpretation to conjunctions of literals—denoted as sets $B(r)$ here); and (ii) the *ultimate approximation* $\mathcal{T}_P^{\text{HEX}}$, which is the most precise approximation of T_P^{HEX} and algebraically definable by (cf. Theorem 1) $\mathcal{T}_P^{\text{HEX}}(I_1, I_2) = (\bigcap T_P^{\text{HEX}}([I_1, I_2]), \bigcup T_P^{\text{HEX}} ([I_1, I_2]))$.

The approximations Φ_P^{HEX} and $\mathcal{T}_P^{\text{HEX}}$ give rise to the Φ_P^{HEX}-*Kripke-Kleene semantics* and the *ultimate Kripke-Kleene semantics*, respectively (cf. Section 2.2).

3.1   Answer-Set Semantics

Recall that given an ordinary normal program P, its 3-valued answer sets are characterized by the fixpoints of the stable revision operator $\Phi_P^{\downarrow\uparrow}$ of the Fitting approximation Φ_P. Moreover, the 2-valued answer sets of P are the fixpoints of $\Phi_P^{\downarrow}$ (cf. [5]).[3]

Likewise, in this section, we extend these definitions to normal HEX programs. In the sequel, let $\mathcal{A}_P^{\text{HEX}} \in \{\Phi_P^{\text{HEX}}, \mathcal{T}_P^{\text{HEX}}\}$. By instantiating the definition in Section 2.2, we define the stable revision operator of $\mathcal{A}_P^{\text{HEX}}$, for every $\mathcal{A}_P^{\text{HEX}}$-reliable $(I_1, I_2) \in \mathcal{I}_P^c$, by $\mathcal{A}_P^{\text{HEX},\downarrow\uparrow}(I_1, I_2) = (\mathcal{A}_P^{\text{HEX},\downarrow}(I_2), \mathcal{A}_P^{\text{HEX},\uparrow}(I_1))$, where $\mathcal{A}_P^{\text{HEX},\downarrow}(I_2) = \text{lfp}(\mathcal{A}_P^{\text{HEX}}(\,.\,, I_2)_1)$ and $\mathcal{A}_P^{\text{HEX},\uparrow}(I_1) = \text{lfp}(\mathcal{A}_P^{\text{HEX}}(I_1, \,.\,)_2)$.

Definition 2 (Answer-Set Semantics). *Let $(I_1, I_2) \in \mathcal{I}_P^c$ be $\mathcal{A}_P^{\text{HEX}}$-reliable, respectively let $I \in \mathcal{I}_P$. We say that*

1. *(I_1, I_2) is a 3-valued Φ_P^{HEX}-answer set (resp., 3-valued ultimate answer set) of P, if (I_1, I_2) is a fixpoint of $\Phi_P^{\text{HEX},\downarrow\uparrow}$ (resp., $\mathcal{T}_P^{\text{HEX},\downarrow\uparrow}$).*
2. *I is a 2-valued Φ_P^{HEX}-answer set (resp., 2-valued ultimate answer set) of P, if I is a fixpoint of T_P^{HEX} and $\Phi_P^{\text{HEX},\downarrow}(I) = I$ (resp., $\mathcal{T}_P^{\text{HEX},\downarrow}(I) = I$).*

Example 2 (Example 1 cont'd). We claim that $I = \{p(a), q(a)\}$ is a 2-valued Φ_P^{HEX}-answer set (and, therefore, an ultimate answer set) of P. First, observe that $T_P^{\text{HEX}}(I) = I$. Second, since $\Phi_P^{\text{HEX}}(\emptyset, I)_1 = \{q(a)\}$, $\Phi_P^{\text{HEX}}(\{q(a)\}, I)_1 = I$, and $\Phi_P^{\text{HEX}}(I, I)_1 = I$, we have $\Phi_P^{\text{HEX},\downarrow}(I) = I$. Hence, I is a 2-valued Φ_P^{HEX}-answer set.

The next Theorem and Example 3 summarize some basic relationships between the standard FLP semantics and the 2-valued Φ_P^{HEX}-answer-set semantics.

Theorem 2. *Let P be a normal HEX program.*

1. *If P is monotone (i.e., contains only monotone external atoms[4]) and negation-free, then $I \in \mathcal{I}_P$ is an Φ_P^{HEX}-answer set iff I is an FLP-answer set of P.*
2. *If $I \in \mathcal{I}_P$ is an Φ_P^{HEX}-answer set, then I is an FLP-answer set of P.*

However, the next example shows that the converse of condition (2) fails in general.

---

[3] Note that $\Phi_P^{\downarrow}(I) = \text{lfp}(T_{P^I})$ where P^I is the Gelfond-Lifschitz reduct [12].
[4] We say that an external atom $f^\#[\mathbf{i}](\mathbf{o})$ is *monotone*, if for every $J, J' \in \mathcal{I}_P$ such that $J \subseteq J'$, $f(J, \mathbf{i}, \mathbf{o}) \leq f(J', \mathbf{i}, \mathbf{o})$, where $\mathbf{f} < \mathbf{t}$.

Example 3. Let $P = \{a \leftarrow f^{\#}[a, b];\ b \leftarrow g^{\#}[a, b]\}$ where f and g are always *true* except for $f(\{a\}, a, b) = \mathbf{f}$ and $g(\{b\}, a, b) = \mathbf{f}$. It is easy to verify that $I = \{a, b\}$ is a minimal model of $fP^I = P$ and, hence, an FLP-answer set of P. In contrast, we have $\mathcal{T}_P^{\text{HEX},\downarrow}(I) = \emptyset$. Consequently, by Proposition 1, I is not an ultimate answer set of P and, hence, by Proposition 2, not an Φ_P^{HEX}-answer set.

Intuitively, the divergence of the 2-valued answer-set semantics based on AFT and the FLP semantics is due to the "non-constructiveness" of the FLP semantics. The intuition behind "constructiveness" is formalized by Fages' *well-supportedness* (adapted to HEX by Shen [18]). Indeed, Theorems 2 and 8 characterize our (2-valued) Φ_P^{HEX}-answer-set semantics as the strict well-supported subclass of the FLP semantics (cf. the discussion in Section 5). While incomparability of ultimate and FLP semantics already follows from the ordinary case [7].

3.2 Well-Founded Semantics

Well-founded semantics play an important role in logic programming and database theory. However, for (normal) HEX programs, to the best of our knowledge, there exist no well-founded semantics up so far. In this section, we define well-founded semantics of normal HEX programs as a special case of 3-valued Φ_P^{HEX}-answer-set semantics, by instantiating the constructions of AFT given in Section 2.2.

Recall from Section 2.2 that every stable revision operator has fixpoints and a least fixpoint. This leads to the following definition.

Definition 3 (Well-Founded Semantics). *Define the $\mathcal{A}_P^{\text{HEX}}$-well-founded model by* $w\left(\mathcal{A}_P^{\text{HEX}}\right) = \text{lfp}(\mathcal{A}_P^{\text{HEX},\downarrow\uparrow})$. *We call the $\mathcal{T}_P^{\text{HEX}}$-well-founded model $w\left(\mathcal{T}_P^{\text{HEX}}\right)$ the ultimate well-founded model of P.*

Example 4. Reconsider the normal HEX program P of Example 3 where we have seen that $I = \{a, b\}$ is an FLP-answer set but not an ultimate answer set of P. Since $\mathcal{T}_P^{\text{HEX},\downarrow\uparrow}(\emptyset, \{a, b\}) = (\emptyset, \{a, b\})$, $w\left(\mathcal{T}_P^{\text{HEX}}\right) = (\emptyset, \{a, b\}) = w\left(\Phi_P^{\text{HEX}}\right)$, i.e., a and b are both *undefined* in the (ultimate) well-founded model.

We can compute $w\left(\mathcal{A}_P^{\text{HEX}}\right)$ by iterating $\mathcal{A}_P^{\text{HEX},\downarrow\uparrow}$, starting at $(\emptyset, At_P)$, until a fixpoint is reached. The $\mathcal{A}_P^{\text{HEX}}$-well-founded model is the least 3-valued $\mathcal{A}_P^{\text{HEX}}$-answer set and approximates every other 3-valued $\mathcal{A}_P^{\text{HEX}}$-answer set, i.e., $w\left(\mathcal{A}_P^{\text{HEX}}\right) \subseteq_p (I_1, I_2)$ for every $\mathcal{A}_P^{\text{HEX}}$-answer set $(I_1, I_2) \in \mathcal{I}_P^c$. In particular, $w\left(\mathcal{A}_P^{\text{HEX}}\right)$ approximates every 2-valued $\mathcal{A}_P^{\text{HEX}}$-answer set; this relation also holds with respect to the FLP semantics.

Theorem 3. *For each FLP-answer set $I \in \mathcal{I}_P$ of P, $w\left(\mathcal{A}_P^{\text{HEX}}\right) \subseteq_p I$.*

Example 5 (Example 4 cont'd). Observe that $w\left(\mathcal{T}_P^{\text{HEX}}\right) = (\emptyset, \{a, b\}) \subseteq_p I$, and that $I = \{a, b\}$ is an FLP-answer set of the normal HEX program P of Example 3.

4   Fixpoint Semantics for Disjunctive HEX Programs

In this section, we extend the 2-valued answer-set semantics [12] to the class of disjunctive HEX programs. For such programs, T_P^{HEX} is no longer a lattice operator; thus

AFT—which studies fixpoints of lattice operators—is not applicable. However, by combining ideas from disjunctive logic programming and AFT, Pelov and Truszczyński [15] extended parts of the AFT to the case of *non-deterministic* operators.

In the sequel, let P be a disjunctive HEX program. First, we define the non-deterministic immediate consequence operator N_P^{HEX}. To this end, we recall the *Smyth ordering* [19] $\sqsubseteq$ on $\mathfrak{P}(\mathcal{I}_P)$, in which $\mathcal{J} \sqsubseteq \mathcal{K}$ if for every $K \in \mathcal{K}$ some $J \in \mathcal{J}$ exists such that $J \subseteq K$. Note that $\sqsubseteq$ is reflexive and transitive, but not anti-symmetric; thus, $\mathfrak{P}(\mathcal{I}_P)$ endowed with $\sqsubseteq$ is not a poset. However, if we consider only the anti-chains in $\mathfrak{P}(\mathcal{I}_P)$ (i.e., only those $\mathcal{J}$ where each $J \in \mathcal{J}$ is minimal w.r.t. $\subseteq$), denoted $\mathfrak{P}_{min}(\mathcal{I}_P)$, then $(\mathfrak{P}_{min}(\mathcal{I}_P), \sqsubseteq)$ is a poset with least element $\{\emptyset\}$ (cf. [13]). Denote for any set D of disjunctions over At_P (i.e., subset of the disjunctive Herbrand base DHB_P of P [13]) by $MM(D)$ the set of the minimal models of D.

Definition 4. *By the* non-deterministic van Emden-Kowalski operator *of P we refer to the operator* $N_P^{\text{HEX}} : \mathcal{I}_P \to \mathfrak{P}_{min}(\mathcal{I}_P)$ *where* $N_P^{\text{HEX}}(I) = MM\left(I \cup T_P^{\text{HEX}}(I)\right)$.

Intuitively, $N_P^{\text{HEX}}(I) = \mathcal{J}$ consists of all interpretations $J \in \mathcal{J}$ representing minimal possible outcomes of P after one step of rule applications; moreover, when applying N_P^{HEX} to I we assume each $a \in I$ to be *true*.

We call $I \in \mathcal{I}_P$ a *fixpoint* of N_P^{HEX}, if $I \in N_P^{\text{HEX}}(I)$, and *pre-fixpoint* of N_P^{HEX}, if $N_P^{\text{HEX}}(I) \sqsubseteq \{I\}$. We denote the set of all minimal fixpoints of N_P^{HEX} by $\text{mfp}(N_P^{\text{HEX}})$.

Proposition 3. *For every $I \in \mathcal{I}_P$, I is a minimal model of P iff $I \in \text{mfp}(N_P^{\text{HEX}})$.*

4.1   Non-deterministic Approximations and Computations

We can consider N_P^{HEX} as an "extension" of T_P^{HEX} to the class of disjunctive HEX programs. However, an important property of T_P^{HEX} which N_P^{HEX} does not enjoy is *iterability*. In this section, we define non-deterministic approximations [15] of N_P^{HEX} and show how to "iterate" them in terms of *computations* [14].

Definition 5. *Define, for each $(I_1, I_2) \in \mathcal{I}_P^c$, (i) the* non-deterministic Fitting approximation *of N_P^{HEX} by* $\mathcal{F}_P^{\text{HEX}}(I_1, I_2) = MM\left(I_1 \cup \Phi_P^{\text{HEX}}(I_1, I_2)_1\right)$, *and (ii) the* non-deterministic ultimate approximation *by* $\mathcal{N}_P^{\text{HEX}}(I_1, I_2) = MM\left(I_1 \cup \mathcal{T}_P^{\text{HEX}}(I_1, I_2)_1\right)$.

In the sequel, let $\mathcal{A}_P^{\text{HEX}} \in \{\mathcal{F}_P^{\text{HEX}}, \mathcal{N}_P^{\text{HEX}}\}$. The next result shows that $\mathcal{F}_P^{\text{HEX}}$ (resp., $\mathcal{N}_P^{\text{HEX}}$) similarly relates to N_P^{HEX} as Φ_P^{HEX} (resp., $\mathcal{T}_P^{\text{HEX}}$) relates to T_P^{HEX}.

Proposition 4. *Let P be a HEX program. Then,*

1. $\mathcal{A}_P^{\text{HEX}}(I, I) = N_P^{\text{HEX}}(I)$, *for every $I \in \mathcal{I}_P$;*
2. $(J_1, J_2) \subseteq_p (I_1, I_2)$ *implies* $\mathcal{A}_P^{\text{HEX}}(J_1, J_2) \sqsubseteq \mathcal{A}_P^{\text{HEX}}(I_1, I_2)$, *for every pair (J_1, J_2) and (I_1, I_2) from $\mathcal{I}_P^c$.*

Like [15], we use computations [14] to formalize the iterated approximation of N_P^{HEX}.

Definition 6. *Let $I \in \mathcal{I}_P$. An $\mathcal{A}_P^{\text{HEX}}$-$I$-computation (in the sense of [14]) is a sequence $J^{I,\uparrow} = (J_i)_{i \geq 0}$, $J_i \in \mathcal{I}_P$, such that $J_0 = \emptyset$ and, for every $n \geq 0$,*
    1. $J_n \subseteq J_{n+1} \subseteq I$,   and   2. $J_{n+1} \in \mathcal{A}_P^{\text{HEX}}(J_n, I)$.
    We call $J^{I,\infty} = \bigcup_{i \geq 0} J_i$ the result of the computation $J^{I,\uparrow}$. Furthermore, $J \subseteq I$ is $\mathcal{A}_P^{\text{HEX}}$-$I$-derivable, if some computation $J^{I,\uparrow}$ with result $J^{I,\infty} = J$ exists.

Example 6. Let $P = \{a \vee b; \ c \leftarrow a, \sim b\}$. We show that $I = \{a, c\}$ is $\mathcal{F}_P^{\text{HEX}}\text{-}I\text{-}$ derivable. Let $J_0 = \emptyset$. We compute $\mathcal{F}_P^{\text{HEX}}(J_0, I) = MM(\{a \vee b\}) = \{\{a\}, \{b\}\}$, and let $J_1 = \{a\}$. For the next iteration, we compute $\mathcal{F}_P^{\text{HEX}}(J_1, I) = MM(\{a, a \vee b, c\}) = \{I\}$, and consider $J_2 = I$. Finally, since we have $\mathcal{F}_P^{\text{HEX}}(J_2, I) = \{I\}$, we set $J_n = I$ for every $n \geq 2$, and obtain a $\mathcal{F}_P^{\text{HEX}}\text{-}I\text{-}$computation $J^{I,\uparrow}$ with result $J^{I,\infty} = I$, which shows that I is $\mathcal{F}_P^{\text{HEX}}\text{-}I\text{-}$derivable.

4.2   Answer-Set Semantics

We now carry the concepts of Section 3.1 over to disjunctive HEX programs and the corresponding non-deterministic operators as follows: (i) instead of considering T_P^{HEX}, we consider the non-deterministic van Emden-Kowalski operator N_P^{HEX} as an appropriate one-step consequence operator; (ii) instead of iterating $\mathcal{A}_P^{\text{HEX}}(.,I)_1$ to the least fixpoint $\mathcal{A}_P^{\text{HEX},\downarrow}(I)$, we "iterate" $\mathcal{A}_P^{\text{HEX}}(.,I)$ in terms of $\mathcal{A}_P^{\text{HEX}}\text{-}I\text{-}$computations to the *minimal* fixpoints $\mathcal{A}_P^{\text{HEX},\downarrow}(I)$; (iii) since disjunctive rules are non-deterministic, we consider minimal instead of least models, and minimal instead of least fixpoints. With these intuitions in place, we now define (2-valued) answer-set semantics (cf. [15]).

Definition 7 (Answer-Set Semantics). *We say that $I \in \mathcal{I}_P$ is an $\mathcal{A}_P^{\text{HEX}}$-answer set, if $I \in \mathcal{A}_P^{\text{HEX},\downarrow}(I) = \text{mfp}\left(\mathcal{A}_P^{\text{HEX}}(.,I)\right)$; it is an* ultimate answer set *of P, if I is an $\mathcal{N}_P^{\text{HEX}}$-answer set of P.*

Example 7. Let $P = \{p(a) \vee q(a); \ \leftarrow \sim p \subseteq^{\#} q\}$, and let $I = \{q(a)\}$. First, we compute $\mathcal{F}_P^{\text{HEX}}(\emptyset, I) = \{\{p(a)\}, I\}$ which shows that $\emptyset$ is not a fixpoint of $\mathcal{F}_P^{\text{HEX}}(.,I)$; second, we compute $\mathcal{F}_P^{\text{HEX}}(I, I) = \{I\}$, that is, I is a minimal fixpoint of $\mathcal{F}_P^{\text{HEX}}(.,I)$ and thus an $\mathcal{F}_P^{\text{HEX}}$-answer set. Since $I' = \{p(a)\}$ violates the constraint and is thus not an $\mathcal{F}_P^{\text{HEX}}$-answer set, I is the only $\mathcal{F}_P^{\text{HEX}}$-answer set.

The next result summarizes some basic relationships between the non-deterministic ultimate and Fitting approximation, e.g., that, as in the normal case, the ultimate approximation $\mathcal{N}_P^{\text{HEX}}$ is "more precise" than the Fitting approximation $\mathcal{F}_P^{\text{HEX}}$.

Proposition 5. *Let P be a HEX program. Then,*

1. *$\mathcal{F}_P^{\text{HEX}}(I_1, I_2) \sqsubseteq \mathcal{N}_P^{\text{HEX}}(I_1, I_2)$, for every $(I_1, I_2) \in \mathcal{I}_P^c$.*
2. *If $I \in \mathcal{I}_P$ is an $\mathcal{F}_P^{\text{HEX}}$-answer set, then I is an ultimate answer set of P.*

A basic requirement for semantics of logic programs is supportedness; in the disjunctive case, we say that an interpretation I is *supported*, if for every atom $a \in I$ there exists some rule $r \in P$ such that $I \models B(r)$, $a \in H(r)$, and $I \not\models H(r) - \{a\}$.

Theorem 4. *Let $I \in \mathcal{I}_P$. If I is an $\mathcal{A}_P^{\text{HEX}}$-answer set, then I is supported.*

Next, we relate our fixpoint-based answer set semantics to the "standard" FLP semantics. Theorem 2 established that all 2-valued Φ_P^{HEX}-answer sets of a normal HEX program are FLP-answer sets. An analogous result holds for disjunctive HEX programs.

Theorem 5. *Let $I \in \mathcal{I}_P$. If I is an $\mathcal{F}_P^{\text{HEX}}$-answer set, then I is an FLP-answer set of P.*

However, the converse of Theorem 5 does not hold in general (cf. Example 3).

Note that $\mathcal{A}_P^{\text{HEX}}$-answer sets as in Definition 7 are non-constructive. However, we can construct every $\mathcal{A}_P^{\text{HEX}}$-answer set bottom-up and identify it with an additional test.

Theorem 6. *Let* $I \in \mathcal{I}_P$. *Then,* I *is an* $\mathcal{A}_P^{\text{HEX}}$-*answer set iff* I *is* $\mathcal{A}_P^{\text{HEX}}$-$I$-*derivable and no* $J \subset I$ *exists such that* $J \in \mathcal{A}_P^{\text{HEX}}(J, I)$.

Example 8. In Example 6, we have seen that the $\mathcal{F}_P^{\text{HEX}}$-answer set $I = \{a, c\}$ of $P = \{a \vee b; \ c \leftarrow a, \sim b\}$ is $\mathcal{F}_P^{\text{HEX}}$-$I$-derivable, and in Example 7 that the $\mathcal{F}_P^{\text{HEX}}$-answer set $I = \{q(a)\}$ of $P = \{p(a) \vee q(a); \ \leftarrow \sim p \subseteq^{\#} q\}$ is $\mathcal{F}_P^{\text{HEX}}$-$I$-derivable. On the other hand, $I = \{p(a), q(a)\}$ is $\mathcal{F}_P^{\text{HEX}}$-$I$-derivable w.r.t. $P = \{p(a) \vee q(a); \ p(a) \leftarrow p \subseteq^{\#} q\}$, while $J = \{p(a)\} \in \mathcal{F}_P^{\text{HEX}}(J, I)$; thus I is not an $\mathcal{F}_P^{\text{HEX}}$-answer set of P.

5   Related Work

Approximation Fixpoint Theory.   AFT [5,7] builds on Fitting's seminal work on bilattices and fixpoint semantics of logic programs. In [5], the framework was introduced upon (symmetric) approximations of $\mathcal{A}$ operating on the bilattice L^2; in logic programming, L^2 corresponds to the set $\mathcal{I}_P^2$ of all *4-valued* interpretations (I_1, I_2) of P. However, as pointed out in [7], under the usual interpretations of logic programs only the *consistent* (i.e., *3-valued*) fragment $\mathcal{I}_P^c$ of $\mathcal{I}_P^2$ has an intuitive meaning. Therefore, [7] advanced AFT for consistent approximations, i.e., the 3-valued case also adopted here.

As demonstrated also by our work, a strength of AFT is its flexibility regarding *language extensions*. Recall from Section 3 that to extend the semantics from ordinary normal programs to normal HEX programs, we just had to extend the 3-valued interpretation to the new language construct (i.e., external atoms). A principled way to cope with language extensions under 4-valued interpretations was recently mentioned in [4]; it hinges on 4-valued immediate consequence operators satisfying certain properties ($\leq_p$-monotonicity and symmetry). It is possible to generalize Definition 1 to this setting.

Pelov and Truszczyński's Computations [15].   We used *non-deterministic* operators to define 2-valued (ultimate) answer sets of *disjunctive* HEX programs, motivated by [15]. However, our approach is not entirely identical to [15]; we elaborate here on differences.

As aggregates can be simulated by external atoms (cf. Section 3.1 in [10]), we translate the definitions in [15] to the language of HEX programs and define, for a disjunctive HEX program P, $N_P^{Sel}(I) = Sel\left(T_P^{\text{HEX}}(I)\right)$ where $Sel : \mathfrak{P}(DHB_P) \to \mathfrak{P}(\mathcal{I}_P)$ is a *selection function*. As we only used the selection function MM in this paper, we focus on N_P^{MM} in the sequel.

Pelov and Truszczyński [15] proposed the notion of *computation* [14] as an appropriate formalization of the process of "iterating" N_P^{MM}. In Section 4, we successfully applied it to non-deterministic (ultimate) approximations, and proved in Theorem 6 that (ultimate) answer sets are *derivable*. The following example shows that the definition of N_P^{MM} as such is not compatible with the notion of computation.

Example 9 ([15], Example 3). Let $P = \{a \vee b \vee c; a \leftarrow b; b \leftarrow c; c \leftarrow a\}$. Observe that $I = \{a, b, c\}$ is the only model of P. By applying N_P^{MM} to $J_0 = \emptyset$, we obtain $N_P^{MM}(J_0) = \{\{a\}, \{b\}, \{c\}\}$. However, since $N_P^{MM}(\{a\}) = \{\{c\}\}$, $N_P^{MM}(\{b\}) = \{\{a\}\}$, and $N_P^{MM}(\{c\}) = \{\{b\}\}$, there is no computation $J^\uparrow$ with result $J^\infty = I$. On the other hand, it is easy to see that I is $\mathcal{A}_P^{\text{HEX}}\text{-}I$-derivable.

Description Logic Programs [9]. Description logic programs[5] (*dl-programs*) [9] are precursors of HEX programs [10] that allow *dl-atoms* (i.e., bi-directional links between a logic program and a description logic ontology) in rule bodies. As shown in [10], we can simulate every dl-program $\mathcal{KB}$ by a normal HEX program $P = P_{\mathcal{KB}}$, which allows us to compare the semantics defined in Section 3 with the strong and weak answer-set semantics and the well-founded semantics defined in [9].

Let $\mathcal{KB}$ be a dl-program and P be the respective normal HEX program; let $At_P^{\#,m}$ be a (fixed) set of all external atoms $a \in At_P^\#$ known to be monotone, and let $At_P^{\#,?} = At_P^\# - At_P^{\#,m}$. Then the *strong Gelfond-Lifschitz reduct* [9] of P relative to $I \in \mathcal{I}_P$ is

$$sP^I = \left\{ H(r) \leftarrow B^+(r) - At_P^{\#,?} : r \in P : I \models B^+(r) \cap At_P^{\#,?}, I \models B^\sim(r) \right\}.$$

Note that sP^I is a negation-free monotone normal HEX program, which implies that $T_{sP^I}^{\text{HEX}}$ has a least fixpoint; we call I a *strong answer set* [9] of P, if $I = \text{lfp}(T_{sP^I}^{\text{HEX}})$.

The next example shows that neither the strong nor the weak answer-set semantics coincides with the (ultimate) answer sets of P.

Example 10. Let $P = \{p(a) \leftarrow\sim (not\ p(a))^\#\}$ where $I \models (not\ p(a))^\#$ if $I \not\models p(a)$.[6] We show that $I = \{p(a)\}$ is a strong answer set of P. As $I \models\sim (not\ p(a))^\#$, sP^I consists of the fact $p(a)$. Hence, I is the least fixpoint of $T_{sP^I}^{\text{HEX}}$ and thus a strong answer set of P. On the other hand $\mathcal{T}_P^{\text{HEX},\downarrow}(I) = \emptyset$, which shows that I is not an ultimate answer set of P and hence not an Φ_P^{HEX}-answer set (cf. Proposition 2). As every strong answer set of P is also a weak answer set [9], the same holds for the weak answer set semantics.

However, the next result shows that for *monotone* dl-programs, the semantics in this paper coincide with the semantics given in [9]; note that well-founded semantics was defined in [9] under restriction to monotone dl-programs using unfounded sets.

Theorem 7. *Let $\mathcal{KB}$ be a monotone dl-program and let $P = P_{\mathcal{KB}}$.*

1. *For each $I \in \mathcal{I}_P$, I is a strong answer set of P iff I is a 2-valued Φ_P^{HEX}-answer set.*
2. *For each $(I_1, I_2) \in \mathcal{I}_P^c$, (I_1, I_2) is the well-founded model of P as defined in [9] iff (I_1, I_2) is the Φ_P^{HEX}-well-founded model.*

Shen's Strongly Well-Supported Semantics [18]. Shen [18] defined (weakly and strongly) well-supported semantics for dl-programs. As the latter can be simulated by normal HEX programs, we rephrase Shen's definition in the HEX-setting.

---

[5] http://sourceforge.net/projects/dlvhex/files/dlvhex-dlplugin/
[6] For readers familiar with dl-programs, note that P amounts to the dl-program $\mathcal{KB} = (\emptyset, \{p(a) \leftarrow\ \sim DL[S \cap p; \neg S](a)\})$ where $DL[S \cap p; \neg S](a)$ is a dl-atom, S is a concept, and $\cap$ is the constraint update operator (cf. [9]).

Given a normal HEX program P and $(I_1, I_2) \in \mathcal{I}_P^c$, Shen's notion "$I_1$ up to I_2 satisfies literal ℓ" is equivalent to our 3-valued evaluation function, in symbols $(I_1, I_2)(\ell) = t$. We thus can characterize Shen's fixpoint operator S_P^{HEX} [18, Definition 5] as follows.

Proposition 6. *For each* $(I_1, I_2) \in \mathcal{I}_P^c$, $S_P^{HEX}(I_1, I_2) = \Phi_P^{HEX}(I_1, I_2)_1$.

Finally, call I an *strongly well-supported answer set* [18] of P, if $I = \mathrm{lfp}\left(S_P^{HEX}(\,.\,, I)\right)$. The following result is an immediate consequence of Proposition 6.

Theorem 8. *Let* $I \in \mathcal{I}_P$. *Then,* I *is a strongly well-supported answer set of* P *iff* I *is a 2-valued* Φ_P^{HEX}-*answer set.*

The equivalences above show that Shen's (strongly) well-supported answer-set semantics is naturally captured within the more general framework of AFT. However the use of AFT allowed us to obtain in addition the whole class of 3-valued (ultimate) answer-set semantics (which contain the well-founded semantics), in a more general (and perhaps more elegant) approach than the one in [18].

6   Discussion and Conclusion

The goal of this paper was to extend the well-founded-, and the (3-valued) answer-set semantics to the class of HEX programs [10] by applying AFT [5,7], and to compare them with the "standard" FLP semantics. This was in particular relevant, because HEX programs constitute a powerful extension of ordinary disjunctive programs, and are able to represent various other formalisms (e.g., dl-programs; see [10]).

As a result of our investigation, we obtained constructive and uniform semantics for a general class of logic programs with nice properties. More precisely, for normal HEX programs, our 2-valued answer-sets based on AFT turned out to be well-supported, which is regarded as a positive feature. Moreover, Shen's strongly well-supported answer set semantics (formulated for dl-programs) coincides with the 2-valued Φ_P^{HEX}-answer set semantics. Furthermore, to the best of our knowledge, the well-founded semantics for normal HEX programs has not been defined before; it coincides on positive programs representing dl-programs with the well-founded semantics in [9], and thus generalizes it *a fortiori* to arbitrary dl-programs. Finally, our 2-valued (ultimate) answer-set semantics of disjunctive HEX programs turned out to be bottom-up computable.

Regarding complexity, assume that checking $I \models f^\#$ is feasible in polynomial time. Then, generalizing ordinary normal logic programs to normal HEX programs does not increase the worst-case complexity of ultimate semantics [7]. Different to the ordinary case, however, computing well-founded and answer set semantics is not easier than ultimate approximation. More specifically, as deciding 3-valued entailment as in Definition 1 is coNP-hard, we obtain that deciding (in the ground case)

- consequence under the (ultimate) well-founded model is Δ_2^P-complete;
- brave consequence is Σ_2^P-complete for 3-valued (ultimate) answer sets; and
- existence of a 2-valued (ultimate) answer-set is Σ_2^P-complete.

Note that for disjunctive HEX programs, despite nondeterministic computations, deciding brave consequence for ultimate answer sets remains Σ_2^P-complete.

Open issues. Some open issues remain. First, in the case of *infinite* HEX programs, the operators defined in this paper all require an *infinite guess*, which makes the notion of *computation* (see Section 4) infeasible. A possible way to tackle this problem consists of three steps: (i) define a HEX program P to be ω-*evaluable*, if there exists an ω-Turing machine [3] M that accepts the answer sets of P; (ii) simulate M by a positive disjunctive HEX program P_M; and (iii) iterate the monotone operator $N_{P_M}^{\text{HEX}}$ in terms of computations as in Definition 6. For example, normal positive HEX programs with finitary external atoms are ω-evaluable, and more generally HEX programs in which atoms depend only on finitely many other atoms [2]; it remains to find further relevant classes of ω-*evaluable infinitary* HEX *programs*.

Second, in the definition of Φ_P^{HEX} (and, consequently, $\mathcal{F}_P^{\text{HEX}}$) the definition of the 3-valued entailment relation plays a crucial role. In a naive realization, evaluating $(I_1, I_2)(f^{\#})$ is exponential, which possibly can be avoided given further knowledge on f (e.g., monotonicity) or relevant interpretations $J \in [I_1, I_2]$ in the context of the program P. Developing respective pruning conditions is interesting and important from a practical perspective. Alternatively, one can imagine to define 3-valued entailment on a substructure of $[I_1, I_2]$, obeying suitable conditions.

Finally, Truszczyński [20] has extended AFT to algebraically capture the notions of strong and uniform equivalence; it is interesting to apply these results to the class of HEX programs by using the results obtained in this paper.

References

1. Antić, C.: Uniform approximation-theoretic semantics for logic programs with external atoms. Master's thesis, TU Vienna (2012),
   http://www.ub.tuwien.ac.at/dipl/2012/AC07814506.pdf
2. Baselice, S., Bonatti, P.A., Criscuolo, G.: On finitely recursive programs. TPLP 9(2), 213–238 (2009)
3. Cohen, R.S., Gold, A.Y.: ω-computations on Turing machines. TCS 6, 1–23 (1978)
4. Denecker, M., Bruynooghe, M., Vennekens, J.: Approximation fixpoint theory and the semantics of logic and answers set programs. In: Erdem, E., Lee, J., Lierler, Y., Pearce, D. (eds.) Correct Reasoning. LNCS, vol. 7265, pp. 178–194. Springer, Heidelberg (2012)
5. Denecker, M., Marek, V., Truszczyński, M.: Approximations, stable operators, well-founded fixpoints and applications in nonmonotonic reasoning. In: Minker, J. (ed.) Logic-Based Artificial Intelligence, pp. 127–144. Kluwer (2000)
6. Denecker, M., Marek, V., Truszczyński, M.: Uniform semantic treatment of default and autoepistemic logics. Artificial Intelligence, 79–122 (2003)
7. Denecker, M., Marek, V., Truszczyński, M.: Ultimate approximation and its application in nonmonotonic knowledge representation systems. Inf. Comp. 192(1), 84–121 (2004)
8. Eiter, T., Brewka, G., Dao-Tran, M., Fink, M., Ianni, G., Krennwallner, T.: Combining nonmonotonic knowledge bases with external sources. In: Ghilardi, S., Sebastiani, R. (eds.) FroCoS 2009. LNCS, vol. 5749, pp. 18–42. Springer, Heidelberg (2009)
9. Eiter, T., Ianni, G., Lukasiewicz, T., Schindlauer, R.: Well-founded semantics for description logic programs in the semantic web. ACM TOCL 12(2), 11:1–11:41 (2011)

10. Eiter, T., Ianni, G., Schindlauer, R., Tompits, H.: A uniform integration of higher-order reasoning and external evaluations in answer-set programming. In: Proc. IJCAI 2005, pp. 90–96 (2005)
11. Faber, W., Leone, N., Pfeifer, G.: Recursive aggregates in disjunctive logic programs: Semantics and complexity. In: Alferes, J.J., Leite, J. (eds.) JELIA 2004. LNCS (LNAI), vol. 3229, pp. 200–212. Springer, Heidelberg (2004)
12. Gelfond, M., Lifschitz, V.: Classical negation in logic programs and disjunctive databases. New Generation Computing 9(3-4), 365–385 (1991)
13. Lobo, J., Minker, J., Rajasekar, A.: Foundations of Disjunctive Logic Programming. MIT Press (1992)
14. Marek, V.W., Niemelä, I., Truszczyński, M.: Logic programs with monotone cardinality atoms. In: Lifschitz, V., Niemelä, I. (eds.) LPNMR 2004. LNCS (LNAI), vol. 2923, pp. 154–166. Springer, Heidelberg (2003)
15. Pelov, N., Truszczyński, M.: Semantics of disjunctive programs with monotone aggreggates - an operator-based approach. In: Proc. NMR 2004, pp. 327–334 (2004)
16. Pelov, N., Denecker, M., Bruynooghe, M.: Well-founded and stable semantics of logic programs with aggregates. Theory and Practice of Logic Programming 7(3), 301–353 (2007)
17. Przymusinski, T.: Well-founded semantics coincides with the three-valued stable semantics. Fundamenta Informaticae 13(4), 445–463 (1990)
18. Shen, Y.D.: Well-supported semantics for description logic programs. In: Proc. IJCAI 2011, pp. 1081–1086. AAAI Press (2011)
19. Smyth, M.B.: Power domains. Journal of Computer and System Sciences 16, 23–36 (1978)
20. Truszczyński, M.: Strong and uniform equivalence of nonmonotonic theories – an algebraic approach. Annals of Mathematics and Artificial Intelligence 48(3-4), 245–265 (2006)
21. Van Gelder, A., Ross, K.A., Schlipf, J.S.: The well-founded semantics for general logic programs. Journal of the ACM 38(3), 619–649 (1991)

Encoding Higher Level Extensions of Petri Nets in Answer Set Programming

Saadat Anwar[1], Chitta Baral[1], and Katsumi Inoue[2]

[1] SCIDSE, Arizona State University, 699 S Mill Ave, Tempe, AZ 85281, USA
[2] Principles of Informatics Research Div., National Institute of Informatics, Japan

Abstract. Answering realistic questions about biological systems and pathways similar to text book questions used for testing students' understanding of such systems is one of our long term research goals. Often these questions require simulation based reasoning. In this paper, we show how higher level extensions of Petri Nets, such as colored tokens can be encoded in Answer Set Programming, thereby providing the right formalisms to model and reason about such questions with relative ease. Our approach can be adapted to other domains.

1 Introduction

One of our long term research objectives is to develop a system that can answer questions similar to the ones given in the biological texts, used to test the understanding of the students. In order to answer such questions, we have to model pathways, add interventions / extensions to them based on the question, simulate them, and reason with the simulation results. We found Petri Nets to be a suitable formalism for modeling biological pathways, as their graphical representation is very close to the biological pathways, and since they can be extended with necessary assumptions and interventions about questions as shown in our prequel paper [1]. We noticed that certain aspects of biological pathways, such as multiple locations with distinct substances, cannot be succinctly represented in a regular Petri Net model. So, here we use Petri Nets with colored tokens.

Fig. 1 shows a Petri Net model of the Electron Transport Chain as given in [2]. Places represent locations, transitions represent processes, $t1-t4$ represent multiprotein complexes, and token color represents substance type. Without colored tokens, this model would become large and cumbersome.

Existing Petri Net modeling and simulation systems either have limited adaptability outside their domain, or limited ease of extension[1]. Also, most systems do not explore all possible state evolutions, allow different firing semantics, or guide search through *way-points* of partial state. We found ease of encoding, extendibility, and reasoning capabilities in Answer Set Programming (ASP).

Previous work on Petri Net to ASP translation is limited to specific classes of Petri Nets. Thus, our main contributions are: ASP encoding of Petri Nets with colored tokens; showing how additional extensions can be encoded via small

---

[1] See full paper for details at: http://arxiv.org/abs/1306.3548

P. Cabalar and T.C. Son (Eds.): LPNMR 2013, LNAI 8148, pp. 116–121, 2013.
© Springer-Verlag Berlin Heidelberg 2013

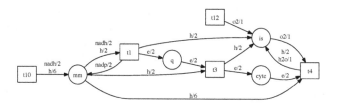

Fig. 1. Petri Net with colored tokens. Colors $= \{e, h, h2o, nadh, nadp, o2\}$. Circles represent places, and rectangles represent transitions. Arc weights such as "$nadh/2, h/2$" specify the multiset of tokens consumed or produced by execution of their respective transitions, e.g. "$nadh/2, h/2$" means 2 tokens of color $nadh$ and 2 tokens of h. Similar notation is used to specify marking on places, when not present, the place is assumed to be empty of tokens.

changes (incl. firing semantics, priority/timed transitions); and showing how our encoding and ASP reasoning can be used to answer biological pathway questions.

2 Fundamentals

For **Answer Set Programming (ASP)** syntax used in this paper, refer to [3].

A **multiset** A over a domain set D is a pair $\langle D, m \rangle$, where $m : D \to \mathbb{N}$ is a function giving the multiplicity of $d \in D$ in A. Given two multsets $A = \langle D, m_A \rangle, B = \langle D, m_B \rangle$, $A \odot B$ if $\forall d \in D : m_A(d) \odot m_B(d)$, where $\odot \in \{<, > , \leq, \geq, =\}$, and $A \neq B$ if $\exists d \in D : m_A(d) \neq m_B(d)$. Multiset sum/difference is defined in the usual way. $d \in A$ represents $m_A(d) > 0$; $A = \emptyset$ represents $\forall d \in D, m(d) = 0$; We use $d/n \in A$ to represent that d appears n-times in A, and drop A when clear from context. See standard texts on multisets for details.

A basic **Petri Net** [4] is a bipartite graph of finite set of place nodes $P = \{p_1, \ldots, p_n\}$, and transition nodes $T = \{t_1, \ldots, t_m\}$ connected through directed arcs $E = E^+ \cup E^-$, where $E^- \subseteq P \times T$ and $E^+ \subseteq T \times P$. A Petri Net's state is given by the number of tokens on each place node, collectively called its *marking* $M = (M(p_1), \ldots, M(p_n)), M(p_i) \in \mathbb{N}$. Arc weights $W : E \to \mathbb{N} \setminus \{0\}$ specify the number of tokens consumed (through E^-) or produced (through E^+) due to firing of the transition at the head or tail of the arc. Modeling capability of basic Petri Nets is enhanced by adding reset, inhibit and read arcs. A **reset arcs** $R : T \to 2^P$ removes all tokens from its input place. An **inhibitor arc** $I : T \to 2^P$ prevents its transition from firing if its source contains tokens. A **read arc** $Q \subseteq P \times T$, $QW : Q \to \mathbb{N} \setminus \{0\}$ prevents its target transition from firing until its source has at least the tokens specified by the arc weight.

Higher level Petri Nets extend the notion of tokens to typed (or colored) tokens. A **Petri Net with Colored Tokens** (with reset, inhibit and read arcs) is a tuple $PN^C = (P, T, E, C, W, R, I, Q, QW)$, where P, T, E, R, I, Q are the same as for basic Petri Nets and its extensions, $C = \{c_1, \ldots, c_l\}$ is a finite set of colors, and arc weights $W : E \to \langle C, m \rangle$, $QW : Q \to \langle C, m \rangle$ are specified as multi-sets of colored tokens over color set C. State (or marking) of place nodes $M(p_i) = \langle C, m \rangle$ is specified as a multiset of colored tokens over set C.

The **initial marking** (M_0) is the initial token assignment of place nodes. Marking at time-step k is written as M_k. The **pre-set** (or input-set) of transition t is •$t = \{p \in P | (p, t) \in E^-\}$, while its **post-set** (or output-set) is $t• = \{p \in P | (t, p) \in E^+\}$. A transition t is **enabled** with respect to marking M_k, $enabled_{M_k}(t)$, if each of its input places p has at least $W(p, t)$ [2] colored-tokens, each of its inhibiting places $p_i \in I(t)$ have zero tokens and each of its read places $p_q : (p_q, t) \in Q$ have at least $QW(p_q, t)$ colored-tokens. A **firing set** is a set $T_k = \{t_{k_1}, \ldots, t_{k_n}\} \subseteq T$ of enabled transitions that fire simultaneously and do not conflict. Execution of a firing set T_k on a marking M_k computes a new marking M_{k+1} by removing tokens consumed by $t \in T_k$ from t's input places and adding tokens produced by $t \in T_k$ to t's output places. A set of enabled transitions is in **conflict** with respect to M_k if firing them will consume more than available tokens at an input place [3]. An **execution sequence** $X = M_0, T_0, M_1, \ldots, T_k, M_{k+1}$ is the simulation of a firing sequence $\sigma = T_1, \ldots, T_k$, $T_i \subseteq T$ is a firing set. It is the transitive closure of executions, where subsequent markings become the initial marking for the next transition set. Thus firing of T_0 at M_0 produces M_1, which becomes initial marking for T_1.

3   Translating Petri Nets with Colored Tokens to ASP

In this section we present an ASP encoding of a Petri Net with colored tokens PN^C (incl. execution behavior), with initial marking M_0 and a simulation length k. This work is an extension of our work or regular Peri Nets in [1].

f1: Facts `place`(p_i) where $p_i \in P$ is a place.

f2: Facts `trans`(t_j) where $t_j \in T$ is a transition.

f3: Facts `col`(c_k) where $c_k \in C$ is a color.

f4: Rules `ptarc`(p_i, t_j, n_c, c, ts_k) :- `time`(ts_k). for each $(p_i, t_j) \in E^-$, $c \in C$, $n_c = m_{W(p_i, t_j)}(c) : n_c > 0$.

f5: Rules `tparc`(t_i, p_j, n_c, c, ts_k) :- `time`(ts_k). for each $(t_i, p_j) \in E^+$, $c \in C$, $n_c = m_{W(t_i, p_j)}(c) : n_c > 0$.

f6: Rules `ptarc`(p_i, t_j, n_c, c, ts_k) :- `holds`(p_i, n_c, c, ts_k), `num`(n_c), n_c>0, `time`(ts_k). for each $(p_i, t_j) : p_i \in R(t_j)$, $c \in C$, $n_c = m_{M_k(p_i)}(c)$.

f7: Rules `iptarc`($p_i, t_j, 1, c, ts_k$) :- `time`(ts_k). for each $(p_i, t_j) : p_i \in I(t_j)$, $c \in C$.

f8: Rules `tptarc`(p_i, t_j, n_c, c, ts_k) :- `time`(ts_k). for each $(p_i, t_j) \in Q$, $c \in C$, $n_c = m_{QW(p_i, t_j)}(c) : n_c > 0$.

i1: Facts `holds`($p_i, n_c, c, 0$). for each place $p_i \in P$, $c \in C$, $n_c = m_{M_0(p_i)}(c)$.

f9: Facts `time`(ts_i) where $0 \le ts_i \le k$ are the discrete simulation time steps.

f10: Facts `num`(n) where $0 \le n \le ntok$ are token quantities [4]

e1: `notenabled(T,TS) :- ptarc(P,T, N,C,TS), holds(P,Q,C,TS), place(P),`
`trans(T), time(TS), num(N), num(Q), col(C), Q<N.`

e2: `notenabled(T,TS) :- iptarc(P,T,N,C,TS), holds(P,Q,C,TS), place(P),`
`trans(T), time(TS), num(N), num(Q), col(C), Q>=N.`

---

[2] We are using $W(p, t), QW(p, t)$ to mean $W(\langle p, t\rangle), QW(\langle p, t\rangle)$ for simplicity.

[3] Our reset arc has modified semantics, which puts token consumption through it in contention with other arcs, but allows us to model elimination of all quantity of a substance as soon as it is produced.

[4] The token count predicate `num`'s limit can be arbitrarily selected to be higher than expected token count. It is there for efficient ASP grounding.

```
e3: notenabled(T,TS) :- tptarc(P,T,N,C,TS), holds(P,Q,C,TS), place(P),
      trans(T), time(TS), num(N), num(Q), col(C), Q<N.
e4: enabled(T,TS) :- trans(T), time(TS), not notenabled(T,TS).
a1: {fires(T,TS)} :- enabled(T,TS), trans(T), time(TS).
r1: add(P,Q,T,C,TS) :- fires(T,TS), tparc(T,P,Q,C,TS), time(TS).
r2: del(P,Q,T,C,TS) :- fires(T,TS), ptarc(P,T,Q,C,TS), time(TS).
r3: tot_incr(P,QQ,C,TS) :- col(C), QQ = #sum[add(P,Q,T,C,TS) = Q : num(Q) :
      trans(T)], time(TS), num(QQ), place(P).
r4: tot_decr(P,QQ,C,TS) :- col(C), QQ = #sum[del(P,Q,T,C,TS) = Q : num(Q) :
      trans(T)], time(TS), num(QQ), place(P).
r5: holds(P,Q,C,TS+1):-place(P),num(Q;Q1;Q2;Q3),time(TS),time(TS+1),col(C),
      holds(P,Q1,C,TS),tot_incr(P,Q2,C,TS),tot_decr(P,Q3,C,TS),Q=Q1+Q2-Q3.
a2: consumesmore(P,TS) :- holds(P,Q,C,TS), tot_decr(P,Q1,C,TS), Q1 > Q.
a3: consumesmore :- consumesmore(P,TS).
a4: :- consumesmore.
```

Proposition 1. *Let PN^C be a Petri Net with colored tokens, reset, inhibit, read arcs, and M_0 be its initial marking. Let $\Pi^3(PN^C, M_0, k)$ be the ASP encoding of PN^C and M_0 over a simulation of length k as defined in Section 3. Then $X^3 = M_0, T_0, M_1, \ldots, T_k$ is an execution sequence of PN^C (with respect to M_0) iff there is an answer-set A of $\Pi^3(PN^C, M_0, k)$ such that: $\{fires(t, j) : t \in T_j, 0 \le j \le k\} = \{fires(t, ts) : fires(t, ts) \in A\}$ and $\{holds(p, q, c, j) : p \in P, c/q \in M_j(p), 0 \le j \le k\} = \{holds(p, q, c, ts) : holds(p, q, c, ts) \in A\}$*

The Petri Net in Fig. 1 with an initial marking of zero tokens is encoded as:[5]:

```
time(0..2). num(0..30). place(mm;is;q;cytc). trans(t1;t3;t4;t10;t12).
col(nadh;h;e;nadp;h2o;o2). holds(mm,0,nadh,0). holds(mm,0,h,0).
tparc(t12,is,1,o2,TS):-time(TS). tparc(t10,mm,6,h,TS):-time(TS).
tparc(t10,mm,2,nadh,TS):-time(TS). ptarc(mm,t1,2,nadh,TS):-time(TS).
```

We get hundreds of answer-sets, for example[6]:

```
fires(t10;t12,0) holds(is,1,o2,1) holds(mm,6,h,1) holds(mm,2,nadh,1)
fires(t1;t10;t12,1) holds(is,2,h,2) holds(is,2,o2,2) holds(mm,10,h,2)
holds(mm,2,nadh,2) holds(mm,2,nadp,2) holds(q,2,e,2) fires(t1;t3;t10;t12,2)
```

4   Extensions

The above code implements a **set** firing semantics, which can produce a large number of answer-sets[7]. In biological domain, it is often preferable to simulate the maximum parallel activity at each step. We accomplish this by enforcing the **maximal firing set** semantics [1] using the following additional rules:

```
a5: could_not_have(T,TS):-enabled(T,TS),not fires(T,TS),ptarc(S,T,Q,C,TS),
      holds(S,QQ,C,TS), tot_decr(S,QQQ,C,TS), Q > QQ - QQQ.
a6: :-not could_not_have(T,TS),time(TS),enabled(T,TS),not fires(T,TS),
      trans(T).
```

---

[5] We show only a few of the `tparc/5`, `ptarc/5`, `holds/4` for illustration.

[6] Only > 0 tokens shown; `fires(t1;...;tm,ts)` $\equiv$ `fires(t1,ts)`,...,`fires(tm,ts)`.

[7] A subset of a firing set can also be fired as a firing set by itself.

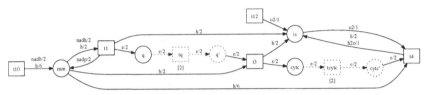

Fig. 2. Extended Petri Net model from Fig. 1 with new timed transitions $tq, tcytc$, modeling decreased fluidity of q and $cytc$. Both $tq, tcytc$ have a duration of 2 each (shown as ("[2]")), others have duration of 1 (assumed).

With maximal firing set, the number of answer-sets reduce to 1.

Other firing semantics can be implemented with similar ease. Let us look at how various Petri Net extensions can be encoded by making small code changes.

Priority transitions favor high priority transitions over lower priority ones, modeling dominant vs. secondary processes. We add the rules $f11, a7, a8$ to encode transition priority [8] and replace $a1, a5, a6$ with $a9, a10, a11$ respectively:

f11: Facts `transpr`(t_i, pr_i) where pr_i is $t_i's$ priority.
a7: `notprenabled(T,TS):-enabled(T;TT,TS),transpr(T,P),transpr(TT,PP),PP<P.`
a8: `prenabled(T,TS)  :- enabled(T,TS), not notprenabled(T,TS).`
a9: `{fires(T,TS)} :- prenabled(T,TS), trans(T), time(TS).`
a10: `could_not_have(T,TS) :- prenabled(TS,TS), not fires(T,TS),`
`    ptarc(S,T,Q,C,TS), holds(S,QQ,C,TS), tot_decr(S,QQQ,C,TS), Q>QQ-QQQ.`
a11: `:- prenabled(tr,TS), not fires(tr,TS), time(TS).`

Timed transitions model the execution time variation of processes. Output from such transitions is produced at the end of their execution duration. We replace $f5$ with $f12$, $r1$ with $r6$, and add $e5$ for non-reentrant transitions:

f12: Rules `tparc`$(t_i, p_j, n_c, c, ts_k, D(t_i))$`:-time`$(ts_k)$`.` for each $(t_i, p_j) \in E^+$, $c \in C$,
$n_c = m_{W(t_i, p_j)}(c) : n_c > 0.$
r6: `add(P,Q,T,C,TSS)  :- fires(T,TS),time(TS;TSS),tparc(T,P,Q,C,TS,D),`
`    TSS=TS+D-1.`
e5: `notenabled(T,TS1):-fires(T,TS0),num(N),TS1>TS0,tparc(T,P,N,C,TS0,D),`
`    col(C), time(TS0), time(TS1), TS1<(TS0+D).`

5   Example Use of Our Encoding and Reasoning Abilities

We illustrate the usefulness of our encoding by applying it to the following simulation based reasoning question[9] from [2]: "Membranes must be fluid to function properly. How would decreased fluidity of the membrane affect the efficiency of the electron transport chain?"

To answer this question, first we build a Petri Net model (see Fig. 1) of the Electron Transport Chain based on [2, Figure 9.15]. We model change in fluidity as an intervention to the Petri Net model. Thus, we add time delay transitions

---

[8] Higher priority numbers signify lower priority
[9] As it appeared in https://sites.google.com/site/2nddeepkrchallenge/

$tq, tcytc$ (with duration 2) to the capture increased time in shuttling electrons (e) from $t1$ to $t3$ and $t3$ to $t4$, with the notion that lower fluidity equals more transport time. The extended model is shown in Fig. 2. We encode both models in ASP based on Sections 3 and 4 and simulate them for a fixed number of time steps (ts) using *maximal firing set* semantics. We compute the chain's efficiency by computing h/ts, where h is the H+ ions moved across the membrane (to is). A plot of H+ ions moved across membrane is shown in Fig. 3. We find that the chain's efficiency decreased from 4.5 to 3 (for reentrant and 2.5 for non-reentrant transitions) due to decreased fluidity, meaning that decreased membrane fluidity leads to lower transport chain efficiency.

ASP's enumeration of the entire simulation evolution allows us to perform additional reasoning not directly possible with Petri Nets. For example, partial state or firing sequence can be encoded (as ASP constraints) as *way-points* to guide the simulation, such as to enumerate answer-sets where a transition t fires when one of its upstream source products S is found to be depleted. The answer-sets are then used to identify another upstream substance responsible for t's firing, by generalization. Our encoding allows various Petri Net dynamic and structural properties to be easily analyzed, as described in our previous work [1].

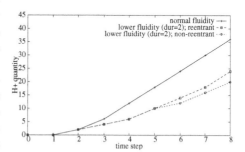

Fig. 3. H+ production in the is over time for the normal fluidity, lower fluidity (reentrant), and lower fluidity (non-reentrant transitions).

Conclusion: In this paper we presented the suitability of using Petri Nets with colored tokens for modeling biological pathways. We showed how such Petri Nets can be intuitively encoded in ASP, simulated, and reasoned with, in order to answer real world questions posed in the biological texts. We showed how our initial encoding can be minimally modified to include extensions. Our encoding has a low specification-implementation gap, it allows enumeration of all possible state evolutions, the ability to guide the search using way-points, and a strong reasoning ability. We showcased the usefulness of our encoding by an example.

References

1. Anwar, S., Baral, C., Inoue, K.: Encoding petri nets in answer set programming for simulation based reasoning (2013), http://arxiv.org/abs/1306.3542
2. Reece, J., Cain, M., Urry, L., Minorsky, P., Wasserman, S.: Campbell Biology. Pearson Benjamin Cummings (2010)
3. Gebser, M., Kaminski, R., Kaufmann, B., Ostrowski, M., Schaub, T., Schneider, M.: Potassco: The Potsdam answer set solving collection. AI Com. 24(2), 105–124 (2011)
4. Peterson, J., et al.: A note on colored petri nets. Information Processing Letters 11(1), 40–43 (1980)

CPLUS2ASP: Computing Action Language $\mathcal{C}+$ in Answer Set Programming

Joseph Babb and Joohyung Lee

School of Computing, Informatics, and Decision Systems Engineering
Arizona State University, Tempe, USA
{Joseph.Babb,joolee}@asu.edu

Abstract. We present Version 2 of system CPLUS2ASP, which implements the definite fragment of action language $\mathcal{C}+$. Its input language is fully compatible with the language of the Causal Calculator Version 2, but the new system is significantly faster thanks to modern answer set solving techniques. The translation implemented in the system is a composition of several recent theoretical results. The system orchestrates a tool chain, consisting of F2LP, CLINGO, ICLINGO, and AS2TRANSITION. Under the incremental execution mode, the system translates a $\mathcal{C}+$ description into the input language of ICLINGO, exploiting its incremental grounding mechanism. The correctness of this execution is justified by the module theorem extended to programs with nested expressions. In addition, the input language of the system has many useful features, such as external atoms by means of Lua calls and the user interactive mode. The system supports extensible multi-modal translations for other action languages, such as $\mathcal{B}$ and $\mathcal{BC}$, as well.

1 Introduction

Action language $\mathcal{C}+$ is a high level language for nonmonotonic causal theories, which allows us to describe transition systems succinctly [1]. The definite fragment of $\mathcal{C}+$ is expressive enough to represent various properties of actions, and was implemented in Version 2 of the Causal Calculator (CCALC)[1]. The system translates an action description in $\mathcal{C}+$ into formulas in propositional logic and calls SAT solvers to compute the models. Though CCALC is not a highly optimized system, it has been used to solve several challenging commonsense reasoning problems, including problems of nontrivial size [2], to provide a group of robots with high-level reasoning [3], to give executable specifications of norm-governed computational societies [4,5], and to automate the analyses of business processes under authorization constraints [6].

An alternative way to compute the definite fragment of Boolean-valued $\mathcal{C}+$ is to translate it into answer set programs as studied in [7,8]. The system reported in [9] and system COALA [10] are implementations of this method and accept the definite fragment of $\mathcal{C}$, a predecessor of language $\mathcal{C}+$. In particular, COALA was

---

[1] http://www.cs.utexas.edu/users/tag/cc

P. Cabalar and T.C. Son (Eds.): LPNMR 2013, LNAI 8148, pp. 122–134, 2013.

shown to be effective for several benchmark problems due to efficiency of ASP solvers.

However, the input language of COALA is missing several important features of $\mathcal{C}+$, such as multi-valued fluents, defined fluents, additive fluents, defeasible causal laws, and syntactically complex formulas. Also, it does not support many useful language constructs allowed in the input language of CCALC, such as user-defined macros, implicit declarations of sorts, and external atoms.

The design aim of system CPLUS2ASP [11] is to utilize the efficient ASP solving techniques as in COALA while supporting the full features of the input language of CCALC. Its design utilizes a standard library with meta-level sorts and meta-level variables, which yields a simple modular and extensible method to represent CCALC input programs in ASP. However, the first version of the system was a prototype implementation for a proof of concept.

This paper presents Version 2 of CPLUS2ASP, which is significantly enhanced in several ways.

- Its input language is fully compatible with the language of CCALC incorporating the features that were missing in CPLUS2ASP v1.
- The system supports extensible multi-modal translations for different action languages. Currently, in addition to $\mathcal{C}+$, the system supports language $\mathcal{B}$ [12], and a recently proposed language $\mathcal{BC}$ [13]. Language $\mathcal{BC}$ combines features of languages $\mathcal{B}$ and $\mathcal{C}$, and allows Prolog-style recursive definitions, which are not allowed in $\mathcal{C}+$.
- The system provides two execution modes: the command line mode and the interactive mode. The interactive mode gives a user-friendly interface for running various commands.
- In CCALC, external atoms are useful for some deterministic computation which is difficult to express directly in $\mathcal{C}+$. For example, they were utilized in [3] for a loose integration of task planning and motion planning. The new version of CPLUS2ASP supports this feature by utilizing Lua calls available in the language of GRINGO.
- The new system provides an incremental computation of action descriptions, which often saves a significant amount of time. Since the translation of action descriptions into answer set programs may contain complex formulas, the justification of this computation uses the module theorem from [14], which extends the module theorem from [15] to first-order formulas under the stable model semantics [16].

In [11], the translation of a definite $\mathcal{C}+$ description into the input language of ASP solvers was explained in multiple steps. A $\mathcal{C}+$ description is first turned into a multi-valued causal theory, and then to a Boolean-valued causal theory by the method described in [17]. The resulting theory is further turned into logic programs with nested expressions by the translation in [8], and then the translation in [18] is applied to turn it into the input language of GRINGO.

In Section 2, we explain the translation in a simpler way by avoiding reference to causal theories but instead by using a recent proposal of multi-valued propositional formulas under the stable model semantics [19]. A $\mathcal{C}+$ description is turned into multi-valued formulas under the stable model semantics, which is further turned into propositional formulas under the stable model semantics [20]. The result is further turned into the input language of GRINGO by the translation described in [18]. Section 3 introduces system CPLUS2ASP v2 and the features of its input language, and Section 4 compares the system with other similar systems. Our experiments show that the new system is significantly faster than the others.

2   From $\mathcal{C}+$ to ASP

2.1   Review: Multi-valued Propositional Formulas

A *(multi-valued propositional) signature* is a set σ of symbols called *constants*, along with a nonempty finite set $Dom(c)$ of symbols, disjoint from σ, assigned to each constant c. $Dom(c)$ is called the *domain* of c. A *Boolean* constant is one whose domain is the set {TRUE, FALSE}. An *atom* of a signature σ is an expression of the form $c=v$ ("the value of c is v") where $c \in \sigma$ and $v \in Dom(c)$. A *(multi-valued propositional) formula* of σ is a propositional combination of atoms. We often write $G \leftarrow F$, in a rule form as in logic programs, to denote the implication $F \rightarrow G$. A finite set of formulas is identified with the conjunction of the formulas in the set.

A *(multi-valued propositional) interpretation* of σ is a function that maps every element of σ to an element in its domain. An interpretation I *satisfies* an atom $c=v$, (symbolically, $I \models c=v$) if $I(c) = v$. The satisfaction relation is extended from atoms to arbitrary formulas according to the usual truth tables for the propositional connectives. I is a *model* of a formula if I satisfies it. We often write an interpretation I with the set of atoms $c=v$ such that $I(c) = v$.

The *stable* models of a multi-valued propositional formula can be defined in terms of a reduct [19]. Let F be a multi-valued propositional formula of signature σ, and let I be a multi-valued propositional interpretation of σ. The reduct F^I of a multi-valued propositional formula F relative to a multi-valued propositional interpretation I is the formula obtained from F by replacing each maximal subformula that is not satisfied by I with $\bot$. I is a *(multi-valued) stable model* of F if I is the unique multi-valued interpretation of σ that satisfies F^I.

Example 1. Assume $\sigma = \{c\}$, and $Dom(c) = \{1, 2, 3\}$. Each of the three interpretations is a model of $c=1 \leftarrow c=1$, but none of them is stable because each reduct has no unique model. Formula $c=1 \leftarrow \neg\neg(c=1)$ has the same models as $c=1 \leftarrow c=1$, but it has one stable model, $\{c=1\}$: the reduct of the formula relative to this interpretation is $c=1 \leftarrow \neg\bot$, and $\{c=1\}$ is its unique model. Similarly, one can check that $(c=1 \leftarrow \neg\neg(c=1)) \wedge (c=2)$ has only one stable model $\{c=2\}$, which illustrates nonmonotonicity of the semantics.

2.2 $\mathcal{C}+$ as Multi-valued Propositional Formulas under SM

Begin with a multi-valued signature partitioned into *fluent* constants and *action* constants. The fluent constants are assumed to be further partitioned into *simple* and *statically determined* fluent constants.

A *fluent formula* is a formula such that all constants occurring in it are fluent constants. An *action formula* is a formula that contains at least one action constant and no fluent constants.

A *static law* is an expression of the form

$$\textbf{caused } F \textbf{ if } G \tag{1}$$

where F and G are fluent formulas. An *action dynamic law* is an expression of the form (1) in which F is an action formula and G is a formula. A *fluent dynamic law* is an expression of the form

$$\textbf{caused } F \textbf{ if } G \textbf{ after } H \tag{2}$$

where F and G are fluent formulas and H is a formula, provided that F does not contain statically determined constants. A *causal law* is a static law, or an action dynamic law, or a fluent dynamic law. An *action description* is a finite set of causal laws.

An action description is called *definite* if F in every causal law (1) and (2) is either an atom or $\bot$.

For any definite action description D and any nonnegative integer m, the multi-valued propositional theory $cplus2mvpf(D, m)$ ("$\mathcal{C}+$ to multi-valued propositional formulas") is defined as follows.[2] The signature of $cplus2mvpf(D, m)$ consists of the pairs $i\!:\!c$ such that

- $i \in \{0, \dots, m\}$ and c is a fluent constant of D, or
- $i \in \{0, \dots, m-1\}$ and c is an action constant of D.

The domain of $i\!:\!c$ is the same as the domain of c. Recall that by $i\!:\!F$ we denote the result of inserting $i\!:$ in front of every occurrence of every constant in a formula F, and similarly for a set of formulas. The rules of $cplus2mvpf(D, m)$ are:

$$i\!:\!F \leftarrow \neg\neg(i\!:\!G) \tag{3}$$

for every static law (1) in D and every $i \in \{0, \dots, m\}$, and for every action dynamic law (1) in D and every $i \in \{0, \dots, m-1\}$;

$$i\!:\!F \leftarrow \neg\neg(i\!:\!G) \wedge (i\!-\!1\!:\!H) \tag{4}$$

for every fluent dynamic law (2) in D and every $i \in \{1, \dots, m\}$;

$$0\!:\!c\!=\!v \leftarrow \neg\neg(0\!:\!c\!=\!v) \tag{5}$$

for every simple fluent constant c and every $v \in Dom(c)$.

---

[2] The translation can be applied to non-definite $\mathcal{C}+$ descriptions as well, but then the semantics does not agree with $\mathcal{C}+$.

Note how the definition of $cplus2mvpf(D, m)$ treats simple fluent constants and statically determined fluent constants in different ways: rules (5) are included only when c is simple.

The translation of $\mathcal{BC}$ into multi-valued propositional formulas is similar. Due to lack of space, we refer the reader to [13, Section 9].

2.3   Translating Multi-valued Propositional Formulas to Propositional Formulas under SM

Note that even when we restrict attention to Boolean constants only, the stable model semantics for multi-valued propositional formulas does not coincide with the stable model semantics for propositional formulas. Syntactically, they are different (one uses expressions of the form $c = \text{TRUE}$ and $c = \text{FALSE}$; the other uses propositional atoms). Semantically, the former relies on the uniqueness of (Boolean)-functions, while the latter relies on the minimization on propositional atoms. Nonetheless there is a simple reduction from the former to the latter.

Begin with a multi-valued propositional signature σ. By σ^{prop} we denote the signature consisting of Boolean constants $c(v)$ for all constants c in σ and all $v \in Dom(c)$. For any multi-valued propositional formula F of σ, by F^{prop} we denote the propositional formula that is obtained from F by replacing each occurrence of a multi-valued atom $c=v$ with $c(v)$. For any constant c with $Dom(c)$, by $UEC(c)$ we denote the existence and uniqueness constraints for c:

$$\bot \leftarrow (c(v) \wedge c(v'))$$

for all $v, v' \in Dom(c)$ such that $v \neq v'$, and

$$\bot \leftarrow \neg \bigvee_{v \in Dom(c)} c(v) \ .$$

By UEC_σ we denote the conjunction of $UEC(c)$ for all $c \in \sigma$.

For any interpretation I of σ, by I^{prop} we denote the interpretation of σ^{prop} that is obtained from I by defining $I^{prop} \models c(v)$ iff $I \models c=v$.

There is a one-to-one correspondence between the stable models of F and the stable models of F^{prop}. The following theorem is a special case of Corollary 1 from [19].

Theorem 1. *Let F be a multi-valued propositional formula of a signature σ such that, for every constant c in σ, $Dom(c)$ has at least two elements. **(I)** An interpretation I of σ is a multi-valued stable model of F iff I^{prop} is a propositional stable model of $F^{prop} \wedge UEC_\sigma$. **(II)** An interpretation J of σ^{prop} is a propositional stable model of $F^{prop} \wedge UEC_\sigma$ iff $J = I^{prop}$ for some multi-valued stable model I of F.*

2.4   Incremental Computation of $\mathcal{C}+$

In answer set planning [21], the length of a plan needs to be specified. When the length is not known in advance, a plan can be found by iteratively increasing the possible plan length. CPLUS2ASP Version 1 calls CLINGO for each such iteration, resulting in redundant computations each time.

Instead, by default, CPLUS2ASP v2 uses ICLINGO, which accepts *incremental logic programs*. Gebser *et al.* [22] define an incremental logic program to be a triple $\langle B, P[t], Q[t] \rangle$, where B is a disjunctive logic program, and $P[t]$, $Q[t]$ are incrementally parameterized disjunctive logic programs. Informally, B is the *base* program component, which describes static knowledge; $P[t]$ is the *cumulative* program component, which contains information regarding every step t that should be accumulated during execution; $Q[t]$ is the *volatile query* program component, containing constraints or information regarding the final step. Conceptually, system ICLINGO computes $B \cup P[1] \cup \cdots \cup P[k] \cup Q[k]$ by increasing k one by one, but avoids reproducing ground rules in each step. Also, previously learned heuristics, conflicts, or loops are reused without having to recompute them. This method turns out to be quite effective. The correctness of this computation assumes that $\langle B, P[t], Q[t] \rangle$ is *acyclic* [14].

Below we show that the translation from $\mathcal{C}+$ described previously can be modified to yield an incremental logic program, which is always acyclic, and thus can be computed by ICLINGO.

For any $\mathcal{C}+$ description D and any formula $F(t)$ (called a *query*) of the same signature as $cplus2mvpf(D, t)$, where t is a parameter denoting a nonnegative integer, we define the corresponding incremental logic program $\langle B, P[t], Q[t] \rangle$ as follows:

- B consists of
  - $0 : UEC(f)$ for every fluent constant f;
  - $0 : c(v) \leftarrow \neg\neg(0 : c(v))$ for every simple fluent c and every $v \in Dom(c)$;
  - $0 : F^{prop} \leftarrow \neg\neg(0 : G^{prop})$ for every static law (1) in D.
- $P[t]$ $(t \geq 1)$ consists of
  - $t : UEC(f)$ for every fluent constant f;
  - $(t-1) : UEC(a)$ for every action constant a;
  - $t : F^{prop} \leftarrow \neg\neg(t : G^{prop})$ for every static law (1) in D;
  - $(t-1) : F^{prop} \leftarrow \neg\neg((t-1) : G^{prop})$ for every action dynamic law (1) in D;
  - $t : F^{prop} \leftarrow \neg\neg(t : G^{prop}) \wedge ((t-1) : H^{prop})$ for every fluent dynamic law (2) in D.
- $Q[t]$ is $\bot \leftarrow \neg(F[t])^{prop}$.

Upon receiving this input and a range of nonnegative integers $[min \ldots max]$, ICLINGO will find an answer set of the module $\mathbf{R}_k$ with $k = min, min + 1, \ldots$ until it finds an answer set, or $k = max$, whichever comes first. In [14], module $\mathbf{R}_k$ is defined from $\langle B, P[t], Q[t] \rangle$ as follows.

$$
\begin{aligned}
\mathbf{P}_0 &= FM(B, \emptyset), \\
\mathbf{P}_i &= \mathbf{P}_{i-1} \sqcup FM(P[i], Out(\mathbf{P}_{i-1})), &(1 \leq i \leq k) \\
\mathbf{R}_k &= \mathbf{P}_k \sqcup FM(Q[k], Out(\mathbf{P}_k)) \,.
\end{aligned}
$$

(Due to lack of space, we refer the reader to [14] for the notations.)

The following theorem states the correctness of incremental execution in CPLUS2ASP.

Theorem 2. *For any definite $\mathcal{C}+$ description D, any non-negative integer k, and any formula $F(k)$ of the same signature as $cplus2mvpf(D, k)$, an interpretation I is a multi-valued stable model of $cplus2mvpf(D, k) \cup \{\bot \leftarrow \neg F(k)\}$ iff I^{prop} is a stable model of $\mathbf{R}_k$. Conversely, an interpretation J is a stable model of $\mathbf{R}_k$ iff $J = I^{prop}$ for some multi-valued stable model of $cplus2mvpf(D, k) \cup \{\bot \leftarrow \neg F(k)\}$.*

Proof. (Sketch) We can check that $\langle B, P[t], Q[t] \rangle$ obtained from the $\mathcal{C}+$ description and a query as above is acyclic according to Definition 12 from [14]. Then the claim follows from Proposition 5 from [14]. ∎

The translation of $\mathcal{BC}$ into an incremental logic program is similar.

3   System CPLUS2ASP v2

System CPLUS2ASP v2 is a re-engineering of the prototypical CPLUS2ASP v1 system [11] and is available under Version 3 of the GNU Public License. Like its predecessor, CPLUS2ASP v2 uses a highly modular architecture that is designed to take advantage of the existing tools, including system F2LP [18] and highly-optimized ASP grounders and solvers in addition to a number of packaged sub-components. Figure 1 shows a high-level conceptualization of the interaction of the sub-components in the CPLUS2ASP v2 architecture.

For a description of the input language of CPLUS2ASP, we refer the reader to the CPLUS2ASP homepage at http://reasoning.eas.asu.edu/cplus2asp or CCALC 2 homepage at http://www.cs.utexas.edu/~tag/ccalc/. A typical run of CPLUS2ASP involves the user interacting with the interactive bridge, a tightly-coupled shell-like interface for CPLUS2ASP, in order to configure the CPLUS2ASP run. CPLUS2ASP.BIN, a translator sub-component, is then called to compile a CCALC 2 input program into a logic program containing complex formulas. Following this, system F2LP further turns the program into the input language of GRINGO. The result of this compilation is given to

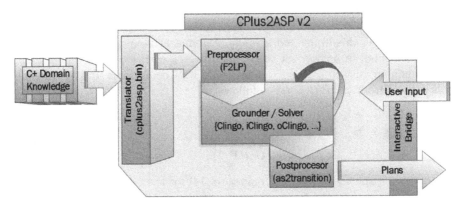

Fig. 1. CPLUS2ASP v2 System Architecture

CLINGO, or a similar answer set solver, and one or more answer sets are calculated. Finally, AS2TRANSITION is invoked in order to format the answer sets into a readable format.

CPLUS2ASP accepts a CCALC 2 style syntax of language $\mathcal{BC}$ as well, for which the user can select a different language mode for running. In addition, CPLUS2ASP is able to provide two target translations, a *static* translation to traditional ASP, and an *incremental* variant, as described in section 2.4.

3.1 Running Modes of System Cplus2ASP v2

In this section we briefly review the usage of CPLUS2ASP v2. For more complete documentation and information on obtaining and installing CPLUS2ASP v2, we invite the reader to visit the CPLUS2ASP homepage.

CPLUS2ASP v2 currently offers two distinct user-interaction methods: command-line and interactive shell. A brief introduction to both modes is provided below.

Using the Command-Line Mode. The command-line mode is designed primarily for interacting with a script or a seasoned CPLUS2ASP user who is familiar with the options available to them. The command-line mode is the default user-interaction mode when a query is provided while calling CPLUS2ASP.

For instance, to run a query labeled "simple" on a $\mathcal{C}+$ description stored in file bw-test, one can run the command:

```
cplus2asp bw-test query=simple
```

In order to run the command under the $\mathcal{BC}$ semantics, the flag --language=bc should be asserted in the command line call.[3]

If more solutions are desired, the number of solutions can be appended to the end of the command-line. As an example, appending 4 to the end of the command will return up to four solutions, while appending all or 0 will return all solutions.

The system provides the following options to write the output of a toolchain component into a file. Below [PROGRAM] may be one of pre-processor, grounder, solver, or post-processor.

--[PROGRAM]-output=[FILE] Writes the output of the toolchain component [PROGRAM] to a persistent output file [FILE].
--to-[PROGRAM] Executes the program toolchain up to and including [PROGRAM]. Similarly, --from-[PROGRAM] selects a program to initiate execution with and continue from.

As an example, if the user wants to run the toolchain up to the preprocessor and store the results for use later, he could use the command

---

[3] The bw-test example program, along with other examples, can be found from the CPLUS2ASP homepage.

```
cplus2asp bw-test --to-pre-processor > bw-test.lp.
```

Later, he could then run the command

```
cplus2asp bw-test.lp --from-grounder query=simple
```

to continue execution.

Using the Interactive Mode. The user-interactive mode provides a shell-like interface which allows the user to perform many of the configurations available from the command line. In general, the user-interactive mode is entered any time the user fails to provide all necessary information within the command-line arguments. As such, the easiest way to enter the user-interactive mode is to neglect to specify a query on the command-line. As an example, the command

```
cplus2asp bw-test
```

will enter the user-interactive mode.

While in the user-interactive mode, the following commands, among others, are available to the user:

help Displays the list of available commands.
config Reveals the currently selected running options.
queries Displays the list of available queries to run.
minstep=[#] Overrides the minimum step to solve for the next query selected.
maxstep=[#] Overrides the maximum step to solve for the next query selected.
sol=[#] Selects the number of solutions to display.
query=[QUERY] Runs the selected query and returns the results.
exit Exits the program.

Following successful execution of a query, the system will return to the inter-active prompt and the process can be repeated. For more information on using CPLUS2ASP v2, we invite the reader to explore the documentation available at http://reasoning.eas.asu.edu/cplus2asp or within the help usage message available by executing cplus2asp --help.

3.2   Lua in System Cplus2ASP v2

System CPLUS2ASP v2 allows for embedding external Lua function calls in the system, which are evaluated at grounding time. These Lua calls allow the user a great deal of flexibility when designing a program and can be used for complex computation that is not easily expressible in logic programs. A Lua function must be encapsulated in #begin_lua ...#end_lua. tags, and, can optionally be included in a separate file ending in .lua. Lua calls occurring within the CPLUS2ASP program are restricted to occurring within the where clause [4] of each rule and must be prefaced with an @ sign.

For example, one can say that moving a block does not always work.[5]

---

[4] The condition in the where clause is evaluated at grounding time.
[5] Note that this is decided at grounding time so this is not truly random.

```
move(B,L) causes loc(B)=L where @roll(1,2).
```

with Lua function defined as

```
#begin_lua
math.randomseed(os.time())
function roll(a,n)--returns 1 with probability a/n
  if(math.random(n) <= a) then return 1
  else return 0
  end
end
#end_lua.
```

A more complete description of the system's Lua functionality and additional examples of its use are available from the CPLUS2ASP homepage.

4   Experiments

In order to compare the performance of the CPLUS2ASP v2 system with its predecessors, we used large variants of several widely known domains [6] and compared the performance of CPLUS2ASP's running modes with the performance of CCALC v2, CPLUS2ASP v1, and the incremental and static running modes of COALA (where applicable). All experiments were performed on an Intel Core 2 Duo 3.00 GHZ CPU with 4 GB RAM running Ubuntu 11.10. The CCALC v2 tests used RELSAT 2.0 as a SAT solver while CPLUS2ASP v1, v2, and COALA tests used the same version of CLINGO, v3.0.5.

The domains tested include a large variant of the Traffic World [2], which models the behavior of cars on a road; a variant of the Blocks World where actions have costs [23]; the Spacecraft Integer [23], which models a spacecraft's movement with multiple independent jets; the Towers of Hanoi; and the Ferryman domain, which involves moving a number of wolves and sheep across a river without allowing the sheep to be eaten. The Towers of Hanoi and Ferryman descriptions are from examples packaged with COALA v1.0.1. In order to run on other systems, we manually turned them into the syntax of CCALC input language.

Table 1 compares the results of the test benchmarks for each of the available configurations. Each measured time includes translation, grounding, and solving for all possible maximum steps between 0 and the horizon (#), as well as the number of atoms and rules produced below each timing. In all test cases CPLUS2ASP's incremental running mode showed a significant performance advantage compared to the other systems, performing roughly 3 times faster than COALA's incremental mode and an order of magnitude faster than its predecessor CPLUS2ASP v1. COALA's incremental running mode comes in the second place in all but one benchmark. CPLUS2ASP v2's static mode tended to outperform its predecessor on the more computation-heavy domains with additive fluents,

---

[6] All benchmark programs are available from the CPLUS2ASP homepage.

Table 1. Benchmarking Results

Domain	steps	CCALC 2	CPLUS2ASP V1	COALA static	incr.	CPLUS2ASP V2 static	incr.
traffic (altmerge)	11	878.59 s + 1 s ^{a} [531552 / 3671940]	95.43 s + 25.95 s [2722247 / 3341068]	$_-$^{b} –	– –	82.16 s + 26.57 s [2262231	14.2 s + 2.6 s / 2766459]
bw-cost (15)^c	8	131.1 s + 5 s [149032 / 624439]	76.16 s + 0.4 s [123517 / 260282]	–	–	17.09 s + 3.16 s [43052	3.47 s + 0.16 s / 526923]
bw-cost (20)	9	52 s + 987 s [374785 / 1584778]	271 s + 9.17 s [279869 / 626496]	–	–	63.26 s + 66.58 s [102426	13.45 s + 2.24 s / 1745166]
spacecraft (15/8)^d	3	173.62 s + 0 s [128262 / 622158]	16.07 s + 2.65 s [146056 / 146056]	–	–	5.57 s + 0.06 s [132918	2.33 s + 0.01 s / 253514]
spacecraft (25/10)	4	*timeout*	208.2 s + 480.24 s [760673 / 1653650]	–	–	67.55 s + 3.42 s [732860	17.46 s + 0.35 s / 1427771]
hanoi (6/3) ^{e}	64	14 s + 1983 s [13710 / 221895]	38.9 s + 137.27 s [37297 / 298047]	1039.15 s + 507.12 s [13798	1.4 s + 51.13 s / 410559]	547.9 s + 47.53 s [10086	0.76 s + 3.5 s / 202694]
towers (8/4)	33	*timeout*	31.19 s + 102.69 s [35041 / 433660]	304.02 s + 3017.87 s [12922	1.51 s + 470.23 s / 655436]	102.81 s + 89.36 s [9074	1.04 s + 14.8 s / 324668]
ferryman (10/4) ^{f}	16	39.45 s + 0 s [55905 / 308909]	8.27 s + 2.98 s [14122 / 120693]	40.85 s + 8.71 s [4973	0.87 s + 1.85 s / 358772]	21.59 s + 2.37 s [12721	0.66 s + 0.25 s / 112912]
ferryman (15/4)	26	1004.26 s + 0 s [256590 / 1452554]	85.21 s + 39.54 s [42687 / 539513]	793.13 s + 169.18 s [15718	6.13 s + 14.73 s / 2275992]	318.4 s + 34.4 s [39536	4.18 s + 2.97 s / 515167]

^{a} preprocessing time + solving time [# atoms / # rules]
^{b} The input language is not expressive enough to represent the domain.
^{c} maximum cost
^{d} domain size ($15 \times 15 \times 15$) / goal position ($8 \times 8 \times 8$)
^{e} # disks / # pegs
^{f} # animals / boat capacity

but was subsequently outmatched in the others. Finally, CCALC 2 and COALA's static mode came in last (with CCALC performing slightly worse in most cases).

Figure 2 shows a more detailed analysis of the execution of the first 100 steps of solving an extreme variant of the ferryman domain consisting of 120 of each animal by graphing the time spent (in seconds) on each step by each configuration. While the static configurations were required to completely re-ground and re-solve the translated answer set program for each maximum step, resulting in an ever-growing amount of work to be performed at each step, CPLUS2ASP v2's incremental running mode is able to avoid this by only grounding the new cumulative ($P[t]$) and volatile ($Q[t]$) components and leveraging heuristics learned from previous iterations. This results in far less time being required for checking each increment.

Although COALA's incremental mode uses the same reasoning engine ICLINGO as CPLUS2ASP v2's incremental mode, system CPLUS2ASP sees a significant overall speed-up over COALA. This is related to a significant reduction in the number of atoms and rules produced during grounding, which also accounts for far fewer conflicts and restarts during solving in all test cases.

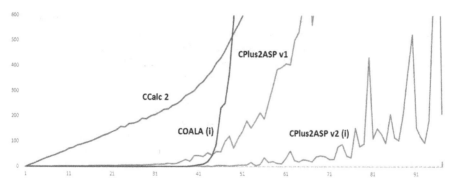

Fig. 2. Ferryman 120/4 Long Horizon Analysis

5 Conclusion

A distinct advantage that CPLUS2ASP v2 has over its prototypical predecessor is that it was re-engineered in order to allow for far greater flexibility and extensibility via a multi-modal execution model. This makes it suitable for use as a base-platform for future input language implementations, input language extensions, or target languages/platforms.

The advances in ASP solving techniques account for the efficiency of system CPLUS2ASP. We expect that the significant speed-up of the system demonstrated by CPLUS2ASP v2, as well as the enhanced expressivity of the input language, will contribute to widening application of action languages in various domains.

Acknowledgements. We are grateful to Michael Bartholomew and the anonymous referees for their useful comments. This work was partially supported by the National Science Foundation under Grant IIS-0916116 and by the South Korea IT R&D program MKE/KIAT 2010-TD-300404-001.

References

1. Giunchiglia, E., Lee, J., Lifschitz, V., McCain, N., Turner, H.: Nonmonotonic causal theories. Artificial Intelligence 153(1-2), 49–104 (2004)
2. Akman, V., Erdoğan, S., Lee, J., Lifschitz, V., Turner, H.: Representing the Zoo World and the Traffic World in the language of the Causal Calculator. Artificial Intelligence 153(1-2), 105–140 (2004)
3. Caldiran, O., Haspalamutgil, K., Ok, A., Palaz, C., Erdem, E., Patoglu, V.: Bridging the gap between high-level reasoning and low-level control. In: Erdem, E., Lin, F., Schaub, T. (eds.) LPNMR 2009. LNCS, vol. 5753, pp. 342–354. Springer, Heidelberg (2009)
4. Artikis, A., Sergot, M., Pitt, J.: Specifying norm-governed computational societies. ACM Transactions on Computational Logic 9(1) (2009)
5. Desai, N., Chopra, A.K., Singh, M.P.: Representing and reasoning about commitments in business processes. In: AAAI, pp. 1328–1333 (2007)

6. Armando, A., Giunchiglia, E., Ponta, S.E.: Formal specification and automatic analysis of business processes under authorization constraints: An action-based approach. In: Fischer-Hübner, S., Lambrinoudakis, C., Pernul, G. (eds.) TrustBus 2009. LNCS, vol. 5695, pp. 63–72. Springer, Heidelberg (2009)

7. McCain, N.: Causality in Commonsense Reasoning about Actions. PhD thesis, University of Texas at Austin (1997)

8. Ferraris, P., Lee, J., Lierler, Y., Lifschitz, V., Yang, F.: Representing first-order causal theories by logic programs. TPLP 12(3), 383–412 (2012)

9. Doğandağ, S., Alpaslan, F.N., Akman, V.: Using stable model semantics (SMODELS) in the Causal Calculator (CCALC). In: Proceedings 10th Turkish Symposium on Artificial Intelligence and Neural Networks, pp. 312–321 (2001)

10. Gebser, M., Grote, T., Schaub, T.: Coala: A compiler from action languages to ASP. In: Janhunen, T., Niemelä, I. (eds.) JELIA 2010. LNCS, vol. 6341, pp. 360–364. Springer, Heidelberg (2010)

11. Casolary, M., Lee, J.: Representing the language of the causal calculator in answer set programming. In: ICLP (Technical Communications), pp. 51–61 (2011)

12. Gelfond, M., Lifschitz, V.: Action languages. Electronic Transactions on Artificial Intelligence 3, 195–210 (1998)

13. Lee, J., Lifschitz, V., Yang, F.: Action language BC: Preliminary report. In: Proc. IJCAI 2013 (to appear, 2013)

14. Babb, J., Lee, J.: Module theorem for the general theory of stable models. TPLP 12(4-5), 719–735 (2012)

15. Janhunen, T., Oikarinen, E., Tompits, H., Woltran, S.: Modularity aspects of disjunctive stable models. Journal of Artificial Intelligence Research 35, 813–857 (2009)

16. Ferraris, P., Lee, J., Lifschitz, V.: Stable models and circumscription. Artificial Intelligence 175, 236–263 (2011)

17. Lee, J.: Automated Reasoning about Actions. PhD thesis, University of Texas at Austin (2005)

18. Lee, J., Palla, R.: System F2LP – computing answer sets of first-order formulas. In: Erdem, E., Lin, F., Schaub, T. (eds.) LPNMR 2009. LNCS, vol. 5753, pp. 515–521. Springer, Heidelberg (2009)

19. Bartholomew, M., Lee, J.: Stable models of formulas with intensional functions. In: Proceedings of International Conference on Principles of Knowledge Representation and Reasoning, KR, pp. 2–12 (2012)

20. Ferraris, P.: Answer sets for propositional theories. In: Baral, C., Greco, G., Leone, N., Terracina, G. (eds.) LPNMR 2005. LNCS (LNAI), vol. 3662, pp. 119–131. Springer, Heidelberg (2005)

21. Lifschitz, V.: Answer set programming and plan generation. Artificial Intelligence 138, 39–54 (2002)

22. Gebser, M., Grote, T., Kaminski, R., Schaub, T.: Reactive answer set programming. In: Delgrande, J.P., Faber, W. (eds.) LPNMR 2011. LNCS, vol. 6645, pp. 54–66. Springer, Heidelberg (2011)

23. Lee, J., Lifschitz, V.: Describing additive fluents in action language $\mathcal{C}+$. In: Proceedings of International Joint Conference on Artificial Intelligence, IJCAI, pp. 1079–1084 (2003)

Towards Answer Set Programming with Sorts

Evgenii Balai, Michael Gelfond, and Yuanlin Zhang

Texas Tech University, USA
{evgenii.balai,michael.gelfond,y.zhang}@ttu.edu

Abstract. Existing ASP languages lack support for conveniently specifying objects, their sorts and the sorts of the parameters of relations in an application domain. However, such support may allow a programmer to better structure the program, to automatically determine some syntax and semantic errors and to avoid thinking about safety of ASP rules — non-declarative conditions on rules required by existing ASP systems. In this paper, we define the syntax and semantics of a knowledge representation language $\mathcal{SPARC}$ which offers explicit constructs to specify objects, relations, and their sorts. The language expands CR-Prolog — an extension of ASP by consistency restoring rules. We introduce an implementation of $\mathcal{SPARC}$ based on its translation to DLV with weak constraints. A syntax checking algorithm helps to avoid errors related to misspellings as well as simple type errors. Another type checking algorithm flags program rules which, due to type conflicts, have no ground instantiations.

1 Introduction

A good knowledge representation methodology should allow one to:

- Identify and describe *sorts* (types, kinds, categories) of objects populating a given domain.
- Identify and classify these objects.
- Identify and precisely define objects *properties* and *relationships* between them.

ASP[1] based knowledge representation languages have powerful means for describing these properties and relationships but lack the means for conveniently specifying objects and their sorts as well as sorts of parameters of the domain relations. There were some attempts to remedy the problem. The #domain statements of lparse [2] — a popular grounder used for a number of ASP systems — define sorts for variables. Even though this device is convenient for simple programs it causes substantial difficulties for medium and large programs. It is especially difficult to put together pieces of programs written at different time and/or by different people. The same variable may be declared as ranging over different sorts by different #domain statements used in different parts of a program. So the process of merging these parts requires renaming of variables. This concern was addressed by Balduccini whose system, RSig[3] , provided an ASP programmer with means for specifying sorts of parameters of the language predicates. RSig is a simple (but very useful) extension of ASP which does not require any shift in perspective and involves only minor changes in existing programs. In this work we further develop the idea of RSIG by introducing a knowledge representation language

P. Cabalar and T.C. Son (Eds.): LPNMR 2013, LNAI 8148, pp. 135–147, 2013.
© Springer-Verlag Berlin Heidelberg 2013

$\mathcal{SPARC}$. In addition to allowing the specification of program relations and their parameters $\mathcal{SPARC}$ provides a programmer with means for defining objects of the program domain and their sorts. This allows *better separation of concerns*. A programmer is encouraged to write rules which express general properties of the domain and do not necessarily refer to particular domain objects. Such rules can be used in conjunction with different collections of objects and/or different placement of objects into sorts. A simple syntax checking algorithm helps a programmer to *avoid errors* related to misspelling as well as simple type errors. (Despite their simplicity such errors are sometimes not easy to identify.) Explicit declaration of sorts allows a programmer to *avoid thinking about safety conditions* in program rules — a feature especially important when $\mathcal{SPARC}$ is used to *teach* declarative programming. Finally a *type checking algorithm* locating rules of the program which, because of the type restrictions on variables, do not have any ground instantiations is useful for determining more subtle potential problems. The paper defines the syntax and semantics of a version of $\mathcal{SPARC}$ defined on top of CR-Prolog — an ASP based language with consistency-restoring rules [4]. It also describes the corresponding syntax and type checking algorithms, and an algorithm for computing answer sets of a $\mathcal{SPARC}$ program based on reduction of such a program to DLV [5] — a language of disjunctive logic programs with weak constraints [6]. The preliminary description of the language and the latter algorithm has been presented in [7] in 2012. The new version of the language however is quite different from that presented in this workshop. The most important improvement is the completely new definition of sorts and domain objects of a program. An implementation of the SPARC system can be found at [8]. The paper is organized as follows. In the next section we define syntax and semantics of $\mathcal{SPARC}$. We then present syntax and type checking algorithms in Sections 3 and 4, and an algorithm for computing answer sets of a $\mathcal{SPARC}$ program in Section 5. Most of the paper can be understood by anyone familiar with logic programming under the answer set semantics. However, full understanding of Section 5 requires knowledge of CR-Prolog.

2   Syntax and Semantics of $\mathcal{SPARC}$

$\mathcal{SPARC}$ vocabulary consists of *variables, sort names, symbolic names, natural numbers*, equality ($=$) and inequality ($!=$) defined on arbitrary terms, *order relations* ($<, \leq$) on numbers and on symbolic names (ordering of symbolic names is lexicographic), and standard *arithmetic functions*. Variables and symbolic names are identifiers which start with capital and lower-case letters respectively; sort name is a symbolic name preceded by $\#$. The vocabulary is used to define $\mathcal{SPARC}$ *terms* which are divided into arithmetic and symbolic. An *arithmetic term* is defined as usual. A *symbolic term* is a symbolic name, or a variable, or a string of the form $f(t_1, \ldots, t_n)$ where f is a symbolic name and $t_1, \ldots, t_n$ are arithmetic or symbolic terms. A term $f(t_1, \ldots, t_n)$ is referred to as a *record* with the *record name* f (of arity n). A term is called *ground* if it contains no variables and no arithmetic operations. A *set expression* of $\mathcal{SPARC}$ is either a sort name, a collection of ground terms $\{t_1, \ldots, t_n\}$, or has the form $(A \odot B)$ where A and B are set expressions and $\odot$ is a set theoretic operation $+, -$ or $\times$. Parentheses can be omitted and standard preference is used to determine the order of operations. We also need two special sorts *dom* and *nat* which belong to every program of $\mathcal{SPARC}$: the former consists of all ground terms from the signature of the program, and the sort *nat* of natural numbers between 0 and $maxint$.

Now we are ready to define the syntax of $\mathcal{SPARC}$. A $\mathcal{SPARC}$ program is constructed from four consecutive parts:

The **first part**, called *directives* consists of a (possibly empty) collection of statements of the form

```
#const <identifier> = <natural_number>.
#maxint = <natural_number>.
```

The **second part** of the program consists of the keyword **sorts** followed by a list of *sort definitions* — statement of the form (1) – (5) below. It is used to define

- objects of the program's domain (often referred to as *domain elements*) and
- sort names and their assignments to non-empty sets of domain elements.

The list consists of statements of the form

```
sort_name = sort_expression
```

where sort expressions are expressions appearing on the right-hand side of statements (1) – (5) below. Each such expression, E, defines a collection $\mathcal{D}(E)$ of ground terms which is assigned to the sort sort_name. In addition, if $\{t_1, \ldots, t_n\}$ occurs in the sort expression on the right then *every t_i together with its subterms* is added to the set, *dom*, of domain elements of the program.
Statement

$$sort\_name = set\_expr \tag{1}$$

defines a sort, *sort_name* using the set expression on the right. For example the sort definition consisting of statements

```
#blocks = {b1,b2}
#locations = #blocks + {table}
```

defines the program domain consisting of elements $\{b1, b2, table\}$; sort #blocks is mapped into $\{b1, b2\}$; and sort #locations mapped into $\{b1, b2, table\}$. The sort definition

```
#names = {name(bob,smith), name{mary,smith}}
```

defines the set names consisting of the two records on the right and expands the set of domain elements by these records and their subterms bob, $mary$, and $smith$.

Statement of the form

$$sort\_name = [n_1..n_2] \tag{2}$$

where n_1 and n_2 are natural numbers and $n_1 \le n_2$ defines the sort $\{n : n_1 \le n \le n_2\}$.

Similarly if id_1 and id_2 are identifiers then the statement

$$sort\_name = [id_1..id_2] \tag{3}$$

defines the sort $\{id : |id_1| \le |id| \le |id_2| \text{ and } id_1 \le id \le id_2\}$ where $\le_l$ is the lexicographic ordering on identifiers.

The next statement has the form

$$sort\_name = f(s_1(var_1), ..., s_n(var_n)) : cond \qquad (4)$$

where f is a new symbolic name, $s_1, \ldots, s_n$ are previously defined sorts and $cond$ has the form $X \diamond Y$, where $X, Y \in \{var_1, \ldots var_n\}$ and $\diamond \in \{<, \leq, =, \neq\}$, or $C_1 \bullet C_2$, where C_1 and C_2 are conditions and $\bullet \in \{\vee, \wedge\}$. Both, the variables and the condition, can be omitted. The new sort is assigned a collection of records of the form $f(t_1, \ldots, t_n)$ where $t_1, \ldots, t_n$ are elements of sorts $s_1, \ldots, s_n$ satisfying condition $cond$. For instance, a statement

```
#actions = put(#blocks,#locations).
```

defines a new sort, $actions$, consisting of records of the form $put(b, l)$ where b is a block and l is a location. Note that, according to this definition, a record $put(b1, b1)$ is an action. Sometimes it is convenient to exclude this possibility. This can be achieved by the following alternative definition of $actions$:

```
#actions = put(#blocks(X),#locations(Y)) : X != Y.
```

Now a record $put(b, l)$ belongs to the sort $actions$ if b is a block, l is a location, and b and l are different. The statement

$$sort\_name = [b\_expr][b\_expr] \ldots [b\_expr] \qquad (5)$$

defines concatenation of $basic$ sorts, i.e., sorts consisting of identifiers and natural numbers; b_expr is the name of a basic sort or a list $t_1, \ldots, t_n$ of natural numbers and symbolic names or expressions of the form $n_1..n_2$ and $id_1..id_2$ where n_1, n_2 are natural numbers and id_1, id_2 are symbolic names. These sort definitions are useful when we want to define large basic sorts, e.g. a sort of blocks $b_1, \ldots, b_{100}$ can be defined as:

```
#blocks = [b][1..100]
```

Definition 1 (Sorts Definitions)
The *list of sort definitions* of a program is a sequence of statements of the form (1)–(5) such that no sort name occurs on the left-hand side of a statement more than once and no sort name occurs on the right-hand side of a statement if it was not previously defined.

The collection of sorts of a program consists of *sorts defined by sort definitions of the program* and sorts *dom* and *nat*.

Definition 2 (Domain Elements)
A ground term t of $\mathcal{SPARC}$ is an *element of the program's domain* if
1. t is a natural number belonging to sort *nat* or
2. t is defined by a sort definition of the form (2)-(5) or
3. there is a sort definition containing an occurrence of $\{.., t, ..\}$ or
4. t is a subterm of a term satisfying one of the above properties.

In the first two cases t belongs to at least one sort defined by the corresponding sort definition. The domain element defined by one of the last two clauses of the definition may or may not have such a sort. In this case it belongs to sort *dom* of the program.

We say that a *record name is defined by a program* Π if it occurs in one of the elements of Π's domain.

The **third part** of the program defines predicate symbols and sorts of their parameters. It starts with a keyword **predicates** and is followed by statements of the form

$$pred(sort\_name, \ldots, sort\_name)$$

where $pred$ is a new identifier and $sort\_name$s are sort names defined by the sort definitions. The statement defines predicate symbol $pred$ and specifies its arity and the sorts of its parameters.

The first three sections of a $\mathcal{SPARC}$ program Π uniquely define the program's signature. To define rules of Π we need the following definitions:

Definition 3 (Program Term)
A $\mathcal{SPARC}$ term t is called a *term of $\mathcal{SPARC}$ program Π* if every ground subterm of t is an element of the program's domain and every record name occurring in t is defined by Π.

Let $p(s_1, \ldots, s_n)$ be a predicate declaration of Π. By $\Sigma(p)$ we denote the sequence $(s_1, \ldots, s_n)$. If p is a sort name, $\Sigma(p)$ is p.

Definition 4 (Program Atom)
A string $p(t_1, \ldots, t_n)$, where p is a predicate symbol or sort of Π and $t_1, \ldots, t_n$ are Π's terms, is an atom of Π if:

- Let $\Sigma(p)$ be $(s_1, s_2, \ldots, s_n)$
- for each $i \in \{1..n\}$:
    - if t_i is a ground symbolic term then t_i belongs to s_i,
    - if t_i is an arithmetic term without variables, s_i must contain the value of t_i (denoted by $val(t_i)$),
    - if t_i is an arithmetic term with variables and at least one arithmetic operation, s_i must contain at least one number.

An atom A of Π or its negation $\neg A$ are called literals of Π.

Example 1 (Program Π_0)
To see some examples consider a program Π_0 containing the following:

```
#const n = 1.
sorts
#s1 = {f(b)}.
#s2 = [0..n].
predicates
p(#s1,#s2).
```

It is easy to see that $\{b, f(b)\}$ where $b \in dom$ and $f(b) \in \#s1$ are non-numerical ground terms of Π_0; $p(X, X)$ is an atom of Π_0, while $p(X, f(b)), p(X, a)$ and $p(0, X)$ are not.

Definition 5 (Program Rules)
A *rule* of a $\mathcal{SPARC}$ program Π is a regular ASP rule

$$l_0 \vee \ldots \vee l_m \leftarrow l_{m+1}, \ldots, l_k, not\ l_{k+1} \ldots not\ l_n \tag{6}$$

or a CR-Prolog rule

$$l_0 \overset{+}{\leftarrow} l_1, \ldots, l_k, not\ l_{k+1} \ldots not\ l_n \tag{7}$$

where l's are literals of Π and l_0 is not formed by a sort name. We say that a rule is *ground* if it is constructed from ground literals.

The **fourth part** of a $\mathcal{SPARC}$ program starts with the keyword **rules** and is followed by a finite collection of rules of Π. *This completes our definition of syntax of $\mathcal{SPARC}$ programs*. In what follows sort definitions, predicate declarations and program rules of Π will be denoted by $\mathcal{S}(\Pi)$, $\mathcal{P}(\Pi)$, and $\mathcal{R}(\Pi)$ respectively.

To define the semantics of $\mathcal{SPARC}$ program Π we will define its *answer sets*. If the rules of Π are ground then answer sets of Π are answer sets of the collection of its ground rules. To define answer sets of a program Π with variables we need some terminology. A *ground instance* of a rule r of Π is a ground rule of Π which is the result of replacing variables of r by properly-sorted elements of the Π's domain; $ground(r)$ is the collection of all such instantiations; $ground(\Pi)$ is the union of $ground(r)$ for all rules of Π.

Definition 6 (Answer Sets)
Answer sets of a $\mathcal{SPARC}$ program Π are answer sets of an unsorted logic program $ground(\Pi)$.

Example 2 (Program Π_0 (continued))
Let us now complete our program Π_0 by adding to it the rules:

```
p.(f(b),0).
p(X,X).
p(f(b),Y+1) :- p(f(b),Y).
```

$ground(\Pi_0)$ consists of ground rules

```
p(f(b),0).
p(f(b),1) :- p(f(b),0).
```

Note that there is no subsitution of X in $p(X, X)$ which respects the sorts of p. Hence, the rule $p(X, X)$ has no ground instantiations; $\{p(f(b), 0), p(f(b), 1)\}$ is the only answer set of $ground(\Pi_0)$ and hence of Π_0. Notice that according to this definition we cannot expand Π_0 by the statement

```
p(X,f(b)).
```

since, according to our sort and predicate declarations, it would not be a rule of the resulting program.

2.1 Discussion

Notice that the above definition of $ground(\Pi)$ involves a non-obvious choice. We do not require the set $ground(r)$ to be non-empty. The alternative would be to prohibit such rules. Under this alternative definition Π_0 would not be a program of $\mathcal{SPARC}$. (Note that $\Pi_0 \cup \{p(X, a)\}$ is not a program under any of these definitions). Our choice is based on the methodology for writing $\mathcal{SPARC}$ programs which attempts to make them elaboration tolerant. We assume that normally programmers will be fully aware of sort, function, and predicate symbols of the program's signature but not necessarily about actual content of the sorts. As an example one may think about a programmer representing "blocks world" domain. He may structure the world in terms of sorts *steps, blocks, locations, actions* and *fluents*, and predicate symbols *holds(fluents, steps)* and *occurs(actions, steps)*, and write causal laws representing the domain, e.g. $holds(on(B, L), I + 1) \leftarrow occurs(put(B, L), I)$. Later he may define the sorts of the program including that of *actions*. Suppose this is done using the first definition of *actions* from page 4. If the programmer later wants to use this knowledge for planning he may decide to exclude generating an action $put(B, B)$ by a constraint

$$\leftarrow occurs(put(B, B), I).$$

After further consideration the definition of *actions* can be changed to the second version, which would leave our constraint without ground instantiations. Should this result in error? Our answer is "no". The rule will simply automatically disappear during the grounding process. We will however have an option of warning the programmer about such an event (see section 4).

3   Checking the Program Syntax

In this section we define a syntax checking algorithm for $\mathcal{SPARC}$ programs. Given program Π, the syntax check of directives and predicate declarations of Π is straightforward. Checking correctness of sort expressions involves checking their syntax, including non-emptiness of the sorts which can be done by a simple recursive algorithm. In the process we also mark all sorts which contain at least one number and create the list of names of all the program records. The rule part of the program is syntactically correct iff each of its rules is correct, i.e. if each rule is properly constructed from program atoms. The main work is performed by functions *IsAtom(A, Π)* and *IsTerm(T, Π)* which return *true* iff A and T are atom and term of Π respectively. Another important function, *ReduceTerm* uses sort definitions of Π, a ground term t and a sort s to construct a formula which evaluates to true iff $t \in s$. To be more precise we need the following definitions:

Definition 7 (Formula)

- $T \in D$, where T is a variable, a ground term or an arithmetic term, and D is a set of ground terms, is a *formula*,
- $t_1 \diamond t_2$, where t_1 and t_2 are terms and $\diamond \in \{=, \neq, \prec, \preceq\}$, is a *formula*, and
- if A and B are formulas then $(A \wedge B)$, $(A \vee B)$, and $\neg A$ are *formulas*.

Formula F is called *ground* if it does not contain variables.

Relation $\prec$ is defined on arbitrary terms; $X \prec Y$ iff X and Y are both symbolic names or both integers and $X < Y$. Otherwise $X \prec Y$ is false. Similarly for $\preceq$.

Definition 8 (Satisfiability). A formula $\mathcal{F}$ is *satisfied* by a substitution θ of variables of $\mathcal{F}$ by ground $\mathcal{SPARC}$ terms if the result, $\mathcal{F}(\theta)$, of this substitution is true.

Now we are ready to describe $IsAtom$ and $IsTerm$:

Algorithm 1. IsAtom

   Input: a string of the form $p(t_1, \ldots, t_n)$, where $t_1, \ldots, t_n$ are $\mathcal{SPARC}$ terms,
         and a $\mathcal{SPARC}$ program Π.
   Output: *true* if $p(t_1, \ldots, t_n)$ is an atom of Π and *false* otherwise.
1 **if** *p is not a sort or a predicate name of Π* **then**
2 | **return** false

3 Let $\Sigma(p)$ be $(s_1, s_2, \ldots, s_n)$
4 **for** each t_i *of* $p(t_1, \ldots, t_n)$ **do**
5   **if** t_i *is a ground term and* $ReduceTerm(t_i, s_i, \Pi)$ *is* false **then**
6 | **return** false

7   **if** t_i *is an arithmetic term without variables and*
      $ReduceTerm(val(t_i), s_i, \Pi)$ *is* false **then**
8 | **return** false

9   **if** t_i *is an arithmetic term with variables and at least one arithmetic*
      operation and s_i *does not contain a number* **then**
10 | **return** false

11   **if** t_i *is not a ground term and* $IsTerm(t_i, \Pi)$ *is* false **then**
12 | **return** false

13 **return** true

Algorithm 2. IsTerm

   Input: a $\mathcal{SPARC}$ term t and a program Π.
   Output: *true* if t is a term of Π and *false* otherwise.
1 **if** *there exists a record name in t that is not defined by Π* **then**
2 | **return** false

3 **for** each *maximum ground subterm u of t* **do**
4   **if** *u is a natural number such that* $u > \#maxint$ **then**
5 | **return** false

6   **if** *u is a symbolic term not occurring in the sort definitions of Π* **then**
7   **if** *there is no sort s such that* $ReduceTerm(u, s, \Pi)$ *is* true **then**
8 | **return** false

9 **return** true

The only thing left is to define function $ReduceTerm(t, s, \Pi)$ mentioned above. Note that for our purpose it is sufficient to define it for a ground term t only. But we introduce a more general algorithm which allows t to be non-ground. This will be useful in the next section.

Algorithm 3. ReduceTerm

Input: a term t and a sort expression E of a $\mathcal{SPARC}$ program Π.
Output: a formula $\mathcal{C}$ which is satisfiable if and only if there exists a substitution
$\qquad \theta$, such that $t\theta \in \mathcal{D}(E)$.

1 **if** E *is a sort name defined by a statement* $E = E_1$ **then**
2 $\quad\lfloor\ \mathcal{C} := ReduceTerm(t, E_1, \Pi)$

3 **else if** E *is of the form* $E_1 \odot E_2$, *where* $\odot \in \{+, -, *\}$ **then**
4 $\quad\bigg\lfloor\ \mathcal{C} := (ReduceTerm(t, E_1, \Pi)) \triangledown (ReduceTerm(t, E_2, \Pi))$, where $A \triangledown B$
$\qquad$ is $A \vee B, A \wedge \neg B$, or $A \wedge B$ when $\odot$ is $+, -$, or $*$ respectively

5 **else if** E *is of the form* $f(s_1[X_1], \ldots, s_n[X_n]) : cond(X_1, \ldots, X_n)$
6 $\qquad$ *where the condition* $cond(X_1, \ldots, X_n)$ *is optional* **then**
7 $\quad\bigg|\ $ **if** t *is not a variable and is not formed by a record name* f **then**
8 $\qquad\lfloor\ $ **return** false

9 $\quad\bigg|\ $ Let $X'_1, \ldots X'_n$ be new variables
10 $\quad\bigg|\ $ **if** t *is of the form* $f(t_1, \ldots t_n)$ **then**
11 $\qquad\lfloor\ \mathcal{C} := (X'_1 = t_1) \wedge \cdots \wedge (X'_n = t_n))$

12 $\quad\bigg|\ $ **else if** t *is a variable* **then**
13 $\qquad\lfloor\ \mathcal{C} := (t = f(X'_1, \ldots X'_n))$

14 $\quad\bigg|\ $ **if** *condition* $cond(X_1, \ldots, X_n)$ *is present in* E **then**
15 $\qquad\bigg|\ \mathcal{C} := \mathcal{C} \wedge cond'(X'_1, \ldots, X'_n)$ where $cond'(X'_1, \ldots, X'_n)$ is obtained
$\qquad\ $ from $cond(X_1, \ldots, X_n)$ by replacing X_i with X'_i and $<, \leq$ with $\prec, \preceq$
$\qquad\lfloor\ $ respectively.

16 $\quad\lfloor\ \mathcal{C} := \mathcal{C} \wedge (ReduceTerm(X'_1, s_1, \Pi)) \wedge \ldots \wedge (ReduceTerm(X'_n, s_n, \Pi))$

17 **else**
18 $\quad\bigg|\ $ **if** t *is not ground term of the form* $f(t_1, \ldots t_n)$ **then**
19 $\qquad\lfloor\ \mathcal{C} = \vee\{(t_1 = t'_1) \wedge \ldots \wedge (t_n = t'_n) | f(t'_1, \ldots, t'_n) \in \mathcal{D}(E)\}^a$

20 $\quad\bigg|\ $ **else**
21 $\qquad\lfloor\ \mathcal{C} = t \in \mathcal{D}(E)$

22 **return** $\mathcal{C}$

---

^{a}empty disjunction is interpreted as `false`

Note that the algorithm *ReduceTerm* comes to line 17 when expression E is of the
form $\{t_1, \ldots, t_n\}$ or is defined by statements of the form 2,3 or 5. In this case the corresponding $\mathcal{D}(E)$ is computed explicitly. The correctness of *ReduceTerm* algorithm is
guaranteed by the following claim:

Claim. Given a $\mathcal{SPARC}$ term t, a sort expression E of a program Π and a substitution
θ, θ is a solution of the formula *ReduceTerm*(t, E, Π) if and only if $t\theta \in \mathcal{D}(E)$.

Example 3 (Tracing the Algorithm)
Now let us trace our syntax checker on a rule p(f(b),Y+1) :- p(f(b),Y) of
program Π_0 from Examples 1 and 2. To check the rule's syntax we use $IsAtom$ to
establish that $p(f(b), Y + 1)$ and $p(f(b), Y)$ are atoms of Π_0. $IsAtom(p(f(b), Y +$
$1), \Pi_0)$ calls $ReduceTerm(f(b), s_1, \Pi_0)$ which returns true (see line 21). After that
we have the following two calls: $IsTerm(Y + 1, \Pi_0)$ and $ReduceTerm(1, s_2, \Pi_0)$.
The latter, and hence the former, return true. Hence, the head of our rule is an atom of
Π_0. Similarly for the body. Therefore, the rule p(f(b),Y+1) :- p(f(b),Y) is
indeed a rule of Π_0.

Now let $\Pi_1 = \Pi_0 \cup \{p(X, f(b)).\}$. This time $IsAtom(p(X, f(b)), \Pi_1)$ will return
false, because $f(b)$ is a ground term which is not an element of corresponding sort $s2$
(therefore, $ReduceTerm(f(b), s2, \Pi_1)$ returns false).

4   Empty Rule Checking

In this section we introduce an algorithm, *IsEmptyRule*, which checks if a rule r of
Π is *empty*, i.e. has no ground instantiations. This is done by applying a standard constraint satisfaction algorithm to a constraint formula over finite domains[9] produced by
function *ReduceRule*.

Algorithm 4. IsEmptyRule

   Input: rule r of a program Π.
   Output: *true* if r is a non-empty rule of Π and *false* otherwise.
1 $C = ReduceRule(r, \Pi)$
2 **return** $satisfiable(C)$

Algorithm 5. ReduceRule

   Input: a rule r and a $\mathcal{SPARC}$ program Π.
   Output: a formula $\mathcal{C}$, which is satisfiable if and only if r is not empty rule of Π.
1 $\mathcal{C} := \texttt{true}$
2 **for each** t_i *of each atom* $p(t_1, \ldots, t_n)$ *occurring in* r **do**
3   Let $(s_1, \ldots, s_n)$ be $\Sigma(p)$
4     $\mathcal{C} := \mathcal{C} \wedge ReduceTerm(t_i, s_i, \Pi)$
5 **return** $\mathcal{C}$

In *ReduceRule*, we extract constraints, using *ReduceTerm*, for every term of every
atom of r and connect them by conjunctions. The function *ReduceTerm* takes a term
t and a sort expression E of a program Π and returns a formula which is satisfiable if
and only if there is an instance of t which belongs to $\mathcal{D}(E)$.

Claim. Given a $\mathcal{SPARC}$ program Π and a program rule r of Π, $IsEmptyRule(\Pi, r)$
returns true if and only if r is not empty.

Example 4 (Empty rule)
Consider the rule p(X,X) of program Π_0. $ReduceRule(r, \Pi_0)$ returns formula $X \in$
$\{f(b)\} \wedge X \in \{1, 2\}$ which is clearly unsatisfiable. Therefore, the rule is an empty rule.

5   Computing Answer Sets of a $\mathcal{SPARC}$ Program

Answer sets of a $\mathcal{SPARC}$ program Π_{sparc} will be computed by translating the program into a program in the language of DLV with weak constraints. First we need some notation: every cr-rule r of Π_{sparc} will be assigned a unique number i. An expression $rn(i, X_1, ..., X_n)$ where $X_1, ..., X_n$ is the list of distinct variables occurring in r will be referred to as the *name of* r. For instance, if rule $p(X, Y) \leftarrow q(Z, X, Y)$ is assigned number 1 then its name is $rn(1, X, Y, Z)$. We also need the following definition:

Definition 9 (DLV counterparts of $\mathcal{SPARC}$ programs). *A DLV program Π_{dlv} is a counterpart of a $\mathcal{SPARC}$ program Π_{sparc} if*

- *the signature of Π_{dlv} is an extension of the signature of Π_{sparc}, and*
- *the answer sets of Π_{sparc} and Π_{dlv} coincide on literals from the language of Π_{sparc}.*

The translation is performed by Algorithm 6. The basic idea is to explicitly add the necessary sorts in the bodies of the DLV rules (which will eliminate possible problems with the safety of variables) and to replace cr-rules by a collection of weak constraints. The latter requires introduction of some new predicate symbols which explains the first requirement in definition 9.

Algorithm 6. Translate

Input: a $\mathcal{SPARC}$ program Π_{sparc}
Output: a DLV counterpart Π_{dlv} of Π_{sparc}.

1 Set variable Π_{dlv} to directives of Π_{sparc}
2 Let appl/1 be a new predicate not occurring in Π_{sparc}
3 **for each** *rule r in Π_{sparc}* **do**
4 $\quad$ $S := \{s(t) \mid$ there exists $p(t_1, ..., t_n)$, occurring in r, such that
5 $\quad\quad$ $p(s_1, ..., s_n) \in \mathcal{P}(\Pi_{sparc});$ for some $i, t = t_i, s = s_i;$ and t is ground$\}$
6 $\quad$ **for each** *distinct sort name s occurring in S* **do**
7 $\quad\quad \lfloor$ $\Pi_{dlv} := \Pi_{dlv} \cup \{s(t).|t \in \mathcal{D}(s)\}$
8 $\quad$ Let rule r' be the result of adding all elements of S to the body of r
9 $\quad$ **if** r' *is a regular rule* **then**
10 $\quad\quad \lfloor$ Add r' to Π_{dlv}
11 $\quad$ **if** r' *is a cr-rule of the form* $q \xleftarrow{+} body$ **then**
12 $\quad\quad$ Add to Π_{dlv} the following rules
$\quad\quad\quad$ `appl`$(rn(i, X_1, ..., X_n)) \vee$ ¬`appl`$(rn(i, X_1, ..., X_n))$ `:- ` *body*.
$\quad\quad\quad$ `:~ appl`$(rn(i, X_1, ..., X_n))$`,` *body*.
$\quad\quad\quad$ q `:- appl`$(rn(i, X_1, ..., X_n))$`,` *body*.
13 $\quad\quad \lfloor$ where $rn(i, X_1, ..., X_n)$ is the name of r

---

The intuitive idea behind the rules added to Π_{dlv} for a cr-rule at line 12 is the following: `appl`$(rn(i, X_1, ..., X_n))$ holds if the cr-rule r is used to obtain an answer set of the $\mathcal{SPARC}$ program; the first of the added rules says that r is either used or not used;

the second rule, a weak constraint, guarantees that r is not used if possible, and the last rule allows the use of r when necessary. The correctness of the algorithm is guaranteed by the following theorem whose complete proof can be found in our technical report [10].

Theorem 1. *A DLV program P_{dlv} obtained from a SPARC program P_{sparc} by the algorithm $Translate$ is a DLV counterpart of P_{sparc}.*

The translation can be used to compute answer set of SPARC program Π_{sparc} by using the DLV solver to compute answer sets of Π_{sparc}'s DLV counterpart and removing from them all auxiliary literals introduced in $Translate$.

Example 5 (Computing answer sets of a SPARC program)
To illustrate the translation and the computation of an answer set of a SPARC program, consider the input program Π_1 obtained from Π_0 by changing the type of one of its rules to consistency restoring:

```
#const n = 1.
sorts
#s1 = f(b).
#s2 = [0..n].
predicates
p(#s1,#s2).
rules
p(f(b),0).
:- not p(f(b),1).
p(X,X).
p(f(b),Y+1) :+ p(f(b),Y).
```

After the execution of the loop at line 3 of algorithm $Translate$, the first three regular program rules will be translated into

```
p(f(b),0).
:- not p(f(b),1).
p(X,X):-s1(X),s2(X).
s2(0). s2(1). s1(f(b)).
```

Assuming the only cr-rule is numbered by 0, it is translated as[1]:

```
appl(rn(0, Y))| -appl(rn(0, Y)):-p(f(b),Y),s2(Y),s2(Y+1).
:~ appl(rn(0, Y)),p(f(b),Y),s2(Y),s2(Y+1).
p(f(b),Y+1):-appl(rn(0, Y)),p(f(b),Y),s2(Y),s2(Y+1).
```

Given the program resulted from $Translate$, DLV solver returns an answer set

$$\{s2(0), s2(1), s1(f(b)), p(f(b), 0), appl(rn(0, 0)), p(f(b), 1)\}.$$

After dropping $appl(rn(0, 0))$, $s2(0)$, $s2(1)$, $s1(f(b))$ from this answer set, we obtain an answer set $\{p(f(b), 0), p(f(b), 1)\}$ for the original program.

---

[1] The actual output result of the implemented version may be different because of variable renaming, change of the order of rules and shifting arithmetic terms.

6   Conclusion

As ASP has been employed to solve more and more problems, we believe constructs are needed to improve the productivity of ASP programmers. Particularly, constructs are needed to allow a programmer to better structure the program, to automatically determine some syntax and semantic errors and to avoid thinking about safety of ASP rules — non-declarative conditions on rules required by existing ASP systems. We define the syntax and semantics of a knowledge representation language $\mathcal{SPARC}$ which offers explicit constructs to specify objects, relations, and their sorts. The new language expands CR-Prolog — an extension of ASP by consistency restoring rules. We introduce an implementation of $\mathcal{SPARC}$ based on its translation to DLV with weak constraints. A simple syntax checking algorithm helps a programmer to avoid errors related to misspelling the names of objects and predicates as well as simple type errors. Another type checking algorithm flags program rules which, due to type conflicts, have no ground instantiations. We hope that the sort related algorithms presented in this paper will be eventually used to make $\mathcal{SPARC}$ a front-end for other ASP based systems (including CR-Prolog system CR-models [11]).

Acknowledgements. This work was partially supported by NSF grant IIS-1018031.

References

1. Gelfond, M., Lifschitz, V.: The stable model semantics for logic programming. In: Proceedings of ICLP 1988, pp. 1070–1080 (1988)
2. Syrjänen, T.: Lparse 1.0 user's manual (2000)
3. Balduccini, M.: Modules and signature declarations for a-prolog: Progress report. In: Software Engineering for Answer Set Programming Workshop, SEA 2007 (2007)
4. Balduccini, M., Gelfond, M.: Logic programs with consistency-restoring rules. In: International Symposium on Logical Formalization of Commonsense Reasoning. AAAI 2003 Spring Symposium Series vol. 102. The AAAI Press (2003)
5. Leone, N., Pfeifer, G., Faber, W., Eiter, T., Gottlob, G., Perri, S., Scarcello, F.: The DLV system for knowledge representation and reasoning. ACM Transactions on Computational Logic (TOCL) 7(3), 499–562 (2006)
6. Buccafurri, F., Leone, N., Rullo, P.: Strong and weak constraints in disjunctive datalog. In: Fuhrbach, U., Dix, J., Nerode, A. (eds.) LPNMR 1997. LNCS, vol. 1265, pp. 2–17. Springer, Heidelberg (1997)
7. Balai, E., Gelfond, M., Zhang, Y.: SPARC – sorted ASP with consistency restoring rules. In: Answer Set Programming and Other Computing Paradigms (2012)
8. SPARC system,
   http://www.depts.ttu.edu/cs/research/krlab/#software
9. Dechter, R.: Constraint Processing. Morgan Kaufmann, San Francisco (2003)
10. Balai, E., Gelfond, M., Zhang, Y.: SPARC – sorted ASP with consistency restoring rules. Technical Report, Texas Tech University, USA (2012),
    http://www.depts.ttu.edu/cs/research/krlab/#papers
11. Balduccini, M.: CR-MODELS: An Inference Engine for CR-Prolog. In: Baral, C., Brewka, G., Schlipf, J. (eds.) LPNMR 2007. LNCS (LNAI), vol. 4483, pp. 18–30. Springer, Heidelberg (2007)

Prolog and ASP Inference under One Roof

Marcello Balduccini[1], Yuliya Lierler[2], and Peter Schüller[3]

[1] Eastman Kodak Company, USA
marcello.balduccini@gmail.com
[2] University of Nebraska at Omaha, USA
ylierler@unomaha.edu
[3] Sabancı University, Turkey
peterschueller@sabanciuniv.edu

Abstract. Answer set programming (ASP) is a declarative programming paradigm stemming from logic programming that has been successfully applied in various domains. Despite amazing advancements in ASP solving, many applications still pose a challenge that is commonly referred to as *grounding bottleneck*. Devising, implementing, and evaluating a method that alleviates this problem for certain application domains is the focus of this paper. The proposed method is based on combining backtracking-based search algorithms employed in answer set solvers with SLDNF resolution from PROLOG. Using PROLOG inference on non-ground portions of a given program, both grounding time and the size of the ground program can be substantially reduced.

Keywords: Answer Set Programming, Prolog, Grounding Bottleneck.

1 Introduction

Answer set programming (ASP) [4] is a declarative programming paradigm stemming from a knowledge representation and reasoning formalism based on the answer set semantics of logic programs. It can be used whenever we want to solve a search problem where the goal is to find solutions among a finite, but potentially very large, number of possibilities. ASP has been successfully applied in different areas of knowledge representation and computer science, including Space Shuttle control [25] and Linux package configuration [13]. Most modern answer set solving tools encapsulate two systems: a grounder, such as LPARSE or GRINGO, and an answer set solver, such as CMODELS or CLASP. A grounder is a software system that takes a logic program *with* variables as an input and produces an equivalent program *without* variables – a ground program. An answer set solver is then invoked on a ground program to generate its answer sets. Answer set solvers typically rely on the enhancements of the Davis-Putnam-Logemann-Loveland procedure [6] – classic backtracking-based search algorithm. Despite amazing advancements in solving technology, many applications still pose a challenge. *Grounding bottleneck* refers to situations where grounding results in programs that are too large for the solving tools to handle effectively. Alleviating grounding bottleneck is the main focus of this work. We describe, implement, and evaluate an approach for combining backtracking-based search algorithms of answer set solvers with SLDNF resolution

P. Cabalar and T.C. Son (Eds.): LPNMR 2013, LNAI 8148, pp. 148–160, 2013.
© Springer-Verlag Berlin Heidelberg 2013

from PROLOG. As a result, the newly implemented approach makes it possible to avoid the grounding of portions of a program by delegating the processing of those parts to a PROLOG system.

The grounding bottleneck has been recognized as a serious issue in recent years. Constraint answer set programming (CASP) [18] is one of the directions of research that has been largely motivated by an attempt to solve the problem. It integrates answer set programming with constraint (logic) programming, which allows applying constraint processing techniques for effective reasoning over non-boolean constructs. CASP introduces a notion of constraint atoms that trigger additional processing by constraint programming tools and at the same time may reduce the size of the grounding. Mellarkod et al. [22] developed one of the earliest CASP languages called *AC*. They also introduced an algorithm for a special class of programs in that language. The primary focus of the work on *AC* was integrating efficient constraint processing capabilities into answer set solving methods. Yet, [22] touched on another crucial aspect of the integration of ASP and CLP: integrating ASP backtracking search with PROLOG SLDNF resolution. In the present paper we resume and expand the investigation on this topic, focusing on a special case of *AC* programs that consist of "standard" (non-constraint) answer set programs.

More on Related Work. Works by Alviano and Faber [1], de Cat et al. [5], Eiter et al. [8,7] are other interesting attempts to alleviate grounding issues. Alviano and Faber propose a *magic sets*-based program rewriting method as a query optimization technique in ASP. This method helps an answer set solver prune the search space by disregarding parts of the program irrelevant to a given query. The goal is achieved by rewriting an original program (if a class of a program permits) in a form that guides the computation by the ASP grounder and solver by taking advantage of information provided by the query. The approach attempts to "mimic" PROLOG-like behavior using ASP technology. The approach advocated here is orthogonal. We propose to take advantage of the PROLOG engine itself when possible. Techniques in the spirit of incremental answer set programming [12] were developed by de Cat et al. [5] and employ a "grounding as needed" approach in solving. The DLVHEX solver [8,7] also provides a possibility for grounding as needed: it uses special *Splitting Sets* to process parts of a program in a sequence, so that the grounding of the *current* part depends on the answer sets of the *previous* parts.

Paper Structure. We start the presentation by a review of preliminary concepts as well as a special case of *AC* programs that are at the center of attention in this work. We then introduce a variant of the *AC* algorithm and describe its implementation within the CASP solver EZCSP [3]. We conclude with a discussion on an experimental analysis that we conducted to assess the introduced technique.

2   Hybrid Programs

A *logic program* is a finite set of rules of the form

$$a_0 \leftarrow a_1, \ldots, a_l, not\ a_{l+1}, \ldots, not\ a_m, not\ not\ a_{m+1}, \ldots, not\ a_n, \tag{1}$$

where a_0 is $\perp$ or an atom, and each a_i $(1 \leq i \leq n)$ is an atom. Atoms may be nonground. We call a rule a *constraint*, if $a_0 = \perp$. This is a special case of programs with nested expressions [21]. We assume that the reader is familiar with the definition of an answer set of such programs and refer to the paper by Lifschitz et al. ([21]) for details. According to [11], a *choice rule* $\{a\}$ of the LPARSE[1] language [23] can be seen as an abbreviation for a rule $a \leftarrow \text{not not } a$. We adopt this abbreviation in the rest of the paper.

The expression a_0 is the *head* of rule (1). If B denotes the *body* of (1), the right hand side of the arrow, we write B^{pos} for the elements occurring in the *positive* part of the body, i.e., $B^{pos} = \{a_1, \ldots, a_l\}$.

To process a logic program, or in other words, to find answer sets of a program or establish some properties about its answer sets, such software systems as answer set solvers and sometimes PROLOG interpreters are used. A sample logic program is:

$$
\begin{aligned}
&down(T) \leftarrow not\ on.\\
&down(0).\ down(1).\ \ldots\ down(3600).\\
&okTime(T) \leftarrow not\ down(T).\\
&\bot \leftarrow occurs(a, 5000),\ not\ okTime(5000).\\
&occurs(a, 5000).\\
&\{on\}.
\end{aligned}
\tag{2}
$$

This program has a unique answer set

$$
\{occurs(a, 5000),\ okTime(5000),\ on,\ down(0), \ldots, down(3600)\}.
\tag{3}
$$

Note that neither answer set solvers nor PROLOG systems can handle such a program. First, program (2) contains a constraint and a choice rule, which makes PROLOG systems inapplicable. Second, (2) contains a rule

$$
okTime(T) \leftarrow not\ down(T),
$$

which violates the common *safety* condition imposed by ASP grounders. A safe rule is such that each variable occurring in its head or its negative part of the body appears in the positive part of the body. Nevertheless, the first three lines of (2) form a logic program that may be processed by PROLOG systems, whereas the last three lines form a program that is acceptable by an answer set solver. In a sense, program (2) is a "hybrid" program that borrows acceptable features from two worlds of logic programming: "classic" PROLOG programming and answer set programming. In this paper we present an algorithm (a family of algorithms) that takes advantage of two inference technologies that are usually used disjointly in logic programming, in PROLOG systems and in answer set solvers. As a result programs such as (2) can be processed by a solver supporting such an algorithm. We implement a variant of this algorithm in the solver EZCSP[2] [3].

In order to treat parts of a program differently (using PROLOG inference in one case, and answer set solver inference in another) we identify a group of program predicates that we use to guide the splitting of the program into two disjoint parts. To make it precise we introduce the following notation.

---

[1] http://www.tcs.hut.fi/Software/smodels/
[2] http://marcy.cjb.net/ezcsp/

For a program Π and a set $\mathbf{p}$ of predicate symbols, the part of Π that consists of all the rules whose heads are atoms formed using predicate symbols from $\mathbf{p}$ is denoted by $\Pi_{\mathbf{p}}$. By $\Pi_{\mathbf{p}}^-$ we denote $\Pi \setminus \Pi_{\mathbf{p}}$. For example, let Π stand for (2) and let $\mathbf{p}_1$ be the set of predicate symbols

$$\{okTime, down\}. \tag{4}$$

Then, $\Pi_{\mathbf{p}_1}$ is:

$$
\begin{aligned}
&down(T) \leftarrow not\ on.\\
&down(0).\ \ldots\ down(3600).\\
&okTime(T) \leftarrow not\ down(T),
\end{aligned}
\tag{5}
$$

whereas

$$
\begin{aligned}
&\bot \leftarrow occurs(a,5000),\ not\ okTime(5000)\\
&occurs(a,5000).\\
&\{on\}.
\end{aligned}
\tag{6}
$$

is $\Pi_{\mathbf{p}_1}^-$. For a program Π, by $ground(\Pi)$ we denote the set of all ground instances of all rules in Π. We say that Π is *semi-ground* w.r.t. a set $\mathbf{p}$ of predicate symbols if $\Pi_{\mathbf{p}}^-$ is a ground program (i.e., contains no variables) and $\Pi_{\mathbf{p}}$ is such that all of its non-ground atoms are formed from predicate symbols in $\mathbf{p}$. For example, program (2) is semi-ground w.r.t. predicate symbols (4).

For any atom $p(\mathbf{t})$, by $p(\mathbf{t})^0$ we denote its predicate symbol p. For any program Π, the *predicate dependency graph* of Π is the directed graph that

– has all predicates occurring in Π as its vertexes, and
– for each rule (1) in Π has an edge from a_0^0 to a_i^0 where $1 \leq i \leq l$.

We say that a program Π is *splittable* w.r.t. predicate symbols $\mathbf{p}$ if each strongly connected component of the predicate dependency graph of Π is either a subset of $\mathbf{p}$ or a disjoint set from $\mathbf{p}$. Program (2) is splittable w.r.t. predicate symbols (4).

The hybrid algorithm that we propose in this note is applicable to splittable programs. To present this algorithm we introduce several concepts.

Given a program Π and a set $\mathbf{p}$ of predicate symbols, a set X of atoms is a *$\mathbf{p}$-input answer set* (or an input answer set w.r.t. $\mathbf{p}$) of Π if X is an answer set of $\Pi \cup X_{\mathbf{p}}^-$ where by $X_{\mathbf{p}}^-$ we denote the set of atoms in X whose predicate symbols are different from those occurring in $\mathbf{p}$. [3] For instance, let X be a set $\{a(1),b(1)\}$ of atoms and let $\mathbf{p}$ be a set $\{a\}$ of predicates, then $X_{\mathbf{p}}^-$ is $\{b(1)\}$. The set X is a $\mathbf{p}$-input answer set of a program $a(1) \leftarrow b(1)$. On the other hand, it is not an input answer set for the same program with respect to a set $\{a,b\}$.

By $At(\Pi)$ we denote the set of all atoms occurring in a program Π.

Proposition 1. *For a program Π and a set $\mathbf{p}$ of predicate symbols, if Π is splittable then a set of atoms A over $At(ground(\Pi))$ is an answer set of Π iff A is an input answer set of $\Pi_{\mathbf{p}}$ w.r.t. $\mathbf{p}$ and A is an input answer set of $\Pi_{\mathbf{p}}^-$ w.r.t. predicate symbols in $\Pi_{\mathbf{p}}^-$ different from $\mathbf{p}$.*

---

[3] Intuitively set $\mathbf{p}$ denotes a set of intentional predicates [10]. The concept of $\mathbf{p}$-input answer sets is closely related to "$\mathbf{p}$-stable models" in [9].

This proposition outlines the basis for our approach. Given a semi-ground and split-table program Π wrt predicate symbols $\mathbf{p}$, we would like to use a PROLOG system for inference over $\Pi_{\mathbf{p}}$ and an answer set solver for inference over $\Pi_{\mathbf{p}}^{-}$. Note that $\Pi_{\mathbf{p}}$ may contain rules that are not ground whereas $\Pi_{\mathbf{p}}^{-}$ is a propositional program so that any answer set solver is applicable to it. Recall that PROLOG is designed to effectively process non-ground programs whereas answer set solvers (without grounders) are able to deal only with propositional programs.

3   Review: Abstract Answer Set Solver

Most state-of-the-art answer set solvers are based on algorithms closely related to the DPLL procedure [6]. Nieuwenhuis et al. described DPLL by means of a transition system that can be viewed as an abstract framework underlying DPLL computation [24]. Our goal is to design a similar framework for describing an algorithm suitable for processing semi-ground splittable programs – QUERY+ASP. As a step in this direction we introduce the graph AS_{Π} that extends the DPLL graph by Nieuwenhuis et al. so that the result can be used to specify an algorithm for finding answer sets of a program.

We frequently identify the body of (1) with the conjunction of its elements (in which *not* is replaced with the classical negation connective $\neg$):

$$a_1 \wedge \cdots \wedge a_l \wedge \neg a_{l+1} \wedge \cdots \wedge \neg a_m \wedge \neg\neg a_{m+1} \wedge \cdots \wedge \neg\neg a_n.$$

Similarly, we often interpret a rule (1) as a clause

$$a_0 \vee \neg a_1 \vee \cdots \vee \neg a_l \vee a_{l+1} \vee \cdots \vee a_m \vee \neg a_{m+1} \vee \cdots \vee \neg a_n \qquad (7)$$

(in the case when $a_0 = \bot$ in (1) a_0 is absent in (7)). Given a program Π, we write Π^{cl} for the set of clauses (7) corresponding to all rules in Π.

For a set σ of atoms, a *record* relative to σ is an ordered set M of literals over σ, some possibly annotated by Δ, which marks them as *decision* literals. A *state* relative to σ is a record relative to σ possibly preceding symbol $\bot$. For instance, some states relative to a singleton set $\{a\}$ of atoms are

$$\emptyset, \quad a, \quad \neg a, \quad a^{\Delta}, \quad a \, \neg a, \quad \bot, \quad a\bot, \quad \neg a\bot, \quad a^{\Delta}\bot, \quad a \, \neg a\bot.$$

We say that a state is inconsistent if either $\bot$ or two complementary literals occur in it. For example, states $a \, \neg a$ and $a\bot$ are inconsistent. Given a state M, we frequently ignore both annotations and order of elements and consider M as a set of literals possibly including the symbol $\bot$.

If neither a literal l nor its complement occur in M, then l is *unassigned* by M.

If C is a disjunction (conjunction) of literals then by $\overline{C}$ we understand the conjunction (disjunction) of the complements of the literals occurring in C. In some situations, we will identify disjunctions and conjunctions of literals with the sets of these literals.

By *Bodies*(Π, a) we denote the set of the bodies of all rules of a ground program Π with the head a. A set U of atoms occurring in a ground program Π is *unfounded* [26,16] on a consistent set M of literals with respect to Π if for every $a \in U$ and every

$B \in Bodies(\Pi, a)$, $M \models \overline{B}$ (where B is identified with the conjunction of its elements), or $U \cap B^{pos} \neq \emptyset$.

Each ground program Π determines its *Answer-Set graph* AS_Π. The set of nodes of AS_Π consists of the states relative to the set of atoms occurring in Π. The edges of the graph AS_Π are specified by the transition rules

Unit Propagate: $M \implies M\,l$ if $C \vee l \in \Pi^{cl}$ and $\overline{C} \subseteq M$

Decide: $M \implies M\,l^\Delta$ if l is unassigned by M

Fail: $M \implies \perp$ if $\begin{cases} M \text{ is inconsistent and different from } \perp, \text{ and} \\ M \text{ contains no decision literals} \end{cases}$

Backtrack: $P\,l^\Delta\,Q \implies P\,\overline{l}$ if $\begin{cases} P\,l^\Delta\,Q \text{ is inconsistent, and} \\ Q \text{ contains no decision literals} \end{cases}$

Unfounded: $M \implies M\,\neg a$ if $a \in U$ for a set U unfounded on M wrt Π.

A node is *terminal* in a graph if no edge leaves this node.

For a set M of literals, by $pos(M)$ and $neg(M)$ we denote the set of positive and negative literals in M respectively. For instance, $pos(\{a, \neg b\}) = \{a\}$ and $neg(\{a, \neg b\}) = \{b\}$.

The graph AS_Π can be used for deciding whether a ground program Π has an answer set by constructing a path from $\emptyset$ to a terminal node. The following proposition serves as a proof of correctness and termination for any procedure that is captured by AS_Π.

Proposition 2. *For any ground program Π,*

(a) graph AS_Π is finite and acyclic,
(b) for any terminal state M of AS_Π other than $\perp$, $pos(M)$ is an answer set of Π,
(c) state $\perp$ is reachable from $\emptyset$ in AS_Π if and only if Π has no answer sets.

Let Π be a program (6). The following is a path in AS_Π, with every edge annotated by the name of a transition rule that justifies the presence of this edge in the graph:

$$\emptyset \stackrel{Unit\ Propagate}{\implies}$$
$$occurs(a, 5000) \stackrel{Unit\ Propagate}{\implies}$$
$$occurs(a, 5000)\ okTime(5000) \stackrel{Unfounded}{\implies}$$
$$occurs(a, 5000)\ okTime(5000)\ \neg okTime(5000) \stackrel{Fail}{\implies}$$
$$\perp$$

Since the last state in the path is terminal and $\perp$, Proposition 3 asserts that this program has no answer sets.

The graph AS_Π is inspired by the graph SM_Π introduced by Lierler [17] for specifying answer set solver SMODELS [23]. The graph SM_Π extends AS_Π by two additional transition rules (in other words, inference rules or propagators): *All Rules Canceled* and *Backchain True*. Lierler and Truszczynski [20] developed a similar framework to model such modern answer set solvers as CMODELS [15], SUP [17], and CLASP [14]. For the simplicity of this presentation, we settle on the AS_Π formalism as a choice for depicting *an* answer set solver. Nevertheless, the procedure described in this paper for combining the inference mechanisms of answer set solving and of PROLOG is not limited to answer set solvers whose algorithm is captured by the AS_Π graph. For example, the procedure can be easily adopted by more sophisticated solvers implementing learning, such as CMODELS or CLASP.

4 Abstract QUERY+ASP

Query, Extensions, and Consequences. For a program Π and a set **p** of predicate symbols, by $At_\mathbf{p}(\Pi)$ we denote a set of atoms occurring in Π whose predicate symbols are in **p**. By $At_\mathbf{p}^-(\Pi)$, we denote a set of atoms in Π whose predicates symbols are not in **p**.

For a semi-ground program Π w.r.t. a set **p** of predicate symbols, a (complete) query Q is a (complete) consistent set of literals over $At_\mathbf{p}^-(\Pi_\mathbf{p}) \cup At_\mathbf{p}(\Pi_\mathbf{p}^-)$. For a query Q of Π, a complete query E is a *satisfying extension* of Q w.r.t. Π if $Q \subseteq E$ and there is an input answer set A of $\Pi_\mathbf{p}$ w.r.t. predicates **p** such that $pos(E) \subseteq A$ and $neg(E) \cap A = \emptyset$.

We say that literal l is a consequence of Π and Q if for every satisfying extension E of Q w.r.t. Π, $l \in E$. By $Cons(\Pi, Q)$, we denote the set of all consequences of Π and Q. If there are no satisfying extensions of Q w.r.t. Π we identify $Cons(\Pi, Q)$ with the singleton $\{\bot\}$.

Let Π be (2) and Q be $\{on\}$. The set $\{on, okTime(5000)\}$ forms a satisfying extension of Q w.r.t. Π. Furthermore, this is the only satisfying extension of Q w.r.t. Π. Consequently, it forms $Cons(\Pi, Q)$. On the other hand, there are no satisfying extensions for a query $Q = \{\neg on, okTime(5000)\}$ so that $\{\bot\}$ corresponds to $Cons(\Pi, Q)$.

The graph $QAS_{\Pi,\mathbf{p}}$. For a program Π and a set **p** of predicate symbols, by Π^c we denote a set of choice rules $\{a\}$ for each atom a in $At_\mathbf{p}(\Pi_p^-)$. For instance, let Π be (2) then Π^c consists of a choice rule

$$\{okTime(5000)\} \tag{8}$$

Let Π be a logic program and **p** a set of predicate symbols. The nodes of the graph $QAS_{\Pi,\mathbf{p}}$ are the states relative to the set of atoms occurring in Π_p^-.

The edges of the graph $QAS_{\Pi,\mathbf{p}}$ include the transition rules of $AS_{\Pi_\mathbf{p}^- \cup \Pi^c}$. Note how these transition rules take into consideration not only a part of program meant to be processed by an answer set solver $\Pi_\mathbf{p}^-$ but also its extension with choice rules for atoms whose predicate symbols are in **p**. For instance, let Π be program (2) and let $\mathbf{p}_1$ be set of predicate symbols (4). The program $\Pi_{\mathbf{p}_1}^- \cup \Pi^c$ contains the rules of (6) extended with choice rule (8).

Another transition rule that concludes the definition of the graph $QAS_{\Pi,\mathbf{p}}$ is called *Query Propagate*. To present this rule we introduce the notion of a query. For a state M of $QAS_{\Pi,\mathbf{p}}$, by $query(M)$ we denote the largest subset of M over $At_\mathbf{p}^-(\Pi_\mathbf{p}) \cup At_\mathbf{p}(\Pi_\mathbf{p}^-)$. Let Π be (2) and M be a state $occurs(a, 5000)\ okTime(5000)\ \neg on^\Delta$, then $query(M)$ is $\{okTime(5000), \neg on\}$.

The transition rule *Query Propagate* follows

$$Query\ Propagate:\ M \Longrightarrow M\ l \ \text{if}\ \ l \in Cons(\Pi, query(M)).$$

The graph $QAS_{\Pi,\mathbf{p}}$ can be used for deciding whether a splittable semi-ground program Π w.r.t. predicate symbols $\mathbf{p}$ has an answer set by constructing a path from $\emptyset$ to a terminal node:

Proposition 3. *For any splittable semi-ground program Π w.r.t. predicate symbols $\mathbf{p}$,*

(a) graph $QAS_{\Pi,p}$ is finite and acyclic,
(b) for any terminal state M of $QAS_{\Pi,p}$ other than $\bot$, $pos(M)$ is a set of all Π_p^- atoms in some answer set of Π,
(c) state $\bot$ is reachable from $\emptyset$ in $QAS_{\Pi,p}$ if and only if Π has no answer sets.

Proposition 3 shows that algorithms, which find a path in the graph $QAS_{\Pi,\mathbf{p}}$ from $\emptyset$ to a terminal node, can be regarded as solvers for splittable semi-ground programs. We call the class of algorithms captured by the graph QUERY+ASP. Let Π be a program (2). The following is a path in $QAS_{\Pi,\mathbf{p}}$, with every edge annotated by the name of a transition rule that justifies the presence of this edge in the graph:

$$\emptyset \overset{\textit{Unit Propagate}}{\Longrightarrow}$$
$$occurs(a,5000) \overset{\textit{Unit Propagate}}{\Longrightarrow}$$
$$occurs(a,5000)\ okTime(5000) \overset{\textit{Decide}}{\Longrightarrow}$$
$$occurs(a,5000)\ okTime(5000)\ \neg on^\Delta \overset{\textit{Query Propagate}}{\Longrightarrow}$$
$$occurs(a,5000)\ okTime(5000)\ \neg on^\Delta\ \bot \overset{\textit{Backtrack}}{\Longrightarrow}$$
$$occurs(a,5000)\ okTime(5000)\ on$$

Since the last state in the path is terminal, Proposition 3 asserts that

$$\{occurs(a,5000),\ okTime(5000),\ on\}$$

is a set of all $\Pi_{\mathbf{p}}^-$ atoms in some answer set of Π. Indeed, recall answer set (3).

We note that the $QAS_{\Pi,\mathbf{p}}$ graph can be seen as a special case of the graph AC_Π introduced in [18] for a more sophisticated class of programs called AC programs.

5   The "blackbox" QUERY+ASP Algorithm

We can view a path in the graph $QAS_{\Pi,\mathbf{p}}$ as a description of a process of search for a set of atoms in some answer set of splittable semi-ground program Π by applying the graph's transition rules. Therefore, we can characterize an algorithm of a solver that utilizes the transition rules of $QAS_{\Pi,\mathbf{p}}$ by describing a strategy for choosing a path in this graph. A strategy can be based, in particular, on assigning priorities to transition rules of $QAS_{\Pi,\mathbf{p}}$, so that a solver never follows a transition due to a rule in a state if a rule with higher priority is applicable.

The priorities

Backtrack, Fail, Unit Propagate, Unfounded, Decide $\gg$ Query Propagate.

describe a *"blackbox"* architecture of a QUERY+ASP system that operates as follows: first, it uses an answer set solver on $\Pi_{\mathbf{p}}^{-} \cup \Pi^{c}$ to find an answer set; then it invokes a procedure to verify whether the *Query Propagate* transition is available; if no such transition is available then the answer set found represents a terminal state of $\text{QAS}_{\Pi,\mathbf{p}}$; otherwise, the answer set solver is instructed to look for another answer set and the process is repeated.

PROLOG **for Implementing** *Query Propagate.* PROLOG systems can be used to implement the *Query Propagate* transition rule for programs satisfying some additional syntactic constraints. We now discuss one class of such programs.

Let Π be a splittable program w.r.t. predicate symbols $\mathbf{p}$. We say that such a program is PROLOG-*friendly* if $\Pi_{\mathbf{p}}$ is in PROLOG syntax (i.e., contains no rules with nested negation) and acyclic [2, Definition 1.4, Corollary 4.3].[4] Recall that an acyclic program (i) has a unique answer set, and (ii) any PROLOG system terminates on it. Thus a PROLOG system can be used to implement the *Query Propagate* transition rule in a situation in which *query*(M) assigns all atoms in $At_{\mathbf{p}}^{-}(\Pi_{\mathbf{p}})$. Indeed, PROLOG can be invoked on (i) a program that consists of $\Pi_{\mathbf{p}}$, and atoms (given as facts) occurring positively in *query*(M) whose predicate symbols are not in $\mathbf{p}$; (ii) a query formed by the literals in *query*(M) whose predicate symbols are in $\mathbf{p}$.

We refer to the variant of QUERY+ASP that implements the "blackbox" approach and uses PROLOG for *Query Propagate* as PROLOG+ASP. It is a direction of future research to find other means for implementing more general settings of QUERY+ASP.

PROLOG+ASP **implementation in** EZCSP: We expect the reader to be familiar with the syntax of the EZCSP language [3] and with the main principles behind this CASP solver. The EZCSP language has been extended to allow a program Π to contain a declaration, $P(\Pi)$, of the form

$$\#begin\_defined. \qquad \Omega \qquad \#end\_defined.$$

where Ω is an acyclic PROLOG program, which intuitively corresponds to $\Pi_{\mathbf{p}}$. All atoms whose predicate symbols are intended to occur in $\Pi_{\mathbf{p}}$ but not in $\mathbf{p}$ must be prefixed by "*prolog_*" (to notify EZCSP that these atoms are relevant to forming a PROLOG program while implementing *Query Propagate*). All atoms whose predicate symbol is in $\mathbf{p}$ are specified as arguments of the special unary relation "*required*" of the language of EZCSP. For instances, logic program (2) in the modified language of EZCSP is:

$$\#begin\_defined.$$
$$down(T) \leftarrow not\ prolog\_on.$$
$$down(0).\ down(1).\ \dots\ down(3600).$$
$$okTime(T) \leftarrow not\ down(T).$$
$$\#end\_defined.$$
$$required(okTime(5000)) \leftarrow occurs(a, 5000).$$
$$occurs(a, 5000) \leftarrow .$$
$$\{prolog\_on\}.$$

---

[4] More general "PROLOG-friendly" syntactic conditions on programs are possible.

The EZCSP algorithm is extended so that, given an EZCSP program Π, it starts by invoking the answer set solver to compute an answer set A of $\Pi \setminus P(\Pi)$. The PROLOG interpreter is then used to determine if the query formed by the atoms of the form $required(\cdot)$ from answer set A holds for the program consisting of Ω and of the "$prolog\_$"-prefixed atoms from A. If the PROLOG interpreter answers positively, then A is returned. Otherwise, the algorithm iterates, instructing the answer set solver to find another answer set.

6   Experimental Domains and Results

In this work we designed an experimental domain called *Emergency Exit* to evaluate the implementation of the PROLOG+ASP procedure in EZCSP. Emergency Exit is a planning problem involving a robot on a grid. Some grid cells are occupied by obstacles and cannot be traversed. One unoccupied cell is selected as a goal cell, and another one as an emergency exit. At every time step, the robot can move along the x or y axis by one cell, as long as the destination cell is unoccupied. The goal of the robot is to reach the goal cell from its initial location in such a way that: (i) doing so takes at most n steps, and (ii) the emergency exit is reachable within k steps from any cell traversed by the

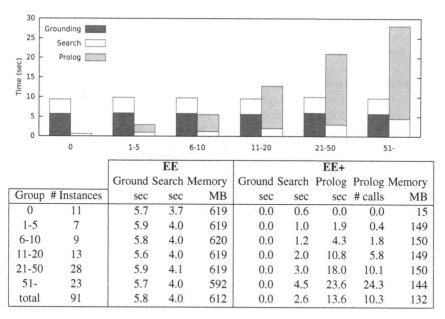

		EE			EE+				
		Ground	Search	Memory	Ground	Search	Prolog	Prolog	Memory
Group	# Instances	sec	sec	MB	sec	sec	sec	# calls	MB
0	11	5.7	3.7	619	0.0	0.6	0.0	0.0	15
1-5	7	5.9	4.0	619	0.0	1.0	1.9	0.4	149
6-10	9	5.8	4.0	620	0.0	1.2	4.3	1.8	150
11-20	13	5.6	4.0	619	0.0	2.0	10.8	5.8	149
21-50	28	5.9	4.1	619	0.0	3.0	18.0	10.1	150
51-	23	5.7	4.0	592	0.0	4.5	23.6	24.3	144
total	91	5.8	4.0	612	0.0	2.6	13.6	10.3	132

Fig. 1. Emergency Exit benchmark results. Instances are grouped by the number of paths from start to goal location. We compare ASP (left stacks) with ASP+Prolog (right stacks) and display the time spent for grounding (dark red), solving (white), and Prolog (light blue) in each stack. Section **EE** in the table shows memory usage, grounding and search time with an encoding in ASP, while **EE+** shows results for PROLOG+ASP and additionally shows the number of calls to PROLOG and the time spent in PROLOG execution.

robot. We also consider a simpler variant in our analysis that we call *Path Finding*. In this problem the task is to find a path that satisfies the requirement (i). It is easy to see that any solution to the Emergency Exit problem is also a solution to the Path Finding problem but not the other way around.

For our experiments we randomly generated 91 instances with a 100×100 grid, $n = 10, k = 194$, cells $(1,6)$, $(6,1)$, and $(100,100)$ marked as a goal, start, and an emergency exit respectively. The instances vary in how obstacles are distributed on a grid: in each case there are between 12 and 25 occupied cells in the part of the grid between $(1,1)$ and $(10,10)$. This randomly varies the number of possible paths from start to goal of length n. Moreover, we selected $k = 194$ and located the emergency exit at $(100,100)$ to ensure that reaching the exit would be possible only for certain paths from start to goal.

In the following presentation, by **PF** we denote an ASP encoding of Path Finding; by **EE** we denote an ASP encoding of Emergency Exit. We constructed **EE** by extending **PF** with an encoding of the reachability requirement (ii). Finally, we constructed variant **EE+** of **EE** in such a way that: (1) the **PF** component is processed by the answer set solver of PROLOG+ASP, whereas (2) **EE+** \ **PF** is processed by the PROLOG interpreter used in the implementation of PROLOG+ASP in EZCSP.

The experiments were run on a Linux server with 32 2.4GHz Intel® E5-2665 CPU cores and 64GB memory. Every run used a single core only. As grounder we used GRINGO 3.0.5. To evaluate PROLOG+ASP on **EE+** we used EZCSP 1.6.20b57 with CMODELS 3.85 (running MINISAT v 1.12b) and BPROLOG 7.8 as backends. As a reference we also present the performance of CMODELS 3.85 (running MINISAT v 1.12b) on **EE**. The supporting files can be found at http://www.mbalduccini.tk/ezcsp/lpnmr2013/.

Figure 1 shows the experimental results. We group the instances according to how many answer sets are found by **PF**. This number serves as an upper bound to the number of invocations of the PROLOG interpreter needed in the PROLOG+ASP algorithm to find a solution or establish the unsatisfiability of a problem in the **EE+** encoding. For each instance group, the histogram reports the grounding time at the bottom, followed by the search time, followed by the PROLOG execution time; the left stacks (with dark red) are for **EE**, the right stacks (with light blue) for **EE+**.

First, we observe that **EE** performs nearly the same for all instance groups, including the ratio between grounding (dark red) and solver (white) effort. The grounding size for **EE** is on average 47MB (not shown in the figure).

EE+ performs quite differently. The number of invocations of the PROLOG interpreter by the algorithm greatly affects the efficiency. Groups of instances with up to 10 plans in **PF** can be computed more efficiently with **EE+**, *exhibiting a difference in order(s) of magnitude*. In the instances that require more iterations, the time spent in the PROLOG interpreter dominates the overall time for solving. As PROLOG is never called in group 0, it has particular low memory usage for **EE+**. The time required for grounding **EE+** is nearly zero, and the average size of grounding is 0.3MB, which is much lower than for **EE**. Overall, we observe that for instances where only few or no plans from start to goal exist, **EE+** is significantly faster than **EE**.

7   Conclusions

In this paper we described a method for alleviating the grounding bottleneck by combining backtracking-based search algorithms employed in answer set solvers with SLDNF resolution from PROLOG. By means of experimental evaluations, we have demonstrated that, for problems where constraints have large groundings, using PROLOG as an inference engine over these constraints *may* save grounding time and memory and *may* lead to significant gains in the performance. However this is only true when the part of a program evaluated by an answer set solver of PROLOG+ASP is such that it produces only few candidates that have to be verified against the constraints evaluated by PROLOG. This conclusion aligns well with an observation reported in [19], where a study was conducted, comparing the solving technology of answer set solvers and of constraint answer set solvers. As in PROLOG+ASP, the answer set solving component of a constraint answer set solver has access only to a portion of all the constraints of the problem. The other constraints are processed separately by a constraint solver. Such separation of concerns may be very fruitful in solving the grounding bottleneck, yet it has to be used with care in order not to undermine the advanced technology of answer set solvers.

Acknowledgments. We are grateful to Yuanlin Zhang, Michael Gelfond, Vladimir Lifschitz, and Mirosław Truszczyński for useful discussions related to the topic of this work. Peter Schüller is supported by TUBITAK 2216 Research Fellowship.

References

1. Alviano, M., Faber, W.: Dynamic magic sets and super-coherent answer set programs. AI Commun. 24(2), 125–145 (2011)
2. Apt, K., Bezem, M.: Acyclic programs. New Generation Computing 9, 335–363 (1991)
3. Balduccini, M.: Representing constraint satisfaction problems in answer set programming. In: Workshop on Answer Set Programming and Other Computing Paradigms, ASPOCP (2009)
4. Brewka, G., Niemelä, I., Truszczyński, M.: Answer set programming at a glance. Communications of the ACM 54(12), 92–103 (2011)
5. de Cat, B., Denecker, M., Stuckey, P.J.: Lazy model expansion by incremental grounding. In: Technical Communications of the International Conference on Logic Programming, ICLP, pp. 201–211 (2012)
6. Davis, M., Logemann, G., Loveland, D.: A machine program for theorem proving. Communications of the ACM 5(7), 394–397 (1962)
7. Eiter, T., Fink, M., Ianni, G., Krennwallner, T., Schüller, P.: Pushing efficient evaluation of HEX programs by modular decomposition. In: Delgrande, J.P., Faber, W. (eds.) LPNMR 2011. LNCS, vol. 6645, pp. 93–106. Springer, Heidelberg (2011)
8. Eiter, T., Ianni, G., Schindlauer, R., Tompits, H.: dlvhex: A Prover for Semantic-Web Reasoning under the Answer-Set Semantics. In: Workshop on Applications of Logic Programming in the Semantic Web and Semantic Web Services, ALPSWS, pp. 33–39. CEUR WS (2006)
9. Ferraris, P., Lee, J., Lifschitz, V.: Stable models and circumscription. Artificial Intelligence 175, 236–263 (2011)

10. Ferraris, P., Lee, J., Lifschitz, V., Palla, R.: Symmetric splitting in the general theory of stable models. In: International Joint Conference on Artificial Intelligence, IJCAI, pp. 797–803 (2009)
11. Ferraris, P., Lifschitz, V.: Weight constraints as nested expressions. Theory and Practice of Logic Programming 5, 45–74 (2005)
12. Gebser, M., Kaminski, R., Kaufmann, B., Ostrowski, M., Schaub, T., Thiele, S.: Engineering an incremental ASP solver. In: Garcia de la Banda, M., Pontelli, E. (eds.) ICLP 2008. LNCS, vol. 5366, pp. 190–205. Springer, Heidelberg (2008)
13. Gebser, M., Kaminski, R., Schaub, T.: aspcud: A linux package configuration tool based on answer set programming. In: International Workshop on Logics for Component Configuration, LoCoCo (2011)
14. Gebser, M., Kaufmann, B., Neumann, A., Schaub, T.: Conflict-driven answer set solving. In: International Joint Conference on Artificial Intelligence, IJCAI, pp. 386–392. MIT Press (2007)
15. Giunchiglia, E., Lierler, Y., Maratea, M.: Answer set programming based on propositional satisfiability. Journal of Automated Reasoning 36, 345–377 (2006)
16. Lee, J.: A model-theoretic counterpart of loop formulas. In: International Joint Conference on Artificial Intelligence, IJCAI, pp. 503–508. Professional Book Center (2005)
17. Lierler, Y.: Abstract answer set solvers. In: Garcia de la Banda, M., Pontelli, E. (eds.) ICLP 2008. LNCS, vol. 5366, pp. 377–391. Springer, Heidelberg (2008)
18. Lierler, Y.: On the relation of constraint answer set programming languages and algorithms. In: AAAI Conference on Artificial Intelligence, AAAI. MIT Press (2012)
19. Lierler, Y., Smith, S., Truszczynski, M., Westlund, A.: Weighted-sequence problem: ASP vs CASP and declarative vs problem-oriented solving. In: Russo, C., Zhou, N.-F. (eds.) PADL 2012. LNCS, vol. 7149, pp. 63–77. Springer, Heidelberg (2012)
20. Lierler, Y., Truszczyński, M.: Transition systems for model generators — a unifying approach. In: Theory and Practice of Logic Programming, International Conference on Logic Programming (ICLP) Special Issue 11(4-5) (2011)
21. Lifschitz, V., Tang, L.R., Turner, H.: Nested expressions in logic programs. Annals of Mathematics and Artificial Intelligence 25, 369–389 (1999)
22. Mellarkod, V.S., Gelfond, M., Zhang, Y.: Integrating answer set programming and constraint logic programming. Annals of Mathematics and Artificial Intelligence (2008)
23. Niemelä, I., Simons, P.: Extending the Smodels system with cardinality and weight constraints. In: Minker, J. (ed.) Logic-Based Artificial Intelligence, pp. 491–521. Kluwer (2000)
24. Nieuwenhuis, R., Oliveras, A., Tinelli, C.: Solving SAT and SAT modulo theories: From an abstract Davis-Putnam-Logemann-Loveland procedure to DPLL(T). Journal of the ACM 53(6), 937–977 (2006)
25. Nogueira, M., Balduccini, M., Gelfond, M., Watson, R., Barry, M.: An A-Prolog decision support system for the Space Shuttle. In: Ramakrishnan, I.V. (ed.) PADL 2001. LNCS, vol. 1990, pp. 169–183. Springer, Heidelberg (2001)
26. Van Gelder, A., Ross, K., Schlipf, J.: The well-founded semantics for general logic programs. Journal of ACM 38(3), 620–650 (1991)

Event-Object Reasoning with Curated Knowledge Bases: Deriving Missing Information

Chitta Baral and Nguyen H. Vo

School of Computing, Informatics, and Decision Systems Engineering,
Arizona State University, Tempe, Arizona, USA

Abstract. The broader goal of our research is to formulate answers to why and how questions with respect to knowledge bases, such as AURA. One issue we face when reasoning with many available knowledge bases is that at times needed information is missing. Examples of this include partially missing information about next sub-event, first sub-event, last sub-event, result of an event, input to an event, destination of an event, and raw material involved in an event. In many cases one can recover part of the missing knowledge through reasoning. In this paper we give a formal definition about how such missing information can be recovered and then give an ASP implementation of it. We then discuss the implication of this with respect to answering why and how questions.

1 Introduction

Our work in this paper is part of two related long terms goals: answering "How", "Why" and "What-if" questions and reasoning with the growing body of available knowledge bases[1], some of which are crowd-sourced. Although answering those questions are important, so far little research has been done on them. Our starting point to address them has been to formulate answers to such questions with respect to abstract knowledge structures obtained from knowledge bases. In particular, in the recent past we considered Event Description Graphs (EDGs) [1] to formulate answers to some "How" and "Why" questions with respect to the Biology knowledge base AURA [2].

Going from the abstract structures to reasoning with real knowledge bases (KBs) we noticed that the KBs often have missing pieces of information, such as properties of an instance (of a class) or relations between two instances. For example, AURA does not encode that *Eukaryotic translation* is the next event of *Synthesis of RNA in eukaryote*; this may be because the two subevents of "Protein synthesis" were encoded independently. The missing pieces make the KB and the Description Graphs constructed from it fragmented and as a result answers obtained with respect to them are not intuitive. Moreover, the KBs like AURA often have two or more names that refer to the same entity. To get intuitive answers they need to be resolved and merged into a single entity.

---

[1] See for example, `http://linkeddata.org/`

P. Cabalar and T.C. Son (Eds.): LPNMR 2013, LNAI 8148, pp. 161–167, 2013.

Such finding of non-identical duplicates in the KB and merging them into one is referred in the literature as entity resolution [3].

In this short paper, we introduce knowledge description graphs (KDGs) as structures obtained from frame based KBs such as AURA. We formulate notions of reasoning with respect to these graphs to obtain certain missing information, present our approach of entity resolution, and use it in recovering additional missing information. After giving an Answer Set Programming (ASP) encoding of our formulation, we conclude with a discussion on the use of the above in answering "why" and "how" questions.

2   Background: Frame-Based Knowledge Bases; ASP

The KB we used in this work is based on AURA [2] and was described in details in [4]. AURA is a frame-based KB manually curated by biology experts; it contains a large amount of frames describing biological entities and events (or processes). The basic class in our KB is *Thing*, which has two children classes: *Entity* and *Event. Entity* is the ancestor of all classes of biological entities; *Event*, of biological events.

Our KB is a set of facts of the form "has(A, slot_name, B)" where A and B are either classes or instances (of classes), *slot_name* is the name of the relation between A and B such as *instance_of, raw_material* or *results*. The statement "*eukaryotic translation* is based on *mRNA*" is represented in our KB as follows.

```
has(e_transl4,instance_of,event). has(mrna6, instance_of, mrna).
has(e_transl4,instance_of,eukaryotic_translation). has(e_transl4,base,mrna6).
```

For the declarative implementation of our formulations, we use ASP [5]. That allows us to use our earlier work [4] on using ASP to reason with frame-based knowledge bases. ASP's strong theoretical foundation [6] and its default negation and recursion are useful in our encoding and in proving results about them.

3   Knowledge Description Graphs

A **Knowledge Description Graph (KDG)** is a structure to represent the facts about instances and classes of events, entities and relationships between them. A KDG is a slight generalization of EDGs in [1] and is constructed from knowledge bases such as AURA. Formal definition of the KDGs is given in the following.

Definition 1. *A KDG is a directed graph with: (i) three types of nodes: event nodes, entity nodes, and class nodes; and (ii) five types of directed edges: compositional edges, class edges, ordering edges, locational edges and participant edges. A KDG has the property that there are no directed cycles within any combination of compositional, locational and participant edges.*

Fig.1 shows the types of edges in a KDG and the corresponding sources and destinations of the edges. For example, compositional edges are from events to events or from entity to entity. We used the slot names in KM [7] and AURA as a guide to categorize the types of edges. Since KDGs can be huge, we usually work on its

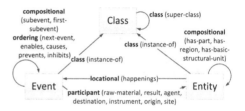

Fig. 1. Types of edges in KDGs

smaller subgraphs that are rooted at an entity or an event. The KDG rooted at Z, denoted as $KDG(Z)$, contains all the accessible nodes from Z (through any edge except ordering) and all the edges between them. Fig. 2 shows the $KDG(Eukaryote)$ where every other nodes can be reached from $Eukaryote$ without going through dashed lines (ordering edges).

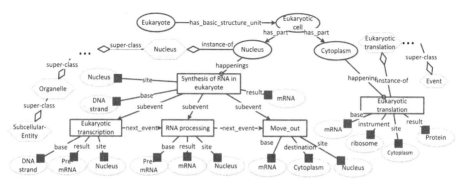

Fig. 2. A KDG rooted at the entity $Eukaryote$. Event, entity, and class nodes are respectively depicted by rectangles, ovals and hexagons. Compositional edges are represented by solid-line arrows; ordering edges by dashed-line arrows; participant edges by lines with a black-square pointed to the entity node; class edges by diamond head arrows and locational edges by lines with a circle pointed to the event node.

4   Reasoning about Missing Info. in KDGs

Two types of Events. In our KB there are two types of events: transport events and operational events. In a transport event, the locations where it happens is changed while the input entity and the output entity are the same. We differentiate two types of events by their ancestor classes: transport events are descendants of the classes *move_through*, *move_into* and *move_out_of*; all other events are operational events.

Input, Output, Input Location, Output Location. We created IO properties of an event based on its specific relations. For examples, input/ output/ input-location/ output-location of a transport event is respectively from object/ object/ {base, origin}/ destination, while those of an operational event is from {object, base, material}/ result/ site/ destination. We can also obtain missing

IO properties of an event from its subevents. For instance, an input of E's first subevent is also an input of E. Similarly, an output of an event can be obtained from its last subevent.

Example: *photosynthesis* has two subevents: *light reaction* and *calvin cycle*, the next event of *light reaction*. *Sunlight* is the raw-material of the *light reaction*, *sugar* is the result of *calvin cycle*. Thus, *sunlight* is the input of *light reaction* as well as *photosynthesis*; *sugar* is the output of both *calvin cycle* and *photosynthesis*.

Similarly, the output location of an operational event is often not defined in the KB but we can use input location as its default value. Using this rule, we can obtain the output locations of three events in Fig. 2: *Synthesis of RNA in eukaryote*, *Eukaryotic translation*, and *Eukaryotic transcription*.

Main Class of an Instance. In our KB, one instance can belong to many classes. For example, *dna_strand19497* - the input of *Eukaryotic transcription* - is an instance of *dna_strand*, *dna_sequence*, *nucleic_acid* and *polymer*[2]. However, to reason about the equality between instances, we need the "main" class(es), which is the most specific class(es) of that instance. We define a main class of E as the minimal element in the set of E's classes w.r.t to subclass ordering. The main classes of *dna_strand19497*, according this definition, are *dna_strand* and *dna_sequence*; the other classes of *dna_strand19497* are ancestors of those two.

5   Entity Resolution and Finding the Possible Next Events

In the KBs such as AURA, especially the ones that are developed using crowd-sourcing, the curation was done in many sessions and probably by many people. The results are, in many cases, (i) two different instance names were used when they are probably the same instance; and (ii) parts of some biological process were encoded as independent events. For example: the input of *Eukaryotic translation* (Fig.2) is *mrna4642* whereas the output of *Move_out* is *mrna22911* while they should be the same; *Synthesis of RNA in eukaryote* and *Eukaryotic translation* should be subevents of "Synthesis of protein in eukaryote" but they are encoded as two separate events. To solve problem (i), we define a match relation. Generally speaking, instance A can match with instance B if A can be safely used in a context where a term of B is expected.

Definition 2. *Let A and B be two instances in $KDG(Z)$. Let $ClassA$ and $ClassB$ be main classes of A and B respectively.*

1. *A matches with B with high confidence[3] if one of the following is true (a) A and B are the same instance; (b)A is cloned from B (Shortcut in AURA to specify that A has all the properties of B); or (c) ClassA is an ancestor of ClassB*

---

[2] For the sake of simplicity, in the previous figures and descriptions, we usually referenced the entities and events by their "main" class(es) and not by the instances' names although our KB and our implementation works on instances' names.

[3] Confidence levels are for greater flexibility in future works.

2. *A matches with B with medium confidence if A and B are both cloned from an instance C.*
3. *A matches with B with low confidence if ClassA = ClassB (A and B are instances of the same main class).*
4. *A matches with B with confidence $min(Conf_1, Conf_2)$ if (a) A matches with C with confidence $Conf_1$; and (b) C matches with B with confidence $Conf_2$.*
5. *Otherwise, A does not match with B.*

Using Def.2, we can match *mrna4642* with *mrna22911* because both have main class *mrna*, but we can not match an instance of *cytoplasm* to an instance of *cytosol*. However when we say *Event A occurs in cytosol*, we can understand that *Event A occurs in cytoplasm*. To overcome this shortcoming, we define the relation *Spatially match* similar to Def.2, except that A can spatially match to B if B is inside A or is a part of A.

Entity resolution can be used in finding the possible next event(s) of a given event. Our approach is that E' is E's next event if E's output matches E''s input and E's output location matches E''s input location. This assumption not only holds in all three consequent events in Fig.2 (i.e. *Eukaryotic transcription, RNA processing* and *Move_out*) but also suggests that *Eukaryotic translation* can be the next event of either *Synthesis of RNA in eukaryote* or *Move_out*. Moreover, we can select the correct event (*Synthesis of RNA in eukaryote*) with an additional constraint to prefer the super class if both A and A's super class are candidates.

6   ASP Encodings

In this section, we give a glimpse of ASP encoding of the formulations in previous section. See [8] for more details.

Inputs and Outputs of Events. $t\_e(E)$ or $o\_e(E)$ is used to indicate a transport event or an operational event, respectively. $event(X)$ indicates that X is an event. We denote the input/output/input location/output location of an event by *input, output, input_loc* and *output_loc* respectively. Rules i1-i5 get the IOs of operational events. IOs of transport events are encoded similarly (rules i6-i10). Rule i11 gets the input of an event from its first subevent. Other rules are encoded in a similar way (rules i13-i24). Rule i25 gets the default output location of an event.

```
ev1: predicates(t_event, move_through; move_into; move_out_of).
ev2: t_e(E) :- has(E, instance_of, Transport_class), predicates(t_event,
     Transport_class), event(E).
ev3: o_e(E) :- event(E), not t_event(E).
i1:input(E,A):-has(E,object,A),o_e(E). i2:input(E,A):-has(E,base,A),o_e(E).
i3:input(E,A):-has(E,raw_material,A),o_e(E).
i4:output(E,A):-has(E,result,A),o_e(E).i5:input_loc(E,A):-has(E,site,A),o_e(E).
i11: has(E, input, A) :- has(SE, input, A), has(E, first_subevent, SE).
i25: has(E, output_location, A) :- not has(E, output_location, A2), has(E,
     input_location, A), entity(A2), event(E), A2 != A.
```

Entity Resolution. *ClassA* is a main class of instance A if *ClassA* is one of A's classes and we do not have *not_main_class(A, ClassA)*. We use predicate *match_with*($A, B, Confidence$) to represent *match with* relation (Def. 2) from instance A to B; *Confidence* can be either *low, medium* or *high*. Rule ma1 encodes the sub-case 2.1.a. The last rule is for Def. 2.4, matching A to B transitively through C. *lowest_confidence*($Conf1, Conf2, Conf$) means $Conf$ is the lowest confidence in $Conf1$ and $Conf2$ (Rules lc1-lc7). Rules for other cases are skipped (ma2-ma5); locational instance matching is encoded in a similar way (rules sma1-sma4).

```
m1:  not_main_class(A, ClassB) :- has(A, instance_of, ClassA), has(A,
       instance_of, ClassB), has(ClassA, ancestorclass, ClassB).
m2:  main_class(A, ClassA) :- has_class(A, ClassA), not not_main_class(A,
       ClassA).
ma1:  match_with(A,B,high):-main_class(A,ClassA),main_class(B,ClassB), A==B.
ma6:  match_with(A,B,Conf) :- match_with(A,C,Conf1), match_with(C,B,Conf2),
       A!=B, A!=C, B!=C, lowest_confidence(Conf1,Conf2,Conf).
```

7   Conclusion and Discussion

In this short paper, we gave a glimpse of several formulations regarding missing knowledge about events and related ASP implementation. One of our formulations was about entity resolution where we resolve multiple entities that may have different names but may refer to the same entity. Our method is different from other methods in the literature [3]. Since each entity resolution method heavily relies on the properties of the database it is working on, and no other system we know of is about AURA or similar event centered knowledge bases we were unable to directly compare our method with the others.

Our approach to use rules (albeit ASP rules) to derive missing information is analogous to use of rules in data cleaning and in improving data quality [9]. However those works do not focus on issues that we discussed in this paper.

Thus, by being able to obtain missing information and enriching the original KDGs one can obtain more accurate and intuitive answers to the various 'why" and "how" questions such as: "How does X occur?", "How does X produce Y?", "How are X and Y related?", "Why X is important to Y?", "How does X participate in process Y?", "How does X do Y?", "Why does X produce Y?" and others. The answer of each question is a subgraph of KDG. For example, the answer of the first question contains only $KDG(X)$ and all the nodes connected to/from X through ordering edges.

References

1. Baral, C., Vo, N.H., Liang, S.: Answering why and how questions with respect to a frame-based knowledge base: a preliminary report. In: Technical Communications of the 28th International Conference on Logic Programming, ICLP 2012, vol. 17, pp. 26–36 (2012)
2. Chaudhri, V.K., Clark, P.E., Mishra, S., Pacheco, J., Spaulding, A., Tien, J.: AURA: capturing knowledge and answering questions on science textbooks. Technical report, SRI International (2009)

3. Getoor, L., Diehl, C.P.: Link mining: a survey. ACM SIGKDD Explorations Newsletter 7(2), 3–12 (2005)
4. Baral, C., Liang, S.: From knowledge represented in frame-based languages to declarative representation and reasoning via ASP. In: 13th International Conference on Principles of Knowledge Representation and Reasoning (2012)
5. Gelfond, M., Lifschitz, V.: The stable model semantics for logic programming. In: Kowalski, R., Bowen, K. (eds.) Logic Programming: Proc. of the Fifth Int'l Conf. and Symp., pp. 1070–1080. MIT Press (1988)
6. Baral, C.: Knowledge representation, reasoning and declarative problem solving. Cambridge University Press (2003)
7. Clark, P., Porter, B., Works, B.: KM: The knowledge machine 2.0: Users manual. Citeseer (2004)
8. Baral, C., Vo, N.H.: Event-object reasoning with curated knowledge bases: Deriving missing information (June 2013), http://arxiv.org/abs/1306.4411
9. Herzog, T.N., Scheuren, F.J., Winkler, W.E.: Data quality and record linkage techniques. Springer (2007)

Towards Query Answering
in Relational Multi-Context Systems*

Rosamaria Barilaro[1], Michael Fink[2], Francesco Ricca[1], and Giorgio Terracina[1]

[1] Dipartimento di Matematica e Informatica, Università della Calabria, Italy
[2] Institute of Information Systems, Vienna University of Technology, Austria

Abstract. We report on preliminary research towards native algorithms for query answering over relational nonmonotonic Multi-Context Systems (MCS), i.e., algorithms that do not rely on computing equilibria. Inspired by techniques for query answering in distributed answer set programming, we identify MCS settings where a generalized query answering algorithm is effective and efficient; confirmed by a preliminary evaluation on a real world application.

1 Introduction

Nonmonotonic Multi-Context Systems (MCS) [2] are a powerful framework for interlinking knowledge of different contexts by so-called bridge rules. Contexts may be represented in heterogeneous knowledge representation formalisms and can be distributed over a network. Recent research outcomes provide effective (distributed) algorithms for implementing the well-established model-based semantics expressed in terms of so-called equilibria. However, these algorithms turned out to be inefficient to serve as a basis for query answering, in particular in the practically relevant case of relational MCS [4] (due to the presence of bridge rules having variables, which in principle can succinctly represent the exchange of large amounts of data).

Addressing the problem, here we consider query answering as the primary task and aim at developing a native query answering algorithm that does not hinge on equilibrium computation. We take inspiration from corresponding techniques in distributed Answer Set Programming (ASP) [1], which are based on both unfolding and weighted hyper-tree decomposition techniques [5]. The idea of unfolding is to express a query directly in terms of the data involved in the query, whereas the goal of hypertree decomposition methods is to compute a decomposition of the original query organized as a join-tree, that can be evaluated efficiently following a bottom up strategy. Aiming at adapting these methods to MCS, our contributions are briefly summarized as follows:
(i) We formally define the problem and characterize a class of MCS that can be adapted.
(ii) We propose an algorithm lifting the approach of [1] to query answering in MCS.
(iii) We report on a preliminary experiment performed on real-world data, considering an MCS modeling a biomedical domain [3] by means of ASP contexts. A comparison with a naïve centralized implementation confirms viability of our approach.

We conclude with some pointers to interesting issues raised for further research.

* This research has been partially supported by the Austrian Science Fund (FWF) grant P24090, the Vienna Science and Technology Fund (WWTF) grant ICT 08-020, the Calabrian Region under project PIA KnowRex POR FESR 2007- 2013 BURC n. 49 s.s. n. 1 16/12/2010, and Italian Ministry for University and Research (MIUR) under project FRAME PON01_02477/4.

P. Cabalar and T.C. Son (Eds.): LPNMR 2013, LNAI 8148, pp. 168–173, 2013.

2 Preliminaries

Relational MCS [4] generalize nonmonotonic MCS [2], as briefly (and, due to space constraints, to some extent informally) follows. MCS represent contextual knowledge by an abstract notion of a so-called *logic* L that is defined in terms of a signature Σ_L, a set of well-formed knowledge bases KB_L, a set of possible belief sets BS_L, and a function $ACC_L : KB_L \rightarrow 2^{BS_L}$ (intuitively representing the semantics of L) that assigns each knowledge-base a set of acceptable sets of beliefs.

While (relational) logics serve the purpose of representing contextual knowledge in terms of knowledge bases, so called *bridge rules* model their interlinking. Intuitively, a relational logic L (sometimes simply called 'logic' subsequently) additionally allows for relational elements in KB_L and BS_L (cf. [4] for details). Given a set of relational logics $\{L_1, \ldots, L_n\}$, let V be a countable set of distinct variable names. We say that a (non-ground) relational element of L_i is of the form $p(t_1, \ldots, t_k)$, where t_j is a term over Σ_{L_i} and V. Then, a *relational bridge rule* is of the form

$$(k{:}s) \leftarrow (c_1{:}p_1), \ldots, (c_j{:}p_j), not(c_{j+1}{:}p_{j+1}), \ldots, not(c_m{:}p_m), \qquad (1)$$

where $1 \leq k \leq n$, and s is an ordinary or relational knowledge base element of L_k, as well as $1 \leq c_\ell \leq n$, and p_ℓ is either an ordinary or relational belief of L_ℓ, for $1 \leq \ell \leq m$. The head belief s is denoted by $hd(r)$, while $head(r) = (k{:}s)$. Moreover, $pos(r) = \{(c_{\ell_1}{:}p_{\ell_1}) \mid 1 \leq \ell_1 \leq j\}$, $neg(r) = \{(c_{\ell_2}{:}p_{\ell_2}) \mid j < \ell_2 \leq m\}$, and $body(r) = pos(r) \cup \{not(c_{\ell_2}{:}p_{\ell_2}) \mid j < \ell_2 \leq m\}$.

A relational MCS consists of a set of contexts each composed of a knowledge base of an associated relational logic and a set of relational bridge rules. More precisely:

Definition 1 (Relational MCS). *A relational MCS $M = (C_1, \ldots, C_n)$ is a collection of contexts $C_i = (L_i, kb_i, br_i, D_i)$, where L_i is a relational logic, kb_i is a knowledge base, br_i is a set of relational bridge rules, and D_i is a collection of import domains $D_{i,\ell}, 1 \leq \ell \leq n$, such that $D_{i,\ell}$ is a subset of the universe of Σ_{L_i}.*

We assume that $D_{i,\ell} = D_\ell^A$, i.e., the active domain of object constants appearing in kb_ℓ or in $hd(r)$, for some $r \in br_\ell$ such that $hd(r)$ is relational. We use br_M to denote $\bigcup_{i=1}^n br_i$. The *import neighborhood of a context* C_k is the set $In(k) = \{c_i \mid (c_i : p_i) \in pos(r) \cup neg(r), r \in br_k\}$. The *import closure of* C_k is $IC(k) = \bigcup_{j \geq 0} IC^j(k)$, where $IC^0(k) = In(k)$, and $IC^{j+1}(k) = \bigcup_{i \in IC^j(k)} In(i)$. The *dependency graph of an MCS* M is the digraph $G = (\{C_1, \cdots, C_n\}, \rightarrow)$, where $C_i \rightarrow C_j$ iff $j \in In(i)$.

The semantics of a relational MCS is defined by grounding bridge rules. The grounding of a relational belief p of L_ℓ is wrt. D_ℓ^A and denoted by $grd(p)$. The set of *ground instances* $grd(r)$ of a relational bridge rule $r \in br_i$ is restricted by admissible substitutions of the variables in r, i.e., such that constants replacing variable X are in the intersection of the domains associated with the relational beliefs of all occurrences of X in r (cf. also [4]). The *grounding* of a relational MCS M, denoted by $grd(M)$, consists of the collection of contexts obtained by replacing br_i with $grd(br_i) = \bigcup_{r \in br_i} grd(r)$.

A belief state $S = (S_1, \ldots, S_n)$ is given by $S_i \in BS_i$. A ground bridge rule r of the form (1) is applicable wrt. S, denoted by $S \models body(r)$, iff $p_\ell \in S_{c_\ell}$ for $1 \leq \ell \leq j$ and $p_\ell \notin S_{c_\ell}$ for $j \leq \ell \leq m$. By $app_i(S)$ we denote the set $\{hd(r) \mid r \in grd(br_i) \wedge S \models r\}$; and $S = (S_1, \ldots, S_n)$ is an equilibrium of MCS M iff $S_i \in ACC_i(kb_i \cup app_i(S))$.

3   Query Answering for MCS

While equilibria provide a declarative model-based semantics for MCS, and thus also a basis for answering queries over MCS, computing equilibria and answering queries on top is for many practical settings not a viable solution. We subsequently consider query answering for relational MCS as a primary task.

We consider *conjunctive queries* $Q_{C_\ell}(\mathbf{t})$ posed to a *query context* C_ℓ of the form:

$$q(\mathbf{t}) \leftarrow (c_1{:}p_1), \ldots, (c_j{:}p_j), not(c_{j+1}{:}p_{j+1}), \ldots, not(c_m{:}p_m),$$

where q is a query predicate, $\mathbf{t} = (t_1, \ldots, t_k)$ is a k-tuple of terms over Σ_{L_ℓ} and V, and every p_i, $1 \le i \le m$, is an ordinary or relational belief of L_{c_i}. We also restrict to queries *compliant with the topology*, i.e., contexts in the body are from the import closure of C_ℓ; formally, $(c_i{:}p_i) \in pos(Q_{C_\ell}(\mathbf{t})) \cup neg(Q_{C_\ell}(\mathbf{t}))$ implies $c_i \in IC(\ell)$.

As usual, query answers are defined in terms of substitutions to the variables in $\mathbf{t}$ (if any) that make the query true. More specifically, given a belief state S we write $S \models Q_{C_\ell}(\mathbf{t})$ if there exists $q_g \in grd(Q_{C_\ell}(\mathbf{t}))$ such that q_g is applicable wrt. S.

Definition 2. *Given a relational MCS M and a conjunctive query $Q_{C_\ell}(\mathbf{t})$, a k-tuple of ground terms $\mathbf{t}' = (t_1', \ldots, t_k')$ is called a) a (possible) query answer to $Q_{C_\ell}(\mathbf{t})$, b) a certain query answer to $Q_{C_\ell}(\mathbf{t})$, if*

(i) $\theta \mathbf{t} = \mathbf{t}'$ for some admissible substitution θ, and
(ii) $S \models \theta Q_{C_\ell}(\mathbf{t})$ holds a) for some, b) for all, equilibria S of M.

The set $Ans(M, Q_{C_\ell}(\mathbf{t}))$ (resp. $Cert(M, Q_{C_\ell}(\mathbf{t}))$) denotes all (certain) query answers.

Towards computing certain answers a first assumption is consistency: an MCS is known to have equilibria. This can often be guaranteed without computing or knowing equilibria (e.g., for totally coherent, acyclic MCS) and yields an important relevance property. For consistent MCS, answers can equivalently be obtained considering the restriction of M to contexts in the import closure $IC(\ell)$ of C_ℓ (denoted by $M|_{IC(\ell)}$).

Theorem 1 (Relevance). *Let M be a consistent MCS. Then, $Ans(M, Q_{C_\ell}(\mathbf{t})) = Ans(M|_{IC(\ell)}, Q_{C_\ell}(\mathbf{t}))$ and $Cert(M, Q_{C_\ell}(\mathbf{t})) = Cert(M|_{IC(\ell)}, Q_{C_\ell}(\mathbf{t}))$, for every $Q_{C_\ell}(\mathbf{t})$ compliant with the topology.*

Another natural restriction that often applies concerns system topology. We say that a relational MCS M is *hierarchical* if its dependency graph is acyclic, i.e. it is a forest. In LP terms they are characterized by stratified and non-recursive bridge rules.

In many relevant scenarios also non-determinism is confined to particular contexts that import deterministic information and provide reasoning capabilities on top. Correspondingly, given a hierarchical MCS M and a (query) context C_ℓ, we say that M is *query-deterministic* for C_ℓ if ACC_{L_j} is deterministic for every $j \in IC(\ell)$ such that $j \ne \ell$. That is, $|ACC_{L_j}(kb)| \le 1$ holds for all $kb \in KB_{L_j}$ and every context $C_j \ne C_\ell$ in the import closure of C_ℓ. Thus, non-determinism is confined to the query context:

Proposition 1. *Let M be a hierarchical and consistent MCS that is query-deterministic for C_ℓ. If $i \ne \ell$, then $S_i^1 = S_i^2$, for any two equilibria S^1 and S^2 of $M|_{IC(\ell)}$.*
Therefore, if C_ℓ is also deterministic, then $M|_{IC(\ell)}$ has a single equilibrium and all respective query answers are certain.

Procedure EvaluateMCSQuery(Query q, Context c, MCS M, Semantics $\mathcal{S}$)

Output: Set of k-tuple of ground terms Res

1 **begin**
2 $\quad$ $Neighbor := \{q\} \cup \bigcup_{c_i \in In(c)} br_{c_i}$; $Ext\_Know := \emptyset$;
3 $\quad$ **foreach** $BridgeRule\ br \in Neighbor$ **do**
4 $\quad\quad$ $Unfold :=$ BridgeUnfold(br,M);
5 $\quad\quad$ **foreach** $Query\ ubr \in Unfold$ **do**
6 $\quad\quad\quad$ **foreach** $(c_o : p_o) \in pos(ubr) \cup neg(ubr)$ s.t. c_o is opaque **do**
7 $\quad\quad\quad\quad$ $Ext\_Know := Ext\_Know \cup$ EvaluateMCSQuery($q_o(\mathbf{t}) \leftarrow (c_o : p_o(\mathbf{t})),c,M$);
8 $\quad\quad$ $Ext\_Know := Ext\_Know \cup$ HT_Evaluation($hd(br),c, Unfold$);
9 $\quad$ $Res :=$ evaluate($q,c,Ext\_Know,\mathcal{S}$);

---

4 Computing Query Answers

Query answering over an MCS exhibiting the conditions outlined in the previous section can be carried out by procedure EvaluateMCSQuery, which takes as input an MCS M and a query q posed to context c. Moreover, it is parameterized depending on whether we are interested in possible or certain answers (i.e., parameter $\mathcal{S}$ can be either "Certain" or "Possible"). The first step identifies the bridge rules of import neighbors of c. All ground instances of corresponding relational elements present in the (unique) equilibrium of $M|_{IC(c)}$ are collected in the set Ext_Know. To compute it, each bridge rule in $Neighbor$ is first unfolded by a call to function BridgeUnfold; this step produces a union of queries (stored in $Unfold$) whose evaluation is done by applying (HT_Evaluation that is) an adaptation of the distributed query evaluation algorithm proposed in [1] to MCS queries as briefly outlined next. The unfolding is carried out by an adaptation to bridge rules of the usual unfolding strategy for Datalog programs. Specifically, each bridge rule is considered a separate query, and the unfolding is carried out by the following head-to-body dependencies among the set of bridge rules of the MCS: whenever possible, elements are recursively substituted in the body rule by their 'definition', as specified within the MCS. We say that a context c is *opaque* if for c an unfolding procedure is not defined. Intuitively, if a context is opaque, it is not possible to 'look inside', i.e., access the definition of its elements (either by the way the logic is defined or by privacy issues), and consequently, it is not possible to unfold through it. Given pair $(c : e)$ occurring in a query, we say that e is *not unfoldable* if either (i) $(c : e)$ appears in the negative part of the query, or (ii) $(c : e)$ is the head of more than one bridge rule in the MCS, or (iii) its context c is *opaque*. If condition (i) or (ii) is satisfied, then the element $(c : e)$ is left unchanged in the query, but the algorithm tries to recursively unfold the defining bridge rules in the MCS, i.e., those with head $(c : e)$. If condition (iii) holds, then $(c : e)$ is interpreted as a query that must be posed to context c, managed by a recursive call to procedure EvaluateMCSQuery. Finally, function *ContextUnfold* applies a context-specific unfolding procedure inside contexts that are not opaque, returning a set of queries. It is thus applicable to context logics admitting unfolding procedures for query answering (e.g., this holds for ASP, DL-Lite, Datalog, etc.). The effect of unfolding is twofold: (i) it restricts the computation to data relevant for answering the query by considering chains of dependencies among bridge rules; and (ii) it rearranges original queries into an equivalent set of queries with longer bodies,

---

Function BridgeUnfold(BridgeRule br, MCS M)

   Output: Set of Bridge rules $Unfold$
1 **begin**
2    $Unfold := \emptyset$;
3   **foreach** $(c : p) \in pos(br) \cup neg(br)$ **do**
4   **if** p *is* unfoldable **then**
5   Let $r_p \in br_M$ s.t. $head(r_p) = (c : p)$;
6          $br := $ Replace$(br, (c : p), body(r_p))$;
7   **if** c *is not* opaque **then**
8   **foreach** $r \in br_M$ s.t. $(c : p) \in head(r)$ **do**
9             $R := $ ContextUnfold(r, c);
10             $Unfold := Unfold \cup \bigcup_{r' \in R}$ BridgeUnfold(r', M);
11    $Unfold := Unfold \cup br$;

---

that thus are more suitable to be analyzed by the subsequent hyper-tree decomposition. Indeed, function HT_Evaluation first applies a weighted hyper-tree decomposition [5] to queries in *Unfold*, and then evaluates them bottom-up according to the body-to-head dependencies. In the mentioned decomposition technique, a query q is associated with a hyper-graph H where hyper-edges represent joins between variables. A hyper-tree decomposition of H decomposes q in sub-queries efficiently bottom-up evaluable. In our setting, body atoms are relational elements defined in some context, and a query has to be executed at a context to compute its ground instances. Moreover, contexts may be distributed on different machines over a network, implying data transfer over the network. Therefore, we a apply a lifting of the hyper-tree weighting function defined in [1] with its execution plan optimization that also evaluates costs in a distributed setting. Finally, once the external knowledge is available at the query context c, function evaluate computes the possible/certain answers to q in context c.

Proposition 2. *Given a relational MCS M and a conjunctive query $Q_{C_\ell}(\mathbf{t})$ posed to context C_ℓ, then both $Ans(M, Q_{C_\ell}(\mathbf{t})) = EvaluateMCSQuery(Q_{C_\ell}(\mathbf{t}), C_\ell, M, Possible)$ and $Cert(M, Q_{C_\ell}(\mathbf{t})) = EvaluateMCSQuery(Q_{C_\ell}(\mathbf{t}), C_\ell, M, Certain)$ hold.*

5   Proof-of-Concept Prototype and Preliminary Evaluation

We developed a proof-of-concept implementation of the approach described in Section 4. Our current implementation restricts to MCSs whose context logics are defined by logic programs under ASP semantics (which are not opaque). Ground relational kb and belief set elements are assumed to reside in DBMSs, locally for each context; they can be possibly accessed by other contexts through ODBC connections. In this setting the implementation of function HT_Evaluation is a straight adaptation of the system [1], having DLV^{DB} as the underlying engine for logic program evaluation. Dealing with ASP contexts allowed to additionally exploit involved optimizations in realizing *ContextUnfold* for query evaluation. Indeed, our implementation also pushes down, whenever possible, constants present in the query through the unfolding process. This significantly reduces data transfers among contexts. We assessed our approach running a preliminary experiment on the real world application presented in [3]. The application domain involves biomedical knowledge resources about genes, drugs and diseases,

Table 1. Experimental results

Query	Our Approach (sec)	Naïve Approach (sec)	Involved Contexts	Involved Tuples	Resulting Tuples
$q1$	38,5	534,3	PHARMGKB,CTD	399.713	41
$q2$	61,1	607,4	PHARMGKB,CTD,SIDER	400.207	535
$q3$	38,9	572,1	PHARMGKB,CTD,BIOGRID	431.528	1
$q4$	381,0	1.847,6	PHARMGKB,CTD	1.375.819	33
$q5$	36,9	345,4	PHARMGKB,CTD,DRUGBANK	261.315	1.726
$q6$	145,7	1.836,3	PHARMGKB,CTD,DRUGBANK,BIOGRID	1.380.398	11

from PHARMGKB, DRUGBANK, BIOGRID, CTD, and SIDER online databases. This scenario has been modeled as an MCS taking each source of knowledge as a context; like in reality, we distributed these contexts, and the corresponding data, on different servers connected by standard Ethernet. Several bridge rules properly represent inter-connections (relations in [3]) and we considered six (deterministic) queries from [3].[1] Given a query q posed to a context c, we executed and compared two algorithms (cf. Table 1): procedure EvaluateMCSQuery and a naïve algorithm, which is a "plain" execution of DLV^{DB} that first transfers to c all the data accessed by its bridge rules, and then locally evaluates q on c. Table 1 shows the promising results obtained. Observe that the number of tuples in the result of the queries considered is much smaller than the involved tuples; this is mainly due to selections present in most queries; these are pushed down to the original data by unfolding.

6 Conclusion

We provided stepping stones towards effective query answering in nonmonotonic MCS, together with experimental confirmation on a real-world application. This opens several interesting issues for future research (partly ongoing), such as generalizing the technique, specifically hyper-tree decomposition, to broader settings, considering language extensions (aggregates), or incorporating sensitivity to privacy issues (beyond import neighborhood). A detailed complexity analysis including data complexity, as well as studying semantic properties (e.g., query containment) are interesting topics in theory.

References

1. Barilaro, R., Ricca, F., Terracina, G.: Optimizing the distributed evaluation of stratified programs via structural analysis. In: Delgrande, J.P., Faber, W. (eds.) LPNMR 2011. LNCS, vol. 6645, pp. 217–222. Springer, Heidelberg (2011)
2. Brewka, G., Eiter, T.: Equilibria in heterogeneous nonmonotonic multi-context systems. In: AAAI, pp. 385–390. AAAI Press (2007)
3. Erdem, E., Erdem, Y., Erdogan, H., Öztok, U.: Finding answers and generating explanations for complex biomedical queries. In: Burgard, W., Roth, D. (eds.) AAAI. AAAI Press (2011)
4. Fink, M., Ghionna, L., Weinzierl, A.: Relational information exchange and aggregation in multi-context systems. In: Delgrande, J.P., Faber, W. (eds.) LPNMR 2011. LNCS, vol. 6645, pp. 120–133. Springer, Heidelberg (2011)
5. Scarcello, F., Greco, G., Leone, N.: Weighted hypertree decompositions and optimal query plans. Journal of Computer and System Sciences 73(3), 475–506 (2007)

---

[1] Note that q1-q6 in our paper correspond to Q6, Q7, Q1, Q8, Q9 and Q2 in [3]; nonetheless, all queries from [3] could be handled within our framework applyng minor syntactic adaptations.

Spectra in Abstract Argumentation:
An Analysis of Minimal Change

Ringo Baumann and Gerhard Brewka

University of Leipzig, Informatics Institute, Germany
{baumann,brewka}@informatik.uni-leipzig.de

Abstract. In this paper we present various new results related to the dynamics of abstract argumentation. Baumann [1] studied the effort needed to enforce a set of arguments E, measured in terms of the minimal number of modifications needed to turn an argumentation framework (AF) $\mathcal{A}$ into a framework $\mathcal{A}^*$ such that $\mathcal{A}^*$ has an extension containing E. This value, called the characteristic, depends on the chosen semantics and the type of admitted modifications. Here we study the inverse problem (called the *spectrum problem*): given a collection of semantics and a modification type, what are the corresponding tuples of characteristics one may obtain for an arbitrary argumentation framework $\mathcal{A}$ and set of arguments E? The set of all these tuples is called the spectrum. We define various properties of spectra and show that the investigation of spectra reveals interesting and surprising insights into the relationship among several semantics.

1 Introduction

Argumentation is the interdisciplinary study of how conclusions can be reached through the construction and evaluation of arguments, that is, structures describing a proposition together with the reasons for accepting it. The field has received growing interest within Artificial Intelligence over the last decades. It covers aspects of knowledge representation and multi-agent systems, but also touches on various philosophical questions (for a very good overview see [2]). Dung's abstract argumentation frameworks (AFs) [3] play a dominant role in the field. In AFs arguments and attacks among them are treated as abstract entities. The focus is on conflict resolution and argument acceptability. Various semantics for AFs have been defined, each of them specifying acceptable sets of arguments, so-called *extensions*, in a particular way.

More recently several problems regarding *dynamic* aspects of abstract argumentation have been addressed in the literature [4–8, 1]. One problem which is relevant to the work presented here concerns the acceptability of certain arguments and is called *enforcing problem* [6]. This is, in brief, the question whether it is possible, given a specific set of allowed operations, to modify a given AF such that a desired set of arguments E is contained in an extension of the modified AF. Several necessary and sufficient conditions under which enforcements are possible were identified.

In addition to clarifying the *possibility* of enforcing certain arguments, a natural further question in this context is concerned with the *effort needed* for the enforcements. This more general problem of *minimal change* [1] can be formulated as follows: what is the minimal number of modifications (additions or removals of attacks) needed to reach

P. Cabalar and T.C. Son (Eds.): LPNMR 2013, LNAI 8148, pp. 174–186, 2013.

an enforcement of E? This value, called *characteristic* in [1], depends on the underlying semantics σ and type of allowed modifications Φ. Quite surprisingly, it was shown that, in case of certain semantics and modification types, there are local criteria to determine the minimal number, although infinitely many possibilities to modify a given AF exist.

In this paper we study a further, closely related question in this context which has some similarity with the famous *Spektralproblem*[1] in model theory [9]. Given a certain semantics σ and a modification type Φ, we study whether there is, for a given natural number n, an AF $\mathcal{A}$ and a set of arguments E such that n is the (σ, Φ)-characteristic of E w.r.t. $\mathcal{A}$. In other words, we want to determine the set of all natural numbers which may occur as (σ, Φ)-characteristics, the so-called (σ, Φ)-*spectrum*. This yields interesting insights into particular semantics. To mention one result, we will show that in case of semi-stable semantics and the addition of weak arguments (arguments which do not attack previous arguments) not each natural number may arise as the minimal effort needed to enforce a certain set D. In particular, the characteristic cannot be 1.

What makes our study even more interesting, as we believe, is the fact that it provides useful and at times surprising new insights into the interrelationships among the studied semantics. To this end, we perform our analysis in parallel for a whole group of semantics which we consider as some of the most important semantics for Dung frameworks. Rather than sets of values, spectra thus become sets of tuples of values. Appropriate properties of the spectra - which we will define in Sect. 3 - will help us to identify such relationships.

The rest of the paper is organized as follows. Sect. 2 reviews the necessary background. Sect. 3 introduces the notion of a spectrum and presents our results for the stable/semi-stable/preferred spectra under various types of modifications. In Sect. 4 we discuss related work and conclude.

2   Background

An *argumentation framework* $\mathcal{F}$ is a pair (A, R), where A is a non-empty finite set whose elements are called *arguments* and $R \subseteq A \times A$ a binary relation, called the *attack relation*. The set of all AFs is denoted by $\mathscr{A}$. If $(a, b) \in R$ holds we say that a *attacks* b, or b is *defeated* by a in $\mathcal{F}$. An argument $a \in A$ is *defended* by a set $A' \subseteq A$ in $\mathcal{F}$ if for each $b \in A$ with $(b, a) \in R$, b is defeated by some $a' \in A'$ in $\mathcal{F}$. Furthermore, we say that a set $A' \subseteq A$ is *conflict-free* in $\mathcal{F}$ if there are no arguments $a, b \in A'$ such that a attacks b. The set of all conflict-free sets of an AF $\mathcal{F}$ is denoted by $cf(\mathcal{F})$. For an AF $\mathcal{F} = (B, S)$ we use $A(\mathcal{F})$ to refer to B and $R(\mathcal{F})$ to refer to S. Finally, we introduce the union of two AFs as usual, namely $\mathcal{F} \cup \mathcal{G} = (A(\mathcal{F}) \cup A(\mathcal{G}), R(\mathcal{F}) \cup R(\mathcal{G}))$.

Semantics determine acceptable sets of arguments for a given AF $\mathcal{F}$, so-called *extensions*. The set of all extensions of $\mathcal{F}$ under semantics σ is denoted by $\mathcal{E}_\sigma(\mathcal{F})$. For two semantics σ, τ we use $\sigma \subseteq \tau$ to indicate that for any $\mathcal{F} \in \mathscr{A}$, $\mathcal{E}_\sigma(\mathcal{F}) \subseteq \mathcal{E}_\tau(\mathcal{F})$. Due to the limited space we consider stable (st), preferred (pr) and semi-stable (ss) semantics only [3, 10].

---

[1] Roughly speaking, Scholz investigated the possible sizes finite models of a first-order sentence may have.

Definition 1 (Semantics). *Given an AF $\mathcal{F} = (A, R)$ and $E \subseteq A$. E is a*

1. *stable extension ($E \in \mathcal{E}_{st}(\mathcal{F})$) iff*
   $E \in cf(\mathcal{F})$ *and each $a \in A \backslash E$ is defeated by some $e \in E$,*
2. *admissible set ($E \in \mathcal{E}_{ad}(\mathcal{F})$) iff*
   $E \in cf(\mathcal{F})$ *and each $e \in E$ is defended by E in $\mathcal{F}$,*
3. *preferred extension ($E \in \mathcal{E}_{pr}(\mathcal{F})$) iff*
   $E \in \mathcal{E}_{ad}(\mathcal{F})$ *and for each $E' \in \mathcal{E}_{ad}(\mathcal{F})$, $E \not\subset E'$ and*
4. *semi-stable extension ($E \in \mathcal{E}_{ss}(\mathcal{F})$) iff*
   $E \in \mathcal{E}_{ad}(\mathcal{F})$ *and for each $E' \in \mathcal{E}_{ad}(\mathcal{F})$, $R_{\mathcal{F}}^{+}(E) \not\subset R_{\mathcal{F}}^{+}(E')$ where $R_{\mathcal{F}}^{+}(E) = E \cup \{b \mid (a,b) \in R, a \in E\}$.*

It is well known that $st \subseteq ss \subseteq pr$. Furthermore, there exist sufficient conditions for the agreement of the considered semantics. In particular, $st = ss$ if $st \neq \varnothing$ [10] and $st = pr$ if the considered AFs are SCC-symmetric and self-loop-free (compare [11]).

Expansions were introduced by [6]. They will be our object of investigation since they represent reasonable types of dynamic argumentation scenarios.

Definition 2 (Expansions). *An AF $\mathcal{F}^*$ is an expansion of AF $\mathcal{F} = (A, R)$ (for short, $\mathcal{F} \leq_E \mathcal{F}^*$) iff $\mathcal{F}^* = (A \cup A^*, R \cup R^*)$ where $A^* \cap A = R^* \cap R = \varnothing$. An expansion is called*

1. *normal ($\mathcal{F} \leq_N \mathcal{F}^*$) iff $\forall ab \; ((a,b) \in R^* \rightarrow a \in A^* \vee b \in A^*)$,*
2. *strong ($\mathcal{F} \leq_S \mathcal{F}^*$) iff $A \leq_N A^*$ and $\forall ab \; ((a,b) \in R^* \rightarrow \neg (a \in A \wedge b \in A^*))$,*
3. *weak ($\mathcal{F} \leq_W \mathcal{F}^*$) iff $A \leq_N A^*$ and $\forall ab \; ((a,b) \in R^* \rightarrow \neg (a \in A^* \wedge b \in A))$.*

For short, normal expansions add new arguments and possibly new attacks which involve at least one of the fresh arguments. Strong (weak) expansions are normal and only add *strong (weak) arguments*, i.e. the added arguments never are attacked by (attack) former arguments. For the purpose of illustration we present the following simple example.

Example 1. The AF $\mathcal{F} = (\{a, b\}, \{(a, b)\})$ is the initial framework. Arbitrary, normal, weak and strong expansions of $\mathcal{F}$ are given by $\mathcal{F}_E$, $\mathcal{F}_N$, $\mathcal{F}_W$ or $\mathcal{F}_S$, respectively.

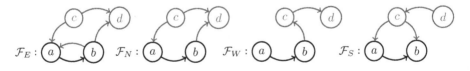

Fig. 1. Notions of Expansions

As usual $\mathcal{F} <_X \mathcal{F}^*$ for $X \in \{E, N, S, W\}$ stands for $\mathcal{F} \leq_X \mathcal{F}^*$ and $\mathcal{F} \neq \mathcal{F}^*$. To simplify notation we will later on often use X to refer to $\leq_X$. Whenever infix notation is used we stick to $\leq_X$, though.

The *minimal change problem* [1] is the problem of determining the minimal effort needed to transform a given argumentation framework, using a particular type of modifications, into a framework that possesses an extension containing a specific set of arguments E. The effort is characterized by the (σ, Φ)-*characteristic*:

Definition 3 (Characteristic). *Given a semantics σ, a binary relation $\Phi \subseteq \mathscr{A} \times \mathscr{A}$ and an AF $\mathcal{F}$. The (σ, Φ)-characteristic of a set $C \subseteq A(\mathcal{F})$ is a natural number or infinity defined by the following function*

$$N_{\sigma,\Phi}^{\mathcal{F}} : \wp(A(\mathcal{F})) \rightarrow \mathbb{N}_\infty$$

$$C \mapsto \begin{cases} 0, & \exists C' : C \subseteq C' \text{ and } C' \in \mathcal{E}_\sigma(\mathcal{F}) \\ k, & k = \min\{d(\mathcal{F}, \mathcal{G}) \mid (\mathcal{F}, \mathcal{G}) \in \Phi, N_{\sigma,\Phi}^{\mathcal{G}}(C) = 0\} \\ \infty, & \text{otherwise.} \end{cases}$$

Here we define $d(\mathcal{F}, \mathcal{G})$ as the number of added or removed attacks needed to transform $\mathcal{F}$ to $\mathcal{G}$. This means $d(\mathcal{F}, \mathcal{G}) = |R(\mathcal{F}) \mathbin{\Delta} R(\mathcal{G})|$ where Δ is the symmetric difference.

3 The Spectrum Problem

Given a semantics and a type of allowed modifications, the characteristic provides information about the effort needed to enforce a set of arguments C starting from an AF $\mathcal{F}$. Here we study the inverse problem, that is, given a particular characteristic, is there an AF $\mathcal{F}$ and a set of arguments C which possess this characteristic. More generally, we will consider n-tuples of semantics and modification types and ask whether some $\mathcal{F}$ and C possess a given n-tuple of characteristics simultaneously. A tuple of characteristics satisfying this condition is called a *fibre*. A fibre is said to be *finite* if all entries are natural numbers. The set of all fibres provides important insights on how close or far apart the characteristics of a set C may be. That's why this set is called the *spectrum*. Here is the formal definition.

Definition 4. *Given n semantics $\sigma_1, ..., \sigma_n$ and n binary relations $\Phi_1, ..., \Phi_n \subseteq \mathscr{A} \times \mathscr{A}$. The $(\sigma_1, \Phi_1, ..., \sigma_n, \Phi_n)$-spectrum is a set of n-tuples (so-called fibres) defined as follows:*

$$\mathcal{S}_{(\sigma_i, \Phi_i)_{i=1}^n} = \{(k_1, ..., k_n) \mid \exists \mathcal{F} \in \mathscr{A} \; \exists C \subseteq A(\mathcal{F}) : N_{\sigma_i, \Phi_i}^{\mathcal{F}}(C) = k_i \text{ for all } i \in \{1, ..., n\}\}.$$

For convenience, if $\Phi_1 = ... = \Phi_n$ we simply write $(\sigma_1, ..., \sigma_n, \Phi)$-spectrum or $\mathcal{S}_{(\sigma_1, ..., \sigma_n, \Phi)}$. These are exactly the types of spectra which we will consider in this paper. Furthermore, we will restrict ourselves to stable (st), semi-stable (ss) and preferred (pr) semantics, arguably the most important semantics for Dung frameworks. The relations we study will be normal, strong, weak and arbitrary expansions.

We first introduce some basic properties spectra may possess.

Definition 5. *A spectrum $\mathcal{S}_{(\sigma_i, \Phi_i)_{i=1}^n}$ is*

1. *m.d.s. iff any finite fibre $(k_1, ..., k_n) \in \mathcal{S}_{(\sigma_i, \Phi_i)_{i=1}^n}$ is a monotonic decreasing sequence,*
2. *m.d.s.-complete iff $\mathcal{S}_{(\sigma_i, \Phi_i)_{i=1}^n}$ is m.d.s. and $\{(k_1, ..., k_n) \in \mathbb{N}^n \mid k_1 \geq ... \geq k_n\} \subseteq \mathcal{S}_{(\sigma_i, \Phi_i)_{i=1}^n}$,*
3. *coherent iff there is no fibre $(k_1, ..., k_n) \in \mathcal{S}_{(\sigma_i, \Phi_i)_{i=1}^n}$, s.t. $k_i = \infty$ and $k_j \neq \infty$ for some indices $1 \leq i, j \leq n$ and*
4. *positive iff any fibre $(k_1, ..., k_n) \in \mathcal{S}_{(\sigma_i, \Phi_i)_{i=1}^n}$ is finite.*

These properties are interesting for the following reasons: if a spectrum for semantics $\sigma_1, \ldots, \sigma_n$ is m.d.s., then we know that whenever enforcing is possible for all of them it is at least as difficult using σ_i as it is using σ_j given that $i < j$. If it is m.d.s.-complete we know in addition that it can in fact be arbitrarily more difficult. Coherence means that whether some C is enforceable or not does not depend on the choice of the considered semantics. Positive means each set C can actually be enforced.

A few relationships among these properties are clear by definition. First, an m.d.s.-complete spectrum is m.d.s. and second, a positive spectrum is coherent. Further interpretations of the introduced properties are given in the following subsections.

3.1   The (st, ss, pr, Φ)-Spectrum ($\Phi \in \{E, N, S\}$)

In this subsection we will characterize the (st, ss, pr)-spectra w.r.t. strong, normal and arbitrary expansions. In [1, Corollary 3] it was shown that the stable (semi-stable) characteristic exceeds the semi-stable (preferred) characteristic w.r.t. any binary relation over the set of all finite AFs. Consequently, the considered spectra are m.d.s.

Quite surprisingly, the following proposition shows that the mentioned spectra are even m.d.s.-complete, i.e. the stable (semi-stable) characteristic may take values which exceed the semi-stable (preferred) characteristic by **any** natural number. In a sense this result is negative as it tells us that information about the characteristic of one semantics does not help in determining the characteristic of the other semantics: even if we know that the characteristic w.r.t. preferred semantics for a certain set D is, say, 1 (i.e., only 1 additional attack is needed), there is no possibility to give an upper bound of the characteristic w.r.t. semi-stable or stable semantics. The result underlines the independence of the considered semantics w.r.t. the minimal change problem. It indicates that the choice of the considered semantics may influence the characteristic dramatically, even though the considered semantics possess many similarities.

Proposition 1. *For any* $\Phi \in \{E, N, S\}$, $\mathcal{S}_{(st,ss,pr,\Phi)}$ *is m.d.s.-complete.*

Proof. Let $\Phi \in \{E, N, S\}$ and $k, l, m \in \mathbb{N}$, s.t. $k \geq l \geq m$. Hence, we may assume that $l = m + n$ and $k = m + n + o$ for some $n, o \in \mathbb{N}$. If we may construct AFs $\mathcal{F}$ and corresponding sets $C \subseteq A(\mathcal{F})$, s.t. $N_{st,\Phi}^{\mathcal{F}}(C) = m + n + o$, $N_{ss,\Phi}^{\mathcal{F}}(C) = m + n$ and $N_{pr,\Phi}^{\mathcal{F}}(C) = m$, then $(k, l, m) \in \mathcal{S}_{(st,ss,pr,\Phi)}$ follows. Thus, $\mathcal{S}_{(st,ss,pr,\Phi)}$ is shown to be m.d.s.-complete. We define the AF $\mathcal{F}_{m,n,o} = (A_{m,n,o}, R_{m,n,o})$ where

$$A_{m,n,o} = \{a\} \cup \{b_j \mid 1 \leq j \leq m\} \cup \{c_j, d_j, e_j \mid 1 \leq j \leq n\} \cup \{f_j \mid 1 \leq j \leq o\} \text{ and}$$

$$\begin{aligned} R_{m,n,o} = \ & \{(b_j, a), (b_j, b_j) \mid 1 \leq j \leq m\} \cup \{(c_j, d_j), (d_j, c_j), (e_j, e_j) \mid 1 \leq j \leq n\} \cup \\ & \{(d_j, e_i) \mid 1 \leq j, i \leq n\} \cup \{(d_j, b_i) \mid 1 \leq j \leq n, 1 \leq i \leq m\} \cup \\ & \{(d_j, d_i) \mid j \neq i, 1 \leq j, i \leq n\} \cup \{(f_j, f_j) \mid 1 \leq j \leq o\} \cup \\ & \{(d_j, f_i) \mid 1 \leq j \leq n, 1 \leq i \leq o\}. \end{aligned}$$

Note that if a subindex equals zero, then there are no corresponding arguments and attacks. For the sake of clarity we present here an instantiation of the presented scheme, namely $\mathcal{F}_{3,2,4}$.

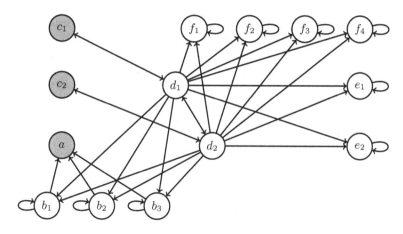

Fig. 2. The AF $\mathcal{F}_{3,2,4}$

The grey highlighted arguments belong to the set $C_2 = \{a, c_1, c_2\}$ which is an instantiation of the scheme $C_n = \{a\} \cup \{c_j \mid 1 \leq j \leq n\}$. We claim that $N_{st,\Phi}^{\mathcal{F}_{m,n,o}}(C_n) = m + n + o$, $N_{ss,\Phi}^{\mathcal{F}_{m,n,o}}(C_n) = m + n$ and $N_{pr,\Phi}^{\mathcal{F}_{m,n,o}}(C_n) = m$. By construction C_n is conflict-free in $\mathcal{F}_{m,n,o}$. Furthermore, C_n does not have proper conflict-free supersets (*). Applying the characterization theorems of [1] (Theorem 9, Def. 8) $N_{pr,\Phi}^{\mathcal{F}_{m,n,o}}(C_n) = V_{ad,S}^{\mathcal{F}_{m,n,o}}(C_n) = \left| R_{\mathcal{F}_{m,n,o}}^-(C_n) \backslash R_{\mathcal{F}_{m,n,o}}^+(C_n) \right| = |\{b_j \mid 1 \leq j \leq m\}| = m$ because these arguments are not counterattacked by C_n. In case of stable semantics $N_{st,\Phi}^{\mathcal{F}_{m,n,o}}(C_n) = V_{st,S}^{\mathcal{F}_{m,n,o}}(C_n) = \left| A(\mathcal{F}_{m,n,o}) \backslash R_{\mathcal{F}_{m,n,o}}^+(C_n) \right| = |\{b_j \mid 1 \leq j \leq m\} \cup \{e_j \mid 1 \leq j \leq n\} \cup \{f_j \mid 1 \leq j \leq o\}| = m + n + o$ since exactly these arguments are not attacked by C_n.

To see that $N_{ss,\Phi}^{\mathcal{F}_{m,n,o}}(C_n) = m + n$ is much more difficult. At first we will show that $N_{ss,E}^{\mathcal{F}_{m,n,o}}(C_n) \geq m+n$ and finally, $N_{ss,S}^{\mathcal{F}_{m,n,o}}(C_n) \leq m+n$. Consequently, $N_{ss,\Phi}^{\mathcal{F}_{m,n,o}}(C_n) = m + n$ for any $\Phi \in \{E, N, S\}$ is proven (Corollary 4 [1]). Consider the n conflict-free sets $S_n^1, ..., S_n^n$ where $S_n^j = \{a\} \cup \{c_i \mid 1 \leq i \leq n\} \backslash \{c_j\} \cup \{d_j\}$. We observe that $C_n \nsubseteq S_n^j$ for $n \geq 1$ and furthermore, $R_{\mathcal{F}_{m,n,o}}^+(C_n) \subset R_{\mathcal{F}_{m,n,o}}^+(S_n^j) = A_{m,n,o}$. Assume now $N_{ss,E}^{\mathcal{F}_{m,n,o}}(C_n) = l' < m + n$. Hence, there is an AF $\mathcal{G}$, s.t. $d(\mathcal{F}_{m,n,o}, \mathcal{G}) = l'$, $\mathcal{F}_{m,n,o} \leq \mathcal{G}$ and furthermore, there is a conflict-free superset C_n' of C_n with the property $C_n' \in \mathcal{E}_{ss}(\mathcal{G})$. In consideration of (*) we deduce that $C_n' = C_n \cup G$ where G is a set of fresh arguments. Since any semi-stable extension is admissible we conclude that each b_j has to be attacked by C_n'. This means at least m additional attacks of $\mathcal{G}$ are required for this task.

Let us consider now the remaining $l'' < n = |\{S_n^j \mid 1 \leq j \leq n\}|$ new attacks. The set S_n^j that we look for satisfies the following conditions: 1. for any $g \in G$, $(d_j, g), (g, d_j) \notin R(\mathcal{G})$, 2. $(d_j, d_j) \notin R(\mathcal{G})$, 3. for any $i \neq j$, $(c_i, d_j), (d_j, c_i) \notin R(\mathcal{G})$ as well as $(a, d_j), (d_j, a) \notin R(\mathcal{G})$ and 4. for any $g \in A(\mathcal{G}) \backslash \{A(\mathcal{F}_{m,n,o}) \cup G\}$, $(c_j, g), (g, d_j) \notin R(\mathcal{G})$. Since any new attack may eliminate at most one potential candidate we deduce that there is indeed such a S_n^j satisfying 1. - 4. We will show now that $S_n^j \cup G \in \mathcal{E}_{ad}(G)$

and $R_{\mathcal{G}}^+(C_n \cup G) \subset R_{\mathcal{G}}^+(S_n^j \cup G)$ contradicting $C'_n \in \mathcal{E}_{ss}(\mathcal{G})$. Let us consider the range $R_{\mathcal{G}}^+(C_n \cup G)$. Obviously, there is an index i, s.t. $e_i \notin R_{\mathcal{G}}^+(C_n \cup G)$ since $l'' < n$ was assumed. Note that $e_i \in R_{\mathcal{G}}^+(S_n^j \cup G)$ by construction of $\mathcal{F}_{m,n,o}$ and S_n^j. Furthermore, in consideration of the first part of condition 4. (c_j does not "reach" further arguments) we immediately conclude that $R_{\mathcal{G}}^+(C_n \cup G) \subseteq R_{\mathcal{G}}^+(S_n^j \cup G)$. Altogether, $R_{\mathcal{G}}^+(C_n \cup G) \subset R_{\mathcal{G}}^+(S_n^j \cup G)$ has to hold. Furthermore, $S_n^j \cup G$ is conflict-free in $\mathcal{G}$ for two reasons, first S_n^j satisfies conditions 1. - 3. and second, $C_n \cup G$ is assumed to be admissible and in particular, conflict-free in $\mathcal{G}$. Assume now that $S_n^j \cup G \notin \mathcal{E}_{ad}(\mathcal{G})$. This means, there is argument $g \in A(\mathcal{G})$ which attacks $S_n^j \cup G$ without being counterattacked. Since conflict-freeness is already shown and $A_{\mathcal{F}_{m,n,o}} \subseteq R_{\mathcal{G}}^+(S_n^j \cup G)$ obviously holds, we deduce $g \in A(\mathcal{G}) \backslash \{A_{m,n,o} \cup G\}$. In consideration of the second part of condition 4. ($(g, d_j) \notin R(\mathcal{G})$) it follows that g attacks some c_i with $i \neq j$ or an argument $g' \in G$. Since $C_n \cup G$ is assumed to be admissible in $\mathcal{G}$ there is an argument $c' \in C_n \cup G$, s.t. $(c', g) \in R(\mathcal{G})$. If $c' \in G$, then obviously $c' \in S_n^j \cup G$. If $c' \in C_n$, then $c' \in S_n^j \cup G$ because the second part of condition 4. ($(c_j, g) \notin R(\mathcal{G})$) guarantees $c_j \neq c'$. This means, under the assumption $N_{ss,E}^{\mathcal{F}_{m,n,o}}(C_n) = l' < m + n$ we derived a contradiction, namely $C'_n \in \mathcal{E}_{ss}(\mathcal{G}) \wedge C'_n \notin \mathcal{E}_{ss}(\mathcal{G})$. Hence, $N_{ss,E}^{\mathcal{F}_{m,n,o}}(C_n) \geq m + n$ is shown.

Let us prove now that $N_{ss,S}^{\mathcal{F}_{m,n,o}}(C_n) \leq m + n$. Consider therefore a fresh argument c and the AF $\mathcal{G}_{m,n} = (A_{m,n,o} \cup \{c\}, R_{m,n,o} \cup \{(c, b_j) \mid 1 \leq j \leq m\} \cup \{(c, d_j) \mid 1 \leq j \leq n\})$. One can easily verify that $C_n \cup \{c\} \in \mathcal{E}_{ss}(\mathcal{G}_{m,n})$ and furthermore, $\mathcal{F} \leq_S \mathcal{G}_{m,n}$. Since $d(\mathcal{F}_{m,n,o}, \mathcal{G}_{m,n}) = m + n$ we conclude $N_{ss,S}^{\mathcal{F}_{m,n,o}}(C_n) \leq m + n$. Finally, $N_{ss,\Phi}^{\mathcal{F}_{m,n,o}}(C_n) = m + n$ for any $\Phi \in \{E, N, S\}$ is proven.

The following proposition shows that the spectrum $\mathcal{S}_{(st,ss,pr,\Phi)}$ is coherent, i.e. any fibre either possesses finite values or all values equal infinity. This means, under the considered semantics it is impossible that a set C may be enforced w.r.t. a semantics σ and simultaneously, C is not enforceable w.r.t. another semantics τ. Furthermore, we show that the considered spectra are not positive, i.e. there are unenforcable sets.

Proposition 2. *For any $\Phi \in \{E, N, S\}$, $\mathcal{S}_{(st,ss,pr,\Phi)}$ is coherent but not positive.*

Proof. Given $\Phi \in \{E, N, S\}$. First, we will prove the coherence of $\mathcal{S}_{(st,ss,pr,\Phi)}$. Since $\mathcal{S}_{(st,ss,pr,\Phi)}$ is already shown to be m.d.s.-complete it suffices to prove that for any fibre $(k, l, m) \in \mathcal{S}_{(st,ss,pr,\Phi)}$, if $m < \infty$, then $l < \infty$ and if $l < \infty$, then $k < \infty$. Let $m < \infty$. Hence there is an AF $\mathcal{F}$ and a set $C \subseteq A(\mathcal{F})$, s.t. $N_{pr,\Phi}^{\mathcal{F}}(C) = m$. This means, C has to be conflict-free in $\mathcal{F}$. Applying Corollary 7 in [1] we deduce $l = N_{ss,S}^{\mathcal{F}}(C) \leq |A(\mathcal{F}) \backslash C| < \infty$. Since $N_{ss,S}^{\mathcal{F}}(C) \geq N_{ss,N}^{\mathcal{F}}(C) \geq N_{ss,E}^{\mathcal{F}}(C)$ (compare Corollary 4 [1]) holds we are done. In the same way one may show that $l < \infty$ implies $k < \infty$.

To prove that $\mathcal{S}_{(st,ss,pr,\Phi)}$ is not positive it suffices to construct a non-finite fibre. Consider therefore $\mathcal{F} = (\{a\}, \{(a, a)\})$ and $C = \{a\}$. Since C does not possess conflict-free supersets we deduce $N_{ad,\Phi}^{\mathcal{F}}(C) = \infty$ (compare Theorem 9, Def. 6 [1]). Furthermore, by Prop. 5 [1] we get $(\infty, \infty, \infty) \in \mathcal{S}_{(st,ss,pr\Phi)}$ concluding the proof.

The following Theorem summarizes the earlier results. Note that the listed properties fully characterize the considered spectra. This means, it is decidable whether an arbitrary fibre belongs to the considered spectra.

Theorem 1. *For any* $\Phi \in \{E, N, S\}$, $\mathcal{S}_{(st,ss,pr,\Phi)}$ *is coherent, m.d.s.-complete but not positive.*

3.2 Properties of the (st, ss, pr, W)-Spectrum

The following example taken from [1] shows some first and notable differences between the coinciding spectra w.r.t. normal, strong and arbitrary expansions and the spectrum w.r.t. weak expansions considered in this section.

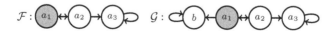

Fig. 3. Non-coherence of the Weak Spectrum

The AFs above exemplify that the (st, ss, pr, W)-spectrum is not coherent since $N^{\mathcal{F}}_{st,W}(\{a_1\}) = \infty$ (unenforceable) and $N^{\mathcal{F}}_{pr,W}(\{a_1\}) = 0$ (already accepted). Furthermore, $1 \leq N^{\mathcal{F}}_{ss,W}(\{a_1\}) \leq 2$ because $\{a_1\}$ and all its proper supersets are not semi-stable in $\mathcal{F}$ but $\{a_1\}$ is semi-stable in $\mathcal{G}$.

Unfortunately, (up to now) there are no characterization theorems for semi-stable semantics. Nevertheless, with the help of the following impossibility result it is shown that $N^{\mathcal{F}}_{ss,W}(\{a_1\}) = 2$ holds. This means, if a desired set of arguments D is not already contained in a semi-stable extension of the initial framework, then the minimal effort needed to enforce D is at least 2 in case of weak expansions.

Proposition 3. $(n, 1, m) \notin \mathcal{S}_{(st,ss,pr,W)}$ *for each* $n, m \in \mathbb{N}_\infty$.

Proof. Since n, m are assumed to be arbitrary natural numbers or ∞ it suffices to prove that $(1) \notin \mathcal{S}_{(ss,W)}$. Assume $(1) \in \mathcal{S}_{(ss,W)}$, i.e. there is an AF $\mathcal{F}$ and a set C with the property $N^{\mathcal{F}}_{ss,W}(C) = 1$. This means there is an AF $\mathcal{G}$, s.t. $\mathcal{F} \leq_W \mathcal{G}$, $d(\mathcal{F}, \mathcal{G}) = 1$ and a set $C' \supseteq C$ with $C' \in \mathcal{E}_{ss}(\mathcal{G})$. W.l.o.g. $C' = D \cup E$ where $C \subseteq D \subseteq A(\mathcal{F})$ and $E \subseteq A(\mathcal{G}) \backslash A(\mathcal{F})$. Since every semi-stable extension is admissible we deduce $D \in \mathcal{E}_{ad}(\mathcal{F})$. Furthermore, $N^{\mathcal{F}}_{ss,\leq_W}(C) = 1 \neq 0$ implies there is an admissible set D' in $\mathcal{F}$, s.t. $R^+_{\mathcal{F}}(D) \subset R^+_{\mathcal{F}}(D')$ (*). We will show now that $D \cup E \notin \mathcal{E}_{ss}(\mathcal{G})$ by proof by cases. Let (d, e) be the new attack. Note that $e \in A(\mathcal{G}) \backslash A(\mathcal{F})$ is implied since $\mathcal{F} \leq_W \mathcal{G}$ is assumed. Furthermore, $d \in D$ and $e \in E$ is impossible since $D \cup E \in cf(\mathcal{G})$.

1^{st}case: Let $d \in D \backslash D'$ and $e \notin E$. We observe $R^+_{\mathcal{G}}(D \cup E) = R^+_{\mathcal{F}}(D) \cup E \cup \{e\}$. Furthermore, E only contains isolated arguments in $\mathcal{G}$ and hence, $D' \cup E \in \mathcal{E}_{ad}(\mathcal{G})$. Because of (*) and $d \in D \backslash D'$ we conclude e is defended by D' in $\mathcal{G}$. Thus, $D' \cup E \cup \{e\} \in \mathcal{E}_{ad}(\mathcal{G})$ and obviously, $R^+_{\mathcal{G}}(D' \cup E \cup \{e\}) = R^+_{\mathcal{F}}(D') \cup E \cup \{e\}$. In consideration of (*) it follows that $R^+_{\mathcal{F}}(D) \cup E \cup \{e\} = R^+_{\mathcal{G}}(D \cup E) \subset R^+_{\mathcal{G}}(D' \cup E \cup \{e\})$ and hence, $D \cup E \notin \mathcal{E}_{ss}(\mathcal{G})$ is shown.

2^{nd}case: Let $d \in D \cap D'$ and $e \notin E$. Consequently, $D' \cup E \in \mathcal{E}_{ad}(\mathcal{G})$ and furthermore, $R^+_{\mathcal{G}}(D \cup E) = R^+_{\mathcal{F}}(D) \cup E \cup \{e\} \subset^{(*)} R^+_{\mathcal{F}}(D') \cup E \cup \{e\} = R^+_{\mathcal{G}}(D' \cup E)$ contradicting $D \cup E \in \mathcal{E}_{ss}(\mathcal{G})$.

3^{rd}case: Let $d \in D'\backslash D$ and $e \notin E$. Again, $D' \cup E \in \mathcal{E}_{ad}(\mathcal{G})$ holds and furthermore, $R_{\mathcal{G}}^+(D \cup E) = R_{\mathcal{F}}^+(D) \cup E \subset^{(*)} R_{\mathcal{F}}^+(D') \cup E \cup \{e\} = R_{\mathcal{G}}^+(D' \cup E)$ in contradiction to $D \cup E \in \mathcal{E}_{ss}(\mathcal{G})$.

4^{th}case: Let $d \in D'\backslash D$ and $e \in E$. Hence, $D' \cup (E\backslash\{e\}) \in \mathcal{E}_{ad}(\mathcal{G})$. Furthermore, $R_{\mathcal{G}}^+(D \cup E) = R_{\mathcal{F}}^+(D) \cup E \subset^{(*)} R_{\mathcal{F}}^+(D') \cup (E\backslash\{e\}) \cup \{e\} = R_{\mathcal{F}}^+(D) \cup E = R_{\mathcal{G}}^+(D' \cup E)$. Consequently, $D \cup E \notin \mathcal{E}_{ss}(\mathcal{G})$ is shown.

5^{th}case: Let $d \in A(\mathcal{F})\backslash(D' \cup D)$ and $e \in E$. Thus, D has to counterattack d in $\mathcal{F}$ since $D \cup E$ is assumed to be admissible in $\mathcal{G}$. Let $d' \in D$ be the counterattacker of d. If $d' \in D'$ we conclude $D' \cup E \in \mathcal{E}_{ad}(\mathcal{G})$. If not, it follows the existence of an argument $d'' \in D'$, s.t. $(d'',d) \in R(\mathcal{F})$ since (*) is assumed. Again, we get $D' \cup E \in \mathcal{E}_{ad}(\mathcal{G})$. In both cases, $R_{\mathcal{F}}^+(D \cup E) = R_{\mathcal{F}}^+(D) \cup E \subset^{(*)} R_{\mathcal{F}}^+(D') \cup E = R_{\mathcal{G}}^+(D' \cup E)$ contradicting $D \cup E \in \mathcal{E}_{ss}(\mathcal{G})$.

6^{th}case: Let $d \in A(\mathcal{F})\backslash(D' \cup D)$ and $e \notin E$. Consequently, $D' \cup E \in \mathcal{E}_{ad}(\mathcal{G})$ and thus, $R_{\mathcal{G}}^+(D \cup E) = R_{\mathcal{F}}^+(D) \cup E \subset^{(*)} R_{\mathcal{F}}^+(D') \cup E = R_{\mathcal{G}}^+(D' \cup E)$ in contradiction to $D \cup E \in \mathcal{E}_{ss}(\mathcal{G})$.

7^{th}case: Let $d,e \in A(\mathcal{G})\backslash A(\mathcal{F})$. Since $d(\mathcal{F},\mathcal{G}) = 1$ it follows that $e \notin E$. Consequently, $D' \cup E \in \mathcal{E}_{ad}(\mathcal{G})$ and furthermore, $R_{\mathcal{G}}^+(D \cup E) = R_{\mathcal{F}}^+(D) \cup R_{\mathcal{G}}^+(E) \subset^{(*)} R_{\mathcal{F}}^+(D') \cup R_{\mathcal{G}}^+(E) = R_{\mathcal{G}}^+(D' \cup E)$ contradicting $D \cup E \in \mathcal{E}_{ss}(\mathcal{G})$.

The proposition above and its usage for the illustrated problem, namely determining the characteristic in a certain argumentation scenario, underline that the investigation of spectra reveals important insights into the minimal change problem. The following impossibility result reveals a further surprising interrelation between the considered semantics, namely that for any $\mathcal{F}$ and any set of arguments E it is impossible that E is already contained in a preferred extension yet unenforceable using semi-stable semantics.

Proposition 4. $(\infty, \infty, 0) \notin \mathcal{S}_{(st,ss,pr,W)}$.

Proof. We will show the stronger result, namely $(\infty, 0) \notin \mathcal{S}_{(ss,pr,W)}$. Assume $(\infty, 0) \in \mathcal{S}_{(ss,pr,W)}$, i.e. there is an AF $\mathcal{F}$ and a set C with the property $N_{pr,\leq_W}^{\mathcal{F}}(C) = 0$ and $N_{ss,\leq_W}^{\mathcal{F}}(C) = \infty$. This means, there exists a set $C' \supseteq C$ with $C' \in \mathcal{E}_{pr}(\mathcal{F})$. Since all considered AFs are assumed to be finite we deduce $C' = \{c_1', ..., c_n'\}$ for some $n \in \mathbb{N}$. Let $D = \{d_1, ..., d_n\}$ be a set of fresh arguments and consider $\mathcal{G} = (A(\mathcal{F}) \cup D, R(\mathcal{F}) \cup \{(d_i, d_i), (c_i', d_i) \mid 1 \leq i \leq n\})$. Obviously, $d(F,G) = 2n$ and $F \leq_W G$. Furthermore, the range of C' in $\mathcal{G}$ includes the set D and obviously, no proper subset of C' possess this property too. Consequently, there is no $C'' \in \mathcal{E}_{ad}(\mathcal{G})$, s.t. $R_{\mathcal{G}}^+(C') \subset R_{\mathcal{G}}^+(C'')$ because C' is also preferred in $\mathcal{G}$. Hence, $C' \in \mathcal{E}_{ss}(\mathcal{G})$ contradicting the assumption.

In the light of Prop. 4 the corresponding question about the fibres (∞, ∞, ∞), $(\infty, 0, 0)$ and $(0, 0, 0)$ arises. The following proposition gives the (positive) answer:

Proposition 5. $\{(\infty, \infty, \infty), (\infty, 0, 0), (0, 0, 0)\} \subseteq \mathcal{S}_{(st,ss,pr,W)}$.

Proof. Consider the AFs $\mathcal{F}_1 = (\{a\}, \{(a,a)\})$, $\mathcal{F}_2 = (\{a,b\}, \{(b,b)\})$ and $\mathcal{F}_3 = (\{a\}, \varnothing)$. In consideration of Theorem 6 and Definition 7 [1] one may easily verify that the set $\{a\}$ possesses the claimed fibres w.r.t. the AFs $\mathcal{F}_1$, $\mathcal{F}_2$ and $\mathcal{F}_3$.

We have already shown that the minimal effort w.r.t. semi-stable semantics and weak expansions needed to enforce a desired set C cannot be 1. This raises the question about other natural numbers lying between 2 and ∞. The following proposition proves that there are infinitely many numbers n between 2 and ∞, s.t. $(\infty, n, 0)$ is a fibre of the (st, ss, pr, W)-spectrum.

Proposition 6. *For any natural number $n \in \mathbb{N}$ there exists $k \in \mathbb{N}$, such that $n \leq k \leq 2n$ and $(\infty, k, 0) \in \mathcal{S}_{(st,ss,pr,W)}$.*

Proof. We define the AF $\mathcal{F}_{\infty,n,0} = (A_{\infty,n,0}, R_{\infty,n,0})$ where

$A_{\infty,n,0} = \{c_j, d_j, e_j \mid 1 \leq j \leq n\}$ and

$R_{\infty,n,0} = \{(c_j, d_j), (d_j, c_j), (d_j, e_j)(e_j, e_j) \mid 1 \leq j \leq n\} \cup \{(d_i, e_j) \mid 1 \leq i, j \leq n\}.$

For the sake of clarity we present here an instantiation of the presented scheme, namely $\mathcal{F}_{\infty,3,0}$.

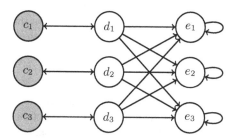

Fig. 4. The AF $\mathcal{F}_{\infty,3,0}$

The grey highlighted arguments belong to the set $C_3 = \{c_1, c_2, c_3\}$ which is an instantiation of the scheme $C_n = \{c_j \mid 1 \leq j \leq n\}$ (D_n, E_n are defined analogously). We claim that $N^{\mathcal{F}_{\infty,n,0}}_{st, \leq_W}(C_n) = \infty$ and $N^{\mathcal{F}_{\infty,n,0}}_{pr, \leq_W}(C_n) = 0$. We observe that no superset of C_n is stable in $\mathcal{F}_{\infty,n,0}$ and furthermore, C_n itself is preferred in $\mathcal{F}_{\infty,n,0}$. Consequently (Theorem 6, Def. 7), the characteristics of C_n in case of stable and preferred semantics hold as claimed.

Consider now the semi-stable semantics. At first we will show that $N^{\mathcal{F}_{\infty,n,0}}_{ss, \leq_W}(C_n) \geq n$. (proof by contradiction) Assume $N^{\mathcal{F}_{\infty,n,0}}_{ss, \leq_W}(C_n) = n' < n$. This means, there is an AF $\mathcal{G}$, s.t. $d(\mathcal{F}_{\infty,n,0}, \mathcal{G}) = n'$, $\mathcal{F}_{\infty,n,0} \leq_W \mathcal{G}$ and furthermore, there is a superset C'_n of C_n, s.t. $C'_n \in \mathcal{E}_{ss}(\mathcal{G})$. We deduce that $C'_n = C_n \cup G$ where G is a set of fresh arguments since we consider weak expansions and furthermore, C_n does not possess proper supersets which are conflict-free in $\mathcal{F}_{\infty,n,0}$.

Since $n' < n$ is assumed it follows that there has to be an index j, s.t. $c_j \in C_n$ does not possesses attacks to arguments in $A(\mathcal{G}) \backslash A_{\infty,n,0}$ (1) and $d_j \in D_n$ does not possesses attacks to arguments in G (2). Consider $S^j_n = \{c_i \mid 1 \leq i \leq n\} \backslash \{c_j\} \cup \{d_j\}$. Obviously, $R^+_{\mathcal{F}_{\infty,n,0}}(C_n) \subset R^+_{\mathcal{F}_{\infty,n,0}}(S^j_n) = A_{\infty,n,0}$ (3). We will show now that $S^j_n \cup G$

is admissible in $\mathcal{G}$ and it possesses a strictly greater range than C_n' in $\mathcal{G}$. Since we assumed $C_n' \in \mathcal{E}_{ss}(\mathcal{G})$, (2) and we are considering weak expansions the conflict-freeness of $S_n^j \cup G$ in $\mathcal{G}$ is implied. Furthermore, admissibility of $S_n^j \cup G$ in $\mathcal{G}$ holds because S_n^j is admissible in $\mathcal{F}_{\infty,n,0}$ and all potential attackers of arguments in G are counterattacked by at least one argument in $S_n^j \cup G$ (d_j counterattacks any e_i, any d_i where $i \neq j$ is counterattacked by c_i, an attacker $g' \in A(\mathcal{G}) \setminus \{A_{\infty,n,0} \cup G\}$ is counterattacked by some $g \in G$ or some $c_i \in S_n^j$ because of the admissibility of C_n' and property (1)). Finally, $R_{\mathcal{G}}^+(C_n \cup G) \subset R_{\mathcal{G}}^+(S_n^j \cup G)$ has to hold because of properties (1) and (3). This contradicts the assumption that $C_n \cup G$ is semi-stable in $\mathcal{G}$.

Let us prove now that $N_{ss,\leq_S}^{\mathcal{F}_{\infty,n,0}}(C_n) \leq 2n$. Let $C_n' = \{c_1', ..., c_n'\}$ a set of fresh arguments and consider $\mathcal{G} = (A_{\infty,n,0} \cup C_n', R_{\infty,n,0} \cup \{(c_i',c_i'),(c_i,c_i') \mid 1 \leq i \leq n\})$. Obviously, $d(F_{\infty,n,0}, G) = 2n$ and $F_{\infty,n,0} \leq_W G$. One can easily verify that $C_n \in \mathcal{E}_{ss}(\mathcal{G})$. Finally, $N_{ss,\leq_W}^{\mathcal{F}_{\infty,n,0}}(C_n) \leq 2n$ is shown.

It is an open question whether each number greater than 1 can appear as the characteristic of semi-stable semantics in a fibre, i.e. whether $\{(\infty, k, 0) \mid 2 \leq k < \infty\} \subseteq \mathcal{S}_{(st,ss,pr,W)}$. We would like to recall that it is already shown that in case of stable and preferred semantics, either a desired set C is already contained in an extension or C is not enforceable [1]. Consequently, an affirmative answer of the open question would imply a complete characterization of the (st, ss, pr, W)-spectrum.

3.3 A Note on the $(st, ss, pr, \mathcal{U})$-Spectrum

We use $\mathcal{U}$ to denote the universal relation among argumentation frameworks. In other words, we allow for arbitrary modifications including deletions of attacks and arguments. What consequences does this have for the corresponding spectrum? In contrast to the other considered spectra the $(st, ss, pr, \mathcal{U})$-spectrum is the first one proven to be positive. This means there are no cases where the enforcing of a certain set D is impossible. Furthermore, the $(st, ss, pr, \mathcal{U})$-spectrum is m.d.s. in analogy to the spectra w.r.t. arbitrary, normal and strong expansions.

Proposition 7. *The spectrum $\mathcal{S}_{(st,ss,pr,\mathcal{U})}$ is positive and m.d.s.*

Proof. Both properties follow immediately by applying Proposition 11 (positive) and Corollary 3 (m.d.s.) in [1]

A detailed analysis of the $(st, ss, pr, \mathcal{U})$-spectrum is part of future work. Due to the multitude of possibilities to modify a certain argumentation scenario if arbitrary modifications are allowed it is a hard task to show further properties. We want to mention that we conjecture that the considered spectrum is m.d.s.-complete (but were unable to find a proof so far).

4   Related Work and Conclusions

In this paper we presented various new results regarding the minimal change problem for Dung's abstract AFs. We introduced the so-called spectra which describe, for a collection of chosen semantics, the range of possible minimal efforts needed to enforce a set of arguments. We focused on stable, semi-stable and preferred semantics and were able to fully characterize the spectra for strong, normal and arbitrary expansions. This analysis revealed the surprising result that, although the three semantics are closely related, it may be arbitrarily more difficult to enforce arguments using stable rather than semi-stable semantics, and also using semi-stable rather than preferred semantics. The analysis of the spectrum for weak expansions turned out to be more difficult. Nevertheless, we were able to prove several useful (im)possibility results.

The presented work continues existing research on the dynamics of abstract argumentation. The paper [4] defines general principles (postulates) individual approaches may satisfy. The principles are illustrated for the special case of the grounded extension. Principles for the multiple extension case are left to further research. The authors of [5] focus on a particular type of change, namely the *addition of a single new argument* which interacts with previous arguments. They study the impact of such additions on the outcome of the argumentation framework, more particularly on the set of its extensions. The closely related paper [7] contains a theoretical study of the impact the *removal* of a single argument may have on the set of extensions of an argumentation framework. The article [8] develops a general method for handling updates of AFs based on a division. In a nutshell, the updated AF is divided into three parts: an unaffected, an affected, and a conditioning part. The status of arguments in the unaffected sub-framework remains unchanged, while the status of the affected arguments is computed in a special AF composed of an affected part and a conditioning part. It is shown that for specific semantics the extensions of the updated framework can be computed by combining the obtained results.

Booth and colleagues [12] investigated several quantitative distance measures for argumentation. In contrast to our work where the focus is on distances among different argumentation frameworks, the distance in that paper measures how far apart two labellings representing two complete extensions of the same argumentation framework are. This has applications in argument-based belief revision (e.g. if an agent is forced to switch to another extension and tries to identify the one closest to his original extension) and in judgement aggregation. Although the goals of this work are different from ours, it remains to be seen whether results from that work can be reused for our purposes.

Baroni et al. [13] introduce so-called input/output argumentation frameworks, an approach to characterize the behavior of an argumentation framework as sort of a black box with a well-defined external interface. The paper defines the notion of semantics decomposability and analyzes complete, stable, grounded and preferred semantics in this regard. It turns out that, under grounded, complete, stable and credulous preferred semantics, input/output argumentation frameworks with the same behavior can be exchanged without affecting the results of the evaluation of other interacting arguments. Since replaceability is one of the main motivations for studying equivalence notions, we plan to explore connections between equivalence and decomposability in the near future.

To the best of our knowledge the kind of questions analyzed in this paper have not been addressed before in argumentation. The analysis of spectra opens a number of new research directions which we want to pursue in the future. As just mentioned, the full characterization of the weak expansion spectrum is still open. Secondly, it would be useful to include further semantics (like grounded or ideal [3, 14]) in the analysis of spectra. Finally, it would be interesting to consider also a stronger form of enforcement where the enforced set of arguments has to be contained in *all* extensions rather than in *some* extension. We also might want to enforce a set of arguments C and at the same time exclude another, disjoint set D, that is, we might be interested in modifications leading to an AF possessing an extension E such that $C \subseteq E$ and $E \cap D = \varnothing$.

References

1. Baumann, R.: What does it take to enforce an argument? Minimal change in abstract argumentation. In: ECAI, pp. 127–132 (2012)
2. Bench-Capon, T.J.M., Dunne, P.E.: Argumentation in artificial intelligence. Artificial Intelligence 171(10-15), 619–641 (2007)
3. Dung, P.M.: On the acceptability of arguments and its fundamental role in nonmonotonic reasoning, logic programming and n-person games. Artificial Intelligence 77(2), 321–357 (1995)
4. Boella, G., Kaci, S., van der Torre, L.: Dynamics in argumentation with single extensions: Abstraction principles and the grounded extension. In: Sossai, C., Chemello, G. (eds.) EC-SQARU 2009. LNCS (LNAI), vol. 5590, pp. 107–118. Springer, Heidelberg (2009)
5. Cayrol, C., Dupin de Saint-Cyr, F., Lagasquie-Schiex, M.C.: Change in abstract argumentation frameworks: adding an argument. Journal of Artificial Intelligence Research 38, 49–84 (2010)
6. Baumann, R., Brewka, G.: Expanding argumentation frameworks: Enforcing and monotonicity results. In: Proc. COMMA 2010, pp. 75–86. IOS Press (2010)
7. Bisquert, P., Cayrol, C., de Saint-Cyr, F.D., Lagasquie-Schiex, M.-C.: Change in argumentation systems: Exploring the interest of removing an argument. In: Benferhat, S., Grant, J. (eds.) SUM 2011. LNCS, vol. 6929, pp. 275–288. Springer, Heidelberg (2011)
8. Liao, B., Jin, L., Koons, R.C.: Dynamics of argumentation systems: A division-based method. Artificial Intelligence 175(11), 1790–1814 (2011)
9. Scholz, H.: Ein ungelöstes Problem in der symbolischen Logik. Journal of Symbolic Logic 17, 160 (1952)
10. Caminada, M.W.: Semi-stable semantics. In: Dunne, P.E., Bench-Capon, T.J. (eds.) Computational Models of Argument. Frontiers in AI and Applications, vol. 144, pp. 121–130. IOS Press (2006)
11. Baroni, P., Giacomin, M.: Characterizing defeat graphs where argumentation semantics agree. In: Simari, G., Torroni, P. (eds.) 1st International Workshop on Argumentation and Non-Monotonic Reasoning, pp. 33–48 (2007)
12. Booth, R., Caminada, M., Podlaszewski, M., Rahwan, I.: Quantifying disagreement in argument-based reasoning. In: Proc. AAMAS 2012, pp. 493–500 (2012)
13. Baroni, P., Boella, G., Cerutti, F., Giacomin, M., van der Torre, L.W.N., Villata, S.: On input/output argumentation frameworks. In: COMMA, pp. 358–365 (2012)
14. Dung, P., Kowalski, R., Toni, F.: Dialectic proof procedures for assumption-based, admissible argumentation. Artificial Intelligence 170(2), 114–159 (2006)

Normalizing Cardinality Rules
Using Merging and Sorting Constructions*

Jori Bomanson and Tomi Janhunen

Helsinki Institute for Information Technology HIIT
Department of Information and Computer Science
Aalto University, FI-00076 AALTO, Finland
{Jori.Bomanson,Tomi.Janhunen}@aalto.fi

Abstract. Answer-set programs become more expressive if extended by cardinality rules. Certain implementation techniques, however, presume the translation
of such rules back into normal rules. This has been previously realized using a
BDD-based transformation which may produce a quadratic number of rules in the
worst case. In this paper, we present two further constructions which are based on
Boolean circuits for merging and sorting and which have been considered, e.g.,
in the context of the propositional satisfiability (SAT) problem and its extensions.
Such circuits can be used to express cardinality constraints in a more compact
way. Thus, in order to normalize cardinality rules, we first develop an ASP encoding of a sorting circuit, on top of which the second translation, one encoding
a selection circuit, is devised. Because sorting is more general than cardinality
checking, we also present ways to prune the resulting sorting and selection programs. The experimental part illustrates the compactness of the new normalizations and points out cases where computational performance is improved.

1 Introduction

Answer-set programming (ASP) [4] is a declarative programming paradigm whose syntax is based on different kinds of rules. The semantics of programs is based on *stable
models* [9], also known as *answer sets*, which are typically in a tight correspondence
with the solutions of a problem being solved by the programmer. The extended rule
types, such as *choice*, *cardinality*, and *weight* rules [15], enable more compact encodings in contrast with *normal* rules which form the basic syntax for ASP. Answer sets are
typically computed using answer set solvers such as CLASP [8] which natively supports
extended rule types in its data structures. However, there are alternative approaches to
compute answer sets, e.g., by translating rules into propositional clauses so that answer sets are captured with satisfying assignments [10,11,14]. Then satisfiability (SAT)
solvers can be used for actual computations. Since clauses stand for simple disjunctive
conditions, it is difficult to support extended rule types in such transformations directly.
One viable approach is to translate such extensions back to normal rules before clausi-
fication. Cardinality rules are also selectively rewritten by native ASP solvers for better

---
* The support from the Finnish Centre of Excellence in Computational Inference Research
  (COIN) funded by the Academy of Finland (under grant #251170) is gratefully acknowledged.

P. Cabalar and T.C. Son (Eds.): LPNMR 2013, LNAI 8148, pp. 187–199, 2013.

performance. In this paper, we call such translation steps *normalization* and concentrate on the normalization of cardinality rules frequently arising in applications.

The body of a cardinality rule involves a *bound* $k \geq 0$ and a list of n literals out of which *at least* k ought to be satisfied.[1] The normalization of this condition is non-trivial except in certain corner cases such as $k \leq 1$ or $k \geq n - 1$. If no new atoms are allowed, a straightforward rewriting yields $\binom{n}{k}$ normal rules which becomes infeasible already for relatively small values of n and in particular when k is close to $n/2$. However, by introducing auxiliary atoms, the cardinality condition in question can be expressed using a polynomial number of rules. For instance, the translation based on *binary decision diagrams* (BDDs) [7] yields of the order of $k \times (n - k)$ normal rules [11]. Again, the worst case behavior of this translation is observed when k and, thus also $n - k$, is close to $n/2$ and the number of normal rules is quadratic in k. In certain application domains, cardinality constraints with thousands of literals are possible which implies that normalization using the BDD scheme would yield millions of rules. The sizes of ground programs used at ASP competitions suggest that such high numbers of rules are hard to deal with efficiently. These observations motivate the main goal of this work, i.e., to find more compact ways to represent cardinality rules in terms of normal rules.

Cardinality constraints have also appeared in other forms such as *pseudo-Boolean constraints* which are typically implemented via translations into SAT [2] or its extensions. The complexities of transformations into pure clauses vary from linear to exponential (see, e.g., [1] for an account). However, as regards exploiting these transformations in the context of ASP, it is highly beneficial if the logical function involved is is monotone, i.e., increasing the number of 1's in input can only increase the number of 1's in output. Then, it is easier to establish substitution properties under stable model semantics. These considerations suggested us to follow the approach of Asín et al. [1] who deploy *merging* and *sorting* circuits as a basis of yet another translation. The translation implements Batcher's *odd-even merge sort* [3] for Boolean values $v_1, \ldots, v_n$ which are sorted by merging the respective odd and even subsequences $v_1, v_3, \ldots$ and $v_2, v_4, \ldots$ first sorted recursively. One may implement a *cardinality check* $k \leq |S|$ for an arbitrary subset S of some base set $B = \{b_1, \ldots, b_n\}$ of interest by sorting the values of membership tests $b_1 \in S, \ldots, b_n \in S$ and by inspecting the kth value in the result. Asín et al. [1] call such circuits *cardinality networks* and show that the number of clauses required to represent them is proportional to $n \times (\log k)^2$. Since cardinality networks are solely based on logical and/or operations, they are inherently monotonic—paving the way for exploiting cardinality networks and their main components, viz. *merging* and *sorting* circuits, in normalization.

The rest of this paper is organized in the following way. First, Section 2 defines the syntax and semantics of cardinality constraint programs. The basic primitives employed in the normalization of cardinality rules, namely *merging* and *sorting* programs, are then introduced in Section 3. Actual normalizations, as to be explained in Section 4, are based on sorting programs and their refinement, viz. *selection* programs. Depending on the bound used in the cardinality rule, only possibly small portions of such programs are needed in practice. We discuss a technique to produce such restricted programs as part of the new normalization procedure proposed in this paper. In Section 5, we present the

---

[1] Analogous upper bounds can be expressed using lower bounds.

correctness arguments for the normalization step. We use the concept of *visible strong equivalence* [12] to establish that the rule being normalized can be replaced by the respective normal rules in any reasonable context. Section 6 is devoted to experimental evaluation where we compare the new method with the one based on BDDs. To this end, we are interested in both the length of normalizations and the execution time of a state-of-the-art ASP solver CLASP. The paper is concluded in Section 7.

2   Preliminaries

In this section, we briefly review the main syntactic fragments of ASP which are interesting for the purposes of this paper, viz. *normal logic programs* (NLPs) and *cardinality constraint programs* (CCPs). For the sake of simplicity, we will restrict the presentation to the *propositional* case although first-order language elements are typically used in ASP. NLPs are finite sets of rules of the form (1) where a, a_i's, b_j's, and c_k's are *propositional atoms* (or *atoms* for short) and not denotes *default negation*. Intuitively, the *head* atom a can be derived whenever the *body* of the rule is satisfied, i.e., when the *positive* body conditions $b_1, \ldots, b_n$ are derivable by the other rules in the program but none of the *negative* body conditions $c_1, \ldots, c_m$ are. A *cardinality rule* (2) is similar but its body is satisfied whenever the number of satisfied body conditions is at least k. CCPs are finite sets of normal and/or cardinality rules. In general, a program P is *positive*, if all of its rules are negation-free, i.e., satisfy $m = 0$ in (1) and (2) below.

$$a \leftarrow b_1, \ldots, b_n, \text{ not } c_1, \ldots, \text{ not } c_m. \tag{1}$$

$$a \leftarrow k \le \{b_1, \ldots, b_n, \text{ not } c_1, \ldots, \text{ not } c_m\}. \tag{2}$$

To define the semantics of programs introduced above, we write $\text{At}(P)$ for the *signature* of a program P, i.e., a finite set of atoms to which all atoms occurring in P belong to. A *positive literal*, i.e., an atom $a \in \text{At}(P)$, is *true* in an *interpretation* $I \subseteq \text{At}(P)$ of P, if $a \in I$, and *false* otherwise ($a \in \text{At}(P) \setminus I$). The body of (1) is satisfied in I, if $\{b_1, \ldots, b_n\} \subseteq I$ and $\{c_1, \ldots, c_m\} \cap I = \emptyset$. Quite similarly, the body of (2) is satisfied in I iff $k \le |\{b_1, \ldots, b_n\} \cap I| + |\{c_1, \ldots, c_m\} \setminus I|$. A rule (1), or alternatively (2), is satisfied in I iff the head a is satisfied by I, i.e., $a \in I$, whenever its body is. An interpretation $I \subseteq \text{At}(P)$ is called a *(classical) model* of P, denoted $I \models P$, iff $I \models r$ for every rule $r \in P$. Moreover, a model $M \models P$ is $\subseteq$-*minimal* iff there is no $M' \models P$ such that $M' \subset M$. A positive program P is guaranteed to have a unique minimal model which coincides with the *least model* of P [9], hereafter denoted by $\text{LM}(P)$.

The semantics of *negative literals* of the form 'not c_i' is set by first evaluating their occurrences in rule bodies. Given a program P and an interpretation $M \subseteq \text{At}(P)$, the *reduct* of P with respect to M, denoted by P^M, contains (i) a rule $a \leftarrow b_1, \ldots, b_n$ for each rule (1) of P such that $\{c_1, \ldots, c_m\} \cap M = \emptyset$ [9] and (ii) a rule $a \leftarrow k' \le \{b_1, \ldots, b_n\}$ for each rule (2) of P where the new lower bound $k' = \max(0, k - |\{c_1, \ldots, c_m\} \setminus M|)$ [15]. The outcome P^M is always a positive program. Thus an interpretation $M \subseteq \text{At}(P)$ of a *program* P is defined as a *stable model* of P iff $M = \text{LM}(P^M)$ [9,15]. The number of stable models, also known as *answer sets*, can vary in general and we let $\text{SM}(P)$ stand for the set of stable models of a program P.

Example 1. Consider a CCP P consisting of the rules $a \leftarrow 1 \leq \{b, \text{not } c\}$; $b \leftarrow 1 \leq \{a, \text{not } c\}$; and $c \leftarrow 1 \leq \{\text{not } a, \text{not } b\}$. Given $M_1 = \{c\}$, the reduct $P^{M_1} = \{a \leftarrow 1 \leq \{b\}; b \leftarrow 1 \leq \{a\}; c \leftarrow 0 \leq \{\}\}$ so that $\text{LM}(P^{M_1}) = \{c\} = M_1$. Thus M_1 is stable. The other stable model $M_2 = \{a, b\}$ of P is easy to verify from the reduct $P^{M_2} = \{a \leftarrow 0 \leq \{b\}; b \leftarrow 0 \leq \{a\}; c \leftarrow 1 \leq \{\}\}$. It can be similarly checked that no other interpretation $M \subseteq \text{At}(P)$ is stable, so that $\text{SM}(P) = \{\{a, b\}, \{c\}\}$.   ■

3   Basic Building Blocks: Mergers and Sorters

Merge sort is a well-known sorting algorithm which splits a sequence of numbers recursively into shorter subsequences and then merges them back together into longer sorted sequences until a fully sorted sequence results. This algorithmic idea lends itself to logical circuits, the main difference being that the goal is to sort sequences consisting of Boolean values 0 and 1 only. *Odd-even mergers* introduced by Batcher [3] enable the construction of more complex merging and sorting circuits as illustrated in Figure 1. The merger on the left side selects the odd and even subsequences 1011 and 1010 of its input sequence 11001110 and merges them together using two smaller mergers (one for the odd subsequence and the other for the even) and a *balanced merger* which combines the two merged subsequences 1110 and 1100 into a final result 11111000.

The idea behind the recursive design is as follows. The mergers for odd and even subsequences produce output sequences which are *balanced*: the numbers of 1's differ at most by two and never to the advantage of the even merger. Such sequences are then easy to merge with the aforementioned balanced merger. The resulting merger forms a building block of sorting circuits, e.g., the merger on the left side of Figure 1 is part of the sorter on the right, which first sorts the input halves 1010 and 0111 into 1100 and 1110, before merging them into 11111000. The circuits demonstrated above have been previously applied when expressing cardinality constraints with propositional clauses [1]. In what follows, we describe merging and sorting circuits in terms of normal rules of the form (1) only. A *merging program* is a NLP, produced by the function Merger to be defined below, having two sequences $\langle l_1, \ldots, l_n \rangle$ and $\langle l'_1, \ldots, l'_m \rangle$ of input literals and a sequence $\langle s_1, \ldots, s_{n+m} \rangle$ of output atoms. Moreover, two further sequences of auxiliary atoms $\langle d_1, \ldots, d_{n_d} \rangle$ and $\langle e_1, \ldots, e_{n_e} \rangle$ where $n_d = \lfloor \frac{n}{2} \rfloor + \lceil \frac{m}{2} \rceil$ and $n_e = \lceil \frac{n}{2} \rceil + \lfloor \frac{m}{2} \rfloor$ are used to represent intermediate merging results. Any recursive invocations of Merger will introduce such atoms, too.

Definition 1. *Given the sequences of input literals $\langle l_1, \ldots, l_n \rangle$, $\langle l'_1, \ldots, l'_m \rangle$, and the sequence of output atoms $\langle s_1, \ldots, s_{n+m} \rangle$ the function* Merger *produces a merging program whose base cases, for $n \leq 1, m \leq 1, 1 \leq n + m$, are defined by*

$$\text{Merger}_{1,0}(\langle l_1 \rangle, \langle \rangle, \langle s_1 \rangle) = \{ s_1 \leftarrow l_1 \},$$
$$\text{Merger}_{0,1}(\langle \rangle, \langle l'_1 \rangle, \langle s_1 \rangle) = \{ s_1 \leftarrow l'_1 \}, \qquad (3)$$
$$\text{Merger}_{1,1}(\langle l_1 \rangle, \langle l'_1 \rangle, \langle s_1, s_2 \rangle) = \{ s_1 \leftarrow l_1; \ s_1 \leftarrow l'_1; \ s_2 \leftarrow l_1, l'_1 \}$$

and whose recursive case, for $n + m > 2$, is

$$\mathsf{Merger}_{n,m}(\langle l_1, \ldots, l_n \rangle, \langle l'_1, \ldots, l'_m \rangle, \langle s_1, \ldots, s_{n+m} \rangle) =$$
$$\mathsf{Merger}_{\lceil \frac{n}{2} \rceil, \lceil \frac{m}{2} \rceil}(\langle l_1, l_3, \ldots, l_{2\lceil \frac{n}{2} \rceil - 1} \rangle, \langle l'_1, l'_3, \ldots, l'_{2\lceil \frac{m}{2} \rceil - 1} \rangle, \langle d_1, \ldots, d_{n_d} \rangle) \cup$$
$$\mathsf{Merger}_{\lfloor \frac{n}{2} \rfloor, \lfloor \frac{m}{2} \rfloor}(\langle l_2, l_4, \ldots, l_{2\lfloor \frac{n}{2} \rfloor} \rangle, \langle l'_2, l'_4, \ldots, l'_{2\lfloor \frac{m}{2} \rfloor} \rangle, \langle e_1, \ldots, e_{n_e} \rangle) \cup \tag{4}$$
$$\mathsf{BalancedMerger}_{n_d, n_e}(\langle d_1, \ldots, d_{n_d} \rangle, \langle e_1, \ldots, e_{n_e} \rangle, \langle s_1, \ldots, s_{n+m} \rangle)$$

where $n_d = \lceil \frac{n}{2} \rceil + \lceil \frac{m}{2} \rceil$ and $n_e = \lfloor \frac{n}{2} \rfloor + \lfloor \frac{m}{2} \rfloor$ as above and BalancedMerger *is defined by*

$$\mathsf{BalancedMerger}_{n_d, n_e}(\langle d_1, \ldots, d_{n_d} \rangle, \langle e_1, \ldots, e_{n_e} \rangle, \langle s_1, \ldots, s_{n+m} \rangle) =$$
$$\{s_1 \leftarrow d_1\} \cup \bigcup_{i=1}^{\min\{n_d - 1, n_e\}} \mathsf{Merger}_{1,1}(\langle d_{i+1} \rangle, \langle e_i \rangle, \langle s_{2i}, s_{2i+1} \rangle) \cup \tag{5}$$
$$\{s_{n+m} \leftarrow e_{n_e} \mid n_d = n_e\} \cup \{s_{n+m} \leftarrow d_{n_d} \mid n_d = n_e + 2\}.$$

The base case $n = m = 1$ of Definition 1 describes the central primitive of merging (sorted sequences) and sorting Boolean values: the first value denoted by s_1 is defined as the disjunction of l_1 and l'_1 whereas the second value denoted by s_2 is defined as their conjunction. Besides the simple "copying" operations involved in (3) and (5), the entire merging program is an amalgamation of instances of this basic primitive.

Example 2. Given two sequences $\langle l_1, l_2 \rangle$ and $\langle l_3, l_4 \rangle$ of input literals and a sequence $\langle s_1, s_2, s_3, s_4 \rangle$ of output atoms, let us expand $\mathsf{Merger}_{2,2}$ according to (4) in Definition 1. The resulting merging program effectively encodes two small mergers for the odd and even subsequences of the input and a balanced merger to combine their output:

$$\begin{array}{lllll} s_1 \leftarrow d_1. & s_2 \leftarrow d_2. & s_2 \leftarrow e_1. & s_3 \leftarrow d_2, e_1. & s_4 \leftarrow e_2. \\ & d_1 \leftarrow l_1. & d_1 \leftarrow l_3. & e_1 \leftarrow l_2. & e_1 \leftarrow l_4. \\ & d_2 \leftarrow l_1, l_3. & & e_2 \leftarrow l_2, l_4. \end{array}$$

In the above, the symbols $l_1, \ldots, l_4$ are still metavariables over literals. In practice, they would be replaced by concrete literals such as a, b, c, and 'not d'. ∎

Definition 2. *Given a set $\{l_1, \ldots, l_{n+m}\}$ of input literals and a sequence $\langle s_1, \ldots, s_{n+m} \rangle$ of output atoms, the function* Sorter *produces a sorting program whose base cases, for $1 \leq n + m \leq 2$, are defined by*

$$\mathsf{Sorter}_1(\{l_1\}, \langle s_1 \rangle) = \mathsf{Merger}_{1,0}(\langle l_1 \rangle, \langle \rangle, \langle s_1 \rangle),$$
$$\mathsf{Sorter}_2(\{l_1, l_2\}, \langle s_1, s_2 \rangle) = \mathsf{Merger}_{1,1}(\langle l_1 \rangle, \langle l_2 \rangle, \langle s_1, s_2 \rangle), \tag{6}$$

and whose recursive case, for $n + m > 2$, is defined by

$$\mathsf{Sorter}_{n+m}(\{l_1, \ldots, l_{n+m}\}, \langle s_1, \ldots, s_{n+m} \rangle) =$$
$$\mathsf{Sorter}_n(\{l_1, \ldots, l_n\}, \langle p_1, \ldots, p_n \rangle) \cup$$
$$\mathsf{Sorter}_m(\{l_{n+1}, \ldots, l_{n+m}\}, \langle q_1, \ldots, q_m \rangle) \cup \tag{7}$$
$$\mathsf{Merger}_{n,m}(\langle p_1, \ldots, p_n \rangle, \langle q_1, \ldots, q_m \rangle, \langle s_1, \ldots, s_{n+m} \rangle)$$

where $\langle p_1, \ldots, p_n \rangle$ and $\langle q_1, \ldots, q_m \rangle$ are sequences of auxiliary atoms.

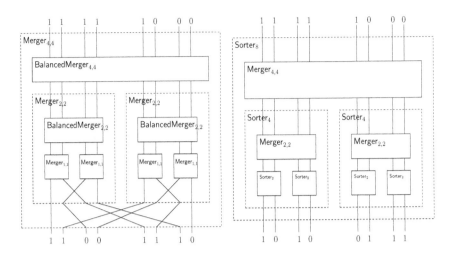

Fig. 1. A merger of size 8 and a sorter of size 8

Example 3. The smallest nontrivial sorting program with $n = m = 1$ is Sorter_2 which by (6) expands into a merging program $\text{Merger}_{1,1}$ having rules $s_1 \leftarrow l_1$; $s_1 \leftarrow l_1$; and $s_2 \leftarrow l_1, l_2$. But as discussed above, these suffice to sort sequences of length two. ∎

4   Normalization of Cardinality Rules

Our next objective is to exploit sorting programs from Section 3 in the normalization of cardinality rules. The key building block is a sorting program corresponding to the body of (2). This program is augmented by a rule $a \leftarrow s_k$ that checks whether at least k literals in the body are satisfied and makes the head a of (2) true accordingly.

Definition 3. *Given a CCP P, the* merge sort normalization $\text{MSN}(P)$ *of P contains each normal rule of P as is and for each cardinality rule of P, the rule $a \leftarrow s_k$ together with the sorting program* $\text{Sorter}_{n+m}(\{b_1, \dots, b_n, \text{not } c_1, \dots, \text{not } c_m\}, \langle s_1, \dots, s_{n+m} \rangle)$.

This definition specifies the outcome when the cardinality rules of a CCP P are substituted by the corresponding sorting programs. Each step in this process is supposed to preserve the meaning of the program subject to normalization under certain practically feasible conditions to be detailed in Section 5. The idea is that the portion of the program that stays unaltered in each step serves as a *context* in which both the rule being normalized and the respective normalization have exactly the same meaning.

Example 4. The programs $\text{Merger}_{2,2}$ and Sorter_2 from Examples 2 and 3 provide us with the subprograms required in the construction of Sorter_4. Figure 2 illustrates such a sorter which, as described above, can be used to normalize any 4-literal cardinality rule (2) when augmented by an appropriate rule $a \leftarrow s_k$ for the bound k. ∎

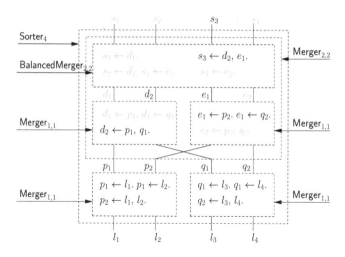

Fig. 2. The program $\mathsf{Sorter}_4(\{l_1, l_2, l_3, l_4\}, \langle s_1, s_2, s_3, s_4 \rangle)$ for encoding a 4-literal cardinality constraint. The rules and auxiliary atoms in gray are irrelevant to the output atom s_3 of interest.

Given the above scheme for normalizing cardinality rules, one may observe that substantial portions of the respective sorting programs might not be needed in practice. Indeed, only the value of s_k is of interest and thus, in a sense, a sorting program represents a *set* of cardinality rules—one for each possible body literal count. In the sequel, we explore two ways to take advantage of this fact. First, we will describe *selection programs* which essentially encode *cardinality networks* [1] whose layout is illustrated in Figure 3. Second, we will introduce symbolic evaluation strategies which are able to determine on the fly which parts of sorting and selection programs are relevant when normalizing a particular cardinality rule. These effects will be studied experimentally in Section 6. Sorting can be replaced by selection for the purpose of cardinality checking with respect to a bound k: it is sufficient to determine the k highest values of membership tests to make a decision. So let k denote the number of requested values. Moreover, given a set L of $n \geq k$ input literals, let $F_1 \sqcup \cdots \sqcup F_m$ partition L into m subsets such that $\lceil k/2 \rceil \leq n_i \leq k$ where $n_i = |F_i|$ holds for each subset F_i of input literals $i = 1, \ldots, m$.

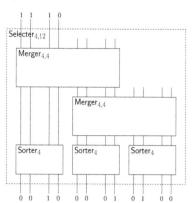

Fig. 3. A selecter that determines the highest four Boolean values of 12 inputs.

Definition 4. *Given a set L of input literals, its partitioning $F_1 \sqcup \cdots \sqcup F_m$ into m subsets, and the sequence $\langle s_{m,1}, \ldots, s_{m,k} \rangle$ of output atoms, the function $\mathsf{Selecter}$ produces, for $m \geq 2$, the selection program defined by $\mathsf{Selecter}_{k,n}(L, \langle s_{m,1}, \ldots, s_{m,k} \rangle) =$*

$$\bigcup_{i=1}^{m} \mathsf{Sorter}_{n_i}(F_i, \langle h_{i,1}, \ldots, h_{i,n_i} \rangle) \cup$$

$$\mathsf{Merger}_{n_1,n_2}(\langle h_{1,1}, \ldots, h_{1,n_1} \rangle, \langle h_{2,1}, \ldots, h_{2,n_2} \rangle, \langle s_{2,1}, \ldots, s_{2,n_1+n_2} \rangle) \cup \quad (8)$$

$$\bigcup_{i=3}^{m} \mathsf{Merger}_{k,n_i}(\langle s_{i-1,1}, \ldots, s_{i-1,k} \rangle, \langle h_{i,1}, \ldots, h_{i,n_i} \rangle, \langle s_{i,1}, \ldots, s_{i,k+n_i} \rangle)$$

where for each $1 \leq i \leq m$, the sequence $\langle h_{i,1}, \ldots, h_{i,n_i} \rangle$ formalizes the result of sorting F_i and the sequence $\langle s_{i,1}, \ldots, s_{i,t} \rangle$ captures an intermediate stage of sorting.

It is worth pointing out that for each intermediate stage $1 \leq i \leq m$, the values of $s_{i,t}$'s with $t > k$ can be discarded since they cannot affect the highest k values. Next, we state an analog of Definition 3 in the case of selection programs.

Definition 5. *Given a CCP P, the selection normalization $\mathrm{SelN}(P)$ of P contains each normal rule of P as is and for each cardinality rule of P, the rule $a \leftarrow s_k$ and the selection program $\mathsf{Selecter}_{k,n+m}(\{b_1, \ldots, b_n, \mathrm{not}\ c_1, \ldots, \mathrm{not}\ c_m\}, \langle s_1, \ldots, s_k \rangle)$.*

As discussed in the context of Figure 2 it is possible to prune sorting programs when only one output atom is of interest. Similar needs arise because the numbers of inputs are typically powers of two for balanced merging and sorting designs whereas cardinality rules (2) can have any dimensions in general. We have considered two strategies to deal with varying dimensions and potential unnecessary structure incurred. The first is based on *partial evaluation* in a bottom-up fashion. For instance, unused inputs can be assigned to 0 and the rest of the circuit can be *symbolically* evaluated to remove unnecessary structure. The second strategy is to proceed top-down and propagate special flags to distinguish which parts of the circuit are really needed and for which rules have to be generated. Due to space limitations, we have to skip a detailed discussion of how such recursive propagation takes place in practice. Nevertheless, the experimental results of Section 6 shed some light on the positive effects of such techniques.

5   Correctness Considerations

Our next objective is to establish that normalization preserves the semantics of cardinality rules (2) in a setting where a rule r is *substituted* by its normalizing program $\mathrm{MSN}(r)$ or $\mathrm{SelN}(r)$ as introduced in Section 4. To this end, we need an appropriate notion of equivalence to address such rule substitutions. *Strong equivalence* [13] was proposed exactly for this purpose, but it does not support auxiliary atoms which were numerously used when describing circuits in Section 3. To circumvent this, we resort to a recent generalization, viz. *visible strong equivalence* [12], which supports both substitutions and auxiliary atoms and has connections to the *relativized* variants of [16].

In what follows, the idea is to *hide* auxiliary atoms when it comes to comparing programs on the basis of their answer sets. Thus, for any program P, the signature $\mathrm{At}(P)$

is partitioned to its visible part $At_v(P)$ and hidden part $At_h(P)$. When comparing two programs P and Q of interest, we insist on $At_v(P) = At_v(Q)$ whereas $At_h(P)$ and $At_h(Q)$ may freely differ. As regards potential *contexts* R of P and Q, we say that P and R *mutually respect each other's hidden atoms* if $At(P) \cap At_h(R) = \emptyset$ and $At_h(R) \cap At(P) = \emptyset$, and analogously for Q and R. Such restriction, however, does not apply to P and Q which means that the hidden atoms of P and Q may overlap.

Definition 6 (Visible Strong Equivalence [12]). *Two programs P and Q are* visibly strongly equivalent, *denoted by $P \equiv_{vs} Q$, iff $At_v(P) = At_v(Q)$ and $SM(P \cup R) =_v SM(Q \cup R)$ for any context R that mutually respects the hidden atoms of P and Q.*

In the above, the relation $=_v$ insists on a strict one-to-one correspondence of models (induced by a bijection f) so that visible projections of models are preserved under f. In analogy to [13,16], there is a model-theoretic characterization of $\equiv_{vs}$ [12]. Given an interpretation $I \subseteq At(P)$, define the visible and hidden projections of I by $I_v = I \cap At_v(P)$ and $I_h = I \cap At_h(P)$, respectively. A model $M \models P$ is $At_h(P)$-minimal iff for no $N \models P$, $N_v = M_v$ and $N_h \subset M_h$, i.e., hidden atoms are *false by default*.

Definition 7 ([12]). *A VSE-model of a program P is a pair $\langle X, Y \rangle$ of interpretations where $X \subseteq Y \subseteq At(P)$ and both X and Y are $At_h(P)$-minimal models of P^Y.*

The set of VSE-models of P is denoted by $VSE(P)$. The intuition of a VSE-model $\langle X, Y \rangle$ is that Y represents a consistent context for P against which the rules of P are reduced (to form P^Y) and X captures a potential closure of P^Y. In this setting, visible atoms are treated classically whereas hidden ones are false by default so that $X_v = Y_v$ implies $X = Y$. In order to *compare* sets S of VSE-models associated with different programs, we define their *second* projections by setting $[S]_2 = \{Y \mid \langle X, Y \rangle \in S\}$.

Definition 8 ([12]). *Given programs P and Q such that $At_v(P) = At_v(Q)$, the respective sets $VSE(P)$ and $VSE(Q)$* visibly match, *denoted $VSE(P) \stackrel{v}{=} VSE(Q)$, if and only if $[VSE(P)]_2 =_v [VSE(Q)]_2$ via a bijection f and for each $Y \in [VSE(P)]_2$, $\{X_v \mid \langle X, Y \rangle \in VSE(P)\} = \{X_v \mid \langle X, f(Y) \rangle \in VSE(Q)\}$.*

Proposition 1 (Characterization of Visible Strong Equivalence [12]). *For programs P and Q with $At_v(P) = At_v(Q)$, $VSE(P) \stackrel{v}{=} VSE(Q)$ implies $P \equiv_{vs} Q$.*

The converse of Proposition 1 is not applicable due to restricted syntax used in this paper. However, the characterization of $\equiv_{vs}$ in Proposition 1 allows one to skip arbitrary context programs when showing P and Q visibly strongly equivalent (cf. Definition 6). This will be our strategy in what follows. First, we illustrate VSE-models with an example and then we provide more general arguments for the soundness of normalization.

Example 5. Let $P = \{a \leftarrow 3 \leq \{b_1, b_2, b_3, \text{not } c\}\}$ be a cardinality constraint program and Q its sorter-based normalization confining to the selected rules of Figure 2 (with literals $l_1, \ldots, l_4$ having been replaced by b_1, b_2, b_3, and not c, respectively):

$$a \leftarrow s_3. \qquad\qquad\qquad s_3 \leftarrow d_2, e_1.$$
$$d_2 \leftarrow p_1, q_1. \qquad\qquad e_1 \leftarrow p_2. \qquad e_1 \leftarrow q_2.$$
$$p_1 \leftarrow b_1. \quad p_1 \leftarrow b_2. \qquad q_1 \leftarrow b_3. \qquad q_1 \leftarrow \text{not } c.$$
$$p_2 \leftarrow b_1, b_2. \qquad\qquad q_2 \leftarrow b_3, \text{not } c.$$

Note that P could be normalized without auxiliary atoms by $a \leftarrow b_1, b_2, b_3$; $a \leftarrow b_2, b_3$, not c; $a \leftarrow b_1, b_3$, not c; and $a \leftarrow b_1, b_2$, not c but this scheme would be quadratic for $k = n + m - 1$ in (2) in general. Now, given $Y = \{a, b_2, b_3\}$, we obtain $P^Y = \{a \leftarrow 2 \leq \{b_1, b_2, b_3\}\}$ and based on these, there are ordinary SE-models of P such as $\langle X, Y \rangle$ with $X = \{b_2\}$ and $\langle Y, Y \rangle$. In total, there are 198 SE-models for P.[2] For Q with $\mathrm{At_v}(Q) = \{a, b_1, b_2, b_3, c\}$, the $\mathrm{At_h}(Q)$-minimal model related with Y is $Y' = \{a, b_2, b_3, p_1, q_1, q_2, d_2, e_1, s_3\}$. Hence the reduct $Q^{Y'}$ is obtained from the rules listed above by removing the occurrences of 'not c' satisfied by Y'. Then $\langle Y', Y' \rangle$ is the VSE-model of Q corresponding to $\langle Y, Y \rangle$: $Y = Y' \cap \mathrm{At}(P)$. For X, the respective $\mathrm{At_h}(Q)$-minimal model of $Q^{Y'}$ is $X' = \{b_2, p_1, q_1, d_2\}$ so that $X = X' \cap \mathrm{At}(P)$. It can be verified (using automated tools as above) that there are 198 VSE-models of Q which are related by a bijection $f(\langle X', Y' \rangle) = \langle X' \cap \mathrm{At}(P), Y' \cap \mathrm{At}(P) \rangle$ from $\mathrm{VSE}(Q)$ to $\mathrm{VSE}(P) = \mathrm{SE}(P)$. Thus $\mathrm{VSE}(P) \stackrel{v}{=} \mathrm{VSE}(Q)$ and $P \equiv_{vs} Q$ by Proposition 1. ∎

Proposition 2. *For a cardinality rule r of the form (2) and its sorting-based normalizations, $\mathrm{VSE}(\{r\}) \stackrel{v}{=} \mathrm{VSE}(\mathrm{MSN}(\{r\}))$ and $\mathrm{VSE}(\{r\}) \stackrel{v}{=} \mathrm{VSE}(\mathrm{SelN}(\{r\}))$.*

Proof. Let r be of the form (2) and define $B = \{b_1, \ldots, b_n\}$ and $C = \{c_1, \ldots, c_m\}$. All atoms in $\mathrm{At}(r) = \{a\} \cup B \cup C$ are assumed to be visible. Moreover, let $P = \mathrm{MSN}(\{r\})$ such that $\mathrm{At_v}(P) = \mathrm{At_v}(r) = \mathrm{At}(r)$, i.e., all auxiliary atoms introduced in the normalization of r are hidden. Let us then consider any $\langle X, Y \rangle \in \mathrm{VSE}(\{r\})$ which implies that $Y \models r$ and $X \models r^Y$ for $X \subseteq Y \subseteq \mathrm{At}(r)$. Thus $k \leq |B \cap Y| + |C \setminus Y|$ implies $a \in Y$. The reduct r^Y is a positive rule $a \leftarrow k' \leq \{b_1, \ldots, b_n\}$ where $k' = \max(0, k - |C \setminus Y|)$ so that $k' \leq |X \cap B|$ implies $a \in X$. Let $l = |B \cap Y| + |C \setminus Y|$ and define an interpretation $Y' = Y \cup H$ where H is picked so that Y' satisfies $\mathrm{Sorter}_{n+m}(\{b_1, \ldots, b_n, \mathrm{not}\, c_1, \ldots, \mathrm{not}\, c_m\}, \langle s_1, \ldots, s_{n+m} \rangle)$ minimally with respect to $\mathrm{At_h}(P)$. Given the structure of the sorting program, hidden atoms occur only positively. This makes H unique. Then suppose that $a \leftarrow s_k$ is not satisfied by Y'. It follows that $a \notin Y'$ and $s_k \in Y'$, i.e., $a \notin Y$ and $s_k \in H$. The rules of the sorting program and the minimal interpretation under H guarantee that $l \geq k$. But then $Y \not\models r$, a contradiction. Hence $Y' \models P$. Taking the reduct $P^{Y'}$ will fix the values of negative literals not $c_1, \ldots,$ not c_m subject to Y' and Y. Thus $X' = X \cup H'$ can be analogously defined so that $H' \subseteq H$, $X' \subseteq Y'$, and $X' \models P^{Y'}$. It follows that $\langle X', Y' \rangle$ is a VSE-model of P and it is a unique extension of $\langle X, Y \rangle$ since H and H' are unique.

Then let $\langle X', Y' \rangle$ be a VSE-model of P and define $\langle X, Y \rangle$ as its projection over $\mathrm{At}(r) = \{a\} \cup B \cup C$. Assuming that $Y \not\models r$ implies that $a \notin Y$ and $l \geq k$ for $l = |B \cap Y| + |C \setminus Y|$. Since $Y' \models P$, we have that $s_k \in Y'$ and $a \in Y'$ as $a \leftarrow s_k$ is in P. Thus $a \in Y$, a contradiction. It can be similarly argued that $X' \models P^{Y'}$ implies $X \models a \leftarrow k' \leq \{b_1, \ldots, b_n\}$, i.e., $X \models r^Y$. Thus $\langle X, Y \rangle$ is an SE-model of $\{r\}$.

It follows that $\mathrm{SE}(\{r\}) \stackrel{v}{=} \mathrm{VSE}(P)$. The proof for $\mathrm{SelN}(r)$ is similar, except that the sorting program is replaced by a selection program (cf. Definition 5). □

The following theorem is obtained as a corollary of Propositions 1 and 2. A simple inductive argument on the number of rule substitutions can be used to show that $P \equiv_{vs} \mathrm{MSN}(P)$ and $P \equiv_{vs} \mathrm{SelN}(P)$ hold for a CCP P in general. Thus $\mathrm{MSN}(P)$ and $\mathrm{SelN}(P)$ can be used to substitute P in contexts which respect their hidden atoms.

---

[2] These were computed using tools CLASSIC (option flag -s for strong equivalence) and CLASP.

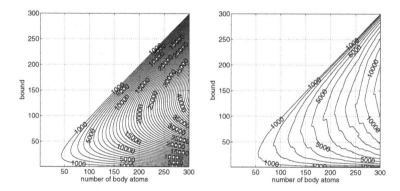

Fig. 4. Level curves describing the resulting number of rules when normalizing cardinality rules using BDDs [7] and selection programs, respectively

Theorem 1. *For a cardinality rule r of the form (2) and its sorting-based normalizations, $\{r\} \equiv_{vs} \mathrm{MSN}(\{r\})$ and $\{r\} \equiv_{vs} \mathrm{SelN}(\{r\})$.*

6 Experiments

In this section we compare normalization schemes based on simplified sorting programs, simplified selection programs, and BDDs [7] with each other. An overview of the sizes of resulting normal programs is illustrated in Figure 4. The size of a translation is measured here as the number of created normal rules. Sorting programs are not shown because their sizes closely matched those of selection. To recap previous results, in [1] it was proven that sorting networks and cardinality networks grow in size proportional to $n \times (\log n)^2$ and $n \times (\log k)^2$, respectively. Since the numbers of clauses are linearly related to the numbers of normal rules in sorting and selection programs, these bounds carry over into normalizations devised in this paper. The contrast with respect to these bounds and our experimental results speaks to the effect of the employed symbolical evaluation strategy. Furthermore, it is evident that the described normalizations fare well when k approaches its extreme values 1 and n. In contrast, when k is close to $n/2$, neither the symbolic evaluation strategy nor the use of selection programs essentially limit the number of produced rules—thus forming the worst case for normalization.

To explore the effects of normalization on solver performance we tried out all three normalization strategies on the NP-complete problem instances of the second answer set programming competition [6] (ASPCOMP-2). Each instance was solved with CLASP, once natively without normalization, and once with each normalization strategy implemented in the tool LP2NORMAL2.[3] The number of problem instances of each NP-complete problem encoding that were solved within 10 minutes and 2.79 GB of memory, matching those used in the competition, are displayed in Table 1. The results indicate that the methods differ only slightly on these problem instances, although the differences

---

[3] Our tools are available at http://research.ics.aalto.fi/software/asp/

Table 1. Numbers of solved ASPCOMP-2 benchmark problems by CLASP (v. 2.1.3) natively and after normalizations based on BDDs, sorting programs, and selection programs

Benchmark	#Inst	Native	BDD	Sorting	Selection
GraphColouring	29	9	8	8	10
WireRouting	23	22	21	23	22
Solitaire	27	19	18	19	20
Labyrinth	29	28	26	27	27
WeightBoundedDominatingSet	29	26	25	29	28
ConnectedDominatingSet	21	20	21	21	21
Other ASPCOMP-2 Problems	358	346	346	346	346
SAT		352	346	354	354
UNSAT		118	119	119	120
Summary	516	470	465	473	474

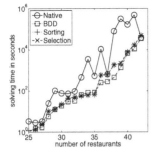

Fig. 5. Performance of CLASP (v. 2.1.3) on new unsatisfiable Fastfood instances   **Fig. 6.** Performance of CLASP in verifying strong visible equivalence (log scale)

are in favor of the new normalizations. The Fastfood benchmark of ASPCOMP-2 was further studied using new unsatisfiable problem instances. Each instance contained one cardinality rule of length equal to the number of *restaurants* in the instance and another that was longer by one. The dimensions were picked so that the bounds of these rules were as close as possible to half of the number of restaurants. The time taken by CLASP to solve these instances without and with different normalization schemes is displayed in Figure 5. The results indicate that any form of normalization is beneficial and, in particular, the one based on BDDs. Moreover, we have carried out automated equivalence checks, as a form of quality control, to ensure the correctness of normalizations produced by LP2NORMAL2 throughout its development process. These verification steps were implemented with the tools CLASSIC, LPEQ, and CLASP. Figure 6 plots with logarithmic level curves the time in seconds taken to prove that a cardinality rule (2) with $(n = m)$ is visibly strongly equivalent with its sorter-based normalization.

7   Conclusions

In this paper, we propose two new techniques to *normalize* cardinality rules frequently arising in ASP applications. In these approaches, two different Boolean circuits that essentially provide a mechanism to sort a sequence of Boolean values are described using normal rules. Such programs encode a cardinality check for bound k, if the kth value in the output is tested additionally. Our experiments indicate that sorting programs lead to a more compact scheme for normalization. Moreover, such normalizations tend to enable faster computation from time to time. We anticipate that this is due to both more concise representation and better propagation properties of sorting networks (cf. [1]).

As regards future work, there is an interesting variant, viz. *pairwise cardinality networks* [5], that are similar to cardinality networks [1]. In the pairwise networks, splitting is done before sorting which could lead to performance differences worth experimenting. In addition, *weight rules* [15] form an important generalization of cardinality rules. Given the translation of [7] based on *adder networks*, it is clear that cardinality rules provide a potential intermediate representation when normalizing such rules.

References

1. Asín, R., Nieuwenhuis, R., Oliveras, A., Rodríguez-Carbonell, E.: Cardinality networks: a theoretical and empirical study. Constraints 16(2), 195–221 (2011)
2. Bailleux, O., Boufkhad, Y., Roussel, O.: A translation of pseudo Boolean constraints to SAT. JSAT 2(1-4), 191–200 (2006)
3. Batcher, K.: Sorting networks and their applications. In: AFIPS Spring Joint Computer Conference, pp. 307–314. ACM (1968)
4. Brewka, G., Eiter, T., Truszczyński, M.: Answer set programming at a glance. Communications of the ACM 54(12), 92–103 (2011)
5. Codish, M., Zazon-Ivry, M.: Pairwise cardinality networks. In: Clarke, E.M., Voronkov, A. (eds.) LPAR-16 2010. LNCS, vol. 6355, pp. 154–172. Springer, Heidelberg (2010)
6. Denecker, M., Vennekens, J., Bond, S., Gebser, M., Truszczyński, M.: The second answer set programming competition. In: Erdem, E., Lin, F., Schaub, T. (eds.) LPNMR 2009. LNCS, vol. 5753, pp. 637–654. Springer, Heidelberg (2009)
7. Eén, N., Sörensson, N.: Translating pseudo-Boolean constraints into SAT. Journal on Satisfiability, Boolean Modeling and Computation 2(1-4), 1–26 (2006)
8. Gebser, M., Kaufmann, B., Schaub, T.: Conflict-driven answer set solving: From theory to practice. Artificial Intelligence 187, 52–89 (2012)
9. Gelfond, M., Lifschitz, V.: The stable model semantics for logic programming. In: Proceedings of ICLP 1988, pp. 1070–1080 (1988)
10. Giunchiglia, E., Lierler, Y., Maratea, M.: Answer set programming based on propositional satisfiability. Journal of Automated Reasoning 36(4), 345–377 (2006)
11. Janhunen, T., Niemelä, I.: Compact translations of non-disjunctive answer set programs to propositional clauses. In: Balduccini, M., Son, T.C. (eds.) Logic Programming, Knowledge Representation, and Nonmonotonic Reasoning. LNCS, vol. 6565, pp. 111–130. Springer, Heidelberg (2011)
12. Janhunen, T., Niemelä, I.: Applying visible strong equivalence in answer-set program transformations. In: Erdem, E., Lee, J., Lierler, Y., Pearce, D. (eds.) Correct Reasoning. LNCS, vol. 7265, pp. 363–379. Springer, Heidelberg (2012)
13. Lifschitz, V., Pearce, D., Valverde, A.: Strongly equivalent logic programs. ACM Transactions on Computational Logic 2(4), 526–541 (2001)
14. Lin, F., Zhao, Y.: ASSAT: computing answer sets of a logic program by SAT solvers. Artificial Intelligence 157(1-2), 115–137 (2004)
15. Simons, P., Niemelä, I., Soininen, T.: Extending and implementing the stable model semantics. Artificial Intelligence 138(1-2), 181–234 (2002)
16. Woltran, S.: A common view on strong, uniform, and other notions of equivalence in answer-set programming. Theory and Practice of Logic Programming 8(2), 217–234 (2008)

Experience Based Nonmonotonic Reasoning

Daniel Borchmann

TU Dresden

Abstract. Within everyday reasoning we often use argumentation patterns that employ the rather vague notion of something being *normally true*. This form of reasoning is usually captured using Reiter's Default Logic. However, in Default Logic one has to make explicit the rules which are to be used for reasoning and which are supposed to be normally true. This is a bit contrary to the everyday situation where people use *experience* to decide what normally follows from particular observations and what not, not using any kind of logical rules at all. To formalize this kind of reasoning we propose an approach which is based on *prior experiences*, using the fact that something follows normally if this is the case for "almost all" of the available experience.

1 Introduction

When we say that "the tram normally is on time," we do so because in most of our previous experiences this has been the case. Of course, when stating this fact, we very well accept the possibility that due to some road accident our tram may not come at all. Even if such a road accident occurs, we may still be of the opinion that the trams are normally on time, because road accidents are "not normal."

This kind of reasoning, which employs the very vague notion of "normality," is rather common to us, and several attempts have been made to formalize this kind of reasoning or embed it into a formal framework. Two of the most common attempts are McCarthy's *Circumscription* [3] and Reiter's *Default Logic* [4]. The former tries to restrict the usual model semantics of first order logic (or propositional logic) to models which are "as normal as possible" by minimizing the amount of abnormality they have. The latter approach adds *justifications* for inferences rules that model normality: a rule is normally applicable, but not if the justification is not valid.

These two approaches have in common that they both start with *knowledge*, expressed using logical formulas, that is assumed to be "normally true", and try to infer new knowledge based on this. However, this is not the case in the situation where we wait for our tram: we do not employ rules like "when there is no construction work nearby, then my tram is usually on time" or similar things.

In fact, what we usually do is that we *compare* our current situation to previous experiences and see what happened in these situations. If we find some occurrence often enough, then we conclude that it "normally occurs" in "situations like this" and say that it should "normally occur now" as well. This form of

P. Cabalar and T.C. Son (Eds.): LPNMR 2013, LNAI 8148, pp. 200–205, 2013.
ⓒ Springer-Verlag Berlin Heidelberg 2013

reasoning does not involve any kind of prior knowledge, but just makes use of previous observations and a certain kind of heuristic which decides when something happened "often enough."

However, note that we can use this comparison with previous situations to *generate* non-monotonic rules. Indeed, when we are waiting for the tram, and no construction work is nearby, then we could conclude from previous experiences that the tram should be on time. So we can extract this rule "no construction work nearby implies tram on time" from our experiences. Such rules could then be used for further reasoning.

The purpose of this work is twofold. First and foremost, we set out to formalize the notion of *normal reasoning based on prior experiences* as discussed above. To this end, we shall make use of the theory of *formal concept analysis* as a framework to model *experiences*. Furthermore, we shall make use of the notion of *confidence*, as it is employed in data mining [1], as a method to formalize the fact that some observation occurred "often enough."

Secondly, we want to foster the discussion of this approach in the non-monotonic reasoning community, and want to ask whether there are close connections to existing approaches.

2   Formal Concept Analysis

For our considerations we require very little from the theory of formal concept analysis [2]. More precisely, we shall introduce the notions of a *formal context* and *contextual derivation*.

A *formal context* is a triple $\mathbb{K} = (G, M, I)$, where G, M are sets and $I \subseteq G \times M$. A popular interpretation of formal contexts is that we think of the set G as of a set of *objects*, and of the set M as of a set of *attributes*. Then an object $g \in G$ *has* and attribute $m \in M$ if and only if $(g, m) \in I$.

Let $A \subseteq G$ be a set of objects. We can ask for the set of *common attributes* of the set A, i. e. for the set

$$A' = \{ m \in M \mid \forall g \in A \colon (g, m) \in I \}.$$

Likewise, for a set $B \subseteq M$ of attributes, we can ask for the set of *satisfying objects*, i. e. for

$$B' = \{ g \in G \mid \forall m \in B \colon (g, m) \in I \}.$$

The mappings $(\cdot)'$ are called the *derivation operators* of $\mathbb{K}$, and the sets A' and B' are called the *derivations* of A and B in $\mathbb{K}$, respectively. A set $B \subseteq M$ of attributes is called an *intent* of $\mathbb{K}$ if and only if $B = B''$.

In formal context, one can ask questions like the following: is it true that every object that has all attributes from A has also all attributes from B? To formalize this notion, we introduce the concept of *implications* as pairs (A, B) of sets $A, B \subseteq M$. To make our intention clearer, we shall often write $A \to B$ instead of (A, B). The set of all implications on a set M is denoted by $\mathrm{Imp}(M)$. The implication $A \to B$ then *holds* in $\mathbb{K}$ if and only if $A' \subseteq B'$.

3 Nonmonotonic Reasoning in Formal Contexts

We shall now make use of the notion of formal context to model non-monotonic reasoning based on prior experiences. For this, let us fix a set M of *relevant attributes*. Then we shall understand an *experience* as a subset $N \subseteq M$. Intuitively, such an experience N consists exactly of all attributes we have observed within this experience. We then collect all such experiences in a formal context $\mathbb{K} = (G, M, I)$, i.e. for each $g \in G$, the set g' is an experience.

Example 1. Suppose (again) we are waiting for our tram at a tram station. It is a sunny day, and we suspect nothing bad. In particular, we do not expect our tram to be late. However, out of the sudden we hear some sirens, which may be due to some road accident that occurred nearby. Because of this extra information, we are not that sure anymore if our tram will arrive at all!

A formal context $\mathbb{K}_{\text{tram}}$ which collects a set of such prior experiences (together with the information whether the tram arrived or not) could be given by

	sunny	sirens	tram on time
Day 1	×		×
Day 2			×
Day 3	×		×
Day 4	×	×	×
Day 5	×	×	
Day 6	×		×

In other words, on day 1, it was sunny and the tram was on time, and on day 5, it was sunny, but there were sirens, and the tram was not on time.

The goal is now to draw conclusions from such a formal context $\mathbb{K} = (G, M, I)$ of experiences. Roughly, we suppose that we are given an *observation* $P \subseteq M$. We then ask for some $m \in M$ whether in "almost all" experiences where P occurred, m occurred as well.

We now formalize this notion of saying that a set $Q \subseteq M$ of attributes "normally follows" from our observation P. For this, we shall introduce the notion of *confidence* $\text{conf}_{\mathbb{K}}(A \to B)$ for implications $(A \to B) \in \text{Imp}(M)$ in the formal context $\mathbb{K}$ as

$$\text{conf}_{\mathbb{K}}(A \to B) = \begin{cases} 1 & A' = \varnothing \\ \frac{|(A \cup B)'|}{|A'|} & \text{otherwise.} \end{cases}$$

In other words, the confidence of the implication $A \to B$ is the relative amount of objects in $\mathbb{K}$ satisfying all attributes in A which also satisfy all attributes in B.

Using the notion of confidence, we say, for some fixed $c \in [0, 1]$, that P *normally (with threshold c) implies* Q in $\mathbb{K}$ if and only if

$$\text{conf}_{\mathbb{K}}(P \to Q) \geq c.$$

Example 2. Let us consider Example 1 again, and let us choose $c = 0.8$. Then on sunny days our tram is normally on time, because

$$\mathrm{conf}_{\mathbb{K}_{\mathrm{tram}}}(\{\text{ sunny }\} \to \{\text{ tram on time }\}) = \frac{4}{5}.$$

However, if we add the extra information that we heard some sirens, then it is not true that our tram will be normally on time, even if it is a sunny day, since

$$\mathrm{conf}_{\mathbb{K}_{\mathrm{tram}}}(\{\text{ sunny, sirens }\} \to \{\text{ tram on time }\}) = \frac{1}{2}.$$

From this example, we already see that this kind of inference is non-monotonic. More precisely, it may happen for some sets $P, Q \subseteq M$ and $p \in M$ that

$$\mathrm{conf}_{\mathbb{K}}(P \to Q) \geqslant c$$
$$\mathrm{conf}_{\mathbb{K}}(P \cup \{p\} \to Q) < c.$$

Furthermore, it is worthwhile to note that the notion of "normally implies" is not transitive in the usual sense.

Example 3. Let us consider the formal context

	a	b	d
1	×		
2	×	×	
3	×	×	×
4		×	×
5			×

and choose $c = \frac{2}{3}$. Then $\{a\}$ normally implies $\{b\}$, and $\{b\}$ normally implies $\{d\}$, but $\{a\}$ does not normally imply $\{d\}$. Even $\{a, b\}$ does not normally imply $\{d\}$.

On the other hand, it is easy to see that

$$\mathrm{conf}_{\mathbb{K}}(A \to C) = \mathrm{conf}_{\mathbb{K}}(A \to B) \cdot \mathrm{conf}_{\mathbb{K}}(B \to C)$$

for $A \subseteq B \subseteq C \subseteq M$.

4   Non-monotonic Rules

Our approach described so far can be used to define a certain kind of non-monotonic rules. More precisely, let us call an implication $A \to B$ a *non-monotonic rule*, and let us say that this rule is *valid* in $\mathbb{K}$ with threshold c if and only if $\mathrm{conf}_{\mathbb{K}}(A \to B) \geqslant c$. Those rules capture all the knowledge we get from non-monotonic reasoning in the formal context $\mathbb{K}$ using the threshold c. The semantics of these rules is obviously model-based.

Let us denote with $R(\mathbb{K}, c)$ the set of all rules of $\mathbb{K}$ using c as threshold. As these rules enjoy a model-based semantics, we can defined the corresponding entailment operator $\models_c$ by

$$\mathcal{L} \models_c (A \to B) \iff (\forall \mathbb{K} : \mathcal{L} \subseteq R(\mathbb{K}, c) \implies (A \to B) \in R(\mathbb{K}, c)),$$

where the formal contexts all have the same set M of attributes.

It is quite easy to see that we do not need all such rules to still be able to do all the reasoning. If we have rules $A \to B$ and $A \to B \cup C$ which are both valid in $\mathbb{K}$ using c, then the latter suffices.

We denote maximal such sets with a special name. Let $P \subseteq M$. We call Q an c-*extension* of P in $\mathbb{K}$ if and only if Q is $\subseteq$-maximal with respect to

$$R(\mathbb{K}, c) \models_c (P \to Q).$$

It thus suffices to know all the rules

$$\{\, P \to Q \mid Q \text{ is a } c\text{-extension of } P \text{ in } \mathbb{K} \,\}.$$

It is also easy to see that for rules $P \to Q$ whose confidence is not 1 in $\mathbb{K}$, it is enough for the sets P and Q to be intents (note that extensions are always intents). In other words, the set

$$\{\, P \to P'' \mid P \subseteq M \,\} \cup$$
$$\{\, P \to Q \mid Q \ c\text{-extension of } P, \mathrm{conf}_{\mathbb{K}}(P \to Q) \neq 1 \text{ and } P, Q \in \mathrm{Int}(\mathbb{K}) \,\}.$$

is complete for $R(\mathbb{K}, c)$. Of course, instead of $\{\, P \to P'' \mid P \subseteq M \,\}$, we could choose any base of $\mathrm{Th}(\mathbb{K})$.

Note that every set P has a c-extension, since $\mathrm{conf}_{\mathbb{K}}(P \to P) = 1$ and $\mathrm{conf}_{\mathbb{K}}(\cdot \to \cdot)$ is antitone in its second argument. More precisely, we have the following characterization of c-extensions.

Proposition 1. *Let $\mathbb{K} = (G, M, I)$ be a finite and non-empty formal context, and let $c \in [0, 1]$. Let $P \subseteq M$. Then a set Q is an c-extension of P in $\mathbb{K}$ if and only if Q is maximal with respect to $\mathrm{conf}_{\mathbb{K}}(P \to Q) \geq c$.*

But note that in contrast to the case $c = 1$, a set P can have multiple c-extensions if $c \neq 1$.

Example 4. Let us consider the formal context

	m	n
1	×	
2	×	×
3	×	×
4	×	×
5		×

and choose $c = 4/5$. Then the set $P = \varnothing$ has two c-extensions, namely $\{\, m \,\}$ and $\{\, n \,\}$.

It is quite easy to see that this example can be generalized to work for every value $c \in [0, 1)$ and for every number of c-extensions.

Also note that in contrast to the classical case, c-extensions are *not* closed under normal entailment.

Example 5. Consider the formal context from Example 4 again, and let $c = 3/4$. Then the c-extensions of $\varnothing$ are $\{m\}, \{n\}$, but both sets on their part have the set $\{m, n\}$ as c-extension.

5 Conclusions and Future Research

We have presented a formalization of evidence based non-monotonic reasoning based on formal concept analysis. To this end, we have used formal contexts to model the set of experiences a person has. Using the notion of confidence, we were able to give a precise formulation of what it means that in "almost all" experiences obtained so far a certain conclusion was correct. Based on this, we have shown that this form of inference indeed yields a non-monotonic formalism.

Albeit the author is quite sure that this approach of combining formal concept analysis and non-monotonic reasoning is original, he is aware of the fact that the general idea underlying this approach is not new. It is thus one of the major next steps in investigating this approach to find connections to existing frameworks for non-monotonic reasoning. Moreover, our formalization yields a connection between formal concept analysis and non-monotonic reasoning. Thus, if we can find that our idea is similar to existing ones, it might be the case that methods from formal concept analysis could be helpful in solving tasks in these existing approaches. Conversely, it is possible that ideas and results from non-monotonic reasoning could be applied to issues of formal concept analysis.

Moreover, as we have already indicated in our considerations above, our formalization could yield a method which allows for the extraction of non-monotonic rules which could be used by other formalisms like default logic. In particular, we have given a first "base" of non-monotonic rules of a formal context, which however might be too large to be practically relevant. A smaller base, maybe comparable to the *canonical base* [2] known in formal concept analysis, maybe of practical interest.

Acknowledgments. The author wants to thank both Bernhard Ganter and Gerd Brewka for their encouraging discussions on this topic.

References

[1] Agrawal, R., Imielinski, T., Swami, A.: Mining Association Rules between Sets of Items in Large Databases. In: Proceedings of the ACM SIGMOD International Conference on Management of Data, pp. 207–216 (1993)

[2] Ganter, B., Wille, R.: Formal Concept Analysis: Mathematical Foundations. Springer, Heidelberg (1999)

[3] McCarthy, J.: Circumscription–A form of non-monotonic reasoning. Artificial Intelligence 13, 1–2 (1980); Special Issue on Non-Monotonic Logic, 27–39

[4] Reiter, R.: A logic for default reasoning. Artificial Intelligence 13, 1–2 (1980); Special Issue on Non-Monotonic Logic, 81–132

An ASP Application in Integrative Biology: Identification of Functional Gene Units

Philippe Bordron[1,2], Damien Eveillard[4], Alejandro Maass[2,3], Anne Siegel[6,5], and Sven Thiele[1,5,6]

[1] INRIA-CIRIC, Rosario Norte 555, Of. 703, Santiago, Chile
[2] CENTER OF MATHEMATICAL MODELING AND CENTER FOR GENOME REGULQTION, Universidad de Chile, Av. Blanco Encalada 2120, Santiago, Chile
[3] DEPARTMENT OF MATHEMATICAL ENGINEERING, Universidad de Chile, Av. Blanco Encalada 2120, Santiago, Chile
[4] ComBi, LINA, Université de Nantes, CNRS UMR 6241, 2 rue de la Houssinière, 44300 Nantes, France
[5] INRIA, Centre Rennes-Bretagne-Atlantique, Projet Dyliss, Campus de Beaulieu, 35042 Rennes cedex, France
[6] CNRS, UMR 6074 IRISA, Campus de Beaulieu, 35042 Rennes, France

Abstract. Integrating heterogeneous knowledge is necessary to elucidate the regulations in biological systems. In particular, such an integration is widely used to identify functional units, that are sets of genes that can be triggered by the same external stimuli, as biological stresses, and that are linked to similar responses of the system. Although several models and algorithms shown great success for detecting functional units on well-known biological species, they fail in identifying them when applied to more exotic species, such as extremophiles, that are by nature unrefined. Indeed, approved methods on unrefined models suffer from an explosion in the number of solutions for functional units, that are merely combinatorial variations of the same set of genes. This paper overcomes this crucial limitation by introducing the concept of "genome segments". As a natural extension of recent studies, we rely on the declarative modeling power of answer set programming (ASP) to encode the identification of shortest genome segments (SGS). This study shows, via experimental evidences, that SGS is a new model of functional units with a predictive power that is comparable to existing methods. We also demonstrate that, contrary to existing methods, SGS are stable in (i) computational time and (ii) ability to predict functional units when one deteriorates the biological knowledge, which simulates cases that occur for exotic species.

1   Introduction

In biological systems one distinguishes several layers of information. One of them represents the set of (bio)chemical reactions that occur within the system. These reactions are called metabolic reactions and form chains of metabolic pathways (i.e. products of reactions are substrates of other reactions) of the whole metabolism. The metabolism is roughly controlled by genes. Indeed, a major

P. Cabalar and T.C. Son (Eds.): LPNMR 2013, LNAI 8148, pp. 206–218, 2013.
© Springer-Verlag Berlin Heidelberg 2013

part of these reactions is catalyzed by enzymes that are encoded by genes. Thus, understanding the link between gene regulation and metabolism is of great interest in biological research. However, despite the interest and efforts, identifying these links remains a difficult task. Recent biological evidence shows that microbial genes present a specific organization. They are grouped together on the DNA strand when they are functionally related [1]. These groups of genes, so called functional units, became the main target of functional biology.

Following this topological assumption several bioinformatics approaches have been proposed to explain functional units [2,3]. All rely on genome scale models integrating genomic information (in particular, genes organization) with metabolic networks. Among these techniques in [4] is proved that shortest paths – namely, *wrr-paths* – in the so called *integrated model* correspond to functional gene units (i.e. called *metabolic operons* in the biological literature). Although shortest wrr-paths show its efficiency to recover functional units when computed in well-curated bacteria like *Escherichia coli*, one observes an explosion in the number of solutions when considering more exotic or less characterized bacteria. This explosion is somehow artifactual, since many wrr-paths in these species are merely combinatorial variations of the same set of genes and can be biologically merged into a single functional gene unit.

In this paper, we propose a modification of the concept of wrr-path in order to overcome the weakness due to the aforementioned combinatorial explosion in wrr-path computation in bad characterized bacteria. We present the concept of *genome segments* which is computable in reasonable time allowing the identification of meaningful functional units on this class of bacteria. Genome segments and wrr-paths are related, they are obtained within the same integrated framework but minimizing different metrics.

To compute functional gene units we formulate the shortest genome segment (SGS) problem by means of Answer Set Programming (ASP) [5] instead of a dedicated algorithm on a graph. This allows a flexible encoding that can be easily adjusted to test different metrics while still being computational efficient. ASP is a declarative problem solving paradigm from the field of logic programming and knowledge representation, that offers a rich modeling language [6] along with highly efficient inference engines [7] based on Boolean constraint solving technology. In the fields of integrative biology ASP applications include the reconstruction of metabolic networks [8], modeling the dynamics of regulatory networks [9], inferring functional dependencies from time-series data [10], and integrating gene expression with pathway information [11].

Finally, we solve the SGS problem on a concrete biological system by taking advantage of ASP's optimization techniques for finding minimal solutions [6,12]. Via further experimental validation, we pinpoint that SGS are a suitable model of functional units with an accurate predictive power. We also demonstrate that the identification of SGS is stable in both computational time and the ability to predict functional units when one deteriorates the biological knowledge, which simulates cases that occur in more exotic species.

2   An Integrated Model: Identifying Functional Gene Units

Metabolic Compounds and Reactions. The metabolism of a given bacterial system is defined by the set C of biological compounds and the set R of metabolic reactions that take place in the system. Each reaction r in R describes the transformation of compounds in C into others, also in C. Consequently, a metabolic compound can be consumed or produced by a metabolic reaction. The compound takes the role of a substrate or a product of the reaction. We define the maps *consume* $: R \rightarrow \mathcal{P}(C)$ and *produce* $: R \rightarrow \mathcal{P}(C)$ (where $\mathcal{P}(C)$ contains subsets of C) to describe the set of substrates and the set of products of a given reaction respectively. The fact that the products of one reaction can be used as substrates by other reactions allows to connect reactions into complex chemical pathways. All possible reactions are usually represented as a metabolic network, which is a graph representation of the metabolism. We use the *reaction graph* representation (R, E) where vertices in R correspond to the set of possible reactions and edges are $E = \{(x, y) \mid x, y \in R, produce(x) \cap consume(y) \neq \emptyset, x \neq y\}$.

Genes, Enzymes, Translation and Catalysis of Reactions. Bacterial genomes are often constituted by only one circular chromosome. Formally, such a genome can be represented as an ordered sequence $\mathcal{G}$ of genes $g_1, \ldots, g_n$ where n is the number of genes in $\mathcal{G}$. The successor and predecessor of a gene $g \in \mathcal{G}$ is naturally defined when looking the genome as a circular word. We denote by G the set of distinct genes that appear in $\mathcal{G}$.

When genes in $\mathcal{G}$ are transcribed and then translated, some proteins catalyzing specific metabolic reactions are produced. The proteins having a catalytic function are called enzymes. Each reaction in R can be catalyzed by one or many enzymes. We define then the map *catalyze* $: G \rightarrow \mathcal{P}(R)$ to describe the set of reactions that a gene can catalyze via the associated enzyme.

The Integrated Gene-reaction Graph. A model that put together all previously described heterogeneous biological knowledge into a weighted directed graph representation is the *integrated gene-reaction graph* [4]. For a genome $\mathcal{G}$, a set of reactions R and a metric between genes $w : G \times G \mapsto \mathbb{R}^+$, we define the integrated graph (V, A) where $V = \{(g, r) \mid g \in G, r \in R, r \in catalyze(g)\}$ is the set of vertices and $A = \{((g, r), (g', r')) \mid (g, r), (g', r') \in V, produce(r) \cap consume(r') \neq \emptyset, r \neq r'\}$ are the edges. Each edge $a = ((g, r), (g', r'))$ of A is weighted by $w(g, g')$. The sum of weights along a path is the path weight. Note that this graph uses the reaction graph as a support which is enriched with genomic information. Using this integrated model allows us to investigate gene regulatory behavior based on the topology of the underlying metabolic reaction network.

Identifying Functional Gene Units. A concept that is used to identify functional gene units is called *without reaction repetition* path (wrr-path) in the integrated gene-reaction graph. Given two reactions r and r' in R. A path $p = v_1 \ldots v_l$ in (V, A) is a wrr-path from r to r' if $v_1 = (g_1, r)$, $v_l = (g_l, r')$ and for all two vertices $v_i = (g_i, r_i)$, $v_j = (g_j, r_j)$ in p with $1 \leq i < j \leq l$ it holds $r_i \neq r_j$. We observe that several wrr-paths can have the same weight. The set of genes that are involved in a wrr-paths can be interpreted as a hint of a functional gene unit. Figure 1(a) illustrates wrr-paths.

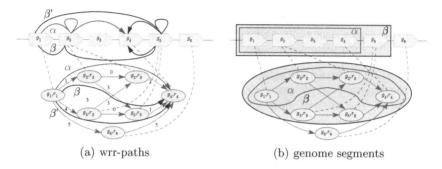

(a) wrr-paths   (b) genome segments

Fig. 1. Difference between wrr-paths and genome segments. On both figures flat arrows represent genes on the genome, and the graph under the genome is the integrated gene-reaction graph. In (a), the shortest wrr-path from r_1 to r_4 is α, and the two second shortest paths are β or β'. In addition, we project in the obvious way those paths on the genome. In (b), the shortest genome segment from r_1 to r_4 is α, whereas the second best is the segment β. The projection of these segments in the integrated gene-reaction graph is depicted.

It has been shown in the well studied organism *Escherichia coli* that the concept of wrr-path is highly useful to determine functional metabolic-gene units [4]. Indeed, besides the fact that this concept was not designed to find operons, the resulting gene units matched up to 45% with known operons which is considered to be a top range technique for identifying operons [13]. Moreover, in [4], only the information of *Escherichia coli* is required, whereas dedicated predictors rely on organisms comparison or/and learning methods [13]. In the case of exotic organisms, the use of information about other organisms may be hurtful.

While computing shortest (in relation to the weight) wrr-paths in a well studied organism like *E. coli* is not problematic, it has certain shortcomings when applied to less refined models. We observed that in unrefined models the number of equally short wrr-paths increases. This is mostly due to the fact that usually gene annotations are less specific. One gene is then often mapped to a multitude of metabolic reactions and the direction of most reactions is unknown and is set to be reversible by default. Thus, the integrated gene-reaction graph of a less well defined model contains naturally more wrr-paths. Most of these paths involve the same genes and therefore belong to the same gene units. These equally short paths are merely combinatoric variants of few gene-reaction pairs. The huge number of alternative paths makes its computation unfeasible on poorly defined models.

3   Genome Segments, an Alternative Approach to Identify Functional Gene Units

We present an alternative approach to identify functional gene units based on gene organization instead of wrr-paths. This allows us to avoid the huge combinatorial problem described above when considering integrated gene-reaction graphs.

Definition 1 (Genome segment). *Given a circular sequence of genes* $\mathcal{G} = g_1 \ldots g_n$, *we define a genome segment of* $\mathcal{G}$ *as a sequence* $\sigma = g_i \ldots g_j$, *where* $1 \leq i \leq j \leq n$, *or a sequence* $\sigma = g_i \ldots g_n g_1 \ldots g_j$, *where* $0 \leq j \leq i \leq n$.

Definition 2 (Induced subgraph). *Let* $\mathcal{G}$ *be a circular sequence of genes and* (R, E) *the reaction graph associated to such sequence. Given a genome segment* σ *of* $\mathcal{G}$, *the induced subgraph is* (R_σ, E_σ) *where* $R_\sigma = \{r \in R \mid \exists\, g \in \sigma,\ r \in catalyze(g)\}$ *and* $E_\sigma = \{(r, r') \in E \mid \exists\, g, g' \in \sigma,\ r \in catalyze(g),\ r' \in catalyze(g')\}$.

The induced subgraph reflects the regulatory influence of the genes in the segment, their corresponding enzymes products and the catalyzed reactions relation of these enzymes. Figure 1(b) illustrates genome segments and the induced subgraphs.

To each wrr-path in an integrated gene-reaction graph one associates the shortest genome segment containing the genes in the path. Thus, two wrr-paths involving the same set of genes are associated to the same genome segment but they may have different weights. In unrefined models, one observes that the number of wrr-paths is huge and that a genome segment can be associated to a big amount of wrr-paths.

Thus, to reduce complexity and in analogy to the wrr-paths problem, where one searches the shortest wrr-paths in terms of their weight, we define the Shortest Genome Segment (SGS) problem as follows.

Definition 3 (Shortest Genome Segment Problem). *The SGS problem receives the following data: a genome sequence* $\mathcal{G}$, *a reaction graph* (R, E), *a map* $catalyze : \mathcal{G} \to \mathcal{P}(R)$ *and two reactions* $r, r' \in R$.

A solution to a SGS problem is the shortest genome segment σ *of* $\mathcal{G}$ *such that the induced subgraph* (R_σ, E_σ) *contains a path from* r *to* r'.

A solution to the SGS problem points to a set of genes that take an active part in the metabolic regulation and form an active gene unit.

Definition 4 (Active gene units). *Let segment* σ *be a solution to the SGS problem with data* $(\mathcal{G}, (R, E), catalyze, r, r')$. *A gene* g *is part of an active gene unit* $AU(\sigma)$ *if and only if* $g \in \sigma$ *and there exists a path* p *in* (R_σ, E_σ) *from* r *to* r' *such that reactions catalyzed by* g, $catalyze(g)$, *intersect the path* p.

These active gene units can be biologically interpreted as functional gene units.

To explore the space of further suboptimal genome segments, we adjust the SGS problem by adding constraints on the minimal length of a segment.

Definition 5 (SGS problem with minimal length). *An SGS problem with minimal length needs the following input data: a genome sequence $\mathcal{G}$, a reaction graph (R, E), a map catalyze : $G \rightarrow \mathcal{P}(R)$, two reactions $r, r' \in R$, and a minimal length min.*

A solution to a SGS problem with minimal length is the shortest genome segment σ of $\mathcal{G}$ with length $l \geq min$, such that the induced subgraph (R_σ, E_σ) contains a path from r to r' and there exists no segment σ' of length $l' < min$ with $AU(\sigma) \subseteq AU(\sigma')$.

Observe that the original SGS problem is a special case of this problem where $min = 1$. It is clear that when min is too large this problem has no solution. On the other hand condition $AU(\sigma) \subseteq AU(\sigma')$ looks to avoid artificial extension of segments with no active genes.

In our application, the identification of functional gene units in exotic organisms, we are especially interested in computing optimal solutions as well as solutions that are close to the optimum. Therefore, we need to solve the following sub-tasks:

- **Problem 1.** Compute the minimal length l of a segment σ with $l \geq min$, such that the induced subgraph contains a path from r to r' and there exists no segment σ' of length $l' < l$ with $AU(\sigma) \subseteq AU(\sigma')$.
- **Problem 2.** Enumerate all segments σ of a given length l such that the induced subgraph contains a path from r to r' and there exists no segment σ' of length $l' < l$ with $AU(\sigma) \subseteq AU(\sigma')$.

In the following, we will show on a real world application that genome segments are a good alternative for the computation of functional gene units.

4   ASP Encoding

We now present our ASP encoding of the SGS problem as defined in Section 3. Additionally the encoding will use an upper bound max representing our knowledge on the maximal length of a genome segment. In some cases can be the length of the genome but typically it is no longer than a few hundreds.

Therefore, an instance of the SGS problem consists of seven components, the sequence of genes $\mathcal{G}$, the reaction graph (R, E), the function $catalyze$ which maps genes to metabolic reactions, metabolic reactions s and e, which represent the start and end of the desired pathway, a lower bound min on the length of the desired genome segment, as well as an upper bound max on the length of the desired genome segment. For our ASP solution, we represent such a problem instance as a set of ground logic facts $\mathcal{F}(\mathcal{G}, (R, E), catalyze, s, e, min, max)$ defined as follows:

$$\mathcal{F}(\mathcal{G}, (R, E), catalyze, s, e, min, max) = \{edge(u, v) \mid (u, v) \in E\}$$
$$\cup \{gene(g) \mid g \in \mathcal{G}\}$$
$$\cup \{cat(g, r) \mid g \in \mathcal{G}, r \in R, r \in catalyze(g)\} \quad (1)$$
$$\cup \{start(s), \ end(e)\}$$
$$\cup \{const \ min, \ max\}.$$

Such a problem instance can then be combined with the logic program in Listing 1.1 to solve the SGS problem.

Listing 1.1. sgs.lp: ASP encoding of shortest genome segments.

```
 1  sgene(G)  :- start(R),cat(G,R).
 2  egene(G)  :- end(R),cat(G,R).
 3
 4  pse(F,L)  :- gene(F;L), F<L,
 5                (L-F)+1 <= max, (L-F)+1 >= min,
 6                sgene(S), F > S-max, L < S+max,
 7                S >=F, S <= L,
 8                egene(E), F > E-max, L < E+max,
 9                E >=F, E <= L.
10
11  1{ se(F,L) : pse(F,L) }1.
12
13  on_segment(G)  :- se(F,L), gene(G), G>=F, G<=L.
14
15  aedge(X,Y)  :- edge(X,Y), cat(G1,X), cat(G2,Y),
16                on_segment(G1;G2).
17
18  from_start(X)  :- start(X), on_segment(G), cat(G,X).
19  from_start(Y)  :- from_start(X), aedge(X,Y).
20  :- not from_start(X), end(X).
21
22  to_end(Y)  :- end(Y), on_segment(G), cat(G,Y).
23  to_end(X)  :- to_end(Y), aedge(X,Y), cat(G,X).
24
25  aunit(G)  :- on_segment(G), cat(G,X), from_start(X),to_end(X).
26  :- se(F,L), not aunit(F).
27  :- se(F,L), not aunit(L).
28
29  length((L-F)+1)  :- F<=L, se(F,L).
30
31  :- length(X), X < min.
32
33  #minimize [ length(L) = L ].
```

This logic program represents a simplified ASP formulation of the SGS problem. We remark that the rules in lines (4-9, 13 and 29) are only handling the case of linear genomes. For the sake of simplicity we omit here the definitions for the circular genome case. The complete encoding is available in the source code[7].

Starting with the rules in lines (1 and 2) the logic program defines the set of genes that are associated to the *start* and *end* reactions via the *cat* predicate. As there can exist more than one gene catalyzing a reaction, there is a set of genes corresponding to them. Each of the genes denoted by the predicate *sgene* can catalyze the *start* reaction, while the predicate *egene* denotes genes which can catalyze the *end* reaction respectively. Note that start and end genes do not necessarily correspond to the beginning and end of a desired genome segment. These genes can occur everywhere and in any order in a genome segment.

The rule in lines (4-9) defines the search space of possible genome segments. A possible genome segment is denoted by the predicate *pse* with two arguments, the first gene F and the last gene L of the segment. The length of the segment is determined by the formula $L - F + 1$ and only segments with $min \leq$ length $\leq max$ are considered. Furthermore, a segment must contain at least one start gene and one end gene respectively. Therefore, it must hold that there exist a start gene S and an end gene E such that $F \leq S \leq L$ and $F \leq E \leq L$ respectively.

Among the possible segments exactly one segment can be chosen. This choice is expressed by the rule in line (11). The genes that lie on the chosen segment are defined by the rule in line (13). These genes induce a set of edges in the reaction graph, the *induced subgraph*, connecting reactions that are catalyzed by genes on the segment. The set of active edges is defined by the rule in lines (15 and 16). An edge (X, Y) is active if X and Y are both catalyzed by genes on the segment.

Given the set of active edges one can test whether there exists a path from the start reaction to the end reaction. The rules in lines (18 and 19) define what is reachable from the start reactions and the integrity constraint in line (20) discards solutions where the end reaction is not reachable. The rules in lines (22 and 23) define nodes that lie on a path of active edges to the *end* reaction.

The genes that catalyze reactions on a pathway from *start* to *end* reaction form the *active gene unit*. They are defined by the rules in line (25). The integrity constraints in lines (26 and 27) discard segments that merely extend shorter segments without extending the active gene unit. A segment is not considered a solution if the first or the last gene is not part of the active gene unit. These integrity constraints are especially important if we look for shortest segments which are bigger than a given size *min*. Without these constraints every solution that is smaller or equal than *min* could easily be extended to a solution of size $min + 1$ by simply prolonging a shorter solution.

Line (29) defines the length of the segment. The integrity constraint in line (36) discards segments whose lengths are shorter than the required minimum *min*.

So far the rules in lines (1 to 31) define all segments that catalyze a reaction pathway from *start* to *end* reaction. To solve the corresponding SGS optimization problem, line (33) declares the objective function via an optimize statement. Preferred solutions are those that minimize the length of the segment.

5   Enumerating Shortest Genome Segments

We provide an application that computes optimal and sub optimal solutions close to the optimum. More precisely, we enumerate all solutions σ of a SGS problem with minimal length from 1 to n until we have at least k distinct active gene units $AU(\sigma)$ or there exist no further solutions.

The intuition is that these active gene units correspond to functional gene units like metabolic operons or regulons. They allow us to investigate the relationship between metabolic pathways and gene localization.

To compute these segments we developed the Python program shogen[1]. It depends on the PyASP[2] library for calling the ASP solvers gringo and clasp and for passing them logic program encoding and problem instances.

shogen takes as input the genome, the metabolic reaction network and information of the catalytic function of the genes. Further, a list of queries start and end reactions for which we want to find functional gene units. In a pre-processing step shogen filters queries that do not have a path in the metabolic reaction network. For the remaining queries shogen computes the shortest genome segments and their active gene units.

The computation is performed in a multi-step process. In a first step, clasp is used to solve **Problem 1**, computing the minimum length of a segment that can catalyze the desired metabolic pathway. Once the optimal length is known, clasp is used with the option --opt-all to solve **Problem 2**, enumerating all solutions that satisfy this optimality criterium. These steps are repeated until at least k segments are computed or no more solutions can be found. The minimum length of a segment is increased whenever all solutions of a given length are computed. The maximum length is fixed and part of the logic problem instance.

The following pseudo-code describes the algorithm.

Algorithm 1. compute the shortest genome segments

Input: A SGS problem instance as facts instance.lp and a parameter k
$Segments \leftarrow \emptyset$;
$min \leftarrow 1$;
while $|Segments| < k$ **do**
$\quad opt \leftarrow$ gringo instance.lp sgs.lp --const min=min | clasp;
$\quad \Sigma_{min} \leftarrow$ gringo instance.lp sgs.lp --const min=min | clasp
$\quad$ --opt-all=opt;
$\quad Segments \leftarrow Segments \cup \Sigma_{min}$;
$\quad min \leftarrow opt+1$;
end
return $Segments$

---

6   Experiments and Results

The functional gene units produced by using the notions of shortest genome segments and shortest wrr-paths were compared. We consider an operational criterium, that is, the computational time needed to obtain them, and also the biological relevance of the results. The benchmark was conduced on the widely studied and well known *Escherichia coli* bacteria. In order to simulate more exotic or less studied organisms, with unrefined models, we create a set of deteriorated models of *E. coli*.

---

[1] https://pypi.python.org/pypi/shogen
[2] http://pypi.python.org/pypi/pyasp

Simulating Exotic Organisms. Although modern genome sequencing techniques allow us to obtain genome data even for little genomes and exotic organisms, the main problem lies in the reconstruction of the metabolic networks for these organisms. While it is often possible to determine the reactions that occur in the metabolic network, a lot of complicated and costly experiments are needed to determine their direction. Thus, the direction of those reactions remains often unknown. *E. coli* is regarded as the best studied organism today and its metabolic network is the most refined existing one. Therefore, we us it as the reference to obtain deteriorated models containing less information about irreversible reactions. Given the set R_i of irreversible reactions known in *E. coli*, we created deteriorated models by taking subsets R_d of R_i and transforming them to reversible reactions. In this way we created models with different deterioration ratios $\frac{|R_d|}{|R_i|}$ ranging from 0 to 1. These deteriorated *E. coli* models aim to simulate the unrefined models of exotic species.

Benchmark and Experimentations. The knowledge about *E. coli* was taken from the `Ecocyc` database (version 16.1). Its genome is composed of one circular chromosome of 4498 genes and its metabolism consists of 2070 distinct reactions separated into 816 reversible and 1254 irreversible ones. We generated four deteriorated models for each of the following deterioration ratios: 0.05, 0.1, 0.2, 0.4, 0.6 and 0.8. We also generated the model with a deterioration ratio of 1.

We confronted each model with a set of queries, a couple of start and end reactions, for which we computed the optimal solutions and the next four levels of suboptimal solutions for the shortest wrr-paths problem and the shortest genome segments problem. We call such solutions a 5-SIP and 5-SGS respectively. The selection of 5 is motivated by computations in [4].

The computations were done on a MacBook Pro 9,2 equipped with an Intel Core i7-3520M processor and 8 Gb of RAM. This computer was running under Mac OS X 10.7.5. The dedicated program `sipper`[3] for computing shortest wrr-paths uses the Java SE Runtime Environment build 1.7.0_13-b20, and the ASP solution for computing shortest genome segments uses the PyASP library including the grounder `gringo` in version 3.0.4 and the ASP solver `clasp` version 2.1.1. Both programs were configured to use only one core. Each computation was repeated twice with a timeout of 24 hours.

Computation Time. Figure 2(a) shows the computation times for the 5-SIP and 5-SGS. When no deterioration is applied, the computation of 5-SIP is faster than the one of 5-SGS. The computation times of both 5-SIP and 5-SGS increase with the deterioration ratio, but the time to compute 5-SIP increases exponentially while the one for 5-SGS increases in a more linear way. For a deterioration ratio over 0.4, the computation of 5-SIP takes more time than the computation of 5-SGS. Moreover, increasing the deterioration ratio even further, the computation of 5-SIP quickly reaches the 24 hours timeout limit.

---

[3] `https://sipper.googlecode.com/`

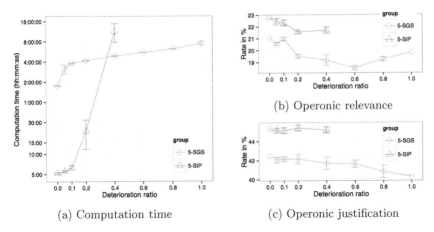

Fig. 2. Results of the computation of 5-SGS and 5-SIP on *E. coli*. The subfigure (a) indicates the computation times according to the deterioration ratio whereas the subfigures (b) and (c) present respectively the operonic relevance and the operonic justification of the sets of gene units.

Biological Relevance. We evaluated the biological relevance of the predicted functional gene units by comparing them against the known metabolic operons in *E. coli*, which are functional units playing a role on both genome regulation and metabolic network studies. We used the set of 278 manually-curated metabolic operons reported in the Ecocyc database version 16.1. We computed the similarity between each predicted gene unit and each operon using the Jaccard measure. As biologists have interest in groups of genes, we do not consider gene units of one gene. A predicted gene unit is said to be similar to an operon when the Jaccard measure between them is greater than or equal to 0.6. We computed then (1) the *operonic relevance* and (2) the *operonic justification* for the set of 5-SGS and the set of 5-SIP. The operonic relevance of the predicted gene units is the proportion of units that are similar to at least one operon. The operonic justification for predicted gene units is the proportion of operons that is similar to at least one predicted gene unit.

Figures 2(b) and 2(c) show the results for operonic relevance and operonic justification, respectively. In both experiments, the deterioration of information leads to predicting gene units with more genes. Thus, we can observe a small decrease of the number of identified operons and a decrease of the operonic justification. Observe also that for 5-SIP a bigger proportion of the predicted gene units is similar to operons and allows to explain a bigger part of the operon than 5-SGS.

7   Discussion

This study demonstrates the biological interest of using SGS as an alternative to shortest wrr-paths, by showing similar capabilities in the identification of confirmed functional gene units. Both concepts were compared using different levels of deteriorated biological knowledge. While using the concept of SGS produced only slightly lower prediction scores for metabolic operons than wrr-paths, it allowed to get meaningful predictions even on unrefined networks with a high deterioration ratio. Therefore, SGS can be considered as an efficient computable alternative to predict functional gene units even on unrefined models.

Beyond the biological evidences for the quality of predictions with SGS, the major outcome of this study remains the efficiency results. Indeed, although wrr-paths is an interesting concept to study organisms for which the biological knowledge is globally complete, this technique drastically fails when applied to incomplete ones, as observed when one studies exotic species. Although the dedicated algorithm to compute the shortest wrr-paths has a better performance on the well refined *E. coli* models, this study emphasizes that the computational time of shortest wrr-paths increases exponentially when the biological knowledge is degraded. This is mainly due to the fact that for one set of genes many equally short wrr-paths can exist, differing only in the order of the involved genes (i.e. a permutation of genes).

SGS does not regard these different permutations. Therefore, the computation times for SGS remain relatively stable, but they are comparatively higher than the dedicated algorithm on the well refined model. This overall high runtime can be explained by the fact the each query represents a separate instance of the ASP problem. Therefore, the times for the problem generation and initialization of data structures, which must be done only once in the dedicated algorithm, is roughly multiplied by the number of queries. A further improvement of the ASP solution could be reached if one can reuse data structures on all problem instances.

Our results confirmed the interest of integrated models and SGS for investigating functional units of exotic species, and the interest of using ASP for deciphering these biological units. From a methodological point of view, ASP allows us to quickly test biological assumptions. In particular, the expressiveness of ASP presents a clear advantage for exploring several hypothesis on biological systems. As a biological perspective, further studies will focus on the extension of the SGS framework to the identification of new functional units. We exploit the flexibility of declarative programming with ASP to create models using more constraint metabolic behavior [8], to explore different metrics based on transcriptomic correlation data instead of a genomic distance, and to identify graph-based units such as CCC (Common Connected Component) [2] or regulons [3], describing co-regulated operons. Therefore, we rely on the versatility of the ASP language and the solving capabilities of ASP solvers to integrate large-scale heterogeneous biological knowledge into computational models.

Acknowledgments. This work was supported by ANR Biotempo (ANR-10-BLANC-0218), Basal-CMM, Fondap-CRG 15090007, INRIA-UChile Integrative-BioChile Associate Team and CIRIC INRIA-Chile.

References

1. Rocha, E.P.C.: The organization of the bacterial genome. Annual Review of Genetics 42, 211–233 (2008)
2. Boyer, F., Morgat, A., Labarre, L., Pothier, J., Viari, A.: Syntons, metabolons and interactons: an exact graph-theoretical approach for exploring neighbourhood between genomic and functional data. Bioinformatics 21, 4209–4215 (2005)
3. Zhang, H.H., Yin, Y.Y., Olman, V.V., Xu, Y.Y.: Genomic arrangement of regulons in bacterial genomes. PLoS ONE 7, e29496–e29496 (2011)
4. Bordron, P., Eveillard, D., Rusu, I.: Integrated analysis of the gene neighbouring impact on bacterial metabolic networks. IET Systems Biology 5, 261–268 (2011)
5. Baral, C.: Knowledge Representation, Reasoning and Declarative Problem Solving. Cambridge University Press (2003)
6. Gebser, M., Kaminski, R., Kaufmann, B., Ostrowski, M., Schaub, T., Thiele, S.: A user's guide to `gringo`, `clasp`, `clingo`, and `iclingo` (2010), http://potassco.sourceforge.net
7. Gebser, M., Kaufmann, B., Neumann, A., Schaub, T.: Conflict-driven answer set solving. In: Veloso, M. (ed.) Proceedings of the Twentieth International Joint Conference on Artificial Intelligence, IJCAI 2007, pp. 386–392. AAAI Press/The MIT Press (2007)
8. Schaub, T., Thiele, S.: Metabolic network expansion with answer set programming. In: Hill, P.M., Warren, D.S. (eds.) ICLP 2009. LNCS, vol. 5649, pp. 312–326. Springer, Heidelberg (2009)
9. Fayruzov, T., Cock, M.D., Cornelis, C., Vermeir, D.: Modeling protein interaction networks with answer set programming. In: IEEE Int. Conf. on Bioinformatics and Biomedicine, BIBM 2009, pp. 99–104 (2009)
10. Durzinsky, M., Marwan, W., Ostrowski, M., Schaub, T., Wagler, A.: Automatic network reconstruction using ASP. Theory and Practice of Logic Programming 11, 749–766 (2011)
11. Papatheodorou, I., Ziehm, M., Wieser, D., Alic, N., Partridge, L., Thornton, J.M.: Using answer set programming to integrate rna expression with signalling pathway information to infer how mutations affect ageing. PLoS ONE 7, e50881 (2012)
12. Gebser, M., Kaminski, R., Schaub, T.: Complex optimization in answer set programming. Theory and Practice of Logic Programming 11, 821–839 (2011)
13. Brouwer, R.W.W., Kuipers, O.P., van Hijum, S.A.F.T.: The relative value of operon predictions. Briefings in Bioinformatics 9, 367–375 (2008)

Evaluating Answer Set Clause Learning
for General Game Playing

Timothy Cerexhe[1], Orkunt Sabuncu[2], and Michael Thielscher[1]

[1] University of New South Wales
{timothyc,mit}@cse.unsw.edu.au
[2] Universität Potsdam
orkunt@cs.uni-potsdam.de

Abstract. In games with imperfect information, the 'information set' is a collection of all possible game histories that are consistent with, or explain, a player's observations. Current game playing systems rely on these best guesses of the true, partially-observable game as the foundation of their decision making, yet finding these information sets is expensive.

We apply reactive Answer Set Programming (ASP) to the problem of sampling information sets in the field of General Game Playing. Furthermore, we use this domain as a test bed for evaluating the effectiveness of oClingo, a reactive answer set solver, in avoiding redundant search by keeping learnt clauses during incremental solving.

1 Introduction

General Game Playing (GGP) research seeks to design systems able to understand the rules of new games and use such descriptions to play those games effectively. These systems must reason their way from the unadorned rules to a strategy capable of defeating adverse opponents under tight time constraints. The recent extension to stochastic games with imperfect information makes this process even harder by requiring players to also reason about knowledge and plan under uncertainty.

In game theory, the *information set* for a specific player is a collection of models (possible histories) of the current state of the game, that are each consistent with all observations made so far, and by extension are indistinguishable for that player [7]. Consider a simple game of 'number guessing' where a player must guess a (random) hidden number by asking a series of 'is the number < n?' questions. Clearly the best strategy is a binary search—by partitioning the search space in half each time we can be guaranteed a logarithmic worst-case. Further, this discovery can be detected in a game-general way by explicitly maintaining every model in the information set. This can be seen as the possible worlds approach. However the size of a typical game is so enormous that maintaining *every* world is impossible.

One response to the limits of a possible worlds approach is to accept a subset of all worlds. Traditional perfect information tree search can then be employed; this is an efficient (and sometimes admissible) substitute for genuinely reasoning about imperfect information [5,10]. In this scenario, a model is 'sampled' from

P. Cabalar and T.C. Son (Eds.): LPNMR 2013, LNAI 8148, pp. 219–232, 2013.
© Springer-Verlag Berlin Heidelberg 2013

the full set, either by progressing (and pruning) all possible worlds up to a fixed size [2], or by re-generating models from the rules. There is evidence that this latter case can be a sufficient approximation in a competition setting [10]. However this generation is expensive.

With this motivation, we seek to expand the current bounds on information set sampling in GGP through a conventional technology—Answer Set Programming (ASP). Specifically, we will benchmark the set sampling problem on Clingo and then compare against the newer oClingo to assess its claims of avoiding redundant search via learnt clauses. We test this problem on three games that have been unplayable at international GGP competitions.[1]

The rest of the paper is organised as follows: first, we formally introduce the Game Description Language, the *gringo* syntax for a logic program, and the *oClingo* extension. In section 4 we explain how to translate GDL to a logic program. Next we describe our experimental setup and present our findings. We conclude with a short discussion.

2   Game Description Language

The science of General Game Playing requires a formal language that allows an arbitrary game to be specified by a complete set of rules. The declarative Game Description Language (GDL) serves this purpose [4]. It uses a logic programming-like syntax and is characterised by the special keywords listed in Table 1.

Table 1. GDL-II keywords

role(?r)	?r is a player
init(?f)	?f holds in the initial position
true(?f)	?f holds in the current position
legal(?r,?m)	?r can do ?m in the current position
does(?r,?m)	player ?r does move ?m
next(?f)	?f holds in the next position
terminal	the current position is terminal
goal(?r,?v)	?r gets payoff ?v
sees(?r,?p)	?r perceives ?p in the next position
random	the random player (aka. Nature)

Originally designed for games with complete information [4], GDL has recently been extended to GDL-II (for: *GDL with incomplete/imperfect information*) by the last two keywords (sees, random) to describe arbitrary (finite) games with randomised moves and imperfect information [13].

Example 1. The GDL-II rules in Fig. 1 formalise a simple game in which a player, whose role name is "guesser", must guess a randomly chosen number

---

[1] 1st Australian Open 2012, see https://wiki.cse.unsw.edu.au/ai2012/GGP

from 1 to 16. The player can ask a series of 'is the number $< n$?' questions before announcing that it is ready to guess.

The intuition behind the rules is as follows.[2] Line 1 introduces the players' names. Lines 3–6 define some basic arithmetic relations as background knowledge. Line 8 defines the two features that comprise the initial game state. The possible moves are specified by the rules for **legal**: in the first round, the **random** player chooses a number (lines 10–11); then the guesser can repeatedly ask "**lessthan**" questions (line 14) until it decides that it is ready to guess (line 15), followed by making a guess (lines 16). The guesser's only percepts are true answers to its yes-no question (lines 18–21). The remaining rules specify the state update (rules for **next**); the conditions for the game to end (rule for **terminal**); and the payoff, which in case of the guesser depends on whether it got the number right and how long it took (rules for **goal**).

GDL-II comes with some syntactic restrictions—for details we must refer to [6,13] due to lack of space—that ensure that every valid game description has a unique interpretation as a state transition system as follows. The **players** in a game are determined by the derivable instances of **role(?r)**. The **initial state** is the set of derivable instances of **init(?f)**. For any state S, the **legal moves** of a player **?r** are determined by the instances of **legal(?r,?m)** that follow from the game rules *augmented by an encoding of the facts in S* using the keyword **true**. Since game play is synchronous in the Game Description Language,[3] states are updated by *joint* moves (containing one move by each player). The **next position** after joint move m is taken in state S is determined by the instances of **next(?f)** that follow from the game rules *augmented by an encoding of m and S* using the keywords **does** and **true**, respectively. The **percepts** (aka. information) a player **?r** gets as a result of joint move m being taken in state S is likewise determined by the derivable instances of **sees(?r,?p)** after encoding m and S using **true** and **does**. Finally, the rules for **terminal** and **goal** determine whether a given state is **terminal** and what the players' **goal values** are in this case.

On this basis, game play in GDL-II follows this protocol:

1. Starting with the initial state, which is completely known to all players, in each state each player selects one of their legal moves. By definition **random** must choose a legal move with uniform probability.
2. The next state is obtained by (synchronously) applying the joint move to the current state. Each role receives their individual percepts resulting from this update.
3. This continues until a terminal state is reached, and then the goal relation determines the result for all players.

---

[2] A word on the syntax: We use infix notation for GDL-II rules as we find this more readable than the usual prefix notation.

[3] Synchronous means that all players move simultaneously. Turn-taking games are modelled by allowing players only one legal move without effect (such as **noop**) if it is not their turn.

```
1 role(guesser).     role(random).
2
3 succ(0,1).   succ(1,2).   ...    succ(15,16).
4 number(?n) <= succ(?m,?n).
5 less(?m,?n) <= succ(?m,?n).
6 less(?m,?n) <= succ(?m,?k), less(?k,?n).
7
8 init(step(0)).     init(starttime).
9
10 legal(random,choosenumber(?n)) <= number(?n), true(starttime).
11 legal(random,noop) <= not true(starttime).
12
13 legal(guesser,noop)            <= true(starttime).
14 legal(guesser,lessthan(?n)) <= number(?n), true(questiontime).
15 legal(guesser,readytoguess) <=  true(questiontime).
16 legal(guesser,guess(?n))     <= number(?n), true(guesstime).
17
18 sees(guesser,yes) <= does(guesser,lessthan(?n)),
19                       true(secretnumber(?m)), less(?m,?n).
20 sees(guesser, no) <= does(guesser,lessthan(?n)),
21                       true(secretnumber(?m)), not less(?m,?n).
22
23 next(secretnumber(?n)) <= does(random,choosenumber(?n)).
24 next(secretnumber(?n)) <= true(secretnumber(?n)).
25
26 next(questiontime) <= true(starttime).
27 next(questiontime) <= true(questiontime), not does(guesser,readytoguess).
28 next(guesstime)    <= does(guesser,readytoguess).
29 next(right)        <= does(guesser,guess(?n)), true(secretnumber(?n)).
30 next(end)          <= does(guesser,guess(?n)).
31 next(step(?n))     <= true(step(?m)), succ(?m,?n).
32
33 terminal <= true(end).
34 terminal <= true(step(16)).
35
36 goal(guesser,100) <= true(right), true(step(?n)), less(?n,8).
37 goal(guesser, 90) <= true(right), true(step(?n)), less(?n,9).
38 ...
39 goal(guesser, 10) <= true(right), true(step(?n)), less(?n,16).
40 goal(guesser,  0) <= not true(right).
41 goal(random,   0).
```

Fig. 1. The GDL-II description of the Number Guessing game at the AI2012 GGP Competition

3   Logic Programming, *gringo*, and reactive ASP

First we recapitulate standard logic programming and answer set programming terminology. Rules are of the form $h_r \leftarrow a_1, \ldots, a_m, not\ a_{m+1}, \ldots, not\ a_n$. where each a_i is an atom of the form $p(t_1, \ldots, t_k)$ and each t_i is a term (constant, variable, or function). The *head* h_r of rule r is either an atom, a *cardinality constraint* of the form $l\{h_1, \ldots, h_k\}u$ in which l, u are integers and $h1, \ldots, h_k$ are atoms, or the special symbol $\perp$. If h_r is a cardinality constraint, we call r a *choice rule*, and an *integrity constraint* if $h_r = \perp$. We denote the atoms occurring in h_r by $head(r)$, ie. $head(r) = \{h_r\}$ if h_r is an atom, $head(r) = \{h_1, \ldots, h_k\}$ if $h_r = l\{h_1, \ldots, h_k\}u$, and $head(r) = \emptyset$ if $h_r = \perp$. The atoms occurring positively and negatively in the body are denoted by $body(r)^+ = \{a_1, \ldots, a_m\}$ and $body(r)^- = \{a_{m+1}, \ldots, a_n\}$. A logic program R is a set of rules; $atom(R)$ denotes the set of atoms occurring in R. $head(R) = \cup_{r \in R} head(r)$ is the collection of all head atoms. The ground program $grd(R)$ is the set of all ground rules constructable from rules $r \in R$ by substituting every variable in r with some element of the Herbrand Universe of R. For further details we recommend [1,11,3].

We now examine Incremental Logic Programs, an extension of logic programming as described above. Incremental programs are constructed from modules, which for the purposes of this paper are effectively subprograms. An Incremental Logic Program $(B, P[t], Q[t])$ is composed of a base module B of time-independent ('rigid') rules, and two parameterised modules: a 'cumulative' module $P[t]$ (instantiated at each successive timestep t and which is accumulated) and a 'volatile' module $Q[t]$ (which is forgotten after each timestep; only one instantiation exists at a time). This is further extended by oClingo to produce an Online Incremental Logic Program. These programs are accompanied by an 'online progression'—a sequence of input atoms for each timestep t. oClingo programs rely on #external directives as domain predicates for grounding rules that rely on these input atoms.

As a final note, a great strength of the Potassco suite of ASP solvers is that clause learning is 'baked in'.

4   Translation

An ASP system is a natural platform for the Game Description Language, due to the finiteness guarantee, uninterpreted functions[4], and the presence of negation-as-failure. Indeed GDL is an extension of Datalog¬ with function symbols, so a syntactic translation is fairly direct [12]. We will now briefly summarise this process, which converts GDL rules to the gringo input language. After this, we will present a modification that produces rules suitable for oClingo as well.

The key aspect of this translation is the 'temporal extension' of the GDL features—GDL has implicit timepoints (initial, current, and next) which must be made explicit for an ASP system. That is, init rules initialise fluents for time zero. Rules for legal or the value of derived fluents are functions of the current time (relative to a state). Fluent update needs to reference the fluent's value at the 'next' timepoint (relative to the current time). This extension is largely achieved by wrapping fluents in binary holds(F,T) relations that tie the fluent F to a given timepoint T. Fluent update is handled by rules for holds with a timepoint one step ahead of the timepoints in the body ($T + 1$ vs T). Derived fluents have the same timepoint in the head and the body.

As noted in the original translation paper [12], this method temporalises all user (derived) rules, even if they are time-independent 'rigids'. This introduces a substantial increase in redundant grounding. As such, we will first formally define the notion of a rigid rule in terms of the dependency graph of the GDL

---

[4] That is, functions have no fixed interpretation and must be specified by other axioms. ASP in contrast typically interprets + as addition (and similarly for other simple arithmetic operators). This means no additional logic needs to be ported along with the GDL when translating to ASP.

rules. Then we present our augmented translation that ensures rigids are left unadorned.

Definition 1. *Construct the dependency graph* $D = (V, E)$ *of a set of GDL rules* G *as follows:*

- *The vertex set* V *contains all predicate symbols found in* G.
- *If predicate symbol* a *appears in the head of some rule* $r \in G$ *and predicate symbol* b *appears in the body of* r, *then* D *has an edge from* b *to* a, *ie.* $(b, a) \in E$.

With the dependency graph, we can now formally define the common notion of rigid rules:

Definition 2. *A rule* $h(a_1, \ldots, a_m)$ <= $b_1, \ldots, b_n$ *is* rigid *wrt a set of GDL rules* G *iff there is no path from* h *to* true *or* does *in the dependency graph for* G.

We now present the main translation:[5]

Definition 3. *Let* G *be a set of GDL rules, then the* temporal extension *of* G, *written* $ext(G)$, *is the set of logic program clauses obtained from* G *as follows. Each occurrence of:*

- init(ϕ) *is replaced by* holds(ϕ,0).
- true(ϕ) *is replaced by* holds(ϕ,T), *and each* next(ϕ) *by* holds(ϕ,T+1).
- sees(R,ϕ) *is replaced by* sees(R,ϕ,T+1).
- distinct(t_1,t_2) *is replaced by* not $t_1 = t_2$.
- $p(t_1, \ldots, t_n)$ *where* p *is keyword* does, legal, terminal, *or* goal *is replaced by* $p(t_1, \ldots, t_n, T)$.
- $p(t_1, \ldots, t_n)$ *where* p *is rigid (by Definition 2) is left unadorned.*[6]

All other atoms $p(t_1, \ldots, t_n)$ *are replaced by* derived($p(t_1, \ldots, t_n), T$) *(or by* derived($p(t_1, \ldots, t_n), 0$) *if they are in the body of an* init *rule).*

In order to produce a valid program, these rules must also be augmented with information about the moves and percepts seen to date, constraints on move selection, and a domain predicate for timepoint variables:

Definition 4. *Given a set of GDL rules* G, *a role name* N, *a round number* $R \geq 1$, *the* move history H *of player* N *(a set of* R does *rules, one for each timepoint) and a set of percepts* P *(of form* observed(S,T) *where* S *is a ground percept and* $T \in [0, R)$ *is the timepoint), construct a logic program* L *containing:*

- *the temporal extension of* G *(by Definition 3).*
- *a time domain predicate* time(0..R-1). *(or* time(0). *if* $R = 1$).
- *our move history* H.

---

[5] Due to space constraints we cannot present a full translation and instead refer to [12].

[6] Note that this includes keyword role due to restrictions in the GDL specification [6].

— *an action 'generator' (choice rule)*
```
{ does(R,A,T) } :- role(R), time(T), legal(R,A,T).
```
— *a unique action constraint*
```
:- not 1 { does(R,A,T) : input(R,A) } 1, role(R), time(T).
```
— *constraints to guarantee correct percepts are generated*
```
:- sees(N,P,T+1), not observed(P,T+1), time(T). and
:- not sees(N,P,T+1), observed(P,T+1), time(T).
```

The logic program produced by Definition 4 is now sufficient to produce a sample of the information set and is the basis for our experiments. Note that we also intend to apply this program to GGP competitions where we only want Clingo to report back the latest game *state*, ie. `holds` statements (since the state, not the history, is the foundation for move selection). This can be achieved with the directives `#hide. #show holds/2.` appended to the rules. Note that our introduction of a `derived` keyword (not present in the original translation) allows us to easily retrieve the complete state if this is preferred.

This translation scheme was conceived for standard ASP systems, but we also wish to employ the newer, reactive *oClingo*—we want to measure the benefit of an incremental logic program to this domain. This introduces two new subtleties: first, the latest timepoint is `t`, so 'next' rules must occupy this time (ie t instead of $T+1$), and 'now' rules must be `t-1` (instead of T)—timepoints will need to be shuffled. A further complexity is that oClingo—for reactive, *incremental* logic programs—has a program that must adhere to module theory, and in particular a firm modularity condition[7].

We first present the alternate temporal extension for an oClingo-compatible domain, and then the game-independent rules that tell oClingo what problem to solve.

Definition 5. *Let G be a set of GDL rules, then the* reactive temporal extension *of G, written oExt(G), is the set of logic program clauses obtained from G as follows. For each rule, adorn the head:*

head	replaced by	time variable in body
$init(\phi)$	$holds(\phi, 0)$	0
$next(\phi)$	$holds(\phi, t)$	$t-1$
$legal(R, A)$	$legal(R, A, t-1)$	$t-1$
$sees(R, P)$	$sees(R, P, t-1)$	$t-1$
$terminal$	$terminal(t)$	t
$goal(R, V)$	$goal(R, V, t)$	t
$p(a_1, \ldots, a_n)$; p is not rigid	$derived(p(a_1, \ldots, a_n), t-1)$	$t-1$
	otherwise the head is unmodified (it and its body are rigid)	

Now update the atoms in the bodies with the appropriate time variable (as determined by the head of the rule):

---

[7] Due to space constraints we must defer this technical detail to [3].

GDL	time variable X (determined by head)
$true\,(\phi)$	$holds\,(\phi, X)$
$does\,(R, A)$	$does\,(R, A, X)$
$distinct\,(t_1, t_2)$	$not\ t_1 = t_2$
$p(a_1, \ldots, a_n)$; p is not rigid	$derived\,(p(a_1, \ldots, a_n), X)$
	otherwise the atom is unmodified (it is rigid)

Definition 6. *Given a set of GDL rules G, a role name N, the move history H of player N (a set of R **does** rules, one for each timepoint) and a set of percepts P (of form **observed**(S, T) where S is a ground percept and T is the timepoint), construct an reactive, incremental logic program L containing:*

- *the reactive temporal extension of G (by Definition 5). Note that the rigid rules go in the base module, all other rules go in the cumulative section.*
- *domain predicates* **input**(R, A) *and* **percepts**(P) *for actions A and percepts P.*
- *#external declarations: #external exec/2. #external observed/2.*
- *an action 'generator' (choice rule)*
  { does(R,A,t-1) } :- role(R), legal(R,A,t-1).
- *a combined uniqueness+liveness constraint*
  :- not 1 { does(R,A,t-1) : input(R,A) } 1, role(R).
- *correct action constraint*
  :- not does(N,A,t-1), exec(A,t-1), input(N,A).
- *constraints to guarantee correct percepts are generated*
  :- sees(N,P,t-1), not observed(P,t-1). and
  :- not sees(N,P,t-1), observed(P,t-1), percepts(P).

And construct an online progression O, as a contiguous sequence of steps of the form:

```
#step X.
exec(A,X-1).
observed(P,X-1).
#endstep.
```

*For each round $X \geq 1$. Note that each step will contain exactly one **exec** statement (player N executed action A at time $X - 1$) and zero or more **observed** statements for the percepts that resulted from that action (as per Definition 4).*

These straight-forward procedures have two additional problems that we have not yet discussed: domain predicates are not always present, and GDL permits a large class of symbols for its identifiers[8]. Obviously the naming issue can be addressed with a simple symbol table. The problem of domain predicates is starting to be mitigated by a growing convention in the GGP community to supply

---

[8] For example hyphens, which ASP systems tend to interpret as a subtraction operator (or classical negation, based on context).

these domains with 'input' and 'base' keywords (for actions and fluents, respectively). However no such keyword has been proposed for percepts. Finding the minimal model of the negation-free program is a reasonably efficient method for grounding these domains on the back-catalog of games without these predicates. Alternatively, more efficient GDL-centric methods have been proposed [12,9], though these are beyond the scope of this paper.

Regarding timepoints. You may note that the choice of actions (does) occurs at time T in Clingo and time $t - 1$ in oClingo. Similarly percepts (sees) occur at time $T + 1$ compared with $t - 1$ between the two versions. The reason for this is historical: the constraints on oClingo are firm[9], but the translation for Clingo was done first (and follows the original translation from [12]). Other variations are possible, however these translations are the ones we tested, and so these are the ones we present.

5 Method

In order to reason about the rules of a game we must first convert them from GDL to an ASP encoding, as presented in Section 4. Next we generate a random play through of the game for each role. This yields a collection of legal (reachable) states, the joint moves that led to those points, and the percepts that each role would see at each step. By replaying one set of moves and percepts for a select player, Clingo (or oClingo) can recreate the state (or find equivalent states subject to its imperfect information). That is, it can sample the information set.

In our experiments we generate 100 random plays for each game for each role[10]. We then ask (o)Clingo to solve for a sample of the game's information set at each round. All times are averaged over three duplicate runs. Experiments were performed on the UNSW cluster to satisfy the time and RAM constraints. Note that individual runs used a single 2.20 GHz Opteron core, but were allocated a complete node (48 processor cores) to eliminate interference from other processes.

We explicitly point out here that our results only measure the time to achieve the *first* model, since we did not have time to repeat our experiments for larger sample sizes. However this is still a useful metric: a single model is enough to start evaluating moves in a game player. Further, the process can be dynamically improved as more models are reported (as in [10]). From this perspective, the time-to-first model is the *most* useful measure of the value of our (ASP) set-sampler, since this is the 'dead time' before the GGP system can start making decisions.

We also ran oClingo with the --ilearnt=forget flag, which disables clause learning between timesteps (ie. clauses learnt in timestep n are thrown away before timestep $n + 1$ begins). Comparing oClingo's performance with and without

---

[9] Facts added 'to the future' are prone to either violating oClingo's modularity condition, or being ignored by the target module parameterisation.

[10] This number was reduced for the larger games due to time constraints.

this feature should demonstrate the value of Incremental Logic Programs for this type of search problem, as well as validate claims regarding the effectiveness of oClingo's clause learning. Finally, by measuring precisely the effect of clause learning between timesteps we can account for how significant its impact is, while controlling for other (smaller) differences between the Clingo and oClingo systems.

Due to the youth of GDL-II and the complexity of games it describes, there is a distinct lack of rules that tax an ASP system under our use-case. Early tests revealed that most games are slow to ground, but their game trees are then fairly simple. For most rounds of most games, both Clingo and oClingo consistently solved the search problem presented in fractions of a second. As such our experiments focus on the role of grounding, and we have chosen three of the hardest domains for the task. These games, taken from past international competitions, are:

Blind Breakthrough. A two-player, zero-sum, turn-taking game played on a chess board. Each player has two rows of pieces against their side, but all pieces are pawns. The winner is the first player to reach the other side of the board ('break through' the opponent's ranks). The 'blind' aspect indicates that a player cannot see the opponent's pieces and is instead informed of the success/failure of attempted moves and the existence of a capturing move. We vary the board size between 6x6, 7x7, and 8x8 squares.

Battleships in Fog. Two navies (on separate, 4x4 grid oceans) can fire at their opponent and are informed of hit/miss. In this variant, players may also sail their single, two-by-one cell ship to an adjacent square, or perform a 'noisy sensing' action that returns three possible opponent locations (one correct, the other two not).

Small Dominion. Players each have a small hand of cards (either money or land) that is filled from a larger, face-down deck. Several low-value cards (eg. copper) can be used to buy a single higher-value card (eg. silver). Doing so allows a player to slowly increase the value of their hand and get more 'victory points' as a result. The game finishes when certain sets of cards are exhausted. These rules yield an interesting alternate strategy where a player buys low-value cards as quickly as possible in order to trigger an early game termination (before the opponent has won by high-value cards).

6   Results

The first and most important observation is that these domains are hard because their search spaces are huge. But these are human-playable games, which suggests some structure must exist on their game trees. This is reflected in our results: grounding remains the most significant factor in the time to find a model, while actually 'solving' is lightning quick. An exception to this case is Blind

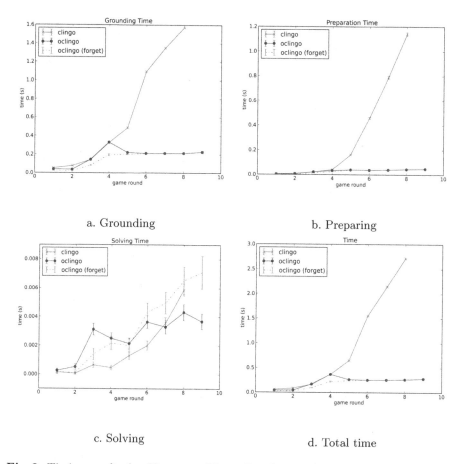

a. Grounding   b. Preparing

c. Solving   d. Total time

Fig. 2. Timing results for Clingo vs oClingo. Results are for Blind Breakthrough and averaged across all board sizes. Error bars indicate a 95% confidence interval.

Breakthrough where solving time can be higher due to the myriad interleavings of moves explaining the same observation.

The second observation is that grounding can be prohibitive in this space and it is necessarily exacerbated by oClingo because it offers (potential) speedups later in a game tree by doing extra work[11] at each step. This was catastrophic for the game of Small Dominion, where the dealer (**random**) chooses three—mostly unused—random values in every round. Obviously this is a poor axiomatisation from an ASP perspective (all these unused values must be ground *before* they can be ignored), but this is the reality of the input GDL, where such encodings are fine for Prolog-based systems. It should be noted that this 'extra work' is clearly

---

[11] Eg. the grounding process needs to account for all the possible external inputs. In contrast, Clingo needs to ground only the inputs it actually receives—a liberty afforded to it since the 'externals' must be provided up-front with the program itself.

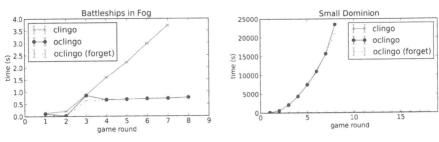

a. Total time in Battleships in Fog   b. Total time in Small Dominion

Fig. 3. Timing in Battleships and Small Dominion. Note that Clingo is dramatically more effective in Small Dominion.

at fault, since straight Clingo was still competitive in this domain. This indicates that there is a cross-over point: small domains are easy for both systems, oClingo has a strong advantage for medium-size domains, but then falls behind as the additional grounding increases and its rewards diminish. That is, after this point oClingo is swamped by its own optimisation.

7   Conclusion

It is clear that oClingo's ability to avoid redundant search and grounding offers an impressive speed-up of over its predecessor Clingo. However this gain is tempered by the size of the target domain; medium-size domains benefit most since they are complex enough to utilise the learnt clauses, but not so large as to grind to a halt whilest grounding. For the field of General Game Playing these features literally increase the horizon of 'solvable' domains. Further, this is achieved within the time constraints of a typical GGP competition—this system is ready for competition play. oClingo is not a silver bullet though, and the largest games are still well beyond the reach of game-general set sampling techniques.

Using an ASP system for a fixed-size sampling of a game's information set is not the only approach to the problem of imperfect-information game play; possible worlds systems store and incrementally update the complete information set. As an efficiency-oriented optimisation, 'particle filter' systems [2] maintain and progressively filter a large subset of the possible worlds—this helps mitigate the capacity demands of storing huge search spaces. Filtering has also been augmented with backtracking in order to avoid pruning all the possible worlds away [10]. This approach excels when successive information sets are local on the game tree, but complex games bring out its exponential complexity. Yet complex games are the interesting ones: games with high branching factors, long periods without percepts, or multiple but very different[12] explanations for the same observations. All of these properties are found in our harder test domains—Blind

---

[12] ie. distant on the game tree

Breakthrough, Battleships in Fog, and Small Dominion—and demonstrate that finding the information set under these constraints is fundamentally a search problem, where an efficient, domain-independent system like an ASP solver is well-suited. Of course other, more efficient methods are also possible when additional assumptions can be made about the domain [8]. A full side-by-side comparison of these methods remains as critical future work.

One notable shortcoming of our approach is the absence of model weights—we find unique histories that describe the current state, but some states are more likely than others. Essentially this means opponent modelling, which is beyond the scope of this paper. Applying soft constraints or gringo's `#maximize` statements to this problem would also be valuable future work.

Acknowledgements. This research was supported under the Go8-DAAD Australia-Germany Joint Research Cooperation scheme and DFG grant SCHA 550/9-1. The third author is the recipient of an ARC Future Fellowship (FT 0991348).

References

1. Baral, C.: Knowledge Representation, Reasoning, and Declarative Problem Solving. Cambridge University Press, New York (2003)
2. Edelkamp, S., Federholzner, T., Kissmann, P.: Searching with partial belief states in general games with incomplete information. In: Glimm, B., Krüger, A. (eds.) KI 2012. LNCS, vol. 7526, pp. 25–36. Springer, Heidelberg (2012)
3. Gebser, M., Kaminski, R., Kaufmann, B., Ostrowski, M., Schaub, T., Thiele, S.: Engineering an incremental ASP solver. In: Garcia de la Banda, M., Pontelli, E. (eds.) ICLP 2008. LNCS, vol. 5366, pp. 190–205. Springer, Heidelberg (2008)
4. Genesereth, M.R., Love, N., Pell, B.: General game playing: Overview of the AAAI competition. AI Magazine 26(2), 62–72 (2005),
http://games.stanford.edu/competition/misc/aaai.pdf
5. Long, J.R., Sturtevant, N.R., Buro, M., Furtak, T.: Understanding the success of perfect information monte carlo sampling in game tree search. In: Fox, M., Poole, D. (eds.) AAAI. AAAI Press (2010)
6. Love, N., Hinrichs, T., Haley, D., Schkufza, E., Genesereth, M.: General game playing: Game description language specification. Tech. Rep. LG–2006–01, Stanford Logic Group (2006)
7. Rasmusen, E.: Games and Information: an Introduction to Game Theory, 4th edn. Blackwell Publishing (2007)
8. Richards, M., Amir, E.: Information set sampling for general imperfect information positional games. In: Proc. IJCAI 2009 Workshop on GGP, GIGA 2009, pp. 59–66 (2009)
9. Saffidine, A., Cazenave, T.: A forward chaining based game description language compiler. In: Proc. IJCAI 2011 Workshop on GGP, GIGA 2011 (July 2011)
10. Schofield, M., Cerexhe, T., Thielscher, M.: Hyperplay: A solution to general game playing with imperfect information. In: Proc. AAAI, Toronto (July 2012)

11. Simons, P., Niemelá, I., Soininen, T.: Extending and implementing the stable model semantics. Artificial Intelligence 138(1-2), 181–234 (2002)
12. Thielscher, M.: Answer set programming for single-player games in general game playing. In: Hill, P.M., Warren, D.S. (eds.) ICLP 2009. LNCS, vol. 5649, pp. 327–341. Springer, Heidelberg (2009)
13. Thielscher, M.: A general game description language for incomplete information games. In: Proc. AAAI, Atlanta, pp. 994–999 (July 2010)

VCWC: A Versioning Competition Workflow Compiler*

Günther Charwat[1], Giovambattista Ianni[2], Thomas Krennwallner[1],
Martin Kronegger[1], Andreas Pfandler[1], Christoph Redl[1], Martin Schwengerer[1],
Lara Katharina Spendier[1], Johannes Peter Wallner[1], and Guohui Xiao[1]

[1] Institute of Information Systems, Vienna University of Technology, 1040 Vienna, Austria
[2] Dipartimento di Matematica e Informatica, Università della Calabria, 87036 Rende (CS), Italy

1 Introduction

System competitions evaluate solvers and compare state-of-the-art implementations on benchmark sets in a dedicated and controlled computing environment, usually comprising of multiple machines. Recent initiatives such as [6] aim at establishing best practices in computer science evaluations, especially identifying measures to be taken for ensuring repeatability, excluding common pitfalls, and introducing appropriate tools. For instance, Asparagus [1] focusses on maintaining benchmarks and instances thereof. Other known tools such as Runlim (http:/fmv.jku.at/runlim/) and Runsolver [12] help to limit resources and measure CPU time and memory usage of solver runs. Other systems are tailored at specific needs of specific communities: the not publicly accessible ASP Competition evaluation platform for the 3rd ASP Competition 2011 [4] implements a framework for running a ASP competition. Another more general platform is StarExec [13], which aims at providing a generic framework for competition maintainers. The last two systems are similar in spirit, but each have restrictions that reduce the possibility of general usage: the StarExec platform does not provide support for generic solver input and has no scripting support, while the ASP Competition evaluation platform has no support for fault-tolerant execution of instance runs. Moreover, benchmark statistics and ranking can only be computed after all solver runs for all benchmark instances have been completed.

A robust job execution platform is a basic requirement for a competition. During benchmark evaluation, several different kinds of failures may happen, mainly **(a)** programming errors in the participant software; **(b)** software bugs in the solution verification programs; or **(c)** hardware failures during a run, which may be local to a machine (e.g., harddisk or memory failure), or global (e.g., when the server room air condition fails).

Moreover, a competition platform must be flexible enough to allow for "late" or updated benchmark and solver submissions. It is not uncommon that anomalies arise during the execution, and changing the course of an evaluation after the platform has started is cumbersome and requires manual effort for the competition maintainers.

A fault-tolerant design helps the competition maintainers to perform all steps and minimizes the action required to come back to a safe state. To address these issues, we introduce the Versioning Competition Workflow Compiler (VCWC) system. VCWC uses a two-step approach: first, a workflow for a competition track is generated; a workflow is a dependency description of jobs that need to be executed in order to come to a

* This research is supported by the Austrian Science Fund (FWF) project P20841 and P24090.

P. Cabalar and T.C. Son (Eds.): LPNMR 2013, LNAI 8148, pp. 233–238, 2013.

ranking of solvers that participate in a competition track. Then, a versatile job scheduling system takes this workflow and executes it. Specifically, VCWC is based on (i) GNU Make and GNU M4 for building the track execution workflow, (ii) the HTCondor [15] high throughput computing platform, which provides flexible means to support the requirements of running a competition, like automated job scheduling on a collection of benchmark servers, and (iii) the Directed Acyclic Graph Manager (DAGMan) [5], a meta-scheduler for HTCondor that maintains the dependencies between jobs and provides facilities for a reliable, fault-tolerant, and self-healing execution of benchmarking workflows. VCWC is open source and implemented using standard UNIX tools, thus it runs on every UNIX-like system that has support for those utilities. VCWC is maintained at https:/github.com/tkren/vcwc, and an extended version of this paper is available at http:/www.kr.tuwien.ac.at/staff/tkren/pub/2013/lpnmr2013-vcwc.pdf.

2   Modeling a Competition

In this section, we describe the basic building blocks of a solver competition. We assume familiarity with the notion of *(computational) problem*, *instance*, and *solution* for a problem; an overview is given, e.g., in [10].

A *benchmark B* is a set of instances I from a well-defined computational problem, where all instances are represented in a standardized format (e.g., as logic programs or as CNF clauses). A *solver S* is an implementation for an algorithm that computes the solution for a given instance I from a benchmark B, where solutions are represented in a standardized format. Given as set of benchmarks $\mathcal{B}$ and a set of solvers $\mathcal{S}$, we define a *track T* as a subset of $\mathcal{B} \times \mathcal{S}$ that is both left-total and right-total, i.e., for each $B \in \mathcal{B}$ there exists an $S \in \mathcal{S}$ such that $(B, S) \in T$, and for every $S \in \mathcal{S}$ there exists a $B \in \mathcal{B}$ such that $(B, S) \in T$. Intuitively, $(B, S) \in T$ means that solver S participates in track T in solving benchmark B. Each track has an associated computation environment $env(T)$ with a fixed number of CPUs, memory size, and available disk space. The set of all participating solvers to a track T is $\mathcal{S}(T) = \{S \mid (S, B) \in T\}$ and the set of all benchmarks is $\mathcal{B}(T) = \{B \mid (S, B) \in T\}$. Then, a *competition* is a collection of tracks. A *run R* of solver S on instance I in track T is the evaluation of S with instance I within the limits of the computation environment $env(T)$. A run has an associated solution $sol(R)$ and performance measurements for evaluation metrics such as runtime and memory usage. In a competition track, every instance is evaluated $k > 1$ times to eliminate outliers and to provide well-founded statistical results.

For example, in the ASP Competition series [3], a *system track T* forms a complete bipartite graph $(\mathcal{B} \cup \mathcal{S}, T)$, i.e., every solver participates in solving all benchmarks. On the other hand, the *model & solve* track does not have this restriction, a participant may choose the benchmarks to solve. Furthermore, tracks are usually classified as *sequential* or *parallel*, which means that their computation environment has exactly one CPU in case of sequential tracks, or more than one CPU in case of parallel tracks.

Several tasks need to be performed in order to evaluate a solver's performance relative to other solvers that participate in a certain track. The outcome of a competition is a ranking of the participating solvers, which should summarize the performance of a solver S on benchmark $\mathcal{B}$ relative to the other solvers that participate in a track.

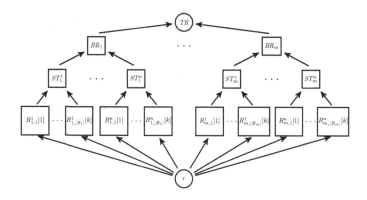

Fig. 1. Competition workflow for a track with m benchmarks and n solvers

A *solution verification* $ver(R)$ of run R is a mapping $ver(R) \in \{0, 1, 2\}$ such that $ver(R) = 0$ whenever $sol(R)$ is not a solution for I, $ver(R) = 1$ for $sol(R)$ being a correct solution for I, and $ver(R) = 2$ otherwise. Note that $ver(R)$ might implement an incomplete verification algorithm, as solution verification could be a computationally hard task. The *solver summary statistics* $sumstat(S, B)$ computes for all runs $R_1, R_2, \ldots$ of solver S on instances I from benchmark B the performance measurements of those runs as summary statistics such as means, median, etc., for all instances $I \in B$. Based on $sumstat(S, B)$, the *benchmark ranking* $bmrank(B)$ of a benchmark B ranks each solver $S \in \mathcal{S}$ based on a predefined benchmark scoring function. Then, the *track ranking* $trackrank(T)$ generates a combined performance evaluation of a track T based on scoring function for $bmrank(B)$ for all benchmarks $B \in \mathcal{B}$.

Modeling the Dependencies in a Competition. As described above, several steps are necessary to generate the outcome $trackrank(T)$ of a competition track T. When combining all the tasks in a dependency graph, where nodes represent tasks and an edges (u, v) represent a dependency between u and v such that u must be executed before v, we get a task model of the competition track, which, when executed in sequence, computes all prerequisite information for each task properly and generates the desired outcome. Such an acyclic dependency graph constitutes a track execution workflow whose tasks can be possibly executed in parallel using proper job scheduling software.

Based on the competition tasks introduced before, we explicitly outline in Fig. 1 the implicit dependencies of the tasks and show a competition workflow that can be used to perform all necessary computational tasks in a competition. Let $n = |\mathcal{S}|$, $m = |\mathcal{B}|$, and k be the number of runs per instance. Nodes $R^u_{v,w}[i]$ stand for the tasks associated with the i-th run, $1 \leq i \leq k$, of solver S_u on instance I_v of benchmark B_w. These tasks are comprehensive of computing the solution and perform the respective verification. The nodes ST^u_w represent the solver summary statistics task of solver S_u in benchmark B_w, i.e., ST^u_w takes all runs executed and verified on S_u that are associated with instances from B_w and creates summary statistics. Then, nodes BR_w represent the benchmark ranking jobs that are connected to all ST^u_w for $1 \leq u \leq n$. The topmost

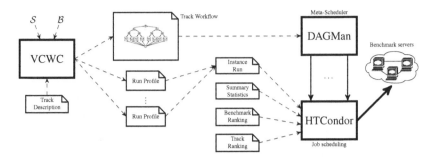

Fig. 2. VCWC System Architecture (dashed lines: data flow, solid lines: call flow)

node TR is the track ranking task in a competition, while the lowest node r gives us the computation root, a unique entry point in the workflow without associated task.

Workflow Versioning. A further benefit of modeling a competition track as a workflow is to have a graph-based representation of tasks that can be easily modified and updated when basic constituents of a track change. To address the problem of late participant submissions or fixing broken benchmark instances or benchmark verification scripts after the competition start, we can introduce a workflow versioning mechanism for incrementally changing the competition execution workflow. Without details, one can add fresh participants, further benchmarks (or instances), or more runs. Additions and removals do not have impact on previously stored executions of the workflow, and statistics will be updated accordingly.

3   Implementation of the VCWC System

The system architecture of VCWC is shown in Fig. 2. The main components are (i) the VCWC compiler, which generates a competition workflow description and profiles for instance parameters; (ii) DAGMan (Directed Acyclic Graph Manager), a meta-scheduler for managing dependencies between jobs built on top of (iii) HTCondor, a job scheduler for building high-throughput computing environments.

VCWC expects two directories as input: a benchmarks directory with all possible benchmarks $\mathcal{B}$ assigned to track T as subdirectories, and a dedicated `participants` directory containing subdirectories for each possible benchmark of a track; participating solvers $\mathcal{S}$ can then choose which benchmark they want to solve. VCWC further takes a track description file as input that records various parameters of a track.

In practice, the VCWC tool consists of a wrapper shell script that invokes GNU Make on a `Makefile`. First, this `Makefile` reads the track description, which references the `benchmarks` and `participants` folders as input, and generates lists of benchmark instances and solvers. Based on this information, the `Makefile` instantiates rules that tell GNU Make how to generate the DAGMan workflow.

For instance, a typical VCWC call generates as output

```
# vcwc trackinfo-t03.mk
Welcome to vcwc 0.1
```

```
generating workflow for track t03 with following setup:
- benchmarks: b01 b02 b04 b05 b06 b07 b08 b09 b10 b11 b12 [...]
- participants: s40 s42 s44 s60 s62 s63
- benchmarks/participants: b18/s40 b18/s60 b18/s42 b18/s63 [...]
- runs: r000 r001 r002
- workflow version: 001
- timestamp: 2013-04-26 14:34:15+02:00
compiling 90 runs for S/t03/b01/s40/001
[...]
compiling 6 participants for B/t03/b01/001
[...]
linking 26 benchmarks for T/t03/001
```

This will generate a DAG workflow file and run profiles for each individual instance run. The generated DAG workflow has always the same shape as Fig. 1. Each node in this DAG encodes the job type, which is an instance run, a solver summary statistics, a benchmark ranking, or the track ranking job. VCWC uses the GNU M4 macro processing language to instantiate workflow templates and run profiles based on the names of benchmarks, solvers, instances, and runs.

Generated workflows can be processed by DAGMan, which submits jobs to HTCondor for execution in the network of benchmark servers. HTCondor is a high-troughput computing framework for distributed computation of computationally intensive tasks. Each task (job) that needs to be executed is first enqueued, and based on priority management and job requirements (such as number of CPUs or memory) it is scheduled to run on one of the target machines that are free for new jobs and fulfill all job requirements. The HTCondor job queue is persistent and make administrator intervention unnecessary in case of a reboot or system crash, as interrupted jobs are automatically rescheduled. The correct topological order of job execution is ensured by DAGMan, which—based on the generated workflow—dispatches, monitors, and keeps track of exit codes of jobs. DAGMan requires human intervention only when no further job can be submitted according to the current topological order, because of a previously failed dependency.

4 Discussion and Conclusions

VCWC has been developed as part of the ASP Competition 2013 evaluation software. A lot of experience had been gained when running the former competition, and the design of VCWC has profit from this. Special care has been given to have a versatile system that allows to address the failure sources (a)–(c) described in Section 1. Even though very unlikely, fatal hardware failures (c) do occur, in fact, during the execution of the ASP Competition 2013, a broken valve actuator prevented to distribute chilled water from the backup cooling system, thus excess heat continued to warm up the data center to an ambient temperature of 45 degrees Celsius, and all server machines had to shut down. After the cooling loop was working again, starting up the benchmark servers automatically re-scheduled all unfinished jobs, and the track workflows continued to run without administrative intervention.

VCWC can easily handle thousands of benchmark runs. With 23 participants among two main tracks and 27 benchmark problems, VCWC has been put under intensive testing: The system track workflow consists of over 18000 jobs, and the size of the DAG file is about 3 MiB. It took about a minute to generate this file, mainly because a lot of small intermediate files had to be written to the harddisk during the compilation. While setting up the competition, the incremental versioning system allowed to make fixes with no impact in the ongoing runs. We got further mileage out of using GNU Make for the implementation of VCWC by using its parallel execution mechanism. In this scenario, we could profit from an immediate 4-fold speedup for compiling the workflows just by turning on parallel make execution on our benchmark servers with two 12-core AMD Opteron Processor 6176 SE processors and 128GiB RAM.

In the ASP community, our VCWC platform follows chronologically and is inspired by the Asparagus Web-based Benchmarking Environment [1] and the (not publicly accessible) 3rd ASP Competition evaluation platform [4]. An attempt at providing a general purpose platform, serving multiple communities and generalizing specific needs is the StarExec platform [13]. Similar efforts in the neighbor communities are the IPC platform [7], the SMT-Exec platform [2] and the TPTP library and associated infrastructure [14]; the QBF-LIB library and evaluation platform [11], and last but not least the SAT Competitions infrastructure [8]. Future versions of VCWC will provide support for more fine-grained instance runs that allow to parametrize solver heuristics, advanced early diagnostics, and database storage facilities.

References

1. Asparagus Web-based Benchmarking Environment, http://asparagus.cs.uni-potsdam.de/
2. Barrett, C., Deters, M., Moura, L., Oliveras, A., Stump, A.: 6 years of SMT-Comp. J. Auto. Reasoning 50(3), 243–277 (2013)
3. Calimeri, F., Ianni, G., Krennwallner, T., Ricca, F.: The Answer Set Programming Competition. AI Mag. 33(4), 114–118 (2012)
4. Calimeri, F., Ianni, G., Ricca, F.: The third open answer set programming competition. Theor. Pract. Log. Prog., FirstView, 1–19 (2012), doi:10.1017/S1471068412000105
5. Couvares, P., Kosar, T., Roy, A., Weber, J., Wenger, K.: Workflow Management in Condor. In: Workflows for e-Science, pp. 357–375. Springer (2007)
6. Collaboratory on Experimental Evaluation of Software and Systems in Computer Science (2012), http://evaluate.inf.usi.ch/
7. The software of the seventh international planning competition (IPC) (2011), http://www.plg.inf.uc3m.es/ipc2011-deterministic/FrontPage/Software
8. Järvisalo, M., Le Berre, D., Roussel, O., Simon, L.: The International SAT Solver Competitions. AI Mag. 33(1), 89–92 (2012)
9. Klebanov, V., Beckert, B., Biere, A., Sutcliffe, G. (eds.): Proceedings 1st Int'l Workshop on Comparative Empirical Evaluation of Reasoning Systems, vol. 873. CEUR-WS.org (2012)
10. Papadimitriou, C.H.: Computational complexity. Addison-Wesley (1994)
11. Peschiera, C., Pulina, L., Tacchella, A.: Designing a solver competition: the QBFEVAL'10 case study. In: Workshop on Evaluation Methods for Solvers, and Quality Metrics for Solutions (EMS+QMS) 2010. EPiC, vol. 6, pp. 19–32. EasyChair (2012)
12. Roussel, O.: Controlling a solver execution with the runsolver tool. J. Sat. 7, 139–144 (2011)
13. Stump, A., Sutcliffe, G., Tinelli, C.: Introducing StarExec: a cross-community infrastructure for logic solving. In: Klebanov, et al. (eds.) [9], p. 2
14. Sutcliffe, G.: The TPTP problem library and associated infrastructure. J. Autom. Reasoning 43(4), 337–362 (2009)
15. Thain, D., Tannenbaum, T., Livny, M.: Distributed Computing in Practice: The Condor Experience. Concurrency Computat. Pract. Exper. 17(2-4), 323–356 (2005)

A Sequential Model for Reasoning about Bargaining in Logic Programs*

Wu Chen[1], Dongmo Zhang[2], and Maonian Wu[3]

[1] College of Computer and Information Science, Southwest University, China
[2] University of Western Sydney, Australia
[3] Guizhou University, China

Abstract. This paper presents a sequential model of bargaining based on abductive reasoning in ASP. We assume that each agent is represented by a logic program that encodes the background knowledge of the agent. Each agent has a set of goals to achieve but these goals are normally unachievable without an agreement from the other agent. We design an alternating-offers procedure that shows how an agreement between two agents can be reached through a reasoning process based on answer set programming and abduction. We prove that the procedure converges to a Nash equilibrium if each player makes rational offer/counter-offer at each round.

Keywords: bargaining theory, logic programming, sequential model.

1 Introduction

Bargaining has been a central research theme in economics for many decades and recently becomes an attractive research topic in artificial intelligence mainly driven by the advance of e-commence and multi-agent systems [1,2]. Different from other disciplines, the research on bargaining in artificial intelligence focuses more on reasoning mechanisms of bargaining process. A number of logical frameworks have been proposed in the literature for modelling different aspects of bargaining reasoning [3,4,5].

There are two different models of bargaining - *cooperative* and *non-cooperative*-that have been proposed in game theory. The cooperative model represents a bargain problem as a one-shot game and specifies the properties of bargaining solutions in an axiomatical system [1]. The noncooperative model of bargaining models a bargaining process as a sequential procedure. To specify bargaining reasoning, both models have been reformulated in logical frameworks. Zhang in [5] has proposed an axiomatic model of bargaining based on propositional logic. Several other authors have also constructed a range of different logic-based sequential models specifying bargaining reasoning, based on either argumentation, propositional logic or logic programming [3,4,6]. However, each of these models has limitation in either reasoning power or game-theoretic properties. The models that describe a bargaining situation in propositional formulas normally treat a formula as a whole therefore either keep a whole formula or drop a formula

---

* This work is supported by the National Natural Science Foundation of China under grants 61003203 and 61262029.

P. Cabalar and T.C. Son (Eds.): LPNMR 2013, LNAI 8148, pp. 239–244, 2013.

(logic is used for consistency maintenance only) [5,6]. The models based on argumentation or logic programs allow break of a formula for bargaining reasoning but the procedures that have been proposed normally lack of game-theoretic properties, such as convergency and pareto optimality [3,4]. This paper proposes a new sequential model of bargaining that specifies procedures of bargaining reasoning in answer set programming. We assume that each agent is represented by a logic program that encodes the background knowledge the agent uses for its bargaining reasoning. Each agent has a set of goals to achieve but these goals are normally unachievable without an agreement from the other agents. We design an alternating-offers procedure that shows how an agreement between two agents can be reached through a reasoning process based on answer set programming and abduction. We prove that the procedure converges to a Nash equilibrium if each player makes rational offer/counter-offer at each round.

The rest of the paper is organised as follows. Section 2 introduces our bargaining model. Section 3 presents the framework of our sequential bargaining model. Section 4 provide a construction of the sequential bargaining procedure and proves its equilibrium properties. The final section concludes the paper.

2   The Bargaining Model

In this section, we introduce a bargaining model in which each agent is equipped with a logic program as its background knowledge for bargaining reasoning and a set of goals to achieve. To make the framework simple we only consider the bargaining situations in which there are only two players.

Assume that $\mathcal{L}$ is a propositional language with finite number of propositional symbols (atoms). A literal can be either a positive atom, say a, or a negative atom, say $\neg a$. a and $\neg a$ are called complementary literals. A set of literals S is *consistent* if it contains no complementary literals; otherwise it is *inconsistent*. A *rule* is a formula

$$L_0 \leftarrow L_1, ..., L_m, not\ L_{m+1}, ..., not\ L_n (0 \leq m \leq n), \tag{1}$$

where each $L_i(0 \leq i \leq n)$ is a literal, *not* is *negation as failure* . We denote its head, positive body and negative body by $Head(r) = \{L_0\}$, $Pos(r) = \{L_1, ..., L_m\}$ and $Neg(r) = \{L_{m+1}, ..., L_n\}$ respectively. r is called a *fact* if $Pos(r) = \emptyset$ and $Neg(r) = \emptyset$. r is a *constraint* if $Head(r) = \emptyset$.

An answer set program is a finite set of *rules*. For a given *logic program* Π, we write $Head(\Pi) = \bigcup_{r \in \Pi} Head(r)$, $Pos(\Pi) = \bigcup_{r \in \Pi} Pos(r)$, $A(\Pi) = Pos(\Pi) \setminus Head(\Pi)$ and $\Pi \cup X = \Pi \cup \{L \leftarrow | L \in X\}$ where X is a set of literals. We use $AS(\Pi)$ to denote the set of consistent answer sets of a *logic program* Π.

In a bargaining situation, an agent might have a number of goals to achieve through the bargaining process. The aim of the agent is to reach an agreement with the other agent so that the other agent agrees the conditions that achieve his goals or some of his goals. If an agent cannot achieve all his goals, the agent might have a preference over these goals. A *model of player* includes an agent's knowledge, bargaining goals and its preference among these goals. The following definition gives such a *model of players*.

Definition 1. *A two-player bargaining game is a tuple* $M = ((\Pi_1, G_1, \leq_1), (\Pi_2, G_2, \leq_2))$, *where, for each* i, Π_i *is a logic program,* G_i *is a set of goals, each goal consisting of a set of literals, and* $\leq_i$ *is a total order over* G_i.

As a convention, we refer the opponent of player i as $-i$ in the sequnt. Given a logic program, a goal is achieved by the logic program if it is in an answer set of the program. However, if it is not achieved, we may wonder what are the conditions that can make the goal true. We call a set of conditions that achieves a goal under a logic program a *support*. In setting of our bargaining model, the concept of supports is important because if a condition that cannot be satisfied by one agent could be satisfied by another agent; an agent may request another agent to satisfy a condition by offering a condition the other agent is needed.

Definition 2. *Given a logic program Π and a set X of literals, we say $\Delta \subseteq A(\Pi)$ is a minimal support for achieving X from Π if it satisfies:*

1. *$X \cap \Delta = \emptyset$.*
2. *There is an answer set $S \in AS(\Pi \cup \Delta)$ such that $X \subseteq S$.*
3. *There is no $\Delta' \subset \Delta$ such that Δ' also satisfies Condition (1) and (2).*

We use $\alpha(\Pi, X)$ to represent the set of all possible minimal supports with respect to Π and X.

Given a bargaining game M, an *offer* of an agent is a pair (D, P), where $D \subseteq Lit$ and $P \subseteq Lit$. The set of all the possible offers is denoted by $\mathcal{O}$. Intuitively, an offer of a player represents the player's demands from the bargaining and the conditions he promises to the other player. D represents the current demands of the player and P denotes the current promises of the player to the other player.

Definition 3. *Let $M = ((\Pi_1, G_1, \leq_1), (\Pi_2, G_2, \leq_2))$ be a bargaining model. An offer $O = (D, P)$ achieves player i's goal $g \in G_i$ if*

1. *$D \in \alpha(\Pi_i, g \cup P)$;*
2. *There is no $g' \in G_i$ such that g' satisfies condition (1) and $g <_i g'$.*

For each player i, let $\mathcal{G}_i : \mathcal{O} \to G_i \cup \{\emptyset\}$ such that for any $O \in \mathcal{O}$, $\mathcal{G}_i(O) = g$ if O of player i *achieves* a goal $g \in G_i$; otherwise, $\mathcal{G}_i(O) = \emptyset$.

For convenience, we assume that for each player i, $\emptyset \notin G_i$, that is, a goal cannot be empty. In addition, we extend the ordering relation $\leq_i$ to $G_i \cup \{\emptyset\}$ such that $\emptyset <_i g$ for all $g \in G_i$.

Definition 4. *For each player i, define an order $\preceq_i$ over $\mathcal{O}$ as follows:*

$$O' \preceq_i O'' \text{ iff } \mathcal{G}_i(O') \leq_i \mathcal{G}_i(O'')$$

We say that O'' *dominates* O' if $O' \preceq_i O''$. Since $\leq_i$ is a total order over $G_i \cup \{\emptyset\}$, it is easy to see that $\preceq_i$ is a total preorder over $\mathcal{O}$.

A player not only has assess each offer he mades to see which goal he can achieve if the offer is accepted but also has to assess the opponent's offer to check if the offer should be accepted. The way of assessing opponent's offers is the following: A goal g of player i is *achievable* with an offer $O_{-i} = (D_{-i}, P_{-i})$ from the opponent of player i if there is a counter-offer $O = (D, P)$ to his opposite such that O achieves g meanwhile $P = D_{-i}$ and $P_{-i} \subseteq D$. We denote $\mathcal{I}_i(O)$ as the maximal goal of player i that is *achievable* with the offer O from player $-i$.

3   Sequential Bargaining Procedures

We design a sequential bargaining procedure as follows. Two players i and $-i$ take actions only at times in the set $T = \{1, 2, \cdots\}$. In each round $t \in T$, one of the players, say i, makes an offer (D^t, P^t) (a member of $\mathcal{O}$), where D^t contains all the items that player i wants the player $-i$ to accept and P^t contains all the items that player i accepts (initially is empty). Then the play passes to round $t + 1$; in this round player $-i$ makes a counter-offer (D^{t+1}, P^{t+1}). A player can terminate the procedure any time either set $D^t = P^{t-1}$ and $P^t = D^{t-1}$, in which case an agreement is reached or say nothing, in which case the game ends with a disagreement. The game continues whenever a player makes a new offer and the play passes to the next round [1].

Following the standard game-theoretical definition of bargaining procedures [1], we define a sequential bargaining procedure as follows. The extensive game of a sequential bargaining is a tuple $(N, H, P, \preceq_i)$ where

1). $N = \{1, 2\}$ is the set of players.

2). H is the set of histories. Each $h \in H$ is a sequence of offers that satisfies the following properties:

2.1). The empty sequence $\emptyset$ is a member of H.

2.2). If $(O^k)_{k=1}^K \in H$ and $L < K$, then $(O^k)_{k=1}^L \in H$.
A history $(O^k)_{k=1}^K \in H$ is *terminal* if there is no O^{K+1} such that $(O^k)_{k=1}^{K+1} \in H$. The set of terminal histories is denoted Z.

3). P is a function that assigns to each nonterminal history a number of N such that $P(h) = 1$ if the length of h is an even number and $P(h) = 2$ if the length of h is an odd number.

4). $\preceq_i$ is a preference relation on Z such that for any two histories $h = (O^k)_{k=1}^K \in Z$ and $h' = (O'^k)_{k=1}^{K'} \in Z$, $h \preceq_i h'$ if and only if $O^K \preceq_i O'^{K'}$.

5). For any $t_1, t_2 \in T$, $O^{t_2} = (D^{t_2}, \emptyset)$ and $O^{t_1} = (D^{t_1}, \emptyset)$ are two offers of player $i(i = 1 \text{ or } 2)$. If $\mathcal{G}_i(O^{t_2}) <_i \mathcal{G}_i(O^{t_1})$, then $t_1 < t_2$. If $t_1 < t_2$, then $\mathcal{G}_i(O^{t_2}) \leq_i \mathcal{G}_i(O^{t_1})$.

6). For any $O^k (k > 1)$, if $O^k = (D^k, \emptyset)$, then $O^{k-1} \neq O^{k+1}$.

Let $A(h) = \{O : (h, O) \in H\}$. We then can define strategies of a player.

Definition 5. *A strategy, s_i, of player $i \in N$ in the extensive game of sequential bargaining is a function that assigns an offer in $A(h)$ to each nonterminal history $h \in H \backslash Z$ for which $P(h) = i$. A pair $s = (s_1, s_2)$ of strategies is called a strategy profile if for each $i \in \{1, 2\}$, s_i is a strategy of player i.*

Definition 6. *A pair of strategies (s_1, s_2) is a Nash equilibrium if, given s_2, no strategy of player 1 results in an outcome that player 1 prefers to the outcome generated by (s_1, s_2), and, given s_1, no strategy of player 2 results in an outcome that player 2 prefers to the outcome generated by (s_1, s_2).*

Nash equilibrium is an important measurement to judge whether a bargaining procedure is designed reasonable or not. The following section will introduce a concrete bargaining procedure and prove that the procedure converges to a Nash equilibrium in finite steps.

4 Construction of Bargaining Procedure

We now give a concrete algorithm to model the bargaining procedure between two players i and $-i$ using abductive reasoning. For convenience, we say g is the best goal of G if $g \in G$ and for any $g' \in G$, $g' \leq g$, which is denoted $B(G)$. We use G_i^t to represent the set of goals of player i at the t round. Let $M = ((\Pi_1, G_1, \leq_1), (\Pi_2, G_2, \leq_2))$ be a bargaining model. Assume that player $-i$ puts forward the first offer $O_{-i}^1 = (D_{-i}^1, P_{-i}^1)$.

Algorithm 1. constructing bargaining procedure with abductive method

---

Input: $\Pi_i (i = 1, 2)$, $G_i (i = 1, 2)$
Output: O_1 and O_2

1 $t := 1$; $G_{-i}^1 := G_{-i}$; $G_i^0 := G_i$;
2 $\mathcal{H}_{-i} := Initialize(\Pi_{-i}, G_{-i}^1)$; $O_{-i}^1 := \mathcal{H}_{-i}.top()$; $\mathcal{H}_i := Initialize(\Pi_i, G_i^0)$;
3 **repeat**
4      $t := t + 1$;
5      $O_i^t := CounterOffer(O_{-i}^{t-1})$;
6      $O_i := O_i^t$;
7   **if** $D_i^t = P_{-i}^{t-1}$ and $D_{-i}^{t-1} = P_i^t$ **then**
8 | break;
9   **end**
10   **if** $P_{-i}^{t-1} = \emptyset$ and $P_i^t = \emptyset$ **then**
11         $\mathcal{H}_{-i} := \mathcal{H}_{-i} \setminus \{\mathcal{H}_{-i}.top()\}$;
12   **if** $\mathcal{H}_{-i} = \emptyset$ **then**
13             $G_{-i}^{t-1} := G_{-i}^{t-1} \setminus \{B(G_{-i}^{t-1})\}$;
14             $\mathcal{H}_{-i} := Initialize(\Pi_{-i}, G_{-i}^{t-1})$;
15   **end**
16   **end**
17   swap i and $-i$;
18 **until** $G_i^t = \emptyset$;

Procedure $Initialize$

Input: Π, G
Output: $\mathcal{H}$

1 **for** $\Delta \in \alpha(\Pi, B(G))$ **do**
2      $O := (\Delta, \emptyset)$;
3      $\mathcal{H}.push(O, G)$;
4 **end**

Given $M = ((\Pi_1, G_1, \leq_1), (\Pi_2, G_2, \leq_2))$ be a sequential bargaining model. The sequential bargaining procedure satisfies the following properties:

Proposition 1. 1. (Mutual commitment) *For any $t \in T$, $P_i^t \subseteq D_{-i}^{t-1}$ and $P_{-i}^{t-1} \subseteq D_i^t$.*
2. (Individual rationality) *For any $t \in T$, if $O_i^t = (D_i^t, P_i^t)$ is a counter-offer of $O_{-i}^{t-1} = (D_{-i}^{t-1}, P_{-i}^{t-1})$, then $(P_{-i}^{t-1}, D_{-i}^{t-1}) \preceq_i O_i^t$.*
3. (Satisfactorily) *For any $t \in T$, if $P_i^t = D_{-i}^{t-1}$ and $D_i^t = P_{-i}^{t-1}$, then $D_{-i}^{t+1} = D_{-i}^{t-1}$.*
4. (Honest) (1)*For any $t_1, t_2 \in T$, let $O^{t_2} = (D_i^{t_2}, \emptyset)$ and $O^{t_1} = (D_i^{t_1}, \emptyset)$. If $\mathcal{G}_i(O^{t_2}) <_i \mathcal{G}_i(O^{t_1})$, then $t_1 < t_2$. If $t_1 < t_2$, then $\mathcal{G}_i(O^{t_2}) \leq_i \mathcal{G}_i(O^{t_1})$. (2)For any $t \in T(t > 1)$, if $O_{-i}^{t-1} = (D_{-i}^{t-1}, \emptyset)$, then $O_i^t \neq O_i^{t-2}$.*

Procedure $CounterOffer$

   Input: O_{-i}^{t-1}
   Output: O_i^t
1 **if** $D_{-i}^{t-1} \neq P_i^{t-2}$ **then**
2   **if** $B(G_i^{t-2}) \leq_i \mathcal{I}_i(O_{-i}^{t-1})$ **then**
3   **foreach** $\Delta \in \alpha(\Pi_i, \mathcal{I}_i(O_{-i}^{t-1}) \cup D_{-i}^{t-1})$ *such that* $P_{-i}^{t-1} \subseteq \Delta$ **do**
4          $G_i^t := G_i^{t-2} \cup \{\mathcal{I}_i(O_{-i}^{t-1})\};$
5          $O_i' := (\Delta, D_{-i}^{t-1});$
6          $\mathcal{H}_i.push(O_i', G_i^t);$
7   **end**
8   **end**
9 **else**
10   **if** $P_{-i}^{t-1} \neq D_i^{t-2}$ **then**
11       $\mathcal{H}_i := \mathcal{H}_i \setminus \{\mathcal{H}_i.top()\};$
12   **if** $\mathcal{H}_i = \emptyset$ **then**
13          $G_i^t := G_i^{t-2} \setminus \{B(G_i^{t-2})\};$
14          $\mathcal{H}_i := Initialize(\Pi_i, G_i^t);$
15   **end**
16   **end**
17 **end**
18 $O_i^t := \mathcal{H}_i.top();$

Theorem 1. *Given any bargaining model, Algorithm 1 generates a strategy profile in finite steps that is a Nash equilibrium under Definition 7.*

5 Conclusion

We have proposed a sequential model of bargaining based on abductive reasoning in ASP and devised a bargaining procedure to demonstrate how two agents reach an agreement through abductive reasoning. We have shown that the sequential bargaining procedure converges a Nash equilibrium. We have also shown that the procedure satisfies a number of desirable properties.

References

1. Osborne, M.J., Rubinstein, A.: Bargaining and Markets. Academic Press (1990)
2. Jennings, N.R., Faratin, P., Lomuscio, A.R., Parsons, S., Sierra, C., Wooldridge, M.: Automated negotiation: prospects, methods and challenges. International Journal of Group Decision and Negotiation 10(2), 199–215 (2001)
3. Kraus, S., Sycara, K., Evenchik, A.: Reaching agreements through argumentation: a logical model and implementation. Artificial Intelligence 104, 1–69 (1998)
4. Son, T.C., Sakama, C.: Negotiation using logic programming with consistency restoring rules. In: Proceedings of the 21st International Joint Conference on Artificial Intelligence, IJCAI 2009, pp. 930–935. Morgan Kaufmann Publishers Inc. (2009)
5. Zhang, D.: A logic-based axiomatic model of bargaining. Artificial Intelligence 174, 1307–1322 (2010)
6. Zhang, D., Zhang, Y.: An ordinal bargaining solution with fixed-point property. Journal of Artificial Intelligence Research 33, 433–464 (2008)

Extending the Metabolic Network of
Ectocarpus Siliculosus Using Answer Set Programming

Guillaume Collet[1,5], Damien Eveillard[2], Martin Gebser[3], Sylvain Prigent[4,5],
Torsten Schaub[3], Anne Siegel[1,5], and Sven Thiele[5,6,1]

[1] CNRS, UMR 6074 IRISA, Campus de Beaulieu, 35042 Rennes, France
[2] Université de Nantes, UMR 6241 LINA, 2 rue de la Houssinière, 44300 Nantes, France
[3] Universität Potsdam, Institut für Informatik, August-Bebel-Str. 89, D-14482, Deutschland
[4] University of Rennes 1, UMR 6074 IRISA, Campus de Beaulieu, 35042 Rennes, France
[5] INRIA, Centre Rennes-Bretagne-Atlantique, Projet Dyliss, Campus de Beaulieu, 35042 Rennes cedex, France
[6] INRIA-CIRIC, Rosario Norte 555, Of. 703, Las Condes, Santiago de Chile, Chile

Abstract. Metabolic network reconstruction is of great biological relevance because it offers a way to investigate the metabolic behavior of organisms. However, reconstruction remains a difficult task at both the biological and computational level. Building on previous work establishing an ASP-based approach to this problem, we present a report from the field resulting in the discovery of new biological knowledge. In fact, for the first time ever, we automatically reconstructed a metabolic network for a macroalgae. We accomplished this by taking advantage of ASP's combined optimization and enumeration capacities. Both computational tasks build on an improved ASP problem representation, incorporating the concept of reversible reactions. Interestingly, optimization greatly benefits from the usage of unsatisfiable cores available in the ASP solver *unclasp*. Applied to *Ectocarpus siliculosus*, only the combination of *unclasp* and *clasp* allowed us to obtain a metabolic network able to produce all recoverable metabolites among the experimentally measured ones. Moreover, 70% of the identified reactions are supported by an homologous enzyme in *Ectocarpus siliculosus*, confirming the quality of the reconstructed network from a biological viewpoint.

1 Introduction

Systems biology is a field at the crossover of biology, computer science, and mathematics, which aims to elucidate the response of a living organism. Among all biological processes occurring in a cell, metabolic networks are in charge of transforming input nutrients into both energy and output nutrients necessary for the functioning of other cells. From an industrial viewpoint, it is crucial to estimate and control the capability of an organism to produce products of interest. Many computational and mathematical methods have been developed to model the response of such systems to external perturbations, and applied to well-studied organisms [1–3].

In the last few years, sequencing technologies have drastically evolved, such that it is now possible to sequence the genome of many less-studied organisms. As a natural follow-up, one needs to estimate the metabolic capability of an "exotic" organism

P. Cabalar and T.C. Son (Eds.): LPNMR 2013, LNAI 8148, pp. 245–256, 2013.

on the basis of its genome, and then apply well-established control methods to the network. The usual strategy consists in checking whether the genome contains known enzymatic "bricks", that is, genomic sequences that appropriately match with genomic sequences of enzymes characterized in other model organisms, such as *Escherichia coli* [4] or *Arabidopsis thaliana* [5], whose genomes and networks have been manually curated over several years [6]. The combination of metabolic reactions associated with the identified enzymes provides a draft of the metabolic network for the studied organism. The integration of the different heterogeneous bio chemical resources leads to inconsistencies and ambiguities in the draft network. Semantic web approaches solve these inconsistencies and rank the retrieved information by exploiting existing ontologies [7]. Nonetheless, genomes are of low quality and the expert community on "exotic" organisms is too small to provide a wide manual curation of this network. Concretely, automatic genome-scale reconstructed networks suffer from substantial incompleteness, and many networks are only partially defined. To overcome this limitation, the next step consists in *filling the gaps* of the draft network. To that end, we rely on reference databases of metabolic reactions and check whether adding such reactions to the network improves its ability to produce metabolite compounds of interest from the growth media of the organism. Several approaches to automatically reconstruct the missing parts of metabolic networks have been proposed. To restore a desired metabolic behavior they propose reactions (picked from reaction databases) that can be added to the network. The reactions are chosen to optimize either graph-based criteria [8] or a linear score modeling the quantitative production of the system [9]. The main limitation of all approaches is the increasing size of the search space, since reaction databases like KEGG[1] or MetaCyc[2] have substantially grown with the availability of high-throughput methods in molecular biology. Other studies propose to overcome this limitation by using sampling heuristics [10], but unfortunately they give little information on the size of solution sets and the quality of the sampling methods.

In previous work [11], we reformulated the gap filling problem as a qualitative combinatorial (optimization) problem, and modeled it using Answer Set Programming (ASP) [12]. The basic idea is that reactions apply only if all their reactants are available, either as nutrients or provided by other metabolic reactions. Starting from given nutrients, referred to as *seeds*, this allows for extending a metabolic network by successively adding operable reactions and their products. The set of metabolites in the resulting network is called the *scope* of the seeds and represents all metabolites that can principally be synthesized from the seeds. In metabolic network completion, we query a database of metabolic reactions looking for minimal sets of reactions that can restore the observed bio-synthetic behavior.

As a follow-up to [11], we attempted to apply the same approach to reconstruct the "exotic" metabolic network of *Ectocarpus siliculosus*, using the MetaCyc database. This organism is a brown algae that belongs to the heterokonts, whose closest relative (diatoms) exhibits a large phylogenetic distance to most other plant model species. Such distinctions make a reconstruction of the metabolic network of *Ectocarpus siliculosus* particularly challenging. In fact, we could not solve the reconstruction problem with the

---

[1] http://www.genome.jp/kegg
[2] http://metacyc.org

original approach that hits its limits with large databases like MetaCyc, which doubled in size over the last four years.

In this work, we push former limits by taking advantage of ASP's combined optimization and enumeration capacities. For one, we introduce an improved ASP problem representation incorporating the concept of reversible reactions. For another, optimization greatly benefits from the usage of unsatisfiable cores available in the ASP solver *unclasp* [13]. Applied to *Ectocarpus siliculosus*, only the combination of *unclasp* and *clasp* [14] allowed us to obtain a metabolic network able to produce all recoverable metabolites among the experimentally measured ones. Moreover, 70% of the identified reactions are supported by an homologous enzyme in *Ectocarpus siliculosus*, confirming the quality of the reconstructed network from a biological viewpoint.

In what follows, we assume some familiarity with ASP, its semantics as well as its basic language constructs. In particular, our encodings are written in the input language of *gringo* 3 [15]. Comprehensive treatments of ASP can be found in [12, 16].

2   Metabolic Network Completion

Metabolism is the sum of all chemical reactions occurring within an organism. As the products of a reaction may be reused as reactants, reactions can be chained to complex chemical pathways. Such complex pathways are described by a metabolic network.

A *metabolic network* is commonly represented as a directed bipartite graph $G = (R \cup M, E)$, where R and M are sets of nodes standing for *reactions* and *metabolites*, respectively. When $(m, r) \in E$ (or $(r, m) \in E$) for $m \in M$ and $r \in R$, the metabolite m is called a *reactant* (or *product*) of reaction r. More formally, for any $r \in R$, define $reac(r) = \{m \in M \mid (m, r) \in E\}$ and $prod(r) = \{m \in M \mid (r, m) \in E\}$.

The biological concept of the synthetic capabilities of a metabolism can be expressed in terms of reachability. Given a metabolic network $(R \cup M, E)$ and a set $S \subseteq M$ of *seed* metabolites, a reaction $r \in R$ is *reachable* from S if all reactants in $reac(r)$ are reachable from S. Moreover, a metabolite $m \in M$ is *reachable* from S if $m \in S$ or if $m \in prod(r)$ for some reaction $r \in R$ that is reachable from S. The *scope* of S, written $\Sigma_{(R \cup M, E)}(S)$, is the closure of metabolites reachable from S.

Given a metabolic network $(R \cup M, E)$, two sets $S, T \subseteq M$ of seed and target metabolites, and a reference network $(R' \cup M', E')$, the *metabolic network completion problem* is to find a set $R'' \subseteq R' \setminus R$ of reactions such that $T \subseteq \Sigma_G(S)$, where

$$G = ((R \cup R'') \cup (M \cup M''), E \cup E''),$$
$$M'' = \{m \in M' \mid r \in R'', m \in reac(r) \cup prod(r)\}, \text{ and}$$
$$E'' = E' \cap ((M'' \times R'') \cup (R'' \times M'')).$$

We call R'' a *completion* of $(R \cup M, E)$ from $(R' \cup M', E')$ wrt (S, T).

For reconstructing *Ectocarpus siliculosus*, we are interested in *cardinality-minimal completions* as well as *necessary reactions* belonging to every completion. Therefore, we need to solve the following sub-tasks:

– **Problem 1**: Compute the minimum size (number of reactions) of a completion.
– **Problem 2**: Enumerate all cardinality-minimal completions.
– **Problem 3**: Compute the intersection of all cardinality-minimal completions.

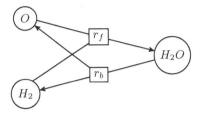

Fig. 1. Example of the first method on $H_2 + O \rightleftharpoons H_2O$

As shown in [17, 18], the reconstruction of metabolic networks and related problems are NP-hard.[3] Problem variants (of higher computational complexity) rely on subset-rather than cardinality-minimal completions. Further refinements may also optimize on the distance between seeds and targets or minimize forbidden side products.

3   Reversible Reactions

Chemical reactions are in essence reversible. However, taking the metabolic context into account (i.e. reactants and products) leads to considering some of them as irreversible in view of energetic cost [19]. In the following, we describe two alternative methods to capture reversible and irreversible reactions.

The first method represents a reversible reaction by two inverse reactions that are separate nodes within the network. For example, given the metabolites H_2, O, and H_2O and the reversible reaction $r = H_2 + O \rightleftharpoons H_2O$, we can construct the metabolic network $(\{H_2, O, H_2O, r_f, r_b\}, \{(H_2, r_f), (O, r_f), (r_f, H_2O), (H_2O, r_b), (r_b, H_2), (r_b, O)\})$, as illustrated in Figure 1. This method allows us to apply the framework presented in [11]. Unfortunately, it also roughly doubles the number of reactions that must be considered when looking for completions.

For an alternative method, let us represent a metabolic network as a graph $G = (R_{rev} \cup R_{irrev} \cup M, E)$, where R_{rev}, R_{irrev}, and M are sets of nodes standing for *reversible reactions*, *irreversible reactions*, and metabolites, respectively. The difference to our previous approach is that we distinguish between nodes for reversible and irreversible reactions. For any reaction $r \in R_{rev} \cup R_{irrev}$, the edges in E describe exactly one direction, that is, $(m, r) \in E$ (or $(r, m) \in E$) expresses that the metabolite $m \in M$ is a reactant (or product) of r. Taking r to be reversible, the network $(\{H_2, O, H_2O, r\}, \{(H_2, r), (O, r), (r, H_2O)\})$ thus captures both reactions displayed in Figure 1.

Given a metabolic network $(R_{rev} \cup R_{irrev} \cup M, E)$ and a set $S \subseteq M$ of seed metabolites, a reaction $r \in R_{rev} \cup R_{irrev}$ is reachable from S if all reactants in $reac(r)$ are reachable from S; when $r \in R_{rev}$ is reversible, r is also *reachable* from S if all products in $prod(r)$ are reachable from S. This reflects that, depending on the direction in which a reversible reaction is applied, the roles of reactants and products may be interchanged. Moreover, a metabolite $m \in M$ is *reachable* from S if $m \in S$ or if $m \in reac(r) \cup prod(r)$ for some reaction $r \in R_{rev} \cup R_{irrev}$ that is reachable from S.

---

[3] That is, the underlying decision problems are NP-hard.

As in the previous section, the scope of S, written $\Sigma_{(R_{rev} \cup R_{irrev} \cup M, E)}(S)$, is the closure of metabolites reachable from S.

Using this alternative representation, the metabolic network completion problem for a network $(R_{rev} \cup R_{irrev} \cup M, E)$, two sets $S, T \subseteq M$ of seed and target metabolites, and a reference network $(R'_{rev} \cup R'_{irrev} \cup M', E')$ is to find a set $R'' \subseteq (R'_{rev} \cup R'_{irrev}) \setminus (R_{rev} \cup R_{irrev})$ of reactions such that $T \subseteq \Sigma_G(S)$, where

$$G = ((R_{rev} \cup R_{irrev} \cup R'') \cup (M \cup M''), E \cup E''),$$
$$M'' = \{m \in M' \mid r \in R'', m \in reac(r) \cup prod(r)\}, \text{ and}$$
$$E'' = E' \cap ((M'' \times R'') \cup (R'' \times M'')).$$

We call R'' a *completion* of $(R_{rev} \cup R_{irrev} \cup M, E)$ from $(R'_{rev} \cup R'_{irrev} \cup M', E')$ wrt (S, T).

Our ASP implementation addresses the alternative representation of reversible reactions by additional facts and rules in comparison to the seminal encoding [11]. In particular, an instance of the network completion problem now contains additional facts *reversible*(r) for reactions $r \in R_{rev} \cup R'_{rev}$, and our new encoding utilizes reversibility information. For instance, the following rules define the scope of a network:

$$scope(M) \leftarrow seed(M)$$
$$scope(M) \leftarrow product(M, R), reaction(R), scope(M') : reactant(M', R)$$
$$scope(M) \leftarrow reactant(M, R), reversible(R), scope(M') : product(M', R)$$

These rules illustrate the changes in our logic program.[4] The first rule states that all metabolites given as seeds are available in an organism, and the second rule derives the products of a reaction whose reactants are available. Moreover, the third rule takes care of interchanged roles of reactants and products in a reversible reaction, where reactants can be derived from available products.

For instance, for implementing the example shown in Figure 1, one may consider the metabolites H_2 and O as seeds as well as H_2O as target. The ground programs obtained with the two alternative methods to represent reversible reactions are given in Listing 1 and 2. Both include similar rules to derive H_2 and O as available in the scope. However, the first program relies on two reactions, r_f and r_b, while the second program addresses the inverse reaction r_b via a rule for reversibility.

The outcomes of the program in Listing 1 are given by the sets $\{r_f\}$ and $\{r_f, r_b\}$ of reactions, the first of which is cardinality-minimal. This tells us that r_f is necessary to produce H_2O from H_2 and O. The unique outcome $\{r\}$ of the program in Listing 2 likewise yields the necessity of applying r, where the actual direction of r needed to produce H_2O from H_2 and O can be inferred easily.

4   Experiments

In order to successfully solve the three problems introduced above, we propose to divide the metabolic network completion into two phases. In the first phase, we compute

---

[4] The full encoding is available at http://pypi.python.org/pypi/meneco.

Listing 1. Ground logic program instance without reversibility.

```
1  seed(H₂). seed(O). target(H₂O).
2
3  { reaction(r_f) }.                      { reaction(r_b) }.
4  reactant(H₂,r_f). reactant(O,r_f).      reactant(H₂O,r_b).
5  product(H₂O,r_f).                       product(H₂,r_b). product(O,r_b).
6
7  scope(H₂)  :- seed(H₂).
8  scope(O)   :- seed(O).
9
10 scope(H₂O) :- product(H₂O,r_f), reaction(r_f), scope(H₂), scope(O).
11
12 scope(H₂)  :- product(H₂,r_b), reaction(r_b), scope(H₂O).
13
14 scope(O)   :- product(O,r_b), reaction(r_b), scope(H₂O).
15
16 :- target(H₂O), not scope(H₂O).
17
18 #minimize{ reaction(r_f), reaction(r_b) }.
```

Listing 2. Ground logic program instance with reversibility.

```
1  seed(H₂). seed(O). target(H₂O).
2
3  { reaction(r) }. reversible(r).
4  reactant(H₂,r). reactant(O,r).
5  product(H₂O,r).
6
7  scope(H₂)  :- seed(H₂).
8  scope(O)   :- seed(O).
9
10 scope(H₂O) :- product(H₂O,r), reaction(r), scope(H₂), scope(O).
11
12 scope(H₂)  :- reactant(H₂,r), reversible(r), scope(H₂O).
13
14 scope(O)   :- reactant(O,r), reversible(r), scope(H₂O).
15
16 :- target(H₂O), not scope(H₂O).
17
18 #minimize{ reaction(r) }.
```

the minimum size of a network completion (**Problem 1**). To this end, ASP provides powerful optimization techniques based on branch-and-bound algorithms. Albeit such techniques can be highly effective, our application pinpoints their current limitations. Hence, we take advantage of *unclasp* (version 0.1), whose usage of unsatisfiable cores is inspired by respective approaches to Maximum Satisfiability (MaxSAT) [20]. In the second phase, we rely on *clasp* (version 2.2.1) to enumerate all minimal completions (**Problem 2**) or to compute the intersection of all minimal completions (**Problem 3**). The experiments were run on a cluster of three machines equipped with 128 to 144 GB RAM and totaling 48 cores, clocked from 2.39 to 2.66 GHz.

Table 1. Ranges of minimum size and number of cardinality-minimal completions for Meta-Cyc subsets The time-outs of *clasp* are also reported with and without the reversibility encoding.

Number of reactions	5000	6000	7000	8000	9000	10000	Full
Minimum completion size	[6,14]	[7,22]	[7,29]	[9,29]	[16,47]	[33,50]	52
clasp time-outs							
with reversibility	0	0	1	3	9	10	10
without reversibility	0	0	0	2	8	10	10
Minimal completions	[4,32]	[6,324]	[6,1728]	[16,3456]	[80,1150]	[180,22800]	2600

4.1   Reconstruction of the Metabolic Network of *Ectocarpus siliculosus*

As a first experiment, we complete a draft metabolic network of the brown algae *Ectocarpus siliculosus* [21] with reactions from MetaCyc. The draft network, produced by merging expert annotations [22] with orthologs in *Arabidopsis thaliana* [23], contained 1210 reactions and 1454 metabolites. Moreover, we consider 44 metabolites as seeds, provided by biological experts, and 51 metabolites, which have been experimentally shown to be natural products of *Ectocarpus siliculosus*, as targets. We checked that the draft network can only produce 23 of the 51 experimentally established targets, which exhibits the insufficiency of the draft network to recover some of the main metabolic capabilities of the brown algae. This also shows that metabolic reconstruction via manual methods is not sufficiently detailed for an "exotic" species like *Ectocarpus siliculosus*.

Applying *unclasp* and *clasp* as described above, we could solve **Problem 1, 2**, and **3** for the draft network. It turns out that at least 52 reactions from the MetaCyc database are required to produce 48 metabolites among the 51 experimentally established targets (**Problem 1**). We checked that the three remaining targets are not producible via reactions from MetaCyc. Moreover, enumeration led to 2600 cardinality-minimal completions (**Problem 2**), whose intersection consists of 45 reactions (**Problem 3**).

The union of all cardinality-minimal completions, 70 reactions, was then added to the draft network to reconstruct the first metabolic network of *Ectocarpus siliculosus*. A comparison of the resulting network, containing 1280 reactions and 1507 metabolites, to sequence information showed that 70% of the reactions are relevant in the brown algae. This suggests that reconstruction by means of ASP is biologically meaningful.

4.2   Study of Scalability

Given that the size of the reaction database constitutes a primary factor regarding the performance of metabolic network reconstruction, we further investigated the scalability of our approach and the benefit of introducing the new model for reversible reactions.

We thus applied our method to the completion of *Ectocarpus siliculosus* relying on databases of different sizes. We created 10 different subsets of MetaCyc, each containing 10000 randomly selected reactions. Starting from them, smaller subsets of size 9000, 8000, 7000, 6000, and 5000 were created by randomly and successively removing reactions, yielding 10 distinguished benchmarks for each size. Each subset includes the same proportion of reversible reactions as the full MetaCyc database ($\approx 42\%$).

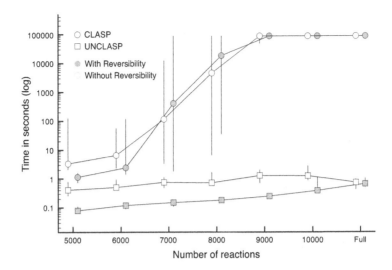

Fig. 2. Runtimes of *clasp* **and** *unclasp* **for computing the minimum size of a completion (Problem 1).** The circles and squares provide the median runtimes of *clasp* and *unclasp*, respectively. In addition, minimum and maximum runtimes are reported as vertical lines.

Table 1 summarizes the minimum network completion sizes, the time-outs of *clasp* upon computing (or proving, respectively) minimum sizes, and the numbers of cardinality-minimal completions for MetaCyc subsets of different sizes. Notably, the minimum sizes of completions recovering producible targets remain relatively small (≤ 50). The small sizes and apparent locality of network completions promote *unclasp*, which turns out to be highly effective upon optimization in the first phase. As the current functionalities of *unclasp* do not include enumeration or intersection computation, the respective experiments are limited to *clasp* in the second phase.

Solving Problem 1. In order to determine the minimum sizes of network completions, we ran *unclasp* in its default configuration as well as *clasp* with the options `--time-limit=86400 --restart-on-model --reset-restarts --local-restarts --opt-heu --save-progress`. The latter configure *clasp*'s sign heuristic to falsify literals subject to minimization and also foster restarts to avoid getting stuck in local minima. However, the runtimes plotted in Figure 2 stay around one second with *unclasp* but grow exponentially with *clasp*. Moreover, the explicit representation of reversible reactions speeds up *unclasp* by factors from 2 to 11, while it leads to more time-outs with *clasp* (cf. Table 1).

Solving Problem 2. For enumerating cardinality-minimal completions, we ran *clasp* with the options `--time-limit=86400 --configuration=handy --opt-all=`*optimum*, where the "handy" configuration is geared towards large

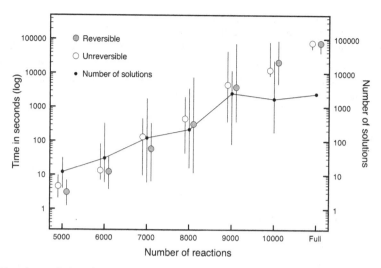

Fig. 3. Runtimes of *clasp* for enumerating all cardinality-minimal completions (Problem 2). The gray and white circles provide the median runtimes of *clasp*; dots indicate the median number of cardinality-minimal completions. Minimum and maximum values are reported as vertical lines.

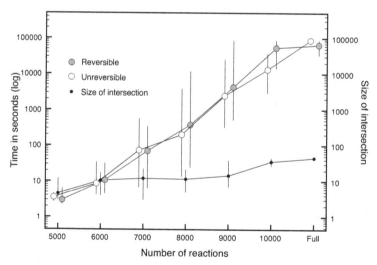

Fig. 4. Runtimes of *clasp* for computing the intersection of all cardinality-minimal completions (Problem 3). The gray and white circles provide the median runtimes of *clasp*; dots indicate the median intersection size. Minimum and maximum values are reported as vertical lines.

problems. The results plotted in Figure 3 show that the runtimes of *clasp* and the numbers of solutions tend to grow exponentially with the size of the reaction database. The number of solutions, however, reaches a plateau from 9000 reactions on, thus exhibiting a correlation with the minimum completion sizes given in Table 1.

Solving Problem 3. Adding the option `--enum-mode=cautious` switches *clasp* from enumeration to computing the intersection of cardinality-minimal completions. The runtimes plotted in Figure 4 still parallel those for enumeration. Unlike this, the intersection size grows much more moderately than the number of cardinality-minimal completions, so that future advancements of ASP solving technology may shrink the efforts of computing consequences below those of enumeration.

5   Conclusions

As a first conclusion, we note that *unclasp* enables the calculation of minimum completion size from an unabridged reaction database, which is necessary to accomplish the metabolic reconstruction of an "exotic" organism like *Ectocarpus siliculosus*. While *clasp* cannot solve this problem for the full MetaCyc database (in allotted time), *unclasp* completes the same task in a few seconds. Moreover, Figure 2 shows that *unclasp* remains almost unaffected by database growth. In fact, the usage of unsatisfiable cores allows for exploiting local problem structure to quickly converge to an optimal solution. Therefore, it appears that *unclasp* is especially well-suited to solve problems with plenty abducibles (>10000 reactions) but rather small optima (about 50 reactions).

As a second conclusion, integrating the reversibility concept into our ASP encoding reduces the runtime of *unclasp* by up to one order of magnitude. Somewhat surprisingly, *clasp* cannot benefit from the improved ASP encoding, even though reversibility reduces the number of candidate reactions from MetaCyc by about one third.

As a third conclusion, only the combination of *unclasp* and *clasp* allowed us to reconstruct the metabolic network of *Ectocarpus siliculosus* because the current functionalities of *unclasp* do not include enumeration or intersection computation. Fortunately, these two tasks can be accomplished by *clasp* when the minimum completion size is known. While enumeration enables an exhaustive exploration of (cardinality-minimal) completions, their intersection yields necessary reactions needed to produce target metabolites. Such information is crucial for the biological post-validation of a metabolic network without manual curation.

In summary, the combination of ASP modeling and solving capacities enabled the successful automatic reconstruction of the first metabolic network of a macroalgae. However, Figure 3 and 4 also indicate that the capabilities of *clasp* to compute all cardinality-minimal completions or their intersection almost hit the limits in view of the current size of the MetaCyc database. Anticipating its future extension, reconstruction tasks will be difficult to address without further advances in ASP solving. To this end, the incorporation of domain knowledge and heuristics to guide the solving process appear to be promising. As a direction for future work, we aim at the development of dedicated heuristics and their employment in the recent ASP solver *hclasp* [24].

Acknowledgments. This work was supported by ANR Biotempo (ANR-10-BLANC-0218), IDEALG (ANR-10-BTBR-04), and DFG (SCHA 550/10-1).

Reference

1. Barabási, A., Oltvai, Z.: Network biology: Understanding the cell's functional organization. Nature Reviews Genetics 5(2), 101–113 (2004)
2. Joyce, A., Palsson, B.: The model organism as a system: Integrating 'omics' data sets. Nature Reviews Molecular Cell Biology 7(3), 198–210 (2006)
3. Yamada, T., Bork, P.: Evolution of biomolecular networks: Lessons from metabolic and protein interactions. Nature Reviews Molecular Cell Biology 10(11), 791–803 (2009)
4. Orth, J., Conrad, T., Na, J., Lerman, J., Nam, H., Feist, A., Palsson, B.: A comprehensive genome-scale reconstruction of Escherichia coli metabolism. Molecular Systems Biology 7, Article 535 (2011)
5. de Oliveira Dal'Molin, C., Quek, L., Palfreyman, R., Brumbley, S., Nielsen, L.: AraGEM, a genome-scale reconstruction of the primary metabolic network in Arabidopsis. Plant Physiology 152(2), 579–589 (2009)
6. Zengler, K., Palsson, B.: A road map for the development of community systems (CoSy) biology. Nature Reviews Microbiology 10(5), 366–372 (2012)
7. Swainston, N., Smallbone, K., Mendes, P., Kell, D., Paton, N.: The subliminal toolbox: automating steps in the reconstruction of metabolic networks. Journal of Integrative Bioinformatics 8, Article 186 (2011)
8. Handorf, T., Ebenhöh, O., Heinrich, R.: Expanding metabolic networks: Scopes of compounds, robustness, and evolution. Journal of Molecular Evolution 61(4), 498–512 (2005)
9. Satish Kumar, V., Dasika, M., Maranas, C.: Optimization based automated curation of metabolic reconstructions. BMC Bioinformatics 8, Article 212 (2007)
10. Christian, N., May, P., Kempa, S., Handorf, T., Ebenhöh, O.: An integrative approach towards completing genome-scale metabolic networks. Molecular BioSystems 5(12), 1889–1903 (2009)
11. Schaub, T., Thiele, S.: Metabolic network expansion with answer set programming. In: Hill, P.M., Warren, D.S. (eds.) ICLP 2009. LNCS, vol. 5649, pp. 312–326. Springer, Heidelberg (2009)
12. Baral, C.: Knowledge Representation, Reasoning and Declarative Problem Solving. Cambridge University Press (2003)
13. Andres, B., Kaufmann, B., Matheis, O., Schaub, T.: Unsatisfiability-based optimization in clasp. In: Dovier, A., Santos Costa, V. (eds.) Technical Communications of the Twenty-Eighth International Conference on Logic Programming, ICLP 2012. Leibniz International Proceedings in Informatics, vol. 17, pp. 212–221. Dagstuhl Publishing (2012)
14. Gebser, M., Kaufmann, B., Schaub, T.: Conflict-driven answer set solving: From theory to practice. Artificial Intelligence 187-188, 52–89 (2012)
15. Gebser, M., Kaminski, R., Kaufmann, B., Ostrowski, M., Schaub, T., Thiele, S.: A user's guide to gringo, clasp, clingo, and iclingo, http://potassco.sourceforge.net
16. Gebser, M., Kaminski, R., Kaufmann, B., Schaub, T.: Answer Set Solving in Practice. Morgan and Claypool Publishers (2012)
17. Nikoloski, Z., Grimbs, S., May, P., Selbig, J.: Metabolic networks are NP-hard to reconstruct. Journal of Theoretical Biology 254(4), 807–816 (2008)
18. Nikoloski, Z., Grimbs, S., Selbig, J., Ebenhöh, O.: Hardness and approximability of the inverse scope problem. In: Crandall, K.A., Lagergren, J. (eds.) WABI 2008. LNCS (LNBI), vol. 5251, pp. 99–112. Springer, Heidelberg (2008)
19. Beard, D., Liang, S., Qian, H.: Energy balance for analysis of complex metabolic networks. Biophysical Journal 83(1), 79–86 (2002)
20. Li, C., Manyà, F.: MaxSAT. In: Biere, A., Heule, M., van Maaren, H., Walsh, T. (eds.) Handbook of Satisfiability. Frontiers in Artificial Intelligence and Applications, vol. 185, pp. 613–631. IOS Press (2009)

21. Tonon, T., Eveillard, D., Prigent, S., Bourdon, J., Potin, P., Boyen, C., Siegel, A.: Toward systems biology in brown algae to explore acclimation and adaptation to the shore environment. Omics: A Journal of Integrative Biology 15(12), 883–892 (2011)
22. Karp, P., Paley, S., Romero, P.: The Pathway Tools software. Bioinformatics 18(suppl. 1), S225–S232 (2002)
23. Loira, N., Dulermo, T., Nicaud, J., Sherman, D.: A genome-scale metabolic model of the lipid-accumulating yeast Yarrowia lipolytica. BMC Systems Biology 6, Article 35 (2012)
24. Gebser, M., Kaufmann, B., Otero, R., Romero, J., Schaub, T., Wanko, P.: Domain-specific heuristics in answer set programming. In: des Jardins, M., Littman, M. (eds.) Proceedings of the Twenty-Seventh National Conference on Artificial Intelligence, AAAI 2013. AAAI Press (to appear, 2013)

Negation as a Resource: A Novel View on Answer Set Semantics[*]

Stefania Costantini[1] and Andrea Formisano[2]

[1] DISIM, Università di L'Aquila, Italy
stefania.costantini@univaq.it
[2] DMI, Università di Perugia, Italy
formis@dmi.unipg.it

Abstract. In recent work, we provided a formulation of ASP programs in terms of linear logic theories. Based on this work, in this paper we propose and discuss a modified Answer Set Semantics, "Resource-based Answer Set Semantics".

Keywords: Answer Set Programming, Linear Logic, Default Negation.

1 Introduction

Answer Set Programming (ASP) is nowadays a well-established programming paradigm, with applications in many areas. RASP [1, 2, 3] is a recent extension of ASP, obtained by explicitly introducing the notion of *resource*. In [4], we proposed a comparison between RASP (and ASP) and linear logic [5]. In defining the correspondence, we introduced a RASP and linear-logic modeling of default negation as understood under the answer set semantics.

In this paper, we show that understanding default negation as a resource may lead to the definition of a generalization of the answers set semantics (for short AS, on which ASP is based), with some potential advantages. We provide a model-theoretic definition of the new semantics, that we call *Resource-based Answer Set Semantics*. In the new setting, there are no inconsistent programs, and basic odd cycles (similarly to basic even cycles in AS) are interpreted as exclusive disjunctions. Constraints must then be represented explicitly (while in ASP they are "simulated" via unary odd cycles). The "practical expressive power" in terms of knowledge representation is improved (as we demonstrate by means of significant examples), though unfortunately also the computational complexity increases.

2 Background on Linear Logic and ASP, and on Their Relationship

We refer the reader to the extensive existing literature about linear logic and ASP, that we are not able to mention for lack of space. We also apologize for the short and incomplete explanation, strictly limited to the features which are of interest here.

---

[*] Research partially funded by INdAM-GNCS-2013 project.

P. Cabalar and T.C. Son (Eds.): LPNMR 2013, LNAI 8148, pp. 257–263, 2013.

Linear logic [5] can be considered as a *resource sensitive* refinement of classical logic: intuitively speaking, in linear logic, two assumptions of a formula P are distinguished from a single assumption of it. Hence, in linear logic contraction and weakening rules are not allowed: while a statement of the form $P \rightarrow P \wedge P$ is valid in classical logic, this is not the case in linear logic.

Linear logic makes a neat distinction between two forms of conjunction. The first one intuitively means "I have both". This is said *multiplicative conjunction* and is written as $\otimes$. The other, the *additive conjunction* means "I have a choice" (and is written as &). Dually, there are two disjunctions. The multiplicative one, written $P \,\mathbin{⅋}\, Q$ can be read as "if not P, then Q", and the additive disjunction $P \oplus Q$, that intuitively stands for the possibility of either P or Q, but we do not know which of the two. That is, it involves "someone else's choice". Linear implication $P \multimap Q$ encodes a form of production process: it can be read as "Q can be derived using P exactly once". (Notice that, in such a process P is "consumed", so it cannot be used again.)

A formal proof system for linear logic can be formulated in terms of a Gentzen-style sequent calculus. A full set of Gentzen-style sequent rules for linear logic can be found, for instance, in [6].

In the answer set semantics (originally named "stable model semantics"), a (logic) program Π (cf., [7]) is a collection of *rules* of the form $H \leftarrow L_1, \ldots, L_n$. where H is an atom, $n \geqslant 0$ and each literal L_i is either an atom A_i or its *default negation* not A_i. Below is the specification of the Answer Set Semantics, reported from [7].

Definition 1. *Let I be a set of atoms and Π a program. A GL-transformation of Π modulo I is a new program Π/I obtained from Π by performing the following two reductions: (1) removing from Π all rules which contain a negative premise not A such that $A \in I$; (2) removing from the remaining rules those negative premises not A such that $A \notin I$. Π/I is a positive logic program, with a unique Least Herbrand Model, denoted as $\Gamma_\Pi(I)$. I is an answer set of Π iff $\Gamma_\Pi(I) = I$.*

Answer sets are minimal supported models, and non-empty answer sets form an antichain with respect to set inclusion. It will be useful in what follows to report from [8] a simple property of Γ_Π: If M is a minimal model of Π, then, $\Gamma_\Pi(M) \subseteq M$.

In [4], we have shown that RASP (and ASP) can be defined as a (propositional) fragment of linear logic by translating programs into a linear logic theory employing as connectives tensor product $\otimes$ (to express concomitant use/production of different resources), linear implication $\multimap$ (to model production processes), and additive conjunction & (to represent alternative/exclusive resource allocation). In well-known terminology, we adopt formulas belonging to the so-called Horn-fragment of linear logic. (We refer the reader to [4] for a detailed treatment.) Briefly, for a positive ASP program Π, each atom q in the body of the j-th rule of a given program is renamed as q^j, where the q^j's are called the standardized-apart versions of q. Since the formalization passes through RASP, which considers atoms as resources, each standardized-apart atom q^j will stand for q^j:1 (In RASP terminology, a writing of the form $q{:}a$ denotes an *amount* a of the *resource* q.) The meaning is that, when using the body of a rule to derive the head, one uses one unit of each atom (seen as a resource) in the body. An &-Horn implication deals with the fact that, in Π, the truth of an atom might be used to prove several consequences (through different rules). To treat the case of full ASP, where occurrences of each negative literals, say *not A*, may appear in the body of the rules of Π, we improve the transformation by making the assumption that *not A* is made available to every rule that intends to adopt it, unless *A* itself is provable. In which case the assumption becomes totally unavailable (as proving *A* consumes the full available quantity of the "resource" *not A*).

It turns out that, if Π is an ASP program, and Σ_Π is the corresponding Linear Logic RASP Theory, and $M = \{A_1, \ldots, A_n\}$ is an answer set for Π, then $A_1 \otimes \ldots \otimes A_n$ is a maximal tensor conjunction provable from Σ_Π. The reverse result does not necessarily hold: this is due (as discussed in [4]) to the lack of relevance of the answer set semantics [9], but also to the locality of a proof-based system such as linear logic.

3   Negation as a Resource: A Novel View on Answer Set Semantics

It is interesting to notice that the linear logic formulation we mentioned in previous section prevents contradictions. Consider for example the program $\Pi_1 = \{p \leftarrow not\ p.\}$. It is transformed into:

$$not\ p^{11}{:}1 \otimes not\ p^{12}{:}1 \multimap p,$$
$$not\ p{:}1,$$
$$(not\ p{:}1 \multimap not\ p^{11}{:}1) \& (not\ p{:}1 \multimap not\ p^{12}{:}1)$$

In the first rule, one occurrence of *not p* corresponds to the one originally present, the other one has been added as for proving *p* it is necessary to "absorb" the whole available quantity of *not p*. We can in fact verify that the singleton tensor conjunction *p* is by no means provable: in fact, it would require two units of *not p*, while just one is available. This does not lead to inconsistency, but simply to the impossibility to prove *p*.

Consider program $\Pi = \{a \leftarrow not\ b.\ b \leftarrow not\ c.\ c \leftarrow not\ a.\}$ which is an "odd cycle" involving three atoms. In our formulation, Σ_Π is the following:

$$not\ a^1{:}1 \otimes not\ b^1{:}1 \multimap a \qquad not\ a{:}1 \qquad (not\ a{:}1 \multimap not\ a^1{:}1) \& (not\ a{:}1 \multimap not\ a^3{:}1)$$
$$not\ c^2{:}1 \otimes not\ b^2{:}1 \multimap b \qquad not\ b{:}1 \qquad (not\ b{:}1 \multimap not\ b^1{:}1) \& (not\ b{:}1 \multimap not\ b^2{:}1)$$
$$not\ a^3{:}1 \otimes not\ c^3{:}1 \multimap c \qquad not\ c{:}1 \qquad (not\ c{:}1 \multimap not\ c^2{:}1) \& (not\ c{:}1 \multimap not\ c^3{:}1)$$

From this linear logic theory we can prove the three maximal tensor conjunctions, namely, *a*, *b* and *c*. Assume, in fact, to try to prove *a* (the cases of *b* and *c* are of course analogous). Proving *a* uses resources $not\ a^1{:}1$ and $not\ b^1{:}1$. Therefore, after proving *a*, *b* cannot be proved because its own negation (i.e., $not\ b^2{:}1$) is not available: in fact, the &-Horn implication related to *b* generates (indifferently) only one of the two items, and has already been requested to produce $not\ b^1{:}1$ for proving *a*. In turn, *c* cannot be proved because $not\ a^3{:}1$ is not available, as the &-Horn implication related to *a* generates (indifferently) only one of the two items, and has already been requested to produce $not\ a^1{:}1$ for proving *a*.

Thus, the 3-atoms odd cycles is interpreted as an exclusive disjunction, exactly like the 2-atoms even cycle (such as $\{q \leftarrow not\ p.\ p \leftarrow not\ q.\}$) in AS. Therefore, in the generate-and-test perspective which is at the basis of the ASP programming methodology, our new view provides a new mean of easily generating the search space.

We call $\{a\}$, $\{b\}$, and $\{c\}$ *resource-based answer sets*, for which we provide below a logic programming characterization. The resource-based answers set for program $\{p \leftarrow not\ p.\}$ is the empty set.

The ternary cycle has many well-known interpretations in terms of knowledge representation, among which the following is an example:

{*beach* ← *not mountain. mountain* ← *not travel. travel* ← *not beach.*}

In our approach we would have exactly one of (indifferently) *beach*, *mountain*, or *travel*. Similarly for the program {*work* ← *not tired. tired* ← *not sleep. sleep* ← *not work.*}.

There are other semantic approaches to managing odd cycles, such as for instance [10, 11] and [12, 13], that can however be distinguished from the present one: in fact, the former proposals basically choose (variants of) the classical models, and the latter ones treat differently the unary and ternary cycles. Below we provide a variation of the answer set semantics that defines resource-based answer sets.

Definition 2. *Let Π be a program and let I be a Π-based minimal model, i.e. a minimal model such that $\forall A \in I$, there exists a rule in Π with head A. M is a resource-based answer set of Π iff $M = \Gamma_\Pi(I)$, where I is a Π-based minimal model of Π.*

It is clear that answer sets are among resource-based answer sets, and that there is a resource-based answer set for each Π-based classical minimal model. Non-empty resource-based answer sets still form an anti-chain w.r.t. set inclusion.

We call the new semantics RAS semantics (Resource-Based Answer Set semantics), w.r.t. AS (Answer Set) semantics. Differently from answer sets, a (possibly empty) resource-based answer set always exists. Complexity of RAS semantics is however higher than complexity of AS semantics: in fact, [14] proves that deciding whether a set of formulas is a minimal model of a propositional theory is co-NP-complete. Clearly, checking whether a minimal model I is Π-based and computing $\Gamma_\Pi(I)$ has polynomial complexity. Then:

Proposition 1. *Given program Π, deciding whether set of atom I is a resource-based answer set of Π is co-NP-complete.*

The result of [4] about the relation with linear logic extends to the new semantics.

It remains to be explained why the new definition models the intuition, and how it applies to practical cases. In particular, given minimal model I of Π, it may be that $\Gamma_\Pi(I) \subset I$, i.e., $\Gamma_\Pi(I)$ is a proper subset of I and thus I is not an answer set, for only one reason. For atom A to belong to a Π-based minimal model I, there exists some rule in Π with head A. For A not to belong to $\Gamma_\Pi(I)$, so that $\Gamma_\Pi(I) \subset I$, each of the rules that could cause A to be in the model, must have been canceled by step (1) of Γ_Π (as it includes a literal *not B*, for $B \in I$, in its body). Atoms belonging to $\Gamma_\Pi(I)$ are therefore those atoms in I that can be derived without such contradictions. As widely discussed in [8, 15], contradictions only arise in program fragments corresponding to *unbounded odd cycles*, i.e., odd cycles where no atom is bounded to be true/false (thus resolving the contradiction) by links with other parts of the program. Starting from Π-based minimal models however, $\Gamma_\Pi(I)$ provides for these cycles the "exclusive or" interpretation that we have proposed above.

In resource-based answer set semantics, there are no inconsistent programs. Nevertheless, the new semantics is useful in knowledge representation not just to fix inconsistencies: rather, it depicts a more general scenario in many reasonable examples.

Consider for instance the variation of the above program (inspired to examples proposed in [10, 11]):

> beach ← not mountain.   passport_ok ← not forgot_renew.
> mountain ← not travel.   forgot_renew ← not passport_ok.
> travel ← not beach, passport_ok.

This program has answer set $M_1 = \{forgot\_renew, mountain\}$, as passport_ok being false forces *travel* to be false, which in turn makes *mountain* true. The answer set semantics cannot cope with the case of the passport being ok, which is in fact excluded as this option determines no answer set. Instead, in resource-based answer set semantics we have, in addition to M_1, three other answer sets stating that, if the passport is ok, any choice is possible, namely we have $M_2 = \{passport\_ok, mountain\}$, $M_3 = \{passport\_ok, beach\}$, and $M_4 = \{passport\_ok, travel\}$. We may notice that the semantics is still a bit strong on this example on the side of the answer set, as one would say that not having *passport_ok* prevents traveling, but any other choice should be possible, while instead the *mountain* choice is forced. A further generalization may be the subject of future work.

In the new semantics, constraints cannot be modeled in terms of odd cycles. Therefore, they have to be modeled explicitly. In particular, let us assume a constraint $\mathscr{C}$ to be of the form $\leftarrow E_1, \ldots, E_n$. where the E_is are atoms. This is with no loss of generality, as a constraint such as, for instance, $\leftarrow A, not\ B$. can be reformulated as the program fragment $\leftarrow A, B'. B' \leftarrow not\ B$. Thus, an overall program $\Pi_\mathscr{C}$ can be seen as composed of answer set program Π plus a set $\{\mathscr{C}_1, \ldots, \mathscr{C}_v\}$ of constraints, and, possibly, an auxiliary program $\Pi_\mathscr{C}$, so that constraints can be defined on atoms belonging to either Π or $\Pi_\mathscr{C}$. We assume however that $\Pi_\mathscr{C}$ is stratified (i.e., it contains no cycles) and that atoms of Π may occur in $\Pi_\mathscr{C}$ only in the body of rules (in the terminology of [8, 16], $\Pi_\mathscr{C}$ is a *top program* of Π).

Consider for instance $\Pi_\mathscr{C}$ to be composed of the following Π:

$\{beach \leftarrow not\ mountain.\ mountain \leftarrow not\ travel.\ travel \leftarrow not\ beach.\ hyperthyroidism.\}$

plus the following $\Pi_\mathscr{C}$:

> $\{unhealthy \leftarrow beach, hyperthyroidism.\}$

plus the constraint $\leftarrow unhealthy$.

The resulting theory will have resource-based answer sets $\{mountain, hyperthyroidism\}$, and $\{travel, hyperthyroidism\}$, while $\{beach, hyperthyroidism, unhealthy\}$ is excluded by the constraint. We now proceed to the formal definition.

Definition 3. *An Answer Set Theory $\mathscr{T}$ is a couple $\langle \Pi_\mathscr{C}, Constr \rangle$, with $\Pi_\mathscr{C} = \Pi \cup \Pi_\mathscr{C}$, where $\Pi_\mathscr{C}$ is a top program for Π, and where Constr is a set $\{\mathscr{C}_1, \ldots, \mathscr{C}_v\}$, $v \geq 0$, of constraints.*

Definition 4. *Given Answer Set Theory $\mathscr{T} = \langle \Pi_\mathscr{C}, Constr \rangle$, a resource-based Answer Set M for Π fulfills the constraints in Constr iff the answer set program Π' is consistent (in the sense of traditional answer set semantics), where Π' is obtained from $\Pi_\mathscr{C}$ by adding all atoms in M as facts, and all constraints in Constr as rules.*

Definition 5. *A* Resource-based Answer Set *M* *of* Answer Set Theory $\mathscr{T} = \langle \Pi_{\mathscr{O}}, Constr \rangle$ *is a resource-based answer set for* Π *that fulfills all constraints in Constr.*

It is easy to see that, in order to check that resource-based Answer Set M for Π fulfills the constraints, one can check consistency of Π' by: (i) computing (in polynomial time) the unique answer set M'' of the stratified program Π'' obtained from $\Pi_{\mathscr{C}}$ by adding all atoms in M as facts, and then (ii) checking constraints on M'' by pattern-matching. Then, for constraints of the above simple form, we can conclude that deciding about the existence of a resource-based answer set is a co-NP-complete problem.

4 Concluding Remarks

In this paper, we have proposed an extension of the answer set semantics where ternary odd cycles are understood as exclusive disjunctions, similarly to binary even cycles. Algorithms underlying answer set solvers do not seem to need substantial modifications in order to cope with the new semantics, that thus might in principle be easily and quickly implemented. In particular, solvers based on SAT appear to be good candidates for extension to the new setting.

References

[1] Costantini, S., Formisano, A.: Answer set programming with resources. Journal of Logic and Computation 20(2), 533–571 (2010)
[2] Costantini, S., Formisano, A.: Modeling preferences and conditional preferences on resource consumption and production in ASP. Journal of Algorithms in Cognition, Informatics and Logic 64(1), 3–15 (2009)
[3] Costantini, S., Formisano, A., Petturiti, D.: Extending and implementing RASP. Fundamenta Informaticae 105(1-2), 1–33 (2010)
[4] Costantini, S., Formisano, A.: RASP and ASP as a fragment of linear logic. Journal of Applied Non-Classical Logics (JANCL) (2013)
[5] Girard, J.Y.: Linear logic. Theoretical Computer Science 50, 1–102 (1987)
[6] Kanovich, M.I.: The complexity of Horn fragments of linear logic. Ann. Pure Appl. Logic 69(2-3), 195–241 (1994)
[7] Gelfond, M., Lifschitz, V.: The stable model semantics for logic programming. In: Kowalski, R., Bowen, K. (eds.) Proc. of the 5th Intl. Conference and Symposium on Logic Programming, pp. 1070–1080. The MIT Press (1988)
[8] Costantini, S.: Contributions to the stable model semantics of logic programs with negation. Theoretical Computer Science 149(2), 231–255 (1995)
[9] Dix, J.: A classification theory of semantics of normal logic programs, I and II. Fundam. Inform. 22(3), 227–255, 257–288 (1995)
[10] Pereira, L.M., Pinto, A.M.: Revised stable models – A semantics for logic programs. In: Bento, C., Cardoso, A., Dias, G. (eds.) EPIA 2005. LNCS (LNAI), vol. 3808, pp. 29–42. Springer, Heidelberg (2005)
[11] Pereira, L.M., Pinto, A.M.: Tight semantics for logic programs. In: Hermenegildo, M.V., Schaub, T. (eds.) Tech. Comm. of the 26th Intl. Conference on Logic Programming, ICLP 2010. LIPIcs, vol. 7, pp. 134–143 (2010)

[12] Osorio, M., López, A.: Expressing the stable semantics in terms of the pstable semantics. In: Proc. of the LoLaCOM 2006 Workshop. CEUR Workshop Proc., vol. 220, CEUR-WS.org (2006)

[13] Osorio, M., Pérez, J.A.N., Ramírez, J.R.A., Macías, V.B.: Logics with common weak completions. J. Log. Comput. 16(6), 867–890 (2006)

[14] Cadoli, M.: The complexity of model checking for circumscriptive formulae. Inf. Process. Lett. 44(3), 113–118 (1992)

[15] Costantini, S.: On the existence of stable models of non-stratified logic programs. Theory and Practice of Logic Programming 6(1-2) (2006)

[16] Lifschitz, V., Turner, H.: Splitting a logic program. In: Proc. of the Intl. Conference on Logic Programming, ICLP 1994, pp. 23–37 (1994)

AGM-Style Belief Revision of Logic Programs under Answer Set Semantics[*]

James Delgrande[1], Pavlos Peppas[2], and Stefan Woltran[3]

[1] School of Computing Science
Simon Fraser University
Burnaby, B.C.
Canada V5A 1S6
[2] Dept of Business Administration
University of Patras
Patras 265 00, Greece
[3] Institut für Informationssysteme
Technische Universität Wien
Favoritenstraße 9–11
A–1040 Vienna, Austria

Abstract. In the past few years, several approaches for revision (and update) of logic programs have been studied. None of these however matched the generality and elegance of the original AGM approach to revision in classical logic. One particular obstacle is the underlying nonmonotonicity of the semantics of logic programs. Recently however, specific revision operators based on the monotonic concept of SE-models (which underlies the answer-set semantics of logic programs) have been proposed. Basing revision of logic programs on sets of SE-models has the drawback that arbitrary sets of SE-models may not necessarily be expressed via a logic program. This situation is similar to the emerging topic of revision in fragments of classical logic. In this paper we show how nonetheless classical AGM-style revision can be extended to various classes of logic programs using the concept of SE-models. That is, we rephrase the AGM postulates in terms of logic programs, provide a semantic construction for revision operators, and then in a representation result show that these approaches coincide. This work is interesting because, on the one hand it shows how the AGM approach can be extended to a seemingly nonmonotonic framework, while on the other hand the formal characterization may provide guiding principles for the development of specific revision operators.

1 Introduction

Answer set programming [5] is an appealing approach for representing problems in knowledge representation and reasoning. It has a conceptually simple theoretical foundation, while at the same time being applicable to a wide range of practical problems. As well, efficient ASP systems have become available. However, a logic program is not

[*] This work was partially supported by a Canadian NSERC Discovery Grant and by the Austrian Science Fund (FWF) under grant P25521-N23.

P. Cabalar and T.C. Son (Eds.): LPNMR 2013, LNAI 8148, pp. 264–276, 2013.

a static object in general, but rather it will evolve and be subject to change, whether as a result of correcting information in the program, adding to the information already present, or in some other fashion modifying the knowledge represented in the program.

In classical logic, the problem of handling knowledge *in flux* has been thoroughly investigated (see [24] for an overview). The seminal AGM approach [1,17], provides a general, elegant, and widely accepted framework for this purpose. Central to this approach are powerful representation theorems, which aim to characterize *all* rational operators satisfying certain postulates. Its generality, emphasizing logical formalization, syntax independence, and minimal change, has made this approach a standard for investigating problems concerned with revision or update of knowledge bases, regardless of the underlying semantics (see, for instance, [16] for AGM-style revision in terms of description logics).

Although there has been very active research in revision (and update) of logic programs [31,2,20,25,14], the generality of the original AGM approach has not been matched yet in any of these approaches. One obstacle is the underlying nonmonotonicity of the semantics of logic programs, which has led to the study of postulates different from the ones in the AGM approach, see e.g. [14,23]. Recently however, specific revision operators based on the monotonic concept of SE-models [28] (which underlies the answer-set semantics of logic programs [21]) have been proposed [12] together with a suitable variant of the AGM postulates; see also [27] for a variant thereof. In recent work [19], the notion of SE-models (and similar concepts) has been put in connection to postulates for belief base change different from AGM.

However, representation theorems are still lacking. One problem is that arbitrary sets of SE-models may not necessarily be expressed via a logic program. (While such a requirement is crucial for a representation theorem, it is not problematic in classical logic, at least in the finite case, since for any set of interpretations there is a set of formulas having these interpretations as its models.) A similar challenge arises in another emerging topic in the area of belief change, namely revision in fragments of classical logic, see e.g. [10,7]. To be more specific, consider the problem of revision in the Horn fragment of classical logic. Since the models of Horn formulas satisfy a certain closure property, the result of a revision requires that this property is obtained in order to be represented in the Horn fragment, too. For representation theorems, it turned out that one needs to suitably integrate this property to the concept of faithful assignments [18] and to add an additional postulate [10].

In this paper we show how classical AGM-style revision can be extended to various classes of logic programs using the concept of SE-models. We give representation theorems for the AGM-style postulates proposed in [12] by exploiting, first, the recent techniques for Horn revision due to [10] and, second, the properties that program classes enjoy in terms of SE-models [13]. This allows us to give representation theorems for the important classes of disjunctive (generalized and ordinary), normal, positive, and Horn logic programs. This work is interesting because, on the one hand it shows how the AGM approach can be extended to a seemingly nonmonotonic framework, while on the other hand the formal characterization may provide guiding principles for the development of specific revision operators beyond the ones suggested in [12].

The remainder of the paper is organized as follows. The next section reviews answer set programming and belief revision, and surveys previous work in belief change in logic programs. The following section examines the problems that arise in a direct application of the AGM approach to answer set programming. Section 4 provides the main formal results, comprising representation theorems for each of the major classes of logic programs. The next section shows that the approach is compatible with iterated revision, while the last section is a summary. Proofs of theorems are omitted due to space considerations but are available on request.

2   Background and Formal Preliminaries

2.1   Answer Set Programming

Let $\mathcal{A}$ be a finite *alphabet* or set of propositional variables. A *(generalised) logic program* (GLP) over an alphabet $\mathcal{A}$ is a finite set of rules of the form

$$a_1; \ldots; a_m; \sim b_1; \ldots; \sim b_n \leftarrow c_1, \ldots, c_j, \sim d_1, \ldots, \sim d_k \qquad (1)$$

where $a_p, b_q, c_r, d_s \in \mathcal{A}$ and $p, q, r, s \geq 0$. Operators ';' and ',' express disjunctive and conjunctive connectives. A *(default) literal* is an atom a or its (default) negation $\sim a$. A rule r as in (1) is a *fact* if $m = 1$ and $n = j = k = 0$, and an *integrity constraint* if $m = n = 0$, yielding an empty disjunction denoted by $\bot$. $\mathcal{LP}$ will denote the set of generalised logic programs. Unless stated otherwise, *logic program* will refer to a GLP.

A rule r as in (1) is called *disjunctive* if $n = 0$; *normal* if $m \leq 1$ and $n = 0$; or *positive* if $n = k = 0$. A program is a *disjunctive logic program* (DLP) if it consists of disjunctive rules only. A program is a *normal logic program* (NLP) if it consists of normal rules only. For completeness, we also consider *positive logic programs* (PLP), consisting of positive rules, and *Horn logic programs* (HLP), consisting of rules that are both positive and normal.

We define $H(r) = \{a_1, \ldots, a_m, \sim b_1, \ldots, \sim b_n\}$ as the *head* of r and $B(r) = \{c_1, \ldots, c_j, \sim d_1, \ldots, \sim d_k\}$ as the *body* of r. Given a set X of literals, $X^+ = \{a \in \mathcal{A} \mid a \in X\}$, $X^- = \{a \in \mathcal{A} \mid \sim a \in X\}$, and $\sim X = \{\sim a \mid a \in X\}$. For simplicity, we sometimes use a set-based notation, expressing a rule as in (1) as $H(r)^+; \sim H(r)^- \leftarrow B(r)^+, \sim B(r)^-$.

An interpretation is represented by the subset of atoms in $\mathcal{A}$ that are true in the interpretation. A *(classical) model* of a program P is an interpretation in which all of the rules in P are true according to the standard definition of truth in propositional logic, and where default negation is treated as classical negation. $Mod(P)$ denotes the set of classical models of P. The *reduct* of a program P with respect to a set of atoms Y, denoted P^Y, is the set of rules:

$$\{H(r)^+ \leftarrow B(r)^+ \mid r \in P, \ H(r)^- \subseteq Y, \ B(r)^- \cap Y = \emptyset\}.$$

Note that the reduct consists of negation-free rules only. An *answer set* Y of a program P is a subset-minimal model of P^Y. The set of all answer sets of a program P is denoted $AS(P)$. For example, the program $P = \{a \leftarrow, \ c; d \leftarrow a, \sim b\}$ has answer sets $AS(P) = \{\{a, c\}, \{a, d\}\}$.

An *SE interpretation* [28] is a pair (X, Y) of interpretations such that $X \subseteq Y \subseteq \mathcal{A}$. The set of all SE interpretations (over $\mathcal{A}$) is denoted by $\mathcal{SE}$. For simplicity, we often drop set-notation within SE interpretations and simply write, e.g., (a, ab) instead of $(\{a\}, \{a, b\})$. An SE interpretation is an *SE model* of a program P if $Y \models P$ and $X \models P^Y$. The set of all SE models of a program P is denoted by $SE(P)$. Note that Y is an answer set of P iff $(Y, Y) \in SE(P)$ and for every $X \subset Y$, $(X, Y) \notin SE(P)$. Also, we have $(Y, Y) \in SE(P)$ iff $Y \in Mod(P)$.

A program P is *satisfiable* just if $SE(P) \neq \emptyset$. Two programs P and Q are *strongly equivalent*, symbolically $P \equiv_s Q$, iff $SE(P) = SE(Q)$. Alternatively, $P \equiv_s Q$ holds iff $AS(P \cup R) = AS(Q \cup R)$, for every program R [21]. We write $P \models_s Q$ iff $SE(P) \subseteq SE(Q)$. This means that P is satisfiable iff $P \not\models_s \bot$.

One feature of SE models is that they contain "more information" than answer sets, which makes them an appealing candidate for problems where programs are examined with respect to further extension (in fact, this is what strong equivalence is about). We illustrate this issue with the following well-known example, involving programs

$$P = \{p; q \leftarrow\} \quad \text{and} \quad Q = \left\{ \begin{matrix} p \leftarrow \sim q \\ q \leftarrow \sim p \end{matrix} \right\}.$$

Here, we have $AS(P) = AS(Q) = \{\{p\}, \{q\}\}$. However, the SE models (we list them for $\mathcal{A} = \{p, q\}$) differ:

$$SE(P) = \{(p, p), (q, q), (p, pq), (q, pq), (pq, pq)\};$$
$$SE(Q) = \{(p, p), (q, q), (p, pq), (q, pq), (pq, pq), (\emptyset, pq)\}.$$

This is to be expected, since P and Q behave differently with respect to program extension, and thus are not strongly equivalent. Consider $R = \{p \leftarrow q, q \leftarrow p\}$. Then $AS(P \cup R) = \{\{p, q\}\}$, while $AS(Q \cup R)$ has no answer set.

Next, we recall several properties the set of SE-models satisfies for certain program classes. These properties when suitably combined characterize a logic program class $\mathcal{C}$ in a necessary (for any program $P \in \mathcal{C}$, $SE(P)$ satisfies certain properties) and sufficient (for each S satisfying these properties, there exists a $P \in \mathcal{C}$, such that $SE(P) = S$) way; see [15,6] and the overview [13]. The properties we require are as follows: A set S of SE interpretations is *well-defined* if, for each $(X, Y) \in S$, also $(Y, Y) \in S$. A well-defined set S of SE interpretations is *complete* if, for each $(X, Y) \in S$, also $(X, Z) \in S$, for any $Y \subseteq Z$ with $(Z, Z) \in S$. A complete set S of SE interpretations is *closed under here-intersection* if, for each $(X, Z), (Y, Z) \in S$ also $(X \cap Y, Z) \in S$. A complete set S of SE interpretations is *positive definable* if, for each $(X, Y) \in S$, also $(X, X) \in S$. Last, a positive definable set S of SE interpretations is *Horn definable* iff $(X_1, Y_1), (X_2, Y_2) \in S$ implies that $(X_1 \cap X_2, Y_1 \cap Y_2) \in S$. Intuitively, these properties capture inherent features that the reducts of program classes enjoy. For instance, for any positive program and any interpretation Y, it holds that $P^Y = P$, as mirrored by the concept of positive definable. We have the following results, c.f. [13].

- For each GLP P, $SE(P)$ is well defined.
- For each DLP P, $SE(P)$ is complete.

- For each NLP P, $SE(P)$ is closed under here-intersection.
- For each PLP P, $SE(P)$ is positive definable.
- For each HLP P, $SE(P)$ is Horn definable.

Moreover, for a set S of SE interpretations:

- if S is well defined, there exists a GLP P such that $SE(P) = S$;
- if S is complete, there exists a DLP P such that $SE(P) = S$;
- if S is closed under here-intersection, there exists a NLP P such that $SE(P) = S$;
- if S is positive definable, there exists a PLP P such that $SE(P) = S$; and
- if S is Horn definable, there exists a HLP P such that $SE(P) = S$.

Consequently, for a set of SE models S, we define $t(S)$ to be a least (with respect to SE models) logic program whose SE models contain S. Note that such a program is unique, up to strong equivalence. We also overload notation and in the case that W is a set of classical interpretations, we define $t(W)$ to be a formula of propositional logic whose models are exactly W. In both cases, since the alphabet is finite, $t(\cdot)$ is guaranteed to exist.

2.2 Belief Revision

The best known work in belief revision is the *AGM approach* [1,17], in which standards for belief *revision* and *contraction* functions are given. In belief revision, a formula is added to a knowledge base such that the resulting knowledge base is consistent (unless the formula to be added is inconsistent). In the AGM approach it is assumed that a knowledge base receives information concerning a static domain. Belief states are modeled by logically closed sets of sentences, called *belief sets*. Thus, a belief set is a set K of sentences which satisfies the constraint

$$\text{if } K \text{ logically entails } \beta, \text{ then } \beta \in K.$$

K can be seen as comprising a partial theory of the world. For belief set K and formula α, $K + \alpha$ is the deductive closure of $K \cup \{\alpha\}$, called the *expansion* of K by α. $K_\perp$ is the inconsistent belief set (i.e., $K_\perp$ is the set of all formulas).

Subsequently, Katsuno and Mendelzon [18] reformulated the AGM approach so that a knowledge base was represented by a formula in some language $\mathcal{L}$. The following postulates comprise Katsuno and Mendelzon's reformulation of the AGM revision postulates, where $*$ is a function from $\mathcal{L} \times \mathcal{L}$ to $\mathcal{L}$:

(R1) $\psi * \mu \vdash \mu$.
(R2) If $\psi \wedge \mu$ is satisfiable, then $\psi * \mu \leftrightarrow \psi \wedge \mu$.
(R3) If μ is satisfiable, then $\psi * \mu$ is also satisfiable.
(R4) If $\psi_1 \leftrightarrow \psi_2$ and $\mu_1 \leftrightarrow \mu_2$, then $\psi_1 * \mu_1 \leftrightarrow \psi_2 * \mu_2$.
(R5) $(\psi * \mu) \wedge \phi \vdash \psi * (\mu \wedge \phi)$.
(R6) If $(\psi * \mu) \wedge \phi$ is satisfiable, then $\psi * (\mu \wedge \phi) \vdash (\psi * \mu) \wedge \phi$.

Katsuno and Mendelzon also show that a necessary and sufficient condition for constructing an AGM revision operator is that there is a function that associates a total preorder on the set of interpretations with any formula ψ, as follows:

Definition 1. *A faithful assignment is a function that maps each formula ψ to a total preorder $\preceq_\psi$ on the set of interpretations $\mathcal{M}$ such that for any interpretations m_1, m_2:*

1. *If $m_1, m_2 \in Mod(\psi)$ then $m_1 \approx_\psi m_2$*
2. *If $m_1 \in Mod(\psi)$ and $m_2 \notin Mod(\psi)$, then $m_1 \prec_\psi m_2$.*
3. *If $\psi \leftrightarrow \mu$ then $\preceq_\psi = \preceq_\mu$.*

The resulting preorder is referred to as a *faithful ranking* associated with ψ. Intuitively, $m_1 \preceq_\psi m_2$ if m_1 is at least as plausible as m_2 with respect to ψ. Katsuno and Mendelzon then provide the following representation result.

Theorem 1 ([18]). *A revision operator * satisfies postulates (R1)–(R6) iff there exists a faithful assignment that maps each formula ψ to a total preorder $\preceq_\psi$ such that*

$$\psi * \mu = t(\min(Mod(\mu), \preceq_\psi)).$$

Thus the revision of ψ by μ is characterized by those models of μ that are most plausible according to the agent.

More recently there has been work in belief revision with respect to subsets of propositional logic. [10] extends the AGM approach to Horn clause knowledge bases while [7] addresses revision in other syntactic restrictions of propositional logic.

2.3 Belief Change in Logic Programming

Most previous work on belief change for logic programs is referred to as *update*. Representative work includes [31,2,20,14,26,30,11]. Strictly speaking, however, such approaches generally do not address "update," at least insofar as the term is understood in the belief revision community, but rather general change to a logic program.

A typical approach (e.g. [14], [30], and [11]) for such updates is to consider a sequence of logic programs $P_1, P_2, \ldots, P_n$, where for P_i, P_j, and $i > j$, the intuition is that P_i has higher priority or precedence over P_j. Given such a sequence, a set of answer sets is determined that in some sense respects the ordering. This may be done by translating the sequence into a single logic program that contains an encoding of the priorities, or by treating the sequence as a prioritized logic program, or by some other appropriate method. The net result, one way or another, is that one obtains a set of answer sets from such a program sequence. In particular, one does not obtain a new program expressed in the language of the original logic programs. Hence, these approaches fall outside the general AGM belief revision paradigm. Such approaches are also clearly syntactic in nature, and fall into the *belief base* category, rather than the *belief set* category.

Several principles have nonetheless been proposed for logic program update. In particular, [14] considers the question of what principles the update of logic programs should satisfy. This is done by re-interpreting different AGM-style postulates for revising or updating classic knowledge bases, as well as introducing new principles. Among the latter, we note the following:

Initialization $\emptyset * P \equiv P$.

Idempotency $(P * P) \equiv P$.
Tautology If Q is tautologous, then $P * Q \equiv P$.
Absorption If $Q = R$, then $((P * Q) * R) \equiv (P * Q)$.
Augmentation If $Q \subseteq R$, then $((P * Q) * R) \equiv (P * R)$.

It can be noted that if $\subseteq$ and $\equiv$ are interpreted in terms of strong equivalence, the first four postulates are implied by the AGM postulates, while the last corresponds to the first of the Darwiche and Pearl iteration postulates [8]. [23] also suggest the following postulate, which is also implied by the AGM approach:

WIS If $Q \equiv_s R$, then $(P * Q) \equiv (P * R)$.

Some work has focussed specifically on *revision* of logic programs. Early work in this direction includes a series of investigations dealing with restoring consistency for programs possessing no answer sets (e.g., [29]). Other work uses logic programs under a variant of the stable semantics to specify database revision, i.e., the revision of knowledge bases given as sets of atomic facts [22]. [12] addresses specific revision (and belief merging) operators based on distances defined in terms of the SE models of the underlying programs. As well, [9] considers the extent to which logic programs per se, are compatible with the AGM approach to revision.

3   Recasting Belief Revision in Terms of Answer Set Programs

The postulates and semantic construction of Section 2 are easily adapted to logic programs; for this, we draw on material from [12,10]. To begin, the *expansion* of logic programs P and Q, $P + Q$, can be defined as a logic program R where $SE(P) \cap SE(Q) = SE(R)$. It can be observed that logic program expansion is unproblematic, since for any programs P, Q of a particular class, $SE(P) \cap SE(Q)$ satisfies the semantic conditions for that class; for example if $SE(P)$ and $SE(Q)$ are complete then so is $SE(P) \cap SE(Q)$.

For the postulates, we have the following, expressed in terms of logic programs. An *(AGM logic program) revision* function $*$ is a function from $\mathcal{LP} \times \mathcal{LP}$ to $\mathcal{LP}$ satisfying the following postulates.

(L0) $P * Q$ is a GLP.
(L1) $P * Q \models_s Q$.
(L2) If $P + Q$ is satisfiable, then $P + Q \equiv_s P * Q$.
(L3) If Q is satisfiable, then $P * Q$ is satisfiable.
(L4) If $P_1 \equiv_s P_2$ and $Q_1 \equiv_s Q_2$, then $P_1 * Q_1 \equiv_s P_2 * Q_2$.
(L5) $(P * Q) + R \models_s P * (Q + R)$.
(L6) If $(P * Q) + R$ is satisfiable, then $P * (Q + R) \models_s (P * Q) + R$.

For later reference we also give a postulate adapted from a similarly-named postulate from Horn revision [10].

(Acyc) If, for $0 \leq i < n$, we have $(P * Q_{i+1}) + Q_i$ is satisfiable and $(P * Q_0) + Q_n$ is satisfiable, then $(P * Q_n) + Q_0$ is satisfiable.

As well, faithful assignments can be defined for logic programs and SE models, basically by changing notation:

Definition 2. *A* faithful assignment *is a function that maps each logic program P to a total preorder $\preceq_P$ on $\mathscr{SE}$ such that for $m_1, m_2 \in \mathscr{SE}$:*

1. *If $m_1, m_2 \in SE(P)$ then $m_1 \approx_P m_2$*
2. *If $m_1 \in SE(P)$ and $m_2 \notin SE(P)$, then $m_1 \prec_P m_2$.*
3. *If $P \equiv_s Q$ then $\preceq_P = \preceq_Q$.*

The resulting preorder is referred to as the *faithful ranking* associated with P. Finally, one can define a function $*$ in terms of a faithful ranking by:

$$P * Q = t(\min(SE(Q), \preceq_P)). \tag{2}$$

The use of $*$ in (2) is suggestive; ideally one would next establish a correspondence between functions that satisfy the postulates and those that can be specified via Definition 2. However, there are two difficulties that arise with the naïve application of AGM revision to logic programs:

1. Some postulates may not be satisfied in a faithful ranking.
2. Necessary logical consequences of (L*0)-(L*6) may not hold in some classes of logic programs.

For the first problem, consider the following example involving normal logic programs:

$$P = \{\bot \leftarrow p, \ \bot \leftarrow q, \ \bot \leftarrow r.\}$$
$$Q = \{\bot \leftarrow \sim p, \ \bot \leftarrow \sim q, \ \bot \leftarrow \sim r, \ r \leftarrow p, \ r \leftarrow q.\}$$
$$R = \{\bot \leftarrow \sim p, \ \bot \leftarrow \sim q, \ \bot \leftarrow \sim r, \ r \leftarrow p, q, \ p \leftarrow q, \ q \leftarrow p.\}$$

We have the corresponding SE models:

$$SE(P) = \{(\emptyset, \emptyset)\}$$
$$SE(Q) = \{(pqr, pqr), (pr, pqr), (qr, pqr), (r, pqr), (\emptyset, pqr)\}$$
$$SE(R) = \{(pqr, pqr), (r, pqr), (\emptyset, pqr)\}$$

Now consider the total preorder over these SE models:

$$(\emptyset, \emptyset) < [(pqr, pqr), (pr, pqr), (qr, pqr)] < (\emptyset, pqr) < (r, pqr) < \langle \text{rest} \rangle$$

It can be verified that $SE((P*Q)+R) = \{(pqr, pqr), (r, pqr)\}$ and $SE(P*(Q+R)) = \{(pqr, prq)\}$. However, this violates (L5).

The second problem is analogous to one that cropped up in [10] with respect to Horn theories: due to the inferential weakness of Horn theories, an operator that satisfied the Horn revision postulates was not strong enough to guarantee the existence of a corresponding faithful ranking; instead the postulate (Acyc) (which is redundant in classical AGM revision) was required. Informally, the problem with respect to Horn theories was that, for two Horn formulas ϕ and ψ, $\phi \vee \phi$ is generally not Horn.

In terms of logic programs, one would require that, for programs P and Q, there is a program with SE models given exactly by $SE(P) \cup SE(Q)$. This is the case for GLPs, but not for any of the other classes of logic programs that we consider. Hence we obtain:

Theorem 2. *For generalised logic programs, (Acyc) is a logical consequence of the postulates (L*0) - (L*6).*

This result does not obtain for other classes of logic programs, and so for these (Acyc) is necessary.

4   Belief Revision of Answer Set Programs

To begin, we need to restrict candidate faithful rankings over SE models to just those that are sensible with respect to a given class of logic programs. The next two definitions serve to eliminate orderings which are incoherent with respect to a class of programs.

Definition 3. *A set of SE models S is GLP (DLP, NLP, PLP, HLP) elementary iff there exists a GLP (DLP, NLP, PLP, HLP) P such that $S = SE(P)$.*

Definition 4. *A faithful ranking on SE models $\preceq_P$ is GLP (DLP, NLP, PLP, HLP) compliant iff for every GLP (DLP, NLP, PLP, HLP) Q, we have that $\min(SE(Q), \preceq_P)$ is GLP (DLP, NLP, PLP, HLP) elementary.*

We have the following conditions on faithful rankings that provide counterparts for the notions of compliance, and that make it easier to work with a given ranking.

Definition 5. *Let $\preceq$ be a faithful ranking on $\mathcal{SE}$ and let $X, Y, Z \subseteq \mathcal{A}$.*
   Then $\preceq$ satisfies:
   $(G{\preceq})$ iff: if $X \subseteq Y$ then $(Y, Y) \preceq (X, Y)$.
   $(D{\preceq})$ iff: $\preceq$ satisfies $(G{\preceq})$ and
         if $X \subseteq Y \subseteq Z$ and $(X, Y) \approx (Z, Z)$ then $(X, Z) \preceq (Z, Z)$.
   $(N{\preceq})$ iff: $\preceq$ satisfies $(D{\preceq})$ and
         if $X, Y \subseteq Z$ and $(X, Z) \approx (Y, Z)$ then $(X \cap Y, Z) \preceq (X, Z)$.
   $(P{\preceq})$ iff: $\preceq$ satisfies $(D{\preceq})$ and
         if $X \subseteq Y$ then $(X, X) \preceq (X, Y)$.
   $(H{\preceq})$ iff: $\preceq$ satisfies $(P{\preceq})$ and
         if $X_1 \subseteq Y_1$ and $X_2 \subseteq Y_2$ then $(X_1 \cap X_1, Y_1 \cap Y_2) \preceq (X_1, Y_1)$.

Theorem 3. *Let $\preceq$ be a faithful ranking on $\mathcal{SE}$.*

1. *$\preceq$ is GLP compliant iff $\preceq$ satisfies $(G{\preceq})$.*
2. *$\preceq$ is DLP compliant iff $\preceq$ satisfies $(D{\preceq})$.*
3. *$\preceq$ is NLP compliant iff $\preceq$ satisfies $(N{\preceq})$.*
4. *$\preceq$ is PLP compliant iff $\preceq$ satisfies $(P{\preceq})$.*
5. *$\preceq$ is HLP compliant iff $\preceq$ satisfies $(H{\preceq})$.*

These notions of compliance on the one hand, and the postulate (Acyc) on the other, prove to be sufficient to extend the AGM approach to capture revision in logic programs. These results are described next.

The next two results (actually, two sets of results) constitute the two parts of a representation theorem. For each, x is a class of logic programs, where x is one of GLP, DLP, NLP, PLP, HLP, and (L0x) is postulate (L0) adjusted for class x. Then we have:

Theorem 4. *Let P be a logic program of class x and $\preceq$ an x-compliant faithful ranking associated with P. Define an operator $* : \mathcal{LP} \times \mathcal{LP} \mapsto \mathcal{LP}$ by $P*Q = t(\min(SE(Q), \preceq))$. Then $*$ satisfies postulates (L0x) - (L6) and (Acyc).*

Theorem 5. *Let $* : \mathcal{LP} \times \mathcal{LP} \mapsto \mathcal{LP}$ be a function satisfying postulates (L0x) - (L6) and (Acyc). Then for fixed program P of class x, there is a faithful ranking $\preceq$ on $\mathcal{SE}$ such that $\preceq$ is x-compliant and for every program Q of class x, $P*Q = t(\min(SE(Q), \preceq))$.*

> **Proof Outline.** Let P be a logic program of type x. For any two SE models m_1 and m_2, $t(\{m_1, m_2\})$ was defined to be a least (with respect to characterizing SE models) logic program of type x containing m_1 and m_2. A binary relation $\preceq'$ over SE models is defined by: $m_1 \preceq' m_2$ iff $m_1 \in SE(P * t(\{m_1, m_2\}))$. (Note that a point of difficulty, and in contrast with the corresponding Katsuno-Mendelzon proof for AGM revision, is that it is possible to have both $m_1 \notin SE(P * t(\{m_1, m_2\}))$ and $m_2 \notin SE(P * t(\{m_1, m_2\})))$.)
> The relation $\preceq'$ is in general not transitive; its transitive closure $\preceq^*$ is, of course, transitive, and moreover for arbitrary logic program Q of type x, the minimal Q SE models in $\preceq^*$ are shown to be the same as the SE models of $P * Q$. Finally, $\preceq^*$ is in general not total. The last step is to show that there is a total preorder on SE models $\preceq$ such that for any program Q of type x, the minimal Q SE models in $\preceq^*$ and $\preceq$ coincide. □

5 Iteration and GLP Belief Revision

The results of the previous section show that the classical AGM postulates can be recast in a logic programming framework. In this section we show that this is also the case for the Darwiche and Pearl postulates for iterated revision [8]. For simplicity and space reasons we just treat GLPs.

The four postulates for iterated revision proposed by Darwiche and Pearl, call them the *DP postulates*, have been characterized by corresponding restrictions on faithful rankings. We express these conditions in terms of logic programs. Let P be a logic program and $\preceq$ a faithful ranking with respect to P, and let us denote by $\preceq_Q$ the total preorder assigned to the logic program $P * Q$ resulting from the revision of P by Q. To save writing two sets of postulates, in the case of a logic program, $SE(\neg Q)$ is understood to mean $\mathcal{SE} \setminus SE(Q)$. In [8] it was shown that the conditions (IL1) - (IL4) below characterize (respectively) the four DP postulates (where, of course P and Q would be formulas of propositional logic):

(IL1) If $w, w' \in SE(Q)$ then $w \prec_Q w'$ iff $w \prec w'$.

(IL2) If $w, w' \in SE(\neg Q)$ then $w \prec_Q w'$ iff $w \prec w'$.

(IL3) If $w \in SE(Q)$ and $w' \in SE(\neg Q)$ then $w \prec w'$ entails $w \prec_Q w'$.

(IL4) If $w \in SE(Q)$ and $w' \in SE(\neg Q)$ then $w \preceq w'$ entails $w \preceq_Q w'$.

The first two postulates assert that following revision by Q, (SE)-models of Q retain their same relative ranking, as do non-models. The next two postulates assert roughly that a non-(SE)-model of Q never becomes more plausible with respect to a model of Q.

Thus to show that the DP postulates are consistent with (L0) - (L6) and (Acyc), it suffices to prove the following result:

Theorem 6. *Let P be a GLP, and $\preceq$ a GLP compliant, faithful ranking with respect to P. Moreover, let $*$ be the GLP revision function induced from $\preceq$ via Definition 2. For every GLP Q, there exists a GLP compliant, total preorder $\preceq_Q$, that is faithful with respect to $P * Q$, and such that (IL1) - (IL4) are satisfied.*

Proof. Let Q be any GLP. If Q is inconsistent, define $\preceq_Q$ to be equal to $\preceq$. Clearly, in this case $\preceq_Q$ satisfies (IL1) - (IL4). Moreover, since $\preceq$ is GLP compliant, so is $\preceq_Q$. Finally for faithfulness, since Q is inconsistent, by (L1), $SE(P * Q) = \emptyset$ and therefore $\preceq_Q$ is trivially faithful with respect to $P * Q$. Hence the theorem is true when Q is inconsistent.

Assume now that Q is consistent. We define $\preceq_Q$ as follows:

$$w \preceq_Q w' \text{ iff } w \in \min(SE(Q), \preceq) \text{ or } w \preceq w' \text{ and } w' \notin \min(SE(Q), \preceq). \quad (3)$$

According to (3), to construct $\preceq_Q$, one starts with $\preceq$ and simply places the minimal Q SE models (with respect to $\preceq$) at the beginning of the ranking; everything else remains the same. This construction is not new. In the propositional setting it was proposed and explored by Boutilier [3,4] in his treatment of iterated revision; it is known to satisfy (IL1) - (IL4).

The ranking $\preceq_Q$ is clearly faithful with respect to $P * Q$. For GLP compliance, let w be a SE model (X, Y) where $X \subset Y$, and let w' be (Y, Y). Since (G$\preceq$) entails GLP compliance, it suffices to show that $w' \preceq_Q w$. We distinguish two cases. First assume that $w' \in \min(SE(Q), \preceq)$. Then $w' \in \min(\mathscr{SE}, \preceq_Q)$, and therefore $w' \preceq_Q w$ as desired. Second assume that $w' \notin \min(SE(Q), \preceq)$. Since $\preceq$ is GLP compliant, by (G$\preceq$) it follows that $w' \preceq w$. Consequently, since $w' \notin \min(SE(Q), \preceq)$, by (3) we obtain $w' \preceq_Q w$.   □

6   Conclusion

In this paper we have shown how classical AGM-style revision may be expressed with respect to the major classes of extended logic programs under the answer-set semantics. That is, on the one hand we rephrased the AGM postulates in terms of logic programs and on the other hand we provided a semantic construction for revision operators analogous to faithful rankings, but with respect to SE models. Except for generalised logic programs, the postulate set had to be augmented by an "acyclicity" postulate; for the ranking on SE models, rankings also have to satisfy a "compliance" condition, specific to the class of logic programs being considered. Since both the new postulate and the compliance conditions are redundant in belief revision with respect to classical logic (as we have shown the additional postulate remains redundant for the most general class of programs, GLPs), our approach in fact extends the AGM approach to logic programs. Given these (postulational and semantic) characterizations, in a representation result we then show that these characterizations capture the same set of revision functions for each class of logic programs.

This work is interesting for several reasons. It shows how the AGM approach can be extended to a seemingly nonmonotonic (and certainly nonclassical) framework. As well, most previous work in logic program change was at the syntax level, in that the results of belief change depended on how a program was expressed. In contract, the approach at hand deals with the semantic level, in which arbitrary syntactic commitments don't play a role. Presumably also, the formal characterization may provide guiding principles for the development of specific revision operators.

References

1. Alchourrón, C., Gärdenfors, P., Makinson, D.: On the logic of theory change: Partial meet functions for contraction and revision. Journal of Symbolic Logic 50(2), 510–530 (1985)
2. Alferes, J., Leite, J., Pereira, L., Przymusinska, H., Przymusinski, T.: Dynamic updates of non-monotonic knowledge bases. Journal of Logic Programming 45(1-3), 43–70 (2000)
3. Boutilier, C.: Revision sequences and nested conditionals. In: Proceedings of the International Joint Conference on Artificial Intelligence, pp. 519–531 (1993)
4. Boutilier, C.: Iterated revision and minimal change of conditional beliefs. Journal of Logic and Computation 25, 262–305 (1996)
5. Brewka, G., Eiter, T., Truszczynski, M.: Answer set programming at a glance. Commun. ACM 54(12), 92–103 (2011)
6. Cabalar, P., Ferraris, P.: Propositional theories are strongly equivalent to logic programs. Theory and Practice of Logic Programming 7(6), 745–759 (2007)
7. Creignou, N., Papini, O., Pichler, R., Woltran, S.: Belief revision within fragments of propositional logic. In: Brewka, G., Eiter, T., McIlraith, S.A. (eds.) Proceedings of the Thirteenth International Conference on the Principles of Knowledge Representation and Reasoning. AAAI Press (2012)
8. Darwiche, A., Pearl, J.: On the logic of iterated belief revision. Artificial Intelligence 89, 1–29 (1997)
9. Delgrande, J.: A program-level approach to revising logic programs under the answer set semantics. In: Theory and Practice of Logic Programming, 26th Int'l. Conference on Logic Programming (ICLP 2010) Special Issue, vol. 10(4-6), pp. 681–696 (July 2010)
10. Delgrande, J., Peppas, P.: Revising Horn theories. In: Proceedings of the International Joint Conference on Artificial Intelligence, Barcelona, Spain, pp. 839–844 (2011)
11. Delgrande, J.P., Schaub, T., Tompits, H.: A preference-based framework for updating logic programs. In: Baral, C., Brewka, G., Schlipf, J. (eds.) LPNMR 2007. LNCS (LNAI), vol. 4483, pp. 71–83. Springer, Heidelberg (2007)
12. Delgrande, J., Schaub, T., Tompits, H., Woltran, S.: A model-theoretic approach to belief change in answer set programming. ACM Transactions on Computational Logic 14(2) (2013)
13. Eiter, T., Fink, M., Pührer, J., Tompits, H., Woltran, S.: Model-based recasting in answer-set programming. Technical Report DBAI-TR-2013-83, Institute of Information Systems 184/2, Vienna University of Technology, Austria (2013)

14. Eiter, T., Fink, M., Sabbatini, G., Tompits, H.: On properties of update sequences based on causal rejection. Theory and Practice of Logic Programming 2(6), 711–767 (2002)

15. Eiter, T., Tompits, H., Woltran, S.: On solution correspondences in answer set programming. In: Proceedings of the 19th International Joint Conference on Artificial Intelligence (IJCAI 2005), pp. 97–102 (2005)

16. Flouris, G., Plexousakis, D., Antoniou, G.: On applying the AGM theory to DLs and OWL. In: Gil, Y., Motta, E., Benjamins, V.R., Musen, M.A. (eds.) ISWC 2005. LNCS, vol. 3729, pp. 216–231. Springer, Heidelberg (2005)

17. Gärdenfors, P.: Knowledge in Flux: Modelling the Dynamics of Epistemic States. The MIT Press, Cambridge (1988)

18. Katsuno, H., Mendelzon, A.: Propositional knowledge base revision and minimal change. Artificial Intelligence 52(3), 263–294 (1991)

19. Krümpelmann, P., Kern-Isberner, G.: Belief base change operations for answer set programming. In: del Cerro, L.F., Herzig, A., Mengin, J. (eds.) JELIA 2012. LNCS, vol. 7519, pp. 294–306. Springer, Heidelberg (2012)

20. Leite, J.: Evolving Knowledge Bases: Specification and Semantics. IOS Press, Amsterdam (2003)

21. Lifschitz, V., Pearce, D., Valverde, A.: Strongly equivalent logic programs. ACM Transactions on Computational Logic 2(4), 526–541 (2001)

22. Marek, V.W., Truszczyński, M.: Revision programming. Theoretical Computer Science 190, 241–277 (1998)

23. Osorio, M., Cuevas, V.: Updates in answer set programming: An approach based on basic structural properties. Theory and Practice of Logic Programming 7(4), 451–479 (2007)

24. Peppas, P.: Belief revision. In: van Harmelen, F., Lifschitz, V., Porter, B. (eds.) Handbook of Knowledge Representation, pp. 317–359. Elsevier Science, San Diego (2008)

25. Sakama, C., Inoue, K.: Updating extended logic programs through abduction. In: Gelfond, M., Leone, N., Pfeifer, G. (eds.) LPNMR 1999. LNCS (LNAI), vol. 1730, pp. 147–161. Springer, Heidelberg (1999)

26. Sakama, C., Inoue, K.: An abductive framework for computing knowledge base updates. Theory and Practice of Logic Programming 3(6), 671–713 (2003)

27. Slota, M., Leite, J.: Robust equivalence models for semantic updates of answer-set programs. In: Brewka, G., Eiter, T., McIlraith, S.A. (eds.) Principles of Knowledge Representation and Reasoning: Proceedings of the Thirteenth International Conference, KR 2012, Rome, Italy, June 10-14. AAAI Press (2012)

28. Turner, H.: Strong equivalence made easy: nested expressions and weight constraints. Theory and Practice of Logic Programming 3(4), 609–622 (2003)

29. Witteveen, C., van der Hoek, W., de Nivelle, H.: Revision of non-monotonic theories: Some postulates and an application to logic programming. In: MacNish, C., Moniz Pereira, L., Pearce, D.J. (eds.) JELIA 1994. LNCS, vol. 838, pp. 137–151. Springer, Heidelberg (1994)

30. Zacarías, F., Osorio, M., Acosta Guadarrama, J.C., Dix, J.: Updates in Answer Set Programming based on structural properties. In: McIlraith, S., Peppas, P., Thielscher, M. (eds.) Proceedings of the 7th International Symposium on Logical Formalizations of Commonsense Reasoning, pp. 213–219. Fakultät für Informatik (May 2005) ISSN 1430-211X

31. Zhang, Y., Foo, N.Y.: Updating logic programs. In: Proceedings of the Thirteenth European Conference on Artificial Intelligence (ECAI 1998), pp. 403–407 (1998)

Efficient Approximation of Well-Founded Justification and Well-Founded Domination

Christian Drescher and Toby Walsh

NICTA and the University of New South Wales

Abstract. Many native ASP solvers exploit unfounded sets to compute consequences of a logic program via some form of well-founded negation, but disregard its contrapositive, well-founded justification (WFJ), due to computational cost. However, we demonstrate that this can hinder propagation of many relevant conditions such as reachability. In order to perform WFJ with low computational cost, we devise a method that approximates its consequences by computing dominators in a flowgraph, a problem for which linear-time algorithms exist. Furthermore, our method allows for additional unfounded set inference, called well-founded domination (WFD). We show that the effect of WFJ and WFD can be simulated for a important classes of logic programs that include reachability. Finally, we take a stand for native ASP solvers and show that unfounded set inference cannot be replaced by logic program transformations or translations into CNF-SAT.

1 Introduction

The task of ASP solving is naturally broken up into a combination of search and propagation. The latter can be viewed in terms of inference operations like unit propagation on the Clark's completion [4] (UP) and unfounded set [22] computation. Unfounded sets characterise atoms in a logic program that might circularly support themselves when they have no external support and are thus not included in any answer set. While propagating consequences from the completion is well studied and implemented[7,10,16], the task of efficiently propagating all information provided by unfounded sets is not yet solved [8]. Instead, native ASP solvers[7,14,20] apply unfounded set propagation asymmetrically via some form of well-founded negation (WFN), e.g., forward loop (FL) inference[8], to exclude atoms that have no external support. However, without their contrapositives, well-founded justification (WFJ) or its restriction, backward loop (BL) inference, as we shall see, propagation of many important conditions may be hindered. An example is given through reachability, which is relevant to a range of real world applications, and for which very natural and efficient ASP encodings exist.

In this paper, we address this deficiency. Our main contribution is a linear-time algorithm that approximates the consequences of WFJ. The approach is based on a novel graph-representation of logic programs, termed the support flowgraph. We show that the problem of finding all dominators in such graph, for which efficient algorithms exist, can be used to approximate WFJ and even simulate BL and WFJ for important classes of logic programs. Our techniques give rise to new forms of ASP inference, well-founded domination (WFD) and loop domination (LD). WFD and LD are the atom counterparts of WFJ and BL, respectively, i.e., they include atoms into an answer set in order to

P. Cabalar and T.C. Son (Eds.): LPNMR 2013, LNAI 8148, pp. 277–289, 2013.
© Springer-Verlag Berlin Heidelberg 2013

guarantee external support to already included atoms. Then, we analyse the ASP inference on reachability. Contrary to the intuition that ASP systems naturally and efficiently handle reachability, we demonstrate that restricting inference to the combination of UP and WFN can hinder its propagation. Additional information, however, can be drawn from unfounded sets. We show that WFD and LD can lead to additional pruning, and that applying UP, FL, BL, and LD on reachability prunes all possible values. Another result shows the expressive power difference between ASP and CNF-SAT. We show that UP hinders the propagation of reachability irrespective of how it is encoded, and that simulating unfounded set inference via UP requires super-polynomial space.

2   Preliminaries

Given a set of atomic propositions $\mathcal{P}$, a *(normal) logic program* Π is a finite set of *rules* r of the form $p_0 \leftarrow p_1, \ldots, p_m, not\ p_{m+1}, \ldots, not\ p_n$ where $p_i \in \mathcal{P}$ are *atoms* $(0 \le i \le n)$ and *not* p_j is the *default negation* of p_j $(m < j \le n)$. The atom $head(r) = p_0$ is referred to as the *head* of r and the set $body(r) = \{p_1, \ldots, p_m, not\ p_{m+1}, \ldots, not\ p_n\}$ as the *body* of r. Let $body(r)^+ = \{p_1, \ldots, p_m\}$ and $body(r)^- = \{p_{m+1}, \ldots, p_n\}$. We denote by $atom(\Pi)$ the set of atoms occurring in Π, and by $body(\Pi)$ the set of bodies in Π. To access bodies sharing the same head p, define $body(p) = \{body(r) \mid r \in \Pi,\ head(r) = p\}$. A set $X \subseteq \mathcal{P}$ is an *answer set* of a logic program Π if X is the least model of the *reduct* $\{head(r) \leftarrow body(r)^+ \mid r \in \Pi,\ body(r)^- \cap X = \emptyset\}$.

Answer sets can be characterised as assignments that assign true to an atom if and only if it is included in the answer set. Extending assignments to bodies in a logic program can greatly reduce proof complexity [8]. Hence, for a logic program Π, we define an *assignment* $\mathbf{A}$ as a set of *literals* of the form $\mathbf{T}x$ or $\mathbf{F}x$ where $x \in atom(\Pi) \cup body(\Pi)$. Intuitively, $\mathbf{T}x$ expresses that x is assigned *true* and $\mathbf{F}x$ that it is *false* in $\mathbf{A}$. The complement of a literal σ is denoted $\overline{\sigma}$. Let $\mathbf{A}^{\mathbf{T}} = \{x \mid \mathbf{T}x \in \mathbf{A}\}$ and $\mathbf{A}^{\mathbf{F}} = \{x \mid \mathbf{F}x \in \mathbf{A}\}$. $\mathbf{A}$ is *conflict-free* if $\mathbf{A}^{\mathbf{T}} \cap \mathbf{A}^{\mathbf{F}} = \emptyset$, and it is *body-saturated* if $\{\beta \in body(\Pi) \mid (\beta^+ \cap \mathbf{A}^{\mathbf{F}}) \cup (\beta^- \cap \mathbf{A}^{\mathbf{T}}) \neq \emptyset\} \subseteq \mathbf{A}^{\mathbf{F}}$, i.e., all bodies containing an atom that is assigned false must be false. Finally, $\mathbf{A}$ is *total* if $atom(\Pi) \cup body(\Pi) = \mathbf{A}^{\mathbf{T}} \cup \mathbf{A}^{\mathbf{F}}$.

Following Lee [13], answer sets of a logic program are given through total, conflict-free assignments that do not violate the conditions induced by the programs *completion* [4], and contain no non-empty unfounded set. We use the concept of nogoods for representing the conditions from a program's completion. Following [7], a *nogood* is a set of literals $\delta = \{\sigma_1, \ldots, \sigma_n\}$, and given a set of nogoods Δ, a total and conflict-free assignment $\mathbf{A}$ is a *solution* if $\delta \not\subseteq \mathbf{A}$ for all $\delta \in \Delta$. In our setting, every nogood is equivalent to a clause in CNF-SAT, e.g., the nogood $\delta = \{\sigma_1, \ldots, \sigma_n\}$ represents the clause $\overline{\sigma_1} \vee \cdots \vee \overline{\sigma_n}$, and vice versa, and a set of nogoods is equivalent to a CNF-SAT formula. To reflect the conditions from a program's completion, for $\beta = \{a_1, \ldots, a_m, not\ a_{m+1}, \ldots, not\ a_n\} \in body(\Pi)$, define

$$\Delta_\beta = \left\{ \begin{array}{l} \{\mathbf{T}a_1, \ldots, \mathbf{T}a_m, \mathbf{F}a_{m+1}, \ldots \mathbf{F}a_n, \mathbf{F}\beta\}, \\ \{\mathbf{F}a_1, \mathbf{T}\beta\}, \ldots, \{\mathbf{F}a_m, \mathbf{T}\beta\}, \{\mathbf{T}a_{m+1}, \mathbf{T}\beta\}, \ldots, \{\mathbf{T}a_n, \mathbf{T}\beta\} \end{array} \right\}$$

and for an atom $p \in atom(\Pi)$ with $body(p) = \{\beta_1, \ldots, \beta_k\}$ define

$$\Delta_p = \big\{ \{\mathbf{T}\beta_1, \mathbf{F}p\}, \ldots, \{\mathbf{T}\beta_k, \mathbf{F}p\}, \{\mathbf{F}\beta_1, \ldots, \mathbf{F}\beta_k, \mathbf{T}p\} \big\}.$$

Intuitively, the nogoods in Δ_β enforce the truth of body β if and only if all its members are satisfied, and the nogoods in Δ_p enforce the truth of an atom p if and only if all at least one of its bodies is satisfied. Let $\Delta_\Pi = \bigcup_{\beta \in body(\Pi)} \Delta_\beta \cup \bigcup_{p \in atom(\Pi)} \Delta_p$. The solutions for Δ_Π correspond to the models of the completion of Π [7].

We now turn to unfounded sets. For a logic program Π and a set $U \subseteq atom(\Pi)$, the *external support* of U is defined as $ES_\Pi(U) = \{body(r) \mid r \in \Pi,\ head(r) \in U,\ body(r)^+ \cap U = \emptyset\}$. Given an assignment $\mathbf{A}$, U is an *unfounded set* [10] of Π w.r.t. $\mathbf{A}$ if $ES_\Pi(U) \subseteq \mathbf{A^F}$. We define $\mathbf{A}$ as *unfounded-free* if $\{p \mid U \subseteq atom(\Pi),\ p \in U,\ ES_\Pi(U) \backslash \mathbf{A^F} = \emptyset\} \subseteq \mathbf{A^F}$, i.e., all atoms from unfounded sets are false. Attention is often restricted to unfounded sets that are subsets of strongly connected components (i.e., loops) in the *dependency graph* of Π given through $DG(\Pi) = (atom(\Pi) \cup body(\Pi), \{(body(r), head(r)) \mid r \in \Pi\} \cup \{(p, body(r)) \mid r \in \Pi,\ p \in body(r)^+\})$. A non-empty set of atoms $U \subseteq atom(\Pi)$ is a *loop* of Π if for any $p, q \in U$ there is a path from p to q in $DG(\Pi)$ such that all atoms in the path belong to U [13]. We denote by $loop(\Pi)$ the set of all loops in Π, and define for $\beta \in body(\Pi)$ the set $scc(\beta)$ as being composed of all atoms that belong to the same strongly connected component as β.

Next, we introduce to propagation in ASP, starting with unit propagation (UP). Given an assignment $\mathbf{A}$, for a nogood δ and $\sigma \in \delta$, if $\delta \backslash \{\sigma\} \subseteq \mathbf{A}$ and $\bar{\sigma} \notin \mathbf{A}$ then δ is *unit* w.r.t. $\mathbf{A}$ and $\bar{\sigma}$ is *unit-resulting*, i.e., only unit-resulting literals can avert $\delta \subseteq \mathbf{A}$. UP is the process of extending an assignment with unit-resulting literals. Formally, we define

$$UP(\Pi, \mathbf{A}) = \begin{cases} \mathbf{A} \cup \{\sigma\} & \text{if } \sigma \text{ is unit-resulting w.r.t. } \mathbf{A} \text{ for some } \delta \in \Delta_\Pi, \\ \mathbf{A} & \text{otherwise.} \end{cases}$$

There might be several choices for σ in general. Therefore, we often consider fixpoints. The effects of UP are determined by the nogoods in Δ_Π, whose intuition was given earlier in this section. In particular, fixpoint operation of UP achieves a body-saturated assignment. Note that the notion of unit nogoods is the nogood-equivalent of unit clauses in CNF-SAT [17]. Hence, application of the unit clause rule (equivalently termed unit propagation) on the CNF-SAT representation of Δ_Π simulates UP on Δ_Π.

An inference operation that aims at unfounded sets is well-founded negation (WFN). WFN is the process of extending an assignment by assigning false to all atoms that are included in an unfounded set. Formally, for sets of atoms $\Omega \subseteq 2^{atom(\Pi)}$ we define

$$WFN[\Omega](\Pi, \mathbf{A}) = \begin{cases} \mathbf{A} \cup \{\mathbf{F}p\} & \text{if } U \in \Omega,\ p \in U,\ ES_\Pi(U) \backslash \mathbf{A^F} = \emptyset, \\ \mathbf{A} & \text{otherwise.} \end{cases}$$

By construction, if $\Omega = 2^{atom(\Pi)}$ then fixpoint operation of $WFN[\Omega](\Pi, \mathbf{A})$ achieves an unfounded-free assignment. In practice, it is enough to consider only unfounded sets that are loops, i.e., $\Omega = loop(\Pi)$, resulting in a restricted from of WFN referred to as forward loop (FL). Fixpoint operation of FL and UP, however, simulates the effect of WFN and UP [8]. FL can be implemented such that it takes $\mathcal{O}(|\Pi|)$ time (cf. [1]). The contrapositive of WFN is well-founded justification (WFJ). It assigns true to the only external support of a set of atoms that contains a true atom.

$$WFJ[\Omega](\Pi, \mathbf{A}) = \begin{cases} \mathbf{A} \cup \{\mathbf{T}\beta\} & \text{if } U \in \Omega,\ p \in U \cap \mathbf{A^T},\ ES_\Pi(U) \backslash \mathbf{A^F} = \{\beta\}, \\ \mathbf{A} & \text{otherwise.} \end{cases}$$

Again, we consider the alternatives $\Omega = 2^{atom(\Pi)}$ (WFJ) and $\Omega = loop(\Pi)$ (backward loop; BL). In general, WFJ propagates more consequences than BL [8]. The time complexity of WFJ is bounded by $\mathcal{O}(|\Pi|^2)$, relatively high computational cost, as it amounts to failed-literal-detection and WFN. Example 1 demonstrates the effect of WFJ.

Example 1. Consider the logic program Π given through the following set of rules.

$$a \leftarrow not\ b \qquad c \leftarrow d \qquad e \leftarrow f \qquad c \leftarrow a$$
$$b \leftarrow not\ a \qquad d \leftarrow c \qquad f \leftarrow e \qquad e \leftarrow not\ a$$

Given the assignment $\mathbf{A} = \{\mathbf{T}c\}$, applying UP to a fixpoint results in the (extended) assignment $\{\mathbf{T}c, \mathbf{T}\{c\}, \mathbf{T}d, \mathbf{T}\{d\}\}$. WFN cannot propagate any additional information. In particular, neither UP nor WFN infer $\mathbf{T}a$ (which is in the only total assignment that contains $\mathbf{A}$ and corresponds to an answer set of Π). However, WFJ yields $\mathbf{T}\{a\}$ since $ES_\Pi(\{c, d\}) \backslash \mathbf{A}^{\mathbf{F}} = \{\{a\}\}$. In turn, repeated application of UP adds $\mathbf{T}a$, $\mathbf{F}\{not\ a\}$ and $\mathbf{F}b$ to the assignment, and WFN yields $\mathbf{F}e$ and $\mathbf{F}f$ since $ES_\Pi(\{e, f\}) \backslash \{\{not\ a\}\} = \emptyset$.

Now that we have established relevance of WFJ, we turn our attention to propagating WFJ. Recall that propagating WFJ can have quadratic costs. In the next section, we will introduce a method that approximates WFJ with only linear costs.

3 Dominators in the Support Flowgraph

We take a look at how support *flows* through a logic program, represented in a flowgraph. In our context, a *flowgraph* is a directed graph with a specially designated *source* node.

Definition 1. *Given a logic program Π and an assignment $\mathbf{A}$. The* support flowgraph *of Π w.r.t. $\mathbf{A}$, denoted $SFG(\Pi, \mathbf{A})$, is a directed graph defined as follows:*

1. *Create a node for each atom in $atom(\Pi)$ and for each body in $body(\Pi)$, labelled with that atom or body, respectively.*
2. *The predecessors of an atom node p are all bodies in $body(p) \backslash \mathbf{A}^{\mathbf{F}}$. The predecessors of a body node β are the set of atoms $\phi(\beta) \backslash \mathbf{A}^{\mathbf{F}}$ where*

$$\phi(\beta) = \begin{cases} \beta^+ & \text{if } scc(\beta) \cap \beta^+ = \emptyset \\ scc(\beta) \cap \beta^+ & \text{otherwise.} \end{cases}$$

   Observe that $\phi(\beta) \subseteq \beta^+$.
3. *Add a special node $\top$ as the predecessor for all body nodes that do not have a predecessor, i.e., bodies from rules $r \in \Pi$ such that $\phi(body(r)) = \emptyset$.*

Nodes corresponding to atoms are referred to as *atom nodes*, and nodes corresponding to bodies are referred to as *body nodes*. We will also identify nodes with the atoms and bodies labelling them. By construction, any predecessor of an atom is always a body, and for every body either all predecessors are atoms it positively depends on or the special node $\top$ is the only predecessor. Note that $SFG(\Pi, \mathbf{A})$ is a flowgraph with source node $\top$. Its size is linear in the size of Π, and its construction can be made incremental w.r.t. the assignment, i.e., edges are removed down any branch of the search tree and re-inserted upon backtracking.

The intuition behind $SFG(\Pi, \mathbf{A})$ is that (1) the node $\top$, representing syntactic truth, provides support to every non-false body that has no positive dependency, (2) every body β potentially provides external support to all atoms that appear in the head of a rule with body β, and in turn, (3) every non-false atom p can provide support to the bodies that are positively dependent on p. The latter is determined by ϕ, according which, bodies in a non-trivial strongly connected component can only receive support from atoms that are in the same component. This design choice is motivated by the desire to restrict the intake of support to atoms in strongly connected components of the logic program.

It is easy to verify that if $\mathbf{A}$ is body-saturated then every body in $body(\Pi) \setminus \mathbf{A^F}$ has a predecessor, and that by design, if $\mathbf{A}$ is unfounded-free then for every atom $p \in atom(\Pi) \setminus \mathbf{A^F}$ there is a path from $\top$ to p.

We use cuts of the support flowgraph to analyse the flow of support. For a directed graph (V, E) a *cut* $c = (S, W)$ is a partition of V into two disjoint subsets S and W. For accessing the nodes in S that have an edge into W, define $front(c) = \{u \in S \mid (u, v) \in E, \ v \in W\}$. Note that, in principle, edges from W to S are allowed. For nodes in W that have an edge into S, define $back(c) = \{u \in W \mid (u, v) \in E, \ v \in S\}$.

Definition 2. *Given a logic program Π and an assignment $\mathbf{A}$. A cut $c = (S, W)$ of $SFG(\Pi, \mathbf{A})$ is a support cut if $\top \in S$, $front(c) \subseteq body(\Pi)$, and $back(c) \subseteq body(\Pi)$.*

In words, for any support cut $c = (S, W)$, the condition $front(c) \subseteq body(\Pi)$ ensures that whenever a body is in W then all its predecessors are in W, and $back(c) \subseteq body(\Pi)$ ensures that whenever an atom is in W then all its successors are in W.

Example 2. Consider the logic program Π given through the following set of rules:

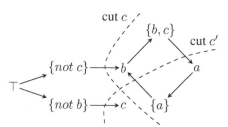

$a \leftarrow b, c \quad b \leftarrow a \quad b \leftarrow not\ c \quad c \leftarrow not\ b$

Note that $c = (\{\top, c, \{not\ b\}, \{not\ c\}\}, \{a, b, \{a\}, \{b, c\}\})$ and $c' = (\{\top, b, c, \{b, c\}, \{not\ b\}, \{not\ c\}\}, \{a, c, \{a\}\})$ both are support cuts of $SFG(\Pi, \emptyset)$. Verify that $ES_\Pi(\{a, b\}) = \{\{not\ c\}\} = front(c)$ and that $ES_\Pi(\{a, c\}) = \{\{not\ b\}\} \subseteq front(c') = \{\{not\ b\}, \{b, c\}\}$.

Observe that $front(c)$ represents external support of $\{a, b\}$, while $front(c')$ approximates (i.e., provides an upper bound of) the external support of $\{a, c\}$.

The following lemma guarantees that every support cut in $SFG(\Pi, \mathbf{A})$ separates a set of atoms from its external support.

Lemma 1. *Given a logic program Π and a body-saturated assignment $\mathbf{A}$. If $c = (S, W)$ is a support cut of $SFG(\Pi, \mathbf{A})$ then $ES_\Pi(W \cap atom(\Pi)) \setminus \mathbf{A^F} \subseteq front(c)$.*

Proof. Let $c = (S, W)$ be a support cut of $SFG(\Pi, \mathbf{A})$. Then, $front(c) \subseteq body(\Pi) \setminus \mathbf{A^F}$. By definition of a support cut, for all $r \in \Pi$ such that $body(r) \notin \mathbf{A^F}$, if $head(r) \in W \setminus \mathbf{A^F}$ then either $body(r) \in front(c)$ or $body(r) \in W$. Since $\mathbf{A}$ is body-saturated, if $body(r) \in W$ then $\phi(body(r)) \subseteq W \cap atom(\Pi)$, and by definition of ϕ, $body(r)^+ \cap W \cap atom(\Pi) \neq \emptyset$. In conclusion, we get $ES_\Pi(W \cap atom(\Pi)) \setminus \mathbf{A^F} \subseteq front(c)$.

Hence, the set of bodies in $front(c)$ provide an upper bound on the external support of the atoms in W. However, we are more interested in finding support cuts that separate a set of atoms from a single external support, i.e., $ES_\Pi(W \cap atom(\Pi)) \backslash \mathbf{A^F} = \{\beta\}$ for some $\beta \in body(\Pi)$. Following from the previous lemma, this single external support is in a domination relationship with the set of atoms it supports. Formally, in a flow-graph (V, E), a node $u \in V$ *dominates* v if every path from the source node to v passes through u. It is easy to verify that a node $v \in S$ dominates all nodes in W if and only if there is a cut $c = (S, W)$ such that $s \in S$ and $front(c) \subseteq \{v\}$.

Theorem 1. *Given a logic program Π and a body-saturated, unfounded-free assignment* **A**. *Let $U \subseteq atom(\Pi)$ such that $U \cap \mathbf{A^T} \neq \emptyset$, and $\beta \in body(\Pi)$. If β dominates all atoms in U in $SFG(\Pi, \mathbf{A})$ then $ES_\Pi(U) \backslash \mathbf{A^F} = \{\beta\}$.*

The previous theorem grants the use of the domination relationship between bodies and atoms to compute consequences from WFJ.

Example 3. Reconsider the logic program from Example 2. The body $\{not\ c\}$ dominates the atom a. Hence, if a is assigned true then WFJ will set $\{not\ c\}$ to true.

A linear-time algorithm for finding all dominators in a flowgraph is provided in [9]. It can be made incremental, i.e., few dominators might be recomputed at any stage during search, subject to removal and re-insertion of edges [21]. This puts our method to approximate WFJ on the same level of computational cost as WFN, resulting in a combined runtime complexity for unfounded set inference of $\mathcal{O}(|\Pi|)$.

The converse of Theorem 1 does not hold in general, but we can provide conditions on logic programs for which our method is guaranteed to compute all consequences from WFJ and BL, respectively.

Definition 3. *A unary logic program is a logic program Π such that for every rule $r \in \Pi$ it holds that $|body(r)^+| \leq 1$. A component-unary logic program is a logic program Π such that for every rule $r \in \Pi$ it holds that $|body(r)^+ \cap scc(body(r))| \leq 1$.*

Observe that every unary logic program is a component-unary logic program, but that component-unary logic programs are much more general. A relevant example from the class of component-unary logic program is discussed in Section 5. In general, any logic program can become (component-) unary as truth values are assigned during search. It is also important to note that for logic programs that are not (component-) unary, our method still simulates WFJ (BL) on the maximal (component-) unary sub-program.

For component-unary logic programs, the domination relationship between body- and atom nodes in the support flowgraph captures BL.

Theorem 2. *Given a component-unary logic program Π, and a body-saturated and unfounded-free assignment* **A**. *Let $L \in loop(\Pi)$ such that $L \cap \mathbf{A^T} \neq \emptyset$, and $\beta \in body(\Pi)$. The body β dominates all atoms in L in $SFG(\Pi, \mathbf{A})$ if and only if $ES_\Pi(L) \backslash \mathbf{A^F} = \{\beta\}$.*

We can guarantee that our method simulates WFJ for unary logic programs.

Theorem 3. *Given a unary logic program Π and a body-saturated, unfounded-free assignment* **A**. *Let $U \subseteq atom(\Pi)$ such that $U \cap \mathbf{A^T} \neq \emptyset$, and $\beta \in body(\Pi)$. If β dominates all atoms in U in $SFG(\Pi, \mathbf{A})$ if and only if $ES_\Pi(U) \backslash \mathbf{A^F} = \{\beta\}$.*

So far, we have restricted our attention to body nodes that dominate a set of atom nodes. In principle, however, any type of node can be a (strict) dominator. We will address dominators that are atom nodes in the next section.

4 Well-Founded Domination

We define an atom-equivalent of WFJ, that is, if a set of atoms U containing at-least one true atom, then any atom that appears positively in all external support of U must likewise be true.

$$WFD[\Omega](\Pi, \mathbf{A}) = \begin{cases} \mathbf{A} \cup \{\mathbf{T}p\} & \text{if } U \in \Omega,\ q \in U \cap \mathbf{A^T}, \text{ and} \\ & ES_\Pi(U) \setminus \mathbf{A^F} \subseteq \{body(r) \mid r \in \Pi,\ p \in body(r)^+\}, \\ \mathbf{A} & \text{otherwise.} \end{cases}$$

As before, we consider the two alternatives $\Omega = 2^{atom(\Pi)}$ (well-founded domination, WFD) and $\Omega = loop(\Pi)$ (loop domination, LD). We reuse the support flowgraph of a logic program and define a new form of cut to approximate consequences of WFD, following the strategy for approximating WFJ from the previous section.

Definition 4. *Given a logic program Π and an assignment* $\mathbf{A}$. *A cut* $c = (S, W)$ *of $SFG(\Pi, \mathbf{A})$ is an* atom cut *if $\top \in S$, $front(c) \subseteq atom(\Pi)$, and $back(c) \subseteq body(\Pi)$.*

The conditions $front(c) \subseteq atom(\Pi)$ and $back(c) \subseteq body(\Pi)$ for an atom cut $c = (S, W)$ ensure that every predecessor and successor of an atom in W is also in W.

Example 4. Consider the logic program Π given through the following set of rules:

$$a \leftarrow b, not\ c \quad a \leftarrow b, not\ d \quad b \leftarrow not\ c$$
$$c \leftarrow not\ d \qquad d \leftarrow not\ c$$

Verify, $c = (\{\top, b, c, d, \{not\ c\}, \{not\ d\}\}, \{a, \{b, not\ c\}, \{b, not\ d\}\})$ is a support cut. Observe that b appears positively in all external support of $\{a\}$, i.e., $ES_\Pi(\{a\}) = \{\{b, not\ c\}, \{b, not\ d\}\} \subseteq \{body(r) \mid r \in \Pi,\ front(c) \in body(r)^+\}$.

The following lemma guarantees that every atom cut in $SFG(\Pi, \mathbf{A})$ separates a set of atoms U from the set of atoms that appear positively in the external support of U.

Lemma 2. *Given a logic program Π and a body-saturated assignment* $\mathbf{A}$. *If $c = (S, W)$ is an atom cut of $SFG(\Pi, \mathbf{A})$ then $ES_\Pi(W \cap atom(\Pi)) \setminus \mathbf{A^F} \subseteq \{body(r) \mid r \in \Pi,\ front(c) \cap body(r)^+ \neq \emptyset\}$.*

Proof. Let $c = (S, W)$ be an atom cut of $SFG(\Pi, \mathbf{A})$. Let $F = W \cap \{body(r) \mid r \in \Pi,\ front(c) \cap \phi(body(r)) \neq \emptyset\}$, the set of bodies in W that have a predecessor in $front(c)$. Construct a cut $c' = (S', W')$ where $S' = S \cup F$ and $W' = W \setminus F$, i.e., all bodies in F are shifted to S. Thus, for all $\beta \in front(c')$ it holds that $front(c) \cap \beta^+ \neq \emptyset$. Next, recall that in a support flowgraph, any predecessor of an atom node is always a body, i.e., no other node has a predecessor in $front(c)$. Hence, we get $front(c') \subseteq body(\Pi)$ and therefore, c' is a support cut of $SFG(\Pi, \mathbf{A})$. By Lemma 1, $ES_\Pi(W' \cap atom(\Pi)) \setminus \mathbf{A^F} \subseteq front(c')$. By construction of c' we have $W \cap atom(\Pi) = W' \cap atom(\Pi)$, and conclude that $front(c) \cap \beta^+ \neq \emptyset$ for every $\beta \in ES_\Pi(W \cap atom(\Pi)) \setminus \mathbf{A^F}$.

Hence, the atoms in $front(c)$ provide an upper bound on the atoms that appear positively in all external support of atoms in W. In order to guarantee that $front(c)$ represents the intersection of all external support, we restrict our attention to atom cuts with a single member in $front(c)$, i.e., dominators. Then, we can approximate WFD.

Theorem 4. *Given a logic program Π and a body-saturated assignment* **A**. *Let $U \subseteq atom(\Pi)$ such that $U \cap \mathbf{A^T} \neq \emptyset$, and $p \in atom(\Pi) \setminus U$. If p dominates all atoms in U in $SFG(\Pi, \mathbf{A})$ then $ES_\Pi(U) \setminus \mathbf{A^F} \subseteq \{body(r) \mid r \in \Pi,\ p \in body(r)^+\}$.*

Example 5. Reconsider the logic program from Example 4, where all external support of $\{a\}$ contains b, and b dominates a. If a is assigned true then WFD will set b to true.

Given a component-unary logic program, the following theorem guarantees that our technique can be used to simulate LD.

Theorem 5. *Given a component-unary logic program Π and a body-saturated assignment* **A**. *Let $L \in loop(\Pi)$ such that $L \cap \mathbf{A^T} \neq \emptyset$, and $p \in atom(\Pi) \setminus L$. The atom node p dominates all atoms in L in $SFG(\Pi, \mathbf{A})$ if and only if $ES_\Pi(L) \setminus \mathbf{A^F} \subseteq \{body(r) \mid r \in \Pi,\ p \in body(r)^+\}$.*

We can even simulate WFD if a unary logic program is given.

Theorem 6. *Given a unary logic program Π and a body-saturated assignment* **A**. *Let $U \subseteq atom(\Pi)$ such that $U \cap \mathbf{A^T} \neq \emptyset$, and $p \in atom(\Pi) \setminus U$. The atom node p dominates all atoms in U in $SFG(\Pi, \mathbf{A})$ if and only if $ES_\Pi(U) \setminus \mathbf{A^F} \subseteq \{body(r) \mid r \in \Pi,\ p \in body(r)^+\}$.*

5   Propagating Reachability in ASP

We want to analyse the impact of propagating ASP inference on the conditions represented by a logic program. These conditions are best studied in terms of constraints over finite domain variables (cf. CSP;[19]). Let V be a finite set of *(domain) variables* where each variable $v \in V$ has an associated finite domain $dom(v)$. A *constraint* c is a k-ary relation on the domains of k variables given by $scope(c) \in V^k$. A *(domain variable) assignment* is a function A that assigns to each variable a value from its domain. For an assignment A, a constraint c is called *domain consistent* if when any $v \in scope(c)$ is assigned any value, there exist values in the domains of the variables in $scope(c) \setminus \{v\}$ such that $A(scope(c)) \in c$, i.e., c is *satisfied*. We will consider variables that represent a directed graph, called *graph variables*, and sets of nodes, called *node set variables*. Following [6], the domain of a graph variable is given via graph inclusion. Graph inclusion defines a partial ordering among graphs, e.g., given two graphs $G = (V, E)$ and $G' = (V', E')$, $G \subseteq G'$ if $V \subseteq V'$ and $E \subseteq E'$. Then, the domain of a graph variable v is defined as the lattice of graphs included between the greatest lower bound $lb(v)$ and the least upper bound $ub(v)$ of the lattice. The domain of a node set variable is bounded by the subsets of nodes in the graph, and we denote the greatest lower bound by $lb(v)$ and the least upper bound by $ub(v)$. If for a domain variable v the associated domain is a singleton, we say that v is *fixed* and simply write v instead of $lb(v)$ or $ub(v)$.

Reachability is a relevant condition in many ASP applications. Given a graph variable G, and node set variables S and N, the constraint $reachable(G, S, N)$ states that N is the set of nodes reachable from some node in S, i.e., the subgraph induced by N is connected. For encoding reachability into ASP we use atoms of the form $edge(Y, X)$, $start(X)$, and $reached(X)$ to capture the membership of edges in G, and nodes in S

and N, respectively. Nodes in G are given implicitly through the edges in G. We denote by REACH$[G, S]$ the following rules:

$$\forall\, X \in ub(S): \qquad reached(X) \leftarrow start(X)$$
$$\forall\, (Y, X) \in ub(G): \qquad reached(X) \leftarrow reached(Y),\, edge(Y, X)$$

We assume that rules for $edge(Y, X)$ and $start(X)$ are provided elsewhere, as we restrict our attention to reachability. It is easy to verify that a node $t \in N$ if and only if $\mathbf{T}reached(t)$ is in an assignment representing an answer set of the resulting program. In the following, we study the impact of propagation on REACH$[G, S]$ in terms of consistency on reachability. We start with the special case where G and S are fixed.

Theorem 7. *If G and S are fixed, then UP and FL on REACH$[G, S]$ achieve domain consistency on reachable(G, S, N).*

Proof. Assume UP and FL reached the fixpoint $\mathbf{A}$, and $\mathbf{A}$ is conflict-free. Let $v \in G$. If $\mathbf{F}reached(v) \notin \mathbf{A}$ then $ES_{\text{REACH}[G,S]}(\{v\}) \setminus \mathbf{A}^{\mathbf{F}} \neq \emptyset$. Hence, either $\mathbf{F}\{start(v)\} \notin \mathbf{A}$ or $\mathbf{F}\{reached(u), edge(u, v)\} \notin \mathbf{A}$ for some $(u, v) \in G$, i.e., either $v \in S$ or v has a predecessor u that is reached. By successively applying the same argument, we obtain loops, each of which concludes in a start node. Hence, there is an assignment with $v \in N$ satisfying the constraint. On the other hand, if $\mathbf{T}reached(v) \notin \mathbf{A}$ then the nogood $\{\mathbf{F}reached(v), \mathbf{T}\{start(v)\}\} \in \Delta_{reached(v)}$ guarantees that $v \notin S$. Similarly, for every $(u, v) \in G$, the nogood $\{\mathbf{F}reached(v), \mathbf{T}\{reached(u), edge(u, v)\}\} \in \Delta_{\text{REACH}[G,S]}$ guarantees that every predecessor u is disconnected. Moreover, the atoms in each loop L starting from v are either disconnected, i.e., we have $ES_{\text{REACH}[G,S]}(L) \setminus \mathbf{A}^{\mathbf{F}} = \emptyset$, or (since the graph is fixed) their subsets are guaranteed external support via a path that does not go through v. Hence, there is an assignment with $v \notin N$ satisfying the constraint. In conclusion, $reachable(G, S, N)$ is domain consistent.

We now turn our attention to another special case of $reachable(G, S, N)$, that is, the value of N is fixed. Then, UP and WFN on REACH$[G, S]$ can hinder propagation, in general, and the construction of a counter example is easy. However, we can guarantee that the addition of WFJ inference prunes all values.

Theorem 8. *If N is fixed then UP and BL on REACH$[G, S]$ achieve domain consistency on reachable(G, S, N).*

Proof. Assume UP and BL result in the fixpoint $\mathbf{A}$, and $\mathbf{A}$ is conflict-free. Let $(u, v) \in ub(G)$. If $\mathbf{T}edge(u, v) \notin \mathbf{A}$ the nogood $\{\mathbf{T}\{reached(u), edge(u, v)\}, \mathbf{F}reached(v)\} \in \Delta_{\text{REACH}[G,S]}$ guarantees that (u, v) does not connect a node that is reached with a disconnected one. Hence, there is an assignment with $(u, v) \in G$ satisfying the constraint. On the other hand, if $\mathbf{F}edge(u, v) \notin \mathbf{A}$ then $ES_{\text{REACH}[G,S]}(\{v\}) \setminus \mathbf{A}^{\mathbf{F}} \neq \{\{reached(u), edge(u, v)\}\}$, i.e., if v is reached then either $v \in S$ or there is some other edge that can connect a reached node to v. By successively applying the same argument, we obtain loops, each of which concludes in a node from S. Hence, there is an assignment with $(u, v) \notin G$ satisfying the constraint. The proof for any $v \in ub(S)$ follows similar arguments. We conclude that $reachable(G, S, N)$ is domain consistent.

If the value of N is not fixed, however, domain consistency is not guaranteed. (Again, the construction of a counter example is easy.) Additional pruning is required. We can show that UP, FL, BL, LD, altogether propagate reachability efficiently.

Theorem 9. *Propagating UP, FL, BL and LD on REACH$[G, S]$ achieves domain consistency on reachable(G, S, N).*

Proof (Sketch). Assume UP, FL, BL and LD reached the fixpoint **A**, and **A** is conflict-free. For any edge $(u, v) \in ub(G)$, the proof follows the one for Theorem 8, i.e., UP ensures that when assigning $(u, v) \in G$ the edge does not connect a node that is reached with a disconnected one, and BL guarantees that when assigning $(u, v) \notin G$ there is some other way to connect to v if $v \in N$. Similarly, for any node $v \in ub(S)$, UP ensures that when assigning $v \in S$ the node v is reached, and BL guarantees that when assigning $v \notin S$ there is some path connecting a start node to v if $v \in N$. Moreover, following the arguments in the proof of Theorem 7, FL removes nodes from $ub(N)$ that cannot be reached in any satisfying domain variable assignment, and for every node $v \in ub(N)$, if $\mathbf{T}reached(v) \notin \mathbf{A}$ then $v \notin lb(S)$ and every predecessor can be disconnected. It remains to show that if $\mathbf{T}reached(v) \notin \mathbf{A}$ then the atoms in each loop L starting from v can be disconnected or reached via some path that does not go through v. This is guaranteed by LD, i.e., we have $ES_\Pi(L) \setminus \mathbf{A^F} \nsubseteq \{body(r) \mid r \in \Pi, \, reached(v) \in body(r)^+\}$. Hence, there is an assignment with $v \notin N$ satisfying the constraint. In conclusion, $reachable(G, S, N)$ is domain consistent.

While previous theorems establish practical relevance of BL and LD, recall that, to our knowledge, existing ASP solvers to not implement BL and LD. However, our efficient approximations of WFJ and WFD can be used to simulate BL and LD, respectively, since REACH$[G, S]$ results in a component-unary logic program. If, in addition, the value of G is fixed such that all atoms in REACH$[G, S]$ of the form $edge(Y, X)$ can be dropped, resulting in a unary logic program, then our method even simulates WFJ and WFD, respectively.

While propagating unfounded sets via FL, UP, and LD on REACH$[G, S]$ prunes all values, we can show that a similar result cannot be simulated by UP. In fact, we show that there is no polynomial size logic program that encodes reachability such that UP achieves domain consistency.

Theorem 10. *There is no polynomial size logic programming encoding of reachability such that UP achieves domain consistency.*

The proof follows from the fact that there is no CNF-SAT encoding.

Proof. Bessiere et al. [2] showed that there is a polynomial size encoding of a constraint into CNF-SAT such that applying unit propagation achieves domain consistency on the constraint if and only if a domain consistency propagator for the constraint can be computed by a polynomial size monotone circuit. Thus, if there exists a polynomial size monotone circuit that computes reachability, we can construct a polynomial size CNF-SAT encoding of reachability such that unit propagation achieves domain consistency. But Karchmer and Wigderson showed that the smallest monotone Boolean circuit for reachability is super-polynomial in the number of vertices in a graph [12]. Since UP on a logic program can be simulated by unit-propagation on a polynomial size CNF-SAT encoding of the program's completion, the smallest encoding of reachability into a logic program such that UP does not hinder propagation is also super-polynomial.

A consequence of previous theorems for SAT-based approaches to ASP solving is that unfounded set inference cannot be simulated by UP using polynomial space. Hence,

Table 1. Experimental Data

Benchmark Class	#N	#S	UP+FL Time	#B	#C	#S	UP+FL+BL Time	#B	#C
Connected Dominating Set	21	20	202	11320.9k	6339.2k	20	3342	6887.4k	3655.0k
Generalised Slitherlink	29	29	3	22.3k	4.7k	29	4	1.3k	0.4k
Graph Partitioning	13	13	147	3159.4k	2344.5k	13	785	1138.5k	810.2k
Hamiltonian Path	29	29	1	44.0k	17.9k	29	8	6.1k	2.9k
Maze Generation	29	26	53	3831.5k	1906.4k	20	1700	1425.8k	880.8k

native ASP solvers can compute more consequences from logic programs than SAT-based solvers. This adds to the study of separating ASP and CNF-SAT in terms of expressive power (see Section 7).

6 Experiments

Implementing Tarjan's linear-time algorithm for finding all dominators in a flowgraph [9] is a challenging engineering exercise as it relies on sophisticated data structures. Hence, for practical reasons, we have integrated BL into the ASP solver *clasp* (2.1.1) via failed-literal-detection and FL. This has high computational costs. To compare with the state-of-the-art, i.e., using only UP and FL, we include the default setting of *clasp* in our analysis. We conducted experiments on search problems that make use of reachability conditions. Our benchmarks stem from the Second ASP Competition [5]. The following definitions apply to Table 1 of results. **UP+FL** denotes *clasp*'s default setting, and **UP+FL+BL** denotes the setting that integrates BL. In each benchmark class, **#N** denotes the total number of instances and **#S** denotes the number of instances for which the program terminated within the allowed time. **Time** denotes the time taken to compute all instances in the class that were solved in both settings. Similarly, **#B** denotes the total number of branches and **#C** denotes the number of conflicts during search, aggregated over all instances in the class that were solved in both settings. Experiments were run on a Linux PC, where each run was limited to 1200s time on a single 2.00 GHz core.

From the results shown in Table 1, it can be concluded that information from BL prunes search dramatically: The additional propagation in **UP+FL+BL** decreases the number of branches and conflicts by roughly one order of magnitude in comparison to **UP+FL**. On the other hand, high computational costs of propagating BL via failed-literal-detection are clearly reflected in the run times of **UP+FL+BL**. These costs, however, can be drastically reduced by using Tarjan's linear-time algorithm, and by making the computation of dominators incremental. In conclusion, our experiments encourage the implementation of our techniques.

7 Related Work

A straightforward way of computing answer sets of logic programs is a reduction to CNF-SAT. This may require the introduction of additional atoms. As shown by Lifschitz and Razborov [15], it is unlikely that, in general, a polynomial-size translation from ASP to CNF-SAT would not require additional atoms. Evidence is provided by the encoding of Lin and Zhao [16] that has exponential space complexity. Another result, shown by Niemelä [18], is that ASP cannot be translated into CNF-SAT in a *faithful* and *modular* way. Reductions based on level-mappings devised in [11] are non-modular but can be computed systematically, using only sub-quadratic space. Our work adds a new result to the study of translating ASP to CNF-SAT. As we have shown in Theorem 10, unfounded set inference cannot be simulated by applying unit propagation on a polynomial-size CNF-SAT encoding, irrespective of the addition of new atoms. To coin a term, reductions from ASP to CNF-SAT cannot *preserve inference*, i.e., in general, every reduction hinders the propagation of consequences from a logic program. Hence, an advantage of native ASP solvers like *clasp* [7], *dlv* [14], and *smodels* [20] over SAT-based systems [10,11,16] is that they can potentially propagate more consequences, e.g., using our techniques.

Formal means for analysing ASP computations in terms of inference were introduced by Gebser and Schaub [8]. According which, *smodels' atmost* and *dlv's greatest unfounded set detection*, both compute WFN and FL, respectively. Similarly, *clasp's unfounded set check* computes FL [1]. Gebser and Schaub also identified the backward propagation operations for unfounded sets, i.e., WFJ and BL. A method that can be used to propagate BL has been proposed by Chen, Ji and Lin [3], but it is inefficient due to high computational costs. We have devised a linear-time approximation of WFJ and shown under which conditions our method simulates WFJ and BL, respectively. Moreover, we have put forward WFD and LD as new forms of inference that can draw additional consequences from unfounded sets. Our approach uses a reduction to the task of finding all dominators in the support flowgraph of a logic program, for which efficient algorithms exist. For instance, Tarjan's algorithm [9] runs in linear time, and computing all dominators can be made incremental [21].

8   Conclusions

Our work is motivated by the desire to understand the effect of propagation in ASP and the diverse modelling choices that arise from logic programming on the process of solving a combinatorial problem. In this paper, we have established that unfounded set inference cannot be simulated by UP on logic program transformations or translations into CNF-SAT. Evidence of practical relevance was given through the problem of reachability. However, as we have seen, even some restricted variants of reachability cannot be efficiently propagated by a combination of UP and WFN. This gap can be closed with WFJ, but existing implementations are inefficient. Our main contribution was a linear-time approximation of WFJ based on a reduction to finding all dominators in a flowgraph representation of the logic program. This gave rise to novel forms of inference, WFD and DL, which can be approximated using the same techniques. We have outlined classes of logic programs for which our approximations simulate WFJ and BL, and WFD and LD, respectively. This includes reachability. Our experimental data encourages the integration of an incremental linear-time algorithm for finding all dominators into an ASP system. Despite our best efforts, efficient algorithms for fully propagating WFJ and WFD remain an open problem.

Acknowledgements. NICTA is funded by the Australian Government as represented by the Department of Broadband, Communications and the Digital Economy and the Australian Research Council through the ICT Centre of Excellence program.

References

1. Anger, C., Gebser, M., Schaub, T.: Approaching the core of unfounded sets. In: Proceedings of NMR 2006, pp. 58–66 (2006)
2. Bessière, C., Katsirelos, G., Narodytska, N., Walsh, T.: Circuit complexity and decompositions of global constraints. In: Proceedings of IJCAI 2009, pp. 412–418 (2009)
3. Chen, X., Ji, J., Lin, F.: Computing loops with at most one external support rule. ACM Trans. Comput. Logic 14(1), 3:1–3:34 (2013)
4. Clark, K.: Negation as failure. In: Logic and Data Bases, pp. 293–322. Plenum Press (1978)
5. Denecker, M., Vennekens, J., Bond, S., Gebser, M., Truszczyński, M.: The second answer set programming competition. In: Erdem, E., Lin, F., Schaub, T. (eds.) LPNMR 2009. LNCS, vol. 5753, pp. 637–654. Springer, Heidelberg (2009)
6. Dooms, G., Deville, Y., Dupont, P.E.: CP(Graph): Introducing a graph computation domain in constraint programming. In: van Beek, P. (ed.) CP 2005. LNCS, vol. 3709, pp. 211–225. Springer, Heidelberg (2005)
7. Gebser, M., Kaufmann, B., Neumann, A., Schaub, T.: Conflict-driven answer set solving. In: Proceedings of IJCAI 2007, pp. 386–392. AAAI Press/MIT Press (2007)
8. Gebser, M., Schaub, T.: Tableau calculi for answer set programming. In: Etalle, S., Truszczyński, M. (eds.) ICLP 2006. LNCS, vol. 4079, pp. 11–25. Springer, Heidelberg (2006)
9. Georgiadis, L., Tarjan, R.E.: Finding dominators revisited. In: Proceedings of SODA 2004, pp. 869–878. SIAM (2004)
10. Giunchiglia, E., Lierler, Y., Maratea, M.: Answer set programming based on propositional satisfiability. Journal of Automated Reasoning 36(4), 345–377 (2006)
11. Janhunen, T., Niemelä, I.: Compact translations of non-disjunctive answer set programs to propositional clauses. In: Balduccini, M., Son, T.C. (eds.) Logic Programming, Knowledge Representation, and Nonmonotonic Reasoning. LNCS, vol. 6565, pp. 111–130. Springer, Heidelberg (2011)
12. Karchmer, M., Wigderson, A.: Monotone circuits for connectivity require super-logarithmic depth. SIAM Journal on Discrete Mathematics 3(2), 255–265 (1990)
13. Lee, J.: A model-theoretic counterpart of loop formulas. In: Proceedings of IJCAI 2005, pp. 503–508. Professional Book Center (2005)
14. Leone, N., Pfeifer, G., Faber, W., Eiter, T., Gottlob, G., Perri, S., Scarcello, F.: The dlv system for knowledge representation and reasoning. ACM Trans. Comput. Log. 7(3), 499–562 (2006)
15. Lifschitz, V., Razborov, A.: Why are there so many loop formulas? ACM Transactions on Computational Logic 7(2), 261–268 (2006)
16. Lin, F., Zhao, Y.: ASSAT: Computing answer sets of a logic program by SAT solvers. In: Proceedings of AAAI 2002, pp. 112–118. AAAI Press/MIT Press (2002)
17. Mitchell, D.: A SAT solver primer. Bulletin of the European Association for Theoretical Computer Science 85, 112–133 (2005)
18. Niemelä, I.: Logic programs with stable model semantics as a constraint programming paradigm. Annals of Mathematics and Artificial Intelligence 25(3-4), 241–273 (1999)
19. Rossi, F., van Beek, P., Walsh, T. (eds.): Handbook of Constraint Programming. Elsevier (2006)
20. Simons, P., Niemelä, I., Soininen, T.: Extending and implementing the stable model semantics. Artificial Intelligence 138(1-2), 181–234 (2002)
21. Sreedhar, V.C., Gao, G.R., Lee, Y.F.: Incremental computation of dominator trees. ACM Trans. Program. Lang. Syst. 19(2), 239–252 (1997)
22. Van Gelder, A., Ross, K.A., Schlipf, J.S.: The well-founded semantics for general logic programs. Journal of the ACM 38(3), 620–650 (1991)

Approximate Epistemic Planning with Postdiction as Answer-Set Programming

Manfred Eppe, Mehul Bhatt, and Frank Dylla

University of Bremen, Germany
{meppe,bhatt,dylla}@informatik.uni-bremen.de

Abstract. We propose a history-based approximation of the Possible Worlds Semantics ($\mathcal{PWS}$) for reasoning about knowledge and action. A respective planning system is implemented by a transformation of the problem domain to an Answer-Set Program. The novelty of our approach is elaboration tolerant support for postdiction under the condition that the plan existence problem is still solvable in NP, as compared to Σ_2^P for non-approximated $\mathcal{PWS}$ of Son and Baral [20]. We demonstrate our planner with standard problems and present its integration in a cognitive robotics framework for high-level control in a smart home.

1 Introduction

Dealing with incomplete knowledge in the presence of abnormalities, unobservable processes, and other real world considerations is a crucial requirement for real-world planning systems. Action-theoretic formalizations for handling incomplete knowledge can be traced back to the Possible Worlds Semantics ($\mathcal{PWS}$) of Moore [15]. Naive formalizations of the $\mathcal{PWS}$ result in search with complete knowledge in an exponential number of possible worlds. The planning complexity for each of these worlds again ranges from polynomial to exponential time [1] (depending on different assumptions and restrictions). Baral et al. [2] show that in case of the action language $\mathcal{A}_k$ the planning problem is Σ_2^P complete (under certain restrictions). This high complexity is a problem for the application of epistemic planning in real-world applications like cognitive robotics or smart environments, where real-time response is needed. One approach to reduce complexity is the approximation of $\mathcal{PWS}$. Son and Baral [20] developed the 0-approximation semantics for $\mathcal{A}_k$ which results in an NP-complete solution for the plan existence problem. However, the application of approximations does not support all kinds of epistemic reasoning, like *postdiction* – a useful inference pattern of knowledge acquisition, e.g., to perform failure diagnosis and abnormality detection. Abnormalities are related to the *qualification problem*: it is not possible to model all conditions under which an action is successful. A partial solution to this is *execution monitoring* (e.g. [18]), i.e. action success is observed by means of specific sensors. If expected effects are not achieved, one can *postdict* about an occurred abnormality.

In Section 3 we present the core contribution of this paper: a 'history' based approximation of the $\mathcal{PWS}$ — called *h-approximation* ($\mathcal{HPX}$) — which supports postdiction. Here, the notion of history is used in an epistemic sense of maintaining and refining knowledge about the past by postdiction and commonsense law of inertia. For instance, if an agent moves trough a door (say at $t = 2$) and later (at some $t' > 2$) comes to know that it is behind the door, then it can postdict that the door must have been open at

P. Cabalar and T.C. Son (Eds.): LPNMR 2013, LNAI 8148, pp. 290–303, 2013.
© Springer-Verlag Berlin Heidelberg 2013

$t = 2$. Solving the plan-existence problem with h-approximation is in NP and finding optimal plans is in Δ_2^P. Despite the low complexity of $\mathcal{HPX}$ compared to $\mathcal{A}_k{}^1$ it is more expressive in the sense that it allows to make propositions about the past. Hence, the relation between $\mathcal{HPX}$ and $\mathcal{A}_k$ is not trivial and deserves a thorough investigation which is provided in Section 4: We extend $\mathcal{A}_k$ and define a *temporal query semantics* ($\mathcal{A}_k{}^{TQS}$) which allows to express knowledge about the past. This allows us to show that $\mathcal{HPX}$ is sound wrt. a temporal possible worlds formalization of action and knowledge.

A planning system for $\mathcal{HPX}$ is developed via its interpretation as an Answer Set Program (ASP). The formalization supports both sequential and (with some restrictions) concurrent planning, and *conditional plans* are generated with off-the-shelf ASP solvers. We provide a case study in a smart home as a proof of concept in Section 5.

2   Related Work

Approximations of the $\mathcal{PWS}$ have been proposed, primarily driven by the need to reduce the complexity of planning with incomplete knowledge vis-a-vis the tradeoff with support for expressiveness and inference capabilities. For such approximations, we are interested in: (i) the extent to which *postdiction* is supported; (ii) whether they are *guaranteed to be epistemically accurate*, (iii) their *tolerance to problem elaboration* [13] and (iv) their *computational complexity*. We identified that many approaches indeed support postdiction, but only in an ad-hoc manner: Domain-dependent postdiction rules and knowledge-level effects of actions are implemented manually and depend on correctness of the manual encoding. For this reason, epistemic accuracy is not guaranteed. Further, even if postdiction rules are implemented epistemically correct wrt. a certain problem, then correctness of these rules may not hold anymore if the problem is elaborated (see Example 1): Hence, ad-hoc formalization of postdiction rules is not elaboration tolerant.

Epistemic Action Formalisms.   Scherl and Levesque [19] provide an epistemic extension and a solution to the frame problem for the Situation Calculus (SC) , and Patkos and Plexousakis [16] as well as Miller et al. [14] provide epistemic theories for the Event Calculus. These approaches are all complete wrt. $\mathcal{PWS}$ and hence suffer from a high computational complexity. Thielscher [21] describes how knowledge is represented in the Fluent Calculus (FC). The implementation in the FC-based framework FLUX is not elaboration-tolerant as it requires manual encoding of knowledge-level effects of actions. Liu and Levesque [11] use a progression operator to approximate $\mathcal{PWS}$. The result is a tractable treatment of the projection problem, but again postdiction is not supported. The PKS planner [17] is able to deal with incomplete knowledge, but postdiction is only supported in an ad-hoc manner. Vlaeminck et al. [24] propose a first order logical framework to approximate $\mathcal{PWS}$. The framework supports reasoning about the past, allows for elaboration tolerant postdiction reasoning, and the projection problem is solvable in polynomial time when using their approximation method. However, the authors do not provide a practical implementation and evaluation and they do not formally relate their approach to other epistemic action languages. To the best of

---

[1] Throughout the paper we usually refer to the full $\mathcal{PWS}$ semantics of $\mathcal{A}_k$. Whenever referring to the 0-approximation semantics this is explicitly stated.

our knowledge, besides [24, 14] there exists no approach which employs a postdiction mechanism that is based on explicit knowledge about the past.

There exist several PDDL-based planners that deal with incomplete knowledge. These planners typically employ some form of $\mathcal{PWS}$ semantics and achieve high performance via practical optimizations such as BDDs [3] or heuristics that build on a relaxed version of the planning problem [8]. The way how states are modeled can also heavily affect performance, as shown by To [22] with the *minimal-DNF* approach. With $\mathcal{HPX}$, we propose another alternative state representation which is based on explicit knowledge about the past.

The $\mathcal{A}$-Family of Languages. The action language $\mathcal{A}$ [7] is originally defined for domains with complete knowledge. Later, epistemic extensions which consider incomplete knowledge and sensing were defined. Our work is strongly influenced by these approaches [12, 20, 23]:   Lobo et al. [12] use epistemic logic programming and formulate a $\mathcal{PWS}$ based epistemic semantics. The original $\mathcal{A}_k$ semantics is based on $\mathcal{PWS}$ and (under some restrictions) is sound and complete wrt. the approaches by Lobo et al. [12] and Scherl and Levesque [19]. Tu et al. [23] introduce $\mathcal{A}_k^c$ and add Static Causal Laws (SCL) to the 0-approximation semantics of $\mathcal{A}_k$. They implement $\mathcal{A}_k^c$ in form of the ASCP planning system which – like $\mathcal{HPX}$ – is based on ASP. The plan-existence problem for $\mathcal{A}_k^c$ is still NP-complete [23]. The authors demonstrate that SCL can be used for an ad-hoc implementation of postdiction. However, we provide the following example to show that an ad-hoc realisation of postdiction is not *elaboration tolerant*:

Example 1. *A robot can drive into a room through a door d. It will be in the room if the door is open:* **causes**($\mathtt{drive}_d, \mathtt{in}, \{\mathtt{open}_d\}$). *An auxiliary fluent* $\mathtt{did\_drive}_d$ *represents that the action has been executed:* **causes**($\mathtt{drive}_d, \mathtt{did\_drive}_d, \emptyset$); *A manually encoded SCL* **if**($\mathtt{open}_d, \{\mathtt{did\_drive}_d, \mathtt{in}\}$) *postdicts that if the robot is in the destination room after driving the door must be open. The robot has a location sensor to determine whether it arrived:* **determines**($\mathtt{sense\_in}, \mathtt{in}$). *Consider an empty initial state* $\delta_{init} = \emptyset$, *a door* $d = 1$ *and a sequence* $\alpha = [\mathtt{drive}_1; \mathtt{sense\_in}]$. *Here* $\mathcal{A}_k^c$ *correctly generates a state* $\delta' \supseteq \{\mathtt{open}_1\}$ *where the door is open if the robot is in the room. Now consider an elaboration of the problem with two doors* ($d \in \{1, 2\}$) *and a sequence* $\alpha = [\mathtt{drive}_1; \mathtt{drive}_2; \mathtt{sense\_in}]$. *By Definitions 4–8 and the closure operator* CL_D *in [23],* $\mathcal{A}_k^c$ *produces a state* $\delta'' \supseteq \{\mathtt{open}_1, \mathtt{open}_2\}$ *where the agent knows that door 1 is open, even though it may actually be closed: this is not sound wrt.* $\mathcal{PWS}$ *semantics.*

Another issue is *concurrent acting and sensing*. Son and Baral [20] (p. 39) describe a modified transition function for the *0-approximation* to support this form of concurrency: they model sensing as determining the value of a fluent after the physical effects are applied. However, this workaround does not support some trivial commonsense inference patterns:

Example 2. *Consider a variation of the Yale shooting scenario where an agent can sense whether the gun was loaded when pulling the trigger because she hears the bang. Without knowing whether the gun was initially loaded, the agent should be able to immediately infer whether or not the turkey is dead depending on the noise. This is not possible with the proposed workaround because it models sensing as the acquisition of a fluent's value after the execution of the sensing: Here the gun is unloaded after executing the shooting, regardless of whether it was loaded before.* $\mathcal{HPX}$ *allows for such inference because here sensing yields knowledge about the value of a fluent at the time sensing is executed.*

3   h-Approximation and Its Translation to ASP

The formalization is based on a foundational theory Γ_{hapx} and on a set of *translation rules* **T** that are applied to a planning domain $\mathcal{P}$. $\mathcal{P}$ is modelled using a PDDL like syntax and consists of the language elements in (1a-1f) as follows: Value propositions ($\mathcal{VP}$) denote initial facts (1a); Oneof constraints ($\mathcal{OC}$) denote exclusive-or knowledge (1b); Goal propositions ($\mathcal{G}$) denote goals[2] (1c); Knowledge propositions ($\mathcal{KP}$) denote sensing (1d); Executability conditions ($\mathcal{EXC}$) denote what an agent must know in order to execute an action (1e); Effect propositions ($\mathcal{EP}$) denote conditional action effects (1f).

$$(\texttt{:init } l^{init}) \quad (1a) \qquad (\texttt{oneof } l^{oo}_1 \ldots l^{oo}_n) \quad (1b) \qquad (\texttt{:goal type (and } l^g_1 \ldots l^g_n)) \quad (1c)$$

$$\begin{array}{c}(\texttt{:action } a \\ \texttt{:observe } f)\end{array} \quad (1d) \qquad \begin{array}{c}(\texttt{:action } a \texttt{ executable} \\ (\texttt{and } l^{ex}_1 \ldots l^{ex}_n))\end{array} \quad (1e) \qquad \begin{array}{c}(\texttt{:action } a \texttt{ :effect} \\ \texttt{when (and } l^c_1 \ldots l^c_n) \; l^e)\end{array} \quad (1f)$$

Formally, a planning domain $\mathcal{P}$ is a tuple $\langle \mathcal{I}, \mathcal{A}, \mathcal{G} \rangle$ where:

- $\mathcal{I}$ is a set of value propositions (1a) and oneof-constraints (1b)
- $\mathcal{A}$ is a set of actions. An action a is a tuple $\langle \mathcal{EP}^a, \mathcal{KP}^a, \mathcal{EXC}^a \rangle$ consisting of a set of effect propositions $\mathcal{EP}^a$ (1f), a set of knowledge propositions $\mathcal{KP}^a$ (1d) and an executability condition $\mathcal{EXC}^a$ (1e).
- $\mathcal{G}$ is a set of goal propositions (1c).

An ASP translation of $\mathcal{P}$, denoted by LP($\mathcal{P}$), consists of a domain-dependent theory and a domain-independent theory:

- Domain-dependent theory (Γ_{world}): It consists of a set of rules Γ_{ini} representing initial knowledge; Γ_{act} representing actions; and Γ_{goals} representing goals.
- Domain-independent theory (Γ_{hapx}): This consists of a set of rules to handle inertia (Γ_{in}); sensing (Γ_{sen}); concurrency (Γ_{conc}), plan verification (Γ_{verify}) as well as plan-generation & optimization (Γ_{plan}).

The resulting Logic Program LP($\mathcal{P}$) is given as:

$$LP(\mathcal{P}) = [\, \Gamma_{in} \cup \Gamma_{sen} \cup \Gamma_{conc} \cup \Gamma_{verify} \cup \Gamma_{plan} \,] \cup [\, \Gamma_{ini} \cup \Gamma_{act} \cup \Gamma_{goal}] \quad (2)$$

Notation. We use the variable symbols A for *action*, EP for *effect proposition*, KP for *knowledge proposition*, T for *time* (or step), BR for *branch*, and F for *fluent*. L denotes *fluent literals* of the form F or ¬F. $\overline{\text{L}}$ denotes the complement of L. For a predicate p(...,L,...) with a literal argument, we denote strong negation "−" with the ¬ symbol as prefix to the fluent. For instance, we denote -knows(F,T,T,BR) by knows(¬ F,T,T,BR). |L| is used to "positify" a literal, i.e. |¬F| = F and |F| = F. Respective small letter symbols denote constants. For example knows(l, t, t', br) denotes that at step t' in branch br it is known that literal l holds at step t.

---

[2] type is either weak or strong. A weak goal must be achieved in only one branch of the conditional plan while a strong goal must be achieved in all branches (see e.g. [3]).

3.1 Translation Rules: $(\mathcal{P} \overset{\text{T1-T8}}{\longmapsto} \Gamma_{world})$

The domain dependent theory Γ_{world} is obtained by applying the set of translation rules $\mathbf{T} = \{T1, \ldots, T8\}$ on a planning domain $\mathcal{P}$.

Actions / Fluents Declarations (T1).   For every fluent f or action a, LP($\mathcal{P}$) contains:
$$fluent(f).\ action(a). \tag{T1}$$

Knowledge $(\mathcal{I} \overset{\text{T2-T3}}{\longmapsto} \Gamma_{ini})$.   Facts Γ_{ini} for initial knowledge are obtained by applying translation rules (T2-T3). For each value proposition (1a) we generate the fact:

$$knows(l^{init}, 0, 0, 0). \tag{T2}$$

For each oneof-constraint (1b) with the set of literals $\mathbf{C} = \{l_1^{oc} \ldots l_n^{oc}\}$ we consider one literal $l_i^{oc} \in \mathbf{C}$. Let $\{l_{i_1}^+, \ldots, l_{i_m}^+\} = \mathbf{C} \setminus l_i^{oc}$ be the subset of literals except l_i^{oc}. Then, for each $l_i^{oc} \in \mathbf{C}$ we generate the LP rule:

$$knows(l_i^{oc}, 0, T, BR) \leftarrow knows(\overline{l_{i_1}^+}, 0, T, BR), \ldots, knows(\overline{l_{i_m}^+}, 0, T, BR). \tag{T3a}$$

$$\begin{aligned} knows(\overline{l_{i_1}^+}, 0, T, BR) &\leftarrow knows(l_i^{oc}, 0, T, BR). \quad \cdots \\ knows(\overline{l_{i_m}^+}, 0, T, BR) &\leftarrow knows(l_i^{oc}, 0, T, BR). \end{aligned} \tag{T3b}$$

(T3a) denotes that if all literals except one are known not to hold, then the remaining one must hold. Rules (T3b) represent that if one literal is known to hold, then all others do not hold. At this stage of our work we only support static causal laws (SCL) to constrain the initial state, because this is the only state in which they do not interfere with the postdiction rules.

Actions $(\mathcal{A} \overset{\text{T4-T7}}{\longmapsto} \Gamma_{act})$.   The generation of rules representing actions covers executability conditions, knowledge-level effects, and knowledge propositions.

Executability Conditions.   These reflect what an agent must know to execute an action. Let $\mathcal{EXC}^a$ of the form (1e) be the executability condition of action a in $\mathcal{P}$. Then LP($\mathcal{P}$) contains the following constraints, where an atom occ(a, t, br) denotes the occurrence of action a at step t in branch br:

$$\begin{aligned} &\leftarrow occ(a, T, BR), not\ knows(l_1^{ex}, T, T, BR). \quad \cdots \\ &\leftarrow occ(a, T, BR), not\ knows(l_n^{ex}, T, T, BR). \end{aligned} \tag{T4}$$

Effect Propositions.   For every effect proposition $ep \in \mathcal{EP}^a$, of the form (when (and $f_1^c \ldots f_{np}^c \neg f_{np+1}^c \ldots \neg f_{nn}^c)\ l^e$), LP($\mathcal{P}$) contains (T5), where hasPC/2 (resp. hasNC/2) represents postive (resp. negative) condition literals, hasEff/2 represents effect literals and hasEP/2 assigns an effect proposition to an action:

$$\begin{aligned} hasEP(a, ep).\ &hasEff(ep, l^e). \\ hasPC(ep, f_1^c).\ &\ldots hasPC(ep, f_{np}^c).\ \ldots \\ hasNC(ep, f_{np+1}^c).\ &\ldots hasNC(ep, f_{nn}^c). \end{aligned} \tag{T5}$$

Knowledge Level Effects of Non-Sensing Actions.   (T6a-T6c)[3]

$$\begin{aligned} knows(l^e, T+1, T1, BR) &\leftarrow apply(ep, T, BR), T1 > T, \\ &knows(l_1^c, T, T1, BR), \ldots, knows(l_n^c, T, T1, BR). \end{aligned} \tag{T6a}$$

---

[3] The frame problem is handled by minimization in the stable model semantics (see e.g. [10]).

$$knows(l_i^c, T, T1, BR) \leftarrow apply(ep, T, BR),$$
$$knows(l^e, T+1, T1, BR), knows(\overline{l^e}, T, T1, BR). \tag{T6b}$$

$$knows(\overline{l_i^{c-}}, T, T1, BR) \leftarrow apply(ep, T, BR), knows(\overline{l^e}, T+1, T1, BR),$$
$$knows(l_{i_1}^{c+}, T, T1, BR), \ldots, knows(l_{i_n}^{c+}, T, T1, BR). \tag{T6c}$$

▶ *Causation* (T6a). If all condition literals l_i^c of an EP (1f) are known to hold at t, and if the action is applied at t, then at $t' > t$, it is known that its effects hold at $t+1$. The atom `apply(ep,t,br)` represents that a with the EP ep happens at t in br.

▶ *Positive postdiction* (T6b). For each condition literal $l_i^c \in \{l_1^c, \ldots, l_k^c\}$ of an effect proposition ep we add a rule (T6b) to the LP. This defines how knowledge about the condition of an effect proposition is postdicted by knowing that the effect holds after the action but did not hold before. For example, if at t' in br it is known that the complement $\overline{l}$ of an effect literal of an EP holds at some $t < t'$ (i.e., `knows`($\overline{l}, t, t', br$)), and if the EP is applied at t, and if it is known that the effect literal holds at $t+1$ (`knows`($l, t+1, t', br$)), then the EP must have set the effect. Therefore one can conclude that the conditions $\{l_1^c, \ldots, l_k^c\}$ of the EP must hold at t.

▶ *Negative postdiction* (T6c). For each potentially unknown condition literal $l_i^{c-} \in \{l_1^c, \ldots, l_n^c\}$ of an effect proposition ep we add one rule (T6c) to the program, where $\{l_{i_1}^{c+}, \ldots, l_{i_n}^{c+}\} = \{l_1^c, \ldots, l_n^c\} \backslash l_i^{c-}$ are the condition literals that are known to hold. This covers the case where we postdict that a condition must be false if the effect is known not to hold after the action and all other conditions are known to hold. For example, if at t' it is known that the complement of an effect literal l holds at some $t+1$ with $t+1 \le t'$, and if the EP is applied at t, and if it is known that all condition literals hold at t, except one literal l_i^{c-} for which it is unknown whether it holds. Then the complement of l_i^{c-} must hold because otherwise the effect literal would hold at $t+1$.

Knowledge Propositions. We assign a KP (1d) to an action a using `hasKP/2`:

$$hasKP(a, f). \tag{T7}$$

Goals ($\mathcal{G} \xrightarrow{\text{T8}} \Gamma_{goal}$). For literals $l_1^{sg}, \ldots, l_n^{sg}$ in a strong goal proposition and $l_1^{wg}, \ldots, l_m^{wg}$ in a weak goal proposition we write:

$$sGoal(T, BR) \leftarrow knows(l_1^{sg}, T, T, BR), \ldots, knows(l_n^{sg}, T, T, BR), s(T), br(BR). \tag{T8a}$$

$$wGoal(T, BR) \leftarrow knows(l_1^{wg}, T, T, BR), \ldots, knows(l_m^{wg}, T, T, BR), s(T), br(BR). \tag{T8b}$$

where an atom `sGoal(t,br)` (resp. `wGoal(t,br)`) represents that the strong (resp. weak) goal is achieved at t in br.

3.2 Γ_{hapx} – Foundational Theory (F1–F5)

The foundational domain-independent $\mathcal{HPX}$-theory is shown in Listing 1. It covers concurrency, inertia, sensing, goals, plan-generation and plan optimization. Line 1 sets the maximal plan length `maxS` and width `maxBr`.

Listing 1. Domain independent theory (Γ_{hapx})

```
1   s(0..maxS). ss(0..maxS-1). br(0..maxBr).
2 ▶ Concurrency (Γconc)
3   apply(EP,T,BR) :- hasEP(A,EP), occ(A,T,BR).
4   contra(EP1,EP) :- hasPC(EP1,F),hasNC(EP,F).
5   :- 2{apply(EP,T,BR):hasEff(EP,F)},br(BR), s(T), fluent(F).
6   :- apply(EP,T,BR), hasEff(EP,F), apply(EP1,T,BR),
        hasEff(EP1,¬F), EP != EP1, not contra(EP1,EP).
7 ▶ Inertia (Γin)
8   initApp(F,T,BR) :- apply(EP,T,BR),hasEff(EP,F).
9   kNotInit(F,T,T1,BR) :- not initApp(F,T,BR),
        uBr(T1,BR), s(T), fluent(F).
10  kNotInit(F,T,T1,BR) :- apply(EP,T,BR), hasPC(EP,F1),
        hasEff(EP,F) ,knows(¬F1,T,T1,BR), T1>=T.
11  knows(F,T+1,T1,BR) :- knows(F,T,T1,BR), kNotTerm(F,T,T1,BR),
        T<T1, s(T).
12  knows(F,T-1,T1,BR) :- knows(F,T,T1,BR),
        kNotInit(F,T-1,T1,BR), T>0, T1>=T, s(T).
13  knows(L,T,T1+1,BR) :- knows(L,T,T1,BR),T1<maxS,s(T1).
14 ▶ Sensing and Branching (Γsen)
15  uBr(0,0). uBr(T+1,BR) :- uBr(T,BR), s(T).
16  kw(F,T,T1,BR):- knows(F,T,T1,BR).
17  kw(F,T,T1,BR):- knows(¬F,T,T1,BR).
18  sOcc(T,BR) :- occ(A,T,BR), hasKP(A,_).
19  leq(BR,BR1) :- BR <= BR1, br(BR), br(BR1).
20  1{nextBr(T,BR,BR1): leq(BR,BR1)}1 :- sOcc(T,BR).
21  :- 2{nextBr(T,BR,BR1) :br(BR):s(T)},br(BR1).
22  uBr(T+1,BR) :- sRes(¬F,T,BR).
23  sRes(F,T,BR) :- occ(A,T,BR),hasKP(A,F),not knows(¬F,T,T,BR).
24  sRes(¬F,T,BR1) :- occ(A,T,BR),hasKP(A,F),not kw(F,T,T,BR),
        nextBr(T,BR,BR1).
25  knows(L,T,T+1,BR) :- sRes(L,T,BR).
26  knows(F1,T,T1,BR1) :- sOcc(T1,BR), nextBr(T1,BR,BR1),
        knows(F1,T,T1,BR), T1>=T.
27  apply(EP,T,BR1) :- sOcc(T1,BR), nextBr(T1,BR,BR1),
        uBr(T1,BR), apply(EP,T,BR), T1>=T.
28  :-2{occ(A,T,BR):hasKP(A,_)}, br(BR), s(T).
29 ▶ Plan verification (Γverify)
30  allWGsAchieved :- uBr(maxS,BR), wGoal(maxS,BR).
31  notAllSGAchieved :- uBr(maxS,BR), not sGoal(maxS,BR).
32  planFound :- allWGsAchieved, not notAllSGAchieved.
33  :- not planFound.
34  notGoal(T,BR) :- not wGoal(T,BR), uBr(T,BR).
35  notGoal(T,BR) :- not sGoal(T,BR), uBr(T,BR).
36 ▶ Plan generation and optimization (Γplan)
37  1{occ(A,T,BR):a(A)}1 :- uBr(T,BR), notGoal(T,BR),
                            br(BR), ss(T). % Sequential planning
38  %1{occ(A,T,BR):a(A)} :- uBr(T,BR), notGoal(T,BR),
                            br(BR), ss(T). % Concurrent planning
39  #minimize {occ(_,_,_) @ 1}    % Optimal planning
```

F1. Concurrency (Γ_{conc}). Line 3 applies all effect propositions of an action a if that action occurs. We need two restrictions regarding concurrency of non-sensing actions: effect similarity and effect contradiction. Two effect propositions are similar if they have the same effect literal. Two EPs are contradictory if they have complementary effect literals and if their conditions do not contradict (*l.* 4). The cardinality constraint *l.* 5 enforces that two similar EPs (with the same effect literal) do not apply concurrently, whereas *l.* 6 restricts similarly for contradictory EPs.

F2. Inertia (Γ_{in}). Inertia is applied in both forward and backward direction similar to [7]. To formalize this, we need a notion on knowing that a fluent is *not* initiated (resp. terminated). This is expressed with the predicates kNotInit/kNotTerm.[4] A fluent could be known to be not initiated for two reasons: (1) if no effect proposition with the respective effect fluent is applied, then this fluent can not be initiated. initApp(f,t,br) (*l.* 8) represents that at t an EP with the effect fluent f is applied in branch br. If initApp(f,t,br) does not hold then f is known not to be initiated at t in br (*l.* 9).

(2) a fluent is known not to be initiated if an effect proposition with that fluent is applied, but one of its conditions is known not to hold (*l.* 10). Note that this requires the concurrency restriction (*l.* 5). Having defined kNotInit/4 and kNotTerm/4 we can formulate forward inertia (*l.* 11) and backward inertia (*l.* 12). Two respective rules for inertia of false fluents are not listed for brevity. We formulate *forward propagation* of knowledge in *l.* 13. That is, if at t' it is known that f was true at t, then this is also known at $t' + 1$.

F3. Sensing and Branching (Γ_{sen}). If sensing occurs, then each possible outcome of the sensing uses one branch. uBr(t,br) denotes that branch br is used at step t. Predicate kw/4 in *ll.* 16-17 is an abbreviation for *knowing whether*. We use sOcc(t,br) to state that a sensing action occurred at t in br (*l.* 18). By leq(br,br') the partial order of branches is precomputed (*l.* 19); it is used in the choice rule *l.* 20 to "pick" a valid child branch when sensing occurs. Two sensing actions are not allowed to pick the same child branch (*l.* 21). Lines 23-24 assign the positive sensing result to the current branch and the negative result to the child branch. Sensing results affect knowledge through *l.* 25. Line 26 represents inheritance: Knowledge and application of EPs is transferred from the original branch to the child branch (*l.* 27). Finally, in line *l.* 28, we make the restriction that two sensing actions cannot occur concurrently.

F4. Plan Verification (Γ_{verify}). Lines 30-33 handle that weak goals must be achieved in only one branch and strong goals in all branches. Information about nodes where goals are not yet achieved (*ll.* 34-35) is used in the plan generation part for pruning.

F5. Plan Generation and Optimization (Γ_{plan}). Line 37 and *l.* 38 implement sequential and concurrent planning respectively. Optimal plans in terms of the number of actions are generated with the optimization statement *l.* 39.

3.3 Plan Extraction from Stable Models

A conditional plan is determined by a set of occ/3, nextBr/3 and sRes/3 atoms.

---

[4] For brevity Listing 1 does only contain rules for kNotInit; the rules for kNotTerm are analogous resp. to *ll.* 8-10.

Definition 1 (Planning as ASP Solving). *Let S be a stable model for the logic program LP(P), then p solves the planning problem* $\mathcal{P}$ *if p is exactly the subset containing all* occ/3, nextBr/3 *and* sRes/3 *atoms of S.*

For example, consider the atoms occ(a_0, t, br), sRes(f, t, br), sRes($\neg f, t, br'$), nextBr(t, br, br'), occ($a_1, t+1, br$) and occ($a_2, t+1, br'$). With a syntax as in [23], this is equivalent to the conditional plan a_0; [if f then a_1 else a_2].

3.4   Complexity of h-Approximation

According to [23], we investigate the complexity for a limited number of sensing actions, and feasible plans. That is, plans with a length that is polynomial wrt. the size of the input problem.

Theorem 1 ((Optimal) Plan Existence). *The plan existence problem for the h-approximation is in NP and finding an optimal plan is in* Δ_2^P.

Proof Sketch: The result emerges directly from the complexity properties of ASP (e.g. [6]).

1. The translation of an input problem via (T1-T8) is polynomial.
2. Grounding the normal logic program is polynomial because the arity of predicates is fixed and maxS and maxBr are bounded due the polynomial plan size.
3. Determining whether there exists a stable model for a normal logic program is NP-complete.
4. Finding an optimal stable model for a normal logic program is Δ_2^P-complete.

3.5   Translation Optimizations

Although optimization of $\mathcal{HPX}$ is not in the focus at this stage of our work we want to note two obvious aspects: (1) By avoiding *unnecessary action execution*, e.g. opening a door if it is already known to be open, search space is pruned significantly. (2) Some domain specificities (e.g., connectivity of rooms) are considered as *static relations*. For these, we modify translation rules (T4) (executability conditions) and (T2) (value propositions), such that knows/4 is replaced by holds/1.

4   A Temporal Query Semantics for $\mathcal{A}_k$

$\mathcal{HPX}$ is not just an approximation to $\mathcal{PWS}$ as implemented in $\mathcal{A}_k$. It is more expressive in the sense that $\mathcal{HPX}$ allows for propositions about the past, e.g. "at step 5 it is known that the door was open at step 3". To find a notion of soundness of $\mathcal{HPX}$ with $\mathcal{A}_k$ (and hence $\mathcal{PWS}$-based approaches in general), we define a *temporal query semantics* ($\mathcal{A}_k^{TQS}$) that allows for reasoning about the past. The syntactical mapping between $\mathcal{A}_k$ and $\mathcal{HPX}$ is presented in the following table:

	$\mathcal{A}_k$	$\mathcal{HPX}$ PDDL dialect
Value prop.	**initially**(l^{init})	(:init l^{init})
Effect prop.	**causes**($a, l^e, \{l_1^c \ldots l_n^c\}$)	(:action a :effect when (and $l_1^c \ldots l_n^c$) l^e)
Executability	**executable**($a, \{l_1^{ex}, \ldots, l_n^{ex}\}$)	(:action a :executable (and $l_1^{ex} \ldots l_n^{ex}$))
Sensing	**determines** ($a, \{f, \neg f\}$)	(:action a :observe f)

An $\mathcal{A}_k$ domain description D can always be mapped to a corresponding $\mathcal{HPX}$ domain specification due to the syntactical similarity. Note that for brevity we do not consider executability conditions in this section. Their implementation and intention is very similar in h-approximation and $\mathcal{A}_k$. Further we restrict the $\mathcal{A}_k$ semantics to allow to sense the value of only one single fluent with one action.

Original $\mathcal{A}_k$ Semantics by Son and Baral [20]. $\mathcal{A}_k$ is based on a transition function which maps an action and a so-called c-state to a c-state. A c-state δ is a tuple $\langle u, \Sigma \rangle$, where u is a state (a set of fluents) and Σ is a k-state (a set of possible belief states). If a fluent is contained in a state, then its value is $true$, and $false$ otherwise. Informally, u represents how the world is and Σ represents the agent's belief. In this work we assume grounded c-states for $\mathcal{A}_k$, i.e. $\delta = \langle u, \Sigma \rangle$ is grounded if $u \in \Sigma$. The transition function for non-sensing actions and without considering executability is:

$$\Phi(a, \langle u, \Sigma \rangle) = \langle Res(a, u), \{Res(a, s') | s' \in \Sigma\} \rangle \text{ where} \tag{3}$$
$$Res(a, s) = s \cup E_a^+(s) \setminus E_a^-(s) \text{ where} \tag{4}$$
$$E_a^+(s) = \{f | f \text{ is the effect literal of an EP and all condition literals hold in } s\}$$
$$E_a^-(s) = \{\neg f | \neg f \text{ is the effect literal of an EP and all condition literals hold in } s\}$$

Res reflects that if all conditions of an effect proposition hold, then the effect holds in the result. The transition function for sensing actions is:

$$\Phi(a, \langle u, \Sigma \rangle) = \langle u, \{s | (s \in \Sigma) \wedge (f \in s \Leftrightarrow f \in u)\} \rangle \tag{5}$$

For convenience we introduce the following notation for a k-state Σ:

$$\Sigma \models f \text{ iff } \forall s \in \Sigma : f \in s \text{ and } \Sigma \models \neg f \text{ iff } \forall s \in \Sigma : f \cap s = \emptyset \tag{6}$$

It reflects that a fluent is known to hold if it holds in all possible worlds s in Σ.

Temporal Query Semantics – $\mathcal{A}_k^{TQS}$.   Our approach is based on a re-evaluation step with a similar intuition as the *update operator* "∘" in [24]: Let $\Sigma_0 = \{s_0^0, \ldots, s_0^{|\Sigma_0|}\}$ be the set of all possible initial states of a (complete) initial c-state of an $\mathcal{A}_k$ domain D. Whenever sensing happens, the transition function will remove some states from the k-state, i.e. $\Phi([a_1; \ldots; a_n], \delta_0) = \langle u_n, \Sigma_n \rangle$, where $\Sigma_n = \{s_n^0, \ldots, s_n^{|\Sigma_n|}\}$ and $|\Sigma_0| \geq |\Sigma_n|$. To reason about the past, we re-evaluate the transition. Here, we do not consider the complete initial state, but only the subset Σ_0^n of initial states which "survived" the transition of a sequence of actions. If a fluent holds in all states of a k-state Σ_t^n, where Σ_t^n is the result of applying $t \leq n$ actions on Σ_0^n, then after the n-th action, it is known that a fluent holds after the t-th action.

Definition 2. *Let $\alpha = [a_1; \ldots; a_n]$ be a sequence of actions and δ_0 be a possible initial state, such that $\Phi([a_1; \ldots; a_n], \delta_0) = \delta_n = \langle u_n, \Sigma_n \rangle$. We define Σ_0^n as the set of initial belief states in Σ_0 which are valid after applying α: $\Sigma_0^n = \{s_0 | s_0 \in \Sigma_0 \wedge Res(a_n, Res(a_{n-1}, \ldots, Res(a_1, s_0) \ldots)) \in \Sigma_n\}$.[5] We say that*

$$\langle l, t \rangle \text{ is known to hold after } \alpha \text{ on } \delta_0$$

if $\Sigma_t^n \models l$ where $\langle u_t, \Sigma_t^n \rangle = \Phi([a_1; \ldots; a_t], \langle u_0, \Sigma_0^n \rangle)$ and $t \leq n$

Soundness wrt. $\mathcal{A}_k^{TQS}$ The following conjecture considers soundness for the projection problem for a sequence of actions:

---

[5] Consider that according to (4) $Res(a, s) = s$ if a is a sensing action.

Conjecture 1. *Let D be a domain specification and $\alpha = [a_1; \ldots; a_n]$ be a sequence of actions. Let $LP(D) = [\Gamma_{in} \cup \Gamma_{sen} \cup \Gamma_{conc} \cup \Gamma_{ini} \cup \Gamma_{act}]$ be a $\mathcal{HPX}$-logic program without rules for plan generation (Γ_{plan}), plan verification (Γ_{verify}) and goal specification (Γ_{goal}). Let Γ_{occ}^n contain rules about action occurrence in valid branches, i.e. $\Gamma_{occ}^n = \{occ(a_0, 0, BR) \leftarrow uBr(0, BR)., \ldots, occ(a_n, n, BR) \leftarrow uBr(n, BR).\}$ Then for all fluents f and all steps t with $0 \leq t \leq n$, there exists a branch br such that:*

$$if\ knows\ (l, t, n, br) \in SM[LP(D) \cup \Gamma_{occ}^n]\ then\ \Sigma_t^n \models l \quad with\ t \leq n. \tag{7}$$

where $SM[LP(D) \cup \Gamma_{occ}^n]$ denotes the stable model of the logic program.

Considerations regarding correctness of the conjecture can be found in an extended version of this paper [5].

5   Evaluation and Case-Study

In order to evaluate practicability of $\mathcal{HPX}$ we compare our implementation with the ASCP planner by Tu et al. [23] and show an integration of our planning system in a smart home assistance system.

Comparison with ASCP.   We implemented three well known benchmark problems for $\mathcal{HPX}$ and the 0-approximation based ASCP planner:[6] *Bomb in the toilet* (e.g. [8]; n potential bombs need to be disarmed in a toilet), *Rings* (e.g. [3]; in n ringlike connected rooms windows need to be closed/locked), and *Sickness* (e.g. [23]; one of n diseases need identified with a paper color test). While $\mathcal{HPX}$ outperforms ASCP for the Rings problem (e.g. $\approx$ 10s to 170s for 3 rooms), ASCP outperforms $\mathcal{HPX}$ for the other domains (e.g. $\approx$ 280s to 140s for 8 bombs and $\approx$ 160s to 1360s for 8 diseases). For the first problem, $\mathcal{HPX}$ benefits from static relations and for the latter two problems ASCP benefits from a simpler knowledge representation and the ability to sense the paper's color with a single action where $\mathcal{HPX}$ needs $n-1$ actions. In both ASCP and $\mathcal{HPX}$ grounding was very fast and the bottleneck was the actual solving of the problems.

Application in a Smart Home.   The $\mathcal{HPX}$ planning system has been integrated within a larger software framework for smart home control in the Bremen Ambient Assisted Living Lab (BAALL) [9]. We present a use-case involving action planning in the presence of abnormalities for an robotic wheelchair: The smart home has (automatic) sliding doors, and sometimes a box or a chair accidentally blocks the door such that it opens only half way. In this case, the planning framework should be able to postdict such an abnormality and to follow an alternative route. The scenario is illustrated in Fig. 1.

Consider the situation where a person instructs a command to the wheelchair (e.g., to reach location; $[S_0]$). An optimal plan to achieve this goal is to pass D1. A more error tolerant plan is: Open D1 and verify if the action succeeded by sensing the door status $[S_1]$; *If* the door is open, drive through the door and approach the user. *Else* there is an abnormality: Open and pass D3 $[S_2]$; drive through the bedroom $[S_3]$; pass D4 and D2 $[S_4]$; and finally approach the sofa $[S_5]$.[7] If it is behind the door then the door was open. For this particular use-case, a sub-problem follows:

---

[6] We used an Intel i5 (2GHz, 6Gb RAM) machine running *clingo* [6] with Windows 7. Tests were performed for a fixed plan length and width.

[7] Abnormalities are considered on the alternative route but skipped here for brevity.

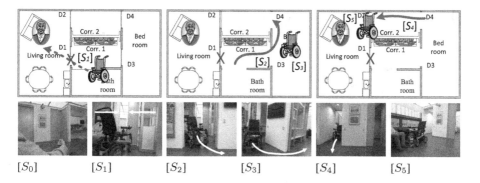

Fig. 1. The wheelchair operating in the smart home BAALL

```
(:action open_door :effect when ¬ab_open open)
(:action drive :executable (and open     ¬in_liv)
                          :effect in_liv)
(:action sense_open :observe open)
(:init ¬in_liv ¬open)     (:goal weak in_liv)
```

The solution to this subproblem is depicted in Fig. 2 (see also state S_1 in Fig. 1). There is an autonomous robotic wheelchair outside the living room ($\neg$in_liv) and the weak goal is that the robot is inside the living room. The robot can open the door (open_door) to the living room. Unfortunately, opening the door does not always work, as the door may be jammed, i.e. there may be an abnormality. However, the robot can perform sensing to verify whether the door is open (sense_open). Figure 2 illustrates our postdiction mechanism. Initially (at $t = 0$ and $br = 0$) it is known that the robot is in the corridor at step 0. The first action is opening the door, i.e. the stable model contains the atom occ(open_door,0,0). Inertia holds for ¬in_liv, because nothing happened that could have initiated ¬in_liv. The rules in *ll*. 8-9 trigger kNotInit(in_liv,0,0,0) and *l.* 13 triggers knows(¬in_liv,0,1,0), such that in turn the forward inertia rule (*l.* 11) causes atom knows(¬in_liv,1,1,0) to hold. Next, sensing happens, i.e. occ(sense_open,1,0). According to the rule in *l.* 23, the positive result is assigned to the original branch and sRes(open,1,0) is produced. According to the rule in *l.* 24, the negative sensing result at step t in branch br is assigned to some child branch br' (denoted by nextBr(t,br,br')) with $br' > br$ (*l.* 20). In the example we have: sRes(¬open,1,1), and due to *l.* 25 we have knows(¬open,1,2,1). This result triggers postdiction rule (T6c) and knowledge about an abnormality is produced: knows(ab_open,0,2,1). Consequently, the wheelchair has to follow another route to achieve the goal. For branch 0, we have knows(open,1,2,0) after the sensing. This result triggers the postdiction rule (T6b): Because knows(¬open,0,2,0) and knows(open,1,2,0) hold, one can postdict that there was no abnormality when open occurred: knows(¬ab_open,0,2,0). Finally, the robot can drive through the door: occ(drive,2,0) and the causation rule (T6a) triggers knowledge that the robot is in the living room at step 3: knows(in_liv,3,3,0).

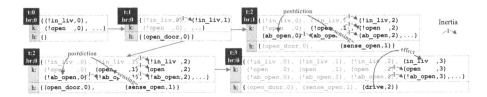

Fig. 2. Abnormality detection as postdiction with *h-approximation*

6   Conclusion

We developed an approximation of the possible worlds semantics with elaboration tolerant support for postdiction, and implemented a planning system by a translation of the approximation to ASP. We show that the plan existence problem in our framework can be solved in NP. We relate our approach to the $\mathcal{PWS}$ semantics of $\mathcal{A}_k$ by extending $\mathcal{A}_k$ semantics to allow for temporal queries. We show that $\mathcal{HPX}$ is sound wrt. this semantics. Finally, we provide a proof of concept for our approach with the case study in Section 5. An extended version of the Case Study will appear in [4]. Further testing revealed the inferiority of the $\mathcal{HPX}$ implementation to dedicated PDDL planners like CFF [8]. This result demands future research concerning the transfer of heuristics used in PDDL-based planners to ASP.

References

[1] Bäckström, C., Nebel, B.: Complexity results for SAS+ planning. Computational Intelligence 11, 625–655 (1995)
[2] Baral, C., Kreinovich, V., Trejo, R.: Computational complexity of planning and approximate planning in the presence of incompleteness. Artificial Intelligence 122 (2000)
[3] Cimatti, A., Pistore, M., Roveri, M., Traverso, P.: Weak, strong, and strong cyclic planning via symbolic model checking. Artificial Intelligence 147, 35–84 (2003)
[4] Eppe, M., Bhatt, M.: Narrative based Postdictive Reasoning for Cognitive Robotics. In: 11th Int'l Symposium on Logical Formalizations of Commonsense Reasoning (2013)
[5] Eppe, M., Bhatt, M., Dylla, F.: h-approximation: History-Based Approximation to Possible World Semantics as ASP. Technical report, arXiv:1304.4925v1 (2013)
[6] Gebser, M., Kaminski, R., Kaufmann, B., Schaub, T.: Answer Set Solving in Practice. Morgan and Claypool (2012)
[7] Gelfond, M., Lifschitz, V.: Representing action and change by logic programs. The Journal of Logic Programming 17, 301–321 (1993)
[8] Hoffmann, J., Brafman, R.I.: Contingent planning via heuristic forward search with implicit belief states. In: ICAPS Proceedings (2005)
[9] Krieg-Brückner, B., Röfer, T., Shi, H., Gersdorf, B.: Mobility Assistance in the Bremen Ambient Assisted Living Lab. Journal of GeroPsyc 23, 121–130 (2010)
[10] Lee, J., Palla, R.: Reformulating the situation calculus and the event calculus in the general theory of stable models and in answer set programming. JAIR 43, 571–620 (2012)
[11] Liu, Y., Levesque, H.J.: Tractable reasoning with incomplete first-order knowledge in dynamic systems with context-dependent actions. In: IJCAI Proceedings (2005)
[12] Lobo, J., Mendez, G., Taylor, S.: Knowledge and the Action Description Language A. Theory and Practice of Logic Programming 1, 129–184 (2001)

[13] McCarthy, J.: Elaboration tolerance. In: Commonsense Reasoning (1998)
[14] Miller, R., Morgenstern, L., Patkos, T.: Reasoning About Knowledge and Action in an Epistemic Event Calculus. In: 11th Int'l Symposium on Logical Formalizations of Commonsense Reasoning (2013)
[15] Moore, R.: A formal theory of knowledge and action. In: Hobbs, J., Moore, R. (eds.) Formal Theories of the Commonsense World, Ablex, Norwood, NJ, pp. 319–358 (1985)
[16] Patkos, T., Plexousakis, D.: Reasoning with Knowledge, Action and Time in Dynamic and Uncertain Domains. In: IJCAI Proceedings, pp. 885–890 (2009)
[17] Petrick, R., Bacchus, F.: Extending the knowledge-based approach to planning with incomplete information and sensing. In: ICAPS Proceedings (2004)
[18] Pettersson, O.: Execution monitoring in robotics: A survey. Robotics and Autonomous Systems 53, 73–88 (2005)
[19] Scherl, R.B., Levesque, H.J.: Knowledge, action, and the frame problem. Artificial Intelligence 144, 1–39 (2003)
[20] Son, T.C., Baral, C.: Formalizing sensing actions - A transition function based approach. Artificial Intelligence 125, 19–91 (2001)
[21] Thielscher, M.: Representing the knowledge of a robot. In: Proc. of KR (2000)
[22] To, S.T.: On the impact of belief state representation in planning under uncertainty. In: IJCAI Proceedings (2011)
[23] Tu, P.H., Son, T.C., Baral, C.: Reasoning and planning with sensing actions, incomplete information, and static causal laws using answer set programming. Theory and Practice of Logic Programming 7, 377–450 (2007)
[24] Vlaeminck, H., Vennekens, J., Denecker, M.: A general representation and approximate inference algorithm for sensing actions. In: Australasian Conference on AI (2012)

Combining Equilibrium Logic and Dynamic Logic

Luis Fariñas del Cerro, Andreas Herzig, and Ezgi Iraz Su*

University of Toulouse
IRIT, CNRS
http://www.irit.fr

Abstract. We extend the language of here-and-there logic by two kinds of atomic programs allowing to minimally update the truth value of a propositional variable here or there, if possible. These atomic programs are combined by the usual dynamic logic program connectives. We investigate the mathematical properties of the resulting extension of equilibrium logic: we prove that the problem of logical consequence in equilibrium models is EXPTIME complete by relating equilibrium logic to dynamic logic of propositional assignments.

Keywords: answer-set programming, here-and-there logic, equilibrium logic, propositional dynamic logic, dynamic logic of propositional assignments.

1 Introduction

Answer Set Programming (ASP) is a successful approach in non-monotonic reasoning. Its efficient implementations became a key technology for declarative problem solving in the AI community [7,8]. In recent years many important results have been obtained from a theoretical point of view, such as the definitions of new comprehensive semantics as equilibrium semantics or the proof of important theorems as strong equivalence theorems [14]. These theoretical and practical results show that ASP is central to various approaches in non-monotonic reasoning.

New applications in AI force us to extend the original language of ASP by some new concepts capable of supporting, for example, the representations of modalities, actions, ontologies or updates. Based on a tradition that was started by Alchourrón, Gärdenfors and Makinson and also by Katsuno and Mendelzon [1,13], several researchers have proposed to extend ASP by operations allowing to update or revise a given ASP program through a new piece of information [3,17,15,16]. The resulting formalisms are quite complex, and we think it is fair to say that it is difficult to grasp what the intuitions should be like under these approaches.

We here propose a different, more modest approach, where the new piece of information is restricted to be atomic. It is based on the update of here-and-there (HT) models. Such models are made up of two sets of propositional variables, H ('here') and T ('there'), such that $H \subseteq T$. We consider two kinds of basic update operations: to set a propositional variable true either here or there according to its truth value in these sets;

---

* We would like to thank the three reviewers of LPNMR 2013 for their helpful comments. This work was partially supported by the French-Spanish *Laboratoire Européen Associé (LEA) "French-Spanish Lab of Advanced Studies in Information Representation and Processing"*.

P. Cabalar and T.C. Son (Eds.): LPNMR 2013, LNAI 8148, pp. 304–316, 2013.

similarly to set it false either here or there, again if possible. From these basic update operations we allow to build update programs by means of the standard dynamic logic program operators of sequential and nondeterministic composition, iteration, and test. We call the result dynamic here-and-there logic (D-HT).

The notions of an equilibrium model and of logical consequence in equilibrium models can then be defined exactly as before. We show that the problem of satisfiability in HT models and of consequence in equilibrium models are both EXPTIME complete. In order to do so, we use dynamic logic of propositional assignments (DL-PA) that was recently studied in [2]. We define a translation tr_1 from the language of D-HT into the language of DL-PA. Our main result says that a formula φ is an equilibrium consequence of a formula χ if and only if the DL-PA formula

$$\langle \pi_1 \rangle \big(tr_1(\chi) \wedge \sim \langle \pi_2 \rangle tr_1(\chi) \supset tr_1(\varphi) \big)$$

is valid, where π_1 and π_2 are DL-PA programs whose length is polynomial in the length of χ and φ. This allows to polynomially embed the problems of D-HT satisfiability and consequence in equilibrium models into DL-PA, and so establishes that they are all in EXPTIME. We moreover show that these upper bounds are tight.

The paper is organized as follows. In Section 2 we introduce dynamic here-and-there logic (D-HT) and define consequence in its equilibrium models. In Section 3 we present dynamic logic of propositional assignments (DL-PA) and establish its complexity. In Section 4 we define translations relating the language of D-HT to the language of DL-PA and vice versa. Section 5 concludes.

2 A Dynamic Extension of HT Logic and of Equilibrium Logic

In this section we propose a dynamic extension of the logic of here-and-there (HT), named D-HT. By means of the standard definition of an equilibrium model, that extension also provides a definition of a non-monotonic consequence relation which is a conservative extension of the standard equilibrium consequence relation.

To begin with, we fix a countable set of *propositional variables* ($\mathbb{P}$) whose elements are noted p, q, etc. The language is produced through adding dynamic modalities to the language of HT. The semantics is based on HT models: an HT model is a couple (H, T) such that $H \subseteq T \subseteq \mathbb{P}$. The sets H and T are respectively called 'here' and 'there'. The constraint that $H \subseteq T$ is the so-called heredity constraint of intuitionistic logic. Each of them is a *valuation*, identified with a subset of $\mathbb{P}$. We write $\mathbb{HT}$ for the set of all HT models. So, $\mathbb{HT} = \{(H, T) \ : \ H \subseteq T \subseteq \mathbb{P}\}$.

2.1 The Language $\mathcal{L}_{\text{D-HT}}$

The language $\mathcal{L}_{\text{D-HT}}$ is defined by the following grammar:

$$\varphi ::= p \mid \perp \mid \varphi \wedge \varphi \mid \varphi \vee \varphi \mid \varphi \to \varphi \mid [\pi]\varphi \mid \langle \pi \rangle \varphi$$
$$\pi ::= +p \mid -p \mid \pi; \pi \mid \pi \cup \pi \mid \pi^* \mid \varphi?$$

where p ranges over $\mathbb{P}$.

We have only two *atomic programs* in the language, namely $+p$ and $-p$. Each of them minimally updates an HT model, if this is possible: in a sense, the former 'upgrades the truth of p' while the latter 'downgrades the truth of p'. More precisely, the program $+p$ makes p true there, but keeps its truth value same here if p is not included there. However, if p exists there, but not here then it makes p true here while keeping its truth value there; otherwise the program $+p$ fails. On the other hand, the program $-p$ sets p false here as it keeps it there if p is contained here. Nevertheless, if p is only contained there, but not here then the program $-p$ excludes p there keeping its truth value same here; or else the program fails.

The operators of sequential composition ("$;$"), nondeterministic composition ("$\cup$"), finite iteration ("(.)*", the so-called Kleene star), and test ("(.)?") are familiar from propositional dynamic logic (PDL).

An *expression* is a formula or a program.

The *length* of a formula φ, noted $|\varphi|$, is the number of symbols used to write down φ, with the exception of [,], $\langle$, $\rangle$, and parentheses. For example, $|p \wedge (q \vee r)| = 1 + 1 + 3 = 5$. The length of a program π, noted $|\pi|$, is defined in the same way. For example, $|([+p]\bot? ; -p)| = 4 + 1 + 2 = 7$.

For a given formula φ, the set of variables occurring in φ is noted $\mathbb{P}_\varphi$. For example, $\mathbb{P}_{[-p](q \vee r)} = \{p, q, r\}$.

The *static fragment* of $\mathcal{L}_{\text{D-HT}}$ is the fragment of $\mathcal{L}_{\text{D-HT}}$ without dynamic operators $[\pi]$ and $\langle\pi\rangle$ for every π, noted $\mathcal{L}_{\text{HT}}$. This is nothing but the language of HT and of equilibrium logic.

Negation of a formula φ, noted $\neg\varphi$, is defined as the abbreviation of $\varphi \rightarrow \bot$. We also use $\top$ as a shorthand for $\bot \rightarrow \bot$.

2.2   Dynamic Here-and-There Logic

We display below the interpretation of formulas and programs together at a time: the interpretation $\|\varphi\|_{\text{D-HT}}$ of a formula φ is a set of HT models, while the interpretation $\|\pi\|_{\text{D-HT}}$ of a program π is a relation on the set of HT models, $\mathbb{HT}$. Note that the interpretation of the dynamic connectives differs from that of usual modal logics because there is a single relation interpreting programs (that therefore does not vary with the models). The definitions are in Table 1.

For instance, $\|\neg p\|_{\text{D-HT}}$ is the set of HT models (H, T) such that $p \notin T$ (and therefore $p \notin H$ by the heredity constraint). Hence, $\|p \vee \neg p\|_{\text{D-HT}}$ is the set of HT models (H, T) such that $p \in H$ or $p \notin T$. $\|\neg\neg p\|_{\text{D-HT}}$ is the set of HT models (H, T) such that $p \in T$. Moreover, $\|\langle +p\rangle\top\|_{\text{D-HT}}$ is the set of HT models (H, T) such that $p \notin H$: when $p \in H$ then p cannot be upgraded and the $+p$ program is inexecutable. Finally, the models of the following formula are all those HT-models (H, T) where T contains p and H does not.

$$\|\langle +p\rangle\top \wedge \langle -p\rangle\top\|_{\text{D-HT}} = \|\neg\neg p\|_{\text{D-HT}} \cap (\mathbb{HT} \setminus \|p\|_{\text{D-HT}})$$
$$= \{(H, T) \ : \ p \notin H \text{ and } p \in T\}$$

Table 1. Interpretation of the D-HT connectives

$$\|p\|_{\text{D-HT}} = \{(H,T) \ : \ p \in H\}$$

$$\|\bot\|_{\text{D-HT}} = \emptyset$$

$$\|\varphi \wedge \psi\|_{\text{D-HT}} = \|\varphi\|_{\text{D-HT}} \cap \|\psi\|_{\text{D-HT}}$$

$$\|\varphi \vee \psi\|_{\text{D-HT}} = \|\varphi\|_{\text{D-HT}} \cup \|\psi\|_{\text{D-HT}}$$

$$\|\varphi \rightarrow \psi\|_{\text{D-HT}} = \{(H,T) \ : \ (H,T),(T,T) \in (\text{HT} \setminus \|\varphi\|_{\text{D-HT}}) \cup \|\psi\|_{\text{D-HT}}\}$$

$$\|[\pi]\varphi\|_{\text{D-HT}} = \{(H,T) \ : \ (H_1,T_1) \in \|\varphi\|_{\text{D-HT}} \text{ for every } ((H,T),(H_1,T_1)) \in \|\pi\|_{\text{D-HT}}\}$$

$$\|\langle\pi\rangle\varphi\|_{\text{D-HT}} = \{(H,T) \ : \ (H_1,T_1) \in \|\varphi\|_{\text{D-HT}} \text{ for some } ((H,T),(H_1,T_1)) \in \|\pi\|_{\text{D-HT}}\}$$

$$\|{+}p\|_{\text{D-HT}} = \{((H_1,T_1),(H_2,T_2)) \ : \ H_2 \setminus H_1 = \{p\} \text{ and } T_2 = T_1, \text{ or } T_2 \setminus T_1 = \{p\} \text{ and } H_2 = H_1\}$$

$$\|{-}p\|_{\text{D-HT}} = \{((H_1,T_1),(H_2,T_2)) \ : \ H_1 \setminus H_2 = \{p\} \text{ and } T_2 = T_1, \text{ or } T_1 \setminus T_2 = \{p\} \text{ and } H_2 = H_1\}$$

$$\|\pi_1;\pi_2\|_{\text{D-HT}} = \|\pi_1\|_{\text{D-HT}} \circ \|\pi_2\|_{\text{D-HT}}$$

$$\|\pi_1 \cup \pi_2\|_{\text{D-HT}} = \|\pi_1\|_{\text{D-HT}} \cup \|\pi_2\|_{\text{D-HT}}$$

$$\|\pi^*\|_{\text{D-HT}} = (\|\pi\|_{\text{D-HT}})^*$$

$$\|\varphi?\|_{\text{D-HT}} = \{((H,T),(H,T)) \ : \ (H,T) \in \|\varphi\|_{\text{D-HT}}\}$$

A formula φ is D-HT *valid* if and only if every HT model is also a model of φ, i.e., $\|\varphi\|_{\text{D-HT}} = \text{HT}$. For example, neither $\langle{+}p\rangle\top$ nor $\langle{-}p\rangle\top$ is valid, but $\langle{+}p \cup {-}p\rangle\top$ is. Moreover, $[{+}p][{+}p]p$, $[{-}p][{-}p]\neg p$, and $[p? \cup \neg p?](p \vee \neg p)$ are all valid. Finally, the following equivalences are valid:

$$[{-}p]\bot \leftrightarrow \neg p$$

$$\langle{-}p\rangle\top \leftrightarrow \neg\neg p$$

$$[{+}p]\bot \leftrightarrow p$$

Therefore $[{-}p]\bot$, $\langle{-}p\rangle\top$ and $[{+}p]\bot$ can all be expressed in $\mathcal{L}_{HT}$. In contrast, $\langle{+}p\rangle\top$ cannot because there is no formula in the static fragment $\mathcal{L}_{HT}$ that conveys that $p \in T \setminus H$. So our extension of HT is more expressive than HT itself.

D-HT logic satisfies the heredity property of intuitionistic logic for atomic formulas: if (H,T) is an HT model of p then (T,T) is also an HT model of p. It is trivially satisfied because for every HT model (H,T), $H \subseteq T$. D-HT logic however fails to satisfy that property for more complex formulas containing dynamic operators. To see this, consider the HT model $(\emptyset, \{p\})$ and the formula $\langle{+}p\rangle\top$: $(\emptyset, \{p\})$ is a model of $\langle{+}p\rangle\top$, while $(\{p\}, \{p\})$ is not.

Our logic D-HT is a particular intuitionistic modal logic. Such logics were studied in the literature [6]. For such logics, duality of the modal operators fails: while $[\pi]\varphi \rightarrow \neg\langle\pi\rangle\neg\varphi$ is valid, the converse is invalid. For example, $(\emptyset, \emptyset)$ is an HT model of $\neg\langle{+}p\rangle\neg p$, but not of $[{+}p]p$.

It follows from the next proposition that we have a finite model property for D-HT: if φ has an HT model then φ has an HT model (H,T) such that T is finite.

Proposition 1. *Let φ be an $\mathcal{L}_{\text{D-HT}}$ formula. Let P be a set of propositional variables such that $P \cap \mathbb{P}_\varphi = \emptyset$, and let $Q \subseteq P$. Then, $(H,T) \in \|\varphi\|_{\text{D-HT}}$ iff $(H \cup Q, T \cup P) \in \|\varphi\|_{\text{D-HT}}$.*

2.3 Dynamic Equilibrium Logic

An *equilibrium model* of an $\mathcal{L}_{\text{D-HT}}$ formula φ is a set of propositional variables $T \subseteq \mathbb{P}$ such that:

1. (T, T) is an HT model of φ;
2. no (H, T) with $H \subset T$ is an HT model of φ.

The valid formulas of D-HT all have exactly one equilibrium model, viz. the empty set. There are formulas that have no equilibrium model, such as $\neg\neg p$. The equilibrium models of equivalent formulas $p \vee \neg p$ and $\neg\neg p \rightarrow p$ are $\emptyset$ and $\{p\}$. The only equilibrium model of $\neg p \rightarrow q$ is $\{q\}$, and of $\langle+p\rangle(\neg p \rightarrow q)$ is $\emptyset$. $\{p\}$ is the only equilibrium model for both $\langle-p\rangle(\neg p \rightarrow q)$, and $\langle+q; +q\rangle(p \wedge q)$. However, $\langle-q\rangle(p \wedge q)$ has no equilibrium model because $\langle-q\rangle(p \wedge q)$ does not even have a D-HT model either.

Let χ and φ be $\mathcal{L}_{\text{D-HT}}$ formulas. φ is a *consequence of χ in equilibrium models*, written $\chi \mathrel{\vert\approx} \varphi$, if and only if for every equilibrium model T of χ, (T, T) is an HT model of φ. For example, $\top \mathrel{\vert\approx} \neg p$, $p \vee q \mathrel{\vert\approx} [\neg p?]q$, and $p \vee q \mathrel{\vert\approx} [\neg p?]\langle+p; +p\rangle(p \wedge q)$.

In our dynamic language we can check not only problems of the form $\chi \mathrel{\vert\approx} [\pi]\varphi$, but also problems of the form $\langle\pi\rangle\chi \mathrel{\vert\approx} \varphi$. The former expresses a hypothetical update of χ: if χ is updated by π then φ follows. The latter may express an actual update of χ, where the program π executes the update 'the other way round': it is the converse of the original update program. For example, suppose we want to update $\chi = p \wedge q$ by $\neg q$. Updates by the latter formula can be implemented by the program $-q; -q$. Now the converse execution of $-q; -q$ is nothing but the execution of the program $\pi = +q; +q$. Therefore, in order to know whether the update of $p \wedge q$ by $\neg q$ results in $p \wedge \neg q$ we have to check whether $\langle+q; +q\rangle(p \wedge q) \mathrel{\vert\approx} p \wedge \neg q$. The latter is indeed the case: we have seen above that the only equilibrium model of $\langle+q; +q\rangle(p \wedge q)$ is $\{p\}$, and $(\{p\}, \{p\})$ is clearly a D-HT model of $p \wedge \neg q$.

3 DL-PA: Dynamic Logic of Propositional Assignments

In this section we define syntax and semantics of dynamic logic of propositional assignments (DL-PA) and state complexity results. The star-free fragment of DL-PA was introduced in [9], where it was shown that it embeds Coalition Logic of Propositional Control [10,11,12]. The full logic with the Kleene star was further studied in [2]. In addition to assignments of propositional variables to true or false, here we allow of assignments to arbitrary formulas as well. We need this extension for the purpose of copying the propositional variables of a valuation and similarly, after some changes to be able to retrieve the initial truth values of that valuation. We will explain these notions later in full detail. However, we keep on calling that logic DL-PA. This is in order because it has the same expressivity and the same complexity as the logic DL-PA of [2].

3.1 Language

The language of DL-PA is defined by the following grammar:

$$\pi ::= p{:=}\varphi \mid \pi; \pi \mid \pi \cup \pi \mid \pi^* \mid \varphi?$$
$$\varphi ::= p \mid \bot \mid \varphi \wedge \varphi \mid \varphi \vee \varphi \mid \varphi \supset \varphi \mid \langle\pi\rangle\varphi$$

where p ranges over a fixed set of propositional variables $\mathbb{P}$. So, an atomic program of the language of DL-PA is a program of the form $p{:=}\varphi$. The program operators of sequential composition (";"), nondeterministic composition ("$\cup$"), finite iteration ("(.)*"), and test ("(.)?") are familiar from Propositional Dynamic Logic (PDL).

The *star-free fragment* of DL-PA is the subset of the language made up of formulas without the Kleene star "(.)*".

We abbreviate the other logical connectives in the usual way; for example, $\sim\varphi$ is defined as $\varphi \supset \bot$. In particular, $\top$ is defined as $\sim\bot = \bot\supset\bot$. Moreover, $[\pi]\varphi$ abbreviates $\sim\langle\pi\rangle\sim\varphi$. The program skip abbreviates $\top$? ("nothing happens"). (Note that it could also be defined by $p{:=}p$, for arbitrary p.) The language of DL-PA allows to express the primitives of standard programming languages. For example, the loop "while φ do π" can be expressed as the DL-PA program $(\varphi?; \pi)^*; \sim\varphi?$.

3.2 Semantics

DL-PA programs are interpreted by means of a (unique) *relation between valuations*: atomic programs $p{:=}\varphi$ update valuations in the obvious way, and complex programs are interpreted just as in PDL by mutual recursion. Table 2 gives the interpretation of the DL-PA connectives.

Table 2. Interpretation of the DL-PA connectives

$\|p{:=}\varphi\|_{\text{DL-PA}} = \{(V, V') : \text{if } V \in \|\varphi\|_{\text{DL-PA}} \text{ then } V' = V\cup\{p\} \text{ and if } V \notin \|\varphi\|_{\text{DL-PA}} \text{ then } V' = V\setminus\{p\}\}$

$\|\pi; \pi'\|_{\text{DL-PA}} = \|\pi\|_{\text{DL-PA}} \circ \|\pi'\|_{\text{DL-PA}}$

$\|\pi\cup\pi'\|_{\text{DL-PA}} = \|\pi\|_{\text{DL-PA}} \cup \|\pi'\|_{\text{DL-PA}}$

$\|\pi^*\|_{\text{DL-PA}} = (\|\pi\|_{\text{D-HT}})^*$

$\|\varphi?\|_{\text{DL-PA}} = \{(V, V) : V \in \|\varphi\|_{\text{DL-PA}}\}$

$\|p\|_{\text{DL-PA}} = \{V : p \in V\}$

$\|\bot\|_{\text{DL-PA}} = \emptyset$

$\|\varphi\wedge\psi\|_{\text{DL-PA}} = \|\varphi\|_{\text{DL-PA}} \cap \|\psi\|_{\text{DL-PA}}$

$\|\varphi\vee\psi\|_{\text{DL-PA}} = \|\varphi\|_{\text{DL-PA}} \cup \|\psi\|_{\text{DL-PA}}$

$\|\varphi\supset\psi\|_{\text{DL-PA}} = (2^{\mathbb{P}} \setminus \|\varphi\|_{\text{DL-PA}}) \cup \|\psi\|_{\text{DL-PA}}$

$\|\langle\pi\rangle\varphi\|_{\text{DL-PA}} = \{V : \text{there is } V' \text{ such that } (V, V') \in \|\pi\|_{\text{DL-PA}} \text{ and } V' \in \|\varphi\|_{\text{DL-PA}}\}$

A formula φ is DL-PA *valid* if $\|\varphi\|_{\text{DL-PA}} = 2^{\mathbb{P}}$, and it is DL-PA *satisfiable* if $\|\varphi\|_{\text{DL-PA}} \neq \emptyset$. For example, the formulas $\langle p{:=}\top\rangle\top$, $\langle p{:=}\top\rangle p$ and $\langle p{:=}\bot\rangle\sim p$ are all valid, as well as $\psi \wedge [\psi?]\varphi \supset \varphi$ and $[p{:=}\top \cup q{:=}\top](p \vee q)$. Moreover, if p does not occur in φ then both $\varphi \supset \langle p{:=}\top\rangle\varphi$ and $\varphi \supset \langle p{:=}\bot\rangle\varphi$ are valid. This is due to the following property that we will use while translating dynamic equilibrium logic into DL-PA.

Proposition 2. *Suppose* $\mathbb{P}_\varphi \cap P = \emptyset$, *i.e., none of the variables of P occurs in* φ. *Then* $V \cup P \in \|\varphi\|_{\mathsf{DL\text{-}PA}}$ *iff* $V \setminus P \in \|\varphi\|_{\mathsf{DL\text{-}PA}}$.

Contrarily to **PDL**, it is shown in [2] that the Kleene star operator can be eliminated in **DL-PA**: for every **DL-PA** program π, there is an equivalent program π' such that no Kleene star occurs in π'. However, the elimination is not polynomial.

3.3   Complexity of the Full Language

It is proved in [2] that both model and satisfiability checking are EXPTIME complete for the fragment of **DL-PA** including the conversion operator and restricting the formulas φ in atomic programs $p:=\varphi$ to either $\top$ or $\bot$. The lower bounds for both problems clearly transfer.

The upper bound for the satisfiability problem is established in [2] by means of a polynomial transformation into the satisfiability problem of **PDL**. An inspection of the proof shows that it generalizes to arbitrary assignments. So, the satisfiability problem of our **DL-PA** has the same complexity as that of **PDL**: it is EXPTIME complete.

The upper bound for the model checking problem can be established just as in [2] by polynomially transforming it into the satisfiability problem: we use that $V \in \|\varphi\|_{\mathsf{DL\text{-}PA}}$ if and only if the formula $\varphi \wedge (\bigwedge_{p \in V} p) \wedge (\bigwedge_{p \notin V} {\sim}p)$ is satisfiable. So, the model checking problem of our **DL-PA** is EXPTIME complete, too.

3.4   Complexity of the Star-Free Fragment

The complexity of the decision problems for the star-free fragment of the language of [2] is established in [9], where it is shown that it is PSPACE complete for both model and satisfiability checking.

As to model checking, the lower bound clearly transfers to our star-free fragment. Furthermore, the PSPACE model checking algorithm of [9] can be extended to our more general star-free fragment without conversion and with general assignment $p:=\varphi$.

As to satisfiability checking, the lower bound of [9] transfers. The upper bound can be proved in the same way as in [9]: given a formula φ, nondeterministically guess a valuation V and model check whether $V \in \|\varphi\|_{\mathsf{DL\text{-}PA}}$. Model checking being in PSPACE, satisfiability checking must therefore be in NPSPACE, and NPSPACE is the same complexity class as PSPACE due to Savitch's theorem.

4   Relating D-HT and DL-PA

In this section we are going to translate **D-HT** and dynamic equilibrium logic into **DL-PA**, and vice versa. The translation is polynomial and allows to check **D-HT** validity and consequence in equilibrium models. This establishes an EXPTIME upper bound for the complexity of the latter problem. We also show that the upper bound is tight.

We start by defining some **DL-PA** programs that will be the building blocks in embedding some notions of **D-HT** into **DL-PA**. Some of these programs require to copy propositional variables.

4.1 Copying Propositional Variables

The translation introduces fresh propositional variables that do not not exist in the formula we translate. Precisely, this requires to suppose a new set of propositional variables: it is the union of the set of 'original' variables $\mathbb{P} = \{p_1, p_2, \ldots\}$ and the set of 'copies' of these variables $\mathbb{P}' = \{p'_1, p'_2, \ldots\}$, where $\mathbb{P}$ and $\mathbb{P}'$ are disjoint. The function $(.)'$ is a bijection between these two sets: for every subset $Q \subseteq \mathbb{P}$ of original variables, the set $Q' = \{p' : p \in Q\} \subseteq \mathbb{P}'$ is its image, and the other way around. We suppose that $(.)'$ is an involution, i.e., it behaves as an identity when applied twice. Now, a DL-PA valuation extends to the form of $X \cup Y'$, where $X \subseteq \mathbb{P}$ and $Y' \subseteq \mathbb{P}'$. As a result, DL-PA validity expands to the power set of $\mathbb{P} \cup \mathbb{P}'$, i.e., $2^{\mathbb{P} \cup \mathbb{P}'}$. In our embedding, X will encode the here-valuation and Y' will encode the there-valuation. Note that in order to respect the heredity constraint hidden in the structure of here-and there models, our translation has to guarantee that X is a subset of Y.

4.2 Useful DL-PA Programs

Table 3 collects some DL-PA programs that are going to be convenient for our enterprise. In that table, $\{p_1, \ldots, p_n\}$ is some finite subset of $\mathbb{P}$ and each p'_i is a copy of p_i as explained above. For $n = 0$ we stipulate that all these programs equal skip.

Table 3. Some useful DL-PA programs

$$\mathtt{mkFalse}^{\geq 0}(\{p_1, \ldots, p_n\}) = (p_1 := \bot \cup \mathtt{skip}); \cdots ; (p_n := \bot \cup \mathtt{skip})$$

$$\mathtt{mkFalse}^{> 0}(\{p_1, \ldots, p_n\}) = (p_1 := \bot \cup \cdots \cup p_n := \bot); \mathtt{mkFalse}^{\geq 0}(P)$$

$$\mathtt{cp}(\{p_1, \ldots, p_n\}) = p'_1 := p_1; \cdots ; p'_n := p_n$$

$$\mathtt{cpBack}(\{p_1, \ldots, p_n\}) = p_1 := p'_1; \cdots ; p_n := p'_n$$

Let $P = \{p_1, \ldots, p_n\}$. The program $\mathtt{mkFalse}^{\geq 0}(P)$ nondeterministically makes some of the variables of P false, possibly none. The program $\mathtt{mkFalse}^{> 0}(P)$ nondeterministically makes false at least one of the variables of P, and possibly more. Its subprogram $p_1 := \bot \cup \cdots \cup p_n := \bot$ makes exactly one of the variables in the valuation P false. The program $\mathtt{cp}(P)$ assigns to each 'fresh' variable p'_i the truth value of p_i, while the program $\mathtt{cpBack}(P)$ assigns to each variable p_i the truth value of p'_i. We shall use the former as a way of storing the truth value of each variable of P before they undergo some changes. That will allow later on to retrieve the original values of the variables in P by means of the $\mathtt{cpBack}(P)$ program. Therefore the sequence $\mathtt{cp}(P); \mathtt{cpBack}(P)$ leaves the variables in P unchanged.

Observe that each program of Table 3 has length linear in the cardinality of P. Observe also that the programs $\mathtt{mkFalse}^{\geq 0}(P)$ and $\mathtt{mkFalse}^{> 0}(P)$ are nondeterministic. In contrast, the programs $\mathtt{cp}(P)$ and $\mathtt{cpBack}(P)$ are deterministic and always executable: $[\mathtt{cp}(P)]\varphi$ and $\langle \mathtt{cp}(P) \rangle \varphi$ are equivalent, as well as $[\mathtt{cpBack}(P)]\varphi$ and $\langle \mathtt{cpBack}(P) \rangle \varphi$.

Lemma 1 (Program Lemma). *Let $P \subseteq \mathbb{P}$ be finite and non-empty. Then*

$$\|\mathtt{mkFalse}^{\geq 0}(P)\|_{\mathsf{DL\text{-}PA}} = \{(V_1, V_2) \ : \ V_2 = V_1 \setminus Q, \text{ for some } Q \subseteq P\}$$

$$\|\mathtt{mkFalse}^{> 0}(P)\|_{\mathsf{DL\text{-}PA}} = \{(V_1, V_2) \ : \ V_2 = V_1 \setminus Q, \text{ for some } Q \subseteq P \text{ such that } Q \neq \emptyset\}$$

$$\|\mathtt{cp}(P)\|_{\mathsf{DL\text{-}PA}} = \{(X_1 \cup Y_1', X_2 \cup Y_2') \ : \ X_2 = X_1 \text{ and } Y_2' = (X_1 \cap P)' \cup (Y_1' \setminus P')\}$$

$$\|\mathtt{cpBack}(P)\|_{\mathsf{DL\text{-}PA}} = \{(X_1 \cup Y_1', X_2 \cup Y_2') \ : \ X_2 = (Y_1' \cap P')' \cup (X_1 \setminus P) \text{ and } Y_2' = Y_1'\}.$$

It follows from the interpretations of $\mathtt{cp}(P)$ and $\mathtt{mkFalse}^{\geq 0}(P)$ that

$$\|\mathtt{cp}(P);\mathtt{mkFalse}^{\geq 0}(P)\|_{\mathsf{DL\text{-}PA}} = \{(X_1 \cup Y_1', X_2 \cup Y_2') \ : X_2 = X_1 \setminus Q \text{ for some } Q \subseteq P$$
$$\text{and } Y_2' = (X_1 \cap P)' \cup (Y_1' \setminus P')\}.$$

4.3   Translating $\mathcal{L}_{\mathsf{D\text{-}HT}}$ to $\mathcal{L}_{\mathsf{DL\text{-}PA}}$

To start with we translate the formulas and programs of the language $\mathcal{L}_{\mathsf{D\text{-}HT}}$ into the language $\mathcal{L}_{\mathsf{DL\text{-}PA}}$. The translation is given in Table 4 in terms of a recursively defined mapping tr_1, where we have omitted the homomorphic cases such as $tr_1([\pi]\varphi) = [tr_1(\pi)]tr_1(\varphi)$ and $tr_1(\varphi?) = (tr_1(\varphi))?$.

Table 4. Translation from $\mathcal{L}_{\mathsf{D\text{-}HT}}$ into DL-PA

$$tr_1(p) = p, \quad \text{for } p \in \mathbb{P}$$
$$tr_1(\varphi \to \psi) = [\mathtt{skip} \cup \mathtt{cpBack}(\mathbb{P}_{\varphi \to \psi})](tr_1(\varphi) \supset tr_1(\psi))$$
$$tr_1(+p) = (\sim\! p'? \, ; p'\!:=\!\top) \cup (\sim\! p \wedge p'? \, ; p\!:=\!\top)$$
$$tr_1(-p) = (p? \, ; p\!:=\!\bot) \cup (\sim\! p \wedge p'? \, ; p'\!:=\!\bot)$$

Observe that tr_1 is polynomial. For example,

$$tr_1(\top) = tr_1(\bot \to \bot) = [\mathtt{skip} \cup \mathtt{skip}]\top$$
$$tr_1(p \vee \neg p) = p \vee [\mathtt{skip} \cup p\!:=\!p']\!\sim\! p$$
$$tr_1(p \to q) = [\mathtt{skip} \cup (p\!:=\!p' \, ; q\!:=\!q')](p \supset q)$$

The first formula is equivalent to $\top$. The second is equivalent to $p \vee (\sim\! p \wedge \sim\! p')$, i.e., to $p \vee \sim\! p'$. The third is equivalent to $(p \supset q) \wedge (p' \supset q')$.

Lemma 2 (Main Lemma). $(H, T) \in \|\varphi\|_{\mathsf{D\text{-}HT}}$ *if and only if* $H \cup T' \in \|tr_1(\varphi)\|_{\mathsf{DL\text{-}PA}}$.

Proof is by induction on the length of expressions (formulas or programs): we show that for every expression ξ,

- if ξ is a formula then $(H, T) \in \|\xi\|_{\mathsf{D\text{-}HT}}$ if and only if $H \cup T' \in \|tr_1(\xi)\|_{\mathsf{DL\text{-}PA}}$, and
- if ξ is a program then
  $((H_1, T_1), (H_2, T_2)) \in \|\xi\|_{\mathsf{D\text{-}HT}}$ if and only if $((H_1 \cup T_1'), (H_2 \cup T_2')) \in \|tr_1(\xi)\|_{\mathsf{DL\text{-}PA}}$.

4.4   From D-HT to DL-PA

We now establish how tr_1 can be used to prove that a given formula φ is D-HT satisfiable. To that end, we prefix the translation by the '$\mathtt{cp}(\mathbb{P}_\varphi)$' program that is followed by the '$\mathtt{mkFalse}^{\geq 0}(\mathbb{P}_\varphi)$' program. The '$\mathtt{cp}(\mathbb{P}_\varphi)$' program produces a 'classical' valuation $T \cup T'$, for some subset T of $\mathbb{P}$ (as far as the variables of φ are concerned), and then '$\mathtt{mkFalse}^{\geq 0}(\mathbb{P}_\varphi)$' program transforms the valuation $T \cup T'$ into a valuation $H \cup T'$ for some H such that $H \subseteq T$.

Theorem 1. *Let φ be an $\mathcal{L}_{\text{D-HT}}$ formula. Then*

- *φ is D-HT satisfiable iff $\langle \mathtt{cp}(\mathbb{P}_\varphi) \rangle \langle \mathtt{mkFalse}^{\geq 0}(\mathbb{P}_\varphi) \rangle tr_1(\varphi)$ is DL-PA satisfiable, and*
- *φ is D-HT valid iff $[\mathtt{cp}(\mathbb{P}_\varphi)][\mathtt{mkFalse}^{\geq 0}(\mathbb{P}_\varphi)]tr_1(\varphi)$ is DL-PA valid.*

This is proved by the Main Lemma and the Program Lemma.

As a result of the theorem above, the formula $[\mathtt{cp}(\{\mathrm{p}\})][\mathtt{mkFalse}^{\geq 0}(\{p\})]tr_1(p \vee \neg p)$ should not be DL-PA valid since we know that $p \vee \neg p$ is not D-HT valid. We indeed have the following sequence of equivalent formulas:

1. $[\mathtt{cp}(\{\mathrm{p}\})][\mathtt{mkFalse}^{\geq 0}(\{p\})]tr_1(p \vee \neg p)$
2. $[p' := p][p := \bot \cup \mathtt{skip}](p \vee [\mathtt{skip} \cup p := p']{\sim}p)$
3. $[p' := p][p := \bot \cup \mathtt{skip}](p \vee {\sim}p')$
4. $[p' := p]([p := \bot](p \vee {\sim}p') \wedge (p \vee {\sim}p'))$
5. $[p' := p]({\sim}p' \wedge (p \vee {\sim}p'))$
6. ${\sim}p \wedge (p \vee {\sim}p)$
7. ${\sim}p$

The last is obviously not DL-PA valid, so the first line is not DL-PA valid either.

4.5   From Dynamic Equilibrium Logic to DL-PA

Having seen how D-HT can be embedded into DL-PA, we now turn to equilibrium logic.

Theorem 2. *For every $\mathcal{L}_{\text{D-HT}}$ formula χ, $T \subseteq \mathbb{P}$ is an equilibrium model of χ if and only if $T \cup T'$ is a DL-PA model of $tr_1(\chi) \wedge {\sim}\langle \mathtt{mkFalse}^{>0}(\mathbb{P}_\chi) \rangle tr_1(\chi)$.*

Proof. $T \cup T'$ is a DL-PA model of $tr_1(\chi) \wedge {\sim}\langle \mathtt{mkFalse}^{>0}(\mathbb{P}_\chi) \rangle tr_1(\chi)$ if and only if

$$T \cup T' \text{ is a DL-PA model of } tr_1(\chi) \tag{1}$$

and

$$T \cup T' \text{ is a DL-PA model of } {\sim}\langle \mathtt{mkFalse}^{>0}(\mathbb{P}_\chi) \rangle tr_1(\chi) \tag{2}$$

By the Main Lemma, (1) is the case if and only if (T, T) is a HT model of χ in D-HT. It remains to prove that (2) is the case if and only if (H, T) is not a HT model of χ, for any set $H \subset T$. We establish this by proving that the following statements are equivalent.

1. $T \cup T'$ is a DL-PA model of ${\sim}\langle \mathtt{mkFalse}^{>0}(\mathbb{P}_\chi) \rangle tr_1(\chi)$
2. $(T \cap \mathbb{P}_\chi) \cup T'$ is not a DL-PA model of $\langle \mathtt{mkFalse}^{>0}(\mathbb{P}_\chi) \rangle tr_1(\chi)$ (Proposition 2)

3. $H \cup T'$ is not a DL-PA model of $tr_1(\chi)$, for any $H \subset T \cap \mathbb{P}_\chi$ (Program Lemma 1)
4. H, T is not a HT model of χ, for any set $H \subset T \cap \mathbb{P}_\chi$ (Main Lemma 2)
5. H, T is not a HT model of χ, for any set $H \subset T$ (Proposition 1).

q.e.d.

Theorem 3. *Let χ and φ be $\mathcal{L}_{\text{D-HT}}$ formulas. Then $\chi \approx \varphi$ if and only if*

$$\langle\mathrm{cp}(\mathbb{P}_\chi \cup \mathbb{P}_\varphi)\rangle\Big((tr_1(\chi) \wedge \sim\langle\mathtt{mkFalse}^{>0}(\mathbb{P}_\chi)\rangle tr_1(\chi)) \supset tr_1(\varphi)\Big)$$

is DL-PA *valid.*

Theorem 3 provides a polynomial embedding of the consequence problem in our dynamic equilibrium logic into DL-PA. Together with the EXPTIME upper bound for the validity problem of DL-PA that we have established in Section 3.3, it follows that the former problem is in EXPTIME. In the next section we establish that the upper bound is tight.

4.6   From DL-PA to D-HT

We establish EXPTIME hardness of the D-HT satisfiability problem by means of a simple translation of the fragment of DL-PA whose atomic assignment programs are restricted to $p:=\top$ and $p:=\bot$ and with the conversion operator: the result follows because it is known that the satisfiability problem for that fragment is already EXPTIME hard [2].

The translation is given in Table 5, where we have omitted the homomorphic cases. In the last two lines, $tr_2(p:=\top)$ makes p true both here and there, while $tr_2(p:=\bot)$ makes p false both here and there. The translation is clearly polynomial.

Table 5. Translation from DL-PA with assignments only to $\top$ and $\bot$ into $\mathcal{L}_{\text{D-HT}}$ (main cases)

$$tr_2(p) = p, \quad \text{for } p \in \mathbb{P}$$
$$tr_2(\varphi \supset \psi) = tr_2(\varphi) \rightarrow tr_2(\psi)$$
$$tr_2(p:=\top) = p? \cup (+p \,;\, +p)$$
$$tr_2(p:=\bot) = \neg p? \cup (-p \,;\, -p)$$

The next lemma is the analog of the Main Lemma adapted to tr_2, and is used in the proof of Theorem 4.

Lemma 3. *Let φ be a* DL-PA *formula. Then,*

$$V \in \|\varphi\|_{\text{DL-PA}} \text{ if and only if } (V, V) \in \|tr_2(\varphi)\|_{\text{D-HT}}.$$

Now, we are ready to show how tr_2 can be used to prove that a given formula φ is DL-PA satisfiable.

Theorem 4. *Let φ be a* DL-PA *formula. Then, φ is* DL-PA *satisfiable if and only if $tr_2(\varphi) \wedge \bigwedge_{p \in \mathbb{P}_\varphi}(p \vee \neg p)$ is satisfiable in* D-HT.

Since the satisfiability problem for the fragment of the language of DL-PA with assignments only to $\top$ and $\bot$ is EXPTIME hard [2], through the theorem above we deduce that the D-HT validity problem is also EXPTIME hard; moreover, Theorem 1 tells us that it is actually EXPTIME complete.

The complexity of the equilibrium consequence problem is at least that of the validity problem in D-HT. Therefore, the consequence problem in dynamic equilibrium logic is EXPTIME hard, too. Moreover, Theorem 3 tells us that it is actually EXPTIME complete.

5 Conclusion

We have defined a simple logic D-HT of atomic change of equilibrium models and have shown that it is strongly related to dynamic logic of propositional assignments (DL-PA). This in particular allows to obtain EXPTIME complexity results both for the D-HT satisfiability and for the consequence in its equilibrium models.

The present paper is part of a line of work aiming at reexamining the logical foundations of equilibrium logic and ASP. In previous works we had analyzed equilibrium logic by means of the concepts of contingency [4] and by means of modal operators quantifying over here-and -there worlds in the definition of an equilibrium model [5]. The present paper adds an analysis of the dynamics by integrating operators of upgrading and downgrading propositional variables.

What about updates by complex programs? Actually we may implement such updates by means of complex D-HT programs. For example, the D-HT program

$$(\neg p \vee q)? \cup (-p; +q)$$

makes the implication $p \rightarrow q$ true, whatever the initial HT model is (although there may be other minimal ways of achieving this). More generally, let us consider that an abstract semantical update operation is a function $f : \mathbb{HT} \longrightarrow 2^{\mathbb{HT}}$ associating to every HT model (H, T) the set of HT models $f(H, T)$ resulting from the update. If the language is finite then for every such f we can design a program π_f such that $\|\pi_f\|_{\text{D-HT}} = f$, viz. the graph of f. This makes use of the fact that in particular we can uniquely (up to logical equivalence) characterize HT models by means of the corresponding formulas. For example, the formula $(\langle +p \rangle \top \wedge \langle -p \rangle \top) \wedge (\bigwedge_{q \neq p} \neg q)$ identifies the HT model $(\emptyset, \{p\})$. Note finally that we cannot express the HT model $(\emptyset, \{p\})$ in the language $\mathcal{L}_{\text{HT}}$, where there is no formula distinguishing that model from the model $(\{p\}, \{p\})$.

References

1. Alchourrón, C., Gärdenfors, P., Makinson, D.: On the logic of theory change: Partial meet contraction and revision functions. J. of Symbolic Logic 50, 510–530 (1985)
2. Balbiani, P., Herzig, A., Troquard, N.: Dynamic logic of propositional assignments: a well-behaved variant of PDL. In: Kupferman, O. (ed.) Logic in Computer Science (LICS), New Orleans, June 25-28. IEEE (2013), http://www.ieee.org/
3. Eiter, T., Fink, M., Sabbatini, G., Tompits, H.: Using methods of declarative logic programming for intelligent information agents. TPLP 2(6), 645–709 (2002)

4. Fariñas del Cerro, L., Herzig, A.: Contingency-based equilibrium logic. In: Delgrande, J.P., Faber, W. (eds.) LPNMR 2011. LNCS, vol. 6645, pp. 223–228. Springer, Heidelberg (2011), http://www.springerlink.com

5. Fariñas del Cerro, L., Herzig, A.: The modal logic of equilibrium models. In: Tinelli, C., Sofronie-Stokkermans, V. (eds.) FroCoS 2011. LNCS, vol. 6989, pp. 135–146. Springer, Heidelberg (2011), http://www.springerlink.com

6. Fischer-Servi, G.: On modal logic with an intuitionistic base. Studia Logica 36(4), 141–149 (1976)

7. Gebser, M., Kaufmann, B., Neumann, A., Schaub, T.: Conflict-driven answer set solving. In: Veloso, M.M. (ed.) IJCAI, pp. 386–392 (2007)

8. Gebser, M., Ostrowski, M., Schaub, T.: Constraint answer set solving. In: Hill, P.M., Warren, D.S. (eds.) ICLP 2009. LNCS, vol. 5649, pp. 235–249. Springer, Heidelberg (2009)

9. Herzig, A., Lorini, E., Moisan, F., Troquard, N.: A dynamic logic of normative systems. In: Walsh, T. (ed.) International Joint Conference on Artificial Intelligence (IJCAI), Barcelona, pp. 228–233. IJCAI/AAAI (2011), Erratum at http://www.irit.fr/~Andreas.Herzig/P/Ijcai11.html

10. van der Hoek, W., Walther, D., Wooldridge, M.: On the logic of cooperation and the transfer of control. J. of AI Research (JAIR) 37, 437–477 (2010)

11. van der Hoek, W., Wooldridge, M.: On the dynamics of delegation, cooperation and control: a logical account. In: Proc. AAMAS 2005 (2005)

12. van der Hoek, W., Wooldridge, M.: On the logic of cooperation and propositional control. Artif. Intell. 164(1-2), 81–119 (2005)

13. Katsuno, H., Mendelzon, A.O.: On the difference between updating a knowledge base and revising it. In: Gärdenfors, P. (ed.) Belief Revision, pp. 183–203. Cambridge University Press (1992); preliminary version in Allen, J.A., Fikes, R., and Sandewall, E. (eds.) Principles of Knowledge Representation and Reasoning: Proc. 2nd Int. Conf., pp. 387–394. Morgan Kaufmann Publishers (1991)

14. Lifschitz, V., Pearce, D., Valverde, A.: Strongly equivalent logic programs. ACM Transactions on Computational Logic 2(4), 526–541 (2001)

15. Slota, M., Leite, J.: Robust equivalence models for semantic updates of answer-set programs. In: Brewka, G., Eiter, T., McIlraith, S.A. (eds.) KR. AAAI Press (2012)

16. Slota, M., Leite, J.: A unifying perspective on knowledge updates. In: del Cerro, L.F., Herzig, A., Mengin, J. (eds.) JELIA 2012. LNCS, vol. 7519, pp. 372–384. Springer, Heidelberg (2012)

17. Zhang, Y., Foo, N.Y.: A unified framework for representing logic program updates. In: Veloso, M.M., Kambhampati, S. (eds.) AAAI, pp. 707–713. AAAI Press/The MIT Press (2005)

ActHEX: Implementing HEX Programs with Action Atoms*

Michael Fink[1], Stefano Germano[2], Giovambattista Ianni[2],
Christoph Redl[1], and Peter Schüller[3]

[1] Institut für Informationssysteme, Technische Universität Wien
[2] Dipartimento di Matematica e Informatica, Università della Calabria
[3] Faculty of Engineering and Natural Sciences, Sabanci University

Abstract. acthex programs are a convenient tool for connecting stateful external environments to logic programs. In the acthex framework, actual actions on an external environment can be declaratively selected, rearranged, scheduled and then executed depending on intelligence specified in an ASP-based language. We report in this paper about recent improvements of the formal and of the operational acthex programming framework. Besides yielding a significant increase in versatility of the framework, we also present illustrative application showcases and a short evaluation thereof exhibiting computational acthex strengths.

1 Introduction

The acthex formalism [1] generalizes HEX programs [4] introducing dedicated action atoms in rule heads. Action atoms can actually operate on and change the state of an *environment*, which can be roughly seen as an abstraction of realms outside the logic program at hand. The acthex framework allows to conveniently design ASP-based applications by properly connecting logic-based decisions to actual effects thereof. We recently advanced the acthex framework wrt. several respects:

– Framework improvements: external atom evaluation has been generalized to take state into account, i.e., the realm of acthex programs has been extended to capture non-deterministic actions and environments. Moreover, support for selecting a single model and a unique corresponding execution schedule has been enhanced, and we developed explicit means for controlling iterative evaluation of logic programs.
– System improvements: we provide a new architecture for the acthex framework efficiently implemented as an extension to the dlvhex system[1].
– Applications: we realized new applications and pursued a preliminary system evaluation exhibiting promising results. In terms of performance, our experiments indicate that, compared to purely declarative approaches, finding problem solutions iteratively may pay off when instances are large. In terms of ease of programming, our approach allows to attach code in arbitrary programming languages to a logic-programming framework. This is dual to other approaches to interoperability of ASP solvers like [5].

* This research has been partially supported by the Austrian Science Fund (FWF) grant P24090, and the Vienna Science and Technology Fund (WWTF) grant ICT 08-020. Peter Schüller is supported by the TUBITAK 2216 Research Fellowship.

[1] Available at http://www.kr.tuwien.ac.at/research/systems/dlvhex/actionplugin.html

P. Cabalar and T.C. Son (Eds.): LPNMR 2013, LNAI 8148, pp. 317–322, 2013.
© Springer-Verlag Berlin Heidelberg 2013

2 Preliminaries

We assume familiarity with ASP and corresponding basic syntactic and semantic notions (atoms, models, etc.). For space reasons, in the following, we also do not present acthex syntax and semantics at full formal detail (for the latter cf. [1,4]).

acthex Syntax. In addition to constants (also used for predicate names) and variables, acthex programs build on external predicate names (prefixed by $\&$) and action predicate names (prefixed by $\#$). An external atom is of the form $\&g[Y_1, \ldots, Y_n](X_1, \ldots, X_m)$, where $Y_1, \ldots, Y_n$ and $X_1, \ldots, X_m$ are lists of terms. An action atom is of the form $\#g[Y_1, \ldots, Y_n]\{o, r\}[w : l]$, where $\#g$ is an action predicate name, $Y_1, \ldots, Y_n$ is a list of input terms of fixed length $in(\#g) = n$. Moreover, attribute $o \in \{b, c, c_p\}$ is called the *action option* that identifies an action as *brave, cautious, or preferred cautious*, while optional integer attributes r, w, and l are called *precedence, weight*, and *level* of $\#g$, respectively. A *rule r* is of the form $\alpha_1 \vee \ldots \vee \alpha_k \leftarrow \beta_1, \ldots, \beta_n, \text{not } \beta_{n+1}, \ldots, \text{not } \beta_m$, where body elements β are (ordinary) atoms or external atoms, and head elements α are (ordinary) atoms or action atoms. An acthex *program* is a finite set of rules.

Example 1. The acthex program $P_1 = \{\#robot[goto, charger]\{b, 1\} \leftarrow \&sensor[bat](low);$ $\#robot[clean, kitchen]\{c, 2\} \leftarrow night; \#robot[clean, bedroom]\{c, 2\} \leftarrow day; night \vee day \leftarrow \}$ uses action atom $\#robot$ to control a robot, and an external atom $\&sensor$ to access sensor data. Intuitively, precedence 1 of action atom $\#robot[goto, charger]\{b, 1\}$ should make the robot recharging its battery, if necessary, before cleaning actions.   □

acthex Semantics. An acthex program P is evaluated wrt. a fixed state (snapshot) of the *external environment* E using the following steps: (i) *answer sets* of P are determined wrt. E, and the set of *best models* is a subset of the answer sets determined by an objective function; (ii) any (best) model originates a set of corresponding *execution schedules* S, i.e., a sequence of actions to execute; (iii) executing the actions of (and sequentially according to) a selected schedule S yields another (not necessarily different) state E' of the environment, called the *observed execution outcome*; finally (iv) the process may be iterated starting at (i), by considering a snapshot E'', which can be different from E' due to exogenous actions (in so-called dynamic environments). Answer Sets are defined similarly to HEX programs [4], i.e., using Herbrand interpretations, the grounding of P wrt. the Herbrand universe, and the FLP reduct; ground action atoms in rule heads are treated like ordinary atoms, see Section 3 for a generalized external atom semantics including the environment E. We denote by $\mathcal{AS}(P, E)$ the collection of all answer sets of P wrt. E. The set of *best models* of P, denoted $\mathcal{BM}(P, E)$, contains those answer sets $I \in \mathcal{AS}(P, E)$ that minimize an objective function over weights and levels of atoms in I (equivalent to the evaluation of weak constraints in [2]). An action $a = \#g[y_1, \ldots, y_n]\{o, r\}[w : l]$ with option o and precedence r is *executable in I wrt. P and E* iff (i) a is brave and $a \in I$, or (ii) a is cautious and $a \in B$ for every $B \in \mathcal{AS}(P, E)$, or (iii) a is preferred cautious and $a \in B$ for every $B \in \mathcal{BM}(P, E)$. An *execution schedule* S_I for a (best) model I is a sequence of all actions executable in I, such that for all pairs of action atoms $a, b \in I$, if $prec(a) < prec(b)$ then a must precede b in S_I, for $prec(c)$ the precedence of an action atom c. Concerning the effects of actually executing actions, as well as corresponding notions of execution outcomes, we also refer to the next section where these notions are generalized compared to definitions in [1].

Example 2. Considering the program of Example 1, if the robot has low battery, then $\mathcal{AS}(P, E) = \mathcal{BM}(P, E)$ contains two models:

$I_1 = \{night, \#robot[clean, kitchen]\{c, 2\}, \#robot[goto, charger]\{b, 1\}\}$, and
$I_2 = \{day, \#robot[clean, bedroom]\{c, 2\}, \#robot[goto, charger]\{b, 1\}\}$.

Both give rise to a single execution schedule S_{I_i}: first charge, then clean.   □

3   Conceptual Improvements to the acthex Framework

The effective implementation of acthex within the dlvhex software, as well as its initial application and preliminary evaluation (cf. Sections 4 and 5), raised practical issues calling for conceptual changes of the acthex framework. Compared to its definition in [1], we incorporated the following improvements.

External Atom and Action Atom Semantics. We *generalize external atom semantics* in order to take the environment into account as follows. With every external predicate name $\&g$ we associate an $(n+m+2)$-ary Boolean function $f_{\&g}$, assigning each tuple $(E, I, y_1, \ldots, y_n, x_1, \ldots, x_m)$ either 0 or 1, where E is an environment state, I an interpretation, and the other parameters are input and output constants of $\&g$, respectively. We say that an interpretation I relative to P is a *model* of a ground external atom $a = \&g[y_1, \ldots, y_n](x_1, \ldots, x_m)$ wrt. environment E, denoted as $I, E \models a$, iff $f_{\&g}(E, I, y_1 \ldots, y_n, x_1, \ldots, x_m) = 1$.

Given a model I, for each action predicate name $\#g$ the *possible effects of executing a ground action* $\#g[y_1, \ldots, y_m]\{o, p\}[w : l]$ on an environment E wrt. I are defined by an associated $(m+2)$-ary function $f_{\#g}$ which returns a set of possible follow-up environment states: $f_{\#g}(E, I, y_1, \ldots, y_m) = \mathcal{E}$. Every $E' \in \mathcal{E}$ thus represents a possible effect. Considering a set of environments rather than a definite effect allows to model nondeterministic actions, and also nondeterministic and/or dynamic environments, where the environment may change without action execution by means of exogenous events.

Model Selection and Execution Schedule Representation. In practice, one usually wants to consider and execute a single execution schedule. This requires the choice of a single best model and a unique corresponding execution schedule. The former is modelled by a *Best Model Selector* function $select_{BM}$ which intuitively decides which model I from $\mathcal{BM}(P, E)$ to use. In our implementation, some simple selection functions (like lexicographic first) are built-in and can be configured. Alternatively, $select_{BM}$ can be provided in terms of user-defined C++ code. The set of all execution schedules of I is given by $\mathcal{ES}_{P,E}(I) =$

$$\{\langle a_1, \ldots, a_n \rangle \mid prec(a_i) \leq prec(a_j), \text{ for } 1 \leq i < j \leq n, \text{ and } \{a_1, \ldots, a_n\} = A_e\}.$$

$\mathcal{ES}_{P,E}(I)$ is in principle as large as $\mathcal{O}(|I|!)$, thus it is of course represented implicitly by its *execution schedule base* $\mathcal{ESB}_{P,E}(I)$, which is defined as a sequence of sets of actions $\mathcal{ESB}_{P,E}(I) = \{\langle A_1, \ldots, A_m \rangle\}$ where $A_i \subseteq A_e$, $1 \leq i \leq m$, and $prec(a) = prec(a')$ for all $a, a' \in A_i$, while $prec(a) < prec(a'')$ holds for all $a \in A_i, a'' \in A_j, 1 \leq j \leq m, i \neq j$. Intuitively, actions in A_i have the same precedence, while the precedence of actions strictly increases along the sequence. Obviously, $\mathcal{ESB}_{P,E}(I)$ has size $\mathcal{O}(|I|)$, and $\mathcal{ES}_{P,E}(I)$ can be recovered from it.

Execution Schedule Selection and Execution Outcomes. Given an execution schedule base, again a particular (customizable) function $build_{ES}$, called *Execution Schedule Builder*, selects a single execution schedule $\langle a_1, \ldots, a_n \rangle \in \mathcal{ES}_{P,E}(I)$ for execution. It defines a strict order over potential schedules, possibly based on general criteria on actions independent from the current execution base. Given an execution schedule $S = \langle a_1, \ldots, a_n \rangle \in \mathcal{ES}_{P,E}(I)$, the set of *possible execution outcomes* of S in environment E wrt. I is defined as $EX(S, I, E) = \{E_n \mid E_0 = E,$ and $E_{i+1} \in f_{\#g}(E_i, I, y_1, \ldots, y_m)\}$, given that a_i is of the form $\#g[y_1, \ldots, y_m]\{o, p\}[w : l]$. Intuitively the initial environment $E_0 = E$ is succeeded by a potential effect of executing each action in S in the given order. Recall that nondeterministic functions $f_{\&g}$ not only capture nondeterministic actions but take into account nondeterministic and/or dynamic environments. Eventually, given acthex program P, environment E, execution schedule builder $build_{ES}$ and best model selector $select_{BM}$, the *observed outcome* of executing P on E is given by some $E_n \in EX(S, I, E)$, where $S = build_{ES}(\mathcal{ESB}_{P,E}(I))$ and $I = select_{BM}(\mathcal{BM}(P, E))$. Unless the environment is static and deterministic, from a modeling perspective, the observed outcome represents a nondeterministic choice. For instance, executing S_{I_1} of Example 2 assuming a static and deterministic environment first yields $\{E_1\} = f_{\#robot}(E, I_1, goto, charger)$, and then $\{E_2\} = f_{\#robot}(E_1, I_1, clean, bedroom)$, where E_2 is the observed execution outcome.

Evaluation Iteration. Another important implementation aspect is an efficient realization of iterative acthex program evaluation. For this purpose we provide support on two aspects. First, in order to capture systems with dynamic environments, the environment state is sensed upon each iteration. This yields the environment $E = E_0$ that is used to evaluate external atoms and upon which the first scheduled action is executed. In general, the environment $E = E'_i$ for evaluation in iteration $i + 1$ can possibly differ from the observed outcome E_i at iteration i. Second, iteration control is provided by dedicated command line options, built-in constants, and specific action atoms. From the command line, and with higher priority by setting built-in constants to true, one can effect iterative evaluation in terms of fixed number of iterations, iteration until a pre-set value of (total) execution time is elapsed (checked after each iteration), and iteration ad infinitum. Special action atoms $\#acthexContinue$ and $\#acthexStop$ have highest priority and provide a declarative means of controlling iteration.

4   System Architecture and Implementation

Figure 1 shows how acthex is implemented within the dlvhex [4] framework, how applications interface with acthex, and the stages of executing an acthex program.

A given acthex program P is first parsed using the dlvhex parser, the acthex-specific parser, and the program rewriter modules. This yields a HEX program P' which contains auxiliary atoms instead of actions. P and P' are such that the sets $\mathcal{AS}(P')$ and $\mathcal{AS}(P, E)$ are in one-to-one correspondence. P' is evaluated using the computational core of dlvhex wrt. *custom* external atoms of an acthex application, then the set of best models $\mathcal{BM}(P, E)$ is computed. One best model I is selected using a Best Model

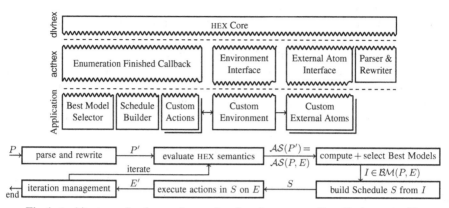

Fig. 1. Architecture of acthex and execution flow for program P on Environment E

Selector module, then an Execution Schedule Builder module creates a unique execution schedule S from actions in I. Both Best Model Selectors and Execution Schedule Builders can be customized, as well as it is possible to program custom action predicates each having its own customizable *Environment* interface. Moreover, the acthex system features iterative evaluation of P'. The iteration process can be controlled as described in Section 3.

5   Application and Evaluation

acthex can be fruitfully used in a variety of contexts, especially when it is expected to take actions which have impact on actual dynamic environments, and which require to repeatedly take new decisions. In this respect, logic-based games are the ideal testbed: we showcase here two pilot applications (addons) we developed using the acthex system.

Sudoku Addon. This addon allows to maintain a Sudoku table of arbitrary size and to perform operations on it. Sudoku tables are seen as stored within the external environment. The addon provides a single action predicate $\#sudoku[A, O_1, O_2, O_3]\{O, P\}$ $[W : L]$, where A is an operation type and O_1, O_2, O_3 are parameters, depending on the operation type. Possible actions are the insertion of a number into a cell, exclusion of a candidate number from the possible values of a cell, and printing the current table in various formats. Other external predicates allow to query the content of the current table. This addon permitted us to experiment with the incremental application of Sudoku inference rules as described in [3]. Large Sudoku tables cannot be solved by pure guess & check strategies: on the other hand, acthex allows to iterate over partially complete tables, and to repeatedly apply a number of deterministic inference strategies depending on the current resolution progress. Our acthex-based iterative player allows to solve Sudoku tables as large as 81×81, which are far out of the performance reach of an ASP-based system using a pure guess & check strategy[2].

---

[2] Detailed results are available at
http://www.kr.tuwien.ac.at/research/systems/dlvhex/actionplugin/SudokuAddon.html#sbench.

Reversi Addon. The Reversi addon allows for playing an online version of the popular board game Reversi. acthex allows to program Reversi heuristic rules using a logic program and to perform actual actions depending on the move of choice. Here the environment includes an external web gaming site[3]; we developed Javascript and Perl scripts in order to access and perform actions on the site, and attached them to the execution of the action atom $\#reversi[A, O_1, O_2]\{O, P\}[W : L]$, where A selects an action type and O_1 and O_2 are parameters, depending on the action type. Possible actions are: setting the game number, logging in, making a move, and waiting until the opponent makes their move. Some external predicates are available for retrieving the current status of the game and the corresponding board. The usage workflow of the Reversi addon is straightforward: after initialization, each iteration extracts the current board state from the Web by means of proper external atoms and performs reasoning about the next move in a logic program, using commonly known heuristic rules for Reversi[4]. The chosen move triggers an action which is executed on the game web site. The iteration progress is then suspended by means of a wait action, which will let a further iteration start when the game opponent replies to the last move. The odering of actions is controlled by the precedence feature of acthex, while the end of the game is detected by means of an external atom, causing to end iteration when a game terminates.

6   Conclusion

In this work we have enriched the acthex semantics by new features and provided an implementation on top of the dlvhex reasoner for HEX-programs. Moreover, an *iteration* framework allows for repeating the evaluation of an acthex program and consequent execution of actions. For evaluation, we applied our system to logic games (Sudoku and Reversi) exhibiting scalability to larger instances and modeling strength. Further work is planned, especially concerning evaluation efficiency. For instance, we are currently considering an incremental evaluation approach similar to iclingo [6], although the latter serves a different purpose, since it does neither address action execution nor maintain arbitrary state information, and hence is less expressive (e.g., for re-planning).

References

1. Basol, S., Erdem, O., Fink, M., Ianni, G.: HEX programs with action atoms. In: International Conference on Logic Programming, Technical Communications, pp. 24–33 (2010)
2. Buccafurri, F., Leone, N., Rullo, P.: Strong and weak constraints in disjunctive datalog. In: Fuhrbach, U., Dix, J., Nerode, A. (eds.) LPNMR 1997. LNCS, vol. 1265, pp. 2–17. Springer, Heidelberg (1997)
3. Calimeri, F., Ianni, G., Perri, S., Zangari, J.: The eternal battle between determinism and nondeterminism: preliminary studies in the sudoku domain. In: RCRA (submitted, 2013)
4. Eiter, T., Ianni, G., Schindlauer, R., Tompits, H.: A Uniform Integration of Higher-Order Reasoning and External Evaluations in Answer-Set Programming. In: IJCAI, pp. 90–96. Professional Book Center (2005)
5. Febbraro, O., Leone, N., Grasso, G., Ricca, F.: Jasp: A framework for integrating answer set programming with java. In: KR (2012)
6. Gebser, M., Kaminski, R., Kaufmann, B., Ostrowski, M., Schaub, T., Thiele, S.: Engineering an incremental ASP solver. In: Garcia de la Banda, M., Pontelli, E. (eds.) ICLP 2008. LNCS, vol. 5366, pp. 190–205. Springer, Heidelberg (2008)

---

[3] "Your Turn My Turn", available at http://www.yourturnmyturn.com

[4] See e.g. the Strategy guide for Reversi at http://www.samsoft.org.uk/reversi/strategy.htm

Debugging Answer-Set Programs with Ouroboros – Extending the SeaLion Plugin[*]

Melanie Frühstück[1], Jörg Pührer[2], and Gerhard Friedrich[3]

[1] Siemens AG Österreich, Corporate Technology, Vienna, Austria
melanie.fruehstueck@siemens.com
[2] Technische Universität Wien,
Institut für Informationssysteme 184/3,
Favoritenstraße 9-11, A–1040 Vienna, Austria
puehrer@kr.tuwien.ac.at
[3] Alpen-Adria Universität, Klagenfurt, Austria
Gerhard.Friedrich@ifit.uni-klu.ac.at

Abstract. In answer-set programming (ASP), there is a lack of debugging tools that are capable of handling programs with variables. Hence, we implemented a tool, called Ouroboros, for debugging non-ground answer-set programs. The system builds on a previous approach based on ASP meta-programming that has been recently extended to cover weight constraints and choice rules. The main debugging question addressed is "given a program P and an interpretation I, why is I not an answer set of P". Our tool gives answers in terms of two categories of explanations: unsatisfied rules and unfounded loops. Ouroboros is a plugin of the SeaLion integrated development environment for ASP that is built on Eclipse. Thereby, Ouroboros complements and profits from SeaLion's Stepping plugin, that implements a different debugging approach for ASP.

1 Introduction

Answer-set programming (ASP) is a well-known declarative problem-solving paradigm [1]. While a great deal of work on ASP implementations has been put into improving solver performance, comparably little effort has been spent on tools that support the development of answer-set programs, in particular, there is a lack of *debugging systems* for ASP. But, in recent years, methods for debugging have been explored theoretically [2–6]. Brain and De Vos [2] discussed what it means for answer-set programs to be incorrect and presented algorithms to locate bugs. Syrjänen [3] proposed to debug contradictory programs by means of ASP meta-programming. Gebser et al. [4] tried to find semantic errors of answer-set programs. The question is why an expected interpretation is not an answer set of a program (spock [7] implements this approach).

[*] This research has been funded by FFG FIT-IT (grant number 825071) within the scope of the RECONCILE project and by the Austrian Science Fund (FWF P21698).

P. Cabalar and T.C. Son (Eds.): LPNMR 2013, LNAI 8148, pp. 323–328, 2013.

However, these approaches are only able to deal with propositional programs which is clearly a limiting factor as far as practical applications are concerned. Therefore, Oetsch et al. [8] developed a meta-program for debugging non-ground programs in ASP. This approach is based on the meta-programming technique of Gebser et al. [4] for propositional programs. Recently, their method has been further extended to cover weight and cardinality constraints [9]. In this paper, we describe the debugging system Ouroboros, which implements this approach and explains why an expected interpretation is not an answer set of a given program. The system gives answers in terms of two categories of explanations: unsatisfied rules and unfounded loops. Intuitively, a rule is unsatisfied if its body is true but all literals in its head are false. Moreover, an unfounded loop is a set of atoms from the interpretation whose truth can only be derived from itself but is not founded in facts. Thus, an unfounded loop is reminiscent of the Ouroboros, a dragon biting its own tail, which our tool is named after. Ouroboros[1] is a plugin of SeaLion [10, 11], an integrated development environment (IDE) for ASP that is based on the Eclipse platform and supports developing answer-set programs using the *Potassco* and DLV solvers [12, 13]. Thereby, Ouroboros complements the *stepping-based debugging mechanism* [14] integrated in SeaLion [11]. It allows the user to interactively build up an interpretation by, stepwise, adding literals derived by a rule whose body is satisfied by the interpretation obtained in the previous step.

While, on the one hand, Ouroboros provides additional debugging functionality for SeaLion, on the other hand, it also profits from the Stepping-plugin which can help in building up the interpretation that is input to our approach. Another possibility to create an interpretation is to use the Kara plugin of SeaLion [15].

2   Backend

As mentioned in the introduction, Ouroboros makes use of ASP meta-programming to find explanations why a given interpretation I is not an answer set of the program P under development. The internal data flow of Ouroboros is depicted in Fig. 1. In a preprocessing step, all cardinality constraints of P are translated into standard rules[2]. Then, P and the expected interpretation I are reified, i.e., P and I are brought onto a meta-level (i.e. a fact person(1) would be presented by rule(r1). head(r1,r1h1). pred(r1h1,person). struct(r1h1,1,const,1).), represented by facts, and joined with rules for identifying the targeted explanations. Finally, the meta-program is fed to an ASP solver and the resulting answer sets get interpreted.

3   Usage and Graphical User Interface

The Ouroboros plugin itself comprises two graphical components, the *Debug Configuration Tab Group* for defining parameters for a debugging session and the

---

[1] The plugin is open source and available from http://www.sealion.at

[2] For a detailed description of these translations we refer the interested reader to a companion paper [9].

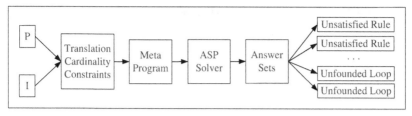

Fig. 1. Data Flow in Ouroboros

Debugging Explanation View that provides the explanations found to the user. In the following two subsections, both will be clarified by means of an example. The example is based on the original house problem [16] that is an abstraction of several configuration problems where entities may be contained in other entities and some additional requirements are defined. We considered a simplification of the modification of this problem [17, 18]. Given a set of cabinets, rooms, persons and objects, the problem consists of assigning objects to cabinets, cabinets to rooms and rooms to persons, such that following constraints hold: cabinets and rooms can contain only a specific number of objects and cabinets, respectively; objects belonging to different persons cannot be placed in the same cabinet; cabinets of different persons cannot be placed in the same room. Fig. 2 depicts the original program and the expected interpretation (in the interpretation view).

3.1   Debug Configurations

Debug configurations are similar to run configurations in Eclipse. They are used to start an application in the debug mode. When clicking on **Ouroboros** in the debug configurations, three tabs occur which the user can select, where one tab is the *Common* Eclipse tab. Let us assume that a user, called Benia, wants to debug the program given in Fig. 2. In the *Input Program/Interpretation* tab Benia selects the program file and the expected interpretation file. When clicking on the *add* button, a window occurs for selecting files from the Eclipse workspace. The currently opened file in the editor is preselected. As Benia wants to check for unsatisfied rules in the program, he selects the explanation type *Unsatisfiability*.

In the *Solver* tab, Benia chooses **Gringo/clasp** as solver configuration. If he had checked for unfounded loops instead unsatisfied rules, the solver would have to be able to deal with disjunctions as they are needed in the meta-program. In that case one can set the first check mark on the bottom of the tab to filter for solver configurations marked as **claspD** configurations. Using some other solver that is able to deal with disjunctions, requires the second check mark to be set.

After all required attributes of the debug configuration are set, Benia can start the debugging process. Now, all steps described in Fig. 1 are run through. When the final answer set is computed, the explanation why the given interpretation is not answer set of the given program is shown in the debugging explanation view.

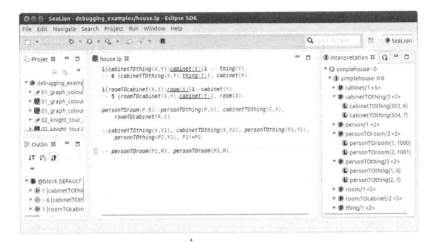

Fig. 2. The given program and the expected interpretation

3.2 Debugging Explanation View

The debugging explanation view consists of two columns. In the first column, the explanation is shown. This is either the rule that is unsatisfied with respect to the interpretation or an unfounded loop. In the second column the meta-programming predicates are shown. This can be either `guessRule/1` and `subst/2` (the former states about the unsatisfied rule and the latter about its substitution) or `inLoop/1` (all literals that form an unfounded loop). Even if the output of the debugger concerns a translation rule of a cardinality constraint, it is mapped back to the cardinality constraint itself. Additionally, all reasons of why the given interpretation is not an answer set of the program are given. In the case of Benia's program, the explanation represents the last constraint given in the program. In particular, the additional condition `P1!=P2` is missing, that means that the overall configuration does not allow a room belonging to two different persons. Thus, one room can only belong to one person. When Benia clicks on the explanation the corresponding rule is highlighted in the editor (see Fig. 3). If the explanation refers to a rule in a specific file, this file is automatically opened.

3.3 Additional Features in the Interpretation View

In addition to the main two components described above, the context menu of the interpretation view of `SeaLion` was extended. In general, the interpretation view provides a tree structure of each answer set of the executed program (cf. Fig. 2). In the context menu of the interpretation some new functionalities were added. The entry *Save as Facts* makes it possible to save the selected interpretation as facts. To do so, the user has to select a project in which the new file is inserted. The predefined name of the new file can be adapted. After finishing this process the file including the facts of the interpretation is opened in the editor.

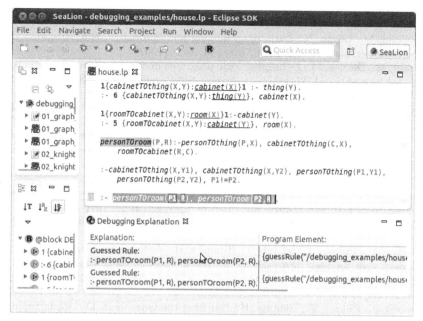

Fig. 3. Debugging Explanation View

Moreover, the user can select *Detect Unsatisfied Rules* if he or she wants to create a new debug configuration where the input interpretation is the one on which the context menu was opened and where the explanation type is automatically set to unsatisfied rules. *Detect Unsatisfied Rules for Launch* lets the user first select an existing debug launch configuration in which the interpretation file is substituted with the interpretation selected. Again, the explanation type is automatically set to unsatisfied rules. The entries *Detect Unfounded Loops* as well as *Detect Unfounded Loops for Launch* work analogously to the just described functionalities, except that the explanation type is set to unfounded loops.

If these four functionalities are used, the user has to be aware of the fact that the selected interpretations are just temporary files. That means, if the user exits Eclipse the interpretation files will get lost. To make them persistent the functionality *Save as Facts* can be used.

4   Conclusion

In this paper we described the debugging extension called Ouroboros in the SeaLion plugin for Eclipse. It provides debugging support of Gringo programs involving variables and cardinality constraints by explaining why a given interpretation is not an answer set of a given program. We concisely presented the the components and functionalities of the Ouroboros plugin and, by means of a debugging example, showed how the user can kick off the debugging process.

References

1. Gelfond, M., Leone, N.: Logic programming and knowledge representation - The A-Prolog perspective. Art. Intell. 138(1-2), 3–38 (2002)
2. Brain, M., De Vos, M.: Debugging logic programs under the answer set semantics. In: 3rd International Workshop on Answer Set Programming (ASP 2005). CEUR Workshop Proceedings, pp. 141–152 (2005)
3. Syrjänen, T.: Debugging inconsistent answer set programs. In: Proc. NMR 2006, pp. 77–83 (2006)
4. Gebser, M., Pührer, J., Schaub, T., Tompits, H.: A meta-programming technique for debugging answer-set programs. In: Proc. AAAI 2008, pp. 448–453. AAAI Press (2008)
5. Caballero, R., García-Ruiz, Y., Sáenz-Pérez, F.: A theoretical framework for the declarative debugging of datalog programs. In: Schewe, K.-D., Thalheim, B. (eds.) SDKB 2008. LNCS, vol. 4925, pp. 143–159. Springer, Heidelberg (2008)
6. Pontelli, E., Son, T.C., Elkhatib, O.: Justifications for logic programs under answer set semantics. TPLP 9(1), 1–56 (2009)
7. Gebser, M., Pührer, J., Schaub, T., Tompits, H., Woltran, S.: spock: A debugging support tool for logic programs under the answer-set semantics. In: Seipel, D., Hanus, M., Wolf, A. (eds.) INAP 2007. LNCS, vol. 5437, pp. 247–252. Springer, Heidelberg (2009)
8. Oetsch, J., Pührer, J., Tompits, H.: Catching the Ouroboros: On debugging non-ground answer-set programs. TPLP 10(4-6), 513–529 (2010)
9. Polleres, A., Frühstück, M., Schenner, G., Friedrich, G.: Debugging non-ground ASP programs with choice rules, cardinality and weight constraints. In: Cabalar, P., Son, T.C. (eds.) LPNMR 2013. LNCS (LNAI), vol. 8148, pp. 452–464. Springer, Heidelberg (2013)
10. Oetsch, J., Pührer, J., Tompits, H.: The SeaLion has landed: An IDE for answer-set programming—Preliminary report. In: Proc. WLP 2011 (2011)
11. Busoniu, P., Oetsch, J., Pührer, J., Skočovský, P., Tompits, H.: SeaLion: An Eclipse-based IDE for answer-set programming with advanced debugging support (submitted draft, 2013)
12. Gebser, M., Kaufmann, B., Kaminski, R., Ostrowski, M., Schaub, T., Schneider, M.T.: Potassco: The Potsdam answer set solving collection. AI Commun. 24(2), 107–124 (2011)
13. Leone, N., Pfeifer, G., Faber, W., Eiter, T., Gottlob, G., Perri, S., Scarcello, F.: The DLV system for knowledge representation and reasoning. ACM Trans. Comput. Logic 7(3), 499–562 (2006)
14. Oetsch, J., Pührer, J., Tompits, H.: Stepping through an answer-set program. In: Delgrande, J.P., Faber, W. (eds.) LPNMR 2011. LNCS (LNAI), vol. 6645, pp. 134–147. Springer, Heidelberg (2011)
15. Kloimüllner, C., Oetsch, J., Pührer, J., Tompits, H.: Kara - A system for visualising and visual editing of interpretations for answer-set programs. In: Proc. WLP 2011, pp. 152–164 (2011)
16. Mayer, W., Bettex, M., Stumptner, M., Falkner, A.: On solving complex rack configuration problems using CSP methods. In: Proc. IJCAI 2009 Workshop on Configuration (2009)
17. Friedrich, G., Ryabokon, A., Falkner, A., Haselböck, A., Schenner, G., Schreiner, H.: (re)configuration using answer set programming. In: Proc. IJCAI 2011 Workshop on Configuration, pp. 17–25 (2011)
18. Aschinger, M., Drescher, C., Vollmer, H.: LoCo—A logic for configuration problems. In: Proc. ECAI 2012, vol. 242, pp. 73–78 (2012)

Game Semantics for Non-monotonic Intensional Logic Programming[*]

Chrysida Galanaki[1], Christos Nomikos[2], and Panos Rondogiannis[1]

[1] Department of Informatics & Telecommunications, University of Athens, Greece
{chrysida,prondo}@di.uoa.gr
[2] Department of Computer Science and Engineering, University of Ioannina, Greece
cnomikos@cs.uoi.gr

Abstract. *Intensional logic programming* is an extension of logic programming based on intensional logic, which includes as special cases both *temporal* and *modal* logic programming. In [OW92], M. Orgun and W. W. Wadge provided a general framework for capturing the semantics of intensional logic programming languages. One key property involved in the construction of [OW92], is the *monotonicity* of intensional operators. In this paper we consider intensional logic programming from a game-theoretic perspective. In particular we define a two-person game and we demonstrate that it is equivalent to the semantics of [OW92]. More importantly, we demonstrate that the game is even applicable to intensional languages with *non-monotonic* operators. In this way we provide the first (to our knowledge) general semantic framework for capturing the semantics of non-monotonic intensional logic programming.

1 Introduction

Intensional Logic is an extension of classical logic that was introduced by R. Montague [Mon74] in order to capture the semantics of natural languages. Roughly speaking, intensional logic was proposed as a formal system for understanding and reasoning about *context-dependent* properties of natural language expressions. In its initial form, intensional logic was a higher-order one, equipped with modal and temporal operators. Nowadays, however, the term "intensional logics" can also be used more loosely in order to describe a large class of logics for reasoning about context-dependent phenomena. In this sense, *temporal logics* and *modal logics* can be viewed as special cases of intensional logic.

Based on this broad interpretation of the term, M. Orgun and W. W. Wadge introduced in [OW92] the notion of *intensional logic programming*, which includes as special cases many non-classical extensions of logic programming (such as *temporal logic programming*, *modal logic programming*, and so on). As pointed

---

[*] This research was supported by the project "Handling Uncertainty in Data Intensive Applications", co-financed by the European Union (European Social Fund) and Greek national funds, through the Operational Program "Education and Lifelong Learning" of the National Strategic Reference Framework (NSRF) - Research Program: THALES, Investing in knowledge society through the European Social Fund.

P. Cabalar and T.C. Son (Eds.): LPNMR 2013, LNAI 8148, pp. 329–341, 2013.

out in [OW92], numerous logic programming languages that have been proposed in the literature can be characterized as "intensional" (such as for example, Chronolog [OW92], Tempura [Mos84], Molog [Far86], and so on). It was therefore natural to wonder whether there exists a common semantic framework for handling all these systems in a uniform way. As it was demonstrated in [OW92], if the intensional operators of the source intensional logic programming language obey some simple semantic properties, then the programs of the language are guaranteed to possess the *minimum model* property. However, all the operators allowed in [OW92] are assumed to satisfy the *monotonicity* property (see Section 2 for a formal definition of this notion), and this excludes many interesting applications that involve *non-monotonicity* (which nowadays is a crucial concept involved in knowledge representation and reasoning).

The purpose of this paper is to extend the framework of [OW92] to allow arbitrary (even non-monotonic) intensional operators in the bodies of program clauses. Our construction is based on *game semantics*, an approach which was initially proposed in [vE86] and has recently been revived and extended in [CLN98, GRW08, Tsou11]. We start by constructing a simple two-person game for the class of intensional logic programs considered in [OW92] and we demonstrate that the outcome of the game coincides with the minimum model semantics obtained in [OW92]. We thus provide an equivalent, purely game-theoretic formulation to the approach of Orgun and Wadge. We then extend the proposed game to handle intensional logic programs that even use non-monotonic operators in the bodies of clauses. In this way we obtain the first (to our knowledge) general semantic framework for non-monotonic intensional logic programming. It should be noted that intensional logic programming, due to its variety of operators, allows a much broader framework for non-monotonicity than the one we are familiar with from classical logic programming (where the main source of non-monotonicity is the operator of negation-as-failure).

2   Preliminaries

2.1   Intensional Logic Programming

For simplicity in our exposition, the programs of our language will be propositional and possibly infinite. This is a common assumption in logic programming which will simplify our presentation. In essence, instead of studying finite first-order programs, we examine the ground instantiations of these programs (which in general are infinite propositional programs).

We assume the existence of an infinite set of intensional operators; programs of our language will use operators of this set. We will use symbols such as $\triangledown$, $\bigcirc$, and so on, to denote arbitrary intensional operators. We assume that an intensional atom consists of a *unique* intensional operator applied to a propositional atom. This assumption does not incur any loss of generality, despite the fact that in most intensional languages an intensional atom is a classical atom prefixed by a finite sequence of intensional operators. For example, in Chronolog [OW92], there exist only two intensional operators, namely first and next, and an intensional atom is a propositional atom prefixed by a finite sequence of these two operators. In our setting, every finite sequence of operators corresponds to a single operator. For example, given the Chronolog intensional atom **next next** A, we will assume that **next next** is a single operator.

A central concept regarding intensional logic programming (and intensional logic in general), is that of its underlying *set of possible worlds*, which is used in order to give semantics to the intensional operators of the language. In the rest of the paper we will use W to denote the set of possible worlds of our language. Following [OW92], the meaning of an intensional operator $\triangledown$ is a function in $2^W \rightarrow 2^W$. We will use $||\triangledown||$ to denote the meaning of $\triangledown$. As discussed in [OW92][page 422], this approach to the semantics of intensional operators is equivalent to the so-called *neighborhood semantics* of Dana Scott [Sco70]. Formally, an intensional language is defined as follows:

Definition 1. *An* intensional language *consists of an infinite set of propositional atoms, the set of usual logical connectives, a possibly infinite set of intensional operators, a set W of* possible worlds *and a set of denotations, ie., functions $2^W \rightarrow 2^W$, one for each intensional operator. An* intensional atom *is a formula $\triangledown A$ where A is a propositional atom and $\triangledown$ is an* intensional operator. *An* intensional clause *is a clause of the form $B_0 \leftarrow B_1, \ldots, B_n$, where $B_0, B_1, \ldots, B_n$ are intensional or propositional atoms. An* intensional (logic) program *is a (possibly infinite) set of intensional clauses.*

The semantics of intensional logic programs is given using the notion of *intensional interpretation* [OW92]. An intensional interpretation I maps each propositional atom to a subset of the set of possible worlds (intuitively, to those possible worlds where this atom is true under I). Notice that, in Section 4 we will treat subsets of W as functions from W to $\{0, 1\}$, i.e., we will identify a subset with its characteristic function.

Definition 2. *An* intensional interpretation I *of an intensional logic program P is a function from the set of propositional atoms of P to 2^W.*

Definition 3. *Let I be an intensional interpretation of a given program P. Then, I can be extended as follows:*

- *$I(\triangledown A) = ||\triangledown||(I(A))$*
- *$I(B_1, \ldots, B_n) = \bigcap_{i=1}^{n} I(B_i)$.*

We will say that I satisfies an intensional clause $B_0 \leftarrow B_1, \ldots, B_n$ if $I(B_0) \supseteq I(B_1, \ldots, B_n)$, and I is a model of P if I satisfies all intensional clauses of P.

We will be interested in intensional operators that possess certain properties that intuitively ensure that the programs of our language are "well-behaved" from a semantic point of view. We follow the terminology introduced in [OW92]:

Definition 4. *Let $\triangledown$ be an intensional operator. We say that $||\triangledown||$ is:*

- monotonic *iff for all $S_1, S_2 \in 2^W$, $S_1 \subseteq S_2$ implies $||\triangledown||(S_1) \subseteq ||\triangledown||(S_2)$.*
- universal *iff for some $S \in 2^W$, $||\triangledown||(S) = W$.*
- conjunctive *iff for all $\{S_i\}_{i \in I} \in 2^{2^W}$, $||\triangledown||(\bigcap_{i \in I} S_i) = \bigcap_{i \in I} ||\triangledown||(S_i)$.*

The above notions can be used in order to state the following theorem (which is a central result established in [OW92]):

Theorem 1. *Let P be an intensional logic program such that the denotations of all intensional operators in the heads of the clauses of P are universal, monotonic and conjunctive, and the denotations of all intensional operators in the bodies of the clauses are monotonic. Then, P has a minimum model M_P.*

Example 1. Consider the following Chronolog program:

```
first p ←
next next p ← p
```

The set of intensional operators of Chronolog consists of all finite sequences of the operators `first` and `next`. The set of possible worlds W is equal to the set $\mathbb{N}$ of natural numbers. Recall now (see [OW92]) that $\|\texttt{next}\|(S) = \{t \in \mathbb{N} \mid t+1 \in S\}$, $\|\texttt{next next}\|(S) = \{t \in \mathbb{N} \mid t + 2 \in S\}$ and $\|\texttt{first}\|(S) = \{t \in \mathbb{N} \mid 0 \in S\}$. One can easily verify that the operators `first` and `next next` are universal, monotonic and conjunctive and therefore the above theorem can be applied. The minimum model of the above program is the interpretation that assigns to p the set of even natural numbers.   □

As we discuss in Section 3, Theorem 1 can also be established in a purely game-theoretic way. Moreover, in Section 4, an extension of this theorem will be obtained for programs with non-monotonic operators in the bodies of clauses.

2.2   Infinite Games of Perfect Information

Infinite games of perfect information [GS53] are games between two players that we will call *Player I* and *Player II*. In such games there does not exist any "hidden information": both players know all the moves that have been played so far, and there are no simultaneous moves. The games are infinite in the sense that they do not terminate at a finite stage and therefore in order to derive the outcome of a play it may be necessary to examine an infinite sequence of moves.

In the following, sequences (finite or infinite in length) will usually be denoted by s or x. A finite sequence of length k will be denoted by $\langle s_0, s_1, \ldots, s_{k-1} \rangle$ and the empty sequence by $\langle \rangle$. Given a set X, an *infinite tree* on X is a set $T_\omega \subseteq X^\omega$ of infinite sequences[1] of members of X.

During a play of a game, the two players exchange moves from a non-empty set X, called the *set of moves*. Initially, Player I chooses some $x_0 \in X$, then Player II chooses $x_1 \in X$, and so on. There also exists a set of *rules* specifying the possible moves of the two players. The rules will usually be defined by putting down (non-blocking) restrictions on the choice of x_n that depend on the preceding moves

---

[1] The definition of an infinite tree as a set of infinite sequences can be justified as follows: the nodes of the tree are the initial segments of the infinite sequences and the root of the tree is the empty sequence $\langle \rangle$. A consequence of this definition is that an infinite tree does not contain terminal nodes (leaves), i.e., it is purely infinite.

$x_0, \ldots, x_{n-1}$. The rules of the game (see for example [Mos80]) implicitly define an infinite tree T_ω on X:

$$\langle x_0, x_1, \ldots \rangle \in T_\omega \Leftrightarrow \text{for each } i \geq 0, \ x_i \text{ is allowed by the rules.}$$

Additionally, we assume the existence of a set D, called the *set of payoffs*, which consists of all possible outcomes of the game. Finally, we consider a function Φ, called the *payoff function*, which calculates the outcome of a play of the game. The above notions are formalized as follows:

Definition 5. *An infinite game of perfect information is a quadruple* $\Gamma = (X, T_\omega, D, \Phi)$, *where:*

- *X is a nonempty set, called the set of moves for Players I and II.*
- *T_ω is an infinite tree on X (i.e., $\subseteq X^\omega$), usually implicitly specified by a set of rules.*
- *D is a linearly ordered set called the set of rewards, with the property that for all $S \subseteq D$, $\mathsf{lub}(S)$ and $\mathsf{glb}(S)$ belong to D.*
- *$\Phi : T_\omega \to D$, is the payoff function of the game.*

Given a game Γ, a *legal sequence of moves* of the game is an initial segment of an infinite path that starts at the root of the tree T_ω of Γ. We define two sets $\mathsf{Strat}^I(\Gamma)$ and $\mathsf{Strat}^{II}(\Gamma)$ which correspond to the set of strategies for Player I and Player II respectively. A strategy $\sigma \in \mathsf{Strat}^I(\Gamma)$ assigns a move to each even length legal sequence of moves; similarly for $\tau \in \mathsf{Strat}^{II}(\Gamma)$ and odd length legal sequences of moves.

Definition 6. *Let $\Gamma = (X, T_\omega, D, \Phi)$ be a game. Let T_n be the set of initial segments of elements of T_ω that have length n. Then, a strategy for Player I is a function $\sigma : (\bigcup_{n<\omega} T_{2n}) \to X$ such that for every $n < \omega$ and for every $\langle x_0, \ldots, x_{2n-1} \rangle \in T_{2n}$, $\langle x_0, \ldots, x_{2n-1}, \sigma(\langle x_0, \ldots, x_{2n-1} \rangle) \rangle \in T_{2n+1}$. Similarly, a strategy for Player II is a function $\tau : (\bigcup_{n<\omega} T_{2n+1}) \to X$ such that for every $n < \omega$ and for every $\langle x_0, \ldots, x_{2n} \rangle \in T_{2n+1}$, $\langle x_0, \ldots, x_{2n}, \tau(\langle x_0, \ldots, x_{2n} \rangle) \rangle \in T_{2n+2}$. We denote by $\mathsf{Strat}^I(\Gamma)$ and by $\mathsf{Strat}^{II}(\Gamma)$ the sets of strategies of Players I and II respectively.*

Two strategies, when played one against the other, define a play of the game:

Definition 7. *Let Γ be a game and let $\sigma \in \mathsf{Strat}^I(\Gamma)$ and $\tau \in \mathsf{Strat}^{II}(\Gamma)$. We define the following sequence:*

$$s_0 = \sigma(\langle \rangle)$$
$$s_{2i} = \sigma(\langle s_0, s_1, \ldots, s_{2i-1} \rangle), \text{ for all } i \geq 1$$
$$s_{2i+1} = \tau(\langle s_0, s_1, \ldots, s_{2i} \rangle), \text{ for all } i \geq 0.$$

The play of the game defined by the strategies σ and τ, *which is denoted by* $\sigma \star \tau$, *is the infinite sequence* $\langle s_0, s_1, s_2, \ldots \rangle$. *The s_i's are the* moves *of the play.*

Until now we have focused on particular plays of a game. We would like to have a notion that gives us the outcome of the whole game provided that Player I tries his best to minimize the result while Player II tries his best to maximize it. Moreover, we would like that during this process, each player can decide for his best strategy, independently of the corresponding choice of the other player. This idea is captured by *determinacy*:

Definition 8 (Determinacy). *Let $\Gamma = (X, T, D, \Phi)$ be a game and let $\mathcal{S} = \mathsf{Strat}^I(\Gamma)$ and $\mathcal{T} = \mathsf{Strat}^{II}(\Gamma)$. Then Γ is* determined *with value $d \in D$ if:*

$$\mathsf{lub}_{\tau \in \mathcal{T}} \, \mathsf{glb}_{\sigma \in \mathcal{S}} \, \Phi(\sigma \star \tau) = \mathsf{glb}_{\sigma \in \mathcal{S}} \, \mathsf{lub}_{\tau \in \mathcal{T}} \, \Phi(\sigma \star \tau) = d.$$

Determinacy is an important notion for game theory. For the games we are considering here, determinacy can easily be established (see Section 4).

3   The Game for Monotonic Intensional Logic Programs

Consider an intensional logic program P that satisfies the requirements of Theorem 1. Let C be a ground intensional or propositional atom and let $w \in W$. Let M_P be the minimum intensional model of P (recall Theorem 1). We introduce a two-player game $\Gamma_P(C, w)$ which (as we are going to see) has the property that $w \in M_P(C)(w)$ if and only if Player II has a winning strategy in $\Gamma_P(C, w)$. In other words, the game will be shown to be equivalent to the minimum model semantics introduced in [OW92].

During $\Gamma_P(C, w)$, Player I has the role of the *Doubter* and Player II the role of the *Believer*. Intuitively, in this game Player I does not believe that program P implies that the atom C is true under the possible world w; on the other hand, Player II believes exactly the opposite. In the following, the moves of the Believer (respectively the Doubter) will be followed by a $+$ (respectively $-$) superscript; this convention will help us distinguish between the two players and avoid confusion (especially in the extended game of the next section).

The infinite tree T_ω of the game $\Gamma_P(C, w)$ consists of all infinite sequences $\langle x_0, x_1, \ldots, x_k, \ldots \rangle$, which satisfy the following restrictions (in which A denotes a propositional atom, B_i denotes an intensional atom, u, v, z denote elements of W and S denotes a subset of W) for each $k \geq 0$:

R$_1$: $x_0 = \langle C, w \rangle^-$.

R$_2$: If $x_k = \langle \nabla A, u \rangle^-$, then $x_{k+1} = \langle A, S \rangle^+$, where $u \in \|\nabla\|(S)$.

R$_3$: If $x_k = \langle A, S \rangle^+$, then $x_{k+1} = \langle A, v \rangle^-$ where $v \in S$.

R$_4$: If $x_k = \langle A, v \rangle^-$, then $x_{k+1} = \langle C, z \rangle^+$ where C is a clause in P of the form $B_0 \leftarrow B_1, \ldots, B_n$, such that either **(i)** $B_0 = A$ and $z = v$ or **(ii)** $B_0 = \nabla A$ and for every S satisfying $z \in \|\nabla\|(S)$ it holds $v \in S$.

R$_5$: If $x_k = \langle B_0 \leftarrow B_1, \ldots, B_n, z \rangle^+$, then $x_{k+1} = \langle B_j, z \rangle^-$, for some j with $1 \leq j \leq n$.

R$_6$: If after x_k has been played, none of the above rules is applicable, then $x_{k+1} = \langle I\text{'ve lost} \rangle$.

R$_7$: If $x_k = \langle I\text{'ve lost} \rangle$, then $x_{k+1} = \langle I\text{'ve won} \rangle$ (and vice-versa).

Some explanations are in order. Suppose that $C = \triangledown A$ (the explanation for $C = A$ is similar). Initially, Player I plays the move $\langle \triangledown A, w \rangle^-$. The intuitive explanation for this move is "*I doubt that $\triangledown A$ is true in world w*". Player II believes the truth of $\triangledown A$ in world w and for this reason he replies to the move of Player I with a pair $\langle A, S \rangle^+$, where $w \in \|\triangledown\|(S)$. The explanation for this move is "*I believe that $\triangledown A$ is true in w; actually, I believe A is true in all the worlds contained in S and this implies that $\triangledown A$ is true in w*". Player I now responds with a pair $\langle A, v \rangle^-$ where $v \in S$. The intuition now is: "*I doubt that A is true in the world v of S (and therefore I continue to believe that $\triangledown A$ is not true in w)*". Player II must now establish that A is true in v. One way to achieve this is to use a clause with head A. A second (less direct) way is to prove that $\triangledown A$ holds at some world z with the property mentioned in Case (ii) of rule $\mathbf{R_4}$; this property guarantees that if $\triangledown A$ holds at z, then A holds at v. Therefore, Player II provides a pair $\langle C, z \rangle^+$, where C is a program clause with head A or $\triangledown A$. The intuition is "*Using this rule and the context z I can establish that A is true in the world v*". Now Player I responds with a pair of the form $\langle B_i, z \rangle^-$, where B_i is one of the intensional atoms in the body of the rule that Player II has just played. The intuition is "*I doubt that B_i is true in world z*". The play of the game then continues along the above lines.

Since we are dealing with *infinite games*, a play continues even if at some point the play of the game has essentially ended in favor of one of the two players; this is achieved using the two moves $\langle I've\ won \rangle$ and $\langle I've\ lost \rangle$. The player who has won the play keeps on playing the move $\langle I've\ won \rangle$, while the other player the move $\langle I've\ lost \rangle$. This way every play is infinite. A play that does not contain $\langle I've\ won \rangle$ and $\langle I've\ lost \rangle$ moves will be called a *genuinely infinite play*.

The set of rewards is $D = \{0, 1\}$. In other words, a play of the game can be assigned the value 0 (this means that Player I has won the play) or the value 1 (Player II has won the play). Finally, the payoff function is defined as follows:

$$\Phi(s) = \begin{cases} 1, & \text{if Player II plays the } \langle I've\ won \rangle \text{ move in } s \\ 0, & \text{otherwise} \end{cases}$$

According to the above definition, Player II wins if he plays the move $\langle I've\ won \rangle$; on the other hand Player I wins if he plays the move $\langle I've\ won \rangle$ or s is a genuinely infinite play. Notice that Player I has an important advantage: he wins if he manages to make the play last for ever (with none of the players winning in a finite number of moves). This completes the formal presentation of the game.

We can now illustrate the game with a simple example:

Example 2. Consider the program of Example 1 and in particular the game $\Gamma_P(\mathtt{next}\ \mathtt{p}, 1)$. A possible play of the game proceeds as follows:

Player I	Player II
$\langle \mathtt{next}\ \mathtt{p}, 1 \rangle^-$	$\langle \mathtt{p}, \{2\} \rangle^+$
$\langle \mathtt{p}, 2 \rangle^-$	$\langle \mathtt{next\ next\ p} \leftarrow \mathtt{p}, 0 \rangle^+$
$\langle \mathtt{p}, 0 \rangle^-$	$\langle \mathtt{first\ p} \leftarrow, 0 \rangle^+$
$\langle I've\ lost \rangle$	$\langle I've\ won \rangle$
$\cdots$	$\cdots$

It is easy to see that Player II actually follows a winning strategy in the above play and therefore the value of the above game is 1.

Consider on the other hand the game $\Gamma_P(\texttt{next p}, 0)$:

Player I	Player II
$\langle \texttt{next p}, 0\rangle^-$	$\langle \texttt{p}, \{1,2,3\}\rangle^+$
$\langle \texttt{p}, 1\rangle^-$	$\langle \textit{I've lost}\rangle$
$\langle \textit{I've won}\rangle$	$\langle \textit{I've lost}\rangle$
$\cdots$	$\cdots$

It can be easily verified that no-matter how Player II plays, Player I can win the game $\Gamma_P(\texttt{next p}, 0)$. Therefore, the value of this game is 0.   □

It can be shown (see Theorem 3) that the game we have defined in this section is determined. Given a program P, we can use the game in order to obtain an intensional interpretation N_P of P as follows:

Definition 9. *Let P be an intensional logic program. We define the game interpretation N_P of P such that for every propositional atom A that appears in P and for every $w \in W$, $w \in N_P(A)$ if and only if the value of the game $\Gamma_P(A, w)$ is equal to 1.*

Recall now that by Theorem 1, every intensional logic program (that satisfies the requirements of the theorem) has a unique minimum model M_P. The following theorem states that actually M_P is identical to the game interpretation N_P:

Theorem 2. *Let P be an intensional logic program and assume that the denotations of all intensional operators in the heads of the clauses are universal, monotonic and conjunctive, and the denotations of all intensional operators that appear in the bodies of the clauses are monotonic. Then, the minimum intensional model M_P and the game interpretation N_P of P coincide.*

The above theorem is a consequence of Theorem 4 of the next section. Notice that by Theorem 4, N_P is a *minimal model* of P; however, by Theorem 1, in the special case of the programs considered in the above theorem, there exists a minimum model. Obviously in this case the minimum model coincides with N_P.

4   The Extended Game

The semantics for intensional logic programs developed in [OW92] as well as the game we presented in the previous section, are restricted to programs that use *monotonic* intensional operators. Non-monotonicity is a central issue in both artificial intelligence and logic programming, and it goes without saying that its study is worthwhile. In logic programming, non-monotonicity is due to the presence of the *negation-as-failure* operator. However, in intensional logic programming there exists a much broader notion of non-monotonicity. In particular, every operator $\nabla : \{0,1\}^W \to \{0,1\}^W$ that does not satisfy the monotonicity property of Definition 4 can be characterized as *non-monotonic*. An obvious such case is the negation operator $\neg$ (mentioned in [OW92]) which for every S returns the complement of S with respect to W (ie., $||\neg||(S) = W - S$). However, there are numerous other interesting cases. For example, when the set of possible worlds W is finite, then we could define the *minority* operator, denoted by $\ominus$:

$$||\ominus||(S) = \begin{cases} W, \text{ if } |S| < \frac{|W|}{2} \\ \emptyset, \text{ otherwise} \end{cases}$$

In the rest of this section we assume that the denotations of the operators that appear in the bodies of clauses are *arbitrary* functions of the form $\nabla : \{0,1\}^W \to \{0,1\}^W$. The key question that arises by allowing the use of arbitrary operators in programs is how can the meaning of such programs be expressed. Our experience from non-monotonic logic programming is that two-valued classical logic is not sufficient in order to properly assign a correct meaning to programs with negation. We adopt exactly the same approach in our case, namely we interpret our programs in a three-valued setting.

Our new truth domain consists of the three truth values $\{0, \frac{1}{2}, 1\}$. As a first step we have to generalize the meaning of the two-valued intensional operators to the three-valued domain. In other words, given an operator $||\nabla|| : \{0,1\}^W \to \{0,1\}^W$, we extend it to an operator $||\nabla||^* : \{0, \frac{1}{2}, 1\}^W \to \{0, \frac{1}{2}, 1\}^W$. Our extension has the property that when $||\nabla||^*$ is applied on a two-valued set, it will give the same output as $||\nabla||$ (i.e., for every $S \in \{0,1\}^W$, $||\nabla||^*(S) = ||\nabla||(S)$). The two following definitions capture the desired extension:

Definition 10. *Let $T \in \{0, \frac{1}{2}, 1\}^W$ and $R \in \{0,1\}^W$. We will say that R is a two-valued extension of T if $R \in \{0,1\}^W$ and for every $w \in W$ either $T(w) = R(w)$ or $T(w) = \frac{1}{2}$.*

Definition 11. *Let $||\nabla|| : \{0,1\}^W \to \{0,1\}^W$ be an intensional operator. We define the operator $||\nabla||^* : \{0, \frac{1}{2}, 1\}^W \to \{0, \frac{1}{2}, 1\}^W$ as follows[2]:*

$$||\nabla||^*(T)(w) = \begin{cases} d, \text{ if } ||\nabla||(R)(w) = d \text{ for all two-valued extensions } R \text{ of } T \\ \frac{1}{2}, \text{ otherwise} \end{cases}$$

The operator $||\nabla||^*$ will be called the three-valued extension *of $||\nabla||$.*

Intensional interpretations can be extended to the three-valued setting:

Definition 12. *A three-valued intensional interpretation I of an intensional logic program P is a function from the set of propositional atoms of P to $\{0, \frac{1}{2}, 1\}^W$.*

We will simply use the term "intensional interpretation" when it is obvious from context whether we are referring to a two-valued or a three-valued one.

Definition 13. *Let I be a three-valued intensional interpretation of a given program P and let $w \in W$. Then, I can be extended as follows:*

- $I(\nabla A)(w) = ||\nabla||^*(I(A))(w)$
- $I(B_1, \ldots, B_n)(w) = min\{I(B_1)(w), \ldots, I(B_n)(w)\}$.

---

[2] Since $||\nabla||^*$ is a function in $\{0, \frac{1}{2}, 1\}^W \to \{0, \frac{1}{2}, 1\}^W$, $||\nabla||^*(T)$ returns an element of $\{0, \frac{1}{2}, 1\}^W$ and therefore $||\nabla||^*(T)(w)$ is simply an element of $\{0, \frac{1}{2}, 1\}$.

We will say that I satisfies an intensional clause $B_0 \leftarrow B_1, \ldots, B_n$ of P if for all $w \in W$, $I(B_0)(w) \geq I(B_1, \ldots, B_n)(w)$ (where $\geq$ is the usual numerical ordering). Moreover, I is a model of P if I satisfies all intensional clauses of P.

Definition 14. Let I, J be three-valued intensional interpretations of a given program P. Then, $I \preceq J$ if for every propositional atom A in P and every $w \in W$, it holds $I(A)(w) \leq J(A)(w)$.

Notice that in the following game, the two players only use the two-valued denotations of intensional operators. However, since the outcome of their game may be a tie, the game interpretation must be three-valued (and therefore the proofs of the theorems involve three-valued denotations of intensional operators).

   The infinite tree T_ω of the game $\Gamma_P(C, w)$ consists of all the infinite sequences $\langle x_0, x_1, \ldots, x_k, \ldots \rangle$ which satisfy the following restrictions (in which A denotes a propositional atom, B_i denotes an intensional atom, u, v, z, y denote elements of W and S, S' denote subsets of W) for each $k \geq 0$:

R_1: $x_0 = \langle C, w \rangle^-$.

R_2: If $x_k = \langle \nabla A, u \rangle^-$, then $x_{k+1} = \langle A, S \rangle^+$, where $u \in ||\nabla||(S)$.

R'_3: If $x_k = \langle A, S \rangle^+$ and $x_{k-1} = \langle \nabla A, u \rangle^-$, then either **(i)** $x_{k+1} = \langle A, v \rangle^-$, where $v \in S$, or **(ii)** $x_{k+1} = \langle A, S, S' \rangle^+$, where $S' \supset S$ and $u \notin ||\nabla||(S')$. A move of type **(ii)** will be called a *role switch*.

R''_3: If $x_k = \langle A, S, S' \rangle^+$, then $x_{k+1} = \langle A, y \rangle^-$, where $y \in (S' - S)$.

R_4: If $x_k = \langle A, v \rangle^-$, then $x_{k+1} = \langle C, z \rangle^+$, where C is a clause in P of the form $B_0 \leftarrow B_1, \ldots, B_n$, such that either **(i)** $B_0 = A$ and $z = v$, or **(ii)** $B_0 = \nabla A$ and for every S satisfying $z \in ||\nabla||(S)$ it holds $v \in S$.

R_5: If $x_k = \langle B_0 \leftarrow B_1, \ldots, B_n, z \rangle^+$, then $x_{k+1} = \langle B_j, z \rangle^-$, for some j with $1 \leq j \leq n$.

R_6: If after the move x_k none of the above rules applies, then $x_{k+1} = \langle I've\ lost \rangle$.

R_7: If $x_k = \langle I've\ lost \rangle$, then $x_{k+1} = \langle I've\ won \rangle$ (and vice-versa).

The explanations of the moves are similar to those of the game of Section 3. The only differences are in rules **R'_3** and **R''_3**. When the game starts, Player I is the Doubter and Player II is the Believer. However, as the game proceeds, the two players may swap roles (ie., the current Believer may become a Doubter and vice-versa). Suppose now that at some point of the game, the current Believer replies to a move of the form $\langle \nabla A, u \rangle^-$ of the Doubter by playing a move $\langle A, S \rangle^+$, where $u \in ||\nabla||(S)$. The Doubter can now respond in two different ways. His first option is to play $\langle A, v \rangle^-$ where $v \in S$. The intuition is: "*I doubt that A is true in the world v of S (and therefore I continue to believe that ∇A is not true in u)*". Alternatively, the Doubter can play $\langle A, S, S' \rangle^+$ where $S' \supset S$ and $u \notin ||\nabla||(S')$. The intuition here is "*I believe that the set of worlds where A is true is $S' \supset S$ and not S (as the Believer just claimed); since $u \notin ||\nabla||(S')$, I was right in my belief that ∇A is not true in u*". This second type of move has made the player that was a Doubter to become a Believer (he believes that the set of worlds where A is true coincides with S') and his opponent to become a Doubter, and so this move is called a *role-switch*. Finally, in move $\mathbf{R}_3''$, the new Doubter plays $\langle A, y \rangle^-$ where $y \in (S' - S)$. The intuition is "*I doubt that the set of worlds where A is true coincides with S'; more specifically, I doubt that A is true in the world y*". The rest of the moves are the same as in the game for monotonic programs.

The set of rewards is $D = \{0, \frac{1}{2}, 1\}$ i.e., a play of the game can be assigned the value 0 (Player I has won the play), value 1 (Player II has won), or the value $\frac{1}{2}$ (the result is a tie). Finally, the payoff function is defined as follows:

$$\Phi(s) = \begin{cases} 1, & \text{if Player II plays the } \langle I've\ won \rangle \text{ move in } s \text{ or } s \text{ is a genuinely} \\ & \text{infinite play that contains an odd number of role-switches} \\ 0, & \text{if Player I plays the } \langle I've\ won \rangle \text{ move in } s \text{ or } s \text{ is a genuinely} \\ & \text{infinite play that contains an even number of role-switches} \\ \frac{1}{2}, & \text{if } s \text{ contains an infinite number of role-switches} \end{cases}$$

According to the above definition, a player wins a play of the game if he manages either to play the $\langle I've\ won \rangle$ move or to remain the doubter after a certain point of the play; otherwise the result of the play is a tie. This completes the formal presentation of the extended game.

The extended game just presented, generalizes the game of Section 3. This is due to the fact that if the operators that appear in a program are all monotonic, then only type (i) of the move that appears in rule $\mathbf{R}_3$ above is applicable: in $\mathbf{R}_3'$ it holds $w \in ||\nabla||(S)$ and since $||\nabla||$ is monotonic, for every $S' \supseteq S$ it will also be the case that $w \in ||\nabla||(S')$. Therefore, if the extended game is applied on a program that contains only monotonic operators, then there are no role-switches and the extended game degenerates to the simpler game of Section 3. We can now illustrate the new game with a simple example.

Example 3. Consider the following intensional program:

```
p ← ⊖p
q ← s,r
▽s ←
△r ←
```

Assume that $W = \{1, 2, 3\}$. Recall that $\ominus$ denotes the minority operator; moreover, the denotations of the operators $\triangle$ and $\triangledown$ are the following:

$$||\triangle||(S) = \begin{cases} W, \text{ if } \{1, 2\} \subseteq S \\ \emptyset, \text{ otherwise} \end{cases} \qquad ||\triangledown||(S) = \begin{cases} W, \text{ if } \{2, 3\} \subseteq S \\ \emptyset, \text{ otherwise} \end{cases}$$

Consider the game $\Gamma_P(\ominus q, 1)$. Intuitively, we expect this game to have value 1: since s is true in the worlds 2 and 3 and r is true in the worlds 1 and 2, then s,r will definitely be true in the world 2. This implies that q is true in the world 2 and since we have no other definite information regarding the truth of q in other worlds, we can deduce that $\ominus q$ is true (in any world). Player II has a winning strategy, which is demonstrated in the following two sample plays:

Player I	Player II
$\langle\ominus q,1\rangle^-$	$\langle q,\{2\}\rangle^+$
$\langle q,2\rangle^-$	$\langle q\leftarrow s,r,2\rangle^+$
$\langle s,2\rangle^-$	$\langle\triangledown s\leftarrow,3\rangle^+$
$\langle\textit{I've lost}\rangle$	$\langle\textit{I've won}\rangle$
$\ldots$	$\ldots$

Player I	Player II
$\langle\ominus q,1\rangle^-$	$\langle q,\{2\}\rangle^+$
$\langle q,\{2\},\{1,2\}\rangle^+$	$\langle q,1\rangle^-$
$\langle q\leftarrow s,r,1\rangle^+$	$\langle s,1\rangle^-$
$\langle\textit{I've lost}\rangle$	$\langle\textit{I've won}\rangle$
$\ldots$	$\ldots$

Consider now the game $\Gamma_P(\ominus p,1)$. Intuitively, we expect the value of this game to be $\frac{1}{2}$ due to the circularity of the first rule, which states that if p is true in a minority of the worlds, then p must be true in every world. This is reminiscent of the circularity of the rule $p \leftarrow$ not p in classical logic programming (which also leads to the value $\frac{1}{2}$). A play of this game is the following:

Player I	Player II
$\langle\ominus p,1\rangle^-$	$\langle p,\emptyset\rangle^+$
$\langle p,\emptyset,\{1,2,3\}\rangle^+$	$\langle p,2\rangle^-$
$\langle p\leftarrow\ominus p,2\rangle^+$	$\langle\ominus p,2\rangle^-$
$\langle p,\emptyset\rangle^+$	$\langle p,\emptyset,\{1,2,3\}\rangle^+$
$\langle p,1\rangle^-$	$\langle p\leftarrow\ominus p,1\rangle^+$
$\langle\ominus p,1\rangle^-$	$\ldots$

In this play, the two players change roles infinitely many times. None of the players has a winning strategy; however both have strategies that may lead the play to a tie. In each of these strategies, the reaction to a move $\langle\ominus p,u\rangle^-$ is $\langle p,\emptyset\rangle^+$, while the reaction to any other move may be any legal choice. $\square$

Regarding the determinacy of the proposed game, we have the following theorem:

Theorem 3. *Let P be a program, C be a propositional or intensional atom and $w \in W$. Then, the game $\Gamma_P(C,w)$ is determined.*

Proof. Following the same basic principles as the proof of the determinacy of the negation game introduced in [GRW08]. $\square$

Given a program P, we can use the game described above in order to obtain an intensional interpretation N_P of P in the same way as we did in Definition 9. Then, the following theorem can be established:

Theorem 4. *Let P be a program and assume that the two-valued denotations of all intensional operators in the heads of the clauses of P are universal, monotonic and conjunctive while the two-valued denotations of intensional operators in the bodies of clauses are arbitrary functions in $\{0,1\}^W \to \{0,1\}^W$. Then, the game interpretation N_P of P is a minimal model of P with respect to $\preceq$.*

Notice that the fact that arbitrary operators are allowed in rule bodies, makes the program to possibly posses many incomparable minimal models and the game identifies a special one of them as the intended meaning of the program. This is the main difference from the monotonic case (where the game identifies the unique minimum model). This state of affairs is similar to the situation in classical logic programming with negation (where the well-founded model is a specially chosen one among the minimal models of the program).

5 Conclusions

We have introduced the first (to our knowledge) general semantic framework for non-monotonic intensional logic programming. The proposed game can be used as a yardstick in order to develop alternative semantical approaches for non-monotonic temporal and modal languages. It would be interesting to broaden our understanding regarding the interplay between logic programming and game-theory, since many recent results ([CLN98, GRW08, Tsou11, DeVos01]) and the present work suggest that this is a fruitful avenue of research. For example, it would be desirable to devise a game semantics for answer set programming.

References

[CLN98]   Di Cosmo, R., Loddo, J.-V., Nicolet, S.: A Game Semantics Foundation for Logic Programming. In: Palamidessi, C., Meinke, K., Glaser, H. (eds.) ALP 1998 and PLILP 1998. LNCS, vol. 1490, pp. 355–373. Springer, Heidelberg (1998)

[Far86]   Fariñas del Cerro, L.: MOLOG: A System that Extends PROLOG with Modal Logic. New Generation Computing 4, 35–50 (1986)

[GRW08]   Galanaki, C., Rondogiannis, P., Wadge, W.W.: An Infinite-Game Semantics for Well-Founded Negation in Logic Programming. Annals of Pure and Applied Logic 151(2-3), 70–88 (2008)

[GS53]   Gale, D., Stewart, F.M.: Infinite Games with Perfect Information. Annals of Mathematical Studies 28, 245–266 (1953)

[Mar75]   Martin, D.A.: Borel Determinacy. Annals of Math. 102, 363–371 (1975)

[Mon74]   Montague, R.: English as a Formal Language. In: Thomason, R.H. (ed.) Formal Philosophy: Selected Papers of Richard Montague, pp. 108–221. Yale University Press (1974)

[Mos80]   Moschovakis, Y.N.: Descriptive Set Theory. North-Holland (1980)

[Mos84]   Moszkowski, B.C.: Executing Temporal Logic Programs. In: Brookes, S.D., Winskel, G., Roscoe, A.W. (eds.) Seminar on Concurrency. LNCS, vol. 197, pp. 111–130. Springer, Heidelberg (1985)

[OW92]   Orgun, M.A.P., Wadge, W.W.: Towards a Unified Theory of Intensional Logic Programming. Journal of Logic Programming 13(4), 413–440 (1992)

[RW05]   Rondogiannis, P., Wadge, W.W.: Minimum Model Semantics for Logic Programs with Negation-as-Failure. ACM Transactions on Computational Logic 6(2), 441–467 (2005)

[Sco70]   Scott, D.: Advice on Modal Logic. In: Lambert, K. (ed.) Philosophical Problems in Logic, pp. 143–173. D. Reidel Publishing Company (1970)

[Tsou11]   Tsouanas, T.: A Game Semantics Approach to Disjunctive Logic Programs. In: GALOP Workshop, Saarbrücken, Germany (2011)

[vE86]   van Emden, M.H.: Quantitative Deduction and its Fixpoint Theory. Journal of Logic Programming 3(1), 37–53 (1986)

[DeVos01]   De Vos, M.: Logic Programming, Decisions and Games. PhD thesis, Vrije Universiteit Brussel (2001)

Matchmaking with Answer Set Programming

Martin Gebser[1], Thomas Glase[1,2], Orkunt Sabuncu[1], and Torsten Schaub[1]

[1] Universität Potsdam
[2] piranha womex AG, Berlin

Abstract. Matchmaking is a form of scheduling that aims at bringing companies or people together that share common interests, services, or products in order to facilitate future business partnerships. We begin by furnishing a formal characterization of the corresponding multi-criteria optimization problem. We then address this problem by Answer Set Programming in order to solve real-world matchmaking instances, which were previously dealt with by special-purpose algorithms.

1 Introduction

Matchmaking is a form of scheduling that aims at bringing companies or people together that share common interests, services, or products in order to facilitate future business partnerships. The matching process usually starts prior to the actual event and is based on a simple search and offer principle. It involves a registration phase in which matchmaking participants declare what they are looking for and what they have to offer. Based on this information, a human matchmaker, equipped with experience in the community and the business, identifies promising matches and makes meeting proposals to the participants, who can then either accept or decline proposed matches.

In this report from the field, we show how we solved the matchmaking problem for several fairs by means of Answer Set Programming (ASP; [1]). The resulting system is used by the company *piranha womex AG*[1] for computing matchmaking schedules for the world music exposition (WOMEX) and other fairs.

We begin by developing a formal characterization of the matchmaking problem in Section 2. We then address this problem by ASP in Section 3. Section 4 provides an empirical analysis showing that our ASP-based approach allows for solving real-world matchmaking instances that were previously dealt with by special-purpose algorithms.

2 Matchmaking Scheduling

Matchmaking events bring companies or people sharing common interests, services, or products together in order to facilitate future business partnerships. Such events usually take place at occasions like technology, business, or music and entertainment fairs. A match is a face-to-face meeting of two parties. While a party may be a representative of a company, it may also be an individual like an artist or a producer. Given that an individual can be regarded as a company with one representative, it is viable to view matches as meetings between company representatives.

---

[1] http://www.piranha.de

P. Cabalar and T.C. Son (Eds.): LPNMR 2013, LNAI 8148, pp. 342–347, 2013.

A matchmaking event starts with an initial phase in which interests and offers of participants are analyzed for potential business partnerships. This phase results in a set of matches recommended to participants, and all accepted recommendations constitute the final set of matches to be scheduled. Scheduling each match for a time slot and a location (e.g. a table) yields the matchmaking schedule of an event. In the following, we formalize these intuitions by defining the matchmaking scheduling problem.

Definition 1. *Let T be a linearly ordered set representing time slots. Locations and companies are represented by sets L and C, respectively. A match is a set consisting of two companies from C. Let M be the set of all matches to be scheduled and P be the set of all persons. Each person $p \in P$ works for a company $w(p) \in C$ and has a time preference $\emptyset \subset t(p) \subseteq T$.*

Matchmaking scheduling is the problem of finding a feasible and optimal schedule $S = \langle S_C, S_P \rangle$, where S_C (resp., S_P) is a relation from M (resp., P) to $T \times L$. A schedule S is feasible if S_C is a total function, i.e. each match is scheduled once, each location hosts at most one match at a time, i.e.

$$\forall t \forall l \forall m_1 \forall m_2 : S_C(m_1, (t, l)) \wedge S_C(m_2, (t, l)) \Rightarrow m_1 = m_2 , \tag{1}$$

and for each company in a match, exactly one employee of the company is scheduled:

$$\forall t \forall l \forall m \forall c \exists! p : S_C(m, (t, l)) \wedge c \in m \Rightarrow S_P(p, (t, l)) \wedge w(p) = c , \tag{2}$$

$$\forall t \forall l \forall p \exists m : S_P(p, (t, l)) \Rightarrow S_C(m, (t, l)) \wedge w(p) \in m . \tag{3}$$

A feasible schedule S is optimal if it is not dominated w.r.t. the following ranked objectives, ordered by their precedence:

- *The number of overlaps, where a person has more than one match at a time slot, should be minimized:*

$$\min \sum_{p \in P, t \in T, \sigma(p,t)>0} \sigma(p, t) - 1 , \text{ where } \sigma(p, t) = |\{l \in L \mid (p, (t, l)) \in S_P\}| . \tag{O}$$

- *The number of matches at unpreferred time slots of persons should be minimized:*

$$\min \sum_{p \in P, t \in T \setminus t(p)} |\{l \in L \mid (p, (t, l)) \in S_P\}| . \tag{P}$$

- *The number of idle time slots between matches of companies should be minimized:*

$$\min \sum_{c \in C} \left| \left\{ t \in T \left| \begin{array}{l} (m_1, (t_1, l_1)) \in S_C, (m_2, (t_2, l_2)) \in S_C, t_1 < t < t_2, \\ c \in m_1, c \in m_2, \{c \in m \mid (m, (t, l)) \in S_C\} = \emptyset \end{array} \right. \right\} \right| . \tag{G}$$

- *The number of location changes between consecutive matches of companies should be minimized:*

$$\min \sum_{c \in C} \left| \left\{ t \in T \left| \begin{array}{l} (m_1, (t, l_1)) \in S_C, (m_2, (s(t), l_2)) \in S_C, l_1 \neq l_2, \\ c \in m_1, c \in m_2, s(t) \text{ is the successor time slot of } t \end{array} \right. \right\} \right| . \tag{T}$$

– *Used resources (time slots and locations) should be minimized:*

$$\min \left(|\{t \in T \mid (m, (t, l)) \in S_{\mathcal{C}}\}| + |\{l \in L \mid (m, (t, l)) \in S_{\mathcal{C}}\}| \right). \qquad \text{(U)}$$

Although a person cannot attend several matches at the same time, we do not impose objective (O) as a hard constraint. The rationale of scheduling all matches and tolerating conflicts is to maximize the chances of future business partnerships. In reality, conflicts may be resolved a posteriori, for instance, by sending more personnel to an event.

Note that the above definition of particular objectives and their precedence reflects practical needs of matchmaking events organized by *piranha womex AG*, and variants as well as extensions (some of which are discussed in [2]) may be useful in other contexts.

3   Matchmaking Scheduling in ASP

In this section, we present our ASP encoding of matchmaking scheduling. A matchmaking instance is given by facts as follows: time(t) and location(l) for each time slot $t \in T$ and location $l \in L$, respectively, works_for(p,c) for each $w(p) = c$, time_pref(p,t) for each $t \in t(p)$, and match(c1,c2) for each $\{c1, c2\} \in M$.

Listing 1 shows our first-order encoding. Following the guess and check methodology of ASP, we first generate a schedule (predicate mm provides $S_{\mathcal{C}}$) that assigns at most one match per time slot and location pair in Line 1. The upper bound of the choice rule neatly encodes the feasibility constraint (1).[2] Given that $S_{\mathcal{C}}$ must be a total function, Line 3–4 force mm to be left-total (all matches have to be scheduled), and Line 6–9 require it to be functional (a match must not be scheduled at multiple time slots or locations). Line 11–13 encode the feasibility constraints (2) and (3) by associating exactly one employee per company involved in a scheduled match with the time slot and location pair of the match (predicate mmperson provides $S_{\mathcal{P}}$).

Line 15–16 implement the overlap objective (O), where an atom overlap(p,t,n) expresses that $n > 0$ for $n = \sigma(p, t) - 1$. The **#minimize** statement in Line 16 further asserts that the sum over all n in overlap(p,t,n) atoms of an answer set ought to be minimal. The time preference objective (P) is encoded in Line 24–25, where nonpref(p,t,l) indicates that person p has a match at an unpreferred time slot $t \notin t(p)$, and the corresponding **#minimize** statement in Line 25 aims at a minimal number of nonpref(p,t,l) atoms in an answer set. Similarly, the objectives (G), (T), and (U) are encoded by the program parts in Line 27–30, 34–35, and 37–38, respectively. Note that we rely on consecutive integers for representing the linearly ordered set T of time slots. Thus, T+1 in Line 34 refers to the successor time slot $s(T)$ of T mentioned in objective (T). Likewise, *gringo*'s built-in comparisons are used in Line 28–29 to check that a company is idle in-between two time slots T1 and T2. Regarding multicriteria optimization, levels provided in the encoding (level "@5" in Line 16 for the most important criterion (O) and then gradually decreasing) represent the precedence of objectives. Moreover, Line 40 confines the visible output to instances of predicates mm and mmperson, which together comprise a schedule S.

---

[2] Although one may likewise generate a time slot and location pair per match, experiments with such an alternative encoding led to performance degradations.

Listing 1. Matchmaking scheduling encoding

```
1    { mm(C1,C2,T,L) : match(C1,C2) } 1 :- time(T), location(L).      % schedule matches

3    scheduled(C1,C2)  :- mm(C1,C2,_,_).                              % all matches must
4    :- match(C1,C2), not scheduled(C1,C2).                           % be scheduled

6    match_at_loc(C1,C2,L)   :- mm(C1,C2,_,L).                        % a match must be
7    match_at_time(C1,C2,T) :- mm(C1,C2,T,_).                         % scheduled once
8    :- match(C1,C2), not { match_at_loc(C1,C2,_) } 1.
9    :- match(C1,C2), not { match_at_time(C1,C2,_) } 1.

11   comp_at_pair(C,T,L) :- mm(C,_,T,L).                              % schedule persons
12   comp_at_pair(C,T,L) :- mm(_,C,T,L).
13   1 { mmperson(P,T,L) : works_for(P,C) } 1 :- comp_at_pair(C,T,L).

15   overlap(P,T,N-1) :- works_for(P,_), time(T), N = { mmperson(P,T,_) }, 1 < N.
16   #minimize[ overlap(P,T,N) = N@5 ].                               % overlap (O)

18   hasoverlap(C) :- overlap(P,_,_), works_for(P,C).
19   matchc(C,MC)   :- MC = { match(C,_), match(_,C) }, company(C).  % match count
20   peoplec(C,PC) :- PC = { works_for(_,C) }, company(C).           % people count
21   timec(TC)       :- TC = { time(_) }.                            % time count
22   :- matchc(C,MC), peoplec(C,PC), timec(TC), PC*TC < MC, not hasoverlap(C).

24   nonpref(P,T,L) :- mmperson(P,T,L), not time_pref(P,T).
25   #minimize{ nonpref(P,T,L)@4 }.                                  % time pref. (P)

27   comp_at_time(C,T) :- comp_at_pair(C,T,_).
28   gap(C,T1,T2-T1-1) :- comp_at_time(C,T1), comp_at_time(C,T2), T1+1 < T2,
29                        not comp_at_time(C,T) : time(T) : T1 < T : T < T2.
30   #minimize[ gap(C,T,N) = N@3 ].                                  % gap (G)

32   :- timec(TC), gap(C,_,N), not { comp_at_time(C,_) } TC-N.

34   tablechange(C,T) :- comp_at_pair(C,T,L1), comp_at_pair(C,T+1,L2), L1 != L2.
35   #minimize{ tablechange(C,T)@2 }.                                % table change (T)

37   usedtime(T) :- comp_at_time(_,T).  usedloc(L) :- match_at_loc(_,_,L).
38   #minimize{ usedtime(T)@1, usedloc(L)@1 }.                       % used res. (U)

40   #hide.  #show mm/4.  #show mmperson/3.
```

Additionally, in Line 18–22 and 32, we make some "redundant" domain knowledge explicit, which is not necessary but may improve solving performance. The integrity constraint in Line 22 states that it is impossible to schedule employees of an overbooking company, which participates in more matches than the number of available time slots multiplied by the man power of the company, without overlap. Furthermore, Line 32 expresses that the length of a gap (a sequence of idle time slots) together with a company's scheduled time slots cannot exceed the total number of available time slots.

4   Experiments

In our experiments, we ran *gringo* (3.0.5) for grounding and *clasp* (2.1.1) for solving. All experiments were performed on a 2.5GHz Intel Core Duo machine with 4GB memory under MacOS X (10.7.5), imposing 3600 seconds as time limit. We configured *clasp* to use the VSIDS decision heuristic (--heuristic=Vsids), which was the best configuration we found so far. We also tried *clasp* options for dedicated multi-criteria optimization [3], but we did not achieve performance improvements with them. Moreover, note that we are unaware of any other freely available matchmaking scheduling system that would be directly comparable with our ASP-based approach.

Table 1. Benchmark results with gradually increasing hierarchy of objectives

Instance	#m	#t	#l	#c	#p	clasp													
						O	P	O	P	G	O	P	G	T	O	P	G	T	U
2on2	2	2	2	3	3	*0	0	*0	0	0	*0	0	0	0	*0	0	0	0	3
3on2	3	2	2	3	3	*1	0	*1	0	0	*1	0	0	1	*1	0	0	1	4
39on14	39	14	20	15	15	*0	0	*0	0	0	0	0	22	23	0	0	16	36	31
180on4	180	4	80	100	100	36	0	88	0	69	84	0	43	134	90	0	52	119	80
ffm11li	13	10	26	9	9	*0	0	*0	0	0	*0	0	0	7	*0	0	0	7	9
ffm11cr	19	10	26	13	16	*1	0	*1	0	0	1	0	0	9	1	0	0	8	14
ffm11mu	24	11	26	14	16	*0	1	*0	1	0	0	1	0	14	0	1	0	14	11
wmx10	77	8	14	26	51	*0	0	*0	0	0	0	0	0	70	0	0	0	72	22
wmx11	69	8	14	26	26	13	3	15	4	1	15	4	13	83	15	5	12	69	22
wmx11e	59	8	14	26	26	0	6	0	6	1	0	6	2	64	0	6	2	70	22
wmx11m	82	8	14	26	26	22	12	22	11	16	22	14	28	89	22	13	20	87	22
wmx11p	69	8	14	26	52	7	32	0	23	25	8	33	34	63	6	29	19	73	22
wmx12	60	7	14	54	54	*1	0	*1	0	0	1	0	0	26	1	0	0	30	19
wmx12m	89	7	14	54	54	5	2	5	2	29	8	2	29	81	6	2	36	82	21
wmx12p	60	7	14	54	75	*0	0	*0	0	0	0	0	0	23	0	0	0	24	19

Table 1 displays benchmark results. The first two columns list instances and their properties, where #m, #t, #l, #c, and #p are numbers of matches, time slots, locations, companies, and persons, respectively. While the first four instances are crafted, the others are real-world instances from Frankfurter Musikmesse 2011 and WOMEX 2010–2012 as well as extended versions of them with additional matches or persons. The columns O, P, G, T, and U show objective values for (O), (P), (G), (T), and (U), respectively, w.r.t. the best answer set found by *clasp* within the allotted time. We started with a relaxed problem using only (O) and (P) as objectives (results shown in the third column) and then gradually added more objectives in the order of precedence, leading to the full problem with all objectives (results shown in the last column). Whenever *clasp* proved a solution to be optimal, the corresponding entry starts with *. For instance, the entry "*1 0 0" for the ffm11cr instance indicates that a schedule with one overlap but no unpreferred time slots or gaps has been found and proven to be optimal by *clasp*. Although more objectives add to the difficulty of multi-criteria optimization, the gradual solutions for instances show only slight degradations. For example, objective values for (P) on the wmx11m instance vary between 11 and 14, and the values are not monotonically increasing with the number of objectives. According to *piranha womex AG*, schedules computed with our ASP-based approach are satisfactory and on some real-world instances even better than previous hand-made schedules.

5   Discussion

We presented an ASP-based approach to matchmaking scheduling, which is a highly combinatorial multi-criteria optimization problem. Our approach allows for solving real-world matchmaking instances that were previously dealt with by special-purpose algorithms. The presented ASP methods are used by *piranha womex AG* for computing matchmaking schedules for the world music exposition (WOMEX) and other fairs.

Although matchmaking scheduling can be regarded as a form of timetabling, to our knowledge, it has not yet received much attention from the timetabling community. Common search methods for timetabling include local search [4], constraint programming [5], and satisfiability solving [6]. Moreover, an extension of disjunctive logic programming by soft constraints was proposed for modeling school timetabling [7]. The commercial matchmaking scheduling system *b2match* [8] supports only gap minimization among the various objectives we considered.

Acknowledgments. This work was partially funded by DFG grant SCHA 550/9-1.

References

1. Baral, C.: Knowledge Representation, Reasoning and Declarative Problem Solving. Cambridge University Press (2003)
2. Glase, T.: Timetabling with answer set programming. Diploma thesis, Institute for Informatics, University of Potsdam (2012)
3. Gebser, M., Kaminski, R., Kaufmann, B., Schaub, T.: Multi-criteria optimization in answer set programming. In: ICLP 2011, pp. 1–10. Dagstuhl Publishing (2011)
4. Cambazard, H., Hebrard, E., O'Sullivan, B., Papadopoulos, A.: Local search and constraint programming for the post enrolment-based course timetabling problem. Annals of Operations Research 194(1), 111–135 (2012)
5. Baptiste, P., Pape, C., Nuijten, W.: Constraint-Based Scheduling. Springer (2001)
6. Achá, R., Nieuwenhuis, R.: Curriculum-based course timetabling with SAT and MaxSAT. Annals of Operations Research (2012) (published online)
7. Faber, W., Leone, N., Pfeifer, G.: Representing school timetabling in a disjunctive logic programming language. In: WLP 1998, pp. 43–52 (1998)
8. b2match, http://www.b2match.com/info/pages/scheduling/

Ricochet Robots: A Transverse ASP Benchmark

Martin Gebser, Holger Jost, Roland Kaminski, Philipp Obermeier, Orkunt Sabuncu,
Torsten Schaub, and Marius Schneider

Universität Potsdam

Abstract. A distinguishing feature of Answer Set Programming is its versatility.
In addition to satisfiability testing, it offers various forms of model enumeration,
intersection or unioning, as well as optimization. Moreover, there is an increasing
interest in incremental and reactive solving due to their applicability to dynamic
domains. However, so far no comparative studies have been conducted, contrasting the respective modeling capacities and their computational impact. To assess the variety of different forms of ASP solving, we propose Alex Randolph's
board game *Ricochet Robots* as a transverse benchmark problem that allows us
to compare various approaches in a uniform setting. To begin with, we consider
alternative ways of encoding ASP planning problems and discuss the underlying
modeling techniques. In turn, we conduct an empirical analysis contrasting traditional solving, optimization, incremental, and reactive approaches. In addition,
we study the impact of some boosting techniques in the realm of our case study.

1 Introduction

A distinguishing feature of Answer Set Programming (ASP; [1]) is its versatility. Its
modeling language and solving technology support various forms of (Boolean) constraint solving that are otherwise restricted to dedicated paradigms. For example, optimization is not supported by standard Satisfiability solvers, and one has to resort to
Maximum Satisfiability solvers to obtain this functionality (cf. [2]). Unlike this, ASP
solvers offer, in addition to satisfiability testing, various forms of model enumeration,
intersection or unioning, as well as (multi-criteria) optimization. Moreover, there is an
increasing interest in incremental and reactive solving due to their applicability to dynamic domains, such as assisted living and cognitive robotics. However, so far no comparative studies have been conducted, contrasting the respective modeling capacities
and their computational impact.

To assess the variety of different forms of ASP solving, we propose the popular board
game *Ricochet Robots* as a transverse benchmark problem that allows us to compare
various approaches in a uniform setting. *Ricochet Robots* is a board game for multiple
players designed by Alex Randolph.[1] A board consists of 16×16 fields arranged in
a grid structure having barriers between various neighboring fields. Four differently
colored robots roam across the board along either horizontally or vertically accessible
fields, respectively. In principle, each robot can thus move in four directions. A robot
cannot stop its move until it hits either a barrier or another robot. Finally, the goal is to

---

[1] http://en.wikipedia.org/wiki/Ricochet_Robot

P. Cabalar and T.C. Son (Eds.): LPNMR 2013, LNAI 8148, pp. 348–360, 2013.

place a particular robot on a target location with a shortest sequence of moves. Often this involves moving several robots to establish temporary barriers. For illustration, consider the reduced board in Figure 1.[2] The red robot can be moved onto the red icon in four steps: down, right, up, and left. The game box offers 96 distinct boards, each of which has sixteen (plus one special) target locations. The overall game is won by the player who wins the majority of individual rounds. (Note that the skill of human players tends to improve over the rounds because they gather knowledge about the board at hand.) *Ricochet Robots* has been studied from the viewpoint of human problem solving [3] and analyzed from a theoretical perspective [4,5,6]. Moreover, it has a large community providing various resources on the web. Among them, there is a collection of fifty-six extensions of the game.[3]

Alex Randolph's *Ricochet Robots* represent a challenging planning problem involving several actors. As such, it allows us to elaborate upon various aspects of ASP. We begin by addressing the corresponding decision problem of whether there is a plan of length smaller or equal than a given horizon. Starting from a plain encoding following traditional ASP planning (cf. [7]), we elaborate upon an alternative encoding featuring several advanced modeling techniques. We further adapt encodings for applying optimization, incremental, and reactive ASP solving techniques. While optimization allows for computing a shortest plan smaller or equal than a given horizon, the incremental and reactive variants do not impose such an upper bound. We use the devised encodings to conduct a comparative empirical analysis addressing the following questions. How do modeling techniques affect the grounding and solving performance of ASP systems? Second, to what extent can algorithm configuration as well as multi-threaded solving speed up the search for an arbitrary or a shortest plan, respectively? Furthermore, how does (bounded) optimization with standard solving techniques compare to (unbounded) optimization using incremental solving? Finally, does reactive ASP solving benefit from proceeding over a series of rounds, rather than tackling them independently?

Last but not least, we provide the visualization tool *robotviz* that, given a stable model, allows us to inspect the corresponding *Ricochet Robots* board on the screen and to interactively trace a plan described by the stable model.

2   Encoding *Ricochet Robots*

In what follows, we present our fact format and two alternative encodings in the input language of the ASP grounder *gringo* 3 [8,9].

2.1   Fact Format

For illustration, consider the reduced board of 8×8 fields in Figure 1 (corresponding to a quarter of an authentic board). Its representation in terms of facts is given in Listing 1. The board size is fixed via the constant `dimension`; its origin (1,1) is in the upper left corner. Barriers are indicated by atoms with predicate `barrier/4`. The first two arguments

---

[2] The enclosed yellow robot is an artifact, given that we took a quarter of an authentic board.

[3] http://www.boardgamegeek.com/boardgame/51/ricochet-robots

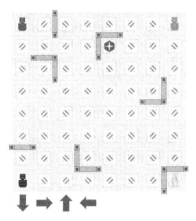

Fig. 1. Visualization of solving through *robotviz*

Listing 1. Example problem instance (board8.lp)

```
#const dimension=8.

barrier(2,1,1,0).          barrier(5,1,0,1).
barrier(2,3,1,0).          barrier(2,2,0,1).
barrier(3,7,1,0).          barrier(7,4,0,1).
barrier(4,2,1,0).          barrier(1,6,0,1).
barrier(7,4,1,0).          barrier(4,7,0,1).
barrier(7,8,1,0).          barrier(8,7,0,1).

position(red,1,1).         position(yellow,dimension,dimension).
position(blue,1,dimension).          position(green,dimension,1).
robot(R) :- position(R,_,_).               target(red,5,2).
```

give the field position and the last two the orientation of the barrier, which is either east
(1,0) or south (0,1). For instance, barrier(2,1,1,0) represents the vertical wall between the fields (2,1) and (3,1), and barrier(5,1,0,1) stands for the horizontal wall
separating (5,1) from (5,2). As specified by atoms with predicate position/3, the four
robots start from board corners. Since each robot has (exactly one) initial position, the
projection robot/1 captures available robots. Finally, target(red,5,2) expresses
that the goal is to move the red robot on the red icon, as displayed in Figure 1.

2.2   Plain Encoding

In Listing 2, we provide an encoding following common practice in ASP planning [7].
That is, sequences of actions are guessed via choice rules (in Line 10), and the respective
successor states are derived via direct effect and frame axioms (in Line 20–22).

In more detail, the first three lines in Listing 2 furnish domain definitions, fixing
the sequence of time steps (time/1),[4] the coordinates of board fields (dim/1), and
two-dimensional representations of the four possible directions (dir/2). The constant
horizon, used to define time/1, is expected to be provided via *gringo*'s command line
option -c (e.g. '-c horizon=10'). Predicate stop/4, which is the symmetric version of

---

[4] The initial time point 0 is handled explicitly.

Listing 2. Plain encoding of ricocheting robots

```
1   time(1..horizon).
2   dim(1..dimension).
3   dir(-1,0;;1,0;;0,-1;;0,1).

5   stop( DX, DY,X,    Y   ) :- barrier(X,Y,DX,DY).
6   stop(-DX,-DY,X+DX,Y+DY) :- stop(DX,DY,X,Y).

8   position(R,X,Y,0) :- position(R,X,Y).

10  1 { move(R,DX,DY,T) : robot(R) : dir(DX,DY) } 1 :- time(T).
11  move(R,T) :- move(R,_,_,T).

13  halt(DX,DY,X-DX,Y-DY,T) :- position(_,X,Y,T), dir(DX,DY), dim(X-DX;Y-DY),
14                             not stop(-DX,-DY,X,Y), T < horizon.

16  goto(R,DX,DY,X,    Y,    T) :- position(R,X,Y,T), dir(DX,DY), T < horizon.
17  goto(R,DX,DY,X+DX,Y+DY,T) :- goto(R,DX,DY,X,Y,T), dim(X+DX;Y+DY),
18                               not stop(DX,DY,X,Y), not halt(DX,DY,X,Y,T).

20  position(R,X,Y,T) :- move(R,DX,DY,T), goto(R,DX,DY,X,Y,T-1),
21                       not goto(R,DX,DY,X+DX,Y+DY,T-1).
22  position(R,X,Y,T) :- position(R,X,Y,T-1), time(T), not move(R,T).

24  :- target(R,X,Y), not position(R,X,Y,horizon).
```

Listing 3. Encoding part for optimization

```
24  goon(T) :- target(R,X,Y), T := 0..horizon, not position(R,X,Y,T).
25  :- goon(horizon).

27  :- move(R,DX,DY,T-1), time(T), not goon(T-1), not move(R,DX,DY,T).

29  #minimize{ goon(_) }.
```

barrier/4 from a problem instance, identifies all blocked field transitions. The initial robot positions are fixed in Line 8.

At each time step, some robot is moved in a direction (cf. Line 10). Such a move can be regarded as the composition of successive field transitions, captured by goto/6 (in Line 16–18). To this end, predicate halt/5 provides temporary barriers due to robots' positions before the move. To be more precise, a robot moving in direction (DX,DY) must halt at field (X-DX,Y-DY) when some (other) robot is located at (X,Y), and an instance of halt(DX,DY,X-DX,Y-DY,T) may provide information relevant to the move at step T+1 if there is no barrier between (X-DX,Y-DY) and (X,Y). Given this, the definition of goto/6 starts at a robot's position (in Line 16) and continues in direction (DX,DY) (in Line 17–18) unless a barrier, a robot, or the board's border is encountered. As this definition tolerates board traversals of length zero, goto/6 yields a successor position for any move of a robot R in direction (DX,DY), so that the rule in Line 20–21 captures the effect of move(R,DX,DY,T). Moreover, the frame axiom in Line 22 preserves the positions of unmoved robots, relying on the projection move/2 (cf. Line 11).

Finally, we stipulate in Line 24 that a robot R must be at its target position (X,Y) at the last time point horizon. Adding directives '**#hide. #show** move/4.' further allows for projecting stable models onto the extension of the move/4 predicate.

The encoding in Listing 2 allows us to decide whether a plan of length horizon exists. For computing a shortest plan, we have two options resting on extended ASP systems. The first alternative is to augment our decision encoding with an optimization directive. This can be accomplished by replacing the integrity constraint in Line 24 in Listing 2 by the encoding part in Listing 3. The new rule in Line 24 indicates whether some goal condition is (not) established at a time point, and compliance with target position(s) at the last time point `horizon` is checked by the integrity constraint in Line 25. Once the goal is established, the additional integrity constraint in Line 27 ensures that it remains satisfied by enforcing that the goal-achieving move is repeated at later steps (without altering robots' positions). Note that the **#minimize** directive in Line 29 aims at few instances of `goon`/1, corresponding to an early establishment of the goal, while further repetitions of the goal-achieving move are ignored. Our extended encoding allows for computing a shortest plan of length bounded by `horizon`. If there is no such plan, the problem can be posed again with an enlarged `horizon`.

For computing a shortest plan in an unbounded fashion, we can take advantage of incremental ASP solving. This allows us to successively explore all bounds from 1 on until a plan is found. An incremental ASP encoding consists of three types of rules: static, cumulative, and volatile ones. Static rules, indicated by **#base**, describe step-independent knowledge. Rules that are accumulated over steps are declared with **#cumulative** (along with a constant standing for the step number), whereas the ones stated as **#volatile** are specific to each time step and discarded upon incrementing the step counter.[5] An incremental ASP encoding is obtained from the one in Listing 2 as follows. First, delete Line 1 and all occurrences of `time(T)`. Second, insert '**#base.**' in Line 1, '**#cumulative t.**' in Line 9, and '**#volatile t.**' in Line 23. Third, replace all occurrences of variable T by the constant t or by t-1 in Line 13–18, respectively. Finally, replace `horizon` in Line 24 by constant t. In doing so, we declare the rules in Line 2–8 to be static, those in Line 10–22 to be cumulative, and Line 24 to be volatile. Given this, an incremental ASP system first grounds the static part, and then it successively grounds (replacing constant t by the current step number) and solves the cumulative and volatile rules, incrementing the step counter until the first plan is found.

The incremental variant of Listing 2 can also be used in a reactive setting [12]. In fact, the only change concerns the way we deal with consecutive target positions. For this purpose, it is sufficient to replace the static target in each problem instance, for example 'target(red,5,2).' in Listing 1, by the following choice rule:

```
1 { target(R,X,Y) : robot(R) : dim(X;Y) } 1.
```

This rule leaves the concrete target position open. Consecutive queries are then posed to a reactive ASP system via a sequence of **#volatile** integrity constraints, like ':- not target(red,5,2).' The volatile nature of each query guarantees that it vanishes after it has been addressed. Note that, for the sake of comparability, we refrain from modifying the initial robot positions. We discuss an experiment that simulates "playing in rounds" in Section 3.

2.3   Advanced Encoding

Next, we present an advanced encoding that differs from the above in two salient ways. Its basic ideas are to guess states rather than actions and to split robots' positions into

---

[5] For a detailed introduction to incremental ASP, the interested reader is referred to [10,11].

Listing 4. Advanced encoding of ricocheting robots

```
1   time(1..horizon).
2   dim(1..dimension).
3   dir(-1,0;;1,0;;0,-1;;0,1).  aux(-1;1).

5   stop( DX,  0,Y,X    ) :- barrier(X,Y,DX,0).
6   stop(  0, DY,X,Y    ) :- barrier(X,Y,0,DY).
7   stop(-DX,-DY,F,L+DX+DY) :- stop(DX,DY,F,L).

9   spot(R, 1,X,0) :- position(R,X,_).
10  spot(R,-1,Y,0) :- position(R,_,Y).

12  same(R,A,RR,T) :- spot(R;RR,A,L,T), R != RR, T < horizon.

14  halt(R,DX,DY,L,       T) :- spot(R,|DY|-|DX|,F,T), stop(DX,DY,F,L), T < horizon.
15  halt(R,DX,DY,L-DX-DY,T) :- same(R,|DY|-|DX|,RR,T), spot(RR,|DX|-|DY|,L,T),
16                             dir(DX,DY), dim(L-DX-DY).

18  goto(R,DX,DY,L,       T) :- spot(R,|DX|-|DY|,L,T), dir(DX,DY), T < horizon.
19  goto(R,DX,DY,L+DX+DY,T) :- goto(R,DX,DY,L,T), dim(L+DX+DY), not halt(R,DX,DY,L,T).

21  goto(R,|DX|-|DY|,L,T) :- goto(R,DX,DY,L,T).
22  halt(R,|DX|-|DY|,L,T) :- halt(R,DX,DY,L,T), dim(L-DX-DY).

24  1 { spot(R,A,L,T) : dim(L) } 1 :- robot(R), aux(A), time(T).
25  :- spot(R,A,L,T), time(T), not goto(R,A,L,T-1).

27  bump(R,A,L,T) :- spot(R,A,L,T), time(T), not spot(R,A,L,T-1).
28  bump(R,A,L,T) :- bump(R,A,L,T-1), time(T), not goon(T-1).
29  bump(R,A,  T) :- bump(R,A,_,T).
30  :- bump(R,A,L,T), dim(L+D) : aux(D), not halt(R,A,L,T-1).
31  :- time(T), not #count{ bump(_,_,_,T) } 1.
32  :- time(T), not bump(R,A,T) : robot(R) : aux(A).
33  :- bump(R,A,T-1;T), goon(T-1).

35  goon(T) :- target(R,X,_), T := 0..horizon, not spot(R, 1,X,T).
36  goon(T) :- target(R,_,Y), T := 0..horizon, not spot(R,-1,Y,T).
37  :- goon(horizon).

39  move(R,DX,DY,T) :- bump(R,|DX|-|DY|,L,T), halt(R,DX,DY,L,T-1) : dim(L+1),
40                     dir(DX,DY), 0 < DX+DY.
41  move(R,DX,DY,T) :- bump(R,|DX|-|DY|,T), not move(R,-DX,-DY,T),
42                     dir(DX,DY), DX+DY < 0.
```

horizontal and vertical coordinates. The first idea is conceptually different from standard encodings in ASP planning, like the one above, and has the advantage that states need not be constructed by effect and frame axioms. The second idea is well-known in automated planning and leads to a significant reduction in the size of ground instantiations. On the other hand, it makes the encoding more complex since robots' positions are not given directly anymore. Further modeling techniques are described on the fly.

Our advanced encoding is given in Listing 4. Apart from two auxiliary atoms, aux(-1) and aux(1), to distinguish horizontal and vertical coordinates, Line 1–3 are as in Listing 2. Moreover, the definition of stop/4 is analogous, except that its format is lined up for one-dimensional movements. In particular, columns X and rows Y are transposed in Line 5 to let stop(DX,DY,F,L) represent that (F,L+DX+DY) is inaccessible from (F,L) (cf. Line 7), where F is a fixed row or column, respectively.

The one-dimensional layout is continued with predicate spot/4 in Line 9, 10, and 24, where atoms spot(R,1,X,T) and spot(R,-1,Y,T) provide the column X and row Y of robot R at time point T. Note that such coordinates are guessed, rather than derived from moves, in Line 24. In order to compensate for the split positions, the rule in Line 12 defines same(R,A,RR,T) (for all but the last time point horizon) to express that distinct robots R and RR are in a common column (A = 1) or row (A = -1).

The one-dimensional counterparts halt/5 and goto/5 of corresponding predicates in Listing 2 capture the effect of moving a robot R in direction (DX,DY) at step T+1, possibly altering its column (|DX|-|DY| = 1) or row (|DX|-|DY| = -1). In fact, barriers and other robots RR sharing the coordinate of R on the orthogonal axis |DY|-|DX| block transitions in direction (DX,DY). This is captured by the definition of halt coordinates for R in direction (DX,DY) in Line 14–16. In turn, starting from its coordinate L on axis |DX|-|DY| (in Line 18), a robot R continues in direction (DX,DY) (in Line 19) unless a halt coordinate or the board's border is encountered.

Given halt/5 and goto/5, the rules in Line 21 and 22 provide the abstractions halt/4 and goto/4, which summarize coordinates affected by horizontal or vertical moves by collapsing directions (DX,DY) to |DX|-|DY|. As a minor optimization, we drop coordinates at the board's border in halt/4 because moves may always halt there.

With predicates providing properties of the predecessor state at hand, we can now constrain successor states guessed in Line 24. First of all, the reachability of successor coordinates along axes is checked by the integrity constraint in Line 25; that is, no robot is allowed to cross barriers or other robots' positions. A new coordinate L for a robot R on axis A at time step T is indicated by deriving bump(R,A,L,T) in Line 27; such atoms point to moves, and the integrity constraint in Line 31 restricts their number to (at most) one per time step. Moreover, the integrity constraint in Line 30 checks that halting at a new coordinate is admissible, which is trivially the case for coordinates at the board's border. The second possibility of deriving bump(R,A,L,T) in Line 28, by which a goal-achieving move at T-1 is (necessarily) repeated at time step T, relies on the absence of goon(T-1) (defined in Line 35–36) for indicating that some goal condition is not yet established. Along with the integrity constraint in Line 31, the rule in Line 28 suppresses any further move after establishing the goal, while still supplying the projection bump/3 (cf. Line 29). The latter is investigated by the integrity constraint in Line 32, stipulating some instance of bump/3 to hold at each time step.[6] Finally, the integrity constraint in Line 33 discards redundant successive moves of a robot R on the same axis A at time steps T-1 and T unless the goal is established at T-1.

As in Listing 3, the integrity constraint in Line 37 checks compliance with target position(s) at the last time point horizon. Moreover, the definition of move/4 in Line 39–42 provides the same format for moves as obtained with the plain encoding in Listing 2, while not carrying relevant information regarding the existence of a plan. Furthermore, note that our advanced encoding can be customized to compute shortest plans, either by adding the **#minimize** directive in Line 29 in Listing 3 or by devising an incremental ASP encoding according to the scheme described in Section 2.2.

2.4   Output Format and Visualization

The *Ricochet Robots* visualization tool *robotviz* allows for displaying the board with barriers, robots, and targets as well as for animating robot moves in a stepwise fashion. *robotviz* is written in C++ and uses *clasp*'s textual output as input. That is, stable models

---

[6] W.l.o.g., we assume that the goal is not readily established at the initial time point 0.

are simply piped into *robotviz*, where they are parsed with a simple string parser. The first stable model is used for interactive visualization. Also, depending on the input, only board and barriers can be displayed or additional robots and a series of moves. For an impression, note that Figure 1 shows a snapshot of *robotviz*. It allows us to visually observe that the yellow robot is trapped in its corner. In addition, the plan of moving the red robot down, right, up, and left is displayed below the board in terms of a sequence of arrows in moved robots' colors, and the steps can further be traced via cursor keys.

The input format is designed for multiple encodings. For this, it is sufficient that certain key atoms belong to the solver's output and are thus declared via appropriate `#show` statements. Barriers are extracted from atoms with predicate `barrier/4`. Analogously, the robots' target and starting positions as well as their moves have to be provided for the interactive step-by-step visualization. If the input lacks `position/3` or `move/4`, only the board is displayed. Finally, the atom `dim(dimension)` must be shown to indicate the board size (otherwise it would not be visible in the solver output).

3   Experimental Case Studies

Our benchmark set is based on an authentic board designed by Alex Randolph of size 16×16. The initial robot positions are in the corners of the board, and the red robot must reach some target position. Given this setting, we obtain a collection of 256 benchmark instances by considering all available fields as target positions. With very few exceptions, the resulting instances are satisfiable when given enough steps, where about twelve steps are required on average. In the following, we first focus on a comparison of encodings as well as solving strategies for decision and optimization tasks. Afterwards, we extend the scope to incremental and reactive ASP systems. All our experiments were run on a Linux machine equipped with two Quad-Core Xeon E5520 2.27GHz processors and 24GB memory, limiting each run to 600 seconds wall-clock time.

3.1   Encodings and Configurations

In an ASP production mode, the most important factor is a scalable encoding. In fact, our advanced encoding in Listing 4 has a clear edge on the plain one in Listing 2 as regards grounding. This can roughly be quantified by a factor of five in terms of time consumption and ground instantiation size. However, space savings due to split positions also incorporate some indirection in referring to the actual fields of robots. Hence, it is interesting to compare search performance relative to encodings in solving decision and optimization tasks. To this end, we fix the constant `horizon` to 20 steps, which (in all but two cases) is sufficient to find plans for satisfiable instances.

In what follows, we investigate the impact of encodings and settings on solver performance. To this end, we consider *clasp* (2.1.3) in its default configuration (including `--heuristic=berkmin`) and the variant with `--heuristic=vsids`, both serving as points of reference. We further contrast these two settings with the following ones:

1. the *clasp* configuration used for the ASP competition in 2013; originally obtained by manual tuning and extensive experimentation; now available via the option `--configuration=handy` in *clasp* (2.1.3),

Table 1. Solving decision and optimization problems with different encodings and *clasp* settings

		Decision problem			Optimization problem		
		Runtime	Timeout	PAR10	Runtime	Timeout	PAR10
plain	*clasp* w berkmin	144	25	671	334	99	2422
	clasp w vsids	136	37	916	299	94	2281
	clasp, manually configured	103	14	398	234	69	1689
	clasp, automatically configured	150	28	741	259	84	2031
	clasp, multi-threaded	62	4	146	173	50	1178
advanced	*clasp* w berkmin	207	63	1536	302	106	2537
	clasp w vsids	140	48	1152	315	114	2720
	clasp, manually configured	65	12	318	192	61	1478
	clasp, automatically configured	44	9	234	136	**35**	**874**
	clasp, multi-threaded	**24**	**3**	**87**	**123**	37	904

2. an automatically generated *clasp* configuration, obtained by means of the algorithm configuration system *smac* (2.02.00; [13]), and
3. a multi-threaded *clasp* configuration using a portfolio of four competitively searching threads (cf. [14]); originally obtained by manual tuning and extensive experimentation; now available via the option --configuration=chatty.

Following common practice in automatic algorithm configuration, we selected the best outcome from ten independent runs of *smac* on a training set of instances relying on a different *Ricochet Robots* board than our benchmarks. Each *smac* run was allotted 100 hours for tuning 94 (discrete and continuous) parameters of *clasp*, using a cutoff of 600 seconds wall-clock time for *clasp*.

Table 1 provides average runtimes in seconds (accounting for timeouts by 600 seconds), absolute numbers of timeouts, and average times in seconds while penalizing timeouts by 6000 seconds (PAR10) over our 256 benchmark instances. We applied the aforementioned *clasp* configurations to solve the decision problem of plan existence as well as the optimization problem of shortest plan computation, relying on the plain encoding in Listing 2 or the advanced encoding in Listing 4, respectively.

Interestingly, the reference configurations with berkmin or vsids, respectively, perform significantly better with the plain than the advanced encoding. Analyzing the search statistics reported by *clasp*, on the one hand, we observed that the five times smaller ground instantiation size with the advanced encoding brings about the same amount of higher raw speed but, on the other hand, leads to roughly one order of magnitude more conflicts upon search. The trade-off between compactness and search efforts shifts towards the advanced encoding for the manually or automatically configured *clasp* settings, although the plain encoding generally still yields fewer conflicts.[7] Also

---

[7] Ground instantiations induce about 800k constraints with the plain encoding and 140k constraints with the advanced encoding. The average number of conflicts reported by *clasp* in its default configuration is 65k with the plain encoding and 510k with the advanced encoding. The latter number reduces to 127k conflicts on average in *smac*'s configuration, while no comparably substantial reductions are achieved with the plain encoding in any configuration.

Table 2. Solving *Ricochet Robots* with different ASP systems

	clasp w berkmin			*clasp* w vsids		
	Runtime	Timeout	PAR10	Runtime	Timeout	PAR10
clasp (decision)	227	71	1725	86	27	655
clasp (optimization)	326	114	2731	224	77	1848
unclasp	574	245	5742	567	242	5671
iclingo	229	83	1980	186	66	1578
oclingo	216	80	1903	179	62	1487

note that automatic configuration via *smac* was accomplished relative to the advanced encoding. Thus, it performs worse than the manually selected configuration with the plain encoding. With the advanced encoding for which it has been tuned, *smac*'s configuration turns out as the best in single-threaded settings. In fact, it even surpasses multi-threaded settings regarding the number of timeouts in optimization, while the parallelism brought by multi-threading exhibits significantly improved robustness otherwise. The success of *smac* (with the advanced encoding) confirms analogous results in ASP [15,16] and related areas [17,18], showing that the burden of solver configuration can and should be taken off the user. Comparing decision and optimization problems, Table 1 further yields consistent relative performances of configurations, suggesting that underlying problem characteristics are quite similar in solving either kind of task.

3.2 ASP Solving Technologies

This section is dedicated to the empirical comparison of different ASP solving technologies. To this end, we concentrate on the advanced encoding given in Section 2.3. As points of reference, we consider *clasp* (1.3.10) for solving decision and optimization problems with a fixed `horizon` of 20 steps.[8] Running this version (rather than 2.1.3) is motivated by its usage in the ASP systems we compare: the *clasp* derivative *unclasp* (0.1; [19]), pursuing an unsatisfiability-based approach to optimization; the incremental ASP system *iclingo* (3.0.5; [10,11]), performing iterative deepening by means of stepwise grounding and solving; and the reactive ASP system *oclingo* (3.0.92; [12]), extending *iclingo* with online capacities to solve sequences of queries. We benchmark all ASP systems in two settings, performing search with *clasp* in its default configuration (including `--heuristic=berkmin`) and the variant with `--heuristic=vsids`.

Table 2 shows experimental results, as before providing average runtimes, absolute numbers of timeouts, and average times penalizing timeouts by 6000 seconds (PAR10). Although *unclasp*'s approach to optimization can be highly effective (cf. [19]), it does not work well for *Ricochet Robots*. In fact, *unclasp* aims at localizing substructures of a problem responsible for penalties within a `#minimize` statement. For the one in Listing 3, this means that atoms with predicate goon/1, indicating that some goal condition is not established at a time point, are gradually admitted to hold. Given that the establishment of goals, as for instance expressed by `target(red,5,2)`, relies on the whole

---

[8] Grounding times of *gringo* are negligible, i.e., less than 0.2 seconds for our 16×16 board.

trajectory from the initial time point 0, reasons for penalties can hardly be subdivided into (independent) local substructures. The inherent causal connection between states in a planning problem thus undermines *unclasp*'s approach to optimization.

The incremental ASP system *iclingo* computes shortest plans in an unbounded fashion by gradually extending the horizon. To be more precise, starting at 1, *iclingo* grounds and solves (the incremental variant of) Listing 4 step by step until the first stable model, corresponding to a shortest plan, is obtained. Accordingly, the performance of *clasp* in optimization constitutes the reference for assessing *iclingo*, and the reduction of timeouts (31 with --heuristic=berkmin and 11 with --heuristic=vsids) shows the success of *iclingo*'s incremental approach. For one, this relies on the fact that stepwise grounding avoids the instantiation of rules for "unnecessary" time points. For another, recorded conflict information may be passed along between successive solving steps given that the solving component of *iclingo* remains in place until a plan is found. That is, grounding as well as solving efforts spent on unsatisfiable (decision) problems with too small horizons still contribute to and potentially foster progress in the sequel.

Going further beyond *iclingo*, the reactive ASP system *oclingo* maintains its solving component for dealing with consecutive target positions. In this way, recorded conflict information can be shared among all benchmark instances, which enables *oclingo* to exploit similarities in solving a series of planning problems. We thus obtain a reduction of timeouts in comparison to *iclingo* (3 with --heuristic=berkmin and 4 with --heuristic=vsids). However, note that *oclingo* does not decrease the planning horizon when a new target is entered since instantiations of rules for time points remain in the system once they have been produced in view of a query. As a consequence, a plan is not guaranteed to be shortest when its target position can be reached without incrementing the step counter. We aim at overcoming this in the future by extending incremental and reactive ASP solving to optimization, via which shortest plans could be addressed without withdrawing any formerly produced ground rules.

An alternative experiment performed with *oclingo* simulates "playing in rounds" by taking robot positions after achieving a goal as initial positions for the next target. Using the same sequence of targets as above, *oclingo* with --heuristic=vsids completed 250 instances in 26 seconds average runtime, thus exhibiting significant improvements over the setting with fixed initial positions. This phenomenon is probably related to the lexicographical order of our sequence of targets, and we aim at further experiments with less regular target sequences in the near future.

4 Discussion

Alex Randolph's board game *Ricochet Robots* offers a rich and versatile benchmark for ASP. As it stands, it represents a simple multi-agent planning problem in which each agent, i.e., robot, has limited sensing capacities (that is, only bumps are detected). This setting leaves room for numerous interesting extensions. For example, we may consider competing or collaborating robots, simultaneous moves, and conceive compelling multi-agent scenarios. Also, the addition of resources like fuel or keys are conceptually interesting extensions, not to mention the plenty variants of the board game available on the web. Moreover, the potential of *Ricochet Robots* is even beyond ASP given that it can be modeled in many other paradigms, like action and planning languages, constraint languages, or in terms of Satisfiability testing.

In this paper, we started by elaborating upon two alternative encodings, one following traditional approaches to ASP planning and another centered around states rather than actions. More disparate encodings will result from the ASP competition in 2013, where *Ricochet Robots* is included in the modeling track. In addition, we provided the graphical tool *robotviz* for visualizing boards as well as solutions to the *Ricochet Robots* problem. The goal of this is to ease acquaintance with the game and to increase its attractiveness, also in view of teaching ASP. The visualization tool *robotviz* along with encodings and instances of *Ricochet Robots* are available at [20].

We illustrated the versatility of the benchmark by conducting two transverse empirical case studies. The first one aimed at assessing the impact of modeling techniques on the performance of ASP systems. This was flanked by an investigation of algorithm configuration and multi-threading as means to speed-up search, which demonstrated the capabilities of automatic solver configuration and parallelism. The second part of our study contrasted distinct ASP solving technologies in a uniform setting. Here, incremental and reactive ASP solving showed to be effective for computing shortest plans. Given that the original *Ricochet Robots* game proceeds in rounds, continuing from the final configuration of the previous round with a new target, automated support of such application scenarios in the future promises a rich source of reactive ASP benchmarks.

Our case studies are of course not sufficient for general claims but show the prospect of having a benchmark for evaluating different aspects in a uniform setting. This is certainly important in view of establishing a production mode for ASP when faced with singular real-world applications. In fact, the development of a robust ASP-based solution to an application problem must account for several interdependent factors and eventually converge to an integrated approach dealing with them. For one, the problem encoding predetermines the prospects of solving methods on problem instances, and its conception thus deserves careful consideration. Second, the application task designates appropriate solving methods, where decision and optimization as well as bounded and unbounded approaches can be distinguished. Finally, algorithm configuration and parallelism are powerful means to improve the efficiency of a solving method.

Acknowledgments. This work was partially funded by DFG grant SCHA 550/9-1.

References

1. Baral, C.: Knowledge Representation, Reasoning and Declarative Problem Solving. Cambridge University Press (2003)
2. Biere, A., Heule, M., van Maaren, H., Walsh, T. (eds.): Handbook of Satisfiability. Frontiers in Artificial Intelligence and Applications, vol. 185. IOS Press (2009)
3. Butko, N., Lehmann, K., Ramenzoni, V.: Ricochet Robots — a case study for human complex problem solving. In: Proceedings of the Annual Santa Fe Institute Summer School on Complex Systems, CSSS 2005 (2005)
4. Engels, B., Kamphans, T.: On the complexity of Randolph's robot game. Research Report 005, Institut für Informatik, Universität Bonn (2005)
5. Engels, B., Kamphans, T.: Randolph's robot game is NP-hard! Electronic Notes in Discrete Mathematics 25, 49–53 (2006)

6. Engels, B., Kamphans, T.: Randolph's robot game is NP-complete! In: Proceedings of the Twenty-second European Workshop on Computational Geometry, EWCG 2006, pp. 157–160 (2006)

7. Lifschitz, V.: Answer set programming and plan generation. Artificial Intelligence 138(1-2), 39–54 (2002)

8. Gebser, M., Kaminski, R., König, A., Schaub, T.: Advances in gringo series 3. In: [21], pp. 345–351

9. Gebser, M., Kaminski, R., Kaufmann, B., Ostrowski, M., Schaub, T., Thiele, S.: A user's guide to gringo, clasp, clingo, and iclingo, http://potassco.sourceforge.net

10. Gebser, M., Kaminski, R., Kaufmann, B., Ostrowski, M., Schaub, T., Thiele, S.: Engineering an incremental ASP solver. In: Garcia de la Banda, M., Pontelli, E. (eds.) ICLP 2008. LNCS, vol. 5366, pp. 190–205. Springer, Heidelberg (2008)

11. Gebser, M., Kaufmann, R., Schaub, T.: Gearing up for effective ASP planning. In: Erdem, E., Lee, J., Lierler, Y., Pearce, D. (eds.) Correct Reasoning. LNCS, vol. 7265, pp. 296–310. Springer, Heidelberg (2012)

12. Gebser, M., Grote, T., Kaminski, R., Schaub, T.: Reactive answer set programming. In: [21], pp. 54–66

13. Hutter, F., Hoos, H.H., Leyton-Brown, K.: Sequential model-based optimization for general algorithm configuration. In: Coello Coello, C.A. (ed.) LION 2011. LNCS, vol. 6683, pp. 507–523. Springer, Heidelberg (2011)

14. Gebser, M., Kaufmann, B., Schaub, T.: Multi-threaded ASP solving with clasp. Theory and Practice of Logic Programming 12(4-5), 525–545 (2012)

15. Gebser, M., Kaminski, R., Kaufmann, B., Schaub, T., Schneider, M., Ziller, S.: A portfolio solver for answer set programming: Preliminary report. In: [21], pp. 352–357

16. Silverthorn, B., Lierler, Y., Schneider, M.: Surviving solver sensitivity: An ASP practitioner's guide. In: [22], pp. 164–175

17. Hutter, F., Babić, D., Hoos, H., Hu, A.: Boosting verification by automatic tuning of decision procedures. In: Proceedings of the Seventh Conference on Formal Methods in Computer-Aided Design, FMCAD 2007, pp. 27–34. IEEE Computer Society Press (2007)

18. Vallati, M., Fawcett, C., Gerevini, A., Hoos, H., Saetti, A.: Generating fast domain-specific planners by automatically configuring a generic parameterised planner. In: Proceedings of the Twenty-First ICAPS Workshop on Planning and Learning, PAL 2011, pp. 21–27 (2011)

19. Andres, B., Kaufmann, B., Matheis, O., Schaub, T.: Unsatisfiability-based optimization in clasp. In: [22], pp. 212–221

20. Potassco Labs, http://potassco.sourceforge.net/labs.html

21. Delgrande, J., Faber, W. (eds.): LPNMR 2011. LNCS, vol. 6645. Springer, Heidelberg (2011)

22. Dovier, A., Santos Costa, V. (eds.): Technical Communications of the Twenty-Eighth International Conference on Logic Programming, ICLP 2012. Leibniz International Proceedings in Informatics, vol. 17. Schloss Dagstuhl (2012)

Decidability and Implementation of Parametrized Logic Programs

Ricardo Gonçalves and José Júlio Alferes*

CENTRIA - Departamento de Informática,
Faculdade de Ciências e Tecnologia, Universidade Nova de Lisboa

Abstract. Parametrized logic programs are very expressive logic programs that generalize normal logic programs under the stable model semantics, by allowing complex formulas of a parameter logic to appear in the body and head of rules. In this paper we study the decidability of these rich programs and propose an implementation that combines, in a modular way, a reasoner for the parameter logic with an answer set solver.

1 Introduction

Parametrized logic programming [9] was introduced as an extension of answer set programming [8] with the motivation of providing a meaning to theories combining both logic programming connectives with other logical connectives, and allowing complex formulas using these connectives to appear in the head and body of a rule. The main idea is to fix a monotonic logic $\mathcal{L}$, called the parameter logic, and build up logic programs using formulas of $\mathcal{L}$ instead of just atoms. The obtained parametrized logic programs have, therefore, the same structure of normal logic programs, the only difference being the fact that atomic symbols are replaced by formulas of $\mathcal{L}$.

When applying this framework, the choice of the parameter logic depends on the domain of the problem to be modeled. As examples, [9] shows how to obtain the answer-set semantics of logic programs with explicit negation, a paraconsistent version of it, and also the semantics of MKNF hybrid knowledge bases [15], using an appropriate choice of the parameter logic. Moreover, [10] introduces deontic logic programs using standard deontic logic [20] as the parameter logic.

Parametrized logic programming can be seen as a framework which allow us to add non-monotonic rule based reasoning on top of an existing (monotonic) language. This view is quite interesting, in particular in those cases where we already have a monotonic logic to model a problem, but we are still lacking some conditional or non-monotonic reasoning. In these situations, parametrized logic programming offers a modular framework for adding such conditional and non-monotonic reasoning, without having to give up on the monotonic logic at

---

* The first author was supported by FCT under the postdoctoral grant SFRH/BPD/47245/2008. The work was partially supported by projects ERRO – PTDC/EIA-CCO/121823/2010, and ASPEN – PTDC/EIA-CCO/110921/2009.

P. Cabalar and T.C. Son (Eds.): LPNMR 2013, LNAI 8148, pp. 361–373, 2013.

hand. One interesting example is the case of MKNF hybrid knowledge bases, where the existing monotonic logics are description logics.

However, in order to make parametrized logic programming usable in practice, we need to prove that this rich combination does not compromise decidability in the case of a decidable parameter logic. Moreover, given a decidable parameter logic, for pragmatic reasons the implementation for a reasoner should make a modular use of an existing reasoner for the parameter logic and an answer set solver. This modularity is extremely important since it allows us to use the large body of successful research done in the area of stable model semantics implementation and answer set programming.

In this paper, after introducing the framework of parametrized logic programs (Section 2), we address the decidability of the stable model entailment of parametrized logic programs and study the implementation of a reasoner for parametrized logic programs, combining a reasoner for the parameter logic and answer set solver in a modular way (Section 3). We also study some interesting examples of parameter logics over a restricted language that have better computational properties than the general case. We end with some conclusions and draw some paths for future research (Section 4).

2   Parametrized Logic Programs

In this section we introduce the syntax and semantics of normal parametrized logic programs [9].

2.1   Language

The syntax of a normal parametrized logic program has the same structure of that of a normal logic program. The only difference is that the atomic symbols of a normal parametrized logic program are replaced by formulas of a parameter logic, which is restricted to be a monotonic logic. Let us start by introducing the necessary concepts related with the notion of (monotonic) logic.

Definition 1. *A (monotonic) logic is a pair $\mathcal{L} = \langle L, \vdash_{\mathcal{L}} \rangle$ where L is a set of formulas and $\vdash_{\mathcal{L}}$ is a Tarskian consequence relation [21] over L, i.e., satisfying the following conditions, for every $T \cup \Phi \cup \{\varphi\} \subseteq L$,*
Reflexivity: *if $\varphi \in T$ then $T \vdash_{\mathcal{L}} \varphi$;*
Cut: *if $T \vdash_{\mathcal{L}} \varphi$ for all $\varphi \in \Phi$, and $\Phi \vdash_{\mathcal{L}} \psi$ then $T \vdash_{\mathcal{L}} \psi$;*
Weakening: *if $T \vdash_{\mathcal{L}} \varphi$ and $T \subseteq \Phi$ then $\Phi \vdash_{\mathcal{L}} \varphi$.*

When clear from the context we write $\vdash$ instead of $\vdash_{\mathcal{L}}$. Let $Th(\mathcal{L})$ be the set of *logical theories* of $\mathcal{L}$, i.e. the set of subsets of L closed under the relation $\vdash_{\mathcal{L}}$. One fundamental characteristic of the above definition is that, for every (monotonic) logic $\mathcal{L}$, the tuple $\langle Th(\mathcal{L}), \subseteq \rangle$ is a complete lattice with smallest element the set $Theo = \{\varphi \in L : \emptyset \vdash \varphi\}$ of theorems of $\mathcal{L}$ and greatest element the set L of all formulas of $\mathcal{L}$. Given a subset A of L we denote by $A^{\vdash_{\mathcal{L}}}$ the smallest logical theory of $\mathcal{L}$ that contains A, and call it *the logical theory generated by A in $\mathcal{L}$*.

In the following we consider fixed a (monotonic) logic $\mathcal{L} = \langle L, \vdash_{\mathcal{L}} \rangle$ and call it the *parameter logic*. The formulas of $\mathcal{L}$ are dubbed *(parametrized) atoms* and a *(parametrized) literal* is either a parametrized atom φ or its negation *not* φ, where *not* denotes default negation. *Default literals* are those of the form *not* φ.

Definition 2. *A normal $\mathcal{L}$ parametrized logic program is a set of rules*

$$\varphi \leftarrow \psi_1, \ldots, \psi_n, not\ \delta_1, \ldots, not\ \delta_m \tag{1}$$

where $\varphi, \psi_1, \ldots, \psi_n, \delta_1, \ldots, \delta_m \in L$.
   A definite *$\mathcal{L}$ parametrized logic program is a set of rules without negations as failure, i.e. of the form $\varphi \leftarrow \psi_1, \ldots, \psi_n$ where $\varphi, \psi_1, \ldots, \psi_n \in L$.*

As usual, the symbol $\leftarrow$ represents rule implication, the symbol "," represents conjunction and the symbol *not* represents default negation. A rule as (1) has the usual reading that φ should hold whenever $\psi_1, \ldots, \psi_n$ hold and $\delta_1, \ldots, \delta_m$ are not known to hold. If $n = 0$ and $m = 0$ then we just write $\varphi \leftarrow$.
   Given a rule r of the form (1), we define $head(r) = \varphi$, $body^+(r) = \{\psi_1, \ldots, \psi_n\}$, $body^-(r) = \{\delta_1, \ldots, \delta_m\}$ and $body(r) = body^+(r) \cup body^-(r)$. Given a parametrized logic program $\mathcal{P}$ we define $form(\mathcal{P})$ to be the set of all formulas of the parameter language L appearing in $\mathcal{P}$, i.e., $form(\mathcal{P}) = \bigcup_{r \in \mathcal{P}} (\{head(r)\} \cup body(r))$. We also define the set $head(\mathcal{P}) = \{head(r) : r \in \mathcal{P}\}$.

2.2   Semantics

The semantics of parametrized logic programs is defined as a generalization of the stable model semantics [8] of normal logic programs.
   In the normal logic programs, an interpretation is just a set of atoms. In a parametrized logic program, since we substitute atoms by formulas of a parameter logic, the first idea is to take sets of formulas of the parameter logic as interpretations. The problem is that, contrary to the case of atoms, the parametrized atoms are not independent of each other. This interdependence is governed by the consequence relation of the parameter logic. For example, if we take classical propositional logic (CPL) as the parameter logic, we have that if the parametrized atom $p \wedge q$ is true then so are the parametrized atoms p and q. If we take, for example, standard deontic logic SDL [20] as parameter, we have that, since $\mathbf{O}(p \vee q), \mathbf{O}(\neg p) \vdash_{SDL} \mathbf{O}(q)$, any SDL logical theory containing both $\mathbf{O}(p \vee q)$ and $\mathbf{O}(\neg p)$ also contains $\mathbf{O}(q)$.
   To account for this interdependence, we use logical theories (sets of formulas closed under the consequence of the logic) as the generalization of interpretations, thus capturing the above mentioned interdependence.

Definition 3. *A (parametrized) interpretation is a logical theory of $\mathcal{L}$.*

Definition 4. *An interpretation T satisfies a rule*

$$\varphi \leftarrow \psi_1, \ldots, \psi_n, not\ \delta_1, \ldots, not\ \delta_m$$

if $\varphi \in T$ whenever $\psi_i \in T$ for every $i \in \{1, \ldots, n\}$ and $\delta_j \notin T$ for every $j \in \{1, \ldots, m\}$.

An interpretation is a model of logic program $\mathcal{P}$ if it satisfies every rule of $\mathcal{P}$. We denote by $Mod_{\mathcal{L}}(P)$ the set of models of $\mathcal{P}$.

The ordering over interpretations is the usual one: If T_1 and T_2 are two interpretations then we say that $T_1 \leq T_2$ if $T_1 \subseteq T_2$. Moreover, given such ordering, minimal and least interpretations may be defined in the usual way.

As in the case of non parametrized programs, we start by assigning semantics to definite parametrized programs. Recall that the stable model of a definite logic program is its least model. In order to generalize this definition to the parametrized case we need to establish that the least parametrized model exists for every definite $\mathcal{L}$ parametrized logic program.

Theorem 1 ([9]). *Every definite $\mathcal{L}$ parametrized logic program has a least model, denoted by $S_P^{\mathcal{L}}$.*

Note that this theorem holds for every choice of the parameter logic $\mathcal{L}$.

The stable model semantics of a normal $\mathcal{L}$ parametrized logic programs is defined using a Gelfond-Lifschitz like operator.

Definition 5. *Let $\mathcal{P}$ be a normal $\mathcal{L}$ parametrized logic program and T an interpretation. The GL-transformation of $\mathcal{P}$ modulo T is the program $\frac{P}{T}$ obtained from P by performing the following operations:*

- *remove from $\mathcal{P}$ all rules which contain a literal not φ such that $T \vdash_{\mathcal{L}} \varphi$;*
- *remove from the remaining rules all default literals.*

Since $\frac{P}{T}$ is a definite $\mathcal{L}$ parametrized program, it has an unique least model J. We define $\Gamma(T) = J$.

Stable models of a parametrized logic program are then defined as fixed points of this Γ operator.

Definition 6. *An interpretation T of an $\mathcal{L}$ parametrized logic program $\mathcal{P}$ is a stable model of $\mathcal{P}$ iff $\Gamma(T) = T$. A formula φ is true under the stable model semantics, denoted by $\mathcal{P} \vDash_{SM} \varphi$ iff it belongs to all stable models of $\mathcal{P}$. We denote by $SM(\mathcal{P})$ the set of all stable models of $\mathcal{L}$.*

2.3 Examples

Parametrized logic programs are very general and flexible, allowing not only to capture well-known extensions of the stable model semantics of normal logic programs, but also to extend them further. In [9] it is shown that normal logic programs and extended logic programs correspond to an appropriate choice of the parameter logic.

One interesting case that already goes beyond the usual extensions of normal logic programs is to use a parameter logic over a full propositional language. Note that this is different, and in fact orthogonal, to the so-called nested logic programs [6]. Nested logic programs are propositional combinations of the logic programming connectives. In the case of parametrized logic programs, propositional nesting only appears at the level of the atoms.

Example 1 (Propositional logic programs). Let us now consider a full propositional language L built over a set $\mathcal{P}$ of propositional symbols using the usual
connectives $(\neg, \vee, \wedge, \Rightarrow)$. Many consequence relations can be defined over this
language. We present three interesting examples: classical logic, Belnap's paraconsistent logic [2] and intuitionistic logic. Consider the following programs:

$$P_1 \begin{cases} p \leftarrow \neg q \\ p \leftarrow q \end{cases} \qquad P_2 \{ p \leftarrow \neg q \vee q \qquad P_3 \begin{cases} q \leftarrow \\ (q \vee s) \Rightarrow p \leftarrow \\ r \leftarrow p \end{cases}$$

$$P_4 \begin{cases} r \leftarrow \\ \neg p \leftarrow \\ (p \vee q) \leftarrow r \\ s \leftarrow q \end{cases} \qquad P_5 \{ p \leftarrow not\ q, not\ \neg q \qquad \begin{array}{l} P_6 \{ p \leftarrow not\ (q \vee \neg q) \\ \\ P_7 \begin{cases} p \leftarrow \\ \neg p \leftarrow \end{cases} \end{array}$$

Let $\mathcal{L} = \langle L, \vdash_{CPL} \rangle$ be Classical Propositional Logic (CPL) over the language
L. Let us study in detail the semantics of P_1. Note that every CPL logical theory
that does not contain neither p nor $\neg p$ satisfies P_1. In particular, the set $Taut$
of tautologies of CPL is a model of P_1. So, $S_{P_1}^{CPL} = Taut$. This means that
$p, \neg p, q, \neg q \notin S_{P_1}^{CPL}$. We also have that $S_{P_2}^{CPL} = \{p\}^{\vDash ails}$. So, in the case of P_2
we have that $p \in S_{P_2}^{CPL}$. Also, we have that $r \in S_{P_3}^{CPL}$ and $s \in S_{P_4}^{CPL}$.

In the case of P_5 its stable models are the CPL logical theories that contain
p and do not contain q nor $\neg q$. Therefore, we have that $p \in S_{P_5}^{CPL}$. In the case of
P_6, since $(p \vee \neg p) \in T$ for every CPL logical theory T we can conclude that the
only stable model of P_6 is the set $Theo$ of theorems of CPL. Therefore $p \notin S_{P_6}^{CPL}$.
Regarding P_7, it is clear that $S_{P_7}^{CPL}$ equals the (inconsistent) set of all formulas.
Note that, like in answer-sets, stable models of parametrized programs can be
inconsistent, this being conceptually different from the case when there are no
answer-sets.

Consider now $\mathcal{L} = \langle L, \vdash_4 \rangle$ the 4-valued Belnap paraconsistent logic *Four*.
Consider the program P_4. Contrarily to the case of CPL, in *Four* it is not the
case that $\neg p, (p \vee q) \vdash_4 q$. Therefore we have that $q, s \notin S_{P_4}^{Four}$.

Let now $\mathcal{L} = \langle L, \vdash_{IPL} \rangle$ be the Intuitionistic Propositional Logic IPL. It is
well-known that $q \vee \neg q$ is not a theorem of IPL. Therefore, considering program
P_2 we have $S_{P_2}^{IPL} = \emptyset^{\vdash IPL}$. So, contrarily to the case of CPL, we have that
$p \notin S_{P_2}^{IPL}$. Using the same idea for program P_6 we can conclude, contrarily to
the case of CPL, that $p \in S_{P_6}^{IPL}$.

Another interesting class of logic that can be taken as parameter are modal
logics [5]. Modal logics are fundamental in many areas of Artificial Intelligence. They are quite flexible, expressive, and quite often decidable. By using
parametrized logic programs with a modal logic as the parameter logic we are
thus adding a non-monotonic layer to an already expressive language.

Example 2 (Modal logic).

Consider modal logic language L_m built over a set $\mathcal{P}$ of propositional symbols
using the usual connectives $\neg, \vee, \wedge, \Rightarrow$ and the modal operators $\Box, \Diamond$. Let $\mathcal{L}_m = \langle L_m, \vdash_m \rangle$ be a modal logic over the language L_m, where the consequence relation
is obtained from usual Kripke style semantics. Of course, the particular modal

logic we obtain depends on the restriction we impose in the Kripke models. Just to mention a few interesting examples, $\mathcal{L}_m$ could be epistemic logic, usually an S_5 modal logic, deontic logic, usually a KD modal logic and doxastic logic, usually a $KD45$ modal logic. Our aim with this example is just to stress that we can choose quite interesting and expressive logics as the parameter logic. Just to give an example, in [10,11] a very rich non-monotonic framework for reasoning about normative systems can be obtained by choosing modal logic KD, also known as Standard Deontic Logic [20], as the parameter logic.

3   Decidability and Implementation

We have seen how general is the construction of logic programs using a parameter logic. The question that naturally arises now is whether this combination of a monotonic logic and a non-monotonic framework preserves decidability. Moreover, even if decidability is preserved, there is still the question of whether we can use existing tools for the parameter logic together with an ASP solver to implement a reasoning tool for the combination. In this section we address both these issues. We first show that decidability is preserved if the parameter logic is decidable and then we also show how to combine an existing reasoner for a given parameter logic with an ASP solver.

We start with an interesting observation: even for logics over a propositional logical language built from a finite number of propositional symbols, the number of logical theories may be infinite. An immediate consequence is that the number of possible stable models of a finite parametrized logic program can be infinite. Interestingly, as we show below, decidability is not necessarily compromised. The key idea is that, given a finite parametrized logic program $\mathcal{P}$, we are able to prove that only those logical theories generated by sets of formulas appearing in $\mathcal{P}$ can be stable models of $\mathcal{P}$, and these are in a finite number.

Theorem 2. *Let $\mathcal{P}$ be a finite parametrized logic program. If T is a stable model of $\mathcal{P}$ then there exists $A \subseteq form(\mathcal{P})$ such that $T = A^{\vdash_\mathcal{L}}$.*

Proof. Let T be a stable model of $\mathcal{P}$. Consider the set $A = T \cap form(\mathcal{P})$, i.e., the restriction of T to the set of formulas appearing in $\mathcal{P}$. Since T is a logical theory of $\mathcal{L}$, $A \subseteq T$ and $\mathcal{L}$ is monotonic, we have that $A^{\vdash_\mathcal{L}} \subseteq T^{\vdash_\mathcal{L}} = T$. We aim to prove that, in fact, $A^{\vdash_\mathcal{L}} = T$. Since A is the restriction of T to the formulas of $\mathcal{P}$ we have that $\frac{\mathcal{P}}{A^{\vdash_\mathcal{L}}} = \frac{\mathcal{P}}{T}$. Then, we have that $A^{\vdash_\mathcal{L}}$ is also a model of $\frac{\mathcal{P}}{T}$. Since T is a stable model of $\mathcal{P}$ it is the minimal model of $\frac{\mathcal{P}}{T}$. Therefore, we can conclude $T \subseteq A^{\vdash_\mathcal{L}}$, which then implies that $T = A^{\vdash_\mathcal{L}}$.   □

The above theorem has as immediate consequence the fact that every finite parametrized logic program has a finite number of stable models.

Corollary 1. *Let $\mathcal{P}$ be a finite parametrized logic program. Then, $\mathcal{P}$ has finitely many stable models.*

With this, we can now prove the decidability result.

Theorem 3. *Let $\mathcal{P}$ be a finite parametrized logic program over a decidable parameter logic $\mathcal{L}$ and φ a formula of $\mathcal{L}$ (not necessarily in $\mathcal{P}$). Then, it is decidable the problem of checking if $\mathcal{P} \vDash_{SM} \varphi$ is the case.*

Proof. First of all, note that we are assuming that $\mathcal{L}$ is a decidable logic, i.e., the problem of checking $\varPhi \vdash_{\mathcal{L}} \varphi$, for a finite set $\varPhi$ of $\mathcal{L}$ formulas, is decidable. Note also that the sets $form(\mathcal{P})$ and its subset $head(\mathcal{P})$ are finite.

Let us now introduce some necessary notation. Given a subset A of $form(\mathcal{P})$ we write $C(A)$ to denote its closure under $\mathcal{L}$ consequence, i.e., $C(A) = A^{\vdash_{\mathcal{L}}} \cap form(\mathcal{P})$. Given a subset A of $form(\mathcal{P})$, we can easily construct $C(A)$ by checking, for each $\psi \in form(\mathcal{P}) \setminus A$, if $A \vdash_{\mathcal{L}} \psi$;

In Fig. 1 we sketch an algorithm showing the decidability of the problem $\mathcal{P} \vDash_{SM} \varphi$. It is based on the Gelfond-Lifschitz transformation with the additional use of an $\mathcal{L}$ oracle. The fundamental tool supporting the algorithm is the result in Theorem 2, since it restricts severely the number of $\mathcal{L}$ theories we need to check. To cut even more the number of theories to be checked we also use the well-known result in the logic programming area: a stable model of a normal logic program is always a subset of the set of heads of rules of the program. □

```
input: finite PLP P and L formula φ
  for each Φ ⊆ head(P) compute C(Φ)
  if Φ = C(Φ) then
     compute P/Φ
     compute least(P/Φ) restricted to form(P):
        define A₀ := C({φ : φ ← ∈ P})
        compute A_{i+1} := C({φ : φ ← ψ₁,...,ψₙ ∈ P and {ψ₁,...,ψₙ} ⊆ A_i})
        until A_k = A_{k+1} for some k
        then set least(P/Φ) := A_k;
     if least(P/Φ) = Φ   /* in this case Φ^{⊢ℒ} is a stable model of P */
        check if Φ ⊢_ℒ φ
  if Φ ⊢_ℒ φ for every Φ ⊆ head(P) such that Φ^{⊢ℒ} is a stable model of P,
     then P ⊨_SM φ is the case.
```

Fig. 1. Decidability algorithm

The algorithm in the proof of Theorem 3 is interesting since it is a (basic) stable model like algorithm which, when necessary, makes queries to an $\mathcal{L}$-oracle. This makes it modular with respect to the $\mathcal{L}$-reasoner and it minimizes the calls to the $\mathcal{L}$-oracle. The algorithm has, nevertheless, a major drawback: it is not modular from the point of view of calculating stable models, in the sense that we cannot use existing ASP solvers to compute the stable models of a parametrized logic program. This modularity is extremely important since it would allow us to use the large body of successful research done in the area of stable model semantics implementation and answer set programming. Our aim is precisely to propose an implementation of a reasoner for parametrized logic programs which modularly combines an ASP solver (such as Clasp [7]) with a reasoner for the parameter logic (such as the KED SDL solver [1] in the case of Standard Deontic Logic [20], or the HermiT [16] reasoner in the case SROIQ description logic [13]).

We start by proving a theorem that sets the ground for the construction of the modular reasoner. Consider a given parametrized logic program $\mathcal{P}$, and construct the following normal logic program $\mathcal{P}^N$ from $\mathcal{P}$:

$$\mathcal{P}^N = \mathcal{P} \cup \{\varphi \leftarrow \psi_1, \dots, \psi_n : \{\psi_1, \dots, \psi_n\} \subseteq form(\mathcal{P}),$$
$$\varphi \in form(\mathcal{P})\backslash\{\psi_1,\dots,\psi_n\},$$
$$\{\psi_1, \dots, \psi_n\} \vdash_{\mathcal{L}} \varphi\}.$$

We call $\mathcal{P}^N$ the normal logic program obtained from $\mathcal{P}$, since the $\mathcal{L}$ formulas appearing in it are to be considered as normal logic programs atoms. The key idea underlying the construction of $\mathcal{P}^N$, in order to enforce the interdependency between the $\mathcal{L}$ formulas (which in $\mathcal{P}^N$ are just atoms), is to enrich $\mathcal{P}$ with rules that represent the possible reasoning in $\mathcal{L}$ occurring with the formulas of $\mathcal{P}$.

Since we are now considering usual normal logic programs, and to distinguish between the set of stable model of a parametrized logic program $\mathcal{P}$ (which is a subset of $2^{Th_{\mathcal{L}}}$) and the set of stable models of $\mathcal{P}$ viewed as a normal logic program (which is a subset of $2^{form(\mathcal{P})}$), we denote the latter by $AS(\mathcal{P})$. Note that $SM(\mathcal{P})$ and $AS(\mathcal{P})$ can be very different since AS does not take into account the interdependency between the (parametrized) atoms. As a simple example let $\mathcal{L}$ be a normal modal logic. Consider $\mathcal{P} = \{\Box p \leftarrow; \Box q \leftarrow \Box(p \vee r)\}$. Then, $SM(\mathcal{P}) = \{\{\Box p, \Box(p \vee r), \Box q\}^{\vdash_{\mathcal{L}}}\}$ but $AS(\mathcal{P}) = \{\{\Box p\}\}$.

Theorem 4. *Given a parametrized logic program $\mathcal{P}$, we have that*

$$SM(\mathcal{P}) = \{A^{\vdash_{\mathcal{L}}} : A \in AS(\mathcal{P}^N)\}$$

Proof. Let us start with some notation and a general comment. We use $\mathcal{P}^{\mathcal{L}}$ to denote the set of rules that are added to $\mathcal{P}$ in the definition of the program $\mathcal{P}^N$, i.e., $\mathcal{P}^N = \mathcal{P} \cup \mathcal{P}^{\mathcal{L}}$. Since the rules in $\mathcal{P}^{\mathcal{L}}$ represent sound consequences in $\mathcal{L}$, and since every logical theory of $\mathcal{L}$ is closed under $\mathcal{L}$ consequence, it follows immediately that every $\mathcal{L}$ logical theory satisfies all the rules in $\mathcal{P}^{\mathcal{L}}$.

We now prove the equality $SM(\mathcal{P}) = \{A^{\vdash_{\mathcal{L}}} : A \in AS(\mathcal{P}^N)\}$ by proving the two inclusions. Let us start by proving the left to right inclusion. Let T be a stable model of $\mathcal{P}$. We aim to prove that there exists a stable model A of $\mathcal{P}^N$ such that $T = A^{\vdash_{\mathcal{L}}}$. Take $A = T \cap form(\mathcal{P})$. We first prove that $A^{\vdash_{\mathcal{L}}} = T$. Since $\mathcal{L}$ is monotone, we have that $A^{\vdash_{\mathcal{L}}} \subseteq T^{\vdash_{\mathcal{L}}} = T$. To prove the reverse inclusion recall that T is the minimal $\mathcal{L}$ logical theory that satisfies $\frac{\mathcal{P}}{T}$. Since $A = T \cap form(\mathcal{P})$, it immediately follows that A satisfies $\frac{\mathcal{P}}{T}$. Therefore, $A^{\vdash_{\mathcal{L}}}$ is an $\mathcal{L}$ logical theory that satisfies $\frac{\mathcal{P}}{T}$. Since T is the minimal one, we have that $T \subseteq A^{\vdash_{\mathcal{L}}}$.

Now that we have proved that $A^{\vdash_{\mathcal{L}}} = T$, we need to prove that A is a stable model of $\mathcal{P}^N$. First of all, observe that $\frac{\mathcal{P}^N}{A} = \frac{\mathcal{P}}{A} \cup \mathcal{P}^{\mathcal{L}} = \frac{\mathcal{P}}{T} \cup \mathcal{P}^{\mathcal{L}}$. Let $B \subseteq form(\mathcal{P})$ be a model of $\frac{\mathcal{P}^N}{A}$. Then, since B satisfies $\mathcal{P}^{\mathcal{L}}$, B is closed under $\mathcal{L}$ consequence.

This in turn implies that $B^{\vdash_{\mathcal{L}}} \cap form(\mathcal{P}) = B$. Clearly $B^{\vdash_{\mathcal{L}}}$ is a model of $\frac{\mathcal{P}}{T}$, and, since T is the minimal $\mathcal{L}$ logical theory satisfying $\frac{\mathcal{P}}{T}$, we can conclude that $T \subseteq B^{\vdash_{\mathcal{L}}}$. But then $A = T \cap form(\mathcal{P}) \subseteq B^{\vdash_{\mathcal{L}}} \cap form(\mathcal{P}) = B$. Since this inclusion is the case for every B model of $\frac{\mathcal{P}^N}{A}$, we can conclude that A is the minimal model of $\frac{\mathcal{P}^N}{A}$, i.e., A is a stable model of $\mathcal{P}^N$.

We now prove the right to left inclusion. Let $A \subseteq form(\mathcal{P})$ be a stable model of $\mathcal{P}^N$. We aim to prove that $A^{\vdash_{\mathcal{L}}}$ is a stable model of $\mathcal{P}$. Since A is a stable model of $\mathcal{P}^N$ we have that A is the minimal model of $\frac{\mathcal{P}^N}{A} = \frac{\mathcal{P}}{A} \cup \mathcal{P}^{\mathcal{L}}$. Since A is a model of $\mathcal{P}^{\mathcal{L}}$ we have that A is closed under $\mathcal{L}$ consequence, i.e., $A^{\vdash_{\mathcal{L}}} \cap form(\mathcal{P}) = A$. We then have that $\frac{\mathcal{P}}{A^{\vdash_{\mathcal{L}}}} = \frac{\mathcal{P}}{A}$. Suppose there exists an $\mathcal{L}$ logical theory T such that $T \subset A^{\vdash_{\mathcal{L}}}$ and T satisfies $\frac{\mathcal{P}}{A^{\vdash_{\mathcal{L}}}}$. In that case, $T \cap form(\mathcal{P}) \subset A^{\vdash_{\mathcal{L}}} \cap form(\mathcal{P}) = A$ and $T \cap form(\mathcal{P})$ satisfies $\frac{\mathcal{P}}{A} \cup \mathcal{P}^{\mathcal{L}} = \frac{\mathcal{P}^N}{A}$. But this contradicts the fact that A is the minimal model of $\frac{\mathcal{P}^N}{A}$.   □

There are some very important consequences of the above theorem. One we already established in Theorem 2: the number of stable models of a finite parametrized logic program is finite. The problem is that each of these stable models is infinite. This is precisely where Theorem 4 gives its fundamental contribution. It presents a finite representation of each of the stable models of $\mathcal{P}$.

Our aim now is to compute the finite representations of the stable models of $\mathcal{P}$. The implicit algorithm in the construction of $\mathcal{P}^N$ is quite basic. It just looks at all possible relations between formulas of the parameter logic appearing in the program. We now develop a more efficient implementation, assuming some mild conditions about the parameter logic. These allow us to prune some search paths in the construction of a normal logic program from $\mathcal{P}$.

Let $\mathcal{L} = \langle L, \vdash_{\mathcal{L}} \rangle$ be a monotonic logic satisfying the following conditions. The first condition, dubbed (Bot) is the existence of a bottom element in the language, i.e., $\bot \in L$ such that for any subset $\Phi \subseteq L$ we have that if $\Phi \vdash_{\mathcal{L}} \bot$ then $\Phi \vdash_{\mathcal{L}} \varphi$ for every $\varphi \in L$. This condition allows to detect an inconsistent set of formulas by checking if it entails $\bot$. The second condition, dubbed (Prop), is that L is built from a set of propositional symbols P and it satisfies: if $propSymb(\Phi) \cap propSymb(\varphi) = \emptyset$ and $\nvdash_{\mathcal{L}} \varphi$ then $\Phi \nvdash_{\mathcal{L}} \varphi$. Intuitively this condition imposes that if a non tautological formula does not have propositional symbols in common with a set of formulas, then it should not be entailed by that set of formulas.

Note that these two conditions are quite mild, and they are satisfied by every example of parameter logic we have shown above. For a parameter logic satisfying these conditions, we can sketch an algorithm, in Fig. 2, that, given a finite parametrized logic program $\mathcal{P}$, returns a normal logic program $\mathcal{P}^{alg}$. This algorithm is an improvement of the one constructing $\mathcal{P}^N$, by pruning several search paths using the conditions imposed on the parameter logic.

We can then prove that the pruned paths do not affect the result of the algorithm, i.e., the constructed program $\mathcal{P}^{alg}$ has the same stable models as $\mathcal{P}^N$. Given these improvements, the algorithm for constructing $\mathcal{P}^{alg}$ does not, in general, return exactly the normal logic program $\mathcal{P}^N$. In fact, one can readily see that $\mathcal{P}^{alg} \subseteq \mathcal{P}^N$. As expected, we can, nevertheless, prove that the extra

```
input: finite parametrized logic program P
  set i = 1; k = lenght(head(P)); P^alg := P ∪ {φ ← : φ ∈ form(P) and ⊢_L φ}
    while i ≤ k
      for each subset A = {δ_1, ..., δ_i} of head(P) of size i
        if A ⊢_L ⊥ then /* A is inconsistent */
          for each φ ∈ form(P) \ A
            add φ ← δ_1, ..., δ_i to P^alg unless
              there is φ ← ψ_1, ..., ψ_n ∈ P^alg with {ψ_1, ..., ψ_n} ⊆ A
        else
          for each φ ∈ form(P) \ A such that propSymb(A) ∩ propSymb(φ) ≠ ∅
            if A ⊢_L φ then
              add φ ← δ_1, ..., δ_i to P^alg unless
                there is φ ← ψ_1, ..., ψ_n ∈ P^alg with {ψ_1, ..., ψ_n} ⊆ A
      i=i+1
return P^alg
```

Fig. 2. Construction of $\mathcal{P}^{alg}$

rules of $\mathcal{P}^N$ are redundant, in the sense that the set of stable model of $\mathcal{P}^N$ and $\mathcal{P}^{alg}$ is the same.

Proposition 1. *Let $\mathcal{L}$ be a monotonic logic satisfying conditions (Bot) and (Prop). Then, for any finite parametrized logic program $\mathcal{P}$ over $\mathcal{L}$, we have that*

$$AS(\mathcal{P}^{alg}) = AS(\mathcal{P}^N).$$

Proof. It follows immediately from the constructions of $\mathcal{P}^{alg}$ and of $\mathcal{P}^N$ that $\mathcal{P}^{alg} \subseteq \mathcal{P}^N$. This implies that $Mod(\mathcal{P}^N) \subseteq Mod(\mathcal{P}^{alg})$. Moreover, given S a subset of $head(\mathcal{P})$, we can readily see that if a rule $r = \varphi \leftarrow \delta_1, \ldots, \delta_n$ is such that $r \in \frac{\mathcal{P}^N}{S}$ but $r \notin \frac{\mathcal{P}^{alg}}{S}$, then there exists $r' = \varphi \leftarrow \psi_1, \ldots, \psi_m \in \frac{\mathcal{P}^{alg}}{S}$ such that $\{\psi_1, \ldots, \psi_m\} \subset \{\delta_1, \ldots, \delta_n\}$. From this observation it follows that $Mod(\frac{\mathcal{P}^N}{S}) = Mod(\frac{\mathcal{P}^{alg}}{S})$. Therefore, S is a stable model of $\mathcal{P}^N$ (minimal model of $\frac{\mathcal{P}^N}{S}$) iff S is a stable model of $\mathcal{P}^{alg}$ (minimal model of $\frac{\mathcal{P}^{alg}}{S}$). □

The above proposition is important since it allows the algorithm of $\mathcal{P}^{alg}$ to actually construct a finite representation of the stable models of a finite parametrized logic program $\mathcal{P}$. This can be done by constructing the normal logic program $\mathcal{P}^{alg}$ from $\mathcal{P}$ and then calculating the stable models of $\mathcal{P}^{alg}$. The latter can be done using any ASP solver. Note that, as we aimed, this construction uses in a modular way a reasoner for the parameter logic and reasoner for the stable model semantics. The reasoner for the parameter logic is only used for the construction of $\mathcal{P}^{alg}$. Then, an ASP solver can be used to obtain the stable models of $\mathcal{P}^{alg}$, which are the finite representations of the stable models of $\mathcal{P}$.

Regarding complexity, it should be clear that the use of parametrized logic programs, with default negation, increases the complexity of the parameter $\mathcal{L}$ alone. This comes from the fact that the stable model semantics, with default negation, adds, as usual, one extra level of non-determinism. From the point of view of logic programming there is also an exponential increasing in the complexity. Recall that in the construction of both $\mathcal{P}^N$ and $\mathcal{P}^{alg}$ we need to query an $\mathcal{L}$-oracle an exponential number of times. Moreover, we then need to compute the stable models of $\mathcal{P}^{alg}$ which, in the extreme case, can have exponentially more rules than the initial program $\mathcal{P}$.

This extra complexity is not surprising given the expressivity of the parametrized logic programs. Recall that a parametrized logic program can have any complex parametrized formula in the head and body of its rules. In some particular applications, however, there is no need for this general expressivity, and we can play the usual game between expressivity and complexity. We end this section with an example showing that we can consider restricted classes of parameter logics that have a more amenable complexity. These restricted languages may well have the necessary expressivity for modeling non-trivial scenarios.

An interesting example is the case of parametrized logic programs over a modal language that only contains literals, the necessity modal operator applied to literals and negations of the necessity operator applied to literals.

Note that we can capture the possibility operator $\lozenge$ since $\lozenge\ell \equiv_m \neg\Box\overline{\ell}$, where $\overline{\ell}$ is the complementary literal of ℓ, i.e., $\overline{\ell} = p$ if $\ell = \neg p$ and $\overline{\ell} = \neg p$ if $\ell = p$. In this restricted language the interaction between modal formulas is limited and, depending on which modal logic axioms the particular logic satisfies, we can construct the normal program $\mathcal{P}^{alg}$ from $\mathcal{P}$ in a simple way.

Proposition 2. *Let $\mathcal{P}$ be a finite parametrized logic program over a modal language only with literals, necessity applied to literals and negations of necessity applied to literals. Consider the following sets*

$$\mathcal{P}_K = \{\varphi \leftarrow \bot : \varphi \in form(\mathcal{P})\} \cup$$
$$\{\bot \leftarrow p, \neg p : \{p, \neg p\} \subseteq head(\mathcal{P})\} \cup$$
$$\{\bot \leftarrow \Box p, \Box\neg p : \{\Box p, \Box\neg p\} \subseteq head(\mathcal{P})\} \cup$$
$$\{\bot \leftarrow \Box\ell, \neg\Box\ell : \{\Box\ell, \neg\Box\ell\} \subseteq head(\mathcal{P})\}.$$
$$\mathcal{P}_D = \{\neg\Box\neg p \leftarrow \Box p : \Box p \in head(\mathcal{P}) \text{ and } \neg\Box\neg p \in form(\mathcal{P})\} \cup$$
$$\{\neg\Box p \leftarrow \Box\neg p : \Box\neg p \in head(\mathcal{P}) \text{ and } \neg\Box p \in form(\mathcal{P})\}.$$
$$\mathcal{P}_T = \{\ell \leftarrow \Box\ell : \Box\ell \in head(\mathcal{P}) \text{ and } \ell \in form(\mathcal{P})\}.$$

Then,

- *if $\mathcal{L}$ is the modal logic K then $\mathcal{P}^{alg} = \mathcal{P} \cup \mathcal{P}_K$;*
- *if $\mathcal{L}$ is the modal logic KD then $\mathcal{P}^{alg} = \mathcal{P} \cup \mathcal{P}_K \cup \mathcal{P}_D$;*
- *if $\mathcal{L}$ is the modal logic KT then $\mathcal{P}^{alg} = \mathcal{P} \cup \mathcal{P}_K \cup \mathcal{P}_T$;*
- *if $\mathcal{L}$ is the modal logic KTD then $\mathcal{P}^{alg} = \mathcal{P} \cup \mathcal{P}_K \cup \mathcal{P}_T \cup \mathcal{P}_D$.*

Proof. The result follows easily from the observation that, for this restricted language, we have that $\Phi \vdash_K \varphi$ iff one of the following conditions holds: for some propositional symbol p, $\{p, \neg p\} \subseteq \Phi$ or $\{\Box p, \Box\neg p\} \subseteq \Phi$; or $\{\Box\ell, \neg\Box\ell\} \subseteq \Phi$

for some literal ℓ. If we add to K the seriality axiom D then we can also entail φ from a set Φ of formulas if $\Box p \in \Phi$ and $\varphi = \neg \Box \neg p$, or when $\Box \neg p \in \Phi$ and $\varphi = \neg \Box p$. In the case of the addition of the transitivity axiom T we can also conclude φ from a set Φ of formulas if $\Box \varphi \in \Phi$.

The above proposition is important because it gives a way to construct $\mathcal{P}^{alg}$ using only syntactical checks, i.e., we do not need to use a modal logic oracle. This is only possible because the interaction between modal formulas in this restricted language is limited and can be clearly described using the above rules. The four rules of $\mathcal{P}_K$ are related to contradictions. The first one refers to the so-called explosion principle: from a contradiction everything follows. The others express how to detect an inconsistency. The rules of $\mathcal{P}_D$ are related to the connection between necessity and possibility: if something is necessary then it is possible, which holds in a modal logic satisfying D. The need for the rules in $\mathcal{P}_T$ comes from the fact that a formula follows from its necessity in a modal logic satisfying the reflexivity axiom T. For lack of space, we did not add several more examples of modal logics to the above proposition. Just to give an example, in the case of doxastic modal logic, which is usually assumed to be a $KD45$ modal logic, we have that $\mathcal{P}^{alg} = \mathcal{P} \cup \mathcal{P}_K \cup \mathcal{P}_D$.

Regarding complexity, it is interesting to note that the maximum number of rules added to $\mathcal{P}^{alg}$ is linear in the number of rules of $\mathcal{P}$.

4   Conclusions and Future Work

In this paper we have proved decidability for parametrized logic programs, assuming the decidability of the parameter logic. We have provided an implementation that combines in modular way a reasoner for a decidable parameter logic with an answer set solver. We have studied examples of modal logics in a restricted language. For those, the construction of a normal logic program $\mathcal{P}^{alg}$ from a given parametrized logic program $\mathcal{P}$ does not need to use a modal logic oracle, and, moreover, the number of rules added to $\mathcal{P}$ in order to obtain $\mathcal{P}^{alg}$ is at most linear in the number of rules of $\mathcal{P}$.

Regarding future work, we want to implement the algorithms presented in this paper in the case of interesting parameter logics. One such example is the case of standard deontic logic, which would then allow us to construct a declarative non-monotonic framework for specifying normative systems [3]. We also want to study in more detail the natural connection between parametrized logic programming and the general approach of multi-context systems [4], along the lines of [12]. Another interesting topic is to investigate belief change in our setting, which would be a challenging problem due to the known difficulties in combining belief change of rules and belief change in classical logic [17], although recent developments have shown a possible unifying view [18,19]. Finally, we would like to study the well-founded semantics for parametrized logic programs along the lines of what is done in [14] for hybrid MKNF.

References

1. Artosi, A., Cattabriga, P., Governatori, G.: Ked: A deontic theorem prover. In: Workshop on Legal Application of Logic Programming, pp. 60–76. IDG (1994)
2. Belnap, N.: A useful four-valued logic. In: Epstein, G., Dunn, M. (eds.) Modern Uses of Multiple-Valued Logic, pp. 7–37. Reidel Publishing Company (1977)
3. Boella, G., van der Torre, L., Verhagen, H.: Introduction to the special issue on normative multiagent systems. JAAMAS 17(1), 1–10 (2008)
4. Brewka, G., Eiter, T.: Equilibria in heterogeneous nonmonotonic multi-context systems. In: AAAI, pp. 385–390. AAAI Press (2007)
5. Chellas, B.: Modal Logic: An Introduction. Cambridge University Press (1980)
6. Ferraris, P.: Logic programs with propositional connectives and aggregates. ACM Trans. Comput. Log. 12(4), 25 (2011)
7. Gebser, M., Kaufmann, B., Neumann, A., Schaub, T.: *Clasp*: A conflict-driven answer set solver. In: Baral, C., Brewka, G., Schlipf, J. (eds.) LPNMR 2007. LNCS (LNAI), vol. 4483, pp. 260–265. Springer, Heidelberg (2007)
8. Gelfond, M., Lifschitz, V.: The stable model semantics for logic programming, pp. 1070–1080. MIT Press (1988)
9. Gonçalves, R., Alferes, J.J.: Parametrized logic programming. In: Janhunen, T., Niemelä, I. (eds.) JELIA 2010. LNCS, vol. 6341, pp. 182–194. Springer, Heidelberg (2010)
10. Gonçalves, R., Alferes, J.J.: An embedding of input-output logic in deontic logic programs. In: Ågotnes, T., Broersen, J., Elgesem, D. (eds.) DEON 2012. LNCS, vol. 7393, pp. 61–75. Springer, Heidelberg (2012)
11. Gonçalves, R., Alferes, J.J.: Specifying and reasoning about normative systems in deontic logic programming. In: van der Hoek, W., Padgham, L., Conitzer, V., Winikoff, M. (eds.) AAMAS, pp. 1423–1424. IFAAMAS (2012)
12. Homola, M., Knorr, M., Leite, J., Slota, M.: MKNF knowledge bases in multi-context systems. In: Fisher, M., van der Torre, L., Dastani, M., Governatori, G. (eds.) CLIMA XIII 2012. LNCS, vol. 7486, pp. 146–162. Springer, Heidelberg (2012)
13. Horrocks, I., Kutz, O., Sattler, U.: The even more irresistible SROIQ. In: KR, pp. 57–67. AAAI Press (2006)
14. Knorr, M., Alferes, J.J., Hitzler, P.: Local closed world reasoning with description logics under the well-founded semantics. Artif. Intell. 175(9-10), 1528–1554 (2011)
15. Motik, B., Rosati, R.: Reconciling description logics and rules. JACM 57(5) (2010)
16. Motik, B., Shearer, R., Horrocks, I.: Hypertableau Reasoning for Description Logics. Journal of Artificial Intelligence Research 36, 165–228 (2009)
17. Slota, M., Leite, J.: On semantic update operators for answer-set programs. In: Coelho, H., Studer, R., Wooldridge, M. (eds.) ECAI. Frontiers in Artificial Intelligence and Applications, vol. 215, pp. 957–962. IOS Press (2010)
18. Slota, M., Leite, J.: A unifying perspective on knowledge updates. In: del Cerro, L.F., Herzig, A., Mengin, J. (eds.) JELIA 2012. LNCS, vol. 7519, pp. 372–384. Springer, Heidelberg (2012)
19. Slota, M., Leite, J.: The rise and fall of semantic rule updates based on SE-models. Theory and Practice of Logic Programming (TPLP) (to appear, 2013)
20. von Wright, G.H.: Deontic logic. Mind 60, 1–15 (1951)
21. Wójcicki, R.: Theory of Logical Calculi. Synthese Library. Kluwer (1988)

Non-monotonic Temporal Goals

Ricardo Gonçalves, Matthias Knorr, João Leite, and Martin Slota

CENTRIA, Universidade Nova de Lisboa, Portugal

Abstract. In this paper we introduce a logic programming based framework which allows the representation of conditional non-monotonic temporal beliefs and goals in a declarative way. We endow it with stable model like semantics that allows us to deal with conflicting goals and generate possible alternatives. We show that our framework satisfies some usual properties on goals and that it allows imposing alternative constraints on the interaction between beliefs and goals. We prove the decidability of the usual reasoning tasks and show how they can be implemented using an ASP solver and an LTL reasoner in a modular way, thus taking advantage of existing LTL reasoners and ASP solvers.

1 Introduction

Mental attitudes such as beliefs, goals, and intentions are well-known to be fundamental for representing autonomous rational agents [4,12,13,5]. Roughly, beliefs represent the agent's knowledge about the state of the world, goals represent states the agent aims at achieving, and intentions are the goals that the agent commits to pursue. We focus on the representation and reasoning about declarative beliefs and goals.

One fundamental ingredient when modeling goals is the notion of time. Goals usually refer to some state of affairs that the agent aims to maintain or achieve sometime in the future. For example, an agent might have the goal to maintain a positive balance on her bank account during the entire month to avoid fines, or to study before the next week's exam. Temporal logic has been shown to be flexible and expressive for representing different goal types [7,10,2], and several works in the literature modeling mental attitudes of agents are based on (extensions of) temporal logic. Namely, [5,10,16,2] are based on Linear Temporal Logic (LTL), while [12] is based on Computational Tree Logic (CTL*).

Another fundamental ingredient is the possibility to model defeasible and conflicting information. As argued in [2], it is quite common that goals have a conditional form and admit exceptions, so the adoption of new beliefs or goals may cause the retraction of some of the agent's current goals. In particular, we may encounter conflicting goals that cannot be pursued at the same time, in which case we have to consider alternative sets of goals. For example, an agent may want to go to Paris for the weekend, but also to London. These goals are conflicting and cannot be achieved together.

Therefore, representation and reasoning about goals would greatly benefit from an approach combining the temporal and non-monotonic aspects. However,

P. Cabalar and T.C. Son (Eds.): LPNMR 2013, LNAI 8148, pp. 374–386, 2013.
© Springer-Verlag Berlin Heidelberg 2013

most of the work on beliefs and goal is monotonic [12,5,10,16], and therefore does not allow us to represent defeasible beliefs and goals. The work in [15] introduces a non-monotonic logic for conditional goals based on default logic, but it does not consider temporal formulas. In [2], a non-monotonic extension of LTL is defined that allows for expressing goals with exceptions, but it does not handle conflicting goals nor does it consider beliefs, therefore not allowing us to model the interaction between these and goals.

In this paper, we bridge the gap between temporal and non-monotonic goal languages by introducing a general non-monotonic goal framework. The language obtained is expressive enough to represent conditional beliefs and goals over complex temporal formulas, and it also allows reasoning about conflicting goals. The semantics defined in the spirit of stable models/answer set programming (ASP) [8] not only endows it with a purely declarative semantics, but also supports a novel perspective on dealing with alternative sets of goals in which each stable model can be seen as a possible consistent set of goals that an agent might adopt. Besides satisfying some usual properties on goals, our framework is also general and flexible enough to represent different constraints on the interaction between beliefs and goals in a simple way. We show decidability and how to implement our framework using existing LTL and ASP solvers.

The paper is structured as follows. In Sect. 2, we introduce the basic language for reasoning about beliefs and goals over a temporal logic. Then, in Sect. 3, we define a non-monotonic framework for representing defeasible beliefs and goals, along with its semantics, and prove some properties of our framework in Sect. 4. We discuss decidability and implementation in Sect. 5, compare with related work in Sect. 6, and conclude in Sect. 7.

2 Logic of Beliefs and Goals

In this section we introduce the language for representing beliefs and goals, which is based on temporal logic. Temporal logic has been shown to be quite flexible and expressive for representing different goal types [7,10,2]. Since our approach is modular on the temporal component, and in order to ease the presentation, we follow [5,10,16,2] and work in this paper with Linear Temporal Logic [11].

2.1 Linear Temporal Logic

Here, we introduce Linear Temporal Logic (LTL). The *language of LTL*, $\mathcal{L}_{\mathrm{LTL}}$, is built from a set of propositional symbols $\mathcal{P}$ using the usual classical connectives $\mathbf{t}, \sim, \sqcap, \sqcup, \Rightarrow$, the unary temporal operator $\bigcirc$ (next) and the binary temporal operator $\mathcal{U}$ (until). Other temporal operators can be defined by abbreviation: $\Diamond\varphi := \mathbf{t}\,\mathcal{U}\varphi$ (eventually), $\Box\varphi := \sim\Diamond\sim\varphi$ (always), $\varphi\mathcal{B}\psi := \sim(\sim\varphi\mathcal{U}\psi)$ (before).

The semantics for $\mathcal{L}_{\mathrm{LTL}}$ is given as interpretation sequences of classical valuations. Formally, an *LTL interpretation* is a sequence $m = (m_i)_{i\in\mathbb{N}}$ where $m_i \subseteq \mathcal{P}$ for each $i \in \mathbb{N}$. The *satisfaction of an LTL formula* by an LTL interpretation $m = (m_i)_{i\in\mathbb{N}}$ at a point i is defined inductively as follows:

- $m, i \Vdash \mathbf{t}$ for every $i \in \mathbb{N}$;
- $m, i \Vdash p$ if $p \in m_i$, for $p \in \mathcal{P}$;
- $m, i \Vdash \sim\varphi$ if $m, i \nVdash \varphi$;
- $m, i \Vdash \varphi_1 \sqcap \varphi_2$ if $m, i \Vdash \varphi_1$ and $m, i \Vdash \varphi_2$;
- $m, i \Vdash \varphi_1 \sqcup \varphi_2$ if $m, i \Vdash \varphi_1$ or $m, i \Vdash \varphi_2$;
- $m, i \Vdash \varphi_1 \Rightarrow \varphi_2$ if $m, i \nVdash \varphi_1$ or $m, i \Vdash \varphi_2$;
- $m, i \Vdash \bigcirc\varphi$ if $m, i+1 \Vdash \varphi$;
- $m, i \Vdash \varphi_1 \mathcal{U} \varphi_2$ if $m, j \Vdash \varphi_2$ for some $j \geq i$ and $m, k \Vdash \varphi_1$ for every $i \leq k < j$.

We say that an LTL interpretation m is a *model* of an LTL formula φ, denoted by $m \Vdash_{\mathrm{LTL}} \varphi$, if $m, 0 \Vdash \varphi$. This is the so-called anchored version of LTL. Given a set Φ of LTL formulas, we denote by $Mod(\Phi)$ the set of LTL models of all formulas in Φ. An LTL formula φ is *valid* if $m \Vdash_{\mathrm{LTL}} \varphi$ for every LTL interpretation m. The consequence relation $\vDash_{\mathrm{LTL}}$ is defined as usual, i.e., $\Phi \vDash_{\mathrm{LTL}} \varphi$ if, for every interpretation m, $m \Vdash \varphi$ whenever $m \Vdash \psi$ for every $\psi \in \Phi$. A set of LTL formulas Φ is an *LTL theory* if, for every $\delta \in \mathcal{L}_{\mathrm{LTL}}$, if $\Phi \vDash \delta$, then $\delta \in \Phi$. We denote by Th_{LTL} the set of all theories over the language $\mathcal{L}_{\mathrm{LTL}}$. Given a set Φ of LTL formulas, we denote by $\Phi^{\vDash_{\mathrm{LTL}}}$ the least LTL theory containing Φ, i.e., the deductive closure of Φ. As it is usual for monotonic logics [19], an important property of Th_{LTL} is the fact that $\langle Th_{\mathrm{LTL}}, \subseteq \rangle$ is a complete lattice, i.e., $\langle Th_{\mathrm{LTL}}, \subseteq \rangle$ is a partial order and for every $\mathcal{A} \subseteq Th_{\mathrm{LTL}}$ the set $\bigcap_{T \in \mathcal{A}} T$ is again a theory over $\mathcal{L}_{\mathrm{LTL}}$.

2.2   Logic of Beliefs and Goals

We now define the language for specifying beliefs and goals. We start by defining the syntax, which is built on top of the LTL language.

Definition 1. *The* language of beliefs and goals, *denoted by $\mathcal{L}_{BG}$, is defined as:*

$$\delta := \mathbf{B}(\varphi) \mid \mathbf{G}(\varphi) \mid \neg\delta \mid \delta \wedge \delta$$

where φ is an LTL formula. A formula of the form $\mathbf{B}(\varphi)$ is called a belief atom *and one of the form $\mathbf{G}(\varphi)$ is called a* goal atom.

Note that the other usual classical connectives can be obtained as abbreviation $\delta_1 \vee \delta_2 := \neg(\neg\delta_1 \wedge \neg\delta_2)$ and $\delta_1 \to \delta_2 := \neg\delta_1 \vee \delta_2$. We denote by $\mathcal{L}_B$ the set of formulas of $\mathcal{L}_{BG}$ that only contain the belief operator $\mathbf{B}$, i.e., those formulas which are built from belief atoms using the classical connectives. We call $\mathcal{L}_B$ the *belief language* and its elements the *belief formulas*. In the same way, we define the *goal language*, $\mathcal{L}_G$, as the set of formulas of $\mathcal{L}_{BG}$ that only contain the goal operator $\mathbf{G}$. Its elements are called *goal formulas*. Also note that, for simplicity, we follow [7,10,15] and do not allow temporal operators outside the scope of a belief or goal operator, nor nesting of belief and goal operators.

Since the language $\mathcal{L}_{BG}$ is built over belief and goal atoms, we define its semantics based on an *interpretation* $T = \langle T_b, T_g \rangle$, i.e., a pair of LTL theories. The first element of the pair is used to interpret belief atoms and the second element to interpret goal atoms. An interpretation $T = \langle T_b, T_g \rangle$ is *consistent* if both T_b and T_g are different from $\mathcal{L}_{LTL}$. We define the satisfaction of a formula in $\mathcal{L}_{BG}$ with respect to a pair $\langle T_b, T_g \rangle$ of LTL theories as follows:

$$\langle T_b, T_g \rangle \Vdash \mathbf{B}(\varphi) \quad \text{if } T_b \models_{LTL} \varphi$$
$$\langle T_b, T_g \rangle \Vdash \mathbf{G}(\varphi) \quad \text{if } T_g \models_{LTL} \varphi$$
$$\langle T_b, T_g \rangle \Vdash \neg\delta \quad \text{if } \langle T_b, T_g \rangle \not\Vdash \delta$$
$$\langle T_b, T_g \rangle \Vdash \delta_1 \wedge \delta_2 \text{ if } \langle T_b, T_g \rangle \Vdash \delta_1 \text{ and } \langle T_b, T_g \rangle \Vdash \delta_2$$

Note that we do not impose any constraints on the relation between the LTL theories T_b and T_g, unlike [7,10] where $T_g \models_{LTL} T_b$ is assumed. Our aim here is to be as general as possible, which is witnessed in Section 4 where we show that our framework is flexible enough to impose such constraints in a simple way.

We can now define the consequence relation over the language $\mathcal{L}_{BG}$.

Definition 2. *Given $\delta \in \mathcal{L}_{BG}$ and $\Gamma \subseteq \mathcal{L}_{BG}$, the consequence relation over $\mathcal{L}_{BG}$ is defined as $\Gamma \models \delta$ iff, for interpretation $\langle T_b, T_g \rangle$, we have that $\langle T_b, T_g \rangle \Vdash \delta$ whenever $\langle T_b, T_g \rangle \Vdash \psi$ for every $\psi \in \Gamma$. We say that δ is valid if $\emptyset \models \delta$.*

3 Non-monotonic Belief and Goal Specification

In this section, we present a non-monotonic framework for specifying conditional non-monotonic temporal beliefs and goals. The framework is based on logic programs built over the language $\mathcal{L}_{BG}$ introduced in the previous section.

3.1 Belief and Goal Bases

Belief bases and goal bases represent the agent's beliefs and goals respectively, and are usually defined as sets of formulas from which we can deduce the beliefs and goals of the agent. For example, in [10], sets of LTL formulas are used.

In this work we generalize this assumption by representing belief and goal bases as sets of non-monotonic rules over $\mathcal{L}_{BG}$, similar to those in Logic Programming, making it possible to represent conditional and defeasible goals.

A *rule* r is of the form

$$\varphi \leftarrow \psi_1, \ldots, \psi_n, not\ \delta_1, \ldots, not\ \delta_m \tag{1}$$

where the *head* of r, φ, and each element of its *body*, $\psi_1, \ldots, \psi_n, \delta_1, \ldots, \delta_m$, is either a goal atom or a belief atom. Like in a logic programming rule, the symbol $\leftarrow$ represents rule implication, the symbol "," represents conjunction and the symbol *not* represents default negation. Thus, r represents that φ holds whenever $\psi_1, \ldots, \psi_n$ hold and $\delta_1, \ldots, \delta_m$ are not known to hold. A rule is called *positive* if it does not contain any occurrence of *not*, and *fact* if its body is empty.

A *belief rule* is a rule of the form (1) where the head φ and all elements of the body are belief atoms. A *goal rule* is a rule of the form (1) where the head φ is a goal atom. A *belief base* $\mathcal{B}$ is a set of belief rules and a *goal base* $\mathcal{G}$ is a set of goal rules. A belief base (goal base) is called *positive* if all its rules are positive.

Definition 3. *An* agent configuration *is a pair* $\mathcal{C} = \langle \mathcal{B}, \mathcal{G} \rangle$ *where* $\mathcal{B}$ *is a belief base and* $\mathcal{G}$ *is a goal base. An agent configuration* $\mathcal{C} = \langle \mathcal{B}, \mathcal{G} \rangle$ *is said to be* positive *if both the belief base* $\mathcal{B}$ *and the goal base* $\mathcal{G}$ *are positive.*

Example 4. Consider the simple example about choosing a means of transport.

$$\mathbf{B}(\Diamond strike) \leftarrow \mathbf{B}(strikeInNews) \tag{2}$$
$$\mathbf{G}(\Diamond work) \leftarrow not\ \mathbf{G}(\Diamond beach) \tag{3}$$
$$\mathbf{G}(\Diamond beach) \leftarrow not\ \mathbf{G}(\Diamond work) \tag{4}$$
$$\mathbf{G}(\Diamond bike) \leftarrow \mathbf{G}(\Diamond beach) \tag{5}$$
$$\mathbf{G}(\Diamond car) \leftarrow \mathbf{G}(\Diamond work), \mathbf{B}(\Diamond strike) \tag{6}$$
$$\mathbf{G}(\Diamond(train \sqcup bus)) \leftarrow \mathbf{G}(\Diamond work), not\ \mathbf{B}(\Diamond strike) \tag{7}$$
$$\mathbf{G}(ticket\ \mathcal{B}\ (train \sqcup bus)) \leftarrow \mathbf{G}(\Diamond(train \sqcup bus)) \tag{8}$$
$$\mathbf{G}(money\ \mathcal{B}\ ticket) \leftarrow \mathbf{G}(\Diamond ticket) \tag{9}$$

Informally, an agent believes that there will be a strike if she sees that in the news (2). Rules (3) and (4) represent conflicting goals, i.e., either the agent has the goal to go to the beach or the goal to go to work, not both. In the former case the agent also has the goal to take the bike (5). In the latter case, depending on whether the agent believes that there will be a strike or not, she has the goal to to go by car (6) or the goal to go by train or by bus (7). Moreover, if the agent has the goal to go by train or by bus, then she also has the goal to buy a ticket before that (8). Finally, if the agent has the goal to buy a ticket, then she has the goal to withdraw money first (9).

3.2   Semantics

The definition of a semantics for belief and goal bases is not straightforward due to their complex language. Recall that belief and goal atoms in the rules may contain arbitrary LTL formulas. Thus, unlike, e.g., first-order atoms, belief or goal atoms may not be independent; for example, the goal atoms $\mathbf{G}(\Box(p \lor q))$, $\mathbf{G}(\Diamond \neg p)$ and $\mathbf{G}(\Diamond q)$ are not. To overcome this difficulty, our notion of interpretation accounts for interdependence between such atoms: since $\{\Box(p \lor q), \Diamond \neg p\} \models_{\text{LTL}} \Diamond q$ and since any LTL theory is closed under logical consequence, any interpretation $\langle T_b, T_g \rangle$ satisfying both $\mathbf{G}(\Box(p \lor q))$ and $\mathbf{G}(\Diamond \neg p)$ must also satisfy $\mathbf{G}(\Diamond q)$.

Satisfaction of rules in interpretations and the notion of model for agent configurations can thus be defined in a standard way.

Definition 5. *An* interpretation $T = \langle T_b, T_g \rangle$ *satisfies a rule of the form (1), if* $T \Vdash \varphi$ *whenever* $T \Vdash \psi_i$ *for every* $i \in \{1, \ldots, n\}$ *and* $T \nVdash \delta_j$ *for every* $j \in \{1, \ldots, m\}$. *An interpretation is a* model *of an agent configuration* $\mathcal{C} = \langle \mathcal{B}, \mathcal{G} \rangle$ *if it satisfies every rule of* $\mathcal{B} \cup \mathcal{G}$. *We denote by* $Mod(\mathcal{C})$ *the set of models of* $\mathcal{C}$.

The ordering over interpretations can easily be defined component-wise: given two interpretations $T = \langle T_b, T_g \rangle$ and $T' = \langle T'_b, T'_g \rangle$ we write $T \leq T'$ if $T_b \subseteq T'_b$ and $T_g \subseteq T'_g$. Using this ordering, the notions of minimal and least interpretations can be defined in the usual way.

We are particularly interested in such minimal interpretations and obtain them in a way similar to the stable model semantics. For that purpose, we start by considering positive agent configurations and adapt a well-known result from logic programs saying that every positive agent configuration has a least model.

Theorem 6. *Every positive agent configuration has a least model.*

Based on the semantics for positive agent configurations, we now define the stable model semantics of an agent configuration that can have default negation.

Definition 7. *Let $C = \langle B, G \rangle$ be an agent configuration and $T = \langle T_b, T_g \rangle$ an interpretation. The agent configuration $\frac{C}{T}$ is obtained from C by:*

- *removing from B and G all rules which contain not φ such that $T \Vdash \varphi$;*
- *removing not φ from the remaining rules of B and G.*

Since $\frac{C}{T}$ is a positive agent configuration, it has a unique least model T'. We define $\Gamma_C(T) = T'$.

An interpretation $T = \langle T_b, T_g \rangle$ is a stable model *of an agent configuration $C = \langle B, G \rangle$ if it is consistent and $\Gamma_C(T) = T$. We denote by $SM(C)$ the set of all stable models of C.*

Each stable model can be thought of as a possible consistent set of goals an agent might adopt, which is why T_b and T_g are required to be consistent. Thus, from the point of view of multi-agent systems, each such stable model represents a possible consistent alternative that the agent can adopt as her set of intentions, i.e., those goals that the agent commits to. We note that choosing a particular set of intentions is out of the scope of this paper since, as common in agent architectures, this is dealt with on a meta-level. Still, the following (well-known) entailment relations can be defined.

Definition 8. *Let C be an agent configuration. A formula δ is true under the* stable model semantics *of C, denoted by $C \vDash_{SM} \delta$, if it is satisfied by every stable model of C. A formula δ is true under the* credulous stable model semantics *of C, denoted by $C \vDash_{CSM} \delta$, if it is satisfied by some stable model of C.*

From the agent's perspective, skeptical entailment $\vDash_{SM}$ represents the goals which she will have independently of the particular set of goals she commits to, while credulous entailment $\vDash_{CSM}$ can be used if an agent needs to know whether it is possible that she might adopt that goal.

Example 9. Recall the agent configuration of Ex. 4. Rules (3) and (4) represent the conflicting goals of going to the beach or going to work. This is captured in the semantics by the existence of two stable models. The first one has as consequences the goals $\mathbf{G}(\lozenge beach)$ and $\mathbf{G}(\lozenge bike)$. The second stable model entails

$\mathbf{G}(\Diamond work)$, $\mathbf{G}(\Diamond(train \sqcup bus))$, $\mathbf{G}(ticket \; \mathcal{B} \; (train \sqcup bus))$ and $\mathbf{G}(money \; \mathcal{B} \; ticket)$. Note the fundamental role of complex temporal reasoning in the calculation of the stable models. For example, in the case of the second stable model, rule (9) only fires because $\mathbf{G}(\Diamond ticket)$ follows from $\mathbf{G}(ticket \; \mathcal{B} \; (train \sqcup bus))$ and $\mathbf{G}(\Diamond(train \sqcup bus))$ together, since $\{ticket \; \mathcal{B} \; (train \sqcup bus), \Diamond(train \sqcup bus)\} \models_{\mathrm{LTL}}$ $\Diamond ticket$. Moreover, since the stable models are closed under consequence and since $\{money \; \mathcal{B} \; ticket, \Diamond ticket\} \models_{\mathrm{LTL}} \Diamond money$ we have that the second stable model also entails $\mathbf{G}(\Diamond money)$.

Adding the fact $\mathbf{B}(strikeInNews) \leftarrow$ to the agent configuration of Ex. 4 does not affect the first stable model, but it affects the second. With this extra rule the goals $\mathbf{G}(\Diamond(train \sqcup bus))$, $\mathbf{G}(ticket \; \mathcal{B} \; (train \sqcup bus))$ and $\mathbf{G}(money \; \mathcal{B} \; ticket)$ no longer follow from the second stable model, but now the goal $\mathbf{G}(\Diamond car)$ follows.

4   Properties

We can find a number of properties in the literature that a logical language modeling goals should exhibit. Some are more consensual than others, but that is not the topic of this paper, we rather point to [18]. The aim of this section is to show that some common properties of beliefs and goals hold in our framework and that other approaches that impose additional conditions on the relation between goals and beliefs can be covered.

A usual property of stable models is that they are minimal.

Proposition 10. *Let $\mathcal{C}$ be an agent configuration. If $T = \langle T_b, T_g \rangle$ is a stable model of $\mathcal{C}$, then there is no stable model $T' = \langle T'_b, T'_g \rangle$ of $\mathcal{C}$ such that $T' < T$.*

Modal logic has been used for modeling beliefs and goals of agents [12,5]. The belief operator is usually described using modal logic $KD45$ and the goal operator using the modal logic KD. It is therefore natural to check if these modal axioms hold in our logic. Note that even though we state the following propositions for $\models_{\mathrm{SM}}$, all of them also hold for $\models_{\mathrm{CSM}}$.

Proposition 11. *Let $\mathcal{C} = \langle \mathcal{B}, \mathcal{G} \rangle$ be an agent configuration. The following holds for all LTL formulas φ and ψ:*

(K_b) $\mathcal{C} \models_{\mathrm{SM}} \mathbf{B}(\varphi \Rightarrow \psi) \to (\mathbf{B}(\varphi) \to \mathbf{B}(\psi))$;
(K_g) $\mathcal{C} \models_{\mathrm{SM}} \mathbf{G}(\varphi \Rightarrow \psi) \to (\mathbf{G}(\varphi) \to \mathbf{G}(\psi))$;
(D_b) $\mathcal{C} \models_{\mathrm{SM}} \mathbf{B}(\varphi) \to \neg\mathbf{B}(\sim\varphi)$;
(D_g) $\mathcal{C} \models_{\mathrm{SM}} \mathbf{G}(\varphi) \to \neg\mathbf{G}(\sim\varphi)$.

The above proposition states that both the axioms K and D hold for both the belief and the goal operator in every agent configuration. Note that we do not consider the modal axioms 4 and 5, which are usually associated with the belief operator. The reason is that these axioms involve formulas with nested beliefs, and therefore cannot be represented in our language.

A property that appears for example in [12] is that beliefs and goals should be closed under implication. Our framework satisfies this property.

Proposition 12. *Let $\mathcal{C} = \langle \mathcal{B}, \mathcal{G} \rangle$ be an agent configuration. The following holds for all LTL formulas φ and ψ:*

- $\mathcal{C} \models_{\text{SM}} (\mathbf{B}(\varphi \Rightarrow \psi) \wedge \mathbf{B}(\varphi)) \rightarrow \mathbf{B}(\psi);$
- $\mathcal{C} \models_{\text{SM}} (\mathbf{G}(\varphi \Rightarrow \psi) \wedge \mathbf{G}(\varphi)) \rightarrow \mathbf{G}(\psi);$

Our notion of agent configuration does not have any built-in constraints on the interaction between beliefs and goals, unlike [12,5]. Our language is designed to be general, in the sense that we do not impose these restrictions, yet expressive enough to allow the representation of such constraints if desired.

A constraint that [5,10] impose is the so-called *realism constraint*. Intuitively this means that an agent should have as goals all her beliefs. Although [12] considers this too restrictive, if we want to impose such a restriction in a given agent configuration $\mathcal{C} = \langle \mathcal{B}, \mathcal{G} \rangle$, we just need to add to $\mathcal{G}$ a rule $\mathbf{G}(\varphi) \leftarrow \mathbf{B}(\varphi)$ for every LTL formula φ appearing in $\mathcal{C}$. Let $\mathcal{C}_{\text{Real}}$ be the resulting agent configuration.

A more or less opposite condition is that an agent should not have a goal that she believes is already the case. In [18] this property is described as goals should be *unachieved* (Un). If we want to impose this restriction in a given agent configuration $\mathcal{C} = \langle \mathcal{B}, \mathcal{G} \rangle$, we just need to substitute every rule $\mathbf{G}(\varphi) \leftarrow body$ of $\mathcal{G}$ by the rule $\mathbf{G}(\varphi) \leftarrow body, not \ \mathbf{B}(\varphi)$. Let $\mathcal{C}_{\text{Un}}$ be the resulting agent configuration.

Another commonly considered constraint, described in [18] as goals should be *possible* (Poss), is that an agent should not have a goal that he believes to be impossible. This restriction can also be applied to an agent configuration $\mathcal{C} = \langle \mathcal{B}, \mathcal{G} \rangle$ by substituting every rule $\mathbf{G}(\varphi) \leftarrow body$ of $\mathcal{G}$ by the rule $\mathbf{G}(\varphi) \leftarrow body, not \ \mathbf{B}(\sim\varphi)$. Denote by $\mathcal{C}_{\text{Poss}}$ the resulting agent configuration.

The following proposition states that the above constructions imply that the desired properties hold in the modified agent configuration.

Proposition 13. *Let $\mathcal{C} = \langle \mathcal{B}, \mathcal{G} \rangle$ be an agent configuration. For every LTL formula φ, we have that*

- $\mathcal{C}_{\text{Real}} \models_{\text{SM}} \mathbf{B}(\varphi) \rightarrow \mathbf{G}(\varphi).$

If φ is an LTL formula such that $head_g(\mathcal{C}) \setminus \{\varphi\} \not\models_{\text{LTL}} \varphi$, then

- $\mathcal{C}_{\text{Un}} \models_{\text{SM}} \mathbf{G}(\varphi) \rightarrow \neg\mathbf{B}(\varphi);$
- $\mathcal{C}_{\text{Poss}} \models_{\text{SM}} \mathbf{G}(\varphi) \rightarrow \neg\mathbf{B}(\sim\varphi);$

where $head_g(\mathcal{C})$ is the set of all LTL formulas occurring in rule heads in $\mathcal{G}$.

The reason why the latter two conditions only hold for formulas φ such that $head_g(\mathcal{C}) \setminus \{\varphi\} \not\models_{\text{LTL}} \varphi$, is that only for such formulas can we guarantee that the rules with head $\mathbf{G}(\varphi)$ are the only responsible for $\mathbf{G}(\varphi)$ being a consequence of $\mathcal{C}$. Otherwise $\mathbf{G}(\varphi)$ could follow from another rule: consider for example $\mathcal{C} = \langle\{\mathbf{B}(p \sqcup q) \leftarrow\}, \{\mathbf{G}(p) \leftarrow\}\rangle$. Then, both $\mathbf{B}(p \sqcup q)$ and $\mathbf{G}(p \sqcup q)$ follow from $\mathcal{C}_{\text{Un}}$.

5   Decidability and Implementation

In this section we discuss the decidability and implementation of the following simple reasoning tasks:

- Given an agent configuration $\mathcal{C}$, does the belief $\mathbf{B}(\varphi)$ follow from $\mathcal{C}$?
- Given an agent configuration $\mathcal{C}$, does the goal $\mathbf{G}(\varphi)$ follow from $\mathcal{C}$?

To answer these queries, we need to compute the stable models of $\mathcal{C}$ and then check if they all entail $\mathbf{B}(\varphi)$ and $\mathbf{G}(\varphi)$, respectively. First of all, we prove that for a finite agent configuration, each of the above problems is decidable.

Note that, even if we restrict to a finite set of propositional symbols (for example those that appear in a finite agent configuration), the number of LTL logical theories over this language is infinite. An immediate consequence is that the number of possible stable models of a finite agent configuration is potentially infinite. Interestingly, as we show below, this is not the case and therefore decidability is not compromised. The key idea is the fact that, given a finite agent configuration $\mathcal{C}$, we are able to prove that only those LTL logical theories generated by sets of LTL formulas appearing in $\mathcal{C}$ can be part of a stable model of $\mathcal{C}$, and there is only a finite number of them. To make this precise consider the following sets. Let $form_b(\mathcal{C})$ be the set of LTL formulas that occur in the agent configuration $\mathcal{C}$ in the scope of the belief operator, and $head_b(\mathcal{C}) \subseteq form_b(\mathcal{C})$ the subset of those that occur in the head of a rule. In the same way we can define $form_g(\mathcal{C})$ to be the set of LTL formulas that occur in $\mathcal{C}$ in the scope of the goal operator, and $head_g(\mathcal{C}) \subseteq form_g(\mathcal{C})$ the subset of those that occur in the head of a rule. Of course, if $\mathcal{C}$ is finite then both $form_b(\mathcal{C})$ and $form_b(\mathcal{C})$ are finite.

Theorem 14. *Let $\mathcal{C} = \langle \mathcal{B}, \mathcal{G} \rangle$ be a finite agent configuration. If $\langle T_b, T_g \rangle$ is a stable model of $\mathcal{C}$, then there exists $A_b \subseteq form_b(\mathcal{C})$ and $A_g \subseteq form_g(\mathcal{C})$ such that $T_b = A_b^{\models_{\mathrm{LTL}}}$ and $T_g = A_g^{\models_{\mathrm{LTL}}}$.*

An immediate consequence of the above theorem is that every finite agent configuration has a finite number of stable models.

Corollary 15. *Every finite agent configuration has finitely many stable models.*

Our aim now is to define an algorithm that modularly combines an LTL reasoner and an ASP solver to compute the answers to the above queries. Recall that the validity problem in LTL is decidable [17]. The advantage of such modular algorithm is that we can leverage existing LTL and ASP reasoners.

Consider a given finite agent configuration $\mathcal{C} = \langle \mathcal{B}, \mathcal{G} \rangle$. We construct the normal logic program $\mathcal{P}^{\mathcal{C}}$ obtained from $\mathcal{C}$ in which belief and goal atoms containing LTL formulas are encoded as normal logic program atoms:

$$
\begin{aligned}
\mathcal{P}^{\mathcal{C}} = \mathcal{B} \cup \mathcal{G} \cup \{\, &\mathbf{B}(\varphi) \leftarrow \mathbf{B}(\psi_1), \ldots, \mathbf{B}(\psi_n) : \{\psi_1, \ldots, \psi_n\} \subseteq head_b(\mathcal{C}), \\
&\varphi \in form_b(\mathcal{C}) \setminus \{\psi_1, \ldots, \psi_n\}, \{\psi_1, \ldots, \psi_n\} \models_{\mathrm{LTL}} \varphi \,\} \cup \\
\cup \{\, &\mathbf{G}(\varphi) \leftarrow \mathbf{G}(\psi_1), \ldots, \mathbf{G}(\psi_n) : \{\psi_1, \ldots, \psi_n\} \subseteq head_g(\mathcal{C}), \\
&\varphi \in form_g(\mathcal{C}) \setminus \{\psi_1, \ldots, \psi_n\}, \{\psi_1, \ldots, \psi_n\} \models_{\mathrm{LTL}} \varphi \,\}
\end{aligned}
$$

To distinguish the set of stable models of an agent configuration C, which is a set of pairs of LTL theories, from the set of stable models of $\mathcal{P}^C$, which is a subset of the set of belief and goal atoms occurring in $\mathcal{P}^C$, we denote the later by $AS(\mathcal{P}^C)$, and by $form(\mathcal{P}^C)$ the set of belief and goal atoms appearing in $\mathcal{P}^C$.

The key idea underlying the construction of $\mathcal{P}^C$ is to enrich the original agent configuration with rules that represent the possible interaction occurring between the formulas of the program in order to enforce the interdependency between temporal formulas appearing in the belief and goal atoms. Note that for a given agent configuration $C = \langle \mathcal{B}, \mathcal{G} \rangle$, the sets $SM(C)$ and $AS(\mathcal{B} \cup \mathcal{G})$ may not be related, since AS does not take into account the logical interdependency between the formulas appearing in C. As a simple example consider the agent configuration $C = \langle \emptyset, \{\mathbf{G}(\Box p) \leftarrow; \ \mathbf{G}(q) \leftarrow \mathbf{G}(\Diamond p)\} \rangle$. Then, $SM(C) = \{\langle \emptyset^{\vDash_{\mathrm{LTL}}}, \{\Box p, \Diamond p, q\}^{\vDash_{\mathrm{LTL}}} \rangle\}$ but $AS(\mathcal{B} \cup \mathcal{G}) = \{\{\mathbf{G}(\Box p)\}\}$. This is the reason why we cannot use an ASP solver directly on the program $\mathcal{B} \cup \mathcal{G}$.

In the case of $\mathcal{P}^C$, we have the following strong relation.

Theorem 16. *Given a finite agent configuration $C = \langle \mathcal{B}, \mathcal{G} \rangle$, we have that*

1. $\{T_b : \langle T_b, T_g \rangle \in SM(C)\} = \{\{\varphi : \mathbf{B}(\varphi) \in A\}^{\vDash_{\mathrm{LTL}}} : A \in AS(\mathcal{P}^C)\}$

2. $\{T_g : \langle T_b, T_g \rangle \in SM(C)\} = \{\{\varphi : \mathbf{G}(\varphi) \in A\}^{\vDash_{\mathrm{LTL}}} : A \in AS(\mathcal{P}^C)\}$

The above theorem presents a finite representation of the stable models of an agent configuration: the stable models of the program $\mathcal{P}^C$. An immediate consequence is that the problems of checking if a belief or a goal atom follows from a finite agent configuration are both decidable.

Corollary 17. *Let C be a finite agent configuration and φ an LTL formula (not necessarily appearing in C). Then, the problems of checking if $C \vDash_{\mathrm{SM}} \mathbf{B}(\varphi)$ and if $C \vDash_{\mathrm{SM}} \mathbf{G}(\varphi)$ are both decidable.*

The decidability of entailment for belief and goal atoms can be extended for complex formulas, since these depend only on the atoms appearing in it.

Corollary 18. *Let C be a finite agent configuration and δ a complex belief or goal formula (not necessarily in C). Then, the problem $C \vDash_{\mathrm{SM}} \delta$ is decidable.*

6   Related Work

There are several approaches in the literature that use temporal logic to model goals [5,12,16,7,10]. The work in [10] uses sets of LTL formulas to define both the belief and the goal bases. These can be easily captured by our non-monotonic framework, as we now show. Formally, in [10] a belief base is a set $\Sigma \subseteq \mathcal{L}_{\mathrm{LTL}}$, a goal base is a set $\Gamma \subseteq \mathcal{L}_{\mathrm{LTL}}$, and a *mental state* is a pair $m = \langle \Sigma, \Gamma \rangle$ such that both Σ and Γ are consistent and $\Gamma \vDash_{\mathrm{LTL}} \Sigma$. The reason for the last condition is to impose the realism principle mentioned in Section 4, i.e., that an agent should have as goals all her beliefs. In fact, the formula $\mathbf{G}(\varphi) \rightarrow \mathbf{B}(\varphi)$ is valid in their logic. Their language of beliefs and goals is the same as our language L and the satisfaction of formulas of $\mathcal{L}_{BG}$ by a mental state m is defined in a similar way as we do for agent configurations. For a given mental state $m = \langle \Sigma, \Gamma \rangle$ we can consider the corresponding (positive) agent configuration $\mathcal{C}_m = \langle \mathcal{B}_m, \mathcal{G}_m \rangle$ where $\mathcal{B}_m = \{\mathbf{B}(\varphi) \leftarrow : \varphi \in \Sigma\}$ and $\mathcal{G}_m = \{\mathbf{G}(\varphi) \leftarrow : \varphi \in \Gamma\}$. It is immediate to check that the unique stable model of $\mathcal{C}_m$ is precisely $\langle \Sigma^{\vDash_{\mathrm{LTL}}}, \Gamma^{\vDash_{\mathrm{LTL}}} \rangle$, and therefore, for every formula δ of $\mathcal{L}_{BG}$, we have that $m \Vdash \delta$ iff $\mathcal{C}_m \vDash_{\mathrm{SM}} \delta$.

The work in [15] defines a framework for conditional goals using a translation to default logic [14]. Although it does not consider temporal goals, it offers an interesting non-monotonic framework for modeling goals. In what follows we briefly sketch the relation between the work in [15] and ours. Let us start with a very brief presentation of their framework. The language for beliefs and goals is a restriction of our language $\mathcal{L}_{BG}$, in the sense that in the scope of a belief or goal operator only propositional formulas without temporal operators are allowed. Conditional goals are defined through *goal inference rules*, which are of the form: $\beta, \kappa^+, \kappa^- \Rightarrow \phi$, where β, κ^+ and κ^- are sets of propositional formulas. In such a goal inference rule, ϕ represents the goal that can be inferred if the beliefs in β are true, the goals in κ^+ are true and the goals in κ^- are not known to be true. A goal base is a set GI of goal inference rules. A belief base σ is just a set of propositional formulas. For beliefs, the semantics is easily defined by $\langle \sigma, GI \rangle \vDash_d \mathbf{B}(\varphi)$ if $\sigma \vDash \varphi$. The semantics of a goal base GI is defined by translating it to a default theory $t(GI)$. This is done by translating each goal inference rule $r = \{\beta_1, \ldots, \beta_k\}, \{\alpha_1, \ldots, \alpha_n\}^+, \{\varphi_1, \ldots, \varphi_m\}^- \Rightarrow \varphi$ in GI such that $\sigma \vDash \{\beta_1, \ldots, \beta_k\}$ to the default rule $t(r) = \alpha_1 \sqcap \cdots \sqcap \alpha_n : \sim\varphi_1, \ldots, \sim\varphi_m, \varphi / \varphi$. The (credulous) entailment for goals is then defined as $\langle \sigma, GI \rangle \vDash_d \mathbf{G}(\varphi)$ if there exists an extension (in the sense of default logic) E of $t(GI)$ such that $E \vDash \varphi$.

Given a goal base GI and a belief base σ consider the agent configuration $\langle \mathcal{B}_\sigma, \mathcal{G}_{GI} \rangle$ such that $\mathcal{B}_\sigma = \{\mathbf{B}(\varphi) \leftarrow : \varphi \in \sigma\}$ and $\mathcal{G}_{GI}$ is obtained by considering, for each rule $r = \{\beta_1, \ldots, \beta_k\}, \{\alpha_1, \ldots, \alpha_n\}^+, \{\varphi_1, \ldots, \varphi_m\}^- \Rightarrow \varphi$ in GI, the rule $\mathbf{G}(\varphi) \leftarrow \mathbf{B}(\beta_1 \wedge \cdots \wedge \beta_k), \mathbf{G}(\alpha_1 \wedge \cdots \wedge \alpha_n), not\ \mathbf{G}(\varphi_1), \ldots, not\ \mathbf{G}(\varphi_m), not\ \mathbf{G}(\neg\varphi)$. Note that we are just using a fragment of our language to embed both σ and GI. In the case of beliefs, we just need to use facts, and in the case of goals, we do not use temporal formulas nor default negated beliefs. We can then prove that $\langle \sigma, GI \rangle \vDash_d \mathbf{B}(\varphi)$ iff $\langle \mathcal{B}_\sigma, \mathcal{G}_{GI} \rangle \vDash_{\mathrm{CSM}} \mathbf{B}(\varphi)$. Also $\langle \sigma, GI \rangle \vDash_d \mathbf{G}(\varphi)$ iff $\langle \mathcal{B}_\sigma, \mathcal{G}_{GI} \rangle \vDash_{\mathrm{CSM}} \mathbf{G}(\varphi)$. Moreover, a similar result holds between the skeptical entailment $\vDash_{dd}$ defined in [15] and our skeptical entailment $\vDash_{\mathrm{SM}}$. In the case of skeptical entailment, the relation is even stronger since it holds for every formula of $\mathcal{L}_{BG}$. The reason why the above proposition does not follow for any complex formula of $\mathcal{L}_{BG}$ is the fact that the entailment $\vDash_d$ of [15] is first defined for belief and goal atoms, and only then extended to complex formulas. In this way, in their framework we may have that $\langle \sigma, GI \rangle \vDash_d \mathbf{G}(p) \wedge \mathbf{G}(\sim p)$, even for consistent σ and GI, since this means that $\mathbf{G}(p)$ is true in one stable model and $\mathbf{G}(\sim p)$ is true in a different stable model. This is not the case in our framework. Although this distinguishes our approach from that in [15], a more fundamental difference is that they adopt a particular entailment over the possible extensions (in the sense of default logic), thus not making use the full richness of the set of stable models. On the contrary, we argue that the set of stable models is fundamental since it can be seen as the set of possible sets of goals an agent can commit to.

Let us now draw some comments on the work of [2]. There, a simple non-monotonic version of temporal logic for specifying goals is defined. This is done by extending the language of temporal logic with two operators to model weak and strong exceptions. Although for lack of space we do not present here the details, it can be shown that the non-monotonic extension N-LTL of LTL presented in [2] can be embedded in our framework. This is not surprising since in our framework the use of default negation allows to model exceptions in a very flexible way.

7   Conclusions and Future Work

In this paper, we have defined a non-monotonic framework for representing temporal beliefs and goals, along with a stable models like semantics for this expressive language. We have argued that an ASP view of the stable models of an agent configuration can bring a novel perspective on dealing with multiple possible sets of goals. We have proven that some usual properties of beliefs and goals hold in our framework. Moreover, we have shown that the problem of checking the entailment of a formula in a given finite agent configuration is decidable, and we presented an implementation that makes a modular use of an LTL reasoner and an ASP solver. In the end, we have briefly hinted on how existing work on the representation of beliefs and goals can be embedded in our framework.

This work raises several interesting directions for future research. Since the temporal operators can only appear in the scope of belief or goal operators, our approach does not deal with the evolution of the belief and goal bases. An interesting idea to cope with such evolution is to use some extension of dynamic logic programs [1]. Also interesting is to study the connection between our framework and the general approach of parametrized logic programming [9].

Additionally, our work could be integrated with existing agent programming languages/architectures, e.g., 2APL [6] and Jason [3], thus increasing their capabilities to represent and reason about goals.

Finally, we also want to extend our work so that we can consider the currently adopted intentions and how they can influence the stable models that encode the new ones to be adopted. This is an interesting problem but rather complex since we typically would like to keep as many of the current intentions as possible, i.e., the problem might be solvable by treating current intentions as beliefs, but it might require a measure of distance and seek for models that encode minimal change between the current intentions and the goals in possible stable models.

Acknowledgments. We would like to thank the anonymous reviewers whose comments helped to improve the paper.

Matthias Knorr, João Leite and Martin Slota were partially supported by FCT under project "ERRO – Efficient Reasoning with Rules and Ontologies" (PTDC/EIA-CCO/121823/2010). Ricardo Gonçalves was supported by FCT grant SFRH/BPD/47245/2008 and Matthias Knorr was also partially supported by FCT grant SFRH/BPD/86970/2012.

References

1. Alferes, J.J., Banti, F., Brogi, A., Leite, J.A.: The refined extension principle for semantics of dynamic logic programming. Studia Logica 79(1), 7–32 (2005)
2. Baral, C., Zhao, J.: Non-monotonic temporal logics for goal specification. In: Veloso, M.M. (ed.) IJCAI, pp. 236–242 (2007)
3. Bordini, R.H., Hübner, J.F., Wooldridge, M.: Programming multi-agent systems in AgentSpeak using Jason. Wiley-Interscience (2007)
4. Bratman, M.: Intention, plans, and practical reason. Harvard University Press, Cambridge (1987)
5. Cohen, P.R., Levesque, H.J.: Intention is choice with commitment. Artif. Intell. 42(2-3), 213–261 (1990)
6. Dastani, M.: 2APL: a practical agent programming language. Autonomous Agents and Multi-Agent Systems 16(3), 214–248 (2008)
7. Dastani, M., van Riemsdijk, M.B., Winikoff, M.: Rich goal types in agent programming. In: Sonenberg, L., Stone, P., Tumer, K., Yolum, P. (eds.) AAMAS, pp. 405–412. IFAAMAS (2011)
8. Gelfond, M., Lifschitz, V.: Classical negation in logic programs and disjunctive databases. New Generation Comput. 9(3-4), 365–385 (1991)
9. Gonçalves, R., Alferes, J.J.: Parametrized logic programming. In: Janhunen, T., Niemelä, I. (eds.) JELIA 2010. LNCS, vol. 6341, pp. 182–194. Springer, Heidelberg (2010)
10. Hindriks, K.V., van der Hoek, W., van Riemsdijk, M.B.: Agent programming with temporally extended goals. In: Sierra, C., Castelfranchi, C., Decker, K.S., Sichman, J.S. (eds.) AAMAS (1), pp. 137–144. IFAAMAS (2009)
11. Pnueli, A.: The temporal logic of programs. In: 18th Annual Symposium on Foundations of Computer Science (Providence, R.I., 1977), pp. 46–57. IEEE Comput. Sci., Long Beach (1977)
12. Rao, A.S., Georgeff, M.P.: Modeling rational agents within a BDI-architecture. In: Allen, J.F., Fikes, R., Sandewall, E. (eds.) KR, pp. 473–484. Morgan Kaufmann (1991)
13. Rao, A.S., Georgeff, M.P.: BDI agents: From theory to practice. In: Lesser, V.R., Gasser, L. (eds.) ICMAS, pp. 312–319. The MIT Press (1995)
14. Reiter, R.: A logic for default reasoning. Artif. Intell. 13(1-2), 81–132 (1980)
15. van Riemsdijk, M.B., Dastani, M., Meyer, J.J.C.: Goals in conflict: semantic foundations of goals in agent programming. Autonomous Agents and Multi-Agent Systems 18(3), 471–500 (2009)
16. van Riemsdijk, M.B., Dastani, M., Winikoff, M.: Goals in agent systems: a unifying framework. In: Padgham, L., Parkes, D.C., Müller, J.P., Parsons, S. (eds.) AAMAS (2), pp. 713–720. IFAAMAS (2008)
17. Sistla, A.P., Clarke, E.M.: The complexity of propositional linear temporal logics. J. ACM 32(3), 733–749 (1985)
18. Winikoff, M., Padgham, L., Harland, J., Thangarajah, J.: Declarative & procedural goals in intelligent agent systems. In: Fensel, D., Giunchiglia, F., McGuinness, D.L., Williams, M.A. (eds.) KR, pp. 470–481. Morgan Kaufmann (2002)
19. Wójcicki, R.: Theory of Logical Calculi. Synthese Library. Kluwer Academic Publishers (1988)

On Equivalent Transformations of Infinitary Formulas under the Stable Model Semantics
(Preliminary Report)

Amelia Harrison[1], Vladimir Lifschitz[1], and Miroslaw Truszczynski[2]

[1] University of Texas, Austin, Texas, USA
{ameliaj,vl}@cs.utexas.edu
[2] University of Kentucky, Lexington, Kentucky, USA
mirek@cs.uky.edu

Abstract. It has been known for a long time that intuitionistically equivalent formulas have the same stable models. We extend this theorem to propositional formulas with infinitely long conjunctions and disjunctions and show how to apply this generalization to proving properties of aggregates in answer set programming.

1 Introduction

This note is about the extension of the stable model semantics to infinitary propositional formulas defined in [6]. One of the reasons why stable models of infinitary formulas are important is that they are closely related to aggregates in answer set programming (ASP). The semantics of aggregates proposed in [1, Section 4.1] treats a ground aggregate as shorthand for a propositional formula. An aggregate with variables has to be grounded before that semantics can be applied to it. For instance, to explain the precise meaning of the expression $1\{p(X)\}$ ("there exists at least one object with the property p") in the body of an ASP rule we first rewrite it as

$$1\{p(t_1), \ldots, p(t_n)\},$$

where $t_1, \ldots, t_n$ are all ground terms in the language of the program, and then turn it into the propositional formula $p(t_1) \vee \cdots \vee p(t_n)$. But this description of the meaning of $1\{p(X)\}$ implicitly assumes that the Herbrand universe of the program is finite. If the program contains function symbols then an infinite disjunction has to be used.[1,2]

---

[1] There is nothing exotic or noncomputable about ASP programs containing both aggregates and function symbols. For instance, the program

$$p(f(a))$$
$$q \leftarrow 1\{p(X)\}$$

has simple intuitive meaning, and its stable model $\{p(f(a)), q\}$ can be computed by existing solvers.

[2] References to grounding in other theories of aggregates suffer from the same problem. For instance, the definition of a ground instance in Section 2.2 of the ASP Core

P. Cabalar and T.C. Son (Eds.): LPNMR 2013, LNAI 8148, pp. 387–394, 2013.

Our goal here is to develop methods for proving that pairs F, G of infinitary formulas have the same stable models. From the results of [5] and [1] we know that in the case of finite propositional formulas it is sufficient to check that the equivalence $F \leftrightarrow G$ is provable intuitionistically. Some extensions of intuitionistic propositional logic, including the logic of here-and-there, can be used as well. In this note we extend these results to deductive systems of infinitary propositional logic.

This goal is closely related to the idea of strong equivalence [4]. The provability of $F \leftrightarrow G$ in the deductive systems of infinitary logic described below guarantees not only that F and G have the same stable models, but also that for any set $\mathcal{H}$ of infinitary formulas, $\mathcal{H} \cup \{F\}$ and $\mathcal{H} \cup \{G\}$ have the same stable models.

We review the stable model semantics of infinitary propositional formulas in Section 2. An infinitary system of natural deduction, similar to propositional intuitionistic logic, is defined in Section 3. Then we discuss the main theorem, which relates this system to stable models (Section 4), and state a few other useful facts (Section 5). In Section 6 this theory is applied to examples involving aggregates.

2   Stable Models of Infinitary Propositional Formulas

The definitions of infinitary formulas and their stable models given below are equivalent to the definitions proposed in [6].

Let σ be a propositional signature, that is, a set of propositional atoms. The sets $\mathcal{F}_0^\sigma$, $\mathcal{F}_1^\sigma$, ... are defined as follows:

- $\mathcal{F}_0^\sigma = \sigma \cup \{\bot\}$,
- $\mathcal{F}_{i+1}^\sigma$ is obtained from $\mathcal{F}_i^\sigma$ by adding expressions $\mathcal{H}^\wedge$ and $\mathcal{H}^\vee$ for all subsets $\mathcal{H}$ of $\mathcal{F}_i^\sigma$, and expressions $F \rightarrow G$ for all $F, G \in \mathcal{F}_i^\sigma$.

The elements of $\bigcup_{i=0}^{\infty} \mathcal{F}_i^\sigma$ are called *(infinitary) formulas* over σ.

Negation and equivalence will be understood as abbreviations: $\neg F$ stands for $F \rightarrow \bot$, and $F \leftrightarrow G$ stands for $(F \rightarrow G) \wedge (G \rightarrow F)$.

We will write $\{F, G\}^\wedge$ as $F \wedge G$, and $\{F, G\}^\vee$ as $F \vee G$. Thus finite propositional formulas over σ can be viewed as a special case of infinitary formulas.

Subsets of a signature σ will be also called its *interpretations*. The satisfaction relation between an interpretation I and a formula F is defined as follows:

- $I \not\models \bot$.
- For every $p \in \sigma$, $I \models p$ if $p \in I$.
- $I \models \mathcal{H}^\vee$ if there is a formula $F \in \mathcal{H}$ such that $I \models F$.
- $I \models \mathcal{H}^\wedge$ if for every formula $F \in \mathcal{H}$, $I \models F$.
- $I \models F \rightarrow G$ if $I \not\models F$ or $I \models G$.

---

document (`https://www.mat.unical.it/aspcomp2013/files/ASP-CORE-2.0.pdf`, Version 2.02) talks about replacing the expression $\{e_1; \ldots; e_n\}$ in a rule with a set denoted by inst($\{e_1; \ldots; e_n\}$). But that set can be infinite.

We say that I satisfies a set $\mathcal{H}$ of formulas if I satisfies all elements of $\mathcal{H}$.

The *reduct* F^I of a formula F with respect to an interpretation I is defined as follows:

- $\perp^I = \perp$.
- For $p \in \sigma$, $p^I = \perp$ if $I \not\models p$; otherwise $p^I = p$.
- $(\mathcal{H}^\wedge)^I = \{G^I \mid G \in \mathcal{H}\}^\wedge$.
- $(\mathcal{H}^\vee)^I = \{G^I \mid G \in \mathcal{H}\}^\vee$.
- $(G \to H)^I = \perp$ if $I \not\models G \to H$; otherwise $(G \to H)^I = G^I \to H^I$.

The *reduct* $\mathcal{H}^I$ of a set $\mathcal{H}$ of formulas is the set consisting of the reducts of the elements of $\mathcal{H}$. An interpretation I is a *stable model* of a set $\mathcal{H}$ of formulas if it is minimal w.r.t. set inclusion among the interpretations satisfying $\mathcal{H}^I$; a stable model of a formula F is a stable model of singleton $\{F\}$. This is a straightforward extension of the definition of a stable model due to Ferraris [1] to infinitary formulas.

3   Basic Infinitary System of Natural Deduction

Inference rules of the deductive system described below are similar to the standard natural deduction rules of propositional logic (see, for instance, [3, Section 1.2.1]). In this system, derivable objects are *(infinitary) sequents*—expressions of the form $\Gamma \Rightarrow F$, where F is an infinitary formula, and Γ is a *finite* set of infinitary formulas ("F under assumptions Γ"). To simplify notation, we will write Γ as a list. We will identify a sequent of the form $\Rightarrow F$ with the formula F.

There is one axiom schema $F \Rightarrow F$. The inference rules are the introduction and elimination rules for the propositional connectives

$$(\wedge I) \;\frac{\Gamma \Rightarrow H \quad \text{for all } H \in \mathcal{H}}{\Gamma \Rightarrow \mathcal{H}^\wedge} \qquad (\wedge E) \;\frac{\Gamma \Rightarrow \mathcal{H}^\wedge}{\Gamma \Rightarrow H} \;\; (H \in \mathcal{H})$$

$$(\vee I) \;\frac{\Gamma \Rightarrow H}{\Gamma \Rightarrow \mathcal{H}^\vee} \;\; (H \in \mathcal{H}) \qquad (\vee E) \;\frac{\Gamma \Rightarrow \mathcal{H}^\vee \quad \Delta, H \Rightarrow F \quad \text{for all } H \in \mathcal{H}}{\Gamma, \Delta \Rightarrow F}$$

$$(\to I) \;\frac{\Gamma, F \Rightarrow G}{\Gamma \Rightarrow F \to G} \qquad\qquad (\to E) \;\frac{\Gamma \Rightarrow F \quad \Delta \Rightarrow F \to G}{\Gamma, \Delta \Rightarrow G}$$

and the contradiction and weakening rules

$$(C) \;\frac{\Gamma \Rightarrow \perp}{\Gamma \Rightarrow F} \qquad (W) \;\frac{\Gamma \Rightarrow F}{\Gamma, \Delta \Rightarrow F} \;.$$

(Note that we did not include the law of the excluded middle in the set of axioms, so that this deductive system is similar to intuitionistic, rather than classical, propositional logic.)

The set of *theorems of the basic system* is the smallest set of sequents that includes the axioms of the system and is closed under the application of its inference rules. We say that formulas F and G are *equivalent in the basic system* if $F \leftrightarrow G$ is a theorem of the basic system. The reason why we are interested in this relation is that formulas equivalent in the basic system have the same stable models, as discussed in Section 4 below.

Example 1. Consider a formula of the form

$$F_0 \wedge \{F_i \to F_{i+1} \mid i \geq 0\}^{\wedge}$$

or, in more compact notation,

$$F_0 \wedge \bigwedge_{i \geq 0} (F_i \to F_{i+1}). \tag{1}$$

Let us check that it is equivalent in the basic system to the formula $\bigwedge_{i \geq 0} F_i$. The sequent

$$F_0 \wedge \bigwedge_{i \geq 0} (F_i \to F_{i+1}) \Rightarrow F_0 \wedge \bigwedge_{i \geq 0} (F_i \to F_{i+1})$$

belongs to the set of theorems of the basic system. Consequently so do the sequents

$$F_0 \wedge \bigwedge_{i \geq 0} (F_i \to F_{i+1}) \Rightarrow F_0$$

and

$$F_0 \wedge \bigwedge_{i \geq 0} (F_i \to F_{i+1}) \Rightarrow F_j \to F_{j+1}$$

for all $j \geq 0$. Consequently the sequents

$$F_0 \wedge \bigwedge_{i \geq 0} (F_i \to F_{i+1}) \Rightarrow F_j$$

for all $j \geq 0$ belong to the set of theorems as well (by induction on j). Consequently so does the sequent

$$F_0 \wedge \bigwedge_{i \geq 0} (F_i \to F_{i+1}) \Rightarrow \bigwedge_{i \geq 0} F_i.$$

A similar argument (except that induction is not needed) shows that the sequent

$$\bigwedge_{i \geq 0} F_i \Rightarrow F_0 \wedge \bigwedge_{i \geq 0} (F_i \to F_{i+1})$$

is a theorem of the basic system also. Consequently so is the sequent

$$\Rightarrow F_0 \wedge \bigwedge_{i \geq 0} (F_i \to F_{i+1}) \leftrightarrow \bigwedge_{i \geq 0} F_i.$$

This argument could be expressed more concisely, without explicit references to the set of theorems of the basic system, as follows. Assume (1). Then F_0 and, for every $i \geq 0$, $F_i \to F_{i+1}$. Then, by induction, F_i for every i. And so forth. This style of presentation is used in the next example.

Example 2. Let $\{F_\alpha\}_{\alpha \in A}$ be a family of formulas from some $\mathcal{F}_i^\sigma$, and let G be a formula. We show that

$$\left(\bigvee_{\alpha \in A} F_\alpha \right) \to G \tag{2}$$

is equivalent in the basic system to the formula

$$\bigwedge_{\alpha \in A} (F_\alpha \to G). \tag{3}$$

Left-to-right: assume (2) and F_α. Then $\bigvee_{\alpha \in A} F_\alpha$, and consequently G. Thus we established $F_\alpha \to G$ under assumption (2) alone for every α, and consequently established (3) under this assumption as well. Right-to-left: assume (3) and $\bigvee_{\alpha \in A} F_\alpha$, and consider the cases corresponding to the disjunctive terms of this disjunction. Assume F_α. From (3), $F_\alpha \to G$, and consequently G. Thus we established G in each case, so that (2) follows from (3) alone.

4 Main Theorem

Main Theorem. *For any set $\mathcal{H}$ of formulas,*

(a) if a formula F is a theorem of the basic system then $\mathcal{H} \cup \{F\}$ has the same stable models as $\mathcal{H}$;

(b) if F is equivalent to G in the basic system then $\mathcal{H} \cup \{F\}$ and $\mathcal{H} \cup \{G\}$ have the same stable models.

The proof of the main theorem relies on the following lemma: *For any theorem $\Gamma \Rightarrow F$ of the basic system and any interpretation I, the sequent $\{G^I \mid G \in \Gamma\} \Rightarrow F^I$ is a theorem of the basic system as well.* To prove the lemma, we show that the set of sequents $\Gamma \Rightarrow F$ such that $\{G^I \mid G \in \Gamma\} \Rightarrow F^I$ is a theorem of the basic system includes the axioms of the basic system and is closed under its inference rules.

The assertion of the theorem will remain true if we add an axiom schema corresponding to an infinitary version of the weak law of the excluded middle $\neg F \vee \neg\neg F$:

$$\bigvee_{\mathcal{I} \subseteq \mathcal{H}} \left(\neg \bigvee_{F \in \mathcal{H} \setminus \mathcal{I}} F \wedge \neg\neg \bigwedge_{F \in \mathcal{I}} F \right), \tag{4}$$

where $\mathcal{H}$ is an arbitrary subset of one of the sets $\mathcal{F}_i$.

5 Some Useful Properties of the Basic System

Let σ and σ' be disjoint signatures. A *substitution* is an arbitrary function from σ' to $\mathcal{F}_i^\sigma$, where i is a nonnegative integer. For any substitution α and any formula F over the signature $\sigma \cup \sigma'$, F^α stands for the formula over σ formed as follows:

- If $F \in \sigma$ or $F = \bot$ then $F^\alpha = F$.
- If $F \in \sigma'$ then $F^\alpha = \alpha(F)$.
- If F is $\mathcal{H}^\wedge$ then $F^\alpha = \{G^\alpha | G \in \mathcal{H}\}^\wedge$.
- If F is $\mathcal{H}^\vee$ then $F^\alpha = \{G^\alpha | G \in \mathcal{H}\}^\vee$.
- If F is $G \to H$ then $F^\alpha = G^\alpha \to H^\alpha$.

Formulas of the form F^α will be called *instances* of F.

Proposition 1. *If F is a theorem of the basic system then every instance of F is a theorem of the basic system also.*

Corollary. *If F is a finite formula provable in intuitionistic propositional logic then every instance of F is a theorem of the basic system.*

Proposition 2. *If for every atom p, $\alpha(p)$ is equivalent to $\beta(p)$ in the basic system then F^α is equivalent to F^β in the basic system.*

6 Examples Involving Aggregates

As discussed in the introduction, infinitary formulas can be used to precisely define the semantics of aggregates in ASP when the Herbrand universe is infinite. In this section, we give three examples demonstrating how the theory described above can be applied to prove equivalences between programs involving aggregates.

Example 3. Intuitively, the rule

$$q(X) \leftarrow 1\{p(X, Y)\} \tag{5}$$

has the same meaning as the rule

$$q(X) \leftarrow p(X, Y). \tag{6}$$

To make this claim precise, consider first the result of grounding rule (5) under the assumption that the Herbrand universe C is finite. In accordance with standard practice in ASP, we treat variable X as global and Y as local. Then the result of grounding (5) is the set of ground rules

$$q(a) \leftarrow 1\{p(a, b) \mid b \in C\}$$

for all $a \in C$. In the spirit of the semantics for aggregates proposed in [1, Section 4.1] these rules have the same meaning as the propositional formulas

$$\left(\bigvee_{b \in C} p(a, b) \right) \to q(a). \tag{7}$$

Likewise, rule (6) can be viewed as shorthand for the set of formulas

$$p(a, b) \to q(a) \tag{8}$$

for all $a, b \in C$. It easy to see that these sets of formulas are intuitionistically equivalent.

How can we lift the assumption that the Herbrand universe is finite? We can treat (7) as an infinitary formula, and show that the conjunction of formulas (7) is equivalent to the conjunction of formulas (8) in the basic system. The fact that the conjunction of formulas (8) for all $b \in C$ is equivalent to (7) in the basic system follows from Example 2 (Section 3).

Example 4. Intuitively,

$$q(X) \leftarrow 2\{p(X, Y)\} \tag{9}$$

has the same meaning as the rule

$$q(X) \leftarrow p(X, Y1), \ p(X, Y2), \ Y1 \neq Y2. \tag{10}$$

To make this claim precise, consider the infinitary formulas corresponding to (9):

$$\left(\bigvee_{b \in C} p(a, b) \wedge \bigwedge_{b \in C} \left(p(a, b) \rightarrow \bigvee_{\substack{c \in C \\ c \neq b}} p(a, c) \right) \right) \rightarrow q(a) \tag{11}$$

$(a \in C)$; see [1, Section 4.1] for details on representing aggregates with propositional formulas. The formulas corresponding to (10) are

$$(p(a, b) \wedge p(a, c)) \rightarrow q(a) \tag{12}$$

$(a, b, c \in C, \ b \neq c)$. Using the propositions stated above, we can show that the conjunction of formulas (11) is equivalent to the conjunction of formulas (12) in the basic system.

Example 5. Intuitively, the cardinality constraint $\{p(X)\}0$ ("the set of true atoms with form $p(X)$ has cardinality at most 0") has the same meaning as the conditional literal $\bot : p(X)$ ("for all X, $p(X)$ is false"). If we represent this conditional literal by the infinitary formula

$$\bigwedge_{a \in C} \neg p(a) \tag{13}$$

then this claim can be made precise by showing that (13) is equivalent to the formula

$$\bigwedge_{\substack{A \subseteq C \\ A \neq \emptyset}} \left(\bigwedge_{a \in A} p(a) \rightarrow \bigvee_{a \in C \setminus A} p(a) \right), \tag{14}$$

which corresponds to $\{p(X)\}0$ in the sense of [1], in the extended system described at the end of Section 4. It is easy to derive (14) from (13) in the basic system. The derivation of (13) from (14) uses the following instance of axiom schema (4):

$$\bigvee_{A \subseteq C} \left(\neg \bigvee_{a \in C \setminus A} p(a) \wedge \neg\neg \bigwedge_{a \in A} p(a) \right). \tag{15}$$

7   Future Work

Two finite propositional formulas are strongly equivalent if and only if they are equivalent in the logic of here-and-there [1, Proposition 2]. The results of this note are similar to the if part of that theorem; we don't know how to extend the only if part to infinitary formulas. Axioms that are stronger than (4) are apparently required (perhaps a generalization of the axiom $F \vee (F \to G) \vee \neg G$ that is known to characterize the logic of here-and-there [2]). Identifying such axioms is a topic for future work.

Acknowledgements. Thanks to Fangkai Yang for comments on a draft of this note.

References

1. Ferraris, P.: Answer sets for propositional theories. In: Baral, C., Greco, G., Leone, N., Terracina, G. (eds.) LPNMR 2005. LNCS (LNAI), vol. 3662, pp. 119–131. Springer, Heidelberg (2005)
2. Hosoi, T.: The axiomatization of the intermediate propositional systems S_n of Gödel. Journal of the Faculty of Science of the University of Tokyo 13, 183–187 (1966)
3. Lifschitz, V., Morgenstern, L., Plaisted, D.: Knowledge representation and classical logic. In: van Harmelen, F., Lifschitz, V., Porter, B. (eds.) Handbook of Knowledge Representation, pp. 3–88. Elsevier (2008)
4. Lifschitz, V., Pearce, D., Valverde, A.: Strongly equivalent logic programs. ACM Transactions on Computational Logic 2, 526–541 (2001)
5. Pearce, D.: A new logical characterization of stable models and answer sets. In: Dix, J., Moniz Pereira, L., Przymusinski, T.C. (eds.) NMELP 1996. LNCS, vol. 1216, pp. 57–70. Springer, Heidelberg (1997)
6. Truszczynski, M.: Connecting first-order ASP and the logic FO(ID) through reducts. In: Erdem, E., Lee, J., Lierler, Y., Pearce, D. (eds.) Correct Reasoning. LNCS, vol. 7265, pp. 543–559. Springer, Heidelberg (2012)

An Application of ASP to the Field of Second Language Acquisition

(Extended Abstract)

Daniela Inclezan

Miami University, Oxford OH 45056, USA
`inclezd@MiamiOH.edu`

Abstract. This paper explores the contributions of Answer Set Programming (ASP) to the study of an established theory from the field of Second Language Acquisition: Input Processing. The theory describes default strategies that learners of a second language use in extracting meaning out of a text, based on their knowledge of the second language and their background knowledge about the world. We formalized this theory in ASP, and as a result we were able to determine opportunities for refining its natural language description, as well as directions for future theory development. We applied our model to automating the prediction of how learners of English would interpret sentences containing the passive voice. We present a system, $PIas$, that uses these predictions to assist language instructors in designing teaching materials.

1 Introduction

This paper extends a relatively new line of research that explores the contributions of Answer Set Programming (ASP) [1–3] to the study and refinement of *qualitative* scientific theories [4, 5]. As pointed out by Balduccini and Girotto [4], qualitative theories tend to be formulated in natural language, often in the form of defaults. Modeling these theories in a precise mathematical language can assist scientists in analyzing their theories, or in designing experiments for testing their predictions. It was shown that ASP is a suitable tool for this task [4, 5], as it provides means for an elegant and accurate representation of defaults, dynamic domains, and incomplete information, among others. In our work, we explore the applicability of ASP to the formalization and analysis of a theory from the field of Second Language Acquisition — a discipline that studies the processes by which people learn a second language.

Our main goal is to illustrate different ways in which modeling the selected theory in ASP can benefit the future development of this theory. In particular, we focus on contributions to (1) the refinement of this theory; (2) the automated testing of its statements; and (3) the development of practical applications for language teaching and testing. A previous version of this work appears in [6].

The theory we consider is Input Processing [7, 8]. We chose it because it is an established theory in the field of Second Language Acquisition, with important

P. Cabalar and T.C. Son (Eds.): LPNMR 2013, LNAI 8148, pp. 395–400, 2013.

consequences on foreign language education. Input Processing (IP) describes the default strategies that second language learners use to get meaning out of text written or spoken in the second language, during comprehension-focused tasks, given the learners' limitations in vocabulary, working memory, or internalized knowledge of grammatical structures. As a result of applying these strategies, even learners with limited grammatical expertise can often, but not always, interpret input sentences correctly. Once grammatical information is internalized, the default strategies are overridden by the always reliable grammatical knowledge. Hence, it can be said that IP describes an example of nonmonotonic reasoning.

IP consists of two principles formulated as defaults and containing sub-principles that represent refinements of, or exceptions to, these defaults. IP predicts, for instance, that beginner learners of English reading the sentence (S_1) *"The cat was bitten by the dog"* would only be able to retrieve the meanings of the words *"cat"*, *"bitten"*, and *"dog"*, and would end up with the sequence of concepts CAT-BITE-DOG. Beginners would generally interpret this sequence incorrectly as *"The cat bit the dog"* because of a hypothesized strategy of assigning agent status to the first noun of a sentence. On the other hand, beginners would correctly interpret the sentence (S_2) *"The shoe was bitten by the dog"* because agent status cannot be assigned to the first noun, as a shoe cannot bite. Similarly, even beginners would interpret correctly (S_3) *"The man was bitten by the dog"* and (S_4) *"Holyfield was bitten by Tyson"* because people are unlikely to bite animals and it is known that Tyson bit Holyfield, respectively. In the case of stories consisting of several sentences, the information extracted from previous sentences conditions the interpretation of latter ones. The second sentence in the paragraph: (S_5) *"The cat killed the dog."* (S_6) *"Then, the dog was pushed by the cat."* would be interpreted correctly by beginners, as a dead dog cannot push.

ASP is a natural choice for modeling the IP theory, first of all because defaults and their exceptions can be represented in ASP in an elegant and precise manner [9]. Moreover, IP takes into consideration the learners' knowledge about the dynamics of the world (e.g., people know under what conditions an action can be performed); in ASP, there is substantial research on how to represent actions and dynamic domains in which change is caused by actions [10, 11].

2   An Analysis of IP Based on Its ASP Model

Logic Form Encoding of a Text. The IP theory assumes that a learner is given a text (called *input* in the enunciation of IP) — a paragraph with one or more sentences. Our logic form encoding, $lp(X)$, of a text X uses relations: $word\_of\_sent(K, S, W)$ (the Kth word of sentence S is W) and $sent\_of\_par(K, P, S)$ (the Kth sentence of paragraph P is S).

Principle 1. Given a sentence and a learner's knowledge of the second language, Principle 1 of IP predicts a possibly partial mapping of words in this sentence into cognitive concepts. Due to limitations in knowledge of the second language or resources in working memory, learners will not map all words into concepts. In our ASP formalization, which we do not present here, the output of Principle

1 is a collection of atoms of the type $map(K, S, Ctg, C)$ – the Kth word of S, belonging to the grammatical category Ctg, was mapped into concept C. By modeling Principle 1, we determined that certain terms in its description (e.g., "sentence initial position") require a more precise definition, and that a different ordering of its sub-principles would facilitate a deeper understanding.

Principle 2. The second principle of IP describes the strategies that learners employ to understand the meaning of a sentence. The input of Principle 2 is the output of Principle 1 for a given sentence (i.e., a mapping of words into concepts), together with the learner's background knowledge about the world. Its output is an event denoting the meaning extracted by the learner from that sentence. This principle is listed in [8, 12] as:

2. *The First Noun Principle (FNP): Learners tend to process the first noun or pronoun they encounter in a sentence as the agent.*

2a. *The Lexical Semantics Principle: Learners may rely on lexical semantics,*[1] *where possible, instead of on word order to interpret sentences.*

2b. *The Event Probabilities Principle: Learners may rely on event probabilities, where possible, instead of on word order to interpret sentences.*

2c. *The Contextual Constraint Principle: Learners may rely less on the First Noun Principle if preceding context constrains the possible interpretation of a clause or sentence.*

2d. *Prior Knowledge: Learners may rely on prior knowledge, where possible, to interpret sentences.*

2e. *Grammatical Cues: Learners will adopt other processing strategies for grammatical role assignment only after their developing system*[2] *has incorporated other cues.*

This principle assumes that learners possess some background knowledge about the world and its dynamics, which we capture using the predicates: $impossible(Ev, I)$ (event Ev is physically impossible to occur at step I of some story); $unlikely(Ev, I)$ (event Ev is unlikely to occur at step I of some story); $hpd(Ev)$ (event Ev is known to have happened in reality). To model the background knowledge base of a learner, we use known methodologies for representing dynamic domains in ASP [10, 11]. Atoms of the type $impossible(Ev, I)$ are derived from axioms specifying preconditions for the execution of actions (i.e., *executability conditions*); $unlikely(Ev, I)$ atoms are obtained from axioms encoding default statements and their exceptions [9]; $hpd(Ev)$ atoms are simply stored as a collection of facts.

When formalizing Principle 2, we assume that each sentence in the input describes exactly one event, and that the Nth sentence of a paragraph describes the Nth occurring event. By the *direct (reverse) meaning* of a sentence we mean the action denoted by the verb of the sentence, and whose agent is the entity denoted by the *first* (*second*) noun appearing in the sentence. For instance, the direct meaning of (S_1) *"The cat was bitten by the dog"* is the event of "the cat

---

[1] *Lexical semantics* refers to the meaning of lexical items.

[2] *Developing system* refers to the representation of grammatical knowledge in the mind of the second language learner, at a certain point in time.

biting the dog," while its reverse meaning is the event of "the dog biting the cat." We use the predicate $dir\_rev\_m(Dir, Rev, S)$ to say that Dir is the direct meaning and Rev is the reverse meaning of sentence S.

Principle 2, also called the First Noun Principle (FNP), is a default statement and its sub-principles express exceptions to it. To encode Principle 2, we use a relation $extr\_m(Ev, S, fnp)$ saying that the learner extracted the meaning Ev from sentence S by applying FNP:

$$extr\_m(Dir, S, fnp) \leftarrow not\ extr\_m(Rev, S, fnp),\ dir\_rev\_m(Dir, Rev, S).$$

The rule says that learners applying the FNP will extract the direct meaning from a sentence, unless they extract the reverse meaning. We represent Principle 2a, 2b, and 2d, respectively, using the axioms:

$$extr\_m(Rev, S, fnp) \leftarrow impossible(Dir, I),\ not\ impossible(Rev, I),$$
$$dir\_rev\_m(Dir, Rev, S),\ sent\_of\_par(I, P, S).$$
$$extr\_m(Rev, S, fnp) \leftarrow not\ impossible(Dir, I),\ unlikely(Dir, I),\ not\ hpd(Dir),$$
$$not\ impossible(Rev, I),\ not\ unlikely(Rev, I),$$
$$dir\_rev\_m(Dir, Rev, S),\ sent\_of\_par(I, P, S).$$
$$extr\_m(Rev, S, fnp) \leftarrow hpd(Rev),$$
$$dir\_rev\_m(Dir, Rev, S),\ sent\_of\_par(I, P, S).$$

Principle 2e is encoded via the rules:

$$extr\_m(Ev, S) \leftarrow extr\_m(Ev, S, grm\_cues).$$
$$extr\_m(Ev, S) \leftarrow extr\_m(Ev, S, fnp),\ not\ extr\_m\_by(S, grm\_cues).$$
$$extr\_m\_by(S, X) \leftarrow extr\_m(Ev, S, X).$$

where $extr\_m(Ev, S)$ says that Ev is the meaning extracted from S; $extr\_m(Ev, S, grm\_cues)$ – the meaning Ev was extracted from S based on grammatical cues (which vary for different grammatical forms); and $extr\_m\_by(S, X)$ – the meaning of S was extracted based on strategy X.

In our formalization of FNP, Principle 2c was embedded in the representation of Principles 2a, 2b, and 2d. The one thing left for contextual constraints is to record the events corresponding to the meaning extracted from previous sentences of the story, assuming the first time step of the story is 1.

$$occurs(Ev, I) \leftarrow extr\_m(Ev, S),\ sent\_of\_par(I, P, S).$$

Principle 2c specifies that preceding sentences in a paragraph constrain the interpretation of latter sentences, but does not mention a possible effect of succeeding sentences on the re-interpretation of earlier sentences that were initially processed incorrectly. This is an interesting direction of research to be addressed.

3   Automating the Predictions of IP

We used our model of the IP theory to generate automated predictions about how sentences like the ones in Section 1 would be interpreted by learners of English.

We exemplify our predictions on Principle 2 and beginner learners that possess limited grammatical knowledge in English. We created a logic program Π by putting together a beginner's knowledge of the second language, his background knowledge about the world, and our formalization of IP. *The answer set(s) of the program $\Pi \cup lp(X)$ corresponds to predictions of the IP theory about how a beginner learner would interpret text X.* Table 3 shows the answer sets of our program for the texts in Section 1. We use terms like $ev(bite, cat, dog)$ to denote events, in this case "a cat biting a dog". Our automated predictions match the ones in Section 1, which suggests that our model of IP is correct.

4   The System *PIas*

We created a system, *PIas*, designed to assist instructors in preparing materials for the passive voice in English. *PIas* follows the guidelines of a successful teaching method called Processing Instruction (PI) [12], developed based on the principles of IP. For a sentence to be *valuable* in this approach, it must lead to an *incorrect* interpretation when grammatical cues are *not* used but the FNP is. S_1 above is a valuable sentence; S_2, S_3, and S_4 are not.

PIas has two functions. The first one is to specify whether sentences and paragraphs created by instructors are valuable or not. *This is relevant because even instructors trained in PI happen to create bad materials.* We define:

$$valuable(S) \leftarrow extr\_m(Ev_1, S, grm\_cues),\ extr\_m(Ev_2, S, fnp),\ Ev_1 \neq Ev_2.$$

We create a module $\mathcal{M}$ containing this definition and its extension to paragraphs. *PIas* takes as an input a text X in natural language, encodes it in its logic form $lp(X)$, and computes the answer sets of a program consisting of Π, $\mathcal{M}$, $lp(X)$ and the grammatical knowledge of an advanced learner. X is valuable if the atom $valuable(X)$ belongs to all answer sets of the resulting program.

The second function of *PIas* is to generate all valuable sentences given a vocabulary and some simple grammar. *This is important because PI requires to expose learners to a large number of valuable sentences.* We add to $\mathcal{M}$ rules for sentence creation. For instance, one particular type of sentence is generated by:

$$word\_of\_sent(1, s(\text{"}The\text{"}, N_1, \text{"}was\text{"}, V, \text{"}by\text{"}, \text{"}the\text{"}, N_2), \text{"}the\text{"}) \leftarrow g(N_1, V, N_2).$$

where $g(N_1, V, N_2)$ is true if N_1 and N_2 are common nouns and V is a verb in the past participle form (e.g., *"bitten"*).

Table 1. Automated Predictions for Principle 2 and Beginner Learners

X	**Answer Set of $\Pi \cup lp(X)$ contains**
S_1	$extr\_m(ev(bite, cat, dog), s_1),\ extr\_m\_by(s_1, fnp)$
S_2	$extr\_m(ev(bite, dog, shoe), s_2),\ impossible(ev(bite, shoe, dog), 0)$
S_3	$extr\_m(ev(bite, dog, man), s_3),\ unlikely(ev(bite, man, dog), 0)$
S_4	$extr\_m(ev(bite, tyson, holyfield), s_4),\ hpd(ev(bite, tyson, holyfield))$
$S_5 \cup S_6$	$extr\_m(ev(kill, cat, dog), s_5),\ extr\_m(ev(push, cat, dog), s_6)$

5   Conclusions

This paper has shown three different directions in which modeling an important theory from the field of Second Language Acquisition can contribute to the development of this theory. First, we identified aspects in the text of the theory description that need refinement, and opportunities for future research. Second, we have shown how our ASP model can be used to automate predictions, which can be beneficial in designing experiments for testing the theory and fine-tuning its parameters. Third, we described a system, *PIas*, that assesses the quality of materials created by language instructors and creates valuable materials. *We hope the application presented here, and its three main contributions, will help promote ASP as a tool for the study of qualitative theories, in different fields.*

Acknowledgments. I warmly thank Michael Gelfond, Marcello Balduccini, and the anonymous reviewers for their valuable comments and suggestions.

References

1. Gelfond, M., Lifschitz, V.: Classical negation in logic programs and disjunctive databases. New Generation Computing 9(3/4), 365–386 (1991)
2. Niemelä, I.: Logic programs with stable model semantics as a constraint programming paradigm. In: Proceedings of the Workshop on Computational Aspects of Nonmonotonic Reasoning, pp. 72–79 (1998)
3. Marek, V.W., Truszczynski, M.: Stable models and an alternative logic programming paradigm. In: The Logic Programming Paradigm: A 25-Year Perspective, pp. 375–398. Springer, Berlin (1999)
4. Balduccini, M., Girotto, S.: Formalization of psychological knowledge in Answer Set Programming and its application. Theory and Practice of Logic Programming 10(4-6), 725–740 (2010)
5. Balduccini, M., Girotto, S.: ASP as a cognitive modeling tool: Short-term memory and long-term memory. In: Balduccini, M., Son, T.C. (eds.) Logic Programming, Knowledge Representation, and Nonmonotonic Reasoning. LNCS, vol. 6565, pp. 377–397. Springer, Heidelberg (2011)
6. Inclezan, D.: Modeling a theory of Second Language Acquisition in ASP. In: Rosati, R., Woltran, S. (eds.) Proceedings of the 14th International Workshop on Non-Monotonic Reasoning, NMR (2012)
7. VanPatten, B.: Learners' comprehension of clitic pronouns: More evidence for a word order strategy. Hispanic Linguistics 1, 57–67 (1984)
8. VanPatten, B.: Input processing in second language acquisition. In: Processing Instruction: Theory, Research, and Commentary, pp. 5–32. Lawrence Erlbaum Associates, Mahwah (2004)
9. Baral, C., Gelfond, M.: Logic programming and knowledge representation. Journal of Logic Programming 19(20), 73–148 (1994)
10. Gelfond, M., Lifschitz, V.: Action languages. Electronic Transactions on Artificial Intelligence 3(16), 193–210 (1998)
11. Balduccini, M., Gelfond, M.: Diagnostic reasoning with A-Prolog. Theory and Practice of Logic Programming 3(4-5), 425–461 (2003)
12. VanPatten, B.: Processing Instruction: An update. Language Learning 52(4), 755–803 (2002)

Turner's Logic of Universal Causation, Propositional Logic, and Logic Programming

Jianmin Ji[1] and Fangzhen Lin[2]

[1] School of Computer Science and Technology
University of Science and Technology of China, Hefei, China
`jianmin@ustc.edu.cn`
[2] Department of Computer Science and Engineering
The Hong Kong University of Science and Technology, Hong Kong
`flin@cs.ust.hk`

Abstract. Turner's logic of universal causation is a general logic for nonmonotonic reasoning. It has its origin in McCain and Turner's causal action theories which have been translated to propositional logic and logic programming with nested expressions. In this paper, we propose to do the same for Turner's logic, and show that Turner's logic can actually be mapped to McCain and Turner's causal theories. These results can be used to construct a system for reasoning in Turner's logic.

1 Introduction

Turner's logic of universal causation [17], called UCL, is a nonmonotonic modal logic that generalizes McCain and Turner's causal action theories [15]. The idea is to use the modal operator $\mathbf{C}$ to specify the statement that a proposition is "caused". For instance, $\psi \supset \mathbf{C}\phi$ says that ϕ is caused whenever ψ obtains.

McCain and Turner's causal action theories have been the basis for the semantics of several expressive action languages, such as $\mathcal{C}$ and $\mathcal{C}+$ [11,5]. They have been translated to propositional logic and logic programming. Ferraris [2] provided a translation from causal theories to disjunctive logic programs. Lee [9] proposed a conversion from causal theories to propositional logic. In this paper, we consider UCL, and show that UCL theories can be converted to propositional theories. We also show that they can be converted to logic programs with nested expressions in polynomial size with polynomial number of new variables. This result improves and generalizes Turner's linear and modular translation from a fragment of UCL to disjunctive logic programs [17]. Furthermore we show that both Ferraris and Lee's translations are special cases of our translations, just as McCain and Turner's causal theories are special theories in UCL. Our motivation for this work is to use the translations to implement a system for computing UCL theories via SAT solvers or ASP solvers, like the system CCalc[1] for causal theories.

---

[1] `http://www.cs.utexas.edu/~tag/ccalc/`.

P. Cabalar and T.C. Son (Eds.): LPNMR 2013, LNAI 8148, pp. 401–413, 2013.

This paper is organized as follows. Section 2 reviews UCL and logic programming. Section 3 shows how Turner's logic can be mapped to propositional logic. Section 4 considers mapping UCL theories to logic programs with nested expressions. Section 5 outlines how the translations here are related to Ferraris and Lee's translations. Finally, Section 6 concludes this paper.

2   Preliminaries

2.1   Propositional Languages

We assume a propositional language with two zero-place logical connectives $\top$ for tautology and $\bot$ for contradiction. We denote by *Atom* the set of atoms, the signature of our language, and *Lit* the set of literals: $Lit = Atom \cup \{\neg a \mid a \in Atom\}$. A set I of literals is called *complete* if for each atom a, exactly one of $\{a, \neg a\}$ is in I. Given a literal l, the *complement* of l, written $\bar{l}$ below, is $\neg a$ if l is a and a if l is $\neg a$, where a is an atom. For a set L of literals, we let $\bar{L} = \{\bar{l} \mid l \in L\}$.

In this paper, we identify an interpretation with a complete set of literals. If I is a complete set of literals, we use it as an interpretation when we say that it is a model of a formula, and we use it as a set of literals when we say that it entails a formula.

2.2   Turner's Logic of Universal Causation

The language of Turner's logic of universal causation (UCL) [17] is a modal propositional language with a modal operator $\mathbf{C}$. UCL *formulas* are propositional formulas with unary modal operator $\mathbf{C}$. A UCL *theory* is a set of UCL formulas.

The semantics of UCL is defined through causally explained interpretations. A UCL *structure* is a pair (I, S) such that I is an interpretation, and S is a set of interpretations to which I belongs. The truth of a UCL sentence in a UCL structure is defined by the standard recursions over the propositional connectives, plus the following two conditions:

$$(I, S) \models a \text{ iff } I \models a \quad \text{(for any atom } a)$$
$$(I, S) \models \mathbf{C}\phi \text{ iff for all } I' \in S, (I', S) \models \phi$$

Given a UCL theory T, we write $(I, S) \models T$ to mean that $(I, S) \models \phi$, for every $\phi \in T$. In this case, we say that (I, S) is a *model* of T. We also say that (I, S) is an *I-model* of T, emphasizing the distinguished interpretation I.

Let T be a UCL theory. An interpretation I is *causally explained* by T if $(I, \{I\})$ is the unique I-model of T.

Note that, if there is a nested occurrence of $\mathbf{C}$, the $\mathbf{C}$ that occurs in the range of another $\mathbf{C}$ can be equivalently[2] removed [17]. In the paper, we only consider UCL formulas with no nested occurrences of $\mathbf{C}$. A formula of the form $\mathbf{C}\phi$, where ϕ is a propositional formula, is called a $\mathbf{C}$-*atom*. Then these UCL formulas are constructed from $\mathbf{C}$-atoms, propositional atoms and connectives.

---

[2] In the sense that, two formulas have the same set of UCL models.

2.3 Logic Programming

A *nested expression* is built from literals using the 0-place connectives $\top$ and $\bot$, the unary connective "*not*" and the binary connective "," and ";".

A *logic program* with nested expressions is a finite set of rules of the form $F \leftarrow G$, where F and G are nested expressions.

The *answer set* of a logic program with nested expressions is defined as in [12]. Given a nested expression F and a set S of literals, we define when S satisfies F, written $S \models F$ below, recursively as follows (l is a literal and G is a nested expression):

- $S \models l$ if $l \in S$,
- $S \models \top$ and $S \not\models \bot$,
- $S \models not\, F$ if $S \not\models F$,
- $S \models F, G$ if $S \models F$ and $S \models G$, and
- $S \models F; G$ if $S \models F$ or $S \models G$.

S satisfies a rule $F \leftarrow G$ if $S \models F$ whenever $S \models G$. S satisfies a logic program P, written $S \models P$, if S satisfies all rules in P.

The *reduct* P^S of P related to S is the result of replacing every maximal subexpression of P that has the form $not\, F$ with $\bot$ if $S \models F$, and with $\top$ otherwise.

Let P be a logic program without *not*, the *answer set* of P is any minimal consistent subset S of Lit that satisfies P. We use $\Gamma_P(S)$ to denote the set of answer sets of P^S. Now a consistent set S of literals is an *answer set* of P iff $S \in \Gamma_P(S)$.

Every logic program with nested expressions can be equivalently translated to disjunctive logic programs with disjunctive rules of the form

$$l_1 \vee \cdots \vee l_k \leftarrow l_{k+1}, \ldots, l_t, not\, l_{t+1}, \ldots, not\, l_m, not\, not\, l_{m+1}, \ldots, not\, not\, l_n,$$

where $n \geq m \geq t \geq k \geq 0$ and $l_1, \ldots, l_n$ are propositional literals. A disjunctive logic program can be computed by disjunctive ASP solvers such as claspD [1], DLV [10], GNT [7] and cmodels [6].

3 From Turner's Logic of Universal Causation to Propositional Logic

Before presenting the translation, we provide some notations. Given a UCL formula F, let $Atom_{\mathbf{C}}(F) = \{ \phi \mid \mathbf{C}\phi$ is a $\mathbf{C}$-atom occurring in $F \}$. Given a UCL theory T, we let $Atom_{\mathbf{C}}(T) = \bigcup_{F \in T} Atom_{\mathbf{C}}(F)$.

We use $tr_p(F)$ to denote the propositional formula obtained from the UCL formula F by replacing each occurrence of a $\mathbf{C}$-atom $\mathbf{C}\phi$ by a new propositional atom a_ϕ w.r.t. ϕ.

Given two propositional formulas ϕ and ψ, we use ϕ^ψ to denote the propositional formula obtained from ϕ by replacing each occurrence of an atom a with a new atom a^ψ w.r.t. ψ.

The following proposition provides a specification of the propositional formula whose models are related to models of a UCL theory.

Proposition 1. *Let T be a UCL theory. A UCL structure $(I, \mathcal{S})$ is a model of T if and only if there exists a model I^* of the propositional formula*

$$\bigwedge_{F \in T} tr_p(F) \wedge \bigwedge_{\phi \in Atom_{\mathbf{C}}(T)} (a_\phi \supset \phi)$$

$$\wedge \bigwedge_{\psi \in Atom_{\mathbf{C}}(T)} \left((\neg a_\psi \wedge \psi) \supset \left(\bigwedge_{\phi \in Atom_{\mathbf{C}}(T)} (a_\phi \supset \phi^\psi) \wedge \neg \psi^\psi \right) \right), \quad (1)$$

such that $I^* \cap Lit = I$ and for each $\phi \in Atom_{\mathbf{C}}(T)$, $a_\phi \in I^*$ iff $\mathcal{S} \models \phi$.

Proof. "$\Rightarrow$" $(I, \mathcal{S})$ is a model of T, then $I \in \mathcal{S}$. If $\mathcal{S} \not\models \psi$ and $I \models \psi$, then there exists another interpretation $I' \in \mathcal{S}$ such that $I' \models \neg \psi$. Thus, we can create an interpretation I^* such that

$$I^* = I \cup \{a_\phi \mid \phi \in Atom_{\mathbf{C}}(T) \text{ and } \mathcal{S} \models \phi\} \cup \{\neg a_\phi \mid \phi \in Atom_{\mathbf{C}}(T) \text{ and } \mathcal{S} \not\models \phi\}$$

$$\cup \bigcup_{\psi \in Atom_{\mathbf{C}}(T), \mathcal{S} \models \psi} \{l^\psi \mid l \in I\} \cup \bigcup_{\psi \in Atom_{\mathbf{C}}(T), \mathcal{S} \not\models \psi, \exists I'. I' \in \mathcal{S}, I' \models \neg \psi} \{l^\psi \mid l \in I'\}.$$

Clearly, $I^* \models (1)$.

"$\Leftarrow$" $I^* \models T$. Let $I = I^* \cap Lit$ and $\mathcal{S} = \{I' \mid$ if $I^* \models a_\phi$ for some $\phi \in Atom_{\mathbf{C}}(T)$, then $I' \models \phi\}$. Note that, $I^* \models \bigwedge_{\phi \in Atom_{\mathbf{C}}(T)} (a_\phi \supset \phi)$, then $I \in \mathcal{S}$. For each $\phi \in Atom_{\mathbf{C}}(T)$, if $I^* \models a_\phi$, then $\mathcal{S} \models \phi$; from (1), if $I^* \models \neg a_\phi$, then there exists an interpretation I' such that $I' \models \neg \psi$ and for each $\psi \in Atom_{\mathbf{C}}(T)$, $I^* \models a_\psi$ implies $I' \models \psi$, thus $I' \in \mathcal{S}$ and $\mathcal{S} \not\models \phi$. Clearly, $(I, \mathcal{S}) \models T$.

Intuitively, the formula

$$\bigwedge_{\psi \in Atom_{\mathbf{C}}(T)} \left((\neg a_\psi \wedge \psi) \supset \left(\bigwedge_{\phi \in Atom_{\mathbf{C}}(T)} (a_\phi \supset \phi^\psi) \wedge \neg \psi^\psi \right) \right)$$

specifies that for each UCL structure $(I, \mathcal{S})$, if $I \models \psi$ and $\mathcal{S} \models \neg \mathbf{C}\psi$, then there exists an interpretation $I' \in \mathcal{S}$ such that $I' \models \neg \psi$.

In the following, we construct propositional formulas whose models are related to causally explained interpretations. First, we consider how to specify the unique model of a propositional formula.

Given a propositional formula ϕ and a nonempty consistent set K of literals, we denote by $\phi|_{K \to \perp}$ the result of replacing each occurrence of an atom a in ϕ by $\perp$ if $a \in K$ and $\top$ if $\neg a \in K$.

Lemma 1. *Let ϕ be a propositional formula, K a nonempty consistent set of literals, and an interpretation $I \supseteq K$. $I \not\models \bigwedge_{l \in K} l \supset \neg \phi|_{K \to \perp}$ if and only if the interpretation $(I \setminus K) \cup \overline{K} \models \phi$.*

Proof. Let $I' = (I \setminus K) \cup \overline{K}$.

"$\Rightarrow$" $I \not\models \bigwedge_{l \in K} l \supset \neg\phi|_{K \to \bot}$, then $I \models \phi|_{K \to \bot}$. Note that, atoms occurring in K do not occur in $\phi|_{K \to \bot}$, then $I' \models \phi|_{K \to \bot}$, furthermore, $I' \models \overline{K}$, thus $I' \models \phi$.

"$\Leftarrow$" $I' \models \phi$ and $I' \models \overline{K}$, then $I' \models \phi|_{K \to \bot}$, thus $I \models \phi|_{K \to \bot}$. Note that $K \subseteq I$, then $I \not\models \bigwedge_{l \in K} l \supset \neg\phi|_{K \to \bot}$.

To avoid influence of auxiliary atoms, we introduce the notion of forgetting provided by Lin and Reiter [14].

Definition 1. *Let ϕ be a propositional formula and S a set of atoms. $forget(\phi; S)$ is the formula inductively defined as follows:*

- $forget(\phi; \emptyset) = \phi$,
- $forget(\phi; \{a\}) = \phi|_{\{a\} \to \bot} \vee \phi|_{\{\neg a\} \to \bot}$,
- $forget(\phi; \{a\} \cup S) = forget(forget(\phi; S), \{a\})$.

Lemma 2 (Theorem 4 in [14]). *Let ϕ be a propositional formula and S a set of atoms. An interpretation $I \models forget(\phi; S)$ if and only if there exists an interpretation $I' \models \phi$ such that $I \setminus S \cup \overline{S} = I' \setminus S \cup \overline{S}$.*

Directly from Lemma 1 and 2, we have the following lemma.

Lemma 3. *Let ϕ be a propositional formula, K a nonempty consistent set of literals, S a set of atoms, and an interpretation $I \supseteq K$. $I \not\models \bigwedge_{l \in K} l \supset \neg forget(\phi; S)|_{K \to \bot}$ if and only if there exists an interpretation $I' \models \phi$ such that $((I \setminus S \cup \overline{S}) \setminus K) \cup \overline{K} = I' \setminus S \cup \overline{S}$.*

Given a propositional formula ϕ, we use $\widehat{\phi}$ to denote the propositional formula obtained from ϕ by replacing each occurrence of an atom a in ϕ by a new atom $\hat{a}$. For a set L of literals, we let $\widehat{L} = \{\hat{l} \mid l \in L\}$. We use Lit_a to denote the set of literals formed from new atoms of the form a_ϕ and $Atom^*$ the set of atoms of the form a^ψ w.r.t. ψ in (1).

Theorem 1. *Let T be a UCL theory. An interpretation I is causally explained by T if and only if there exists a model I^* of the propositional formula*

$$(1) \wedge \bigwedge_{\phi \in Atom_{\mathbf{C}}(T)} (a_\phi \supset \widehat{\phi})$$

$$\wedge \bigwedge_{\substack{A \subseteq Lit_a \\ A \text{ is nonempty and consistent}}} \left(\bigwedge_{l_a \in A} l_a \supset \neg\, forget((1); Atom^*)\Big|_{A \to \bot} \right)$$

$$\wedge \bigwedge_{\substack{K \subseteq Lit \\ K \text{ is nonempty and consistent}}} \left(\bigwedge_{l \in K} \hat{l} \supset \neg \bigwedge_{\phi \in Atom_{\mathbf{C}}(T)} (a_\phi \supset \widehat{\phi})\Big|_{\widehat{K} \to \bot} \right), \quad (2)$$

such that $I^* \cap Lit = I$.

Proof. "⇒" I is causally explained by T, then $(I, \{I\})$ is the unique I-model of T. We can create an interpretation I^* such that

$$I^* = I \cup \{a_\phi \mid \phi \in Atom_{\mathbf{C}}(T) \text{ and } I \models \phi\} \cup \{\neg a_\phi \mid \phi \in Atom_{\mathbf{C}}(T) \text{ and } I \not\models \phi\}$$

$$\cup \bigcup_{\psi \in Atom_{\mathbf{C}}(T)} \{l^\psi \mid l \in I\} \cup \{\hat{l} \mid l \in I\}.$$

From Proposition 1, $I^* \models (1) \wedge \bigwedge_{\phi \in Atom_{\mathbf{C}}(T)} (a_\phi \supset \widehat{\phi})$.

If $I^* \not\models \bigwedge_{l_a \in A} l_a \supset \neg\, forget((1); Atom^*)\big|_{A \to \perp}$ for some nonempty consistent set $A \subseteq Lit_a$, similar to the proof of Lemma 3, then these exists another interpretation $I^{*\prime}$ such that $((I^* \setminus Atom^* \cup \overline{Atom^*}) \setminus A) \cup \overline{A} = I^{*\prime} \setminus Atom^* \cup \overline{Atom^*}$ and $I^{*\prime} \models (1)$. From Proposition 1, there exists another set $\mathcal{S}'$ of interpretations such that $\mathcal{S}' \neq \{I\}$, $I \in \mathcal{S}'$ and $(I, \mathcal{S}')$ is an I-model of T, which conflicts to the condition that $(I, \{I\})$ is the unique I-model of T.

If $I^* \not\models \bigwedge_{l \in K} \hat{l} \supset \neg \bigwedge_{\phi \in Atom_{\mathbf{C}}(T)} (a_\phi \supset \widehat{\phi})\big|_{\widehat{K} \to \perp}$ for some nonempty consistent set $K \subseteq Lit$, similar to the proof of Lemma 1, then there exists another interpretation $I^{*\prime}$ such that $I^{*\prime} = (I^* \setminus \widehat{K}) \cup \overline{\widehat{K}}$ and $I^{*\prime} \models (1) \wedge \bigwedge_{\phi \in Atom_{\mathbf{C}}(T)} (a_\phi \supset \widehat{\phi})$. From Proposition 1, there exists another interpretation I' such that $(I, \{I, I'\}) \models T$, which conflicts to the condition that $(I, \{I\})$ is the unique I-model of T, thus $I^* \models (2)$.

"⇐" $I^* \models (2)$. Let $I = I^* \cap Lit$, if there exists another UCL structure $(I, \mathcal{S})$ such that $(I, \mathcal{S}) \models T$ and $\mathcal{S} \neq \{I\}$, then there are two cases: 1. there exists $\phi \in Atom_{\mathbf{C}}(T)$ such that $I \models \phi$ and $\mathcal{S} \not\models \phi$; 2. for each $\phi \in Atom_{\mathbf{C}}(T)$, $I \models \phi$ if and only if $\mathcal{S} \models \phi$.

For case 1, let $A = \{a_\phi \mid \phi \in Atom_{\mathbf{C}}(T), I \models \phi, \mathcal{S} \not\models \phi\}$, then $I^* \models \bigwedge_{l_a \in A} l_a$ and $I^* \models forget((1); Atom^*)\big|_{A \to \perp}$, which conflicts to the condition that $I^* \models (2)$, thus it is impossible.

For case 2, let $I' \in \mathcal{S}$ and $I' \neq I$, then for each $\phi \in Atom_{\mathbf{C}}(T)$, $I^* \models a_\phi$ implies $I' \models \phi$, thus there exists $K = I \setminus I'$ such that $I^* \models \bigwedge_{l \in K} \hat{l}$ and $I^* \models \bigwedge_{\phi \in Atom_{\mathbf{C}}(T)} (a_\phi \supset \widehat{\phi})\big|_{\widehat{K} \to \perp}$, which conflicts to the condition that $I^* \models (2)$. So I is the only interpretation that satisfies $\{\phi \in Atom_{\mathbf{C}}(T) \mid I^* \models a_\phi\}$, then $(I, \{I\})$ is the unique I-model of T.

Note that, the size of formula (2) is exponential increased from T, as the number of all possible nonempty consistent sets of literals is 3^n, where n is the number of atoms. In fact, we only need to consider a subset of these sets. Details are proposed in Section 5.3.

As a simple example, given the UCL theory $T = \{\mathbf{C}(p \vee q), \mathbf{C}p \supset \mathbf{C}q, \mathbf{C}q \supset \mathbf{C}p\}$, from the definition of (1), we obtain the following propositional formula:

$$a_{p \vee q} \wedge (a_p \equiv a_q) \wedge (p \vee q) \wedge (a_p \supset p) \wedge (a_q \supset q) \wedge$$
$$\big((\neg a_p \wedge p) \supset (\neg p^2 \wedge q^2)\big) \wedge \big((\neg a_q \wedge q) \supset (\neg q^3 \wedge p^3)\big) \quad (3)$$

From the definition of (2), we obtain the following formula[3]

$$(3) \wedge \left(a_{p \vee q} \supset (\hat{p} \vee \hat{q})\right) \wedge (a_p \supset \hat{p}) \wedge (a_q \supset \hat{q})$$
$$\wedge \left(a_p \supset \neg(\neg a_q)\right) \wedge \left(a_q \supset \neg(\neg a_p)\right) \wedge \left((a_p \wedge a_q) \supset \bot\right)$$
$$\wedge \left(\neg a_p \supset \neg(a_q \wedge p \wedge q)\right) \wedge \left(\neg a_q \supset \neg(a_p \wedge p \wedge q)\right) \wedge \left(\neg a_p \wedge \neg a_q \supset \neg(p \wedge q)\right)$$
$$\wedge \left(\hat{p} \supset \neg((a_{p \vee q} \supset \hat{q}) \wedge \neg a_p \wedge (a_q \supset \hat{q}))\right) \wedge \left(\hat{q} \supset \neg((a_{p \vee q} \supset \hat{p}) \wedge (a_p \supset \hat{p}) \wedge \neg a_q)\right)$$
$$\wedge \left(\neg \hat{p} \supset \neg(a_q \supset \hat{q})\right) \wedge \left(\neg \hat{q} \supset \neg(a_p \supset \hat{p})\right)$$

where $\hat{p}$ and $\hat{q}$ are new atoms. The formula implies that

$$a_{p \vee q} \wedge \hat{p} \wedge \hat{q} \wedge a_p \wedge a_q \wedge p \wedge q \wedge ((a_p \wedge a_q) \supset \bot)$$

which is inconsistent. From Theorem 1, there does not exist an interpretation I such that I is causally explained by the UCL theory T.

4   From Turner's Logic of Universal Causation to Logic Programming

Formula (2) in propositional logic is complex, as it needs to include constraints to make it satisfied by a "unique model". The problem becomes easier when we consider logic programming. Based on the propositional formula (1), we can translate a UCL theory T to a logic program with nested expressions.

Note that, every propositional formula ϕ can be equivalently translated to CNF as

$$(l_1^1 \vee \cdots \vee l_{n^1}^1) \wedge \cdots \wedge (l_1^m \vee \cdots \vee l_{n^m}^m), \tag{4}$$

where $l_1^1, \ldots, l_{n^m}^m$ are literals.

For any propositional formula, we can convert it to the nested expression by replacing each $\wedge$ with a comma, each $\vee$ with a semicolon and $\neg$ with not.

Given a UCL theory T, we use $tr_{ne}(T)$ to denote the nested expression obtained from (1). We use $Atom'$ to denote the set of atoms that occur in (1) but not in $Atom$. Now we define $tr_{lp}(T)$ to be the logic program containing $\bot \leftarrow not\, tr_{ne}(T)$, the following rules for each $\phi \in Atom_{\mathbf{C}}(T)$ whose CNF is in the form of (4)

$$l_1^1; \ldots; l_{n^1}^1 \leftarrow not\, not\, a_\phi, (\bar{l}_1^1; not\, \bar{l}_1^1), \ldots, (\bar{l}_{n^1}^1; not\, \bar{l}_{n^1}^1),$$
$$\cdots$$
$$l_1^m; \ldots; l_{n^m}^m \leftarrow not\, not\, a_\phi, (\bar{l}_1^m; not\, \bar{l}_1^m), \ldots, (\bar{l}_{n^m}^m; not\, \bar{l}_{n^m}^m),$$

and

$$a'; \neg a' \leftarrow \top, \qquad \text{(for each } a' \in Atom').$$

---

[3] The formula is simplified due to Theorem 5 in Section 5.3.

Lemma 4. *Let T be a UCL theory and I and J two interpretations. $(I, \{I, J\}) \models T$ if and only if there exists a set S of literals occurring in $tr_{lp}(T)$ such that $S \models (tr_{lp}(T))^{S \cup I}$ and $S \cap Lit = I \cap J$.*

Proof. "$\Rightarrow$" $(I, \{I, J\}) \models T$. Similar to the proof of Proposition 1, we can create an interpretation I^* such that $I^* \cap Lit = I$ and $I^* \models (1)$. Note that, $(tr_{lp}(T))^{I^*}$ contains rules of the form

$$l_1; \ldots; l_n \leftarrow \underset{l \in \{l_1, \ldots, l_n\}, \bar{l} \in I}{,} \bar{l}, \tag{5}$$

where $I^* \models a_\phi$ for corresponding $\phi \in Atom_{\mathbf{C}}(T)$.

Note that, $\{I, J\} \models \phi$, $I \models l_1 \vee \cdots \vee l_n$ and $J \models l_1 \vee \cdots \vee l_n$. Consider the case, for each literal $l \in \{l_1, \ldots, l_n\}$, $\bar{l} \in I$ implies $\bar{l} \in I \cap J$, then there exists literal $l \in \{l_1, \ldots, l_n\}$ and $l \in J$ such that $l \in I$ (if not, $\bar{l} \in I$ which implies $\bar{l} \in J$), thus $(I \cap J) \models (5)$.

We denote $S = (I^* \setminus I) \cup (I \cap J)$. Clearly, $S \models (tr_{lp}(T))^{S \cup I}$.

"$\Leftarrow$" $S \models (tr_{lp}(T))^{S \cup I}$ and $S \cap Lit = I \cap J$. $(tr_{lp}(T))^{S \cup I}$ contains rules of (5) and $S \models a_\phi$ for corresponding $\phi \in Atom_{\mathbf{C}}(T)$.

Note that $I \cap J \models (5)$, then $I \cap J \models l_1 \vee \cdots \vee l_n$ whenever $\bar{l} \in I \cap J$ for all $\bar{l} \in I$ and $l \in \{l_1, \ldots, l_n\}$. If $J \not\models l_1 \vee \cdots \vee l_n$, then there exists $\bar{l} \in I$ and $\bar{l} \notin I \cap J$, thus $\bar{l} \notin J$ and $l \in J$ which conflicts to $J \not\models l_1 \vee \cdots \vee l_n$. So $J \models \phi$, from Proposition 1, $(I, \{I, J\}) \models T$.

Theorem 2. *Let T be a UCL theory. An interpretation I is causally explained by T if and only if there exists an answer set S of the logic program $tr_{lp}(T) \cup \{\bot \leftarrow not\, a, not\, \neg a \mid a \in Atom\}$, such that $S \cap Lit = I$.*

Proof. I is causally explained by T means that $(I, \{I\})$ is the unique I-model of T. From Lemma 4, this is equivalent to the condition, for every set S of literals occurring in $tr_{lp}(T)$ and interpretation J such that $S \models (tr_{lp}(T))^{S \cup I}$, $S \cap Lit = I \cap J$ iff $J = I$. This means that there exists an answer set S of $tr_{lp}(T) \cup \{\bot \leftarrow not\, a, not\, \neg a \mid a \in Atom\}$ such that $S \cap Lit = I$.

5   Related Work

5.1   Turner's Conversion from a Fragment of UCL to Disjunctive Logic Programming

Turner [17] proposed a simple translation from a subset of UCL theories to disjunctive logic programs [3] via disjunctive default logic [4].

Turner's translation considers the UCL formula of the form

$$\mathbf{C}(l_1 \wedge \cdots \wedge l_k) \wedge l_{k+1} \wedge \cdots \wedge l_m \supset \mathbf{C}l_{m+1} \vee \cdots \mathbf{C}l_n, \tag{6}$$

where $l_1, \ldots, l_n$ are literals.

A UCL formula of the form (6) is translated to the disjunctive rule

$$l_{m+1} \vee \cdots \vee l_n \leftarrow l_1, \ldots, l_k, not\, \bar{l}_{k+1}, \ldots, not\, \bar{l}_m.$$

It has been proved that, given a set T of UCL formulas in the form (6), an interpretation I is an answer set of the corresponding disjunctive logic program if and only if I is causally explained by T.

When every formula in range of $\mathbf{C}$ is a literal, our translation seems more complex than Turner's translation. However, some steps in the translation can also be simplified. Consider the following proposition proposed in [2].

Proposition 2 (Proposition 1 in [2]). *For any literal l and any nested expression F, the one-rule logic program*

$$l \leftarrow F, (\bar{l}; not \, \bar{l})$$

is strongly equivalent to $l \leftarrow F$.

5.2   Ferraris's Translation from Causal Theories to Logic Programs

Ferraris [2] proposed a translation from McCain and Turner's causal theories [15] to logic programs with nested expressions. As causal theories can be easily converted into UCL, we show that Ferraris's translation is a special case of our translation proposed in Section 4. First, we briefly review causal theories and Ferraris's translation, then we consider the relation to our translation.

A causal theory according to McCain and Turner [15] is a set of *causal laws* of the following form

$$\psi \Rightarrow \phi, \tag{7}$$

where ϕ and ψ are propositional formulas.

Ferraris's translation converts the causal law

$$\psi \Rightarrow l_1 \vee \cdots \vee l_n, \tag{8}$$

to the rule

$$l_1; \ldots; l_n \leftarrow not \, not \, \psi^{ne}, (\bar{l}_1; not \, \bar{l}_1), \ldots, (\bar{l}_n; not \, \bar{l}_n),$$

where ψ^{ne} stands for the nested expression of ψ. Theorem 1 in [2] proved that models of a set of causal laws in the form (8) are identical to complete answer sets of the corresponding logic programs.

According to Turner [17], a causal law of the form (7) can be translated to his logic as

$$\psi \supset \mathbf{C}\phi. \tag{9}$$

Thus given our translation from Turner's logic to logic programming, we have a translation from McCain and Turner's causal theory to logic programming as well.

A UCL formula of the form (9) is called *regular*. A *regular* UCL theory is a set of regular UCL formulas.

Note that, when T is a regular UCL theory, formula (1) in Proposition 1 can be simplified to

$$\bigwedge_{F \in T} tr_p(F) \wedge \bigwedge_{\phi \in Atom_{\mathbf{C}}(T)} (a_\phi \supset \phi). \tag{10}$$

Proposition 3. *Let T be a regular UCL theory. A UCL structure $(I, \mathcal{S})$ is a model of T if and only if there exists a model I^* of formula (10) such that $I^* \cap Lit = I$ and for each $\phi \in Atom_{\mathbf{C}}(T)$, $a_\phi \in I^*$ iff $\mathcal{S} \models \phi$.*

Based on Proposition 3, the translation in Section 4 can also be simplified. Given a regular UCL theory T, we use $tr'_{ne}(T)$ to denote the nested expression obtained from (10). We define $tr'_{lp}(T)$ the same as $tr_{lp}(T)$ except $tr_{ne}(T)$ is replaced by $tr'_{ne}(T)$.

Theorem 3. *Let T be a regular UCL theory. An interpretation I is causally explained by T if and only if there exists an answer set S of the logic program $tr'_{lp}(T) \cup \{\bot \leftarrow not\ a, not\ \neg a \mid a \in Atom\}$, such that $S \cap Lit = I$.*

It is easy to find out that, for regular UCL theory T, $tr'_{lp}(T)$ is equivalent to the result of Ferraris's translation.

Our translation in Section 4 can also be specified by Ferraris's translation. First, a UCL theory can be converted to a regular UCL theory.

Theorem 4. *Let T be a UCL theory. An interpretation I is causally explained by T if and only if I is causally explained by the regular UCL theory with following formulas*

(1),

$a_\phi \supset \mathbf{C}\phi$,   *(for each $\phi \in Atom_{\mathbf{C}}(T)$)*

$a' \supset \mathbf{C}a'$, $\neg a' \supset \mathbf{C}\neg a'$.   *(for each $a' \in Atom'$)*

Then we can use Ferraris's translation turning the regular UCL theory in the above theorem to a logic program with nested expressions.

5.3   Lee's Translation from Causal Theories to Propositional Theories with Loop Formulas

Lee [9] proposed a translation from McCain and Turner's causal theories to propositional theories with loop formulas. In this section, we show that we can also define so called "loop formulas" for our translation in Section 3 and Lee's translation would be a special case.

Given a set Π of propositional clauses, i.e. disjunctions of literals, the *dependency graph* of Π is the directed graph G_Π such that

- the vertices of G_Π are literals in Π, and
- for any two vertices l_1, l_2, there is an edge from l_1 to l_2 if there is a clause $C \in \Pi$ such that l_1 and $\overline{l_2}$ are in C.

A nonempty consistent set L of literals is called a *loop* of Π if for any literals l_1 and l_2 in L, there is a path from l_1 to l_2 in G_Π such that all the vertices in the path are in L, i.e. the L-induced subgraph of G_Π is strongly connected. Specially, the singleton set $\{l\}$ for every literal $l \in Lit$ is a loop. We use $Loop(\Pi)$ to denote the set of all loops of Π.

The *loop formula* associated with a loop L under a set Π of propositional clauses, denoted by $LF(\Pi, L)$, is a sentence of the form:

$$\bigwedge_{l \in L} l \supset \neg \bigwedge_{C \in \Pi} C|_{L \to \bot}.$$

We can simplify the translation from UCL to propositional logic by loops.

Proposition 4. *Let Π be a set of propositional clauses,*

$$\bigwedge_{C \in \Pi} C \wedge \bigwedge_{L \in Loop(\Pi)} LF(\Pi, L) \supset \bigwedge_{C \in \Pi} C \wedge \bigwedge_{\substack{K \subseteq Lit \\ K \text{ is nonempty} \\ \text{and consistent}}} \left(\bigwedge_{l \in K} l \supset \neg \bigwedge_{C \in \Pi} C|_{K \to \bot} \right).$$

Proof. Let $L \in Loop(\Pi)$, K a nonempty consistent set of literals s.t. $L \subseteq K$, and there does not exist an edge of G_Π from a literal in L to a literal in $K \setminus L$.

There does not exist an edge of G_Π from a literal in L to a literal in $K \setminus L$, then there does not exist a clause C in Π of the form

$$l_1 \vee \cdots \vee l_n$$

such that $l_i \in L$ and $l_j \in K \setminus L$ for some $1 \leq i, j \leq n$. Thus, if $C \in \Pi$ and $C \cap L \neq \emptyset$, then $C \cap (K \setminus L) = \emptyset$.

For each clause $C \in \Pi$, as $L \subseteq K$, there are three different cases.

Case 1, $L \cap C = \emptyset$. If $L \cap \overline{C} \neq \emptyset$, then $K \cap \overline{C} \neq \emptyset$, thus $\neg C|_{L \to \bot} \equiv \neg C|_{K \to \bot} \equiv \bot$. If $L \cap \overline{C} = \emptyset$, then $\neg C|_{L \to \bot} \equiv \neg C$, thus $C \wedge LF(\{C\}, L) \supset C \wedge (\bigwedge_{l \in K} l \supset \neg C|_{K \to \bot})$.

Case 2, $L \cap C \neq \emptyset$, $L \cap \overline{C} = \emptyset$, and $K \cap \overline{C} = \emptyset$. From the above condition, $C \cap (K \setminus L) = \emptyset$, then $\neg C|_{L \to \bot} \equiv \neg C|_{K \to \bot}$.

Case 3, $K \cap \overline{C} \neq \emptyset$. Then $\neg C|_{K \to \bot} \equiv \bot$, thus $C \wedge LF(\{C\}, L) \supset C \wedge (\bigwedge_{l \in K} l \supset \neg C|_{K \to \bot})$.

Based on the above results,

$$\bigwedge_{C \in \Pi} C \wedge LF(\Pi, L) \supset \bigwedge_{C \in \Pi} C \wedge \left(\bigwedge_{l \in K} l \supset \neg \bigwedge_{C \in \Pi} C|_{K \to \bot} \right).$$

In addition, for every nonempty consistent set K of literals, there always exists a loop $L \subseteq K$ such that there does not exist an edge of G_Π from a formula in L to a formula in $K \setminus L$. So the proposition is proved.

Given a UCL theory T, with a slight abuse of notations, we use $Loop(T)$ to the set of loops of the set of clauses which are in CNF of $\phi \in Atom_C(T)$. Similarly, we use $Loop_a(T)$ to the set of loops of the set of clauses which are in CNF of (1).

Theorem 5. *Let T be a UCL theory. An interpretation I is causally explained by T if and only if there exists a model I^* of the propositional formula*

$$(1) \wedge \bigwedge_{\phi \in Atom_{\mathbf{C}}(T)} (a_\phi \supset \widehat{\phi})$$

$$\wedge \bigwedge_{A \subseteq Lit_a, A \in Loop_a(T)} \left(\bigwedge_{l_a \in A} l_a \supset \neg \ forget((1); Atom^*)\big|_{A \to \perp} \right)$$

$$\wedge \bigwedge_{L \in Loop(T)} \left(\bigwedge_{l \in L} \widehat{l} \supset \neg \bigwedge_{\phi \in Atom_{\mathbf{C}}(T)} (a_\phi \supset \widehat{\phi})\big|_{\widehat{L} \to \perp} \right), \quad (11)$$

such that $I^* \cap Lit = I$.

Similar to the discussion in the previous section, when T is a regular UCL theory, the above theorem can be simplified.

Theorem 6. *Let T be a regular UCL theory. An interpretation I is causally explained by T if and only if there exists a model I^* of the propositional formula*

$$\bigwedge_{F \in T} tr_p(F) \wedge \bigwedge_{\phi \in Atom_{\mathbf{C}}(T)} (a_\phi \supset \phi) \wedge \bigwedge_{\phi \in Atom_{\mathbf{C}}(T)} (a_\phi \supset \widehat{\phi})$$

$$\wedge \bigwedge_{L \in Loop(T)} \left(\bigwedge_{l \in L} \widehat{l} \supset \neg \bigwedge_{\phi \in Atom_{\mathbf{C}}(T)} (a_\phi \supset \widehat{\phi})\big|_{\widehat{L} \to \perp} \right), \quad (12)$$

such that $I^* \cap Lit = I$.

When each formula in the range of $\mathbf{C}$ is a clause in the regular UCL theory T, comparing the above theorem with Theorem 1 in [9], it is easy to find out that formula (12) corresponds to $DR(T) \cup CLC(T)$ in Lee's Theorem.

6   Conclusion

We have provided translations from Turner's logic of universal causation to propositional logic and logic programming. These translations generalize the respective translations by Ferraris and Lee for McCain and Turner's causal theories. Our next step is to use these results to implement Turner's logic using SAT and ASP solvers.

It is worth mentioning here that our results in this paper can also be used to map Turner's logic to fixed-point nonmonotonic logics such as default logic [16] and Lin and Shoham's logic of GK [13,8].

Acknowledgments. This work had been supported by the National Hi-Tech Project of China under grant 2008AA01Z150, the Natural Science Foundation of China under grant 60745002 and 61175057, the USTC Key Direction Project, the Fundamental Research Funds for the Central Universities, the Youth Innovation Fund of USTC, and HK RGC GRF 616909 . We thank the anonymous reviewers for their valuable comments on an earlier version of the paper.

References

1. Drescher, C., Gebser, M., Grote, T., Kaufmann, B., König, A., Ostrowski, M., Schaub, T.: Conflict-Driven Disjunctive Answer Set Solving. In: Brewka, G., Lang, J. (eds.) Proceedings of the 11th International Conference on Principles of Knowledge Representation and Reasoning, KR 2008, pp. 422–432. AAAI Press, Menlo Park (2008)
2. Ferraris, P.: A logic program characterization of causal theories. In: Proceedings of the 20th International Joint Conference on Artificial Intelligence, IJCAI 2007, pp. 366–371 (2007)
3. Gelfond, M., Lifschitz, V.: Classical negation in logic programs and disjunctive databases. New Generation Comput. 9(3/4), 365–386 (1991)
4. Gelfond, M., Lifschitz, V., Przymusińska, H., Truszczyński, M.: Disjunctive Defaults. In: Allen, J.F., Fikes, R., Sandewall, E. (eds.) Proceedings of the 2nd International Conference on Principles of Knowledge Representation and Reasoning, KR 1991, pp. 230–237. Morgan Kaufmann, San Fransisco (1991)
5. Giunchiglia, E., Lee, J., Lifschitz, V., McCain, N., Turner, H.: Nonmonotonic causal theories. Artificial Intelligence 153(1-2), 49–104 (2004)
6. Giunchiglia, E., Lierler, Y., Maratea, M.: Answer Set Programming Based on Propositional Satisfiability. J. Autom. Reasoning 36(4), 345–377 (2006)
7. Janhunen, T., Niemelä, I.: GNT — A Solver for Disjunctive Logic Programs. In: Lifschitz, V., Niemelä, I. (eds.) LPNMR 2004. LNCS (LNAI), vol. 2923, pp. 331–335. Springer, Heidelberg (2003)
8. Ji, J., Lin, F.: From Turner's Logic of Universal Causation to the Logic of GK. In: Erdem, E., Lee, J., Lierler, Y., Pearce, D. (eds.) Correct Reasoning. LNCS, vol. 7265, pp. 380–385. Springer, Heidelberg (2012)
9. Lee, J.: Nondefinite vs. definite causal theories. In: Lifschitz, V., Niemelä, I. (eds.) LPNMR 2004. LNCS (LNAI), vol. 2923, pp. 141–153. Springer, Heidelberg (2003)
10. Leone, N., Pfeifer, G., Faber, W., Eiter, T., Gottlob, G., Perri, S., Scarcello, F.: The DLV system for knowledge representation and reasoning. ACM Transactions on Computational Logic 7(3), 499–562 (2006)
11. Lifschitz, V.: Action languages, answer sets and planning. In: The Logic Programming Paradigm: a 25-Year Perspective, pp. 357–373. Springer (1999)
12. Lifschitz, V., Tang, L.R., Turner, H.: Nested expressions in logic programs. Annals of Mathematics and Artificial Intelligence 25(3-4), 369–389 (1999)
13. Lin, F., Shoham, Y.: A logic of knowledge and justified assumptions. Artificial Intelligence 57(2-3), 271–289 (1992)
14. Lin, F., Reiter, R.: Forget it. In: Working Notes of AAAI Fall Symposium on Relevance, pp. 154–159 (1994)
15. McCain, N., Turner, H.: Causal theories of action and change. In: Proceedings of the 14th National Conference on Artificial Intelligence, AAAI 1997, pp. 460–465 (1997)
16. Reiter, R.: A logic for default reasoning. Artificial Intelligence 13, 81–132 (1980)
17. Turner, H.: Logic of universal causation. Artificial Intelligence 113(1), 87–123 (1999)

Concrete Results on Abstract Rules

Markus Krötzsch, Despoina Magka, and Ian Horrocks

Department of Computer Science, University of Oxford, UK

Abstract. There are many different notions of "rule" in the literature. A key feature and main intuition of any such notion is that rules can be "applied" to derive conclusions from certain premises. More formally, a rule is viewed as a function that, when invoked on a set of known facts, can produce new facts. In this paper, we show that this extreme simplification is still sufficient to obtain a number of useful results in concrete cases. We define abstract rules as a certain kind of functions, provide them with a semantics in terms of (abstract) stable models, and explain how concrete normal logic programming rules can be viewed as abstract rules in a variety of ways. We further analyse dependencies between abstract rules to recognise classes of logic programs for which stable models are guaranteed to be unique.

1 Introduction

A large variety of different types of "rules" are considered in logic programming, knowledge representation, production rule systems, and databases. While many rules have a common background in predicate logic, there are still important differences between, say, normal logic programs [13], existential rules [3], and database dependencies [1]. It is, however, highly desirable to transfer concrete results and insights between these domains.

This goal is best illustrated by considering a concrete example. In a recent publication, the authors analyse *nonmonotonic existential rules* under a stable model semantics [14]. The work identifies syntactic conditions to guarantee finiteness and uniqueness of stable models, and shows how this can be applied to improve the performance of reasoning over real-world data. To fully exploit these ideas, this approach can be further extended in at least two ways: (1) other types of logic rules, e.g., normal logic programs, could be considered; (2) extend the scope of the approach for relevant special forms of programs, e.g., for equality and datatype reasoning.

In terms of (2), the authors already extended their results by considering integrity constraints [14]. Unfortunately, even this relatively small change required laborious extensions of all previous correctness proofs. While the structure of arguments remains similar, each individual step now needs to take constraints into account. Following this pedestrian approach, repeated effort is required for each modification in the underlying language. With the continued elaboration of rule-based languages, important ideas often remain confined to one sub-area and are rather reinvented than being transferred. For instance, the notion of rule dependency that was extended to nonmonotonic existential rules in [14] has first been proposed for conceptual graph rules in 2004 [2] and rediscovered for databases in 2008 [4].

P. Cabalar and T.C. Son (Eds.): LPNMR 2013, LNAI 8148, pp. 414–426, 2013.

To address this issue, we propose to take a more abstract view on "rules" that can be instantiated in many different cases. Our main intuition is that many kinds of rules can be "applied" in certain sense to derive conclusions from premises. We formalise this by viewing a rule as a function that, when invoked on a set of known facts, can produce new facts. This is a rather natural view on rules. Our main contribution is to show that this extreme simplification is still sufficient to derive interesting results that are easy to instantiate in concrete cases. Our contributions are organised as follows.

- We define *abstract rules* and provide them with a semantics in terms of *abstract stable models* that agrees with the standard notion of stable model semantics in concrete cases (Section 2).
- We present *operations for constructing abstract rules* (Section 3) and establish several strong equivalence results to show that these operations preserve our semantics (Section 4).
- We reformulate the condition of *R-stratification* from [14] for abstract rules (Section 5) and show that this condition leads to a *unique stable model* even in the abstract case (Section 6).
- We apply our results to *equality reasoning* in normal logic programs and propose a new approach for stratifying programs with equality using constraints (Section 7).

Proofs that are not included here can be found in an accompanying report [10].

2   Abstract Rules and Models

We consider a countable set B of basic logical expressions, which we think of as facts that might be true or false in a given situation; subsets $F \subseteq B$ will therefore be called *sets of (abstract) facts*. B contains a distinguished element $\bot$, which denotes a contradiction.

Definition 1. *An* abstract rule *r is a function $r : 2^B \to 2^B$ with the following properties.*

- *r is* extensive: *$r(F) \supseteq F$ for every $F \subseteq B$*
- *r is* compact *(or* finitary*): for every $F \subseteq B$ and $f \in r(F)$, there are finite sets $B^+, B^- \subseteq B$ with $B^+ \subseteq F$ and $B^- \cap F = \emptyset$, such that $f \in r(F')$ for every set F' with $B^+ \subseteq F'$ and $B^- \cap F' = \emptyset$.*

If $f \in r(F)$, we say that r derives *f from F. A pair of minimal sets $\langle B^+, B^- \rangle$ that witnesses the compactness property for the derivation $f \in r(F)$ is called an* abstract body *of r with respect to f and F.*

An abstract program *is a countable set of abstract rules, denoted P, possibly with subscripts or primes. Uncountable rule sets are not considered herein.*

A rule is *monotone* if $F_1 \subseteq F_2$ implies $r(F_1) \subseteq r(F_2)$ for all sets of facts $F_1, F_2 \subseteq B$, but this is not required by our definition. For monotone rules, every derivation has a body $\langle B^+, B^- \rangle$ with $B^- = \emptyset$, as one would expect.

Example 1. A propositional logic programming rule is an expression of form $H \leftarrow B_1, \ldots, B_n, \mathbf{not}\ B_{n+1}, \ldots, \mathbf{not}\ B_m$, where $B_1, \ldots, B_m$ and H are propositional letters. It can be viewed as an abstract rule r: let B be the set of propositional letters and define $r(F) := F \cup \{H\}$ if $B_1, \ldots, B_n \in F$ and $B_{n+1}, \ldots, B_m \notin F$, and $r(F) := F$ otherwise. An abstract body is given by the sets $B^+ = \{B_1, \ldots, B_n\}$ and $B^- = \{B_{n+1}, \ldots, B_m\}$.

Note that this approach does not define a one-to-one correspondence of concrete rules and abstract rules. For example, the rules $A \leftarrow A$ or $A \leftarrow A, B$ both lead to the same abstract rule with $r(F) = F$. Thus our abstraction cannot capture any logic programming semantics where the presence of such rules is relevant.

Example 2. A normal logic programming rule is an expression of form $H \leftarrow B_1, \ldots, B_n$, **not** $B_{n+1}, \ldots, $ **not** B_m, where $B_1, \ldots, B_m$ and H are predicate-logic atoms over some logical signature. It can be viewed as an abstract rule r: let B be the set of all ground atoms over the signature (the so-called *Herbrand base*), let G_r be the set of all ground instantiations of r, and define $r(F) := \bigcup_{r_g \in G_r} r_g(F)$, where $r_g(F)$ is defined as for propositional rules in Example 1. An abstract body for a ground fact H_g is obtained as the abstract body for any ground instance r_g that can be used to derive H_g from F.

Example 2 illustrates that abstract bodies may not be unique, since several ground instantiations of a logic programming rule may have the same head but different bodies.

Example 3. Consider the Herbrand base $\mathsf{B} = \{n(a), n(s(a)), n(s(s(a))), \ldots\}$, and let $\mathsf{B}^{\geq 1}$ denote the set $\mathsf{B} \setminus \{n(a)\}$. Consider the following functions:

$$r_1(F) := F \cup \{n(a)\} \quad \text{if } \mathsf{B}^{\geq 1} \subseteq F; \qquad r_1(F) := F \quad \text{otherwise} \qquad (1)$$

$$r_2(F) := F \cup \{n(a)\} \quad \text{if } \mathsf{B}^{\geq 1} \not\subseteq F; \qquad r_2(F) := F \quad \text{otherwise} \qquad (2)$$

$$r_3(F) := F \cup \{n(a)\} \quad \text{if } F \text{ is finite}; \qquad r_3(F) := F \quad \text{otherwise} \qquad (3)$$

Each of these functions is extensive. Function r_1 is not an abstract rule: it is monotone but depends on an infinite set $\mathsf{B}^{\geq 1}$ of premises, hence is not compact. Function r_2 is an abstract rule: for any fact $f \in \mathsf{B}^{\geq 1}$ such that $f \notin F$, the sets $B^+ = \emptyset$ and $B^- = \{f\}$ form a body of the derivation $n(a) \in r_2(F)$. A concrete representation of this rule along the lines of Example 2 is $n(a) \leftarrow$ **not** $n(s(X))$, although this may not be syntactically allowed in all logic programming approaches, since X occurs in negated atoms only. Function r_3 is not an abstract rule: it is nonmonotonic but there is no finite negative body B^- for any derivation (for finite F, we would get $B^+ = \emptyset$ and $B^- = \mathsf{B} \setminus F$).

We can define the consequence operator T_P for abstract rules as usual.

Definition 2. *For a set of rules P and a set of facts F, we define $T_P(F) := \bigcup_{r \in P} r(F)$. Moreover, we set $T_P^0(F) := F$, $T_P^{i+1}(F) := T_P(T_P^i(F))$, and $T_P^\infty(F) := \bigcup_{i \geq 0} T_P^i(F)$.*

Since we do not assume rules to be monotone, different orders of rule applications might lead to very different sets of derived sets of facts, and in particular $T_P^\infty(F)$ may not capture the desired semantics of the program. The next definition describes types of derived sets of facts that are more suitable for defining the semantics of abstract rules.

Definition 3. *Consider an abstract program P and a set of facts F_0. A set of facts F is well-supported for P and F_0 if there is a well-founded partial order $\prec$ on F such that, for every fact $f \in F \setminus F_0$, there is a rule $r_f \in P$ with body $\langle B_f^+, B_f^- \rangle$ for f and F, and $f' \prec f$ for all $f' \in B_f^+$. We assume that the choice of r_f and $\langle B_f^+, B_f^- \rangle$ is part of each well-supported set (there might be other choices, leading to other well-supported sets).*

A set of facts F is an abstract model for a set of rules P and a set of input facts F_0 if $\perp \notin F$, $F_0 \subseteq F$, and $r(F) \subseteq F$ for every rule $r \in P$. An abstract stable model is a well-supported abstract model.

For the case of normal logic programming rules, our definition of well-supported model agrees with that of Fages [7]. Our terminology is justified by Fages's result that well-supported models are exactly the (classical) stable models [7, Theorem 2.1]. We obtain the following theorem as a corollary.

Theorem 1. *The abstract stable models of a normal logic program P, viewed as a set of abstract rules as in Example 2, are exactly the classical stable models of P.*

Like in the classical case, T_P can be used to compute models, which, however, may not be well-supported.

Proposition 1. $T_P^\infty(F)$ *is an abstract model for P and F.*

Proof. Suppose for a contradiction that the claim does not hold. Then there is a rule $r \in P$ and a fact $f \in r(T_P^\infty(F))$ with $f \notin T_P^\infty(F)$. By compactness, there is a finite set $F' \subseteq T_P^\infty(F)$ such that $f \in r(F'')$ for every $F' \subseteq F'' \subseteq T_P^\infty(F)$. By construction of $T_P^\infty(F)$, there is some i such that $F' \subseteq T_P^i(F)$. But then $f \in r(T_P^i(F))$ by compactness, and hence $f \in T_P^{i+1}(F) \subseteq T_P^\infty(F)$—a contradiction. $\square$

The well-founded order $\prec$ in Definition 3 leaves a lot of flexibility for establishing that a set is well-supported. Due to compactness, however, it is generally enough to order facts by a finite rank that can be expressed as a natural number.

Proposition 2. *If F is well-supported for P and F_0, then this is witnessed by an order $\prec$ such that $\langle F, \prec \rangle$ has an order-preserving injection into the natural numbers $\langle \mathbb{N}, < \rangle$.*

Proof. Since F is well-supported, there is an order $\prec_0$ as in Definition 3. For each fact $f \in F \setminus F_0$, there is a rule $r_f \in P$ and a body $\langle B_f^+, B_f^- \rangle$ as in Definition 3. The order $\prec_1$ on F is the transitive closure of the set $\{f' \prec_1 f \mid f' \in B_f^+\}$. Clearly, $\prec_1 \subseteq \prec_0$, so $\prec_1$ is well-founded. By definition, $\prec_1$ can be used to show that F is well-supported, using the same choice of rules r_f and bodies B_f^+ as for $\prec_0$.

We claim that for every fact $f \in F$, the set $f\!\downarrow := \{f' \mid f' \prec_1 f\}$ is finite (*). Indeed, by construction, $f\!\downarrow = \bigcup_{f' \in B_f^+} f'\!\downarrow$. The claim follows by well-founded induction: if $f'\!\downarrow$ is finite for all $f' \prec_1 f$, then $f'\!\downarrow$ is finite for all $f' \in B_f^+$, and thus $f\!\downarrow$ is finite, too.

To construct a total well-founded order $\prec$ as required in the claim, we re-order the elements of F as a sequence. Since F is countable, there is an injective mapping $\iota : F \to \mathbb{N}$ from F to natural numbers. We recursively construct a (possibly infinite) sequence $f_1, f_2, \ldots$ of facts from F as follows. To select f_i, consider the set of $\prec_1$-minimal elements M_i in the set $F \setminus \{f_1, \ldots, f_{i-1}\}$. If $M_i = \emptyset$ then there is no f_i and the construction terminates with a finite sequence. Otherwise, define $f_i \in M_i$ to be the ι-smallest element of M_i, i.e., $\iota(f_i) \leq \iota(f)$ for all $f \in M_i$.

We claim that the constructed sequence $S = \{f_1, f_2, \ldots\}$ contains exactly the elements of F. For a contradiction, suppose that there is an element $f \in F \setminus S$. By (*), the set $\{f\} \cup f\!\downarrow$ is finite. Consider the set $M := \{f\} \cup f\!\downarrow \setminus S$. By our assumptions, $f \in M$, so M is finite but not empty. Thus there is a $\prec_1$-minimal element f' in M, which is also a $\prec_1$-minimal element of $F \setminus S$. As there are only finitely many elements in F that are $\prec_1$-smaller than f', there is a finite index j such that f' is a $\prec_1$-minimal element of

$F \setminus \{f_1, \ldots, f_j\}$. There are at most $\iota(f') - 1$ many elements that can be added to S after f_j, before f' must also be added. Thus $f' \in S$, which contradicts our assumptions.

We define the order $\prec$ on F by setting $f_i \prec f_j$ for all $i < j$. This makes $\prec$ a suborder of $\langle \mathbb{N}, < \rangle$, and thus well-founded. Moreover, $\prec_1 \subseteq \prec$, so $\prec$ can be used to show that F is well-supported. $\square$

3   Constructing Abstract Rules

The examples given so far mainly show that abstract rules can capture normal logic programs. In this section, we show that they are significantly more general, even on a base set of facts B that is the Herbrand base of a predicate logic signature. For this purpose, we introduce various operations for constructing new abstract rules from existing ones, and show that these operation preserve stable model semantics (Section 4). The basic operations we consider are union, composition and saturation of rules.

Definition 4. *Let P be a program. The* union $\bigcup P$ *of P is defined by setting* $(\bigcup P)(F) := \bigcup_{r \in P} r(F)$ *if $P \neq \emptyset$. For $P = \emptyset$, we define* $(\bigcup \emptyset)(F) := F$.
   The intersection $\bigcap P$ *of P is defined as* $(\bigcap P)(F) := \bigcap_{r \in P} r(F)$ *if $P \neq \emptyset$. For $P = \emptyset$, we define* $(\bigcap \emptyset)(F) := \mathsf{B}$.

Example 4. The abstract rule induced by a normal logic programming rule as in Example 2 is the infinite union of the abstract rules obtained from its ground instantiations. Likewise, the one-step T_P operator of Definition 2 is the abstract rule $\bigcup P$. Intersections of rules are the abstract counterpart to conjunctions in rule bodies. For example, the intersection of the rules $q \leftarrow p_1$ and $q \leftarrow p_2, \mathbf{not}\ p_3$ can be expressed as $q \leftarrow p_1, p_2, \mathbf{not}\ p_3$.

Intersections of abstract rules do not always result in abstract rules. For example, the function in (1), which is not compact, can be viewed as the intersection of the infinite set of all rules $n(a) \leftarrow n(s^i(a))$ with $i \geq 1$. However, abstract rules are closed under infinite unions and finite intersections, as shown next.

Theorem 2. *The union $\bigcup P$ of an abstract program P is an abstract rule. If P is finite, then the intersection $\bigcap P$ is also an abstract rule.*

Proof. First consider $\bigcup P$. For all $r \in P$ we find $F \subseteq r(F)$ by extensiveness; hence $F \subseteq \bigcup_{r \in P} r(F) = \bigcup P(F)$ and $\bigcup P$ is extensive. If $P = \emptyset$ then $(\bigcup P)(F) = F$, so for any derivation $f \in (\bigcup P)(F)$ the sets $B^+ = \{f\}$ and $B^- = \emptyset$ show compactness of $(\bigcup P)$. If $P \neq \emptyset$ then, for any fact $f \in (\bigcup P)(F)$, there is a rule $r \in P$ such that $f \in r(F)$; this implies the existence of suitable sets B^+ and B^- to show compactness of $\bigcup P$.

Now assume that P is finite and consider $\bigcap P$. Extensiveness of $\bigcap P$ is again immediate from the extensiveness of rules in P. If $P = \emptyset$ then $(\bigcap P)(F) = \mathsf{B}$, so for any derivation $f \in (\bigcap P)(F)$ the sets $B^+ = B^- = \emptyset$ show compactness of $(\bigcap P)$. If $P \neq \emptyset$ then, for any fact $f \in (\bigcap P)(F)$ and any rule $r_i \in P$, we find sets B^+ and B^- by compactness of the derivation $f \in r_i(F)$. Thus, the sets $\bigcup_{r_i \in P} B_i^+$ and $\bigcup_{r_i \in P} B_i^-$ show compactness of $(\bigcap P)$. In particular, these sets are finite since P is. $\square$

Another interesting type of operations is based on functional composition.

Definition 5. *The composition $r_2 \circ r_1$ of r_1 and r_2 is the function with $(r_2 \circ r_1)(F) := r_2(r_1(F))$. The n-iteration r^n of a rule r is the n-fold composition with itself, i.e., r^0 is the identity function and $r^{i+1} = r \circ r^i$. The saturation r^∞ of r is the union of all its n-iterations, i.e., $r^\infty := \bigcup \{ r^i \mid i \geq 0 \}$.*

Example 5. Iterations of the T_P operator of Definition 2 are equivalent to iterations of abstract rules: $T_P^i = (\bigcup P)^i$ and $T_P^\infty = (\bigcup P)^\infty$.

Example 6. In general, composition does not preserve compactness. Consider the rules $r_1 : q \leftarrow p(X)$ and $r_2 : r \leftarrow \textbf{not } q$ over the infinite base $B = \{q, r, p(a), p(s(a)), \ldots\}$. The composition $r_2 \circ r_1$ can be described as

$$(r_2 \circ r_1)(F) = r_1(F) \cup \begin{cases} \{r\} & \text{if } p(s^n(a)) \notin F \text{ for all } n \geq 1 \\ \emptyset & \text{otherwise.} \end{cases}$$

Thus, for every choice of finite sets $\langle B^+, B^- \rangle$, there is a set F' with $B^+ \subseteq F'$ and $B^- \cap F' = \emptyset$ such that $r \notin (r_2 \circ r_1)(F')$. The function $r_2 \circ r_1$ is not compact.

Theorem 3. *Let r_1 and r_2 be abstract rules. If r_2 is monotone, then $r_2 \circ r_1$, r_2^n for all $n \geq 0$, and r_2^∞ are abstract rules.*

Proof. Consider $r_2 \circ r_1$. Extensiveness of r_1 and r_2 yields $F \subseteq r_1(F) \subseteq r_2(r_1(F)) = (r_2 \circ r_1)(F)$. For compactness, consider some $f \in (r_2 \circ r_1)(F)$. By compactness of r_2, we find finite sets B_2^+ and B_2^- for deriving f from $r_1(F)$. Since r_2 is monotone, we can assume without loss of generality that $B_2^- = \emptyset$. As B_2^+ is finite, it has the form $\{f_1, \ldots, f_m\}$. For each $f_i \in B_2^+$, there are sets B_{1i}^+ and B_{1i}^- that show compactness of the derivation $f_i \in r_1(F)$. The sets $B^+ = \bigcup_{i=1}^m B_{1i}^+$ and $B^- = \bigcup_{i=1}^m B_{1i}^-$ show the compactness of the derivation $f \in (r_2 \circ r_1)(F)$.

The claim for r^n follows by induction: the result is clear for r^0, and the induction step follows from the result for composition. The claim for saturation follows by combining the results for n-iteration and Theorem 2. □

4 Strong Equivalence of Abstract Programs

Logic programs P_1 and P_2 are strongly equivalent if the programs $P \cup P_1$ and $P \cup P_2$ have exactly the same stable models for any program P and set of facts F_0 [12,16]. We apply the same definition to abstract logic programs.

Theorem 4. *Every abstract logic program P is strongly equivalent to $\{\bigcup P\}$.*

Proof. To simplify the proof, we use the following auxiliary definition. An abstract program P_1 *is subsumed by* an abstract program P_2, written $P_1 \sqsubseteq P_2$, if the following holds: for every rule $r_1 \in P_1$ and derivation $f \in r_1(F)$ with a body $\langle B^+, B^- \rangle$, there is a rule $r_2 \in P_2$ and derivation $f \in r_2(F)$ for which $\langle B^+, B^- \rangle$ is also a body. Clearly, $P \sqsubseteq \{\bigcup P\}$ and $\{\bigcup P\} \sqsubseteq P$.

To complete the proof, we show some general properties of subsumption. Consider a set of facts F_0 and abstract programs P_1 and P_2.

1. If $P_1 \sqsubseteq P_2$, then every well-supported set for P_1, F_0 is well-supported for P_2, F_0.
2. If $P_2 \sqsubseteq P_1$, then every model of P_1, F_0 is a model of P_2, F_0.
3. If $P_1 \sqsubseteq P_2$ and $P_2 \sqsubseteq P_1$, then P_1 and P_2 are strongly equivalent.

The overall claim thus is an immediate consequence of the last item.

Assume that $P_1 \sqsubseteq P_2$ and that F is well-supported for P_1, F_0 using the order $\prec$. Then for every $f \in F$ there is a rule $r_1 \in P_1$ such that $f \in r_1(F)$ has a body $\langle B^+, B^- \rangle$ in F with $f' \prec f$ for all $f' \in B^+$. Since $P_1 \sqsubseteq P_2$, there is a rule $r_2 \in P_2$ with $f \in r_2(F)$ and the same body.

Assume that $P_2 \sqsubseteq P_1$ and that F is a model for P_1, F_0. Suppose for a contradiction that F is not a model of P_2, F_0. Then there is a rule $r_2 \in P_2$ and a fact $f \in r_2(F) \setminus F$. By $P_2 \sqsubseteq P_1$, there is a rule $r_1 \in P_1$ with $f \in r_1(F)$. This contradicts the assumptions that F is a model of P_1, F_0.

Assume that $P_1 \sqsubseteq P_2$ and $P_2 \sqsubseteq P_1$, and let P be an arbitrary abstract program. Clearly, $P \cup P_1 \sqsubseteq P \cup P_2$ and $P \cup P_2 \sqsubseteq P \cup P_1$. Thus, by the first two properties, every stable model of $P \cup P_1$ and F_0 is also a stable model of $P \cup P_2$ and F_0, and vice versa. $\square$

It is easy to see that intersection $\bigcap P$ does not lead to strong equivalence. However, we can establish relevant results for composition, iteration, and saturation. The proof of the following result uses Proposition 2 to construct the well-founded order that is needed to show that a model is stable.

Proposition 3. *For monotone rules r_1 and r_2, $\{r_1, r_2\}$ is strongly equivalent to $\{r_2 \circ r_1\}$.*

Theorem 5. *For a monotone rule r, all of the programs $\{r\}$, $\{r^n\}$ for $n \geq 2$, and $\{r^\infty\}$ are pairwise strongly equivalent.*

Proof. The strong equivalence of $\{r\}$ and $\{r^n\}$ for any $n \geq 2$ is shown by induction. By Proposition 3, $\{r^{n+1}\}$ is strongly equivalent to $\{r, r^n\}$. By induction $\{r^n\}$ is strongly equivalent to $\{r\}$, so that $\{r, r^n\}$ is strongly equivalent to $\{r, r\} = \{r\}$ as required.

For the limit $\{r^\infty\}$, note that $\{r^\infty\} = \bigcup\{r^n \mid n \geq 1\}$ is strongly equivalent to $\{r^n \mid n \geq 1\}$ by Theorem 4. The result follows as each $\{r^n\}$ is strongly equivalent to $\{r\}$. $\square$

5   Reliances and Stratifications

Stable models can not always be computed by applying rules in a bottom-up fashion. Due to nonmonotonicity, a rule that was applicable initially may no longer be applicable after further facts have been derived. Conversely, it can also happen that one rule is applicable only after another rule has been applied. Both types of relationships between rules are useful to gain insights about the stable models of a given program and to guide the computation of stable models.

We are interested in two types of dependencies, which we call *reliances* to avoid confusion with existing notions: negative reliance (the application of a rule may inhibit the application of another rule) and positive reliance (the application of a rule may enable the application of another rule). In both cases we ask whether this interaction of rules can occur during a normal derivation, i.e., when considering some set of (already derived) facts. We could just consider arbitrary sets of facts here, but we can obtain stronger results if we restrict attention to fact sets which can actually occur during the derivation of a stable model. For the next definition, recall that the notation r_f and $\langle B_f^+, B_f^- \rangle$ was introduced for well-supported sets in Definition 3.

Definition 6. *Given a rule r and finite sets B^+ and B^-, we say that f follows from $\langle B^+, B^- \rangle$ by r if $f \in r(F)$ for every set $F \subseteq B$ with $B^+ \subseteq F$ and $B^- \cap F = \emptyset$.*

Let $\mathscr{D} \subseteq 2^B$ be a set of sets of facts that are admissible as input. *A set $F \subseteq B$ is* derivable *from $\mathscr{D}$ and P if there is a set $F_0 \in \mathscr{D}$ such that F is well-supported for P and F_0, and for all $f \in F \setminus F_0$ we have: $f' \in r(F')$ for all f' that follow from $\langle B_f^+, B_f^- \rangle$ by r_f.*

Intuitively, derivable sets are well-supported sets that contain all the facts that must certainly follow when from the rule applications that establish well-supportedness. The use of $\mathscr{D}$ allows us to consider all sets of facts as admissible inputs (if $\mathscr{D} = 2^B$) or to restrict attention to a single input F_0 (if $\mathscr{D} = \{F_0\}$). A common restriction in Datalog rules is that some "intensional" predicate symbols are not allowed in the input, while in existential rules one does not allow function symbols in input facts, although (skolem) functions may occur in derivations. When irrelevant or clear from the context, we speak of derivable sets without mentioning $\mathscr{D}$ and P explicitly.

Definition 7. *A rule r_2 positively relies on a rule r_1, written $r_1 \overset{+}{\rightarrow} r_2$, if there is a derivable set of facts F with $\bot \notin F$ such that there is a fact $f_2 \in F$ with $r_{f_2} = r_2$, and a fact $f_1 \in B_{f_2}^+$ with $r_{f_1} = r_1$.*

A rule r_2 negatively relies on a rule r_1, written $r_1 \overset{-}{\rightarrow} r_2$, if there is a derivable set of facts F, a derivation $f_2 \in r_2(F)$ with body $\langle B_2^+, B_2^- \rangle$, and a derivation $f_1 \in r_1(F) \cap B_2^-$ with body $\langle B_1^+, B_1^- \rangle$, such that $\bot$ does not follow from $\langle B_1^+, B_1^- \rangle$ by r_1.

In both cases, $\bot$ is taken into account to exclude situations where the application of r_1 leads to an inconsistency.

In practice, it may not always be possible to compute $\overset{+}{\rightarrow}$ and $\overset{-}{\rightarrow}$ exactly. For example, it may be difficult to determine if a certain set is derivable (based on a given choice of $\mathscr{D}$). However, all of our results remain correct when working with larger relations instead of $\overset{+}{\rightarrow}$ and $\overset{-}{\rightarrow}$. Therefore, a practical algorithm may overestimate reliances without putting correctness at risk.

Example 7. A very simple overestimation of reliances on normal logic programs is related to the classical notion of stratification. Consider logic programming rules r_1 and r_2. We write $r_1 \rightsquigarrow^+ r_2$ if a predicate symbol that occurs in the head of r_1 occurs in a non-negated body atom of r_2, and $r_1 \rightsquigarrow^- r_2$ if a head predicate symbol of r_1 occurs in a negated body atom of r_2. It is easy to see that $\overset{+}{\rightarrow} \subseteq \rightsquigarrow^+$ and $\overset{-}{\rightarrow} \subseteq \rightsquigarrow^-$.

Example 8. A more elaborate notion of reliance was recently developed for *existential rules* by the authors [14]. Existential rules are first skolemised, which leads to normal logic programs where each function symbol occurs in the head of exactly one rule and in no rule bodies. Functions are not allowed in input fact sets either. In this special case, one can find all positive and negative reliances by considering only sets of facts F that contain no function symbols. It has been shown that, for programs that do not use $\bot$, checking $r_1 \overset{+}{\rightarrow} r_2$ is NP-complete, while $r_1 \overset{-}{\rightarrow} r_2$ can be checked in polynomial time [14]. This is an exact computation of the relations of Definition 7 for this specific case, not an overestimation.

Definition 8. *Consider a program P and a (finite or countably infinite) sequence of disjoint sets* $\mathbf{P} = \langle P_1, P_2, \ldots \rangle$ *with* $\bigcup_{P_i \in \mathbf{P}} P_i = P$. $\mathbf{P}$ *is an R-stratification of P if, for all rules* $r_1 \in P_i$ *and* $r_2 \in P_j$,

- *if* $r_1 \xrightarrow{\pm} r_2$ *then* $i \leq j$;
- *if* $r_1 \xrightarrow{-} r_2$ *then* $i < j$.

If P has an R-stratification then it is called R-stratifiable.

It should be noted how reliances interact with unions of rules. Any R-stratification of $P \cup \{\bigcup P'\}$ gives rise to an R-stratification of $P \cup P'$, while the converse is not true in general. This is analogous to the relationship of classical stratification (considering normal rules as unions of their groundings as in Example 4) to *local stratification* (considering stratification on the infinitely many ground instances [15]). It also illustrates that our approach can capture (and extend) both of these ideas.

6   Computing Stable Models of Stratified Rule Sets

We now show that R-stratified abstract programs have at most one stable model, which can be obtained by deterministic computation. For programs that have a finite stratification, this leads to a semi-decision procedure for entailment, provided that the given abstract rules are computable functions.

Definition 9. *Given a stratification* $\mathbf{P} = \langle P_1, P_2, \ldots \rangle$ *of P, we define*

$$S_{\mathbf{P}}^0(F) := F, \qquad S_{\mathbf{P}}^{i+1}(F) := T_{P_{i+1}}^\infty(S_{\mathbf{P}}^i(F)), \qquad S_{\mathbf{P}}^\infty(F) := \bigcup_{P_i \in \mathbf{P}} S_{\mathbf{P}}^i(F).$$

For the remainder of this section, let P denote an R-stratified program with R-stratification $\mathbf{P} = \langle P_1, P_2, \ldots \rangle$, and let F denote an admissible set of facts, i.e., $F \in \mathscr{D}$. We use the abbreviations $P_1^m := \bigcup_{i=1}^m P_i$, $P_1^0 := \emptyset$, and $S_P^i := S_P^i(F)$. The main result that we will show in this section is the following.

Theorem 6. *If* $\bot \notin S_P^\infty$ *then* S_P^∞ *is the unique stable model of P and F. Otherwise, P and F do not have a stable model.*

Lemma 1. *For every* $P_j \in \mathbf{P}$ *and* $\ell \geq 0$, *if* $\bot \notin T_{P_j}^\ell(S_P^{j-1})$ *then* $T_{P_j}^\ell(S_P^{j-1})$ *is derivable.*

Proof. For any $i \geq 1$ and $k \geq 0$, we use the abbreviation $T_i^k := T_{P_i}^k(S_P^{i-1})$. We define a well-founded partial order $\prec$ on S_P^∞ by setting $f_1 \prec f_2$ for facts $f_1, f_2 \in S_P^\infty$ iff there are numbers $i, k \geq 0$ such that $f_1 \in T_{i+1}^k$ and $f_2 \notin T_{i+1}^k$.

We proceed by induction over the derivation steps, i.e., we assume that T_i^k is derivable for all i, k such that $i < j$, or $i = j$ and $k < \ell$. Consider an arbitrary fact $f \in T_j^\ell$. Let $P_i \in \mathbf{P}$ and $k \geq 0$ be such that f was first derived in T_i^k. If $k = 0$, then $i = 1$ and $f \in F$ (since all facts in T_i^0 for $i > 0$ do already occur in an earlier iteration T_{i-1}^m for some $m \geq 1$); in this case, f clearly satisfies the conditions of derivability. If $k > 0$, then there is a rule $r \in P_i$ such that $f \in r(T_i^{k-1})$. Let $\langle B^+, B^- \rangle$ be a body for this derivation with respect to T_i^{k-1}. We claim that $\langle B^+, B^- \rangle$ is also a body for $f \in r(T_j^\ell)$.

Clearly, $B^+ \subseteq T_i^{k-1} \subseteq T_j^\ell$, and thus $\hat{f} \prec f$ for every $\hat{f} \in B^+$. Moreover, we show that $B^- \cap T_j^\ell = \emptyset$. By definition, $B^- \cap T_i^k = \emptyset$. Now suppose for a contradiction that $B^- \cap T_j^\ell \neq \emptyset$. Then there is a rule $r' \in P_{i'}$ and a number k' such that $B^- \cap r'(T_{i'}^{k'}) \neq \emptyset$, where either $i < i'$, or $i = i'$ and $k \leq k'$, and also $i' < j$, or $i' = j$ and $k' < \ell$. By induction hypothesis, $T_{i'}^{k'}$ is derivable. Since $\bot \notin T_j^\ell$, we also have $\bot \notin r'(T_{i'}^{k'})$. Hence $r' \twoheadrightarrow r$; together with $i \leq i'$ this contradicts the assumed stratification. Hence $\langle B^+, B^- \rangle$ is a body for $f \in r(T_j^\ell)$. This establishes the conditions for well-supportedness of f. The remaining conditions for derivability are immediate by construction, since $r(T_i^{k-1}) \subseteq T_j^\ell$. □

Lemma 2. *Consider numbers $i \leq j$ with $P_i, P_j \in \mathbf{P}$, and a rule $r \in P_i$. If $\bot \notin S_P^j$ then $r(S_P^j) \subseteq S_P^j$.*

Proof. By Proposition 1, $r(S_P^i) \subseteq S_P^i$. Now consider $j > i$. Suppose for a contradiction that $r(S_P^i) \not\subseteq S_P^j$. There is $k > i$ and $\ell \geq 0$ with $T_k^\ell := T_{P_k}^\ell(S_P^{k-1})$ such that $r(T_k^\ell) \subseteq S_P^j$ and $r(T_k^{\ell+1}) \not\subseteq S_P^j$. Thus, there is a fact $f \in r(T_k^{\ell+1}) \setminus S_P^j$. Let $\langle B^+, B^- \rangle$ be a body for this derivation. We have $B^+ \not\subseteq T_k^\ell$. Thus there is a fact $f' \in B^+ \setminus T_k^\ell$ that is derived by a rule $r' \in P_k$ from T_k^ℓ. By Lemma 1, $T_k^{\ell+1}$ is derivable, where $r_{f'} = r'$. Since $\bot \notin S_P^j$ and $T_k^{\ell+1} \subseteq S_P^j$, also $\bot \notin T_k^{\ell+1}$. Hence $r' \twoheadrightarrow r$. Together with $i < j$ this contradicts the assumed stratification. □

Proposition 4. *If $\bot \notin S_P^\infty$ then S_P^∞ is a stable model of P and F.*

Proof. For every $r \in P$ and every derivation $f \in r(S_P^\infty)$, there is a body $\langle B^+, B^- \rangle$ by compactness. Since B^+ is finite, there is some $n \geq 0$ such that $B^+ \subseteq S_P^n$. By Lemma 2, $f \in S_P^n$. Hence S_P^∞ is a model of P and F.

For every $n \geq 0$, S_P^n is well-supported by Lemma 1. Let $\prec_n$ be an according well-founded order that is a suborder of $\langle \mathbb{N}, < \rangle$, which exists by Proposition 2. We construct a suitable order $\prec$ to show well-supportedness of S_P^∞ as follows. For every $n \geq 1$, let L_n be the set $S_P^n \setminus S_P^{n-1}$, ordered by the well-founded order $\prec_n$ (restricted from S_P^n to L_n). We now define $\prec$ to be the transitive closure of the following set:

$$\{f_1 \prec f_2 \mid f_1, f_2 \in L_n, f_1 \prec_n f_2\} \cup \{f_1 \prec f_2 \mid f_1 \in L_n, f_2 \in L_m, n < m\}.$$

This order is well-founded (it can clearly be embedded into the ordinal ω^2, since every $\prec_i$ can be embedded into ω). If $f_1, f_2 \in S_P^n$ and $f_1 \prec_n f_2$, then $f_1 \prec f_2$. Therefore, the bodies used to show well-foundedness of a fact $f \in L_n$ can be used to show well-foundedness of $f \in S_P^\infty$. □

The final ingredient to the proof of Theorem 6 is the following lemma. In the classical case, an analogous result was shown by taking advantage of the Gelfond-Lifschitz reduct of P [14]. Since this is not available for abstract rules, we need to take a very different approach, using an induction over the sets of facts in M for which well-foundedness is established by a rule from stratum P_k or below.

Lemma 3. *If M is a stable model of P and F, then $S_P^\infty = M$.*

Proof (of Theorem 6). If $\bot \notin S_P^\infty$ then by Proposition 4, S_P^∞ is a stable model of P and F. Together with Lemma 3 this implies that S_P^∞ is the unique stable model of P and F.

If $\bot \in S_P^\infty$ suppose for a contradiction that M is a stable model of P and F. Then by Lemma 3, $S_P^\infty = M$, which contradicts the fact that $\bot \notin M$.   □

7 Stratifying Programs with Equality

In this section, we apply the previous results to show how a normal logic program with equality may be stratified. Classical logic programming engines support syntactic (term) equality that is easy to handle: it may only occur in the body of a rule. In contrast, *equality generating dependencies* in databases are rules that may infer new equalities between domain elements [1]. Inferred equality also plays a major role in ontology languages, which can be processed with answer set programming engines [6]. Fortunately, the special characteristics of equality can be fully expressed by logic programming rules, using the following well-known equality theory:

$$X \approx X \leftarrow \tag{4}$$

$$X \approx Y \leftarrow Y \approx X \tag{5}$$

$$X \approx Z \leftarrow X \approx Y, Y \approx Z \tag{6}$$

$$p(X_1, \ldots, Y, \ldots, X_n) \leftarrow X \approx Y, p(X_1, \ldots, X, \ldots, X_n) \tag{7}$$

where a rule of the form (7) is required for every n-ary predicate p (in a given program P), and every position of X within that predicate. We call this logic program $P_\approx$. While this approach allows logic programs to support equality without defining a special semantics, it has severe effects on stratification.

Example 9. Consider the program P that consists of the following rules

$$\mathsf{human}(X) \leftarrow \mathsf{biped}(X), \mathbf{not}\ \mathsf{bird}(X) \tag{8}$$

$$Y \approx Z \leftarrow \mathsf{human}(X), \mathsf{birthplace}(X,Y), \mathsf{birthplace}(X,Z) \tag{9}$$

together with a suitable equality theory $P_\approx$ for the predicates used therein. Rule $r_{(9)}$ states that each human has at most one birthplace. Let r_{bird} denote the version of rule (7) in $P_\approx$ for predicate bird. Now P cannot be R-stratified: if all set of ground facts are allowed as input, we have $r_{(8)} \overset{+}{\rightarrow} r_{(9)} \overset{+}{\rightarrow} r_{\mathsf{bird}} \overset{-}{\rightarrow} r_{(8)}$.

The previous example illustrates the fact that the equality theory leads to almost arbitrary reliances between otherwise unrelated rules, thus preventing stratification. This potential interaction is hardly desirable in this case, since no bird can ever be a birthplace. In [14] the authors have proposed the use of constraints to reduce the amount of reliances. We can obtain a similar effect using abstract rules.

Example 10. Consider the constraint $r_\bot : \bot \leftarrow \mathsf{bird}(X), \mathsf{birthplace}(Y,X)$ and the program P of Example 9. Define the program $P' := \{r_\bot \circ r \mid r \in P\}$, which immediately applies the constraint after each rule application. Instead of $r_{(9)} \overset{+}{\rightarrow} r_{\mathsf{bird}}$, we now find $(r_\bot \circ r_{(9)}) \overset{+}{\nrightarrow} (r_\bot \circ r_{\mathsf{bird}})$ since $\bot$ is derived in all cases where the reliance could occur. Unfortunately, this approach still fails to make P' R-stratifiable, since we still find $(r_\bot \circ r_{(8)}) \overset{+}{\rightarrow} (r_\bot \circ r_{(9)}) \overset{+}{\rightarrow} (r_\bot \circ r_{(5)}) \overset{+}{\rightarrow} (r_\bot \circ r_{\mathsf{bird}}) \overset{-}{\rightarrow} (r_\bot \circ r_{(8)})$.

The symmetry rule (5) is used in the previous example to ensure that every reliance can be shown without violating the constraint. This is unfortunate since the program has only at most one unique stable model: in all situations where the chain of reliances of the example is mirrored by an actual chain of rule applications, the constraint $r_\perp$ must be violated. This problem can be overcome by incorporating equality reasoning into each rule application as follows.

Example 11. Let $P_\approx$ denote the equality theory for rules (8) and (9). Define a rule $\hat{r} := r_\perp \circ (\bigcup P_\approx)^\infty$, and rules $r_1 := \hat{r} \circ r_{(8)}$ and $r_2 := \hat{r} \circ r_{(9)}$. Note that these are indeed abstract rules by Theorem 3. Admissible input sets $\mathscr{D}$ are defined to be all models of $\hat{r}$. Then the program $\{r_1, r_2\}$ is R-stratified, and the only reliance is $r_1 \overset{+}{\to} r_2$. By Theorems 4 and 5, as well as Proposition 3, the stable models of $\{r_1, r_2\}$ are identical to the stable models of P' from Example 10. By Theorem 6, P' thus has a unique stable model whenever it is satisfiable.

The previous example outlines an interesting general approach to analyse the effects of equality. More important, however, is the fact that we have defined this extension and verified its key properties in a few lines. In contrast, the extension with constraints sketched in Example 10 originally required several pages of correctness proofs [14]. A major goal of our abstract framework is to extract the common ideas of such proofs, to provide an easy-to-use toolbox for establishing similar properties for many different scenarios and kinds of rules.

8   Conclusions

In this work, we proposed an abstract framework for studying logic programs, where rules are simply viewed as functions over an abstract set of derivable facts. We have shown that recent results on stratification and stable models can be lifted to this general case, and how these results can be instantiated to obtain a new approach for dealing with equality in logic programs. This result also takes advantage of a variety of semantic-preserving algebraic operations that we have introduced to construct abstract rules.

The purpose of this work is to demonstrate how abstract rules can serve as a powerful framework for establishing universal results about rule-based reasoning, which can readily be instantiated in concrete cases. There are numerous directions into which this research can be extended next. An obvious step is to define abstract notions of other semantics, such as the well-founded semantics, and to lift other relevant conditions, such as order consistency [8] and acyclicity [9,4,11]. In each case, new concrete applications of these results should be established, thus bridging gaps between different areas where rule languages are considered. Finally, even our basic notion of abstract rule may still be extended further, e.g., by allowing rules to retract facts. Moreover, our approach does not cover disjunctive rules, although these can often be simulated using nonmonotonic negation by rewriting $p \to q_1 \vee q_2$ as $p, \mathbf{not}\ q_1 \to q_2$ and $p, \mathbf{not}\ q_2 \to q_1$ [5].

Besides extensions of the abstract rules framework, it is also worthwhile to explore further application areas for these notions. All examples given herein are based on normal logic programs. Our treatment of equality shows that this already allows interesting applications, but it would also be interesting to consider different notions of rules. Possible candidates are rules that support datatype reasoning and data-related constraints.

Acknowledgements. This work was supported by the Royal Society, the Seventh Framework Program (FP7) of the European Commission under Grant Agreement 318338, 'Optique', and the EPSRC projects ExODA, Score! and MaSI3.

References

1. Abiteboul, S., Hull, R., Vianu, V.: Foundations of Databases. Addison Wesley (1994)
2. Baget, J.F.: Improving the forward chaining algorithm for conceptual graphs rules. In: Dubois, D., Welty, C.A., Williams, M.A. (eds.) Proc. 9th Int. Conf. on Principles of Knowledge Representation and Reasoning, KR 2004, pp. 407–414. AAAI Press (2004)
3. Baget, J.F., Leclère, M., Mugnier, M.L., Salvat, E.: On rules with existential variables: Walking the decidability line. Artificial Intelligence 175(9-10), 1620–1654 (2011)
4. Deutsch, A., Nash, A., Remmel, J.B.: The chase revisited. In: Lenzerini, M., Lembo, D. (eds.) Proc. 27th Symposium on Principles of Database Systems, PODS 2008, pp. 149–158. ACM (2008)
5. Eiter, T., Fink, M., Tompits, H., Woltran, S.: On eliminating disjunctions in stable logic programming. In: Dubois, D., Welty, C.A., Williams, M.A. (eds.) Proc. 9th Int. Conf. on Principles of Knowledge Representation and Reasoning, KR 2004, pp. 447–458. AAAI Press (2004)
6. Eiter, T., Krennwallner, T., Schneider, P., Xiao, G.: Uniform evaluation of nonmonotonic DL-programs. In: Lukasiewicz, T., Sali, A. (eds.) FoIKS 2012. LNCS, vol. 7153, pp. 1–22. Springer, Heidelberg (2012)
7. Fages, F.: A new fixpoint semantics for general logic programs compared with the well-founded and the stable model semantics. New Generation Comput. 9(3/4), 425–444 (1991)
8. Fages, F.: Consistency of Clark's completion and existence of stable models. Meth. of Logic in CS 1(1), 51–60 (1994)
9. Fagin, R., Kolaitis, P.G., Miller, R.J., Popa, L.: Data exchange: semantics and query answering. Theoretical Computer Science 336(1), 89–124 (2005)
10. Krötzsch, M., Magka, D., Horrocks, I.: Concrete results on abstract rules. Tech. rep., University of Oxford (2013), http://korrekt.org/page/Publications
11. Lierler, Y., Lifschitz, V.: One more decidable class of finitely ground programs. In: Hill, P.M., Warren, D.S. (eds.) ICLP 2009. LNCS, vol. 5649, pp. 489–493. Springer, Heidelberg (2009)
12. Lifschitz, V., Pearce, D., Valverde, A.: Strongly equivalent logic programs. ACM Trans. Comput. Logic 2(4), 526–541 (2001)
13. Lloyd, J.W.: Foundations of Logic Programming. Springer (1988)
14. Magka, D., Krötzsch, M., Horrocks, I.: Computing stable models for nonmonotonic existential rules. In: Proc. 23rd Int. Joint Conf. on Artificial Intelligence, IJCAI 2013. AAAI Press/IJCAI (to appear, 2013)
15. Przymusinski, T.C.: On the declarative and procedural semantics of logic programs. J. Autom. Reasoning 5(2), 167–205 (1989)
16. Turner, H.: Strong equivalence made easy: nested expressions and weight constraints. TPLP 3(4-5), 609–622 (2003)

Linear Logic Programming for Narrative Generation

Chris Martens[1], Anne-Gwenn Bosser[2], João F. Ferreira[2], and Marc Cavazza[2]

[1] Carnegie Mellon University
[2] Teesside University

Abstract. In this paper, we explore the use of Linear Logic programming for story generation. We use the language Celf to represent narrative knowledge, and its own querying mechanism to generate story instances, through a number of proof terms. Each proof term obtained is used, through a resource-flow analysis, to build a directed graph where nodes are narrative actions and edges represent inferred causality relationships. Such graphs represent narrative plots structured by narrative causality. This approach is a candidate technique for narrative generation which unifies declarative representations and generation via query and deduction mechanisms.

Keywords: Linear Logic Programming, Narrative Modelling, Celf.

1 Introduction

Linear Logic [5] has recently been proposed as a suitable representational model for narratives [2]: its resource-sensitive nature allows to naturally reason about narrative actions and the changes they cause in the environment. In this paper, we explore Linear Logic programming as a tool for narrative representation and narrative generation. We describe how initial circumstances and narrative actions can be declared in the Linear Logic programming language Celf [11] and how using Celf's search mechanism allows the generation of proof terms which can be interpreted as causally structured narrative plots. To improve narrative analysis, we developed a prototype front-end to Celf. We illustrate how to use story material to program and generate a variety of plots using the novel *Madame Bovary* [4]: its narrative causal structure has been emphasized in Flaubert's working material [8]. Preliminary results are encouraging, allowing the generation of story variants through a methodical programming approach.

2 Related Works

Narratives have always been an important topic for research in AI for their role as knowledge structures [12] and recent years have seen the widespread adoption of planning techniques for the construction of narrative generation systems [14], mostly because they support the representation of causality. Linear

P. Cabalar and T.C. Son (Eds.): LPNMR 2013, LNAI 8148, pp. 427–432, 2013.

Logic provides an expressive model of action and change (information revision is dealt with at the level of the logical rules through the linear implication) which has led previous work to explore its suitability for narrative representation using a story-as-proof analogy [2]. The intractability of proof search in expressive fragments of Linear Logic has led to the use of a proof assistant [3] for story generation and for evidencing properties transcending all narratives in a semi-automated manner. Support of narrative causality at the logical level is also an advantage when compared with standard logic programming approaches to narrative generation [13]. LolliMon [9] and Celf [11] are recent systems that have extended Lolli [7] (which follows a goal-directed backward proof-search interpretation in the intuitionistic fragment of Linear Logic) and where forward and backward chaining phases may be controlled by the programmer using a monad. We refer the reader to [10] for an overview and application survey.

3   Programming a Narrative

3.1   A Celf Program Describing a Narrative

Celf[1] [11] uses dependent types for the representation of logical predicates; this approach to logic programming means that the result of a query is a *term* of the corresponding *type*, which can be analysed as a computational artefact. Celf programs are normally divided into two main parts: **a signature**, which is a declaration of type and terms constants describing data and transitions, and **query directives**, defining the problem for which Celf will try to find solutions (proof terms showing that a given type is inhabited).

The technique used by Celf to compute proof terms is called focusing, based on the foundations of *Focused Linear Logic* [1] interpreted as Monadic Concurrent Logic Programming [9]: Celf gives the programmer control over when to enter a forward-chaining phase, which may use synchronous connectives, through the use of a monad (denoted using curly brackets {...}). The search triggered by a query in Celf begins in a backward-chaining phase using the query type as its goal, and if that type includes a monadic expression, it will enter a forward-chaining phase. This phase is implemented with a committed choice semantics, backtracking over the selection of a rule only when its antecedents cannot be met—effectively inducing a random choice between all fireable rules on each forward chaining step. This built-in nondeterminism lets us go automatically from a *specification* of a narrative structure to the automatic generation of stories.

3.2   Identification of Narrative Elements

The process of programming a narrative is that of describing circumstances that can, by execution of the program, generate one or many stories. Following a widespread paradigm in narrative generation research, we use an existing, linear, baseline story to support our experiments. Identifying the circumstances

---

[1] The Celf system can be obtained from `https://github.com/clf/celf`

```
1  emma : type.
2  emmaCharlesMarried : type.
3  <.............>
4  arsenic : type.
5  emmaIsDead : type.
6  emmaSpendsYearsInConvent : type = emma * convent -o {!novels * !grace * !
        education * @emma}.
7  emmaMarriesCharles : type = emma * escapism * grace * charles -o {
        emmaIsBored * @emma *!emmaCharlesMarried}.
8  emmaDoesNotGoToBall : type = emma * ball -o {emmaIsBored * @emma}.
9  < ......>
10 emmaContractsDebts : type = emma * emmaIsBored -o {@debt * @emma}.
11 emmaGetsSick : type = emma * emmaIsDespaired -o {@debt * @debt * @debt *
        @debt * !charlesIsConcerned * @emma}.
12 emmaJumpsThroughWindow : type = emma * emmaIsDespaired * emmaRebels -o {
        @emmaIsDead}.
13 emmaLearnsBovaryFatherDeath : type = emma * leonEmmaTogether *
        charlesIsConcerned * homais -o {@arsenic * @inheritance *
        @leonEmmaTogether * @emma}.
14 emmaCommitsSuicide : type = emma * ruin * arsenic * emmaRebels -o {
        @emmaIsDead}.
15 init : type =
16   { convent * @emma * @leonIsBored * !charles * !rodolphePastLoveLife * !
        homais
17           * @emmaSpendsYearsInConvent
18           * @(emmaGoesToBall & emmaDoesNotGoToBall)
19           <.......>
20           * !emmaContractsDebts
21 }.
22 #query * * * 100 (init -o {emmaIsDead}).
```

Fig. 1. Celf excerpt for a fragment of *Madame Bovary*. Atomic types (narrative resources) are followed by types describing narrative actions, the initial environment declaration and a query of 100 attempts to generate stories ending with Emma's death. The complete file (105 lines of code) is available on https://github.com/jff/TeLLer.)

within a static story such as Madame Bovary [4] is a human activity that can be assisted by companion works [8]. Figure 1 shows an example of a Celf program representative of the form we use to model narratives, and composed of a signature and a query (line 22).

The narrative elements we identify and model fall into two main categories. **Narrative resources** are available story elements (including characters) as well as states of the story, which may be related to characters and motives. In the present example, we model them using **atomic types** (lines 1–5). **Narrative actions** are transforming events occurring in the narrative. We model the impact they have on the narrative, in terms of resource creation and consumption using asynchronous types (lines 6–14), here linear implication formulae.

The type `init` on line 15 describes the initial narrative environment. Resources can be introduced as a) linear (default): there is one copy in the initial environment, and it will need to be consumed for any computation to terminate successfully. Emma's boredom is modelled as linear, since one of the driving force for her actions in the story is to escape this state; b) affine (using @): there is initially one copy in the initial environment and it may or may not be consumed by a successful computation. Because Emma may die in the story, the corresponding resource is introduced as persistent; c) persistent, (using !): there are arbitrarily many copies in the initial environment and any number of them may be consumed by a successful computation. We use this to denote immutable facts and hard rules, such as Emma and Charles married status for instance.

In addition to the author's notes [8] for filtering through story events irrelevant for the modelled narrative structure, we proceed iteratively, and lazily model a new resource when we model a narrative action involving it. The narrative action corresponding to Emma taking arsenic to poison herself illustrates this process: Emma (returning late from a date with Leon) learns about her father's death from Homais because Charles is afraid to upset her. She learns about inheritance. We first model:

```
emmaLearnsBovaryFatherDeath : type = emma * leonEmmaTogether *
    charlesIsConcerned * homais -o {@inheritance * @leonEmmaTogether *
    @emma}
```

During the same event, a side conversation occurs between two of the characters present during which Emma incidentally learns where to find arsenic. The importance of this knowledge becomes apparent only when we model the narrative action corresponding to Emma's death. We then modify the code so that the action adds the corresponding resource to the environment and obtain the code on lines 13 and 14.

Mutually exclusive narrative actions can be explicitly suggested using the choice connective & in the declaration of the initial conditions. These can be used to encode key turning points in the narrative that are broadly recognized as such, which is frequently the case when using existing stories as a baseline. We use this connective to model Emma's choice to attend the ball (see line 18).

Once the narrative modelled, it is run using Celf and the proof-terms obtained are post-processed for causality analysis. Following a long tradition of analysing causality via graphs [6], we developed a prototype tool, *CelfToGraph*[2], that automatically transforms proof terms generated by Celf into directed acyclic graphs. Such graphs represent narrative plots, structured by narrative causality, where nodes are narrative actions and edges represent inferred causality relationships.

One advantage of modelling narratives using a programming language is the ability to iteratively fine tune the model: a programmer alternates between coding and testing phases, which is facilitated by the frontend that we developed: in addition to the generation of causal graphs representing narratives, *CelfToGraph* queries can exhibit plots with specific characteristics. One can also verify if the generated set has a varied output (differing significantly from the original plot), test the impact of more **narrative drive** on the generation (for instance by comparing the effect of affine vs. linear models of narrative actions), or fine-tune resource threshold quantities.

3.3   Generated Plots

The entire code corresponding to the excerpt on Figure 1 consists of a total of 105 lines, including 31 narrative action descriptions. As we have only explicitly

---

[2] *CelfToGraph* requires Celf v2.9 and is available at `https://github.com/jff/TeLLer`

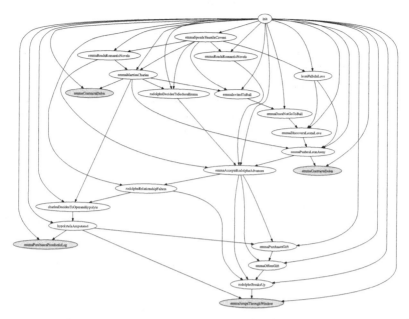

Fig. 2. One of 41 causally structured generated plots exhibited using *CelfToGraph*. In this variant, Emma does not attend the ball and defenestrates when left by Rodolphe.

encoded one branching choice, the variety of outputs is due to the linear semantics of narrative actions (producing resources that may be contended) and forward chaining variability.

The code described allows to **generate** 72 different narrative sequences for 100 attempts. After an automatic comparison of the corresponding plots using *CelfToGraph*, , we can exhibit 41 different plots (characterised by different generated causal structures), meaning that a number of different narrative sequences share the same causal structures. This allows the characterisation of classes of true **story variants**. Figure 2 shows a story variant among those generated, which has been exhibited by the tool: in this story, Emma jumps through the window following the departure of Rodolphe. If we look at the code Figure 2 l. 11 and l. 12 two narrative actions consume the resource `emmaIsDespaired`. When the first is triggered by the forward chaining mechanism, we obtain a story ending with Emma jumping through the window. When requesting 1000 query attempts, we obtain 747 solutions, among which 697 are different narrative sequences, and 226 true story variants.

4   Conclusion

There has been much interest in the use of Linear Logic to represent natural language semantics and the semantics of action and change. Narrative structures are based on the integration of the above phenomena, and Linear Logic programming provides a direct mechanism to operationalize these descriptions.

Our first results reported here are clearly encouraging, offering all the benefits of a declarative representation. This opens perspectives for applications such as Interactive Storytelling, where narrative generation is a default interaction paradigm, allowing narratives to adapt to changes in the environment.

In future work, we intend to develop this approach with the definition of an interaction paradigm using Linear Logic's choice connectives and on-the-fly environment modifications. Another interesting line of inquiry would be to explore the possible definition of normal forms for stories generated.

References

1. Andreoli, J.: Logic programming with focusing proofs in Linear Logic. Journal of Logic and Computation 2, 297–347 (1992)
2. Bosser, A.G., Cavazza, M., Champagnat, R.: Linear Logic for non-linear storytelling. In: ECAI 2010. Frontiers in Artificial Intelligence and Applications, vol. 215. IOS Press (2010)
3. Bosser, A.-G., Courtieu, P., Forest, J., Cavazza, M.: Structural analysis of narratives with the Coq proof assistant. In: van Eekelen, M., Geuvers, H., Schmaltz, J., Wiedijk, F. (eds.) ITP 2011. LNCS, vol. 6898, pp. 55–70. Springer, Heidelberg (2011)
4. Flaubert, G.: Madame Bovary. Revue de Paris (1857), edition 2001 Collection Folio Classiques, ISBN 9782070413119
5. Girard, J.Y.: Linear Logic. Theoretical Computer Science 50(1), 1–102 (1987)
6. Greenland, S., Pearl, J., Robins, J.: Causal diagrams for epidemiologic research. Epidemiology, 37–48 (1999)
7. Hodas, J.S., Miller, D.: Logic programming in a fragment of Intuitionistic Linear Logic. Information and Computation 110(2), 327–365 (1994)
8. Leclerc, Y.: Flaubert, Plans et Scénarios de Madame Bovary, Zuma, Cadeilhan (1995)
9. López, P., Pfenning, F., Polakow, J., Watkins, K.: Monadic concurrent Linear Logic programming. In: Proceedings of the 7th International ACM SIGPLAN Conference on Principles and Practice of Declarative Programming (2005)
10. Miller, D.: Overview of Linear Logic programming. Linear Logic in Computer Science 316, 119–150 (2004)
11. Schack-Nielsen, A., Schürmann, C.: Celf – A logical framework for deductive and concurrent systems (System description). In: Armando, A., Baumgartner, P., Dowek, G. (eds.) IJCAR 2008. LNCS (LNAI), vol. 5195, pp. 320–326. Springer, Heidelberg (2008)
12. Schank, R., Abelson, R.: Scripts, plans, goals and understanding: An inquiry into human knowledge structures. Psychology Press (1977)
13. Schroeder, M.: How to tell a logical story. In: Narrative Intelligence: Papers from the AAAI Fall Symposium. AAAI Press (1999)
14. Young, R.M.: Notes on the use of plan structures in the creation of interactive plot. In: Narrative Intelligence: Papers from the AAAI Fall Symposium. AAAI Press (1999)

Implementing Informal Semantics of ASP

Artur Mikitiuk[1] and Miroslaw Truszczynski[2]

[1] Mathematics & Natural Sciences, Cardinal Stefan Wyszyński University, Warsaw, Poland
[2] Department of Computer Science, University of Kentucky, Lexington, KY 40506-0633, USA

Abstract. We describe a system that, given a theory of an answer-set programming (ASP) system *psgrnd*, generates its informal reading in natural language. That reading helps understand the *psgrnd* theory, and verify its correctness or identify programming errors. Similar tools can be developed for other ASP formalisms. To this end, the basic language used by the system has to be extended to allow the programmer provide (minimal) additional information on how to understand atomic concepts, of which the theory (program) is built.

1 Introduction

Declarative programming rests on the idea that coding simply consists of writing down problem specifications given in natural language as formal expressions in some logic (declarative programming language). It requires a close and unambiguous syntactic connection between informal statements of the natural language and formal statements of the target declarative programming language. The connection must tie the *intended informal meaning* of natural text expressions with the *formal meaning* of formal statements (programs or theories). This connection exists in the case of the first-order (FO) logic and its extension, the logic FO(ID), that was proposed as a computational knowledge representation formalism [2]. It is in fact the main reason why FO logic and its extensions gained so much attention by the knowledge representation community.

Our goal in this work is to show that the informal semantics can be "implemented," that is, one can generate an informal reading of a formal declarative program (knowledge base) representing a problem. We describe an implementation of the informal semantics of an ASP system *psgrnd* [4,3]. That system extends the FO logic with cardinality and weight constraints, and with Horn logic program rules to represent monotone definitions. It can be seen as a fragment of the computational knowledge representation system FO(ID) [2,1], when the latter is restricted to the class of Herbrand models.

Our approach follows our earlier work in this direction [6]. The main extensions the present system *pspbdb* offers consist of the support to handle definitions and aggregates (neither issue was addressed before), and in redesigned natural language templates.

The ASP community has been addressing the issues of testing and debugging of ASP programs [7,9], and program development environments [5,8]. We believe tools producing informal readings of answer-set programs will further promote the acceptance of ASP.

2 The Implementation of the Program *pspbdb*

The language PS^{pb} has been proposed in [3] for modeling search problems specified in terms of boolean combinations of pseudo-boolean constraints (pb-constraints for short).

P. Cabalar and T.C. Son (Eds.): LPNMR 2013, LNAI 8148, pp. 433–438, 2013.

This language extends FO logic with the ability to model explicitly pb-constraints and to define Horn predicates. A formal grammar describing the syntax of the language PS^{pb} is available at http://www.cs.uky.edu/psgrnd/.

Program *pspbdb* is a tool translating theories in the language PS^{pb} into English sentences that represent the informal meaning (semantics) of a PS^{pb} theory.[1] The translation is based on comments for the debugger placed in the theory being translated next to data atom and weight function definitions and next to predicate declarations (cannot be placed before such definitions or declarations). Here are some examples.

```
vtx(1..200). %%vtx(X) means #vertex X
edge(86,163). %%edge(A,B) means #(A,B) is an edge
pred hc(vtx,vtx):edge.
        %%hc(A,B) means #edge(A,B) belongs to Ham cycle
```

A comment for the debugger starts with a double percent sign followed by a predicate or weight function name with names of its arguments, followed by the word *means* and a hash sign. The text after the hash sign until the end of the line is an explanation. This explanation is used by *pspbdb* during translation of program or data atoms and weight function calls appearing in program rules. The quality of the translation depends on the accuracy of the provided comments.

Program *pspbdb* has been written in C++. It follows our earlier work [6] and reuses large parts of the *psgrnd* code [3]. The formal grammar for the language PS^{pb} has been extended with additional productions including syntax of comments for the debugger. The lexical analyzer from *psgrnd* had to be modified to recognize additional tokens (double percent sign, keyword means, hash sign). The *psgrnd* code related to parsing had to be modified by adding new functions needed to process comments for the debugger and to translate rules in the processed PS^{pb} theory into English sentences.

Translation of Atoms and Weight Function Calls. The translation of a program atom is done by replacing symbols (formal parameters) provided in the comment following the predicate declaration with actual arguments. For example, given the declaration

```
pred color(vtx,clr). %%color(A,B) means #vertex A is colored B
```

atom $color(X, red)$ is translated as *vertex X is colored red*. The weight function calls are translated similarly as atoms (but comments explaining such calls should be written in plural). Next, if a program atom has a list of local variables, it is translated as *for some value of . . .* or *for some tuple* $(\ldots)$ depending on whether the list contains one or more variables.

A cardinality constraint of the form $l\{A_1, \ldots, A_k\}u$ is in general case translated as

$$\text{there are at least } l \text{ but but not more than } u \text{ atoms such that}$$
$$\text{"translation of } A_1 \text{" OR } \ldots \text{ OR "translation of } A_k \text{"}$$

Our program recognizes several special cases: when the lower bound l and the upper bound u are equal, when one or both of the bounds are equal to 1, when one of the bounds is missing. In such cases the translation is modified accordingly.

The translation of pb-constraints (weighted atoms) is similar to that of cardinality constraints. Given the following explanations of weight function *size* and predicate *in*

---

[1] Download from http://www.cs.uky.edu/ai/software/pspbdb.tar.gz.

```
size(5)=13. %%size(A) means #the sizes of A
pred in(item). %%in(A) means #A that are in the knapsack
```

a pb-constraint "$\{in(X) = size(X)[X]\}15$." is translated as

```
The sum of the sizes of X FOR THOSE X
that are in the knapsack is not more than 15
```

Translation of Rules. A rule $A_1, A_2, \ldots, A_s \rightarrow B_1|B_2|\ldots|B_t$, is translated into a sentence of one of the following two forms depending on whether the body is present.

(a) IF   "translation of A_1" (b)
      AND   "translation of A_2"

      . . .

      AND   "translation of A_s"
      THEN   "translation of B_1"   "translation of B_1"
      OR   "translation of B_2"   OR   "translation of B_2"

      . . .   . . .

      OR   "translation of B_t"   OR   "translation of B_t"

The sentence may be preceded by the statement "For every X_1 and $\ldots$ and X_n," to indicate which variables in the formula are universally quantified. For example, if we provide the following meaning for predicate in

```
pred in(time,pos,pos,entry). %%in(A,B,C,D) means
    #at time A entry D is in position (B,C)
```

the rule "$in(T, X, Y, A)[A]$." is translated as

```
For every T and X and Y,
    at time T entry A is in position (X,Y) for some value of A
```

When the head is missing, the clause specifies a constraint. We do not present translations of constraints due to space restrictions.

Horn rules in PS^{pb} are used to *define* Horn predicates. Such definitions can be explained under two restrictions: (1) all rules defining the same Horn predicate are not separated with other rules; (2) the predicate in the heads of all its defining rules has the same list of arguments. Then

                                      DEFINING

$A :\!- B_{1,1}, \ldots, B_{1,k_1}$ "Translation of atom A"
    $\ldots$ translates   (case 1)   IF "translation of the 1st body"
                          to   . . .
$A :\!- B_{m,1}, \ldots, B_{m,k_m}$ (case m)   IF "translation of the mth body"
                                     AND in no other case

Dealing with Symbolic Constants. Program *pspbdb* works in two stages due to the way *psgrnd* works. The grounder accepts theories and data sets containing some symbolic constants (integer parameters). For example, when a graph has n vertices, one can define data predicate $vtx(1..n)$. When the program is invoked, this value has to be provided in the command line. Since *pspbdb* reuses large parts of code from *psgrnd*, it also requires the value of n to be provided in the command line. In the first stage *pspbdb* creates a temporary file with translation containing numerical values of such parameters.

In the second stage the temporary file is scanned for values of command line parameters. However, a number may appear in the text as the value of a parameter but it may also appear because the number occurs in the translated rule. If the parameter does not appear in the rule, we leave the number. If the number does not appear in the rule, we replace it with the name of the parameter.

The problem is what to do when both the parameter and the number occur in the original rule. For example, one could have a cardinality constraint of the form $2\{\ldots\}n$ when the value of n is 2. We could also have two parameters with the same value. In such cases *pspbdb* leaves the number in the translation (we plan to work on this case in the future). To avoid such situations, the user could give each parameter a different value (this value does not matter for *pspbdb*) and avoid giving small values (0, 1, 2) because such numbers often appear in the rules. If one notices in the translation a number instead of a parameter, one can re-run the program with another value of this parameter.

Program Usage. Program is invoked from a UNIX command line in the following way:

$$pspbdb \ [dataFileList] \ ruleFile \ [constantList] \ [-o \ output]$$

An optional *dataFileList* consists of one or more files specifying the "data" component of the problem description while *ruleFile* contains the "program" component. A *constantList* is a list of value assignments for parameters appearing in these files. If the names of constants are $n1, \ldots, nM$ and the corresponding values are $v1, \ldots, vM$, the *constantList* has form $n1 = v1 \ \ldots \ nM = vM$. If -*o* option is not specified, the default output file is *out.pspbdb*.

3   Results

Due to space restriction, we will present in this section only two examples of PS^{pb} theories with added comments for the debugger and the results produced by *pspbdb*. The examples illustrate in particular the new features of the program: the cardinality constraints and Horn rules. Our first example is a program for the *n queens* problem.

```
num(1..n).
pred q(num,num). %%q(A,B) means #queen in row B and col A
var num C,R,I.
1{q(C,R)[C]}1. %exactly one queen in every row
1{q(C,R)[R]}1. %exactly one queen in every col
q(C,R), q(C+I,R+I) -> . %no two queens on
q(C,R), q(C+I,R-I) -> . %the same diagonal
```

Processing this theory gives the following result (to save space we skip empty lines):

```
LINE 4:
1{q(C,R)[C]}1. %exactly one queen in every row
For every R,
```

```
      queen in row R and col C for exactly one value of C
LINE 5:
1{q(C,R)[R]}1. %exactly one queen in every col
For every C,
      queen in row R and col C for exactly one value of R
LINE 6:
q(C,R), q(C+I,R+I) -> . %no two queens on
The following properties must not occur together:
      queen in row R and col C
      queen in row R + I and col C + I
LINE 7:
q(C,R), q(C+I,R-I) -> . %the same diagonal
The following properties must not occur together:
      queen in row R and col C
      queen in row R - I and col C + I
```

The second example is a program encoding the *Hamiltonian cycle* problem.

```
%%edge(A,B) means #(A,B) is an edge
start(1). %%start(X) means #X is a start vertex
pred hc(vtx,vtx):edge.
%%hc(A,B) means #edge(A,B) belongs to Ham cycle
pred visit(vtx). %%visit(A) means
#vertex A can be reached from start via Ham cycle edges
var vtx X,Y.
1{hc(X,Y)[Y]}1.
1{hc(X,Y)[X]}1.
visit(Y) :- visit(X),hc(X,Y).
visit(Y) :- start(Y).
visit(X).
```

The translation produced by *pspbdb* is given below.

```
LINE 1263:
1{hc(X,Y)[Y]}1.
For every X,
   edge(X,Y) belongs to Ham cycle for exactly one value of Y
LINE 1264:
1{hc(X,Y)[X]}1.
For every Y,
   edge(X,Y) belongs to Ham cycle for exactly one value of X
LINE 1265:
visit(Y) :- visit(X),hc(X,Y).
LINE 1266:
visit(Y) :- start(Y).
DEFINING
 vertex Y can be reached from start via Ham cycle edges
  (case 1) IF vertex X can be reached from start via Ham cycle
             edges
           AND edge(X,Y) belongs to Ham cycle
  (case 2) IF Y is a start vertex
```

```
AND in no other case
LINE 1267:
visit(X).
For every X,
    vertex X can be reached from start via Ham cycle edges
```

4   Discussion and Future Work

The examples presented above show that if a PS^{pb} program is annotated with the instructions how to render predicates, the tool *pspbdb* generates text that well captures the intended meaning of the program. The annotations are minimal and not a burden on the programmer. They simply represent associations between formal symbols and their intended reading the programmer established in the program design phase.

Our work suggests several uses for programs such as *pspbdb*. In ASP they can assist the programmer in developing correct programs and they can help in understanding programs developed by others (an onerous task). They can also help in teaching formal logics. When students are asked to represent a statement in logic, a program translating logical expressions into English could help them recognize and correct errors.

There are several possible improvements to *pspbdb* concerning replacing numeric values with symbolic constants, relaxing restrictions on Horn rules, and an interactive operation of the tool. Importantly, the tool can be extended to other flavors of answer set programming that are based on the syntax of (disjunctive) logic programs.

References

1. Denecker, M.: A knowledge base system project for FO(.). In: Hill, P.M., Warren, D.S. (eds.) ICLP 2009. LNCS, vol. 5649, p. 22. Springer, Heidelberg (2009)
2. Denecker, M., Ternovska, E.: A logic for non-monotone inductive definitions. ACM Transactions on Computational Logic 9(2) (2008)
3. East, D., Iakhiaev, M., Mikitiuk, A., Truszczyński, M.: Tools for modeling and solving search problems. AI Communications 19(4), 301–312 (2006)
4. East, D., Truszczyński, M.: Predicate-calculus based logics for modeling and solving search problems. ACM Transactions on Computational Logic 7, 38–83 (2006)
5. Febbraro, O., Reale, K., Ricca, F.: ASPIDE: Integrated development environment for answer set programming. In: Delgrande, J.P., Faber, W. (eds.) LPNMR 2011. LNCS, vol. 6645, pp. 317–330. Springer, Heidelberg (2011)
6. Mikitiuk, A., Moseley, E., Truszczynski, M.: Towards debugging of answer-set programs in the language pspb. In: Proceedings of the 2007 International Conference on Artificial Intelligence, ICAI 2007, pp. 635–640 (2007)
7. Oetsch, J., Prischink, M., Pührer, J., Schwengerer, M., Tompits, H.: On the small-scope hypothesis for testing answer-set programs. In: Brewka, G., Eiter, T., McIlraith, S.A. (eds.) Proceedings of the 13th International Conference on Principles of Knowledge Representation and Reasoning, KR 2012. AAAI Press (2012)
8. Oetsch, J., Pührer, J., Tompits, H.: The sealion has landed: An ide for answer-set programming—preliminary report. CoRR abs/1109.3989 (2011)
9. Vos, M.D., Kisa, D.G., Oetsch, J., Pührer, J., Tompits, H.: Annotating answer-set programs in lana. Theory and Practice of Logic Programming 12(4-5), 619–637 (2012)

Implementing Belief Change in the Situation Calculus and an Application

Maurice Pagnucco [1], David Rajaratnam [1],
Hannes Strass [2], and Michael Thielscher [1]

[1] School of Computer Science and Engineering, UNSW, Australia
{morri,daver,mit}@cse.unsw.edu.au
[2] Leipzig University, Leipzig, Germany
strass@informatik.uni-leipzig.de

Abstract. Accounts of belief and knowledge in the Situation Calculus
have been developed and discussed for some time yet there is no extant
implementation. We develop a practical implementation of belief and belief change in the Situation Calculus based on default logic for which we
have an implemented solver. After establishing the mapping with default
logic we demonstrate how belief change in the Situation Calculus can be
used to solve an interesting problem in robotics – reasoning with misleading information. Motivated by a challenge in the RoboCup@Home
competition, we give a solution to the problem of planning robustly in
cases where operators provide the robot with misleading or incorrect
information.

1 Introduction

Several accounts of belief and knowledge in the Situation Calculus have been developed and discussed for some time [1–3] yet there is no extant implementation.
In this paper, we show how belief and belief change in the Situation Calculus
according to the account of [1, 4] can be implemented. We do so by mapping this
formalisation of belief change in the Situation Calculus to an account of default
logic for which we have an implemented solver [5].

As we establish this mapping into default logic, we demonstrate how belief
change in the Situation Calculus can be used to solve an interesting problem in
robotics – reasoning with misleading information. Motivated by a challenge in
the RoboCup@Home competition, we give a solution to the problem of planning
robustly in cases where operators provide the robot with misleading or incorrect
information. What, for example, should a robot given the task of returning
with the red cup from the kitchen table do when it arrives in the kitchen to
find no red cup but instead notices a blue cup and a red plate on the table?
In RoboCup@Home, the best course of action is not to return empty-handed
but to attempt to salvage the situation by applying a form of commonsense
preferences to return with one of the objects available. Our results pave the way
for a practical and efficient solution to such problems.

P. Cabalar and T.C. Son (Eds.): LPNMR 2013, LNAI 8148, pp. 439–451, 2013.

The rest of the paper proceeds as follows. We first provide the technical background to understand the paper. Then we describe a formal specification of belief change in the Situation Calculus. A motivating example based on RoboCup@Home is introduced in order to demonstrate a challenging problem that can be solved using this account of belief change in the Situation Calculus. Next, we present another solution, that is based on an implementable fragment of prioritised default logic and show how belief change in the Situation Calculus can be translated into this logic. Finally, we show that the two solutions yield the same results, discuss our findings in a broader context and conclude.

2 Technical Preliminaries

2.1 Situation Calculus

The Situation Calculus provides a formal language based on that of classical first-order logic in which to describe dynamic domains [6, 7]. Three types of terms are distinguished: *situations* representing a snapshot of the world; *fluents* denoting domain properties that may change as a result of actions; and *actions* that can be performed by the reasoner. We use the predicate $Holds(f, s)$ to specify that a fluent f holds at a particular situation. As a matter of convention a short form is adopted such that for any n-ary fluent $f(x_1, \ldots, x_n)$, writing $f(x_1, \ldots, x_n, s)$ is a short form for $Holds(f(x_1, \ldots, x_n), s)$. A special function $do(a, s)$ represents the situation that results from performing action a at situation s. S_0 denotes the initial situation where no actions have taken place. For each action we need to specify preconditions $Poss(a, s)$ specifying the conditions under which action a is possible in situation s and effect axioms that specify how the value of a fluent changes when an action is performed.[1] For a more comprehensive formulation of what is required of a Situation Calculus basic action theory (BAT), the reader is referred to [7].

2.2 Iterated Belief Revision in the Situation Calculus

A request to an agent to achieve a goal affects its beliefs. For instance, when the agent is asked to collect the red cup from the kitchen table, it is reasonable for the agent to believe that there is in fact a red cup located on the kitchen table. We therefore adopt an extension to the Situation Calculus capable of representing beliefs. Several accounts exist [1–3] however we use that of Shapiro et al. [1]. It is based on the ideas of Moore and extended by Cohen and Levesque [8] who introduced knowledge into the Situation Calculus by reifying the accessibility relation in modal semantics for knowledge. Two types of actions are distinguished: *physical actions* which alter the world (and hence fluent values) when performed; and, *sensing actions* that are associated with a sensor and determine the value of a fluent (e.g., a vision system used to determine whether a red cup is on a

---

[1] In fact, we compile effect axioms into *successor state axioms* (SSAs) [7].

table). Sensing actions are also referred to as *knowledge producing* actions as they inform the reasoner about fluent values but do not alter the world state.

Scherl and Levesque [9] introduced the relation $B(s', s)$ denoting that if the agent were in situation s, it would consider s' to be possible.[2] This is adopted by Shapiro et al. [1]. The successor state axiom for the B relation is given in the table below as Axiom (B1) and states that s'' is possible at the situation resulting from performing action a at situation s whenever the sensing action associated with a agrees on its value at s and s'. $SF(a, s)$ is a predicate that is true whenever the sensing action a returns the sensing value 1 at s and was introduced by Levesque [10]. The innovation of Shapiro et al. [1] is to associate a plausibility with situations. Plausibility values are introduced, in decreasing order with a value of 0 being the most plausible, for initial situations and these values remain the same for all successor situations as expressed in Axiom (B2) below. This is critical for preserving the introspection properties for belief. The plausibility values themselves are not important, only the ordering over situations that they induce. Axioms (B3) and (B4) define the situations s' that are most plausible (MP) and most plausible situations that are possible – i.e., B-related – (MPB) at s, respectively. In Axiom (B5) we define sentence ϕ to be believed in situation s ($Bel(\phi, s)$) whenever it is true at all the most plausible situations that are possible at s. Finally, Axiom (B6) specifies that any situations B-related to an initial situation are also initial situations. The distinguished predicate $Init(s)$ indicates that s is an initial situation.

B1. $B(s'', do(a, s)) \equiv \exists s'[B(s', s) \wedge s'' = do(a, s') \wedge SF(a, s') \equiv SF(a, s))]$

B2. $pl(do(a, s)) = pl(s)$

B3. $MP(s', s) \stackrel{\text{def}}{=} \forall s''.B(s'', s) \supset pl(s') \leq pl(s'')$

B4. $MPB(s', s) \stackrel{\text{def}}{=} B(s', s) \wedge MP(s', s)$

B5. $Bel(\phi, s) \stackrel{\text{def}}{=} \forall s'.MPB(s', s) \supset \phi[s']$

B6. $Init(s) \wedge B(s', s) \supset Init(s')$

3   Formalisation in the Situation Calculus

The formalisation of our approach is based on the iterated belief revision extension to the Situation Calculus. Notably, the problem is specified in terms of *primitive fluents, primitive actions, sensing actions,* an *initial state, precondition axioms,* and *successor state axioms.* In order to deal with the potential for defeasible information we introduce a number of restrictions to this formalism. In essence these restrictions are designed to exploit the way in which abstract logical names can be anchored to the perception of actual objects in the environment.

Objects. We require a fixed set I of individual objects, to which a unique names assumption is applied. Intuitively, they identify the items that a robot is trained to recognise. We introduce the fluent $Same(x, y)$ to express that two

---

[2] Note the order of the arguments as it differs from that commonly used in modal semantics of knowledge.

names refer to the same real object, and allow a set of additional names $N = \{O_1, \ldots, O_n\}$, ensuring that these names only refer to existing objects in the domain:

$$\bigvee_{A \in I} Same(O_i, A, s), \quad \text{for } 1 \leq i \leq n$$

Same is required to be reflexive, symmetric and transitive and is further axiomatised using "substitutivity" axioms to enforce that identical objects agree on all fluent properties F of the domain

$$Same(x, y, s) \supset (F(\bar{z}, s)[z_i/x] \equiv F(\bar{z}, s)[z_i/y])$$

and the SSA $Same(x, y, do(a, s)) \equiv Same(x, y, s)$. The trivial successor state axiom of this fluent reflects the intuition that hypotheses about names referring to objects will only be affected by knowledge-producing actions.

Informing the Robot. Informing the robot about the operator's belief in the state of the world is formalised outside of the underlying action calculus at the meta-level and is subsequently compiled into the initial state axioms.

Let f be a fluent literal. Then $Told(f, S_0)$, which we abbreviate as $Told(f)$, represents the act of the operator informing the robot about the operator's understanding of the initial state of the world. Additionally, object references in f must consist only of the names in N, reflecting the intuition that the operator may only ever refer to objects on the basis of their properties, but not by using their names. Finally, a set of operator commands T is *consistent* provided there is no fluent f such that $Told(f) \in T$ and $Told(\neg f) \in T$.

Setting Goals. Requests from the user for the robot to perform a task are required to be of the form $Goal(\exists s.\phi(s))$ where $\phi(s)$ is a sentence expressing the goal to be achieved. As with the operator commands, all objects referenced in $\phi(s)$ must be referred to only by the names in N.

Motivating Example: Dealing with Misleading Information

The following example will be used to illustrate our approach. It represents a reasonably practical example of moderate sophistication sufficient for the space available. Moreover it is of interest as it represents an instance of goal revision which can be innovatively handled by the account of belief change in the Situation Calculus that we adopt here.

The Robocup@Home (`robocupathome.org`) competition is an international initiative to foster research into domestic robots. Effective domestic robots must be able to perform tasks in response to user commands and to behave *robustly* if the information provided is in some way erroneous. This is demonstrated in the "General Purpose Service Robot" challenge of the Robocup@Home 2010 Competition,[3] and the following scenario is based on an example from this challenge.

---

[3] `http://www.robocupathome.org/documents/rulebook2010_FINAL_VERSION.pdf`

Scenario 1. *The robot is in the living room of the home. The home has a kitchen with a table in the middle. The robot is told to fetch the red cup from the kitchen table. However, there is no red cup on the kitchen table and the robot only discovers this fact once it arrives in the kitchen and looks for the cup on the table.*

We highlight two separate cases. In the base case there is only a blue cup on the table. In the extended case there is a blue cup and a red plate on the table.

While a robot cannot know the precise intentions of the human operator, it can nevertheless apply commonsense knowledge in its responses. In the first case, faced with no alternatives, it might simply fetch the blue cup. In the second case, it might assume that the user is more interested in the type of object than its colour and so would prefer the blue cup over the red plate.

For simplicity of presentation we provide a compact encoding of this scenario. In particular for binary properties we adopt only one of each binary pair, with the intuition that the negation of the given property implies that its pair must hold. For example, if an object is not a cup then it must be a plate.

Objects. The Robocup@Home challenge deals with a fixed set of household objects that are determined at the start of the competition. This allows the teams time to train their vision systems to be able to detect and distinguish between these objects. In our example scenario, there are two cups, one red and one blue, and a red plate: $I = \{C_R, C_B, P_R\}$.

Primitive Fluents. The primitive fluents in our domain and their meanings are as follows. *InKitchen*: the robot is in the kitchen, *Holding*(o): the robot is holding an object, *OnTable*(o): the object is on the kitchen table, *Cup*(o): the object is a cup, *Red*(o): the object is red.

Primitive Actions *SwitchRoom*: moving from the living room to the kitchen and vice-versa; *PickUp*(o): pick up an object from the kitchen table.

Sensing. The robot is trained to recognise the pre-determined set of objects I. The main sensing task is then to detect whether or not these specific objects are located on the kitchen table. This is encapsulated by the sensing action *SenseOT*(o) that senses if object $o \in I$ is on the table. The $SF(a, s)$ predicate, introduced in the previous section, is used to axiomatise the act of sensing:

$$SF(PickUp(o), s) \equiv true$$
$$SF(SwitchRoom, s) \equiv true$$
$$InKitchen(s) \supset (SF(SenseOT(o), s) \equiv OnTable(o, s))$$

Initial State. In the initial state the robot is in the living room (i.e., not in the kitchen) and is not holding anything: $\neg InKitchen(S_0) \wedge (\forall x)(\neg Holding(x, S_0))$.

Informing the Robot. The robot is told that there is a red cup on the table: $Told(Cup(O_1))$, $Told(Red(O_1))$, $Told(OnTable(O_1))$.

Precondition Axioms. The robot can only pick up an item when it is not already holding an object and the item in question is on the kitchen table:

$Poss(PickUp(o), s) \equiv (\forall x)(\neg Holding(x, s)) \wedge InKitchen(s) \wedge OnTable(o, s)$; the robot can always switch locations: $Poss(SwitchRoom, s) \equiv true$.

Successor State Axioms. If the robot wasn't already in the kitchen then it will be as a result of switching rooms: $InKitchen(do(a, s)) \equiv (\neg InKitchen(s) \wedge a = SwitchRoom) \vee (InKitchen(s) \wedge a \neq SwitchRoom)$; an item will be on the table only if it was previously on the table and has not been picked up: $OnTable(o, do(a, s)) \equiv OnTable(o, s) \wedge a \neq PickUp(o)$; the robot will be holding an object if it picks it up or was already holding the object: $Holding(o, do(a, s)) \equiv a = PickUp(o) \vee Holding(o, s)$; object type is persistent: $Cup(o, do(a, s)) \equiv Cup(o, s)$; colour is persistent: $Red(o, do(a, s)) \equiv Red(o, s)$.

Preferences. In order to use the framework for belief change in the Situation Calculus to deal with misleading information we proceed as follows. Every planning problem (i.e., request to achieve a goal) is considered a new reasoning problem.[4] The statements, $Told(f(\bar{x}))$ and $Goal(\exists s.\phi(s))$, are used to ascribe initial beliefs and a goal to achieve. They are interpreted at the meta-level and are not part of the object language. In our example scenario, the request $Told(Cup(O_1))$, $Told(Red(O_1))$, $Told(OnTable(O_1))$, $Goal(\exists s.Holding(O_1, s))$ asks the agent to collect a red cup from the table. This results in the specification of a reasoning about action problem in the Situation Calculus extended with beliefs. In particular, the request specifies what should be believed in the initial situation S_0 and as such partially restricts the plausibility relation $pl()$. However, our beliefs may be mistaken – there is no red cup on the table – and as a result we need to formulate an alternative course of action to get the best out of the situation at hand. Which alternative course of action to take is determined by preferences that are specified using a meta-level preference relation $<_C$. These preferences further restrict the plausibility of situations $pl()$.

Preferences reflect the robot's commonsense knowledge. In our scenario, for example, the robot may prefer to fetch an object that is of the same type as requested but of a different colour, and most of all prefer to find an object in the room to which it was sent.

$$OnTable <_C Cup <_C Red \tag{1}$$

It is of course possible to conceive of a scenario in which the above preference for, say, non-red cups in the kitchen over red non-cups elsewhere is reversed. The operator may be a child building a colour collage and therefore assign greater importance to the colour of the object than its type.

In reality, determining the best set of preferences would be a complex task requiring the robot to combine subtleties of natural language processing with specific knowledge about the operator. Such considerations are beyond the scope of this paper, and so we just presuppose a given *commonsense preference ordering*, represented by a partial order among fluent names.

Next, we directly compile the $Told()$ statements plus an ordering like (1) into a plausibility ordering over all the initial situations. Here, the initial situations

---

[4] This is not crucial to our approach but considerably simplifies the notation and formal machinery required and, in any case, is not central to the main contributions.

encode all possible hypotheses of what the operator might have meant by their commands. The commonsense preference is then used to rank these hypotheses according to their plausibility. In order to relate this preference ordering to the $Told()$ statements we introduce the notation $\langle \cdot \rangle$ to extract the fluent name from a fluent literal (e.g., $\langle \neg Cup(O_1) \rangle = Cup$).

Definition 1. *Let Σ be a Situation Calculus BAT, B the axioms for iterated belief revision in the Situation Calculus, I be the set of domain objects, N be a set of additional names, T be a set of consistent operator commands and $<_C$ be a commonsense preference ordering. Then $(\Sigma \cup B, T, <_C)$ is a Situation Calculus BAT extended with belief and commonsense preferences such that:*

1. *The initial situations are created by the axioms*

$$(\exists s)\left(B(s, S_0) \wedge \bigwedge_{O \in N} Same(O, \sigma(O), s) \right) \tag{2}$$

   for all functions $\sigma : N \to I$.
2. *For every pair of initial situations s_1, s_2, we define*

$$pl(s_1) < pl(s_2)$$

   iff both
   (a) there is some $Told(f(\bar{x})) \in T$ such that $\Sigma \cup \{(2)\} \models f(\bar{x}, s_1)$ and $\Sigma \cup \{(2)\} \models \neg f(\bar{x}, s_2)$; and,
   (b) for every $Told(f(\bar{x}_1)) \in T$ such that

$$\Sigma \cup \{(2)\} \models \neg f(\bar{x}_1, s_1) \text{ and } \Sigma \cup \{(2)\} \models f(\bar{x}_1, s_2)$$

   there is a $Told(g(\bar{x}_2)) \in T$ such that $\langle g \rangle <_C \langle f \rangle$,

$$\Sigma \cup \{(2)\} \models g(\bar{x}_2, s_1) \text{ and } \Sigma \cup \{(2)\} \models \neg g(\bar{x}_2, s_2)$$

Part 1 creates all the initial situations. Intuitively, the function σ says which names are assigned to which real object; so $\sigma_1(O_1) = \sigma_1(O_3) = C_R$ means that O_1 and O_3 are considered the same as the red cup C_R. The number of axioms thus generated is polynomial in the number of domain objects, but exponential in the number of additional names.[5] This is one of the main reasons why a direct implementation of Situation Calculus with belief change would be practically infeasible for our problem at hand and why we are interested in developing a more practical implementation based on default logic.

Part 2 restricts the plausibility relation over initial situations. Initial situation s_1 is preferred to s_2 whenever s_1 assigns the value true to a fluent preferred under the preference ordering $<_C$ and s_2 assigns the value false; additionally, for all fluents f where this is the other way around (f is true in s_2 and false in s_1), there must be a preferred fluent g which holds in s_1 but not in s_2.

From this formalisation of the scenario, we can establish the fact that the robot will initially believe what it is told.

---

[5] Recall that the number of functions $\sigma : N \to I$ is $|I|^{|N|}$.

Proposition 1. *Let Σ be a Situation Calculus BAT, B the axioms for iterated belief revision in the Situation Calculus, T be a set of consistent operator commands, and $<_C$ be a commonsense preference ordering. Then $(\Sigma \cup B, T, <)$ is a Situation Calculus BAT extended with belief and commonsense preferences such that for all $Told(f(\bar{x})) \in T$ we have $(\Sigma \cup B, T, <_C) \models Bel(f(\bar{x}), S_0)$.*

Plan Execution. This formalism allows the robot to change its beliefs about what it is told. In this paper we assume that the robot has determined a plan and begun its execution. We can therefore consider the robot's changing beliefs with regards to satisfying its goal of holding object O_1 by considering the situation[6]

$$do([SwitchRoom, SenseOT(C_R), SenseOT(C_B), SenseOT(P_R), PickUp(O_1)], S_0)$$

Initially the robot believes that the object O_1 refers to the red cup C_R . However when the robot arrives in the kitchen it finds that there is only a blue cup on the table. Consequently the robot changes its belief about O_1 to now refer to the blue cup C_B . This scenario is visualised by Figure 1 showing the possible situations based on the robot's beliefs and the plausibility relation.

In the extended example the robot arrives in the kitchen to find both a blue cup and red plate on the table. It therefore has a choice, which it resolves based on its preference for object type over colour (1), consequently modifying its belief about O_1 to again refer to the blue cup.

4   A Default Logic Approach

The Situation Calculus with beliefs provides an expressive formalism for tackling the problem of agents receiving erroneous information and expected to use some basic commonsense reasoning under these circumstances. Next we address the problem of turning the theory into a practical implementation. To this end we adapt a recently developed extension of action logics with default reasoning [11], which can be effectively implemented using Answer Set Programming [12]. The idea is to treat potentially erroneous information as something that is considered true by default but can always be retracted should the agent make observations to the contrary. We extend the existing approach by *prioritised* defaults that allow us to provide our robot with preferences among different ways of remedying a situation in which it has been misled.

Supernormal Defaults. To begin with, we instantiate the general framework of [11] to the Situation Calculus and to a restricted form of default rules. Each operator command $Told([\neg]f(\bar{x}), s)$ is translated into a *supernormal* default of the form

$$\frac{: f(\bar{x}, s)}{f(\bar{x}, s)} \quad \text{or} \quad \frac{: \neg f(\bar{x}, s)}{\neg f(\bar{x}, s)}$$

With these rules the robot will believe, by default, everything it is told. For our running example we thus obtain these three defaults about the initial situation:

$$\delta_{Cup} = \frac{: Cup(O_1, S_0)}{Cup(O_1, S_0)} \quad \delta_{Red} = \frac{: Red(O_1, S_0)}{Red(O_1, S_0)} \quad \delta_{OnTable} = \frac{: OnTable(O_1, S_0)}{OnTable(O_1, S_0)}$$

---

[6] $do([a_1, \ldots, a_n], s)$ abbreviates $do(a_n, \ldots, do(a_2, do(a_1, s)) \ldots)$.

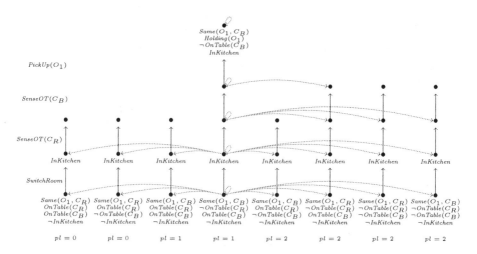

Fig. 1. The robot is told to pick up the red cup from the table, but finds only a blue cup. For succinctness, details of the red plate and the status of the persistent fluents *Cup* and *Red* are omitted. Furthermore, only the accessibility relations (dotted lines) for the actual situation (fourth from the left) are shown. The transition of situations based on actions are indicated by the solid vertical lines. Values for the plausibility relation are assigned to the initial situations based on the preferences. The initial situations in which the robot believes that it is going to pick up the red cup on the table are the most preferred ($pl = 0$). Next are those in which the robot believes that it is going to pick up the blue cup on the table ($pl = 1$). Finally, the least preferred are situations where the robot believes that the item to pick up is not on the table ($pl = 2$).

A *Situation Calculus default theory* is a pair (Σ, Δ) where Σ is a Situation Calculus BAT as above and Δ is a set of default rules.

Priorities. In a *prioritised default theory* [13], the default rules are partially ordered by $\prec$, where $\delta_1 \prec \delta_2$ means that the application of default δ_1 is preferred over the application of default δ_2. For our purpose, we can map a given commonsense preference ordering among fluent names directly into a partial ordering among the defaults from above. For example, with the ordering given by (1) we obtain $\delta_{OnTable} \prec \delta_{Cup} \prec \delta_{Red}$. A *prioritised Situation Calculus default theory* is a triple $(\Sigma, \Delta, \prec)$ where (Σ, Δ) is as above and $\prec$ is a partial ordering on Δ.

Extensions. Reasoning with default theories is based on the concept of so-called *extensions*, which can be seen as a way of assuming as many defaults as possible without creating inconsistencies [13, 14].

Definition 2. *Consider a prioritised Situation Calculus default theory* $(\Sigma, \Delta, \prec)$. *Let E be a set of formulas and define $E_0 := Th(\Sigma)$ and, for $i \geq 0$,*

$$E_{i+1} := Th(E_i \cup \{\gamma \mid \tfrac{:\neg\gamma}{\gamma} \in \Delta, \; \neg\gamma \notin E\})$$

Then E is an extension *of* $(\Sigma, \Delta, \prec)$ *iff* $E = \bigcup_{i \geq 0} E_i$.

Let a partial ordering be defined as $E_1 \lessdot E_2$ iff both

(a) there is $\frac{:\gamma}{\gamma}$ in Δ such that $\gamma \in E_1$ but $\gamma \notin E_2$; and,

(b) for every $\frac{:\gamma_1}{\gamma_1}$ such that $\gamma_1 \notin E_1$ but $\gamma_1 \in E_2$ there is $\frac{:\gamma_2}{\gamma_2} \prec \frac{:\gamma_1}{\gamma_1}$ in Δ such that $\gamma_2 \in E_1$ but $\gamma_2 \notin E_2$.

Extension E is a preferred extension of $(\Sigma, \Delta, \prec)$ iff there is no E' such that $E' \lessdot E$. Entailment $(\Sigma, \Delta, \prec) \approx \phi$ is defined as ϕ being true in all preferred extensions.

In our running example, when initially the robot has no information to the contrary it can consistently apply all defaults, resulting in a unique preferred extension that entails $Cup(O_1, S_0) \wedge Red(O_1, S_0) \wedge OnTable(O_1, S_0)$. Based on these default conclusions the Situation Calculus axioms entail the same plans for a given goal as those for the Situation Calculus extended with belief and commonsense preferences. But suppose that the robot enters the kitchen and observes what is indicated in Figure 1, that is,

$$Same(O_1, C_R, S) \vee Same(O_1, C_B, S)$$
$$Cup(C_R, S) \wedge Red(C_R, S)$$
$$Cup(C_B, S) \wedge \neg Red(C_B, S)$$
$$\neg OnTable(C_R, S) \wedge OnTable(C_B, S)$$

where S is the situation after $SwitchRoom$ followed by $SenseOT(C_R)$ and $SenseOT(C_B)$. Disregarding priorities for now, there are two extensions, characterised by

$$\{Same(O_1, C_R, S), \neg OnTable(O_1, S)\} \subseteq E_1$$
$$\{Same(O_1, C_B, S), \neg Red(O_1, S)\} \qquad \subseteq E_2$$

However, given the priorities from above, only E_2 is a preferred extension, triggering the robot to pick up the blue cup.

In the second case of the scenario, the robot further senses that there is also a red plate on the table. In this case there will be a third extension E_3 such that $\{Same(O_1, P_B, S), \neg Cup(O_1, S)\} \subseteq E_3$. However, as with the first case, E_2 is still the only preferred extension and therefore the robot selects the blue cup.

Implementation. Answer Set Programming (ASP) [12] is well-suited for efficiently implementing nonmonotonic reasoning formalisms. Extended logic programs can be seen as special kinds of default theories [15] and this correspondence can be used to transform a default theory into an answer set program. Entailment of a formula by the default theory can then be determined by querying the answer set program. This transformation technique has been developed in [5]. In the following, we outline this technique (steps 2 to 4) and extend it to cover preferences (step 1). This allows the transformation of a sufficiently restricted prioritised Situation Calculus default theory $(\Sigma, \Delta, \prec)$ into an answer set program $P_{\Sigma, \Delta, \prec}$.

Step 1. We transform the *prioritised* Situation Calculus default theory $(\Sigma, \Delta, \prec)$ into a Situation Calculus default theory $(\Sigma^\prec, \Delta^\prec)$ where the preferences have been encoded at the object-level [16]. This is done by explicitly keeping track of default δ's meta-level applicability $ok(\delta)$ and whether it was applied ($ap(\delta)$) or blocked ($bl(\delta)$). For example, δ_{Cup} and δ_{Red} are transformed into

$$\frac{ok(\delta_{Cup}) : Holds(Cup(O_1), S_0)}{Holds(Cup(O_1), S_0) \wedge ap(\delta_{Cup})} \qquad \frac{ok(\delta_{Cup}) \wedge \neg Holds(Cup(O_1), S_0) :}{bl(\delta_{Cup})}$$

$$\frac{ok(\delta_{Red}) : Holds(Red(O_1), S_0)}{Holds(Red(O_1), S_0) \wedge ap(\delta_{Red})} \qquad \frac{ok(\delta_{Red}) \wedge \neg Holds(Red(O_1), S_0) :}{bl(\delta_{Red})}$$

The preference between the defaults is enforced by statements like $(ap(\delta_{Cup}) \vee bl(\delta_{Cup})) \supset ok(\delta_{Red})$, effectively saying that δ_{Red} can only be applied once it is clear whether the more preferred default δ_{Cup} has been "processed".

Step 2. We instantiate the defaults from $\Delta^{\prec}$ and the axioms from $\Sigma^{\prec}$ for the given Situation Calculus signature. This yields a propositional default theory.

Step 3. We rewrite the ground instantiation of $\Sigma^{\prec}$ into a set $P_{\Sigma^{\prec}}$ of extended logic program rules.

Step 4. We map $\Delta^{\prec}$ into a set of logic program rules. A default of the form $\frac{p:q}{r_1 \wedge r_2}$ becomes $r_i \leftarrow p$, not $-q$ for $i = 1, 2$; a rule $\frac{p \wedge q:}{r}$ is turned into $r \leftarrow p, q$. Here not is the usual nonmonotonic negation of normal logic programs; $-q$ is a new predicate symbol standing for the (classical) negation of q [15]. The resulting rules together with $P_{\Sigma^{\prec}}$ now form the corresponding answer set program $P_{\Sigma, \Delta, \prec}$ of the initial prioritised Situation Calculus default theory $(\Sigma, \Delta, \prec)$.

5   Equivalence of the Two Approaches

We are now in a position to state the central result of this paper, which says that our prioritised Situation Calculus default theories are suitable approximations of the Situation Calculus extended with belief and commonsense preferences. Unfortunately, lack of space prevents us from giving a rigorously formal account. Generally speaking, the latter is more expressive for two reasons. First, it allows to infer meta-statements about beliefs, as in the formula $Bel(Bel(Red(O_1), S_0), do(SwitchRoom, S_0))$. Second, all possible situations are ranked according to $pl()$, thus allowing to draw conclusions about their relative ordering, whereas in prioritised default logics the non-preferred extensions are not considered for entailment. However, neither of these two features is relevant for the problem at hand, and we can prove the following.

Theorem 1. *Let Σ be a Situation Calculus BAT, B the axioms for iterated belief revision in the Situation Calculus, T a set of consistent operator commands, $<_C$ a commonsense preference ordering, $\Delta, \prec$ a set of default rules and an ordering as explained above, $a_1, \ldots, a_n$ a sequence of actions, and $SF_n := \{[\neg]SF(a_1, S_0), \ldots, [\neg]SF(a_n, do([a_1, \ldots, a_{n-1}], S_0))\}$ a set of literals describing a particular sequence of sensing results. Then for any objective formula ϕ (that is, any formula ϕ without Bel) we find*

$$(\Sigma \cup B \cup SF_n, T, <_C) \models Bel(\phi, do([a_1, \ldots, a_n], S_0))$$
$$\text{iff } (\Sigma \cup SF_n, \Delta, \prec) \approx Holds(\phi, do([a_1, \ldots, a_n], S_0))$$

Proof (sketch): By induction on the number of actions n. If $n = 0$, by Proposition 1 the robot believes all operator commands; in a similar way it can be shown that there is a unique preferred extension which entails the exact same statements about S_0 that are true in all most plausible initial situations. For the induction step, if a_{n+1} is a physical action the claim follows from the fact that both axiomatisations share the same basic action theory. If a_{n+1} is a sensing action, then any possible situation in $do([a_1, \ldots, a_n], S_0)$ that contradicts $[\neg]SF(a_{n+1}, do([a_1, \ldots, a_n, a_{n+1}], S_0)$ is no longer possible in $do([a_1, \ldots, a_n, a_{n+1}], S_0)$; likewise, any extension of $(\Sigma \cup SF_n, \Delta, \prec)$ that contradicts this sensing literal is no longer an extension of $(\Sigma \cup SF_{n+1}, \Delta, \prec)$. The claim follows from the structural equivalence of the construction of the plausibility ordering in Def. 1 (Item 2) and of preferred extensions in Def. 2.   □

6   Conclusions

We developed an effective implementation of a well established approach to belief change in the Situation Calculus [1, 4]. This was achieved by mapping a problem instance expressed using this particular approach to belief change in the situation calculus into a default logic theory for which an ASP based implementation exists [11]. We illustrated our approach using an example inspired by the RoboCup@Home rulebook. This example innovatively solves the problem of how a reasoner faced with an unachievable goal should nevertheless do its best to salvage the situation by relying on its preferences.

It is important to observe that while our example scenario encodes a user request to fetch a single item, the formalism allows for more complex cases, such as conjunctive and disjunctive goals. However care must be taken when formulating requests. For example, a disjunctive request to fetch a fork or a spoon should be encoded as a request to fetch one of two distinct objects, a spoon object or a fork object. The alternative, and less intuitive, encoding would be to fetch a single object for which the operator is unsure if it is a fork or a spoon. This latter encoding is not possible due to restrictions on the *Told* statements.

We formalised our solution using an extension of the Situation Calculus to handle beliefs and mapped this solution into a solvable default logic theory. An alternative approach to tackling the example we presented would have been to consider goal revision [17]. However note that proposals like this one modify goals at the explicit request of an agent and do not consider that the goals themselves may be unachievable. In our approach, the goal cannot be achieved and we argue that this is more accurately dealt with by reasoning about the robot's beliefs (i.e., expectations about the world).

In related work, Lee and Palla [18] implement the situation calculus in ASP. However, adding the belief axioms of our paper to their approach would entail explicitly representing all possible alternative situations (since their plausibilities matter). Our approach avoids this technical problem by using default logic where only preferred extensions are considered for entailment.

Acknowledgements. This research was supported under Australian Research Council's (ARC) *Discovery Projects* funding scheme (project DP 120102144). The fourth author is the recipient of an ARC Future Fellowship (FT 0991348) and is also affiliated with the University of Western Sydney.

References

1. Shapiro, S., Pagnucco, M., Lespérance, Y., Levesque, H.: Iterated belief change in the situation calculus. AIJ 175(1), 165–192 (2011)
2. Demolombe, R., del Pilar Pozos Parra, M.: A simple and tractable extension of situation calculus to epistemic logic. In: Ohsuga, S., Raś, Z.W. (eds.) ISMIS 2000. LNCS (LNAI), vol. 1932, pp. 515–524. Springer, Heidelberg (2000)
3. Demolombe, R., Pozos-Parra, M.P.: Belief change in the situation calculus: A new proposal without plausibility levels. In: Proc. of the Workshop on Belief Revision and Dynamic Logic at ESSLLI (2005)
4. Shapiro, S., Pagnucco, M., Lespérance, Y., Levesque, H.: Iterated belief change in the situation calculus. In: KR, pp. 527–538 (2000)
5. Strass, H.: The draculasp system: Default reasoning about actions and change using logic and answer set programming. In: NMR (2012)
6. McCarthy, J.: Situations, actions and causal laws. Stanford AI Project Memo 2 (1963)
7. Reiter, R.: Knowledge in Action: Logical Foundations for Specifying and Implementing Dynamical Systems. The MIT Press (2001)
8. Cohen, P., Levesque, H.: Rational interaction as the basis for communication. In: Cohen, P., Morgan, J., Pollack, M. (eds.) Intentions in Communication, pp. 221–256. MIT Press (1990)
9. Scherl, R., Levesque, H.: Knowledge, action, and the frame problem. AIJ 144(1-2), 1–39 (2003)
10. Levesque, H.: What is planning in the presence of sensing? In: AAAI, pp. 1139–1146 (1996)
11. Baumann, R., Brewka, G., Strass, H., Thielscher, M., Zaslawski, V.: State defaults and ramifications in the unifying action calculus. In: KR, pp. 435–444 (2010)
12. Gelfond, M.: Answer Sets. In: Handbook of KR, pp. 285–316 (2008)
13. Brewka, G.: Adding priorities and specificity to default logic. In: MacNish, C., Moniz Pereira, L., Pearce, D.J. (eds.) JELIA 1994. LNCS (LNAI), vol. 838, pp. 247–260. Springer, Heidelberg (1994)
14. Reiter, R.: A logic for default reasoning. AIJ 13, 81–132 (1980)
15. Gelfond, M., Lifschitz, V.: Classical Negation in Logic Programs and Disjunctive Databases. New Gen. Comp. 9, 365–385 (1991)
16. Delgrande, J., Schaub, T.: Expressing preferences in default logic. AIJ 123(1-2), 41–87 (2000)
17. Shapiro, S., Lespérance, Y., Levesque, H.: Goal change. In: IJCAI 2005, pp. 582–588 (2005)
18. Lee, J., Palla, R.: Situation Calculus as Answer Set Programming. In: Proceedings of the Twenty-Fourth Conference on Artificial Intelligence, AAAI 2010, pp. 309–314 (July 2010)

Debugging Non-ground ASP Programs with Choice Rules, Cardinality and Weight Constraints

Axel Polleres[1], Melanie Frühstück[1],
Gottfried Schenner[1], and Gerhard Friedrich[2]

[1] Siemens AG Österreich, Siemensstraße 90, 1210 Vienna, Austria
[2] Alpen-Adria Universität, Klagenfurt, Austria

Abstract. When deploying Answer Set Programming (ASP) in an industrial context, for instance for (re-)configuration [5], knowledge engineers need debugging support on non-ground programs. Current approaches to ASP debugging, however, do not cover extended modeling features of ASP, such as choice rules, conditional literals, cardinality and weight constraints [13]. To this end, we encode non-ground ASP programs using extended modeling features into normal logic progams; this encoding extends existing encodings for the case of ground programs [4, 10, 11] to the non-ground case. We subsequently deploy this translation on top of an existing ASP debugging approach for non-ground normal logic programs [14]. We have implemented and tested the approach and provide evaluation results.

1 Introduction

Answer Set Programming (ASP), with its intuitive and declarative modeling features – offering the possibility to model knowledge base constraints concisely in the form of non-ground programs plus advanced modeling feature such as choice rules, cardinality constraints and weight constraints [13] – has become an attractive tool for knowledge engineers also in an industrial context. For instance, within the RECONCILE project[1] we deploy ASP for modeling and solving configuration and (re-)configuration problems [5] occurring in practical settings such as in large-scale projects in the railway automation domain.

While in such a context the advanced features in Answer Set Programming (ASP) significantly increase the declarative modeling capabilities of the language, debugging tools that support the full language of ASP are still missing: most approaches for debugging are only able to deal with propositional programs [1, 2, 9, 16, 18], with the exception of Oetsch et al. [14], who developed a meta-program for debugging normal logic programs. Still, this latter approach does not support debugging in the presence of features such as choice rules, cardinality constraints and weight constraints.

Notably, as shown in earlier works, these language constructs do not raise expressivity beyond normal logic programs: Ferraris and Lifschitz [4] have shown how weight

---

[1] https://www.cee.siemens.com/web/at/de/corporate/portal/
Innovation/InnovationStories/Pages/Reconcile.aspx

P. Cabalar and T.C. Son (Eds.): LPNMR 2013, LNAI 8148, pp. 452–464, 2013.

constraints can be encoded as nested expressions, while Janhunen and Niemelä [11] provide a translation of choice rules, cardinality and weight constraints into normal logic programs. In one step of the translation process they show how to transform SMODELS programs to normal programs. Another representation of the translation of choice and cardinality rules to normal logic programs can be found in [10]; their approach is based on [11] introducing more intermediate steps in the translation.

However, all the above mentioned literature focus on propositional ASP programs. In this paper we describe a transformation of non-ground choice rules, as well as cardinality and weight constraints with conditions into non-ground logic programs. Our proposal is mainly based on the structure of rules of Gebser and Schaub [10]. Eventually, we show how to deploy this non-ground embedding for debugging programs using advanced ASP features: based on the non-ground debugging approach by Oetsch et al. [14] for normal logic programs, after applying our translation process, ASP debugging also becomes feasible for programs using more advanced ASP features.

We first introduce the ASP language used herein in Section 2. Then, we extend the translation of [10] to the non-ground case (Section 3). We present an evaluation of this translation, comparing our non-ground embedding to the propositional embedding from [11] in Section 4. Finally, in Section 5 we illustrate how our translation can be embedded into the debugging approach of [14], before we conclude in Section 6.

2 Preliminaries

Syntax. A *literal* is an atom that is possibly preceded by the *strong negation* symbol ¬. We define a *normal* (non-ground) rule r as

$$h(\overline{x_h}) \leftarrow Body(\overline{X_{Body}}). \tag{1}$$

where $h(\overline{x_h})$ defines the *head* of the rule, i.e. a literal including its vector of parameters $\overline{x_h}$ (variables and constants).[2] The *body* of a rule $Body(\overline{X_{Body}}) = Body^+(\overline{X_{Body}}) \cup Body^-(\overline{X_{Body^-}})$ consists of $Body^+(\overline{X_{Body}})$, the set of all positive body literals, and $Body^-(\overline{X_{Body^-}})$, a set of default-negated body literals (i.e. literals preceded by *not*). X_{Body} denotes the set of all variables occurring in body literals; note that since we assume *safety* all these variables also occur in positive body literals, i.e. more precisely: a rule r of the form (1) is called *safe* if $X_h \subseteq X_{Body}$ and $X_{Body^-} \subseteq X_{Body}$; if the head is omitted then r is called a *constraint*; if the body is empty then r is called a *fact*.

Additionally to normal rules of the form (1), we consider programs expanded with choice rules and cardinality rules, as for instance supported by Potsdam Answer set Solving Collection (Potassco) [7]. Choice rules have the form

$$\{h_1(\overline{x_{h_1}}) : Cond(\overline{Y_{h_1}}), \ldots, h_n(\overline{x_{h_n}}) : Cond(\overline{Y_{h_n}})\} \leftarrow Body(\overline{X_{Body}}). \tag{2}$$

---

[2] As usual, we denote variables by upper case letters and constants by alphanumeric strings starting with a lower case letter. We further denote mixed vectors of constants and variables by overlined lower case letters (such as $\overline{x}$) whereas, accordingly, we denote the corresponding vector of all variables occurring in $\overline{x}$ by $\overline{X}$ (preserving order) and by X we denote the respective corresponding (unordered) set of variables.

where $h_i(\overline{x_{h_i}}) : Cond(\overline{Y_{h_i}})$ is called a *conditional literal*. The *condition* $Cond(\overline{Y_{h_i}})$ consists of positive literals (with variables in Y_{h_i}) separated by further colons, read as a conjunction, and can be possibly empty.[3]

Further, *cardinality constraints* $l\{h_1(\overline{x_{h_1}}) : Cond(\overline{Y_{h_1}}), \ldots, h_n(\overline{x_{h_n}}) : Cond(\overline{Y_{h_n}})\}u$ where l, u are either numeric constants or variables representing lower and upper bounds are allowed in rule heads and bodies, i.e. w.l.o.g. in rules of the following forms:

$$h(\overline{x_h}) \leftarrow l\{b_1(\overline{x_{b_1}}) : Cond(\overline{Y_{b_1}}), \ldots, b_n(\overline{x_{b_n}}) : Cond(\overline{Y_{b_n}})\}u, Body(\overline{X_{Body}}). \quad (3)$$

$$h(\overline{x_h}) \leftarrow not\ l\{b_1(\overline{x_{b_1}}) : Cond(\overline{Y_{b_1}}), \ldots, b_n(\overline{x_{b_n}}) : Cond(\overline{Y_{b_n}})\}u, Body(\overline{X_{Body}}). \quad (4)$$

$$l\{h_1(\overline{x_{h_1}}) : Cond(\overline{Y_{h_1}}), \ldots, h_n(\overline{x_{h_n}}) : Cond(\overline{Y_{h_n}})\}u \leftarrow Body(\overline{X_{Body}}). \quad (5)$$

We call a set P of *safe* rules of the forms (1)–(5) a *program*: here, we extend the standard notion of safety for rules of the form (1) to rules (2)–(5) as follows: a conditional literal $h_i(\overline{x_{h_i}}) : Cond(\overline{Y_{h_i}})$ within a rule r is *safe* if for all $1 \leq i \leq n$ it holds that $X_{h_i} \subseteq X_{Body} \cup Y_{h_i}$. Accordingly, rules of the forms (2)–(5) are *safe* if (i) they are safe in the standard sense (see above), (ii) all conditional literals are safe, and (iii) bounds l, u are either constants or variables from X_{Body}.[4]

Semantics. The *Herbrand universe* HU_P of a program P is the set of all constants appearing in P and the *Herbrand base* HB_P is the set of all ground atoms constructed by predicate symbols in P using constants from HU_P.[5]

As usual in ASP, we define the semantics of a program P in terms of its grounding; the *grounding* of a rule r, $ground(r)$, is defined by the set of ground rules obtained from (i) taking the set of all its ground instantiations, and (ii) replacing each conditional literal $h_i(\overline{x_{h_i}}) : Cond(\overline{Y_{h_i}})$ (within a choice or a cardinality constraint) with the **set** of all possible ground conditional literals obtained from substituting variables with constants from HU_P. Accordingly, we call $ground(P) = \bigcup_{r \in P} ground(r)$, the *grounding* of program P. Note that this procedure covers the two-step instantiation described in [17]: i.e. what they call "global" variables are replaced through step (i) and "local" variables during the expansion in step (ii).

An interpretation $I \subseteq HB_P$ satisfies a ground literal b, written $I \models b$, if $b \in I$. Analogously, $I \models b$ for a ground cardinality constraint $b = l\{h_1 : c_{1,1} : \ldots : c_{1,m_1}, \ldots, h_n : c_{n,1} : \ldots : c_{n,m_n}\}u$, if

$$l \leq |\ \{h_i\ |\ \{h_i, c_{i,1}, \ldots, c_{i,m_i}\} \subseteq I\}\ | \leq u$$

---

[3] For simplicity we only consider positive conditional literals and conditions herein; tools of Potassco also allow default negation within conditional literals which we leave to future work. In our formal definitions we also exclude built-ins in conditions (which we allow though in our implementation, cf. Section 5).

[4] Note that we leave out the form where cardinality constraints can be used to assign values to unsafe variables; tools of Potassco also allow cardinality constraints of the form $X = \{h_1(\overline{x_{h_1}}) : Cond(\overline{Y_{h_1}}), \ldots, h_n(\overline{x_{h_n}}) : Cond(\overline{Y_{h_n}})\}$ which assign the cardinality to an (unsafe) variable; we leave this extension to future work.

[5] Note that we assume no "overloading", i.e. each predicate symbol has a fixed arity. This restriction, which is not made in current ASP tools (like those of Potassco), can be easily lifted in a preprocessing step where you replace predicate names occurring in different arities with new unique predicate names per arity, e.g. $p(X), p(X, Y)$ become $p_{/2}(X, Y), p_{/1}(X)$, or alike.

Next, we define the *reduct* r^I of a rule r wrt. $I \subseteq HB_P$ as a set of rules as follows
- if $r : h \leftarrow Body$ is a ground rule where h is a literal, then

$$r^I = \begin{cases} \{h \leftarrow Body^+\} & \text{if there is no } not\ b \in Body^- \text{ with } I \models b. \\ \emptyset & \text{otherwise} \end{cases}$$

- if $r : \{h_1 : c_{1,1} : ... : c_{1,m_1}, ..., h_n : c_{n,1} : ... : c_{n,m_n}\} \leftarrow Body$ is a ground choice rule, then r^I is a *set* containing, for each $h_i \in I \cap \{h_1, ..., h_n\}$, the rule

$$r^I_{h_i} = \begin{cases} h_i \leftarrow c_{i,1}, ..., c_{i,m_i}, Body^+ & \text{if there is no } not\ b \in Body^- \text{ with } I \models b. \\ \emptyset & \text{otherwise} \end{cases}$$

Any consistent interpretation I, such that I is a (subset-)minimal model of $P^I = \bigcup_{r \in P} r^I$ is called an *answer set*; likewise, answer sets for a non-ground program P are defined as the answer sets of $ground(P)$.

As in [17], we view rules of the form (5) as syntactic sugar not treated separately in the semantics; we will get back to these in the next section.

As a further extension, *weighted conditional literals* which assign a weight w_i (either a numeric constant or a safe variable, cf. footnote 4) to a conditional literal are allowed in so called *weight constraints* of the form

$$l[h_1(\overline{x_{h_1}}) : Cond(\overline{Y_{h_1}}) = w_1, ..., h_n(\overline{x_{h_n}}) : Cond(\overline{Y_{h_n}}) = w_n]u \qquad (6)$$

These distinguish from cardinality constraints in that values are summed in a multi-set semantics, i.e. weights w_i, w_j of satisfying ground instances for each conditional literal i and j count separately even if $h_i = h_j$, i.e. when replacing cardinality constraints within rules of the forms (3)–(5) with their weighted counterparts, the upper and lower bounds mean to indicate bounds for sums of *weights* of satisfied instances, rather than counting distinct instances. The semantics of weight constraints extends the semantics for cardinality constraints straightforwardly, formal details of which we omit here for space limitations.

3   Translation to Normal Rules

We translate rules of forms (2)–(5) successively to normal rules in several steps.

Step 1. We first consider choice rules of the form (2). A choice rule can be translated into following rules:

$$h_1(\overline{x_{h_1}}) \leftarrow Body(\overline{X_{Body}}), Cond(\overline{Y_{h_1}}), not\ h'_{r,1}(\overline{x_{h_1}}).$$
$$h'_{r,1}(\overline{x_{h_1}}) \leftarrow Body(\overline{X_{Body}}), Cond(\overline{Y_{h_1}}), not\ h_1(\overline{x_{h_1}}).$$
$$... \qquad (7)$$
$$h_n(\overline{x_{h_n}}) \leftarrow Body(\overline{X_{Body}}), Cond(\overline{Y_{h_n}}), not\ h'_{r,n}(\overline{x_{h_n}}).$$
$$h'_{r,n}(\overline{x_{h_n}}) \leftarrow Body(\overline{X_{Body}}), Cond(\overline{Y_{h_n}}), not\ h_n(\overline{x_{h_n}}).$$

where the $h'_{r,i}$ are new predicate symbols, unique to the rule r they appear in (to avoid interferences between the translations of several choice rules).

Step 2. Next, we reduce any rules with cardinality constraints to the form of (3), that is, we rewrite rules of the forms (4)+(5) in such a way that cardinality constraints only appear positively in rule bodies: a cardinality rule r of the form (5) is replaced by

(i) its unconstrained variant, i.e. the translation (according to Step 1) of the choice rule obtained by removing upper and lower bounds l and u; and
(ii) the following pair of rules (where, c_r is a "fresh" predicate symbol):

$$c_r(\overline{X_{Body}}) \leftarrow l\{h_1(\overline{x_{h_1}}) : Cond(\overline{Y_{h_1}}), ..., h_n(\overline{x_{h_n}}) : Cond(\overline{Y_{h_n}})\}u, Body(\overline{X_{Body}}). \quad (8)$$
$$\leftarrow not\ c_r(\overline{X_{Body}}), Body(\overline{X_{Body}}). \quad (9)$$

Similarly, cardinality rules of the form (4) are replaced by the pair of rules

$$c_r(\overline{X_{Body}}) \leftarrow l\{b_1(\overline{x_{b_1}}) : Cond(\overline{Y_{b_1}}), ..., b_n(\overline{x_{b_n}}) : Cond(\overline{Y_{b_n}})\}u, Body(\overline{X_{Body}}). \quad (10)$$
$$h(\overline{x_h}) \leftarrow not\ c_r(\overline{X_{Body}}), Body(\overline{X_{Body}}). \quad (11)$$

Step 3. Finally, cardinality rules of the form (3) – including those of the forms (8)+(10) obtained in the previous step – are translated as follows.

(i) First, we translate the body cardinality constraint to a variant with only lower bounds as follows

$$h(\overline{x_h}) \leftarrow l_r(\overline{X_{Body}}), not\ u_r(\overline{X_{Body}}), Body(\overline{X_{Body}}). \quad (12)$$
$$l_r(\overline{X_{Body}}) \leftarrow l\{b_1(\overline{x_{b_1}}) : Cond(\overline{Y_{b_1}}), ..., b_n(\overline{x_{b_n}}) : Cond(\overline{Y_{b_n}})\}, Body(\overline{X_{Body}}). \quad (13)$$
$$u_r(\overline{X_{Body}}) \leftarrow u + 1\{b_1(\overline{x_{b_1}}) : Cond(\overline{Y_{b_1}}), ..., b_n(\overline{x_{b_n}}) : Cond(\overline{Y_{b_n}})\}, Body(\overline{X_{Body}}). \quad (14)$$

where l_r, u_r are new predicate symbols. Note that $Body(\overline{X_{Body}})$ in rule (12) is not strictly necessary when both a lower and an upper bound are given, but, since both $l_r(\overline{X_{Body}})$ and $not\ u_r(\overline{X_{Body}})$ are optional in this rule, it is necessary to guarentee safety in the absence of the latter. Likewise, rule (13) (and (14), resp.) is only needed in case a lower (or upper, resp.) bound is given.
(ii) Next, we translate rules with a body cardinality constraint with only lower bounds, i.e. rules of the form

$$h(\overline{x_h}) \leftarrow l\{b_1(\overline{x_{b_1}}) : Cond(\overline{Y_{b_1}}), ..., b_n(\overline{x_{b_n}}) : Cond(\overline{Y_{b_n}})\}, Body(\overline{X_{Body}}). \quad (15)$$

are translated to

$$h(\overline{x_h}) \leftarrow cnt_r(\overline{X_{Body}}, C), Body(\overline{X_{Body}}), C \geq l. \quad (16)$$

The definition of the new predicate cnt_r is given as follows. We assume a built-in predicate "$<$" defining a total, lexical order for pairs of constants in HU_P. Further, for sequence $\overline{x_{b_i}}$, let $\overline{x_{b_i}}'$ denote the sequence obtained from replacing each variable x occurring in $\overline{x_{b_i}}$ by a fresh variable x'. Lastly, let $\overline{x_{r,i}^{\cup}} = (\overline{x_{b_i}}, \overline{X_{Body}})$, i.e. the concatenation of the two vectors $\overline{x_{b_i}}$ and $\overline{X_{Body}}$ and $\overline{x_{r,i}^{\cup}}' = (\overline{x_{b_i}}', \overline{X_{Body}})$.

We now define the predicate cnt_r by the following auxiliary rules, for each $i \in \{1, \ldots, n\}$

$$val_{r,b_i}(\overline{x_{r,i}^\cup}) \leftarrow b_i(\overline{x_{b_i}}), Cond(\overline{Y_{b_i}}), Body(\overline{X_{Body}}). \tag{17}$$

$$exists_{r,b_i}(\overline{X_{Body}}) \leftarrow Body(\overline{X_{Body}}), val_{r,b_i}(\overline{x_{r,i}^\cup}). \tag{18}$$

$$exists^<_{r,b_i}(\overline{x_{r,i}^\cup}) \leftarrow val_{r,b_i}(\overline{x_{r,i}^\cup}), val_{r,b_i}(\overline{x_{r,i}^\cup}'), \overline{x_{b_i}}' <_{|x_{b_i}|} \overline{x_{b_i}}. \tag{19}$$

$$exists^>_{r,b_i}(\overline{x_{r,i}^\cup}) \leftarrow val_{r,b_i}(\overline{x_{r,i}^\cup}), val_{r,b_i}(\overline{x_{r,i}^\cup}'), \overline{x_{b_i}} <_{|x_{b_i}|} \overline{x_{b_i}}'. \tag{20}$$

$$next_{r,b_i}(\overline{x_{r,i}^\cup}, \overline{x_{r,i}^\cup}') \leftarrow val_{r,b_i}(\overline{x_{r,i}^\cup}), val_{r,b_i}(\overline{x_{r,i}^\cup}'), \overline{x_{b_i}} <_{|x_{b_i}|} \overline{x_{b_i}}', \tag{21}$$

$$not\ between_{r,b_i}(\overline{x_{r,i}^\cup}, \overline{x_{r,i}^\cup}').$$

$$between_{r,b_i}(\overline{x_{r,i}^\cup}, \overline{x_{r,i}^\cup}'') \leftarrow val_{r,b_i}(\overline{x_{r,i}^\cup}), val_{r,b_i}(\overline{x_{r,i}^\cup}'), val_{r,b_i}(\overline{x_{r,i}^\cup}''), \tag{22}$$

$$\overline{x_{b_i}} <_{|x_{b_i}|} \overline{x_{b_i}}', \overline{x_{b_i}}' <_{|x_{b_i}|} \overline{x_{b_i}}''.$$

$$cnt_{r,b_i}(\overline{x_{r,i}^\cup}, 1) \leftarrow val_{r,b_i}(\overline{x_{r,i}^\cup}), not\ exists^<_{r,b_i}(\overline{x_{r,i}^\cup}). \tag{23}$$

$$cnt_{r,b_i}(\overline{x_{r,i}^\cup}', N+1) \leftarrow next_{r,b_i}(\overline{x_{r,i}^\cup}, \overline{x_{r,i}^\cup}'), cnt_{r,b_i}(\overline{x_{r,i}^\cup}, N). \tag{24}$$

$$cnt'_{r,b_i}(\overline{X_{Body}}, N) \leftarrow cnt_{r,b_i}(\overline{x_{r,i}^\cup}, N), not\ exists^>_{r,b_i}(\overline{x_{r,i}^\cup}) \tag{25}$$

$$cnt'_{r,b_i}(\overline{X_{Body}}, 0) \leftarrow Body(\overline{X_{Body}}), not\ exists_{r,b_i}(\overline{X_{Body}}). \tag{26}$$

where $<_n$ is an auxiliary predicate of arity $2n$ which determines whether the first of two vectors of the same length n is lexicographically smaller than the latter. For $n > 0$, the predicate $<_n$ can be easily defined recursively over the built-in predicate "$<$" in the rules (27)+(28) as follows:

$$(X_1, \ldots, X_k) <_k (Y_1, \ldots, Y_k) \leftarrow X_1 < Y_1. \qquad \forall 1 \le k \le n \tag{27}$$

$$(X_1, X_2, \ldots, X_k) <_k (X_1, Y_2, \ldots, Y_k) \leftarrow (X_2, \ldots, X_k) <_{k-1} (Y_2, \ldots, Y_k). \quad \forall 1 < k \le n \tag{28}$$

Rule (17) "collects" all possible bindings ("values") for variables that make a particular conditional atom b_i true, dependent on a particular body instantiation. The auxiliary rule (18) determines whether a value exists at all for a particular body instantiation; existence of a smaller, or greater, resp., than a prticular value is computed in the auxiliary rules (19)+(20). Rules (21) and (22) define a total order over values, defining a successor predicate ($next$) via the auxiliary information that no value lies in between two consecutive values. The cnt_{r,b_i} predicate then counts all the instantiations that belong to a particular conditional atom b_i, cf. rules (23)+(24). Rule (25) collects, for each b_i and body instantiation, the maximum count in the auxiliary predicates cnt'_{r,b_i}, where rule (26) sets this predicate to 0, in case no actual value exists for the conditional atom b_i. Finally, cnt_r is defined by the following rule which simply sums up all the maximum counts for the respective b_i's.

$$cnt_r(\overline{X_{Body}}, N) \leftarrow cnt'_{r,b_1}(\overline{X_{Body}}, N_1), \ldots, cnt'_{r,b_m}(\overline{X_{Body}}, N_m), N = N_1 + \ldots + N_m. \tag{29}$$

where $\{b_1, \ldots, b_m\}$ is the set of distinct predicate names occurring in $\{b_1, \ldots, b_n\}$.

Proposition 1. *The answer sets of a program P and its translation obtained from Steps 1-3 outlined above are in 1-to-1 correspondence.*

While we omit a full proof, we argue that the translation steps outlined above "emulate" semantics as described in Section 2 on non-ground programs, when assuming that HU_P contains apart from explicitly mentioned constants, integers from 0 to a finitely computable upper bound for instantiating and evaluating N in rules (24),(25), and (29) correctly; state-of-the-art ASP solvers like Potassco deal with such arithmetics appropriately out-of-the-box, which is our main concern when deploying the translation within our debugging use case (cf. Section 5 below).

As a possible optimization, which reduces the number and size of non-ground rules, note that it is possible to equivalently replace rules (19)–(25) with the following rules

$$cnt_{r,b_i}(\overline{x_{r,i}^{\cup}}, 1) \leftarrow val_{r,b_i}(\overline{x_{r,i}^{\cup}}). \tag{23'}$$

$$cnt_{r,b_i}(\overline{x_{r,i}^{\cup}}', N+1) \leftarrow val_{r,b_i}(\overline{x_{r,i}^{\cup}}'), \overline{x_{b_i}} <_{|X_{b_i}|} \overline{x_{b_i}}', cnt_{r,b_i}(\overline{x_{r,i}^{\cup}}, N). \tag{24'}$$

$$cnt_{r,b_i}'(\overline{X_{Body}}, N) \leftarrow cnt_{r,b_i}(\overline{x_{r,i}^{\cup}}, N), not\ nmax_{r,b_i}(\overline{X_{Body}}, N). \tag{25'}$$

$$nmax_{r,b_i}(\overline{X_{Body}}, N-1) \leftarrow cnt_{r,b_i}(\overline{x_{r,i}^{\cup}}, N). \tag{25''}$$

The idea behind this optimization is that, despite getting potentially various derivations for each N per body instance in rules (23')+(24'), there is only one unique *maximum* N derived per body instance, cf. rule (25'), which is the only relevant fact for rule (29), and in consequence for rule (16). Here, the new auxiliary rule (25'') is needed to assess that a certain value N is not the maximum count.

Taking this further, the instances of $<_{|X_{b_i}|}$ above can be replaced by a custom comparison predicate $smaller_{r,b_i}$ for each conditional atom b_i. Let k_i denote the arity of b_i, then $smaller_{r,b_i}$ is defined by the following set of rules:

$$smaller_{r,b_i}(X_1, \ldots, X_{k_i}, Y_1, \ldots, Y_{k_i}) \leftarrow X_1 < Y_1,$$
$$val_{r,b_i}(X_1, \ldots, X_{k_i}, \overline{X_{Body}}), val_{r,b_i}(Y_1, \ldots, Y_{k_i}, \overline{X_{Body}}).$$
$$smaller_{r,b_i}(X_1, X_2, \ldots, X_{k_i}, X_1, Y_2, \ldots, Y_{k_i}) \leftarrow X_2 < Y_2,$$
$$val_{r,b_i}(X_1, X_2, \ldots, X_{k_i}, \overline{X_{Body}}), val_{r,b_i}(X_1, Y_2, \ldots, Y_{k_i}, \overline{X_{Body}}). \tag{30}$$
$$\vdots$$
$$smaller_{r,b_i}(X_1, \ldots, X_{k_i}, X_1, \ldots, X_{k_i-1}, Y_{k_i}) \leftarrow X_{k_i} < Y_{k_i},$$
$$val_{r,b_i}(X_1, \ldots, X_{k_i}, \overline{X_{Body}}), val_{r,b_i}(X_1, \ldots, X_{k_i-1}, Y_{k_i}, \overline{X_{Body}}).$$

The idea of this definition is that the $smaller_{r,b_i}$ predicate really only compares values relevant for the particular b_i, instead of defining a generic smaller relation between *any* tuples in HU_P^n, which potentially narrows down the size of the grounding.

3.1 Extending the Translation by Weights

So far, we have only treated "pure" cardinality constraints, involving only conditional atoms with the default weight 1. It is not hard to extend the translation above to arbitrary weight constraints involving weighted conditional literals of the form (6). Firstly, we redefine $\overline{x_{r,i}^{\cup}}$ as follows

$$\overline{x_{r,i}^{\cup}} = (\overline{x_{b_i}}, \overline{X_{Body}}, w_i)$$

i.e. we carry over weights as an additional parameter in our auxiliary predicates. Apart from this change, rules (17)–(22) are modified with respect to the predicate names val_{r,b_i}, $exists_{r,b_i}$, $exists^<_{r,b_i}$, $exists^>_{r,b_i}$, $next_{r,b_i}$, and $between_{r,b_i}$ which are now replaced with $val_{r,i}$, $first_{r,i}$, $exists_{r,i}$, $exists^<_{r,i}$, $exists^>_{r,i}$, $next_{r,i}$, and $between_{r,i}$, respectively. I.e. values are no longer collected "per predicate" b_i, but separately for each weighted conditional literal at position $1 \leq i \leq n$, in order to cater for the multiset semantics of weight constraints. Secondly, we need to replace the counting rules (23)–(26) and (29) by rules that do summation instead; we use, in analogy to the cnt and cnt' predicates from above a new predicates sum and sum' here:

$$sum_{r,i}(\overline{x^\cup_{r,i}}, w_i) \leftarrow val_{r,i}(\overline{x^\cup_{r,i}}), not\ exists^<_{r,i}(\overline{x^\cup_{r,i}}). \tag{31}$$

$$sum_{r,i}(\overline{x^\cup_{r,i}}, W + w_i) \leftarrow next_{r,i}(\overline{x^\cup_{r,i}}', \overline{x^\cup_{r,i}}), sum_{r,i}(\overline{x^\cup_{r,i}}', W). \tag{32}$$

$$sum'_{r,i}(\overline{X_{Body}}, W) \leftarrow sum_{r,i}(\overline{x^\cup_{r,i}}, W), not\ exists^>_{r,i}(\overline{x^\cup_{r,i}}). \tag{33}$$

$$sum'_{r,i}(\overline{x^{\cup,0}_{r,i}}, 0) \leftarrow Body(\overline{X_{Body}}), not\ exists_{r,i}(\overline{X_{Body}}). \tag{34}$$

$$sum_r(\overline{X_{Body}}, W) \leftarrow sum'_{r,1}(\overline{X_{Body}}, W_1), ..., sum'_{r,n}(\overline{X_{Body}}, W_n), \tag{35}$$
$$W = W_1 + ... + W_n.$$

Similar to the predicates cnt and cnt' before, the unique total sum value over all values is collected in the $sum'_{r,i}$ predicates for each i, whereas the $sum_{r,i}$ predicates collect the respective intermediate sums. Note that, due to negative weights, sums are not necessarily monotonically increasing over all values; this prevents, on the one hand, the same optimization as for cnt (cf. rules (23')+(25")) to be applied in the case of weight constraints. On the other hand, the resulting encoding can – assuming that respective arithmetic is supported – deal with negative weights out-of-the-box, i.e. negative weights do not need to be eliminated as in [13].

Finally, rule (16) is analogously replaced by

$$h(\overline{X_h}) \leftarrow sum_r(\overline{X_{Body}}, W), Body(\overline{X_{Body}}), W \geq l. \tag{36}$$

4   Evaluation

Obviously, the additional machinery added in our translation comes at a cost. In order to evaluate how much it affects program size and performance in state of the art solvers, we chose some benchmark problems from the second Answer Set Programming Competition [3] involving cardinality constraints and choices: we took 8 different instances for *graph colouring*, *knight tour*, *hanoi* and *partner units*.

For grounding and solving we used gringo (v. 3.0.5) and clasp (v. 2.1.1) from Potassco. Results are reported in Table 1: each column reports size of the non-ground program ($\#ng$), size of the program after grounding ($\#g$), and evaluation time (t) in seconds (including grounding, translation and solving). We report results for grounding and evaluating the original program ($orig$), our naïve translation (tr), the optimized translation (tr_{opt}) using rules (23')–(30). Additionally, as a reference, we compare our results to first grounding the original program and then applying a ground transformation $tr_{\texttt{lp2normal}}$ to normal programs, using the tool lp2normal by Janhunen and Niemelä [11].

Table 1. Total times (in seconds)

Program	Instance*	$orig$	tr	tr_{opt}	$tr_{lp2normal}$
		$\#ng/\#g/t$	$\#ng/\#g/t$	$\#ng/\#g/t$	$\#ng/\#g/t$
Graph	1 − 125	1672/6903/1.11	1690/23753/20.48	1687/20503/2.02	1672/10235/0.2
Colouring	11 − 130	1757/7243/ > 900	1775/22778/ > 900	1772/19653/ > 900	1757/9780/ > 900
	21 − 135	1986/8087/ > 900	2004/25232/ > 900	2001/21857/ > 900	1986/11194/ > 900
	30 − 135	1794/7415/14.24	1812/24560/24.71	1809/21185/4.51	1794/10522/13.44
	31 − 140	2039/8315/419.13	2057/26095/ > 900	2054/22595/ > 900	2039/11537/283.05
	40 − 140	2219/8945/ > 900	2237/26725/ > 900	2234/23225/ > 900	2219/12167/ > 900
	41 − 145	2262/9138/ > 900	2280/27553/ > 900	2277/23928/ > 900	2262/12475/ > 900
	51 − 150	2405/9681/ > 900	2423/28731/ > 900	2420/24981/ > 900	2405/13133/ > 900
Knight	01 − 8	21/1852/0	61/23078/0.45	55/17794/0.2	21/5043/0.01
Tour	03 − 12	22/4526/0.01	62/68044/6.34	56/50975/1.08	22/13386/0.04
	05 − 16	21/8388/0.03	61/136810/44.62	55/101314/5.01	21/25822/0.08
	06 − 20	21/13432/0.05	61/229346/232.84	55/168760/15.46	21/42233/0.12
	07 − 30	21/31222/0.14	61/564694/ > 900	55/412272/181.31	21/100752/0.43
	08 − 40	21/56412/0.34	61/1048642/ > 900	55/762774/758.63	21/184230/0.85
	09 − 46	21/75078/0.42	61/1410338/ > 900	55/1024440/ > 900	21/246336/1.23
	10 − 50	22/89000/0.88	62/1681185/ > 900	56/1220267/ > 900	22/292706/1.58
Hanoi	09 − 28	104/37323/1.56	168/3279898/ > 900	156/1745347/51.24	104/52445/4.39
	11 − 30	106/40041/13.62	170/3514328/ > 900	158/1870177/51.42	106/56243/5.74
	15 − 34	110/45477/51.56	174/3983116/ > 900	162/2119757/441.48	110/63839/31.71
	16 − 40	100/31886/1.56	164/2811325/ > 900	152/1496006/19.15	100/44848/2.07
	22 − 60	102/33314/0.82	166/3175997/633.51	154/1683463/26.99	102/46472/1.37
	36 − 80	106/40041/1.19	170/3514364/600.88	158/1870217/30.89	106/56243/1.11
	41 − 100	104/37322/0.48	168/3279933/321.64	156/1745386/22.45	104/52444/0.98
	47 − 120	99/30527/1.9	163/2694686/845.48	151/1434231/16.28	99/42949/0.75
Partner	176 − 24	68/13213/0.61	146/162007/40.98	131/102667/7.1	68/19347/1.07
Units	29 − 40	108/61777/0.13	186/1068887/ > 900	171/631427/ > 900	108/79679/6.98
	23 − 30	117/40332/0.13	195/451513/ > 900	180/277733/8.51	117/51127/0.49
	207 − 58	136/162537/0.61	214/4857458/ > 900	199/2730134/ > 900	136/203258/1.48
	204 − 67	141/223931/1.54	219/7672436/ > 900	204/4285390/ > 900	141/276403/1.86
	175 − 75	290/689087/20.78	368/15446057/ > 900	353/8611453/ > 900	290/762711/27.44
	52 − 100	254/963749/ > 900	332/36843565/ > 900	317/20137215/ > 900	254/1082289/ > 900
	115 − 100	254/963806/ > 900	332/37214669/ > 900	317/20328419/ > 900	254/1082942/ > 900

*) For the instance naming convention, please refer to http://dtai.cs.kuleuven.be/events/ASP-competition/index.shtml.

As expected, the results in Table 1 show that the time that evaluation time rises significantly, which is mainly due to a blowup during grounding. There are certain exceptions, as in our selection one particular graph colouring example where our optimized encoding even outperforms all others. Solving the instances with pre-grounding the original program and using `lp2normal` on the ground instantiated program shows better results, however we couldn't use this approach in our use case of debugging, described in the next section.

5   Debugging with Ouroboros

In our project, we deploy ASP programs for encoding (re-)configuration problems [5], where debugging of the resulting (non-ground) programs became a significant issue in practical use cases. We base our debugger on the approach of Oetsch et al. [14], who developed a meta-program for debugging non-ground programs in ASP. The basic idea of this debugging method is to reify a program P as well as the fully expected interpretation I. Reification means that the program and interpretation are brought onto a meta-level. Finally, the meta program and meta interpretation are fed to an ASP solver. The obtained answer sets explain why I is not an answer set of P.

There are two main explanation classes why an interpretation I is not answer set of P. First, some atoms of the interpretation can form an unfounded loop. A non-empty set

L of ground literals is a loop of P iff for each pair $(a, b) \in L$ there is a path from a to b in the positive dependency graph. The length of the path from a to b can be equal to or greater than 0. Additionally, let I, J be interpretations. J is supported by P wrt. I if the grounding of P contains some rule r whose body is satisfied by I and some head atoms of r are included in J, but all head atoms of r that are not included in J are false under I. Moreover, this support is ensured to be external, that means without any reference to the set J itself [14]. If J is not externally supported by P wrt. I, J is called *unfounded* by P with respect to I. In particular, if there is a loop in P that is contained in I but this loop is not externally supported (unfounded) by P with respect to I then I is not an answer set and the debugger program returns the unfounded (sub)set of I. The second type of explanation are unsatisfied rules, that means that instantiations of rules in P are not satisfied by I, in this case the debugger returns the (non-ground) unsatisfied rule(s).

The original meta-program debugger was written for DLV System [12] and can handle non-ground (even disjunctive) logic programs, integer arithmetic $(+, *)$, comparison predicates $(=, \neq, \leq, <, \geq, >)$ and strong negation. We made some minor adaptions to use Potassco, which we deployed throughout our project, where we do not need disjunction but make heavy use of other extended constructs such as choices, cardinality and weight constraints: As a first step, we transformed the meta-program in such a way that the usage of Potassco system [8] for debugging was facilitated. Debugging of programs containing choices, cardinality and weight constraints was enabled straightforwardly by (i) applying our presented translation from Section 3 above to the input, whereas we translate back debugging results from the meta-program, such that they refer back to the original rules with choices and cardinality constraints, whenever a rule occurring from our translation is identified as "buggy". Minor additional adaptions of the meta-program included support for extended integer arithmetic (e.g. to support use of $-$ in rule (25")). To support the debugging process including the translation of cardinality constraints, we extended the SeaLion Eclipse plugin [15] (an integrated development environment (IDE) for Answer Set Programming) by the Ouroboros plugin[6]. Our extended Ouroboros plugin can handle rules with cardinality constraints (and has not yet implemented the translation of weight constraints).

The plugin, including the new transformed and extended meta-program debugger based on [14], can be found at https://mmdasp.svn.sourceforge.net/svnroot/ mmdasp/sealion/trunk/org.mmdasp.sealion.ouroboros/. To illustrate a simple debugging scenario consider the following example from constraint-based configuration. ASP programmer Lilian wants to assign each thing to exactly one cabinet with the constraint that there should not be more than two things in one cabinet. Her program, P_1, looks as follows:

```
thing(th1). thing(th2). thing(th3).
cabinet(c1). cabinet(c2). cabinet(c3).
1 {cabinetToThing(X, Y) : cabinet(X)} 1 :- thing(Y).
:- 2 {cabinetToThing(X, Y) : thing(Y)}, cabinet(X).
```

Executing P_1, Lilian gets six answer sets. However, she wonders why there are only answer sets where in each of them one cabinet has exactly one thing. Normally, there

---
[6] Details on the Ouroboros plugin can be found in a companion system description [6].

should be answer sets where e.g. cabinet $c2$ has two things. So she decides to save the following interpretation – I_1 – as facts, where she replaced cabinetToThing(c3, th2) with cabinetToThing(c2, th2) to check why it is not an answer set:

```
thing(th1). thing(th2). thing(th3).
cabinet(c1). cabinet(c2). cabinet(c3).
cabinetToThing(c1, th3).
cabinetToThing(c2, th1).
cabinetToThing(c2, th2).
```

Now she creates a debug configuration and selects the program file as well as the adapted interpretation file and chooses the explanation type *Unsatisfiability*. After debugging the explanation says *Guessed rule: :- 2 {cabinetToThing(X, Y) : thing(Y)}, cabinet(X)*. Indeed, investigation of this rule reveals that the lower bound was set wrongly and should be 3 instead of 2.

In the background of this debugging process, the following happens: Let us denote the set of rules containing cardinality constraints or choices from a given program P as $P_{cc} = \{r_{cc_1}, \ldots, r_{cc_n}\}$. Moreover, let $tr(r_{cc_i})$ be the translation of a resp. rule r_{cc_i} according to Section 3. In our case, the two cardinality constraint rules of $P_{1,cc}$ are translated as follows:

```
cabinetTOthing(X, Y) :- thing(Y), cabinet(X), not -cabinetTOthing(X, Y).
-cabinetTOthing(X, Y) :- thing(Y), cabinet(X), not cabinetTOthing(X, Y).
:- not lowerUpperOK_1(Y), thing(Y).
lowerUpperOK_1(Y) :- not upper_1(Y), lower_1(Y), thing(Y).
lower_1(Y) :- cnt_1(Y, CounterC), CounterC >= 1, thing(Y).
upper_1(Y) :- CounterC > 1, cnt_1(Y, CounterC), thing(Y).
val_1_0(X, Y, Y) :- cabinet(X), cabinetTOthing(X, Y), thing(Y).
exists_1_0(Y) :- thing(Y), val_1_0(X, Y, Y).
smaller_1_0(X, Y, X1, Y1) :- val_1_0(X, Y, YBody), val_1_0(X1, Y1, YBody), X < X1.
smaller_1_0(X, Y, X, Y1) :- val_1_0(X, Y, YBody), val_1_0(X, Y1, YBody), Y < Y1.
cnt_1_0(X, Y, YBody, 1) :- val_1_0(X, Y, YBody).
cnt_1_0(X1, Y1, YBody, Ncounter1) :- val_1_0(X1, Y1, YBody),
    smaller_1_0(X, Y, X1, Y1), cnt_1_0(X, Y, YBody, Ncounter), Ncounter1 = Ncounter+1.
cntPrime_1_0(YBody, Ncounter) :- cnt_1_0(X, Y, YBody, Ncounter),
    not nmax_1_0(YBody, Ncounter).
cntPrime_1_0(Y, 0) :- thing(Y), not exists_1_0(Y).
nmax_1_0(YBody, Ncounter1) :- cnt_1_0(X, Y, YBody, Ncounter), Ncounter1 = Ncounter-1.
cnt_1(YBody, Ncounter0) :- cntPrime_1_0(YBody, Ncounter0).
:- lower_2(X), cabinet(X).
lower_2(X) :- cnt_2(X, CounterC), CounterC >= 2, cabinet(X).
val_2_0(X, Y, X) :- thing(Y), cabinetTOthing(X, Y), cabinet(X).
exists_2_0(X) :- cabinet(X), val_2_0(X, Y, X).
smaller_2_0(X, Y, X1, Y1) :- val_2_0(X, Y, XBody), val_2_0(X1, Y1, XBody), X < X1.
smaller_2_0(X, Y, X, Y1) :- val_2_0(X, Y, XBody), val_2_0(X, Y1, XBody), Y < Y1.
cnt_2_0(X, Y, XBody, 1) :- val_2_0(X, Y, XBody).
cnt_2_0(X1, Y1, XBody, Ncounter1) :- val_2_0(X1, Y1, XBody),
    smaller_2_0(X, Y, X1, Y1), cnt_2_0(X, Y, XBody, Ncounter), Ncounter1 = Ncounter+1.
cntPrime_2_0(XBody, Ncounter) :- cnt_2_0(X, Y, XBody, Ncounter),
    not nmax_2_0(XBody, Ncounter).
cntPrime_2_0(X, 0) :- cabinet(X), not exists_2_0(X).
nmax_2_0(XBody, Ncounter1) :- cnt_2_0(X, Y, XBody, Ncounter), Ncounter1 = Ncounter-1.
cnt_2(XBody, Ncounter0) :- cntPrime_2_0(XBody, Ncounter0).
```

Since the debugging approach requires a complete interpretation, we first have to extend the interpretation I given for debugging by the newly derivable auxiliary literals introduced in the translation. For this purpose a distinction must be made between satisfied and unsatisfied (wrt. I) cardinality constraints: if r_{cc_i} involves a cardinality constraint with bounds, then $tr(r_{cc_i})$ contains an integrity constraint (either the rule is a constraint, then see rule (8) or (10) or otherwise see rule (9)); now, if the cardinality constraint is satisfied under the interpretation at hand, solving $tr(r_{cc_i}) \cup I$ yields one answer set that contains the additionally required literals. If a cardinality constraint P_i is not satisfied under the interpretation, then solving $tr(r_{cc_i}) \cup I$ yields no answer set at all. In this case, the original interpretation I is used. Thus, if a cardinality constraint is unsatisfied under I the debugger meta-program will state that rule (17) and rule (26) are unsatisfied.

As another case, some atoms of the interpretation can also form an unfounded loop. Let us consider Lilian's program just with the first cardinality constraint, denoted as P_2:

```
thing(th1). thing(th2).
cabinet(c1). cabinet(c2).
1 {cabinetTOthing(X, Y) : cabinet(X)} 1 :- thing(Y).
```

This program has four answer sets. However, Lilian expects to have some answer sets something like cabinetTOthing(c3,th3), i.e. expects interpretation I_2 to be an desired answer set:

```
thing(th1). thing(th2).
cabinet(c1). cabinet(c2).
cabinetTOthing(c2, th1).
cabinetTOthing(c1, th2).
cabinetTOthing(c3, th3).
```

In this case, the debugging output explains that cabinetTOthing(c3, th3). forms an unfounded loop. In particular, there is neither a fact thing(th3) nor a fact cabinet(c3).

We emphasize that both these kinds of errors – wrong cardinalities, missing facts – occurred in practice in the encodings of our practical configuration settings.

6   Conclusions

We have presented a non-ground embedding of advanced ASP constructs (choices, cardinality and weight constraints) into normal logic programs and demonstrated how this embedding can be used to debug non-ground ASP programs using these constructs in the domain of configuration. While the non-ground embedding allowed us to extend an existing debugging approach for normal non-ground progams [15] relatively straightforwardly, our preliminary evaluation of the non-ground transformation shows that it cannot compete directly with non-ground embeddings as of yet. An investigation of further optimizations, or the possibility to use more efficient ground transformations directly in our debugger are on our agenda for future work.

Acknowledgements. The authors would like to thank Tomi Janhunen for providing advice on how to use the tools from [11] in our evaluation and Jörg Pührer for supporting and giving advice regarding the Ouroboros plugin. This work was funded by FFG FIT-IT within the scope of the project RECONCILE (grant number 825071).

References

1. Brain, M., De Vos, M.: Debugging logic programs under the answer set semantics. In: 3rd International Workshop on Answer Set Programming, ASP 2005. CEUR Workshop Proceedings, pp. 141–152 (2005)
2. Caballero, R., García-Ruiz, Y., Sáenz-Pérez, F.: A theoretical framework for the declarative debugging of datalog programs. In: Schewe, K.-D., Thalheim, B. (eds.) SDKB 2008. LNCS, vol. 4925, pp. 143–159. Springer, Heidelberg (2008)
3. Denecker, M., Vennekens, J., Bond, S., Gebser, M., Truszczyński, M.: The second answer set programming competition. In: Erdem, E., Lin, F., Schaub, T. (eds.) LPNMR 2009. LNCS, vol. 5753, pp. 637–654. Springer, Heidelberg (2009)
4. Ferraris, P., Lifschitz, V.: Weight constraints as nested expressions. Theory Pract. Log. Program. 5(1-2), 45–74 (2005)
5. Friedrich, G., Ryabokon, A., Falkner, A.A., Haselböck, A., Schenner, G., Schreiner, H.: (Re)configuration using Answer Set Programming. In: IJCAI 2011 Workshop on Configuration, pp. 17–25 (2011)
6. Frühstück, M., Pührer, J., Friedrich, G.: Debugging answer-set programs with Ouroboros – extending the SeaLion plugin. In: Cabalar, P., Son, T.C. (eds.) LPNMR 2013. LNCS (LNAI), vol. 8148, pp. 323–328. Springer, Heidelberg (2013)
7. Gebser, M., Kaminski, R., Kaufmann, B., Ostrowski, M., Schaub, T., Thiele, S.: A user's guide to gringo, clasp, clingo, and iclingo (2010)
8. Gebser, M., Kaufmann, B., Kaminski, R., Ostrowski, M., Schaub, T., Schneider, M.: Potassco: The potsdam answer set solving collection. AI Commun. 24(2), 107–124 (2011)
9. Gebser, M., Pührer, J., Schaub, T., Tompits, H.: A meta-programming technique for debugging answer-set programs. In: Proceedings of the 23rd National Conference on Artificial Intelligence, AAAI 2008, vol. 1, pp. 448–453. AAAI Press (2008)
10. Gebser, M., Schaub, T.: Answer set solving in practice (2011), http://www.cs.uni-potsdam.de/~torsten/ijcai11tutorial/asp.pdf (visited on October 18, 2012)
11. Janhunen, T., Niemelä, I.: Compact translations of non-disjunctive answer set programs to propositional clauses. In: Balduccini, M., Son, T.C. (eds.) Logic Programming, Knowledge Representation, and Nonmonotonic Reasoning. LNCS, vol. 6565, pp. 111–130. Springer, Heidelberg (2011)
12. Leone, N., Pfeifer, G., Faber, W., Eiter, T., Gottlob, G., Perri, S., Scarcello, F.: The dlv system for knowledge representation and reasoning. ACM Trans. Comput. Logic 7(3), 499–562 (2006)
13. Niemelä, I., Simons, P., Soininen, T.: Stable model semantics of weight constraint rules. In: Gelfond, M., Leone, N., Pfeifer, G. (eds.) LPNMR 1999. LNCS (LNAI), vol. 1730, pp. 317–331. Springer, Heidelberg (1999)
14. Oetsch, J., Pührer, J., Tompits, H.: Catching the ouroboros: On debugging non-ground answer-set programs. Theory Pract. Log. Program. 10(4-6), 513–529 (2010)
15. Oetsch, J., Pührer, J., Tompits, H.: The sealion has landed: An IDE for answer-set programming. In: 25th Workshop on Logic Programming, WLP (2011)
16. Pontelli, E., Son, T.C., Elkhatib, O.: Justifications for logic programs under answer set semantics. Theory Pract. Log. Program. 9(1), 1–56 (2009)
17. Syrjänen, T.: Cardinality constraint programs. In: Alferes, J.J., Leite, J. (eds.) JELIA 2004. LNCS (LNAI), vol. 3229, pp. 187–199. Springer, Heidelberg (2004)
18. Syrjänen, T.: Debugging inconsistent answer set programs. In: Proc. NMR, vol. 6, pp. 77–83 (2006)

Conflict-Based Program Rewriting for Solving Configuration Problems

Anna Ryabokon[1], Gerhard Friedrich[1], and Andreas A. Falkner[2]

[1] Universitaet Klagenfurt, Austria
`firstname.lastname@aau.at`
[2] Siemens AG Österreich, Vienna, Austria
`firstname.{middleinitial.}lastname@siemens.com`

Abstract. Many real-world design problems such as product configuration require a flexible number of components and thus rely on tuple generating dependencies in order to express relations between entities. Often, such problems are subject to optimization, preferring models which include a minimal number of constants substituted in existentially quantified formulas.

In this paper we propose an approach based on automated program rewriting which avoids such substitutions of existentially quantified variables that would lead to a contradiction. While preserving all solutions, the method significantly reduces runtime and solves instances of a class of real-world configuration problems which could not be efficiently solved by current techniques.

1 Introduction

A number of important real-world applications require knowledge representation (KR) languages that are able to express the existence of certain objects. For instance, a computer configuration system [20] might require the existence of a compatible CPU for each motherboard. The rules with existentially quantified heads used to express these relations are often referred to as tuple generating dependencies (TGDs). Modern knowledge representation (KR) formalisms such as Datalog+/-[4] or Description Logic [2] are able to represent TGDs. They are used for query answering and allow to verify whether a given set of facts is a problem solution. However, in some applications such as knowledge-based configuration the problem solutions are unknown a priori and have to be generated, e.g. by computing (subsets) of preferred logical models.

General languages for configuration problems such as LoCo [1] allow to model conditional inclusion of components by means of TGDs. Since in general case a knowledge base containing TGDs might have infinite models (configurations), a language must ensure the finiteness of models by bounding the number of generated components depending on the user input. Namely, we have to verify whether the set of user-defined input components of a configuration problem, given as facts, suffice to make the configuration problem finite. Assuring the finiteness of models is desirable not only for guaranteeing decidability, but is also obvious for practical reasons such as realizability of a configuration since infinite configurations cannot be manufactured. After the finite bounds on the number of required components are computed, configurations are found by the means of model construction.

P. Cabalar and T.C. Son (Eds.): LPNMR 2013, LNAI 8148, pp. 465–478, 2013.
© Springer-Verlag Berlin Heidelberg 2013

In many cases, computation of precise bounds is impossible since current methods [7,1,8] do not consider additional constraints given in a problem description. In this case one generates a number of components corresponding to an upper bound, which usually is more than required to solve a configuration problem instance. In order to obtain a solution with the minimal number of components, a system requires a set of preference criteria to be defined by a user. These criteria are then provided to a solver which returns preferred models. The computation of a(n optimal) model can be done by translating a general logical description of a problem instance to Answer Set Programming (ASP) [13,3].

In practice finding an optimal solution might take unacceptable time because of the large number of constants (components) and the presence of symmetric models. They can be obtained from other models by interchanging constants substituted in existentially quantified variables. As practice shows, the larger the number of existing symmetric models is the worse is solving performance.

There are two general ways to overcome this problem: extend the knowledge base with additional symmetry breaking constraints [5] and/or reduce the sets of generated constants used in the rules approximating TGDs (domain filtering [9,14]). However, these techniques have only been developed for specific KR languages, e.g. ASP or constraint programming, and are not supported by general languages such as [1]. Moreover, the evaluation presented in [18] shows that the application of the known symmetry-breaking approach for ASP [5] works well for the pigeonhole problem, but does not improve solving performance when applied to such problems as rack configuration.

In this paper we propose a novel *TGD rewriting* approach which is applicable to general knowledge representation languages. This method improves the elimination of existential quantifiers based on or-terms. The basic idea of eliminating existential quantification [1] is to compute a sufficiently large set of fresh constants (called domain) for each existential quantifier and replace the existentially quantified variables in the TGDs by an or-term. Roughly speaking, the or-term contains an atom for every combination of elements of the domains of the existential quantifiers. We exploit conflicting variable substitutions in order to reduce the length of these or-terms thus reducing the number of choice points in the search space. To the best of our knowledge there are no previous proposals for an automatic rewriting of existential rules. The algorithm is evaluated on a set of industrial configuration problem instances corresponding to combinatorial problems which include several optimization criteria. These problems require selection and assignment of hardware modules depending on the user and system requirements and occur frequently during configuration of technical systems produced by Siemens, such as telephone switching systems, electronic railway interlocking system, automation systems, etc.

The experimental results show that the algorithm can find (optimal) solutions for problems occurring in practice within an acceptable time. For the set of reference industrial problems, the method was able to find solutions which are up to 525% better with respect to the specified preference criteria. For the largest problem instances the standard solver was not able to find a solution with optimal costs using the original program in 3 hours, whereas the rewritten program required at most 11 minutes.

In Section 2 we present an approach for rewriting TGDs that allows the search to be performed efficiently. In Section 3 the implementation details are provided and in Section 4 the evaluation results are analyzed. Finally, in Section 5 we conclude and discuss future work.

2 Rewriting of Existential Rules

In this section, after an introduction of some preliminaries, we describe the rewriting algorithm that uses binary constraints to limit the number of generated constants (nulls) and, thus, accelerates evaluation of programs containing TGDs.

2.1 Preliminaries

In configuration languages applied by Siemens in practice variants of TGDs are used to express relations between two components. Formally, a TGD σ of such a language is a first order formula of the form

$$\forall X \in S^X \ \Phi(X) \to \exists_l^u Z \in S^Z \ \Psi(X, Z) \tag{1}$$

where $\Phi(X)$ is an atom and $\Psi(X, Z)$ is a conjunction of atoms. An *atom* is an expression of the form $r_i(t_1, \ldots, t_n)$, where $t_1, \ldots, t_n$ are terms and r_i is an element of a finite set of relation names (predicates) $\mathcal{R} = \{r_1, \ldots, r_n\}$. Each *term* can be either a variable or a constant or a null (Skolem constant). The infinite countable domains of the terms are denoted by Δ_V, Δ_C and Δ_N respectively and the union of these domains by Δ. An atom $r_i(t_1, \ldots, t_n)$ is called *ground* if all terms in the tuple $\langle t_1, \ldots, t_n \rangle$ are elements of the set $\Delta_C \cup \Delta_N$. The set $Hbase(\Delta_C \cup \Delta_N)$ (Herbrand base) contains all ground atoms that can be generated using predicates in $\mathcal{R}$ and terms in $\Delta_C \cup \Delta_N$.

Formula (1) includes two extensions w.r.t. classical first-order formulas, namely, counting existential quantifiers and sorts $S \subseteq \Delta_C \cup \Delta_N$. The latter is required by the fact that all components of one type in a component catalog must have a unique identifier. Consequently, configuration languages employed in industry and proposed in [1] allow the definition of a *sort* of identifiers for each component type of a problem. All sorts defined in a problem description are required to be mutually disjoint, i.e. $S_i \cap S_j = \emptyset$. As it is shown in [1] a program including rules of the form (1) can be reduced to a classical first-order program.

An example of a typical configuration TGD is a binary relation between two components C_1 and C_2 that relates at least l and at most u instances of C_2 with each C_1:

$$\forall X \in S^X \ C_1(X) \to \exists_l^u Z \in S^Z \ C_2(Z) \wedge C_1 2 C_2(X, Z)$$

where sorts S^X and S^Z for variables X and Z contain identifiers of component instance available in a problem description.

An atom $\Phi(X)$ is often called the *body* of a rule and the conjunction $\Psi(X, Z)$ is the *head*. A rule with an empty body $body(r) = \emptyset$ is usually referred to as a *fact*. Facts can be of the two types *existential* and *ground* depending on whether they contain some existentially quantified variables or not. Note that, an existential fact can be a conjunction of atoms including one existentially quantified variable. A program Σ is a finite set of rules. It contains rules of the form (1) as well as first-order rules with only universally quantified variables. In addition, $edb(\Sigma)$ denotes a set of all ground facts of the program Σ and by $idb(\Sigma)$ all other rules.

A *substitution* is a homomorphism $\theta : \Delta \mapsto \Delta_C \cup \Delta_N$ that maps elements of Δ_C and Δ_N to themselves. In order to simplify the presentation we denote a substitution of the variables in an atom at by $\theta(at)$. For the sorted variables the substitution function is defined as $\theta : \Delta \mapsto S$, where $S \subseteq \Delta_C \cup \Delta_N$.

Since sorts and counting existential quantifies can be reduced to classical first-order logic we use the following semantic of a program Σ. Let $\mathcal{A}$ be a set of atoms. Given an atom at the set $\mathcal{A}$ entails at ($\mathcal{A} \models at$) if there is a substitution θ such that $\theta(at) \in \mathcal{A}$. A set of ground atoms $\mathcal{M} \subset Hbase(\Delta_C \cup \Delta_N)$ is a *model* of the program Σ if for every rule $\sigma \in \Sigma$ there exists a substitution $\theta(body(\sigma)) \subseteq \mathcal{M}$ such that $\mathcal{M} \models \theta'(head(\sigma))$, where θ' contains all and only substitutions for existentially quantified variables Z, see [6,4,15] for more details. In configuration solutions often correspond to a minimal set of atoms in order to reduce the costs. Therefore, in practice we can focus on the computation of minimal models. The computation of finite models containing the optimal solutions (configurations) can be accomplished by using ASP encodings and solvers.

Note, that in case of configuration the number of nulls used in a solution must be finite, since infinite solutions are of no practical interest. Therefore, configuration languages use methods as in [8] to compute the required number of nulls for each sort or to determine that there is no finite solution. Consequently, we assume that the sorts are fixed. The bounds can be determined for the number of components by methods presented in [7,1,8]. For the practical configuration problems of Siemens these bounds can be computed in polynomial time.

2.2   Conflict-Based Program Rewriting

Given a program Σ, a reasoning algorithm, e.g. chase [6,4], usually starts from the set of rules $edb(\Sigma)$ and iteratively extends it by searching any applying substitutions to the rules in $idb(\Sigma)$. In case of a TGD σ with $\theta(body(\sigma)) \subseteq edb(\Sigma)$ the reasoning algorithm first rewrites it as an existential fact of the form

$$\exists_l^u Z \in S^Z \ \Psi(\theta(X), Z) \tag{2}$$

where $\theta(X)$ maps variable X to some constant in S^X. Next, the algorithm searches for an *extension* θ' of the substitution θ on $X \cup Z$ associating the variable Z with some element of S^Z. The resulting ground facts are then added to $edb(\Sigma)$. Usually reasoning algorithms use a variant of an extension function θ' that associates a *fresh* null with each variable in order to obtain a universal solution $\mathcal{M}$, i.e. such solution that any other solutions, say $\mathcal{M}'$, can be obtained from $\mathcal{M}$ by a homomorphism $h : \Delta_N \cup \Delta_C \rightarrow \Delta_N \cup \Delta_C$ that maps elements of Δ_C to themselves [6,15].

Example 1. Consider the following program capturing a frequent case in technical configuration which includes two component types and a typical $\exists_1^1$ relation between them.

Given a set of things, store each of them in exactly one cabinet taking into account that things t_1 and t_2 cannot be placed in the same cabinet. The domain of cabinets is defined as $S^Z = \{\varphi_1, \varphi_2, \varphi_3\}$.

$$r_1 : \quad thing(t_1) \wedge thing(t_2) \wedge thing(t_3)$$
$$r_2 : \quad \forall X \, thing(X) \rightarrow \exists_1^1 Z \in S^Z \, t2c(X, Z)$$
$$r_3 : \quad \forall X \, t2c(t_1, X) \wedge t2c(t_2, X) \rightarrow$$

The reasoning algorithm starts with an $edb(\Sigma) = \{thing(t_1), thing(t_2), thing(t_3)\}$ and finds a substitution $\theta_1 = \{X/t_1\}$. This substitution allows to rewrite the second rule as an existential fact $\exists_1^1 Z \in S^Z \, t2c(t_1, Z)$. The second substitution $\theta_2 = \{X/t_2\}$ results in the fact $\exists_1^1 Z \in S^Z \, t2c(t_2, Z)$ and the third $\theta_3 = \{X/t_3\}$ in $\exists_1^1 Z \in S^Z \, t2c(t_3, Z)$. An extension function might introduce different mappings. For instance, it can map every variable to a different null and obtain three grounded atoms $t2c(t_1, \varphi_1)$, $t2c(t_2, \varphi_2)$, and $t2c(t_3, \varphi_3)$, where each φ_i corresponds to a cabinet. $edb(\Sigma)$ extended with these facts is a model of the program above. However, an extension resulting in $t2c(t_1, \varphi_1)$, $t2c(t_2, \varphi_1)$, and $t2c(t_3, \varphi_1)$ does not allow to obtain a model of the program because of the constraint (rule r_3).

Definition 1 (Conflict set). *Let* $\Psi := \{\Psi(\theta_1(X), Z), \ldots, \Psi(\theta_n(X), Z)\}$ *be the set of sorted existential facts* Ψ *of a program* Σ *in which variable* Z *range over the same sort* S^Z. *A pair of facts* $CS \subseteq \Psi$ *is a* conflict set *iff* $\Sigma \cup \{\exists Z \bigwedge_{a \in CS} a\}$ *is inconsistent.*

A set **CS** of conflict sets can be computed using this definition as shown in Section 3. Given the set **CS**, we can rewrite TGDs in Σ in a way that the reasoning algorithm will extend substitutions for any pair of conflicting atoms with different constants (nulls).

Definition 2 (TGD rewriting problem). *Let* **CS** *be a set of conflict sets and* Σ *be a program including TGDs of the form (1). The TGD rewriting problem is to find such domain restrictions for the substitution function* θ *for each TGD in* Σ *such that:*

a) each set Ψ *of sorted existential facts do not contain conflict sets* $CS \in$ **CS***;*
b) all possible models are preserved up to symmetric ones (renaming of nulls).

In order to rewrite the original TGD, first, we use Algorithm 1 which finds for each existential fact $\Psi_i \in \Psi$ a tuple $\langle D_i^X, D_i^Z \rangle$, where the set $D_i^X \subseteq S^X$ contains all constants that are substituted by θ_i to the variable X in Formula (1) and the set $D_i^Z \subseteq S^Z$ contains all nulls which are used to substitute Z. Next, given the resulting set of all tuples $\mathcal{SN} = \{\langle D_1^X, D_1^Z \rangle, \ldots, \langle D_n^X, D_n^Z \rangle\}$ we rewrite the TGD (1) as follows:

$$\forall X \in D_i^X \, \Phi(X) \rightarrow \exists_l^u Z \in D_i^Z \, \Psi(X, Z) \tag{3}$$

Assuming that the initial program Σ is consistent, none of the rewritten TGDs (3) in the program Σ' can result in generation of a conflict set $CS \in$ **CS**.

Our algorithm uses three functions: GETSUBSTITUTIONCONSTANTS, GETNULLS, and GETCONSTANTS. The first function retrieves all constants that were substituted for the universally quantified variable X. The injective function GETNULLS : $\Psi \mapsto \mathcal{P}(S \cap \Delta_N)$

Algorithm 1. GenerateExtensions

> **input** : A set **CS** of conflict sets, a set Ψ of existential facts, a domain S, program Σ
> **output**: A set of substitutions associated with a set of nulls $\mathcal{SN}$

1 $\mathcal{SN} \leftarrow \emptyset$;
2 **for** $i \leftarrow 1$ **to** $|\Psi|$ **do**
3     $D^X \leftarrow$ GETSUBSTITUTIONCONSTANTS(Ψ_i);
4     $D^Z \leftarrow$ GETNULLS(Ψ_i, S);
5     $D^Z \leftarrow D^Z \cup$ GETCONSTANTS(Σ, S);
6   **for** $j \leftarrow 1$ **to** $i - 1$ **do**
7   **if** $\forall CS \in$ **CS** $CS \nsubseteq \{\Psi_i, \Psi_j\}$ **then**
8           $\left\lfloor D^Z \leftarrow D^Z \cup \text{GETNULLS}(\Psi_j, S); \right.$
9     $\mathcal{SN} \leftarrow \mathcal{SN} \cup \{\langle D^X, D^Z \rangle\}$;
10 **return** $\mathcal{SN}$;

---

associates a set of nulls $\varphi_i \subseteq S \cap \Delta_N$ with every $\Psi_i \in \Psi$, such that φ_i includes u fresh nulls for the variable Z, which are not used in any of the previous substitutions. GETCONSTANTS returns all constants from the set $S \cap \Delta_C$ contained in Σ. The sort S is initialized prior to the solving and includes a sufficient number of fresh nulls (see Section 2.1) and additional constants contained in the original program. Such constants should always be included in every domain D^Z to ensure that all models are preserved.

For each atom $\Psi_i \in \Psi$ Algorithm 1 initializes the set D^Z with a set of fresh nulls φ_i corresponding to the fact Ψ_i and all constants of S appearing in Σ. Next, D^Z is extended with the nulls corresponding to $\Psi_j \in \Psi$ non-conflicting with Ψ_i. If there are multiple sets of existential facts $\Psi = \{\Psi_1, \ldots, \Psi_i\}$ ranging over mutually disjoint sorts $S_1^Z, \ldots, S_i^Z$ then the algorithm is applied to each $\Psi_k \in \Psi$ separately.

The rewriting approach modifies the range of the substitution function, defined as a homomorphism $\theta : \Delta \mapsto S$ by replacing S with its subset D. All other elements of the structure such as predicates remain constant. Such modification guarantees that only conflicting pairs of nulls, i.e. $\{\varphi_i, \varphi_j\} \nsubseteq D^Z$, are not substituted for existentially quantified variables of a TGD. Assuming that the instantiations of universally quantified variables is complete based on the grounded facts, we argue that our rewriting is model preserving (up to renaming of nulls) because (a) we consider all constants of the corresponding sort from the program, (b) we add u fresh nulls, and (c) eliminate nulls which are conflicting w.r.t. a given conflict set **CS**. The case where additional deductions extend the set of grounded facts is described in Section 2.3.

Of course, the method does not ensure the elimination of all possible conflicts, but only those given in **CS**. In addition, we also do not require **CS** to contain all conflicts. In particular, in our implementation we focus on conflicts which can be efficiently detected by using the Horn fragment of Σ, where consistency can be checked in polynomial time. Similar to filtering methods used in constraint satisfaction problem solving, we focus on the elimination of the binary conflicts in order to increase efficiency of the algorithm. Our evaluation shows that the suggested algorithm allows to achieve significant reduction of computation time for real-world problems.

Note that Algorithm 1 prevents the generation of many symmetric solutions by generating only a lower triangle of the matrix of nulls that can be substituted for existentially quantified variables. Each row of this matrix corresponds to an element of D^X and each column to one of the nulls in D^Z. For instance, if atoms $t2c(t_1, Z)$ and $t2c(t_2, Z)$ are not conflicting then the algorithm generates two tuples $\langle \{t_1\}, \{\varphi_1\} \rangle$ and $\langle \{t_2\}, \{\varphi_1, \varphi_2\} \rangle$, thus, avoiding a symmetric solution $\{t2c(t_1, \varphi_2), t2c(t_2, \varphi_2)\}$.

Example 2 (continue Example 1). Let $edb(\Sigma)$ be extended with facts defining two more things $\{thing(t_4), thing(t_5)\}$ and $S^Z = \{\varphi_1, \ldots, \varphi_5\}$. Also in $idb(\Sigma)$ we replace the constraint r_3 with a set of constraints that do not allow to assign any of the things $\{t_1, t_2\}$ and $\{t_3, t_4, t_5\}$ to the same cabinet. Assume that for the set of facts

$$\Psi = \{t2c(t_1, Z), t2c(t_2, Z), t2c(t_3, Z), t2c(t_4, Z), t2c(t_5, Z)\}$$

all constraints in $idb(\Sigma)$ result in the following set of conflict sets:

$$\mathbf{CS} = \{ \{t2c(t_1, Z), t2c(t_3, Z)\}, \{t2c(t_1, Z), t2c(t_4, Z)\}, \{t2c(t_1, Z), t2c(t_5, Z)\}, \\ \{t2c(t_2, Z), t2c(t_3, Z)\}, \{t2c(t_2, Z), t2c(t_4, Z)\}, \{t2c(t_2, Z), t2c(t_5, Z)\}\}$$

Application of Algorithm 1 to the given sets Ψ, $\mathbf{CS}$ and the domain of nulls $S^Z = \{\varphi_1, \ldots, \varphi_5\}$ results in the following set of extensions $\mathcal{SN}$:

$$t_1 : \{\varphi_1\} \quad t_2 : \{\varphi_1, \varphi_2\} \quad t_3 : \{\varphi_3\} \quad t_4 : \{\varphi_3, \varphi_4\} \quad t_5 : \{\varphi_3, \varphi_4, \varphi_5\}$$

Given the set $\mathcal{SN}$ we use the suggested transformation to rewrite the TGD in Example 2 as follows:

$$thing(t_1) \rightarrow \exists_1^1 Z \in \{\varphi_1\} \, t2c(t_1, Z) \qquad thing(t_3) \rightarrow \exists_1^1 Z \in \{\varphi_3\} \, t2c(t_3, Z)$$
$$thing(t_2) \rightarrow \exists_1^1 Z \in \{\varphi_1, \varphi_2\} \, t2c(t_2, Z) \quad thing(t_4) \rightarrow \exists_1^1 Z \in \{\varphi_3, \varphi_4\} \, t2c(t_4, Z)$$
$$thing(t_5) \rightarrow \exists_1^1 Z \in \{\varphi_3, \varphi_4, \varphi_5\} \, t2c(t_5, Z)$$

The resulting program reduces the number of possible extensions significantly. In the Example 1 the standard elimination of existential quantification would result in a set of five nulls $\{\varphi_1, \ldots, \varphi_5\}$, where each existential quantified variable can be substituted by an element of this set. Such a set of nulls allows $5^5 = 3125$ possible combinations of substitutions of nulls, whereas the rewritten program allows only 12 combinations which is 260 times smaller than using the common algorithm. In this example we preserve all possible models, up to symmetric ones, because the algorithm eliminated only those nulls from the sets D_i^Z whose substitution results in an inconsistency.

2.3   Rewriting of Multiple TGDs

Descriptions of complex problem instances might include multiple TGDs such that some of them depend on the others, i.e. the body $\Phi(X)$ of a TGD σ_1 contains an atom $at \in \Phi(X)$ which can be derived from some atom in the head $\Psi(X, Z)$ of a TGD σ_2.

Recursive Algorithm 2 starts with computation of a partial solution F using GET-NEXTSOLUTION function. This function returns a set of grounded facts computed for

---

Algorithm 2. Solve

---

input : A program Σ

output: A set of atoms representing the best solution of a problem instance

1 $Sol \leftarrow \emptyset$;

2 **loop**

3 $\quad\quad F \leftarrow$ GETNEXTSOLUTION(Σ);

4 $\quad\quad$ **if** $F = \emptyset$ **then break** $EF \leftarrow$ GETEXISTENTIALFACTS$(\Sigma \cup F)$;

5 $\quad\quad$ **if** $EF \neq \emptyset$ **then**

6 $\quad\quad\quad\quad \mathcal{SN} \leftarrow \emptyset$;

7 $\quad\quad\quad\quad$ **foreach** $\langle \mathbf{\Psi}_i, S_i \rangle \in EF$ **do**

8 $\quad\quad\quad\quad\quad\quad \mathbf{CS} \leftarrow$ GETCONFLICTSETS$(\mathbf{\Psi}_i, \Sigma \cup F)$;

9 $\quad\quad\quad\quad\quad\quad \mathcal{SN} \leftarrow \mathcal{SN} \cup$ GENERATEEXTENSIONS$(\mathbf{CS}, \mathbf{\Psi}_i, S_i, \Sigma)$;

10 $\quad\quad\quad\quad \Sigma' \leftarrow$ REWRITE$(\mathcal{SN}, \Sigma \cup F)$;

11 $\quad\quad\quad\quad PS \leftarrow$ SOLVE(Σ');

12 $\quad\quad\quad\quad Sol \leftarrow Sol \cup \{PS \cup F\}$;

13 $\quad\quad$ **else** $Sol \leftarrow Sol \cup \{F\}$

14 **return** GETBESTSOLUTION(Sol);

---

a (rewritten) program excluding all TGDs of form (1) with existential quantifiers. In the first iteration the function returns all facts that are derivable from $edb(\Sigma)$ without TGDs including existential quantification. In the next iteration, it returns a set of grounded facts $\{\Psi(\theta'_1(X), \theta'_1(Z)), \ldots, \Psi(\theta'_n(X), \theta'_n(Z))\}$ computed for TGDs justified by the facts derived in the previous iteration, etc. If the next partial solution F is empty, then the algorithm exits the loop and returns the best of the solutions stored in Sol. Otherwise, GETEXISTENTIALFACTS retrieves a set EF of tuples $\langle \mathbf{\Psi}_i, S_i \rangle$ from Σ, where all $\mathbf{\Psi}_i$ are justified by the facts in F. In addition, the function selects only those existential facts Ψ_i that are generated by new substitutions of the universal variables. By "new" we mean those which were not used for rewriting in previous iterations.

For each tuple $\langle \mathbf{\Psi}_i, S_i \rangle$, comprising a set of existential facts $\mathbf{\Psi}_i$ and a sort S_i of existentially quantified variables occurring in $\mathbf{\Psi}_i$, we compute a set of conflicts according to the Definition 1. Given the set of conflicts $\mathbf{CS}$ Algorithm 1 generates the set of extensions $\mathcal{SN}$ and the target TGDs are then rewritten to obtain Σ', as described above. Next, the algorithm calls itself providing the extended program as an input. The facts F, added on the rewriting step (REWRITE), justify the bodies of TDGs that were not justified on the previous iteration, thus allowing us to apply the rewriting one more time. The program continues the search in a depth-first order until all solutions are enumerated and returns the best one in case preference criteria are defined by a user.

When the rewriting method is applied to recursive TGDs we guarantee its termination because the lower and upper bounds on the number of constants corresponding to each component are finite and determined prior to execution of the Algorithm 2. To rewrite a recursive TDG we have to take into account all constants of a sort that are already appearing in a partial solution. These constants must be added to the initial set D^Z by the function GETCONSTANTS (see Algorithm 1).

The algorithm allows the identification of a solution by executing its fast variant in which we investigate only the left-most branch in the search tree, i.e. it selects one of the optimal partial solutions on each step and continues until the whole model is generated. This greedy approach is important in practice since users often want to test the model they are developing and require a solving method which responds quickly.

3   Implementation

The method suggested in the paper can be applied to knowledge bases defined using knowledge representation languages such as LoCo [1], thus allowing application of modern solvers of combinatorial search problems like SAT, answer set programming or constraints. We ensure the finiteness of the logical models by bounding the number of generated nulls depending on the user input. The rewriting approach is implemented using the translation to ASP as suggested in [10]. Following this methodology each TGD is translated to a choice rule [19] of the form:

$$lower \, \{\Psi(X, Z) : domain(Z)\} \, upper : - \, \Phi(X). \tag{4}$$

where $domain(Z)$ is a domain for the existentially quantified variables, which contains as many constants as the corresponding set D^Z. An operator ":" generates a set of atoms by substituting all possible constants appearing in the atoms with the $domain$ predicate. The ASP semantics [12] ensures that at least $lower$ and at most $upper$ atoms from the set of atoms defined in the head of a choice rule are elements of each answer set. The latter is a set of atoms justified by a program with respect to the ASP semantics [3]. The implementation is done using Potassco system [11] which is the winner of the last ASP competition.

Example 3. The sample program Σ presented in Example 1 is encoded as follows:

$thing(t1).   thing(t2).   thing(t3).   domain(c1).   domain(c2).   domain(c3).$
$1 \, \{t2c(X, Z) : domain(Z)\} \, 1 : - thing(X).$
$: - t2c(t1, X), t2c(t2, X).$

Application of Gringo, which is the Potassco grounder, expands the choice rule and the constraints into three pairs of rules such as:

$$1 \, \{t2c(t1, c1), t2c(t1, c2), t2c(t1, c3)\} \, 1 : - thing(t1).$$
$$: - t2c(t1, c1), t2c(t2, c1).$$

A sample answer set justified by the grounded program includes all facts given in $edb(\Sigma)$ and a set of atoms $\{t2c(t1, c1), t2c(t2, c2), t2c(t3, c1)\}$. A solver identifies 18 possible answer sets for this problem.

Minimization of the number of nulls, i.e. constants defined in the domain, used in the solution is very important. In problems like configuration this corresponds to the minimization of the number of components used in the system, thus reducing the costs of a product. In our case, minimization can be done using the optimization algorithms available in ASP systems. We can add the following rules to model this in Potassco:

$$usedNull(Z) :\text{-} t2c(X, Z).$$
$$\#minimize \{usedNull(Z)\}.$$

The latter rule specifies a preference criterion which forces a solver to return only answer sets including a minimal number of atoms from the set. In our example, the preferred answer set is the one which includes a minimal number of elements from the set $\{usedNull(c1), usedNull(c2), usedNull(c3)\}$. Given such preference criteria a solver returns only 12 answer sets.

The rewriting program executes the algorithms presented in Section 2. The identification of conflict sets is implemented using Definition 1. We remove from the grounded program all choice rules and add pairs of facts which represent a potential conflict set. If the resulting program is unsatisfiable then the pair is a conflict set. Note, that the translated ASP program without choice rules includes only Horn clauses (clauses with at most one positive literal). Completeness of the set of conflicts **CS** is not required. Taking into account that there are at most $n(n-1)/2$ possible pairs of atoms, the identification of conflicts can be done in polynomial time, i.e. $O(n^2)$.

In addition, performance of the solver can be improved by introduction of the ordering on the sets of fresh nulls. For the original program we specify a set of rules defining lexicographical ordering of all constants in a domain. Applied to Example 1 the ordering rules force a solver to use cabinets corresponding to lexicographically smaller constants first. We define ordering rules for nulls which are used in the rewritten program associated with each set of non-conflicting atoms, i.e. such atoms of any facts Ψ_i and Ψ_j for which $D_i^Z \cap D_j^Z \neq \emptyset$. The order is defined by the number of occurrences of a null in tuples of the set SN. The more frequently the null is used the higher is its order. In Example 2 for set $\{t_1, t_2\}$ the set of nulls is ordered as $\{\varphi_1, \varphi_2\}$ since φ_1 is used twice and φ_2 only once. Such ordering requires a solver to substitute φ_1 first.

4   Evaluation

We evaluated our approach on a slightly modified version of the House configuration problem proposed in [17]. The problem is an abstraction of various configuration problems occurring in practice of Siemens, where entities may be contained in other entities and several requirements and constraints in and between entities must be fulfilled. Namely, we considered the modification of this problem studied and evaluated in [10,1]. The benchmarks[1] correspond to a reconfiguration problem which includes knowledge describing a legacy solution (previous solution which has to be reconciled with new requirements) used for finding a reconfiguration solution. It is impossible to compare our results with the results provided in [10] since the presence of a legacy solution in some cases simplifies the solving, e.g. in problem instances *Newroom*. Ignoring the legacy knowledge we get a set of configuration problem instances which were evaluated in this paper using the plain ASP encoding without the rewriting and the rewriting

---

[1] http://proserver3-iwas.uni-klu.ac.at/reconcile/index.php/benchmarks

program. Moreover, from four types of configuration problems described in [10] we present the evaluation results for *Empty*, *Long* and *Newroom* instances. As it turned out during the experiments, *Swap* does not contain the binary conflicts necessary for an efficient application of the conflict-based rewriting and only the common encoding can be applied. The overhead for the rewriting of the hardest *Swap* instance is about 3 minutes which is rather small in comparison to the overall solution time.

A knowledge base for a House problem instance comprises objects such as person, thing, cabinet, and room, where things can be long or short and cabinets high or small. The number of cabinets and rooms generated for each problem equals the number of things such that a fresh constant can be used in each relation. The configuration problem is to place these things into cabinets and the cabinets into rooms such that the following configuration requirements are satisfied:

- each thing must be stored in exactly one cabinet;
- a cabinet can contain at most 5 things;
- every cabinet must be placed in exactly one room;
- a room can contain at most 4 cabinets;
- a person can own any number of rooms;
- each room belongs to a person;
- a room may only contain cabinets storing things of the owner of the room;
- a long thing can only be put into a high cabinet;
- a small cabinet occupies 1 and a high cabinet 2 of 4 positions available in a room;
- a solution with minimal number of high cabinets, cabinets and rooms is preferred.

The evaluation experiments were performed using the grounder Gringo v. 3.0.4 and the solver Clasp v. 2.0.6 (default portfolio options) from Potassco ASP collection[2] [11] on a system with Intel i7-3930K CPU (3.20GHz), 32Gb of RAM and running Ubuntu 11.10. Note that the method is not only restricted to the mentioned ASP system. Usage of systems such as DLV [16] is possible with some modifications in the translation, since choice rules are not supported by its knowledge representation language. The algorithms presented in Section 2 as well as supporting methods, such as finding of justified choice rules or their modifications, were implemented in Java. In our evaluation we tested the algorithm allowing the backtracking only in a situation when a set partial solution cannot be extended. That is, any extension results in an unsatisfiable program due to non-binary constraints. The algorithm quits as soon as the first solution is found.

The set of benchmark problems comprised 24 configuration instances including 8 instances of each type. The name of an instance indicates the number of persons and things declared in an EDB. The size of the grounded program depending on an EDB varies from 0.6 to 620 Mb with an average of 133 Mb (Lparse format). The ASP program corresponding to the House problem included a set of rules expressing the requirements given above. The costs of a solution were determined as a number of generated components used in a solution.

We compared the suggested rewriting approach with an ASP program without rewriting referred to as *Rewriting* and *Original* respectively. We measured execution time and

---

[2] http://potassco.sourceforge.net/

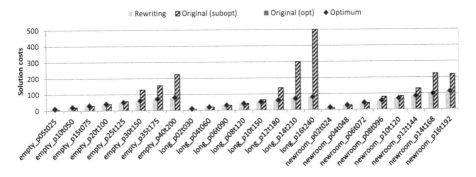

Fig. 1. Quality of solutions identified by Original and Rewriting in 900 seconds

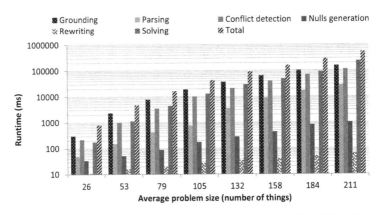

Fig. 2. Time to find a solution using three optimization criteria and Rewriting

optimality of the solution found within 900 seconds. In both cases we used the ordering rules as defined in Section 2.

The evaluation showed that the solver Clasp was able to find solutions with the minimal costs, i.e. minimum number of used nulls (high cabinets/cabinets/rooms) for the 10 smallest cases presented in Fig. 1 using *Original*. Among them it was able to prove the optimality only for three smallest instances `empty_p05t025`, `long_p02t030`, `newroom_p02t024` in 0.3, 1.2 and 103 seconds respectively which are presented by *Original (opt)*. Generally, the solver had a problem with proving optimality of a solution given *Original* encoding and in most of the cases the timeout was reached – *Original (subopt)*. In opposite, *Rewriting* was able to find a solution with optimal costs for all considered cases requiring 640 seconds for the hardest instance `long_p16t240`. Additionally, we did an experiment running the solver on the instance `long_p16t240` for 3 hours with *Original* and the solver still failed to improve the costs of the preferred solution. Thus, *Rewriting* finds significantly better solutions in less time than *Original*.

Fig. 2 presents the overall runtime of the *Rewriting* solving the House configuration problem including three optimization criteria, i.e. minimizing the number of high cabinets, cabinets and rooms. The total runtime was divided into the following processing stages: grounding, parsing, conflict detection, nulls generation, rewriting and solving.

All Empty, Long and Newroom instances are grouped depending on the average number of things. We took the number of things to represent the instances and their average time to illustrate the results. Here we provide time segments for all instances and a total time using *Rewriting*. As the diagram shows, the hardest instance was solved in about 640 seconds including 225 seconds for grounding, 170 seconds for the conflict detection and 205 seconds for solving. The remaining 40 seconds were used for parsing, generation of nulls and rewriting. The rewriting takes very little time, 79 milliseconds for the mentioned test instance. This makes it invisible on the diagram compared with other stages for the smallest test instances.

5   Conclusions and Future Work

The evaluation results demonstrate the convincing superiority of the rewriting approach over the common (original) encoding using state-of-the art solvers for instances relevant to practice. Modifying the existential rules by elimination of the problems' binary conflicts allows to speed up computations and strongly enhance the quality of returned solutions. However, a tight integration of the method into an ASP solver will improve the performance even more. Nevertheless, we demonstrated a significant speed up for a typical class of configuration problems occurring in practice. From the application point of view we are going to extend our method to a reconfiguration case, when a legacy configuration has to be reconciled to meet new requirements.

Acknowledgments. This research has been funded by the Austrian Research Promotion Agency (grant numbers: 825071 and 840242).The authors would like to thank all anonymous reviewers for their comments and especially Conrad Drescher for the discussions regarding LoCo.

References

1. Aschinger, M., Drescher, C., Vollmer, H.: LoCo – A Logic for Configuration Problems. In: Proceedings of the 20th European Conference on Artificial Intelligence, pp. 73–78 (2012)
2. Baader, F., Calvanese, D., McGuinness, D., Nardi, D., Patel-Schneider, P. (eds.): The Description Logic Handbook. Cambridge University Press (2010)
3. Brewka, G., Eiter, T., Truszczynski, M.: Answer set programming at a glance. Communications of the ACM 54(12), 92–103 (2011)
4. Calì, A., Gottlob, G., Lukasiewicz, T., Pieris, A.: Datalog+/-: A family of languages for ontology querying. In: Proceedings of the 25th Annual IEEE Symposium on Logic in Computer Science, pp. 351–368 (2010)
5. Drescher, C., Tifrea, O., Walsh, T.: Symmetry-breaking Answer Set Solving. AI Communications 24(2), 177–194 (2011)
6. Fagin, R., Kolaitis, P.G., Popa, L.: Data exchange: getting to the core. ACM Transactions on Database Systems 30(1), 174–210 (2005)
7. Falkner, A., Feinerer, I., Salzer, G., Schenner, G.: Computing Product Configurations via UML and Integer Linear Programming. Journal of Mass Customisation 3(4), 351–367 (2010)
8. Feinerer, I.: Efficient large-scale configuration via integer linear programming. AI for Engineering Design, Analysis and Manufacturing 27, 37–49 (2013)

9. Freuder, E.C.: Eliminating interchangeable values in constraint satisfaction problems. In: Proceedings of the 9th National Conference on Artificial Intelligence, pp. 227–233 (1991)
10. Friedrich, G., Ryabokon, A., Falkner, A.A., Haselböck, A., Schenner, G., Schreiner, H.: (Re) configuration based on model generation. In: Proceedings of the Second Workshop on Logics for Component Configuration, LoCoCo, vol. 65, pp. 26–35 (2011)
11. Gebser, M., Kaufmann, B., Kaminski, R., Ostrowski, M., Schaub, T., Schneider, M.T.: Potassco: The Potsdam Answer Set Solving Collection. AI Communications 24(2), 107–124 (2011)
12. Gebser, M., Kaufmann, B., Schaub, T.: Conflict-driven answer set solving: From theory to practice. Artificial Intelligence 187-188, 52–89 (2012)
13. Gelfond, M., Lifschitz, V.: The stable model semantics for logic programming. In: 5th International Conference and Symposium on Logic Programming, pp. 1070–1080 (1988)
14. Haselböck, A.: Exploiting interchangeabilities in constraint-satisfaction problems. In: Proceedings of the 13th International Joint Conference on Artificial Intelligence, pp. 282–289 (1993)
15. Leone, N., Manna, M., Terracina, G., Veltri, P.: Efficiently Computable Datalog∃ Programs. In: Proceedings of the 13th International Conference on Principles of Knowledge Representation and Reasoning, pp. 13–23 (2012)
16. Leone, N., Pfeifer, G., Faber, W., Eiter, T., Gottlob, G., Perri, S., Scarcello, F.: The DLV system for knowledge representation and reasoning. ACM Transactions on Computational Logic (TOCL) 7(3), 499–562 (2006)
17. Mayer, W., Bettex, M., Stumptner, M., Falkner, A.: On solving complex rack configuration problems using CSP methods. In: Proceedings of the Workshop on Configuration (2009)
18. Ryabokon, A.: Study: Symmetry breaking for ASP. CoRR arXiv:1212.2657 (2012)
19. Simons, P., Niemelä, I., Soininen, T.: Extending and implementing the stable model semantics. Artificial Intelligence 138, 181–234 (2002)
20. Stumptner, M.: An overview of knowledge-based configuration. AI Communications 10(2), 111–125 (1997)

Program Updating by Incremental and Answer Subsumption Tabling

Ari Saptawijaya* and Luís Moniz Pereira

Centro de Inteligência Artificial (CENTRIA), Departamento de Informática
Faculdade de Ciências e Tecnologia, Univ. Nova de Lisboa, 2829-516 Caparica, Portugal
ar.saptawijaya@campus.fct.unl.pt, lmp@fct.unl.pt

Abstract. We propose a novel conceptual approach to program updates implementation that exploits two features of tabling in logic programming (in XSB
Prolog): incremental and answer subsumption tabling. Our approach, EVOLP/R,
is based on the constructs of Evolving Logic Programs (EVOLP), but simplifies it at first by restricting updates to fluents only. Rule updates are nevertheless
achieved via the mechanism of rule name fluents, placed in rules' bodies, permitting to turn rules on or off, through assertions or retractions of their corresponding
unique name fluents. Incremental tabling of fluents allows to automatically maintain – at engine level – the consistency of program states. Answer subsumption of
fluents addresses the frame problem – at engine level – by automatically keeping
track of their latest assertion or retraction. The implementation is detailed here to
the extent that it may be exported to other logic programming tabling systems.

Keywords: logic program updates, incremental tabling, answer subsumption
tabling.

1 Introduction

In this paper we explore the use of state-of-the-art logic programming implementation
techniques to exploit their use in addressing a classical non-monotonic reasoning problem, that of logic program updates, with incidence on representing change, i.e. internal
or self and external or world changes. Our approach, EVOLP/R, follows the paradigm
of Evolving Logic Programs (EVOLP) [1], by adapting its syntax and semantics, but
simplifies it at first by restricting updates to fluents only. This restriction nevertheless
permits rule updates to take place, as long as we know the rules beforehand, i.e. ones not
constructed, learnt, or externally given. To update the program with such known-from-
the-start rules, special fluents that serve as names of rules and identify rules uniquely
are introduced. Such a rule name fluent is placed in the body of a rule to turn the rule
on and off (cf. [2]), this being achieved by asserting or retracting the rule name fluent.

We foster a novel implementation technique to program updates by exploiting Prolog tabling mechanisms, notably two features of XSB Prolog: incremental and answer
subsumption tabling. Incremental tabling of fluents allows to automatically maintain
the consistency of program states, analogously to assumption based truth-maintenance

---

* Affiliated with Fakultas Ilmu Komputer at Universitas Indonesia, Depok, Indonesia.

P. Cabalar and T.C. Son (Eds.): LPNMR 2013, LNAI 8148, pp. 479–484, 2013.

system, due to assertion and retraction of fluents. On the other hand, answer subsumption of fluents allows to address the frame problem by automatically keeping track of their latest assertion or retraction, whether obtained as updated facts or concluded by rules. The employment of these tabling features has profound consequences in modeling agents. It permits separating higher-level declarative representation and reasoning, as a mechanism pertinent to agents, from a world's inbuilt reactive laws of operation. The latter, being of no operational concern to the problem representation level, are relegated to engine-level enacted tabling features. EVOLP/R is realized using a program transformation plus a library of system predicates. The transformation adds some extra information, e.g. timestamps, for internal processing. Rule name fluents are system generated and also added in the transform. System predicates are defined to operate on the transform by combining the usage of incremental and answer subsumption tabling.

We describe the constructs of EVOLP/R (Section 2), detail the implementation technique (Section 3), and discuss related work along with concluding remarks (Section 4).

2   The EVOLP/R Language

For convenience, we represent EVOLP/R programs as propositional Horn theories, by simply adapting EVOLP definitions [1]. Let $\mathcal{K}$ be an arbitrary set of propositional variables. We denote $\tilde{\mathcal{K}}$ as the *extension* of $\mathcal{K}$, and is defined as $\tilde{\mathcal{K}} = \{A : A \in \mathcal{K}\} \cup \{\sim A : A \in \mathcal{K}\}$. Atoms $A \in \mathcal{K}$ and $\sim A$ are called *positive fluents* and *negative fluents*, respectively. As in EVOLP, program updates are enacted by having the reserved predicate $assert/1$ in the head of a rule. We define now the EVOLP/R language and program.

Definition 1. *Let $\tilde{\mathcal{K}}$ be the extension of a set $\mathcal{K}$ of propositional variables. The* EVOLP/R *language $\mathcal{L}$ is defined inductively as follows:*

1. *All propositional atoms in $\tilde{\mathcal{K}}$ are propositional atoms in $\mathcal{L}$.*
2. *If A is a propositional atom in $\mathcal{L}$, then $assert(A)$ is a propositional atom in $\mathcal{L}$.*
3. *If A is a propositional atom in $\mathcal{L}$, then $\sim assert(A)$ is a propositional atom in $\mathcal{L}$.*
4. *If A_0 is a propositional atom in $\mathcal{L}$ and $A_1, \ldots, A_n$, with $n \geq 0$, are literals in $\mathcal{L}$ (i.e. a propositional atom A, or its default negation not A), then $A_0 \leftarrow A_1, \ldots, A_n$ is a rule in $\mathcal{L}$.*
5. *Nothing else is a propositional atom in $\mathcal{L}$.*

An EVOLP/R *program over a language $\mathcal{L}$ is a (possibly infinite) set of rules in $\mathcal{L}$.*

We extend the notion of positive and negative fluents in $\tilde{\mathcal{K}}$ to propositional atoms A and $\sim A$ in $\mathcal{L}$, respectively. They are said to be *complement* each other. When it is clear from the context, we refer both of them as fluents. Retraction of fluent A (or $\sim A$), making it false, is achieved by asserting its complement $\sim A$ (or A, respectively). I.e., no reserved predicate for retraction is needed. Non-monotonicity of a fluent can thus be admitted by asserting its complement, so as to let the latter supervene the former. Observe that the syntax permits embedded assertions of literals, e.g. $assert(assert(a))$, $\sim assert(assert(a))$; the latter being the complement of the former.

By Definition 1, EVOLP/R programs are not generalized logic programs (like in EVOLP), but they nevertheless permit negative fluents in the rules' heads. Indeed, one may view negative fluents as explicit negations, and due to the coherence principle [3], that explicit negation entails default negation, negative fluents obey the principle. Therefore, the two forms of rules' heads, i.e. $assert(not\ A)$ in EVOLP and $assert(\sim A)$ in EVOLP/R, can be treated equivalently. This justification allows the semantics of EVOLP/R to be safely based on that of EVOLP, as long as the paraconsistency of simultaneously having A and $\sim A$ is duly detected and handled, say with integrity constraints or preferences. Note that EVOLP/R restricts updates to fluents only. Nevertheless, rule updates (like in EVOLP) can be achieved, via the mechanism of rule name fluents, placed in rules' bodies, allowing to turn rules on or off, through assertions or retractions of their corresponding unique name fluents.

Like EVOLP, besides the self-evolution of a program, EVOLP/R also allows influence from the outside, either as an observation of fluents that are perceived at some state, or assertion orders of fluents on the evolving program. Different from EVOLP, the outside influence in EVOLP/R, referred as *external updates*, persist by inertia as long as they do not conflict with the more recent values for them. Nevertheless, we may easily define external updates that do not persist by inertia, called *events* in EVOLP, by defining for every atomic event E the rule: $assert(\sim E) \leftarrow E$, i.e. if event E is imposed at some state i, then it is no longer assumed from the next state, i.e. $(i + 1)$, onwards. In other words, E holds at state i only.

3   Implementing EVOLP/R in Tabled Logic Programming

Tabling in logic programming affords reuse of solutions, rather than recomputing them, by maintaining subgoals and their answers (obtained in query evaluation) in a table. In implementing EVOLP/R, we exploit in combination two features of tabling in XSB Prolog [4]: (1) Incremental tabling, which ensures the consistency of answers in tables with all dynamic facts and rules upon which the tables depend, and (2) Answer subsumption, which allows tables to retain only answers that subsume others with respect to some partial order relation. The reader is referred to [5] for the definitions, options, examples and details of both features.

The EVOLP/R implementation consists of a compiled program transformation plus a library of system predicates. The transformation adds information to program clauses: (1) *Timestamp* includes two extra arguments of fluents, i.e. *holds time* (the time when a fluent is true) and *query time* (the time when it is queried), (2) *Rule name* as a special fluent $\$rule(p/n, id_i)$, which identifies rule of predicate p with arity n by its unique name identity id_i, and is introduced in its body, for checking that the rule still holds.

Transformation. Example 1 illustrates the transformation technique and how the extra information figures in the transform (predicates $\$rule$ and $assert$ are written as $\$r$ and as, respectively). In EVOLP/R, the initial timestamp is set at 1, when a program is inserted. Fluent predicates can be defined as facts (extensional) or by rules (intensional).

Extensional fluent instances, like a, are translated into a rule which inertially constrains its validity from its holds time up to query time Q. In Example 1, a holds at the initial time 1. This validity may become superseded by that of the fluent's complement. For rule regulated intensional fluent instances, like b and $as(\sim a)$, unique rule name fluents are introduced and translated just like for extensional fluents (lines 2, 4, 6).

Line 3 shows the translation of rule $b \leftarrow a$. The extra arguments in its head are holds time H of fluent b and the query time Q. Calls to the goals in the body are translated into calls to the system predicate $holds/3$ (defined later). In the transform of $b \leftarrow a$ (line 3), the first goal in its body verifies whether the unique rule name fluent $\$r(b/0, id_1)$ holds within query time Q, in which case its latest holds time (i.e. the latest time up to Q this rule was turned on) H_r is returned. The next goal verifies whether a holds at Q by returning its latest holds time H_a. The validity of b at Q, with its holds time H ($\leq Q$), is thus obtained from the maximum of H_r and H_a (i.e. H is determined by which inertial fluent in its body holds latest), via $max/2$ system predicate.

Rule $as(\sim a) \leftarrow b$ is transformed into two rules: the transform in line 5 is similar to that of rule $b \leftarrow a$, whereas the one in line 7 is derived as the effect of asserting $\sim a$. I.e., the validity of $\sim a$, being queried at time Q, depends on the latest time when its rule was turned on (H_r in 1st goal in the body) and when $as(\sim a)$ took place (H_{as} in 4th goal in the body). The latter goal is considered at a query time Q_{as}, where $1 \leq Q_{as} \leq Q - 1$ (generated recursively via $gen/2$ system predicate), i.e. existential H_{as} is obtained by querying at a time point Q_{as} within $Q - 1$, just before $\sim a$ is queried (at Q). The holds time H ($\leq Q$) of $\sim a$ is thus determined, via $max/2$, between H_r and $H_{as} + 1$ (rather than H_{as}, because $\sim a$ is actually asserted one time step from the time $as(\sim a)$ holds).

Example 1. Program: a. $b \leftarrow a$. $as(\sim a) \leftarrow b$.   transforms into:

1. $a(1, Q)$ $\leftarrow 1 \leq Q$.
2. $\$r(b/0, id_1, 1, Q)$ $\leftarrow 1 \leq Q$.
3. $b(H, Q)$ $\leftarrow holds(\$r(b/0, id_1), H_r, Q),\ holds(a, H_a, Q),$
                                        $max([H_r, H_a], H),\ H \leq Q$.

4. $\$r(as(\sim a/0), id_1, 1, Q) \leftarrow 1 \leq Q$.
5. $as(\sim a, H, Q)$ $\leftarrow holds(\$r(as(\sim a/0), id_1), H_r, Q),\ holds(b, H_b, Q),$
                                        $max([H_r, H_b], H),\ H \leq Q$.

6. $\$r(\sim a/0, id_1, 1, Q)$ $\leftarrow 1 \leq Q$.
7. $\sim a(H, Q)$ $\leftarrow holds(\$r(\sim a/0, id_1), H_r, Q),\ Q'$ is $Q - 1,$
                                        $gen(Q_{as}, Q'),\ holds(as(\sim a), H_{as}, Q_{as}),$
                                        H'_a is $H_{as} + 1,\ max([H_r, H'_a], H),\ H \leq Q$.

Since any fluents occurring in the program may be updated, all fluents and their complements should be declared as dynamic and incremental (in order to benefit from incremental tabling), e.g. `:- dynamic a/2, '~a'/2 as incremental`. Their incremental assertions may influence program states, notably the latest time when they are true, which is maintained in conjunction with answer subsumption tabling.

System Predicates. We first introduce predicate $fluent/3$, i.e. given query time Qt, $fluent(F, Ht, Qt)$ looks for (dynamic) definitions of fluent F, and returns the one with the latest holds time Ht. It makes good combined use of tabling features: (1) Since $fluent/3$ aims at returning only the latest holds time of F, $fluent/3$ can be tabled using answer subsumption on its second argument; and (2) Predicate $fluent/3$ depends on dynamic fluent definitions of F, and this dependency indicates that $fluent/3$ can be tabled incrementally, to avoid abolishing the table each time a Prolog assertion is made and then recomputing from scratch. Consequently, predicate $fluent/3$ is declared as `:- table fluent(_,po('>'/2),_) as incremental`. It is defined as:

$$fluent(F, Ht, Qt) \leftarrow extend(F, [Ht, Qt], F'), \; call(F').$$

where $extend(F, Args, F')$ extends the arguments of fluent F with those in list $Args$ to obtain F'. Since $fluent/3$ enjoys incremental and answer subsumption tabling, it cannot also be dynamic [5]; the latter being delegated to F'.

Example 1 describes how predicate $holds(F, Ht, Qt)$ should be interpreted, i.e. it verifies whether fluent F is true in a given query time Qt, in which case its *latest* holds time Ht is returned. It suggests that $holds/3$ can be defined using $fluent/3$, which provides such latest holds time. But additionally, $holds/3$ has to make sure its fluent complement $\sim F$ does not hold after Ht, in which case F will fail to hold. I.e.,

$$holds(F, Ht, Qt) \leftarrow compl(F, F'), fluent(F, Ht, Qt), fluent(F', Ht', Qt),$$
$$(Ht \neq 0 \rightarrow Ht \geq Ht' \; ; \; fail).$$

where $compl(F, F')$ obtains the fluent complement F' from F. The last goal in the body, i.e. $(Ht \neq 0 \rightarrow Ht \geq Ht' \; ; \; fail)$, specifies the condition for F to successfully hold. Observe that this condition requires every fluent and its complement to be defined at time 0 (zero), i.e. they are set to true in that special (vacuum) moment in time. This aims to prevent $holds/3$ to fail prematurely in calls to $fluent/3$, which may happen when a fluent or its complement is not defined yet. The condition reads quite straightforward, where only positive timestamps are countenanced, i.e. $Ht \neq 0$ (as they reflect actual time after 0 when a fluent is true): F holds lastly at Ht with respect to query time Qt only if Ht is at least the same as the latest holds time Ht' of $\sim F$. Note that the condition also implicitly covers the case when $\sim F$ is never asserted (i.e. $Ht' = 0$). It also allows paraconsistency (in case $Ht = Ht'$), to be dealt by the user as desired.

Example 2. Recall Example 1, which is loaded initially at time 1. It is easy to verify that query $holds(a, H, 1)$ succeeds with $H = 1$, whereas $holds(a, H, 2)$ fails, but $holds(\sim a, H, 2)$ succeeds with $H = 2$; the latter two persist by inertia. Suppose at time 3, an external update $\{a, \sim\$r(b/0, id_1)\}$ is given. Now, $holds(a, H, 3)$ no longer fails, but succeeds with $H = 3$, because $fluent(a, H, 3)$ succeeds, now with $H = 3$ (instead of with $H = 1$), thanks to incremental tabling (triggered by the external update a) and answer subsumption, whereas $fluent(\sim a, H', 3)$ succeeds with $H' = 2$, and $H \geq H'$. Moreover, due to the external update $\sim\$r(b/0, id_1)$, rule $b \leftarrow a$ is turned off at time 3; consequently $holds(b, H, 3)$ fails (so do $holds(as(\sim a), H, 3)$ and $holds(\sim a, H, 4)$). Thus, a continues to hold at time 4, i.e $holds(a, H, 4)$ succeeds with $H = 3$, onwards.

4   Concluding Remarks

We have proposed EVOLP/R as a simplified EVOLP, by restricting updates to fluents only, for the moment. Rule updates can nevertheless be enacted by introducing a unique rule name fluent to each rule, placed in its body, functioning as a switch to turn the rule on and off. We also showed how incremental tabling is useful to facilitate fluent updates incrementally in dynamic environments and evolving systems (in line with the goals of introducing incremental tabling [6]), and in conjunction with answer subsumption, to avoid recursing through the frame axiom but instead allow direct access to the latest time when a fluent is true.

As a distinct but somewhat similar and complementary approach, we should mention the recent Logic-based Production System with abduction [7], and its successive installments [8], aiming at defining a new encompassing logic-based framework for computing, for knowledge representation and reasoning. It relies on the fundamental role of state transition systems in computing, and involving fluent updates by destructive assignment. It is implemented in LPA Prolog [9], but no details are given about it. In future, we intend to learn from their results and evolve EVOLP/R towards enabling their higher level constructs and compare implementations. Their approach differs from ours in that it defines a new language and an operational semantics, rather than taking an existing one, and implements it on a commercial Prolog system with no underlying tabling mechanisms.

It is our purpose to combine EVOLP/R with tabled abduction [10], so as to jointly afford abduction and updating in one integrated XSB system by exploiting its tabling features, and to apply the integrated system to abductive moral reasoning (cf. [11, 12]), with updating and argumentation, as a sequel to our ongoing approach to this type of non-monotonic reasoning.

Acknowledgements. We thank David S. Warren for elucidating features of tabling. AS acknowledges the support of FCT/MEC Portugal, grant SFRH/BD/72795/2010.

References

1. Alferes, J.J., Brogi, A., Leite, J.A., Pereira, L.M.: Evolving logic programs. In: Flesca, S., Greco, S., Leone, N., Ianni, G. (eds.) JELIA 2002. LNCS (LNAI), vol. 2424, pp. 50–61. Springer, Heidelberg (2002)
2. Poole, D.L.: A logical framework for default reasoning. Artificial Intelligence 36(1), 27–47 (1988)
3. Alferes, J.J., Pereira, L.M.: Reasoning with Logic Programming. LNCS (LNAI), vol. 1111. Springer, Heidelberg (1996)
4. Swift, T., Warren, D.S.: XSB: Extending Prolog with tabled logic programming. Theory and Practice of Logic Programming 12(1-2), 157–187 (2012)
5. Swift, T., Warren, D.S., Sagonas, K., Freire, J., Rao, P., Cui, B., Johnson, E., de Castro, L., Marques, R.F., Saha, D., Dawson, S., Kifer, M.: The XSB System Version 3.3.x Volume 1: Programmer's Manual (2012)
6. Saha, D.: Incremental Evaluation of Tabled Logic Programs. PhD thesis, SUNY Stony Brook (2006)
7. Kowalski, R., Sadri, F.: Abductive logic programming agents with destructive databases. Annals of Mathematics and Artificial Intelligence 62(1), 129–158 (2011)
8. Kowalski, R., Sadri, F.: Towards a logic-based unifying framework for computing (2013), http://www.doc.ic.ac.uk/~rak/papers/TUF.pdf
9. Logic Programming Associates Ltd.: LPA prolog, http://www.lpa.co.uk/
10. Saptawijaya, A., Pereira, L.M.: Tabled abduction in logic programs. Accepted as Technical Communication at ICLP 2013 (2013), http://centria.di.fct.unl.pt/~lmp/publications/online-papers/tabdual_lp.pdf
11. Pereira, L.M., Saptawijaya, A.: Modelling Morality with Prospective Logic. In: Anderson, M., Anderson, S.L. (eds.) Machine Ethics, pp. 398–421. Cambridge U. P. (2011)
12. Han, T.A., Saptawijaya, A., Pereira, L.M.: Moral reasoning under uncertainty. In: Bjørner, N., Voronkov, A. (eds.) LPAR-18. LNCS, vol. 7180, pp. 212–227. Springer, Heidelberg (2012)

Characterization Theorems for Revision of Logic Programs

Nicolas Schwind and Katsumi Inoue

National Institute of Informatics
2-1-2, Hitotsubashi, Chiyoda-ku, Tokyo 101-8430, Japan
{schwind,inoue}@nii.ac.jp

Abstract. We address the problem of belief revision of logic programs, i.e., how to incorporate to a logic program $\mathcal{P}$ a new logic program $\mathcal{Q}$. Based on the structure of SE interpretations, Delgrande *et al.* [5] adapted the AGM postulates to identify the rational behavior of *generalized* logic program (GLP) revision operators and introduced some specific operators. In this paper, a constructive characterization of *all* rational GLP revision operators is given in terms of an ordering among propositional interpretations with some further conditions specific to SE interpretations. It provides an intuitive, complete procedure for the construction of all rational GLP revision operators and makes easier the comprehension of their semantic properties. In particular, we show that every rational GLP revision operator is derived from a propositional revision operator satisfying the original AGM postulates. Taking advantage of our characterization, we embed the GLP revision operators into structures of Boolean lattices, that allow us to bring to light some potential weaknesses in the adapted AGM postulates. To illustrate our claim, we introduce and characterize axiomatically two specific classes of (rational) GLP revision operators which arguably have a drastic behavior.

1 Introduction

Logic programs (LPs) are well-suited for modeling problems which involve common sense reasoning (e.g., biological networks, diagnosis, planning, etc.) Due to the dynamic nature of our environment, beliefs represented through an LP $\mathcal{P}$ are subject to change, i.e., because one wants to incorporate to it a new LP $\mathcal{Q}$. Since there is no unique, consensual procedure to revise a set of beliefs Alchourrón, Gärdenfors and Makinson [1] introduced a set of desirable principles w.r.t. belief change called *AGM postulates*. Katsuno and Mendelzon [14] adapted them for propositional belief revision and distinguished two kind of change operations, i.e., *revision* and *update* [13] characterized for each one of these change operations by a set of so-called *KM postulates*. Revision consists in incorporating a new information into a database that represents a static world, i.e., new and old beliefs describe the same situation but the new ones are more reliable. In the case of update, the underlying world evolves by the occurence of some events i.e., new and old beliefs describe two different states of the world.

Our interests focus here on the problem of revision of logic programs. Most of works dealing with belief change in logic programming are concerned with update

P. Cabalar and T.C. Son (Eds.): LPNMR 2013, LNAI 8148, pp. 485–498, 2013.

[20,2,8], and they do not lie into the AGM framework, particularly due to their syntactic, rule-based essence. Indeed, given the nonmonotonic nature of LPs the AGM/KM postulates can not be directly applied to logic programs. Still, the notion of *SE interpretations* [19] - initially introduced to characterize the strong equivalence between logic programs [16] - provide a monotonic semantical characterization of LPs. Then, based on these structures, Delgrande *et al.* [5,7] adapted the AGM/KM postulates in the context of logic programming. They proposed several revision operators and investigated their properties w.r.t. the adapted postulates. Their work covered a serious drawback in the field of belief revision in logic programming. However the constructive characterization of *all* rational belief revision operators remains an open issue.

In this paper, we consider the revision of *generalized* logic programs (GLPs), which is a very general form of programs. We provide a characterization theorem for the GLP revision operators, that is, a sound and complete model-theoretic construction of the rational LP revision operators (i.e., those which fully satisfy the adaptation of AGM postulates to LPs). Interestingly, our result shows that every rational LP revision operator is derived from a rational propositional revision operator (i.e., satisfying the KM postulates in the propositional setting). Our characterization makes easier the refined analysis of LP revision operators. Indeed, we can embed the GLP revision operators into structures of Boolean lattices, that allows us to bring out some potential weaknesses in the original postulates and pave the way for the discrimination of some rational GLP revision operators.

The next section introduces some preliminaries about belief revision in propositional logic and some necessary background on answer-set programs. Section 3 introduces the LP revision operators and some preliminary results. Section 4 provides the characterization theorem for GLP revision operators. In Section 5 we partition the class of GLP revision operators into subclasses of Boolean lattices, then we introduce and characterize axiomatically two specific classes of (rational) GLP revision operators, i.e., the *skeptical* and *brave* GLP revision operators. We conclude in Section 6 and propose some perspectives for further work. For space reasons, only proof sketches of some propositions are provided in an appendix.

2   Preliminaries

We consider a propositional language $\mathcal{L}$ defined from a finite set of propositional variables (also called *atoms*) $\mathcal{A}$ and the usual connectives. $\bot$ (resp. $\top$) is the Boolean constant always false (resp. true). An *interpretation* over $\mathcal{A}$ is a total function from $\mathcal{A}$ to $\{0,1\}$. To avoid heavy expressions, an interpretation I is also viewed as the subset of atoms from $\mathcal{A}$ that are true in I. For instance, if $\mathcal{A} = \{p,q\}$, then the interpretation over $\mathcal{A}$ such that $I(p) = 1$ and $I(q) = 0$ is also represented as the set $\{p\}$. The set of all interpretations is denoted Ω. An interpretation I is a *model* of a formula $\phi \in \mathcal{L}$ iff it makes it true in the usual truth functional way. A *consistent* formula is a formula that admits a model. $mod(\phi)$ denotes the set of models of formula ϕ, i.e., $mod(\phi) = \{I \in \Omega \mid I \models \phi\}$.

2.1   Belief Revision in Propositional Logic

This section introduces some background on propositional belief revision. Basically, a revision operator $\circ$ is a mapping associating two formulae ϕ, ψ with a new formula, denoted $\phi \circ \psi$. The AGM framework [1] describes the standard principles for belief revision (e.g., consistency preservation and minimality of change), which capture changes occuring in a static domain. Katsuno and Mendelzon [13] equivalently rephrased the AGM postulates as follows:

Definition 1 (KM revision operator). *A KM revision operator $\circ$ is a propositional revision operator that satisfies the following postulates, for all formulae $\phi, \phi_1, \phi_2, \psi, \psi_1, \psi_2$:*

(R1) $\phi \circ \psi \models \psi$;
(R2) *If $\phi \wedge \psi$ is consistent, then $\phi \circ \psi \equiv \phi \wedge \psi$;*
(R3) *If ψ is consistent, then $\phi \circ \psi$ is consistent;*
(R4) *If $\phi_1 \equiv \phi_2$ and $\psi_1 \equiv \psi_2$, then $\phi_1 \circ \psi_1 \equiv \phi_2 \circ \psi_2$;*
(R5) $(\phi \circ \psi_1) \wedge \psi_2 \models \phi \circ (\psi_1 \wedge \psi_2)$;
(R6) *If $(\phi \circ \psi_1) \wedge \psi_2$ is consistent, then $\phi \circ (\psi_1 \wedge \psi_2) \models (\phi \circ \psi_1) \wedge \psi_2$.*

These so-called *KM postulates* capture the desired behavior of a revision operator, e.g., in terms of consistency preservation and minimality of change.

KM revision operators can be characterized in terms of total preorders over interpretations. Indeed, each KM revision operator corresponds to a faithful assignment [13]:

Definition 2 (Faithful assignment). *A faithful assignment is a mapping which associates with every formula ϕ a preorder $\leq_\phi$ over interpretations[1] such that for all interpretations I, J and all formulae ϕ, ϕ_1, ϕ_2, the following conditions hold:*

(a) *If $I \models \phi$ and $J \models \phi$, then $I \simeq_\phi J$;*
(b) *If $I \models \phi$ and $J \not\models \phi$, then $I <_\phi J$;*
(c) *If $\phi_1 \equiv \phi_2$, then $\leq_{\phi_1} = \leq_{\phi_2}$.*

Theorem 1 ([14]). *A revision operator $\circ$ is a KM revision operator if and only if there exists a faithful assignment associating every formula ϕ with a total preorder $\leq_\phi$ such that for all formulae ϕ, ψ, $mod(\phi \circ \psi) = \min(mod(\psi), \leq_\phi)$.*

KM revision operators include the class of distance-based revision operators (see, for instance, [4]), i.e., those operators characterized by a distance between interpretations:

Definition 3 (Distance-based revision operator). *Let d be a distance between interpretations[2], extended to a distance between every interpretation I and*

---

[1] For each preorder $\leq_\phi$, $\simeq_\phi$ denotes the corresponding indifference relation and $<_\phi$ the corresponding strict ordering.

[2] Actually, a pseudo-distance is enough, i.e., triangular inequality is not mandatory.

every formula ϕ by $d(I, \phi) = min\{d(I, J) \mid J \models \phi\}$ if ϕ is consistent, 0 otherwise. The distance-based revision operator $\circ^d$ is defined for all formulae ϕ, ψ by $mod(\phi \circ^d \psi) = min(mod(\psi), \leq^d_\phi)$ where the preorder $\leq^d_\phi$ induced by ϕ is defined for all interpretations I, J by $I \leq^d_\phi J$ iff $d(I, \phi) \leq d(J, \phi)$.

Theorem 2. *Every distance-based revision operator is a KM revision operator, i.e., it satisfies the postulates (R1 - R6).*

Usual distances are d_D, the drastic distance ($d_D(I, J) = 1$ iff $I \neq J$), and d_H the Hamming distance ($d_H(I, J) = n$ if I and J differ on n variables). Noteworthy, the faithful assignment corresponding to the revision operator based on the drastic distance d_D (so-called drastic revision operator) associates with every formula a (unique) two-level preorder:

Definition 4 (Drastic revision operator). *The drastic revision operator, denoted $\circ_D$, is the revision operator based on the drastic distance.*

Likewise, the revision operator based on Hamming distance corresponds to the well-known Dalal revision operator [4]:

Definition 5 (Dalal revision operator). *The Dalal revision operator, denoted $\circ_{Dal}$, is the revision operator based on the Hamming distance.*

2.2 Logic Programming

In this section, we define the syntax and semantics of generalized logic programs. We use the same notations as in [5]. A *generalized logic program* (GLP) is a finite set of rules of the form

$$a_1; \ldots; a_k; \sim b_1; \ldots; \sim b_l \leftarrow c_1, \ldots, c_m, \sim d_1, \ldots, \sim d_n,$$

where $k, l, m, n \geq 0$.

Each a_i, b_i, c_i, d_i is either one of the constant symbols $\bot, \top$, or an atom from $\mathcal{A}$; $\sim$ is the negation by failure; "$;$" is the disjunctive connective, "$,$" is the conjunctive connective of atoms. The right-hand and left-hand sides of r are respectively called the head and body of r. For each rule r, we define $H(r)^+ = \{a_1, \ldots, a_k\}$, $H(r)^- = \{b_1, \ldots, b_l\}$, $B(r)^+ = \{c_1, \ldots, c_m\}$, and $B(r)^- = \{d_1, \ldots, d_n\}$. For the sake of simplicity, a rule r is also expressed as follows:

$$H(r)^+; \sim H(r)^- \leftarrow B(r)^+, \sim B(r)^-.$$

A logic program is interpreted through its preferred models based on the answer set semantics. A *(classical) model* X of a GLP $\mathcal{P}$ (written $X \models \mathcal{P}$) is an interpretation from Ω that satisfies all rules from $\mathcal{P}$ according to the classical definition of truth in propositional logic. $mod(\mathcal{P})$ will denote the set of all models of a GLP $\mathcal{P}$. An *answer set* X of a GLP $\mathcal{P}$ is a minimal (w.r.t. set inclusion) set of atoms from $\mathcal{A}$ that is a model of the program $\mathcal{P}^X$, where $\mathcal{P}^X$ is called the *reduct* of $\mathcal{P}$ relative to X and is defined as $\mathcal{P}^X = \{H(r)^+ \leftarrow B(r)^+ \mid r \in \mathcal{P}, H(r)^- \subseteq X, B(r)^- \cap X = \emptyset\}$. The classical notion of equivalence between programs corresponds to the correspondence of their answer sets.

SE interpretations are semantic structures characterizing *strong equivalence* between logic programs [19], they provide a monotonic semantic foundation of logic programs under answer set semantics. An SE interpretation over $\mathcal{A}$ is a pair (X, Y) of interpretations over $\mathcal{A}$ such that $X \subseteq Y$. An *SE model* (X, Y) of a logic program $\mathcal{P}$ is an SE interpretation over $\mathcal{A}$ that satisfies $Y \models \mathcal{P}$ and $X \models \mathcal{P}^Y$, where $\mathcal{P}^Y$ is the reduct of $\mathcal{P}$ relative to Y. For the sake of simplicity, set-notations will be dropped within SE interpretations, e.g., the SE interpretation $(\{p\}, \{p, q\})$ will be simply denoted (p, pq). Through their SE models, logic programs are semantically described in a stronger way than through their answer sets, as shown in the following example which belongs to [5]:

Example 1. Let $\mathcal{P} = \{p; q \leftarrow \top\}$ and $\mathcal{Q} = \{p \leftarrow\sim q, q \leftarrow\sim p\}$. Then $AS(\mathcal{P}) = AS(\mathcal{Q}) = \{\{p\}, \{q\}\}$, that is, they admit the same answer sets, however their SE models differ: $SE(\mathcal{P}) = \{(p, p), (q, q), (p, pq), (q, pq), (pq, pq)\}$, while $SE(\mathcal{Q}) = \{(p, p), (q, q), (p, pq), (q, pq), (pq, pq), (\emptyset, pq)\}$.

A program $\mathcal{P}$ is *consistent* if $SE(P) \neq \emptyset$. Two programs $\mathcal{P}$ and $\mathcal{Q}$ are said to be *strongly equivalent*, denoted $\mathcal{P} \equiv_s \mathcal{Q}$, whenever $SE(\mathcal{P}) = SE(\mathcal{Q})$. We also write $\mathcal{P} \subseteq_s \mathcal{Q}$ if $SE(\mathcal{P}) \subseteq SE(\mathcal{Q})$. Two programs are equivalent if they are strongly equivalent, but the other direction does not hold in general. Note that Y is an answer set of $\mathcal{P}$ iff $(Y, Y) \in SE(\mathcal{P})$ and no $(X, Y) \in SE(\mathcal{P})$ with $X \subset Y$ exists. We also have $(Y, Y) \in SE(\mathcal{P})$ iff $Y \in mod(\mathcal{P})$. A set of SE interpretations S is *well-defined* if for every interpretation X, Y with $X \subseteq Y$, if $(X, Y) \in S$ then $(Y, Y) \in S$. Every GLP has a well-defined set of SE models. Moreoover, from every well-defined set S of SE models, one can build a GLP P such that $SE(P) = S$ [10,3].

3   Logic Program Revision Operators

Given the nonmonotonic nature of answer-set programs, Delgrande *et al.* [5] pointed out that the rational behavior of revision operators for logic programs cannot be expressed using the original KM postulates (cf. Definition 1). Therefore, they proposed an adaptation of these postulates in the context of logic programming using the characterization of logic programs through their SE models. To this end, they first defined the operation of *expansion* of two logic programs:

Definition 6 (Expansion operator [5]). *Given two programs $\mathcal{P}, \mathcal{Q}$, the expansion of $\mathcal{P}$ by $\mathcal{Q}$, denoted $\mathcal{P} + \mathcal{Q}$ is any program $\mathcal{R}$ such that $SE(\mathcal{R}) = SE(\mathcal{P}) \cap SE(\mathcal{Q})$.*

Though the expansion of logic programs trivializes the result whenever the two input logic programs admit no common SE models, this operation is of interest in its own right. Indeed, it has be shown that if $\mathcal{P}$ and $\mathcal{Q}$ are GLPs then there exists a construction of a logic program $\mathcal{P} + \mathcal{Q}$ that is also a GLP [6].

Expansion of programs corresponds to the model-theoretical definition of expansion expressed through KM postulates. Delgrande *et al.* rephrased the full set of KM postulates in the context of GLPs. Beforehand, we define a logic program revision operator as a simple function, that considers two GLPs (the original one and the new one) and returns a revised GLP:

Definition 7 (LP revision operator). *A* LP revision operator $\star$ *is a mapping associating two GLPs* $\mathcal{P}, \mathcal{Q}$ *with a new GLP, denoted* $\mathcal{P} \star \mathcal{Q}$.

Definition 8 (GLP revision operator [5]). *A* GLP revision operator $*$ *is an LP revision operator that satisfies the following postulates, for all GLPs* $\mathcal{P}, \mathcal{P}_1, \mathcal{P}_2, \mathcal{Q}, \mathcal{Q}_1, \mathcal{Q}_2, \mathcal{R}$:

(RA1) $\mathcal{P} * \mathcal{Q} \subseteq_s \mathcal{Q}$;
(RA2) *If* $\mathcal{P} + \mathcal{Q}$ *is consistent, then* $\mathcal{P} * \mathcal{Q} \equiv_s \mathcal{P} + \mathcal{Q}$;
(RA3) *If* $\mathcal{Q}$ *is consistent, then* $\mathcal{P} * \mathcal{Q}$ *is consistent;*
(RA4) *If* $\mathcal{P}_1 \equiv_s \mathcal{P}_2$ *and* $\mathcal{Q}_1 \equiv_s \mathcal{Q}_2$, *then* $\mathcal{P}_1 * \mathcal{Q}_1 \equiv \mathcal{P}_2 * \mathcal{Q}_2$;
(RA5) $(\mathcal{P} * \mathcal{Q}) + \mathcal{R} \subseteq_s \mathcal{P} * (\mathcal{Q} + \mathcal{R})$;
(RA6) *If* $(\mathcal{P} * \mathcal{Q}) + \mathcal{R}$ *is consistent, then* $\mathcal{P} * (\mathcal{Q} + \mathcal{R}) \subseteq_s (\mathcal{P} * \mathcal{Q}) + \mathcal{R}$.

Delgrande *et al.* proposed in [5] a specific revision operator that is inspired from Satoh's propositional revision operator [18], i.e., it is based on the set containment of SE interpretations. This operator satisfies postulates (RA1 - RA5). Though it seems to have a good behavior on some instances, this operator does not satisfy (RA6), so that it does not fully respect the principle of minimality of change (see [12], Section 3.1 for details on this postulate). However, the whole set of postulates is consistent, as they later introduce the so-called *cardinality-based revision operator* [6] that reduces to the Dalal revision operator over propositional models[3], and that satisfies all the postulates (RA1 - RA6):

Definition 9 (Cardinality-based revision operator). *Given a program* $\mathcal{P}$, *let* $\phi_{\mathcal{P}}, \psi_{\mathcal{P}}, \psi_{\mathcal{Q}}, \alpha_{(\mathcal{P},\mathcal{Q})}$ *be propositional formulae satisfying* $mod(\phi_{\mathcal{P}}) = \{X \mid (X,Y) \in SE(\mathcal{P})\}$, $mod(\psi_{\mathcal{P}}) = mod(\mathcal{P})$, $mod(\psi_{\mathcal{Q}}) = mod(\mathcal{Q})$ *and* $mod(\alpha_{(\mathcal{P},\mathcal{Q})}) = \{X \mid (X,Y) \in SE(\mathcal{Q}), Y \models \psi_{\mathcal{P}} \circ_{Dal} \psi_{\mathcal{Q}}\}$. *The cardinality-based operator, denoted* $\star_c$, *is defined for all programs* $\mathcal{P}, \mathcal{Q}$ *by* $SE(\mathcal{P} \star_c \mathcal{Q}) = \{(X,Y) \mid Y \models \psi_{\mathcal{P}} \circ_{Dal} \psi_{\mathcal{Q}}, X \models \phi_{\mathcal{P}} \circ_{Dal} \alpha_{(\mathcal{P},\mathcal{Q})}\}\}$.

Theorem 3 ([6]). $\star_c$ *is a GLP revision operator.*

In addition, we introduce below a simple LP revision operator which also satisfies the whole set of postulates (RA1 - RA6):

Definition 10 (Drastic LP revision operator). *The* drastic GLP revision operator $*_D$ *is defined for all GLPs* $\mathcal{P}, \mathcal{Q}$ *as* $\mathcal{P} *_D \mathcal{Q} = \mathcal{P} + \mathcal{Q}$ *if* $\mathcal{P} + \mathcal{Q}$ *is consistent, otherwise* $\mathcal{P} *_D \mathcal{Q} = \mathcal{Q}$.

Proposition 1. $*_D$ *is a GLP revision operator.*

---

[3] This definition is equivalent to the original one introduced in [6], reformulated here for space reasons.

Theorem 3 and Proposition 1 show that postulates (RA1 - RA6) form a consistent set of properties, but it is not known whether there exist more GLP revision operators than the cardinality-based and the drastic LP revision operators. Moreoever, the cardinality-based revision operator has a parsimonious behavior compared to the drastic LP revision operator; however, both are fully satisfactory in terms of revision principles; this raises the problem on how to discard some rational operators from others.

In the next section, we fill the gap and we give a constructive, full characterization of GLP revision operators. This allows us to get a clear, complete picture of the class of GLP revision operators.

4   Characterization of GLP Revision Operators

We now provide the main result of our paper, i.e., a characterization theorem for GLP revision operators. That is, we show that each GLP revision operator (i.e., each LP revision operator satisfying the postulates (RA1 - RA6)) can be characterized in terms of preorders over the set of all classical interpretations, with some further conditions specific to SE interpretations.

Definition 11 (LP faithful assignment). *A LP faithful assignment is a mapping which associates with every GLP $\mathcal{P}$ a preorder $\leq_{\mathcal{P}}$ over interpretations such that for every GLP $\mathcal{P}, \mathcal{Q}$ and every interpretation Y, Y', the following conditions hold:*

(1) *If $Y \models \mathcal{P}$ and $Y' \models \mathcal{P}$, then $Y \simeq_{\mathcal{P}} Y'$;*
(2) *If $Y \models \mathcal{P}$ and $Y' \not\models \mathcal{P}$, then $Y <_{\mathcal{P}} Y'$;*
(3) *If $\mathcal{P} \equiv_s \mathcal{Q}$, then $\leq_{\mathcal{P}} = \leq_{\mathcal{Q}}$.*

Definition 12 (Well-defined assignment). *A well-defined assignment is a pair (Φ, Ψ), where Φ is an LP faithful assignment and Ψ is a mapping which associates with every GLP $\mathcal{P}$ and every interpretation Y a set of interpretations $\Psi(\mathcal{P}, Y)$ (simply denoted $\mathcal{P}(Y)$), such that for all GLPs $\mathcal{P}, \mathcal{Q}$ and all interpretations X, Y, the following conditions hold:*

(a) *$Y \in \mathcal{P}(Y)$;*
(b) *If $X \in \mathcal{P}(Y)$, then $X \subseteq Y$;*
(c) *If $(X, Y) \in SE(\mathcal{P})$, then $X \in \mathcal{P}(Y)$;*
(d) *If $(X, Y) \notin SE(\mathcal{P})$ and $Y \models \mathcal{P}$, then $X \notin \mathcal{P}(Y)$;*
(e) *If $\mathcal{P} \equiv_s \mathcal{Q}$, then $\mathcal{P}(Y) = \mathcal{Q}(Y)$.*

We are ready to bring to light our main result:

Proposition 2. *An operator $\star$ is a GLP revision operator iff there exists a well-defined assignment (Φ, Ψ), where Φ associates with every GLP $\mathcal{P}$ a total preorder $\leq_{\mathcal{P}}$, Ψ associates with every GLP $\mathcal{P}$ and every interpretation Y a set of interpretations $\mathcal{P}(Y)$, such that for all GLPs $\mathcal{P}, \mathcal{Q}$, $SE(\mathcal{P} \star \mathcal{Q}) = \{(X, Y) \mid (X, Y) \in SE(\mathcal{Q}), \forall Y' \models \mathcal{Q}\ Y \leq_{\mathcal{P}} Y', X \in \mathcal{P}(Y)\}$.*

Note that there is no relationship between the mappings Φ, Ψ induced from a well-defined assignment, that is, each one of them can be defined in a completely independent way. Therefore, an interesting consequence from Theorem 1 and Proposition 2 is that every GLP revision operator is an extension of a (propositional) KM revision operator:

Definition 13 (Propositional-based LP revision operator). *Given a program* $\mathcal{P}$, *let* $\psi_{\mathcal{P}}$ *be any propositional formula such that* $mod(\psi_{\mathcal{P}}) = mod(\mathcal{P})$. *Let* $\circ$ *be a propositional revision operator and* f *be a mapping from* Ω *to* 2^{Ω} *such that for every interpretation* Y, $Y \in f(Y)$ *and if* $X \in f(Y)$ *then* $X \subseteq Y$. *The propositional-based LP revision operator w.r.t.* $\circ$ *and* f, *denoted* $\star^{\circ,f}$, *is defined for all GLPs* $\mathcal{P}, \mathcal{Q}$ *by* $SE(\mathcal{P} \star^{\circ,f} \mathcal{Q}) = SE(\mathcal{P} + \mathcal{Q})$ *if* $\mathcal{P} + \mathcal{Q}$ *is consistent, otherwise* $SE(\mathcal{P} \star^{\circ,f} \mathcal{Q}) = \{(X, Y) \mid (X, Y) \in SE(\mathcal{Q}), Y \models \psi_{\mathcal{P}} \circ \psi_{\mathcal{Q}}, X \in f(Y)\}$.
  $\star^{\circ,f}$ *is said to be a propositional-based GLP revision operator if* $\circ$ *is a KM revision operator (i.e., satisfying postulates (R1 - R6)).*

Proposition 3. *The classes of GLP revision operators and propositional-based GLP revision operators coincide.*

For every propositional revision operator $\circ$, let $GLP(\circ)$ denote the set of all propositional-based LP revision operators w.r.t. $\circ$. From Definition 13, it is easy to see that each propositional-based LP revision operator is built from a *unique* propositional revision operator, that is, for all propositional revision operators $\circ_1, \circ_2$, we have $\circ_1 \neq \circ_2$ if and only if $GLP(\circ_1) \cap GLP(\circ_2) = \emptyset$. Therefore, a direct consequence of Proposition 3 is that the class of GLP revision operators can be viewed as the partition $\{GLP(\circ) \mid \circ \text{ is a KM revision operator}\}$. Similarly, for each propositional revision operator $\circ$, for all propositional-based LP revision operators $\star^{\circ,f_1}, \star_2^{\circ,f_2}$, we have $\star^{\circ,f_1} \neq \star_2^{\circ,f_2}$ if and only if $f_1 \neq f_2$.

Note that the cardinality-based revision operator $\star_c$ (cf. Definition 9) corresponds to the propositional-based GLP revision operator $\star^{\circ_{Dal},f}$, where $\circ_{Dal}$ is the Dalal revision operator (cf. Definition 5) and f is defined for every interpretation Y as $f(Y) = \{X \mid X \subseteq Y, \exists Z \models \psi_{\mathcal{P}} \circ_{Dal} \psi_{\mathcal{Q}}, \forall X' \subseteq Y, \forall Z' \models \psi_{\mathcal{P}} \circ_{Dal} \psi_{\mathcal{Q}}, d_H(X, Z) \leq d_H(X', Z')\}$. In addition, the drastic GLP revision operator (cf. Definition 10) corresponds to the propositional-based GLP revision operator $\star^{\circ_D,f}$, where $\circ_D$ is the drastic revision operator (cf. Definition 4) and f is defined for every interpretation Y as $f(Y) = 2^Y$.

Remark that in the case where $\mathcal{P}$ and $\mathcal{Q}$ have no common SE models, then a propositional-based GLP revision operator $\star^{\circ,f}$ gives preference to the second component of SE interpretations, that is driven by the choice of the underlying propositional revision operator $\circ$. However, one can directly see from Definition 13 that the first element of SE interpretations (that is specified using f) is totally unconstrained. We will show in the next section that this "freedom" on the choice of the first component of SE interpretations raises some issues for some subclasses of fully rational LP revision operators.

Our characterization theorem provides an intuitive construction of GLP revision operators and aids the analysis of their semantic properties, as it is illustrated in the next section.

5 GLP Revision Operators Embedded into Boolean Lattices

We now take a closer look to the set of GLP revision operators associated with each given KM revision operator. The characterization theorem provided in the previous section allows us to embed the subclass $GLP(\circ)$, for each KM revision operator $\circ$, into a structure of Boolean lattice[4].

Definition 14. *Let $\circ$ be a propositional revision operator. We define the binary relation $\preceq_\circ$ over $GLP(\circ)$ as follows: for all propositional-based LP revision operators $\star^{\circ,f_1}, \star^{\circ,f_2}$, $\star^{\circ,f_1} \preceq_\circ \star^{\circ,f_2}$ if and only for every interpretation Y, we have $f_2(Y) \subseteq f_1(Y)$.*

It can be easily checked that for each propositional revision operator $\circ$, $(GLP(\circ), \preceq_\circ)$ forms a Boolean lattice, that corresponds to the product of the Boolean lattices $\{(\mathbb{B}_Y, \subseteq) \mid Y \in \Omega\}$, where $\mathbb{B}_Y = \{Z \cup \{Y\} \mid Z \in 2^{2^Y \setminus Y}\}$. The following result shows that this lattice structure can be used to analyse the relative semantic behavior of GLP revision operators from $(GLP(\circ), \preceq_\circ)$.

Proposition 4. *Let $\circ$ be a KM revision operator. Then for all GLP revision operators $\star_1, \star_2 \in GLP(\circ)$, $\star_1 \preceq_\circ \star_2$ if and only if for all GLPs $\mathcal{P}, \mathcal{Q}$, we have $AS(\mathcal{P} \star_1 \mathcal{Q}) \subseteq AS(\mathcal{P} \star_2 \mathcal{Q})$.*

This result paves the way for the choice of a specific GLP revision operator depending on the desired "amount of information" provided by the revised GLP in terms of number of its answer sets. We illustrate this notion by considering two specific classes of GLP revision operators that correspond respectively to the suprema and infima of lattices $(GLP(\circ), \preceq_\circ)$ for all KM revision operators $\circ$.

Definition 15 (Skeptical GLP revision operators). *The skeptical GLP revision operators, denoted $\star_S^\circ$ are the propositional-based GLP revision operators $\star^{\circ,f}$ where f is defined for every interpretation Y by $f(Y) = 2^Y$.*

Definition 16 (Brave GLP revision operators). *The brave GLP revision operators, denoted $\star_B^\circ$ are the propositional-based GLP revision operators $\star^{\circ,f}$ where f is defined for every interpretation Y by $f(Y) = \emptyset$.*

For each propositional revision operator $\circ$, we have $\star_S^\circ = inf(GLP(\circ), \preceq_\circ)$ and $\star_B^\circ = sup(GLP(\circ), \preceq_\circ)$. We now illustrate how much the behavior of skeptical and brave GLP revision operators diverge through the following representative example:

Example 2. Consider $\circ_D$ the propositional drastic revision operator. Let $\mathcal{P} = \{p \leftarrow \top, q \leftarrow \top, \bot \leftarrow r\}$ and $\mathcal{Q} = \{\bot \leftarrow p, q, \sim r\}$. We have $AS(\mathcal{P}) = \{p, q\}$, $AS(\mathcal{Q}) = \{\emptyset\}$, $AS(\mathcal{P} \star_S^{\circ_D} \mathcal{Q}) = \{\emptyset\}$ and $AS(\mathcal{P} \star_B^{\circ_D} \mathcal{Q}) = \{\emptyset, \{p\}, \{q\}, \{r\}, \{pr\}, \{qr\}, \{pqr\}\}$.

---

[4] A Boolean lattice is a partially ordered set $(E, \leq_E)$ which is isomorphic to the set of subsets of some set F together with the usual set-inclusion operation, i.e., $(2^F, \subseteq)$.

We provide an axiomatic characterization of each one of these two subclasses of GLP revision operators in order to get a clearer view of their general behavior. Each characterization theorem below is given in terms of answer sets of the revised program.

Proposition 5. *The skeptical GLP revision operators are the only GLP revision operators $\star$ such that for all GLPs $\mathcal{P}, \mathcal{Q}$, whenever $\mathcal{P} + \mathcal{Q}$ is inconsistent, we have $AS(\mathcal{P} \star \mathcal{Q}) \subseteq AS(\mathcal{Q})$.*

Proposition 6. *Given a program $\mathcal{P}$, let $\psi_{\mathcal{P}}$ be any propositional formula such that $mod(\psi_{\mathcal{P}}) = mod(\mathcal{P})$. The brave GLP revision operators are the only GLP revision operators $\star^{\circ, f}$ such that for all GLPs $\mathcal{P}, \mathcal{Q}$, whenever $\mathcal{P} + \mathcal{Q}$ is inconsistent, we have $AS(\mathcal{P} \star^{\circ, f} \mathcal{Q}) = mod(\psi_{\mathcal{P}} \circ \psi_{\mathcal{Q}})$.*

Remark that the drastic GLP revision operator (cf. Definition 10), i.e., the skeptical GLP revision operator based on the propositional drastic revision operator, is a specific case from the result given in Proposition 5 where $AS(\mathcal{P} \star_S^{\circ_D} \mathcal{Q}) = AS(\mathcal{Q})$ whenever $\mathcal{P} + \mathcal{Q}$ is inconsistent. In addition, the brave GLP revision operator based on the propositional drastic revision operator satisfies $AS(\mathcal{P} \star_B^{\circ_D} \mathcal{Q}) = mod(\mathcal{Q})$ whenever $\mathcal{P} + \mathcal{Q}$ is inconsistent.

Though they are rational LP revision operators w.r.t. the postulates (RA1 - RA6), skeptical and brave operators have a rather trivial, thus undesirable behavior. On the one hand, consider where p is believed to be true, then learned to be false. That is, $\{\bot \leftarrow p\} \subseteq \mathcal{P}$ and $\mathcal{Q} = \{p \leftarrow \top\}$. Then one obtains that $AS(\mathcal{P} \star_S^{\circ} \mathcal{Q}) \subseteq AS(\mathcal{Q})$, that is, $AS(\mathcal{P} \star_S^{\circ} \mathcal{Q}) = \{\{p\}\}$, i.e., for any such program $\mathcal{P}$, on learning that p is true the revision states that *only* p is true. On the other hand, brave operators only focus on classical models of logic programs $\mathcal{P}, \mathcal{Q}$ to compute $\mathcal{P} \star_B^{\circ} \mathcal{Q}$ (whenever $\mathcal{P} + \mathcal{Q}$ is inconsistent), thus they do not take into consideration the inherent, non-monotonic behavior of logic programs. As a consequence, programs $\mathcal{P} \star_B^{\circ} \mathcal{Q}$ will often admit many answer sets that are actually irrelevant to the input programs $\mathcal{P}$ and $\mathcal{Q}$. Stated otherwise, skeptical and brave GLP revision operators are dual sides of a "drastic" behavior for the revision. These operators are representative examples that provide some "bounds" of the complete picture of GLP revision operators $GLP(\circ)$, for each KM revision operator $\circ$. Discarding such drastic behaviors may call for additional postulates in order to capture more parsimonious revision procedures in logic programming, as for instance the cardinality-based revision operator (cf. Definition 9) which is neither brave nor skeptical. Stated otherwise, it seems necessary to refine the existing properties that every rational revision operator should satisfy so that the answer sets of the revised program $\mathcal{P} \star^{\circ, f} \mathcal{Q}$ fall "between" these two extremes (i.e., between $AS(\mathcal{Q})$ and $mod(\mathcal{P} \circ \mathcal{Q})$) in the sense of set inclusion.

6   Conclusion and Perspectives

In this paper, we pursued some previous work on revision of logic programs, where the adopted approach is based on a monotonic characterization of logic programs using SE interpretations. We considered the revision of generalized logic programs (GLPs) and characterized the class of rational GLP revision operators in terms of an ordering among classical interpretations with some further conditions specific to SE interpretations. The constructive characterization we provide facilitates the comprehension of their semantic properties by drawing a clear, complete picture of GLP revision operators. Interestingly, we showed that a GLP revision operator is an extension of a rational propositional revision operator, that is, each propositional revision operator corresponds to a specific subclass of GLP revision operators. Moreover, we showed that each one of these subclasses can be embedded into a Boolean lattice, which infimum and supremum, the so-called *skeptical* and *brave* GLP revision operators, have some drastic behavior.

This work can be extended into several directions in belief change theory for logic programming. Our results make easier the improvement of the current AGM framework in the context of logic programming. Indeed, though the subclasses of skeptical and brave GLP revision operators are fully satisfactory w.r.t. the AGM revision principles, their behavior is shown to be rather trivial. This may call for additional postulates which would aim to capture more parsimonious, balanced classes of GLP revision operators. Additionally, we will investigate the case of logic program *merging* operators (merging can be viewed as a multi-source generalization of belief revision). Indeed it is not even known whether there exists a fully rational merging operator, i.e., that satisfies the whole set of postulates proposed by Delgrande *et al.* [7] for logic programs merging operators.

References

1. Alchourrón, C.E., Gärdenfors, P., Makinson, D.: On the logic of theory change: Partial meet contraction and revision functions. Journal of Symbolic Logic 50(2), 510–530 (1985)
2. Alferes, J.J., Leite, J.A., Pereira, L.M., Przymusinska, H., Przymusinski, T.C.: Dynamic updates of non-monotonic knowledge bases. Journal of Logic Programming 45(1-3), 43–70 (2000)
3. Cabalar, P., Ferraris, P.: Propositional theories are strongly equivalent to logic programs. Theory and Practice of Logic Programming 7(6) (2007)
4. Dalal, M.: Investigations into a theory of knowledge base revision: Preliminary report. In: Proc. of the 7th National Conference on Artificial Intelligence, AAAI 1988, pp. 475–479 (1988)
5. Delgrande, J.P., Schaub, T., Tompits, H., Woltran, S.: Belief revision of logic programs under answer set semantics. In: Proc. of the 11th International Conference on Principles of Knowledge Representation and Reasoning, KR 2008, pp. 411–421 (2008)
6. Delgrande, J.P., Schaub, T., Tompits, H., Woltran, S.: A general approach to belief change in answer set programming. CoRR, abs/0912.5511 (2009)
7. Delgrande, J., Schaub, T., Tompits, H., Woltran, S.: Merging logic programs under answer set semantics. In: Hill, P.M., Warren, D.S. (eds.) ICLP 2009. LNCS, vol. 5649, pp. 160–174. Springer, Heidelberg (2009)

8. Eiter, T., Fink, M., Sabbatini, G., Tompits, H.: On properties of update sequences based on causal rejection. Theory and Practice of Logic Programming 2(6), 711–767 (2002)

9. Eiter, T., Fink, M., Tompits, H., Woltran, S.: On eliminating disjunctions in stable logic programming. In: Proc. of the 9th International Conference on Principles of Knowledge Representation and Reasoning, KR 2004, pp. 447–458 (2003)

10. Eiter, T., Tompits, H., Woltran, S.: On solution correspondences in answer set programming. In: Proc. of the 19th International Joint Conference on Artificial Intelligence, IJCAI 2005, pp. 97–102 (2005)

11. Inoue, K., Sakama, C.: Negation as failure in the head. Journal of Logic Programming 35(1), 39–78 (1998)

12. Katsuno, H., Mendelzon, A.O.: A unified view of propositional knowledge base updates. In: Proc. of the 11th International Joint Conference on Artificial Intelligence, IJCAI 1989, pp. 1413–1419 (1989)

13. Katsuno, H., Mendelzon, A.O.: On the difference between updating a knowledge base and revising it. In: Proc. of the 2nd International Conference on Principles of Knowledge Representation and Reasoning, KR 1991, pp. 387–394 (1991)

14. Katsuno, H., Mendelzon, A.O.: Propositional knowledge base revision and minimal change. Artificial Intelligence 52(3), 263–294 (1992)

15. Konieczny, S., Pino Pérez, R.: Merging information under constraints: a logical framework. Journal of Logic and Computation 12(5), 773–808 (2002)

16. Lifschitz, V., Pearce, D., Valverde, A.: Strongly equivalent logic programs. ACM Transactions on Computational Logic 2(4), 526–541 (2001)

17. Lifschitz, V., Woo, T.Y.C.: Answer sets in general nonmonotonic reasoning (preliminary report). In: Proc. of the 3rd International Conference on Principles of Knowledge Representation and Reasoning, KR 1992, pp. 603–614 (1992)

18. Satoh, K.: Nonmonotonic reasoning by minimal belief revision. In: Proc. of the International Conference on Fifth Generation Computer Systems, FGCS 1988, pp. 455–462 (1988)

19. Turner, H.: Strong equivalence made easy: nested expressions and weight constraints. Theory and Practice of Logic Programming 3(4-5), 609–622 (2003)

20. Zhang, Y., Foo, N.Y.: Towards generalized rule-based updates. In: Proc. of the 15th International Joint Conference on Artificial Intelligence, IJCAI 1997, pp. 82–88 (1997)

Appendix: Proof Sketches

Proposition 2

(Only if part) In this proof, for every well-defined set of SE interpretations S, $\mathsf{glp}(S)$ denotes any GLP $\mathcal{P}$ such that $SE(\mathcal{P}) = S$. Let $*$ be a GLP revision operator. For every GLP $\mathcal{P}$, define the relation $\leq_{\mathcal{P}}$ over interpretations such that $\forall Y, Y' \in \Omega$, $Y \leq_{\mathcal{P}} Y'$ iff $Y \models \mathcal{P} * \mathsf{glp}(\{(Y,Y),(Y',Y')\})$. Moreover, for every GLP $\mathcal{P}$, $\forall Y \in \Omega$, let $\mathcal{P}(Y) = \{X \subseteq Y \mid (X,Y) \in SE(\mathcal{P} * \mathsf{glp}(\{(X,Y),(Y,Y)\}))\}$. We claim that $\leq_{\mathcal{P}}$ is a total preorder (this part of proof is similar to the one given for Theorem 11 in [15][5]).

---

[5] In the proof of Theorem 11 in [15], propositional merging operators are considered. Multi-sets of formulae (so-called *profiles*) are merged under a certain integrity constraint represented by a formula. This part of our proof is similar if one restricts ourselves to singleton profiles.

Now we show that $SE(\mathcal{P} * \mathcal{Q}) = \{(X,Y) \mid (X,Y) \in SE(\mathcal{Q}), \forall Y' \models \mathcal{Q}, Y \leq_{\mathcal{P}} Y', X \in \mathcal{P}(Y)\}$. Let us denote S the latter set and first show the first inclusion $SE(\mathcal{P} * \mathcal{Q}) \subseteq_s$ S. Let $(X,Y) \in SE(\mathcal{P} * \mathcal{Q})$ and let us show that *(i)* $(X,Y) \in SE(\mathcal{Q})$, *(ii)* $\forall Y' \models \mathcal{Q}, Y \leq_{\mathcal{P}} Y'$ and that *(iii)* $X \in \mathcal{P}(Y)$. (i) is direct from (RA1). For (ii), let $Y' \models \mathcal{Q}$. Since $*$ returns a GLP, $SE(\mathcal{P} * \mathcal{Q})$ is well-defined. That is, since $(X,Y) \in SE(\mathcal{P} * \mathcal{Q})$, we also have $(Y,Y) \in SE(\mathcal{P} * \mathcal{Q})$. Therefore, $\mathcal{P} * \mathcal{Q} + \mathsf{glp}(\{(Y,Y),(Y',Y')\})$ is consistent. So by (RA5) and (RA6), $(Y,Y) \in \mathcal{P} * \mathsf{glp}(\{(Y,Y),(Y',Y')\})$. Hence, $Y \leq_{\mathcal{P}} Y'$. For (iii), since $(X,Y) \in SE(\mathcal{P} * \mathcal{Q})$, $SE(\mathcal{P} * \mathcal{Q}) + \mathsf{glp}(\{(X,Y),(Y,Y)\})$ is consistent, we have $(X,Y) \in SE(\mathcal{P} * \mathsf{glp}(\{(X,Y),(Y,Y)\}))$ by (RA5) and (RA6); hence, $X \in \mathcal{P}(Y)$. Let us now show the other inclusion S $\subseteq_s SE(\mathcal{P} * \mathcal{Q})$. Assume $(X,Y) \in$ S. So $\forall Y' \models \mathcal{Q}$, $Y \leq_{\mathcal{P}} Y'$ and $X \in \mathcal{P}(Y)$. First, from the definition of $\mathcal{P}(Y)$ and by (RA1) and (RA3), we have $Y \in \mathcal{P}(Y)$. So $Y \in$ S. Since S $\neq \emptyset$, $\mathcal{Q}$ is consistent, thus by (RA1) and (RA3) $\exists Y_* \models \mathcal{Q}, Y_* \in SE(\mathcal{P} * \mathcal{Q})$. Let $\mathcal{R}_{\#} = \mathsf{glp}(\{(X,Y),(Y,Y),(Y_*,Y_*)\})$. So $\mathcal{P} * \mathcal{Q} + \mathcal{R}_{\#}$ is consistent. Then by (RA5) and (RA6), $\mathcal{P} * \mathcal{Q} + \mathcal{R}_{\#} = \mathcal{P} * \mathcal{R}_{\#}$. We have to show that $(X,Y) \in \mathcal{P} * \mathcal{R}_{\#}$. Assume towards a contradiction that $(X,Y) \notin \mathcal{P} * \mathcal{R}_{\#}$. By (RA1) and (RA3) and since $(Y_*,Y_*) \in \mathcal{P} * \mathcal{R}_{\#}$, we have two remaining cases: (i) $\mathcal{P} * \mathcal{R}_{\#} = \mathsf{glp}(\{(Y_*,Y_*)\})$. Since $\mathcal{P} * \mathcal{R}_{\#} + \mathsf{glp}(\{(Y,Y),(Y_*,Y_*)\})$ is consistent, by (RA5) and (RA6) we get that $\mathcal{P} * \mathsf{glp}(\{(Y,Y),(Y_*,Y_*)\}) = \mathsf{glp}(\{(Y_*,Y_*)\})$. This contradicts $Y \leq_{\mathcal{P}} Y'$. (ii) $\mathcal{P} * \mathcal{R}_{\#} = \mathsf{glp}(\{(Y,Y),(Y_*,Y_*)\})$. Since $\mathcal{P} * \mathcal{R}_{\#} + \mathsf{glp}(\{(X,Y),(Y,Y)\})$ is consistent, by (RA5) and (RA6) we get that $\mathcal{P} * \mathsf{glp}(\{(X,Y),(Y,Y)\}) = \mathsf{glp}(\{(Y,Y)\})$. This contradicts $X \in \mathcal{P}_Y$.

It is harmless to verify that all conditions (1 - 3) of the faithful assignment and conditions (a - e) of the well-defined assignment are satisfied: conditions (a) and (b) are direct from the definition of $\mathcal{P}(Y)$, conditions (1), (2), (c) and (d) come from (RA2), and conditions (3) and (e) are derived from (RA4).

(If part) We consider a faithful assignment that associates with every GLP $\mathcal{P}$ a total preorder $\leq_{\mathcal{P}}$ and a well-defined assignment that associates with every GLP $\mathcal{P}$ and every interpretation Y a set $\mathcal{P}(Y) \subseteq \Omega$, such that $\forall \mathcal{P}, \mathcal{Q}, SE(\mathcal{P} * \mathcal{Q}) = \{(X,Y) \mid (X,Y) \in SE(\mathcal{Q}), \forall Y' \models \mathcal{Q}, Y \leq_{\mathcal{P}} Y', X \in \mathcal{P}(Y)\}$. Let $\mathcal{P}, \mathcal{Q}$ be two GLPs and $X, Y \in \Omega$. We have to show that $SE(P * Q)$ is well-defined. Let $(X,Y) \in SE(\mathcal{P} * \mathcal{Q})$. Since $SE(P * Q)$ is a set of SE interpretations, $X \subseteq Y$. Moreover, by condition (a) of the well-defined assignment, $Y \in \mathcal{P}(Y)$, so $Y \in SE(\mathcal{P} * \mathcal{Q})$. Hence, $SE(\mathcal{P} * \mathcal{Q})$ is well-defined.

It is harmless to verify that postulates (RA1 - RA6) are satisfied: (RA1) and (RA3) are obvious from the definition of $SE(\mathcal{P} * \mathcal{Q})$, (RA2) comes from conditions (1), (2), (c) and (d), (RA4) is derived from conditions (3) and (e), and (RA5) and (RA6) hold by definition.   ∎

Proposition 3

Consider beforehand that $\mathcal{P} + \mathcal{Q}$ is inconsistent (the other case is trivial from Proposition 2 and postulate (RA2)). When reducing the SE interpretations to their second components, the fact that the set of all classical models of $\mathcal{P} \star^{\circ,f} \mathcal{Q}$

corresponds to the models of $\psi_{\mathcal{P}} \circ \psi_{\mathcal{Q}}$ comes from the similarities between an LP faithful assignment (cf. Definition 11) and a faithful assignment (cf. Definition 2), and from Proposition 2 and Theorem 1. Regarding all first components of SE interpretations, the correspondence between f (cf. Definition 13) and $\mathcal{P}(Y)$ (cf. conditions (a - e) of the well-defined assignment in Definition 12) can be easily seen.   ∎

Flexible Combinatory Categorial Grammar Parsing Using the CYK Algorithm and Answer Set Programming

Peter Schüller

Cognitive Robotics Laboratory, Sabancı University, Turkey
`peterschueller@sabanciuniv.edu`

Abstract. Combinatory Categorial Grammar (CCG) is a grammar formalism used for natural language parsing. CCG assigns structured lexical categories to words and uses a small set of combinatory rules to combine these categories in order to parse sentences. In this work we describe and implement a new approach to CCG parsing that relies on Answer Set Programming (ASP) — a declarative programming paradigm. Different from previous work, we present an encoding that is inspired by the algorithm due to Cocke, Younger, and Kasami (CYK). We also show encoding extensions for parse tree normalization and best-effort parsing and outline possible future extensions which are possible due to the usage of ASP as computational mechanism. We analyze performance of our approach on a part of the Brown corpus and discuss lessons learned during experiments with the ASP tools dlv, gringo, and clasp. The new approach is available in the open source CCG parsing toolkit AspCcgTk which uses the C&C supertagger as a preprocessor to achieve wide-coverage natural language parsing.

1   Introduction

Parsing is the task of recovering the structure of sentences which is an important task in natural language processing (NLP). Contemporary NLP systems often process input in a 'pipeline' consisting of sequential steps of chunking, part-of-speech tagging, parsing, semantical annotation, and further steps. A widely-used technique in such pipelines is to statistically select a single best result of one stage and feed it to the next one.

However many natural language effects cannot be handled satisfactorily with such an approach, because natural language ambiguities can emerge in various levels of processing and some of them can only be resolved on other levels. For example the sentence

$$\text{"John saw the astronomer with the telescope."} \tag{1}$$

admits two structures, intuitively one where John used a telescope to see the astronomer, and one where John saw an astronomer who had a telescope. On the other hand,

$$\text{"John saw the astronomer with the dog."} \tag{2}$$

cannot only have the structure such that John saw an astronomer who had a dog. These sentences both have a syntactic ambiguity: whether the with-clause modifies 'saw' or

P. Cabalar and T.C. Son (Eds.): LPNMR 2013, LNAI 8148, pp. 499–511, 2013.

'the astronomer'. In (2) semantic information about "dog", i.e., that it can (usually) not be used as a "tool for seeing", rules out the structure where 'John performed the action of seeing by means of the dog'. On the other hand (1) can only be disambiguated using contextual information from the world or from previous or following sentences.

These examples show that, to make sense of natural language, a bidirectional integration of natural language processing modules is necessary. Answer Set Programming (ASP) [3,7] is a declarative logic programming formalism which is well-suited to serve as computational formalism for NLP tasks: ASP programs can contain (i) guesses, which support modeling ambiguities of any kind; (ii) definitions of auxiliary concepts, which support modeling processes of natural language (in particular compositionality), and (iii) constraints which support modeling of linguistic constraints on all phenomenological levels.

In this work we describe an efficient encoding for parsing Combinatory Categorial Grammar (CCG) using ASP. CCG is a popular grammar formalism used in natural language parsing, which assigns structured categories to words of a sentence and uses a set of combinatory rules to combine these categories and to parse the sentence. Different from previous work [22] which modeled CCG parsing as action planning we here propose an encoding that is inspired by the CYK algorithm [12, 19, 33] and performs the major computational effort already within instantiation of the program. The combinatorial power of ASP is used for reasoning about parse tree shapes, parse tree normalizations [9, 15, 32], best-effort parsing and further possible extensions, e.g., [23].

Our main contributions are:

- we describe an adaptation of CYK algorithm for CCG parsing and give an encoding which builds a CYK chart during instantiation of the ASP encoding;
- we provide an encoding for enumerating parse trees based on the above encoding;
- we show how normalizing constraints for CCG can be realized as an additional ASP program module;
- we describe an extension of our encoding that supports best-effort parsing, i.e., providing maximal coverage of the input if no full parse tree is possible;
- we report on experiments which show that our approach provides reasonable parsing times and compare the new encoding to the previous approach for CCG parsing with ASP [22] and to the C&C parser;
- we discuss several lessons learned and interesting observations gained from this application of ASP.

An extension of the encodings presented in this work are released as version 0.4 of the open source CCG parsing toolkit ASPCCGTK[1] which uses the C&C supertagger [10] to achieve wide coverage and can visualize multiple CCG parse trees [22].

2 Preliminaries

CCG. A *Combinatory Categorial Grammar* (CCG) [29] is a tuple $G = (\Sigma, N, S, f, R)$ with Σ a finite set of *terminal symbols*, N a finite set of *atomic categories*, $S \in N$ the

---

[1] http://www.kr.tuwien.ac.at/staff/former_staff/ps/aspccgtk/

start category, f a function mapping from terminal symbols to complex categories, and R a finite set of *combinatory rules*. *Complex categories* are defined as follows: every atomic category is a complex category; given complex categories A and B, A/B and $A\backslash B$ are complex categories; nothing else is a complex category.

A *combinatory rule* (also called *combinator*) is of the form

$$\frac{X_1 \quad \cdots \quad X_n}{C} \; \mathfrak{A} \tag{3}$$

where $\mathfrak{A}$ is a symbol indicating the name of the rule and C, $X_1, \ldots, X_n$ are categories: we call $X_1, \ldots, X_n$ the precondition categories of $\mathfrak{A}$ and C the result category of $\mathfrak{A}$.

English can be parsed with the following combinators [29]: forward and backward application ($>$ and $<$, respectively), forward and backward *composition* ($>$**B** and $<$**B**), forward and backward *type raising* ($>$**T** and $<$**T**), backward *cross composition*, backward *cross substitution*, and *coordination*.

We limit the presentation of our work to the following set of combinators.

$$\frac{A/B \quad B}{A} > \qquad \frac{A/B \quad B/C}{A/C} >\text{B} \qquad \frac{A}{B/(B\backslash A)} >\text{T}$$

$$\frac{B \quad A\backslash B}{A} < \qquad \frac{B\backslash C \quad A\backslash B}{A\backslash C} <\text{B} \qquad \frac{A}{B\backslash(B/A)} <\text{T}$$

These combinatory rules are rule schemas with one or two preconditions, called unary and binary combinators, respectively, in the following. In this work we only consider the syntactic side of these combinators and disregard the semantic operations (following principles of combinatory logic) which are associated with combinators.

We use A, B, C, ... to denote variables in CCG rule schemas. CCG derivation is defined in terms of a function f which maps terminal symbols to CCG categories, and in terms of substitution of adjacent CCG categories by instantiations of combinators. Formally the *CCG derivation relation* $\Rightarrow$ contains for all $\alpha, \beta \in (N \cup \Sigma)^\star$

(i) $\alpha C \beta \Rightarrow \alpha c \beta$ for terminal symbol $c \in \Sigma$ and category $C \in f(c)$ i.e., the terminal symbol c is mapped to category C by f, furthermore
(ii) $\alpha C \beta \Rightarrow \alpha X_1 \cdots X_n \beta$ for an instantiation of a combinatory rule $\mathfrak{A} \in R$ of form (3).

The language generated by a CCG G is the set $\{\alpha \in \Sigma^\star \mid S \Rightarrow^\star \alpha\}$ where $\Rightarrow^\star$ is the transitive closure of $\Rightarrow$.

A derivation $S \Rightarrow^\star \alpha$ can be considered as a set of parse trees: "S" is the root of a tree, a tree node is either the category C on the left side of (i) or (ii) above; a tree node generated by (i) has one child which is terminal c; a node generated by (ii) has an ordered sequence of children $X_1, \ldots, X_n$; in-order traversal of the tree leafs yields α.

The problem of *enumerating parse trees* is to obtain all distinct parse trees given α.

In the following we will call a natural language input a 'sentence' and the terminal symbols of such an input we will call 'tokens'.[2]

Example 1. The sentence "The dog bit John" with f such that $f(\text{"}The\text{"}) = \{NP/N\}$, $f(\text{"}dog\text{"}) = \{N\}$, $f(\text{"}bit\text{"}) = \{S\backslash NP, (S\backslash NP)/NP\}$, $f(\text{"}John\text{"}) = \{NP\}$ can be derived using combinators $>$ and $<$ as follows:

---

[2] Using the term 'word' can be misleading: what we call 'sentence' is sometimes called 'word', words such as "it's" can become multiple tokens, and punctuation symbols are tokens as well.

$$\cfrac{\cfrac{The}{NP/N} \quad \cfrac{dog}{N}}{NP} > \quad \cfrac{\cfrac{bit}{(S\backslash NP)/NP} \quad \cfrac{John}{NP}}{S\backslash NP} >$$
$$\cfrac{}{S} <$$

where lines below tokens of the sentence show derivations of type (i), i.e., mappings using f; lines below show derivations of type (ii), i.e., instantiations of combinators. Note that f provides two categories for "bit" corresponding to the ambiguity between the intransitive and transitive reading of the verb "to bite".

ASP. Answer set programming (ASP) [3, 7, 25, 27] is a declarative programming formalism based on the answer set semantics of logic programs [18]. The idea of ASP is to represent a given computational problem by a program whose answer sets correspond to solutions, and then use an answer set solver to generate answer sets for this program. A common methodology in ASP is called GENERATE-DEFINE-TEST [24]: the GENER-ATE part of a program describes a collection of answer set candidates; the TEST part consists of constraints that eliminate candidates that do not correspond to solutions; the DEFINE part defines concepts in terms of other concepts.

We will present this work using the ASP-Core-2 language [8] of which we introduce a subset on the following example: given as a set of facts of form $edge(X, Y)$ encoding a graph, a typical logic program for solving the 3-colorability problem looks as follows:

$$vertex(X) \leftarrow edge(X, \_).$$
$$vertex(Y) \leftarrow edge(\_, Y).$$
$$1 \leq \{ \, color(X, red), color(X, green), color(X, blue) \, \} \leq 1 \leftarrow vertex(X).$$
$$\leftarrow color(X_1, C), color(X_2, C), edge(X_1, X_2).$$

The first two rules DEFINE $vertex$ in terms of $edge$ ('$\_$' symbols are anonymous variables); the third rule GENERATEs one color for each vertex; the fourth rule performs a TEST: it is a constraint which eliminates candidate solutions where two adjacent vertexes have the same color.

Variables are universally quantified over rules; a logic program is generally evaluated by (i) grounding it, i.e., instantiating all variables with terms that contain no variables, and (ii) searching for an answer set using methods related to SAT solving [17]. In this work we will use uninterpreted function symbols, e.g., in addition to constants programs can contain function terms. We use this to represent non-atomic CCG categories: $r(\text{"S"}, \text{"S"})$ and $l(r(\text{"S"}, \text{"NP"}), \text{"NP"})$ represent S/S and $(S/NP)\backslash NP$, respectively.

Additionally we use count aggregates as literals in rule bodies: intuitively the literal $2 \leq \#count \, \{ \, X : pred(X, Y) \, \} \leq 4$ is true iff the set of substitutions for variable X such that $pred(X, Y)$ is true in the answer set candidate has cardinality 2, 3, or 4.

For a detailed description of ASP-Core-2 and for semantics of ASP we refer to [8].

3   Realizing CCG Parsing with CYK in Answer Set Programming

A sentence α with n tokens is presented to our CCG parser encoding as a set of facts of the form $catFor(C, P)$ where C specifies the CCG category and P the token position: $P \in \{1, \ldots, n\}$. Intuitively these categories are obtained from statistical tagging; a parser must select a single category at each position for generating a parse tree. Given a sentence α we denote by $inp(\alpha)$ its encoding in terms of token categories.

Example 2 (ctd.). "The dog bit John" has $n = 4$ tokens and $inp(\text{"The dog bit John"}) =$
$$= \{ catFor(l(\text{"NP"}, \text{"N"}), 1), catFor(\text{"N"}, 2), catFor(l(\text{"S"}, \text{"NP"}), 3),$$
$$catFor(r(l(\text{"S"}, \text{"NP"}), \text{"NP"}), 3), catFor(\text{"NP"}, 4) \}$$

CYK for CCG. The CYK algorithm was originally proposed for parsing Context Free Grammar in Chomsky Normal Form [12, 19, 33], i.e., grammars with rules of the form $A \Rightarrow BC$ (corresponding to binary CCG combinators) and $A \Rightarrow c$ (corresponding to application of the function f) only, where A, B, and C are nonterminals, and c is a terminal. An adaptation for parsing grammars which also contain rules of the form $A \Rightarrow B$ (corresponding to CCG type raising) has been discussed in [21], in [30] specific problems of parsing CCG with CYK are discussed.

Algorithm 1 shows an adaptation of CYK to parse CCG in the spirit of [21]; Figure 1 visualizes a CYK chart and possible combinator applications for $n = 3$ tokens.

3.1 Building a CYK Chart via ASP Grounding

The ASP encoding we present in the following realizes Algorithm 1 such that a CYK chart for the given input is computed during program instantiation. To that end we present a non-ground encoding where identifiers starting with a capital letter denote variables and '_' denotes anonymous variables.

We represent contents of chart cells $C \in \mathcal{T}_{I,J}$ using predicate $grid(I, J, C)$. Corresponding to line 2 of Alg. 1 we fill diagonal cells $\mathcal{T}_{D,D}$ with $f(a_D)$ using rule

$$grid(D, D, C) \leftarrow catFor(C, D). \qquad (4)$$

We track applicability of unary (lines 2 and 10) and binary (line 8) combinators using predicates $applicableU$ and $applicableB$, resp.: $applicableU(R, I, J, C', C)$ represents that combinator R can be applied to category C in cell $\mathcal{T}_{I,J}$ and yields category C' in the same cell; $applicableB(R, I, J, H, C', X, Y)$ represents that combinator R can be applied to categories X and Y in cells $\mathcal{T}_{H,J}$ and $\mathcal{T}_{I,H+1}$, resp., and yields category C' in $\mathcal{T}_{I,J}$ (see also Fig. 1). We next give examples for encoding combinators $>\mathbf{T}$ and $>$:

$$applicableU(\text{">\mathbf{T}"}, I, J, r(B, l(B, A)), A) \leftarrow \qquad \left[\frac{A}{B/(B \backslash A)} >\mathbf{T} \right] \qquad (5)$$
$$grid(I, J, A), raiseCategory(A, B).$$

$$applicableB(\text{">"}, I, J, H, A, r(A, B), B) \leftarrow \qquad \left[\frac{A/B \quad B}{A} > \right] \qquad (6)$$
$$grid(H, J, r(A, B)), grid(I, H+1, B).$$

We define categories resulting from applicable rules to be part of the chart using rules

$$grid(I, J, C) \leftarrow applicableU(\_, I, J, C, \_).$$
$$grid(I, J, C) \leftarrow applicableB(\_, I, J, \_, C, \_, \_). \qquad (7)$$

This concludes the deterministic non-ground program Π_{CYK} which consists of rules (4) to (7). Intuitively this encoding defines applicability from chart cells as lines 2, 8 and 10

Algorithm 1. CYK adapted for CCG Parsing

Input: CCG $G=(\Sigma, N, S, f, R)$; token sequence $\alpha = a_1, \ldots, a_n \in \Sigma^*$ with $n \geq 1$

1 **for** $d = 1, \ldots, n$ **do** // initialize diagonal cells (d, d) using f and unary combinators in R

2 $\quad \mathcal{T}_{d,d} := f(a_d) \cup \{\mathcal{B}' \mid \mathcal{A}' \in f(a_d) \text{ and } \frac{\mathcal{A}'}{\mathcal{B}'} \text{ is an instantiation of a combinator } \frac{\mathcal{A}}{\mathcal{B}} \text{ in } R\}$

3 **for** $i = 2, \ldots, n$ **do** // iterate columns i from left to right

4 $\quad$ **for** $j = i - 1, \ldots, 1$ **do** // iterate rows j from bottom to top

5 $\quad\quad \mathcal{T}_{i,j} := \emptyset$

6 $\quad\quad$ **for** $h = j, \ldots, i - 1$ **do** // iterate distance of source cells from left and from top

7 $\quad\quad\quad$ **foreach** *combinator* $\frac{\mathcal{B} \quad \mathcal{C}}{\mathcal{A}}$ *in* R **do**

8 $\quad\quad\quad\quad$ **if** $\mathcal{B}' \in \mathcal{T}_{h,j}$ and $\mathcal{C}' \in \mathcal{T}_{i,h+1}$ and $\mathcal{A}', \mathcal{B}', \mathcal{C}'$ can instantiate $\mathcal{A}, \mathcal{B}, \mathcal{C}$ **then**

9 $\quad\quad\quad\quad\quad \mathcal{T}_{i,j} := \mathcal{T}_{i,j} \cup \{\mathcal{A}\}$

10 $\quad\quad \mathcal{T}_{i,j} := \mathcal{T}_{i,j} \cup \{\mathcal{B}' \mid \mathcal{A}' \in \mathcal{T}_{i,j} \text{ and } \frac{\mathcal{A}'}{\mathcal{B}'} \text{ is an instantiation of a combinator } \frac{\mathcal{A}}{\mathcal{B}} \text{ in } R\}$

11 **if** $S \in \mathcal{T}_{1,n}$ **then return** *yes* **else return** *no*

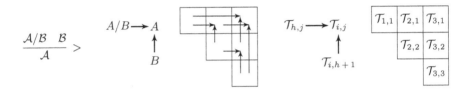

Fig. 1. CYK chart visualization for input sentence with 3 tokens and for $>$ combinator

of Alg. 1 do, furthermore it defines that categories that result from applying combinators are again part of chart cells.

Given a sentence α, program $inp(\alpha) \cup \Pi_{CYK}$ has a single answer set I which represents the CYK chart for α as produced by lines 1 to 10 of Alg. 1.

For space reasons we here do not present type conversion (e.g., $N \Rightarrow NP$) and punctuation rules (e.g., to handle commas) as described in [11, Appendix A]; these features are necessary for wide-coverage parsing and they are implemented in ASPCCGTK.

Language Membership. We can check whether an input is part of the language by checking whether cell $(n, 1)$ contains category S, e.g., using the following constraint:

$$\leftarrow not\ grid(n, 1, \text{``}S\text{''}).\tag{8}$$

or by performing an ASP query (see Section 4 on Magic Set experiments).

3.2 Enumerating Parse Trees

We next describe an encoding for enumerating all parse trees of a given input. We achieve this by (i) guessing which applicable combinators are applied and (ii) restricting the guess to a tree such that the chosen combinators form edges, input tokens are leaves, and category "S" in cell $(n, 1)$ is the root node.

For each applicable combinator we guess whether it is applied or not:

$$0 \leq \{\ applyB(R, I, J, H, X, Y, Z)\ \} \leq 1 \leftarrow applicableB(R, I, J, H, X, Y, Z).$$
$$0 \leq \{\ applyU(R, I, J, X, Y)\ \} \leq 1 \leftarrow applicableU(R, I, J, X, Y). \tag{9}$$

To ensure that the above guess induces a parse tree, we first define reachability of categories in cells from other categories via applied combinators.

$$reach(H, J, C_{left}) \leftarrow reach(I, J, C), applyB(\_, I, J, H, C, C_{left}, \_).$$
$$reach(I, H+1, C_{down}) \leftarrow reach(I, J, C), applyB(\_, I, J, H, C, \_, C_{down}). \tag{10}$$
$$reach(I, J, C_{sameCell}) \leftarrow reach(I, J, C), applyU(\_, I, J, C_{sameCell}, C)$$

We ensure that the guess is restricted to parse using the following rules

$$reach(n, 1, \text{``S''}) \leftarrow \tag{11}$$
$$\leftarrow valid(I, I), \#count\{\ C : reach(I, I, C), catFor(C, I)\ \} \leq 0 \tag{12}$$
$$\leftarrow applyB(\_, I, J, \_, C, \_, \_), not\ reach(I, J, C)$$
$$\leftarrow applyU(\_, I, J, C, \_), not\ reach(I, J, C) \tag{13}$$
$$\leftarrow valid(I, J), 2 \leq \#count\{\ C : applyU(\_, I, J, C, \_)\ \}$$
$$\leftarrow valid(I, J), 2 \leq \#count\{\ C : applyB(\_, I, J, \_, C, \_, \_)\ \} \tag{14}$$

These rules define the root category "S" to be reachable (11), require that each word is reachable (12), disallow unreachable combinators to be applied (13), and disallow more than one binary (resp., unary) combinator application in one cell (14).

By Π_{Tree} we denote rules (9) to (14). Π_{Tree} follows the classical GENERATE-DEFINE-TEST approach: it guesses a subset of applicable combinators (9), defines a notion of reachability (10)-(11), and restricts the guess to certain trees (12)-(14).

With Π_{Tree} we can enumerate parse trees: given an input sentence α the answer sets of program $inp(\alpha) \cup \Pi_{CYK} \cup \Pi_{Tree}$ correspond 1-1 to the CCG parse trees of α.

Note that we do not use f in our encoding, instead we use f to generate $inp(\alpha)$; this is because f corresponds to statistical tagging which is handled outside of our encoding.

Parse Tree Normalization. CCG generates *spurious* parse trees which are not of interest because they provably lead to the same linguistic interpretation as other parse trees; this can only be avoided by *normalization* of parse trees [9, 15, 32] which is performed by constraining the shape of CCG parse trees depending on the type of combinator used to create each tree node. For example, using the category resulting from $>\mathbf{T}$ as the first prerequisite of $>$ can always be replaced by a single application of $<$. Fortunately such constraints can be represented easily in ASP, the above normalization is encoded as

$$\leftarrow applyB(\text{``>''}, I, J, H, \_, L, \_), applyU(\text{``>T''}, I, H, L, \_). \tag{15}$$

Such normalizations can eliminate an exponential number of spurious parse trees [15]. Thanks to the modularity of our encoding we can maintain multiple sets of normalizing constraints (e.g., to normalize towards left- or right-branching) and simply add them to our program when needed without changing rules in the encoding.

Best-Effort CCG Parsing. If a sentence has no parse tree it can be useful to obtain a best-effort parse forest. This can be done with respect to various optimization criteria, e.g., finding a minimum of root nodes for a set of parse trees which contains all tokens of an input or ignoring a minimum of input tokens. Our parser encoding allows us to perform such best-effort parsing with only small modifications. For example we can create a parser that enumerates (i) complete parse trees if one exists, otherwise (ii) partial parse trees with a minimum number of root nodes such that all input tokens are reachable. This is achieved by replacing (11) with the following set of rules:

$$0 \leq \{ \ guessReach(I,J,C) \ \} \leq 1 \leftarrow grid(I,J,C).$$
$$reach(I,J,C) \leftarrow guessReacch(I,J,C). \qquad (16)$$
$$\#minimize\{1 : guessReach(I,J,C)\}.$$

4 Experimental Evaluation

For performance evaluation we used Section A of the Brown corpus [16] which is a freely available English language corpus. We selected Section A because it contains newspaper articles and the C&C supertagger we use for tagging the input is trained on a newspaper corpus. Section A contains 4611 sentences in total, Table 1 groups these sentences in terms of their length, e.g., the corpus contains 684 sentences with a length between 11 and 15 words (inclusive) where the average sentence length is 13.

Our experiments were performed similar as the C&C parser operates when parsing a sentence: we obtain tags of probability class $\beta = 0.075, 0.03$, and 0.001 from the C&C supertagger, then we run our encoding on categories obtained with $\beta \geq 0.075$, if this does not yield a parse tree we retry with $\beta \geq 0.03$, then with $\beta \geq 0.001$, and we register failure if even this does not yield a parse trees.

Where not otherwise indicated we used GRINGO version 3.0.5 and CLASP[3] version 2.1.1 for experiments; some experiments were performed with DLV[4] version 2012-12-17. We used a timeout of 300 seconds and enumerated up to 100 parse trees for each sentence, this was done in single threaded mode on a Linux server with 32 2.4GHz Intel® E5-2665 CPU cores and 64GB memory.

Table 1 reports the number of CCG tags required to parse a sentence (if no parse was found the value for $\beta = 0.001$), the number of parse trees obtained for each sentence, and the percentage of sentences where a parse tree was found. For example, sentences with 11-15 words required 25 tags for parsing on average, each sentence yielded on average 38 parse trees, and 84.9% of sentences yielded at least one parse tree.

Comparison with Planning Approach.. We compare our approach (CYK+ASP) with the ASP formulation for CCG parsing that uses planning [22]. The CYK algorithm is a (dynamic programming) approach, therefore in a CYK chart partial parse trees can be reused between complete parse trees. This is not possible in the planning approach which requires the notion of time to define an order of combinator applications.

Table 1 reports performance of the planning approach: experiments show that the CYK approach scales much better than planning, especially for larger sentences. Performance is similar only for the shortest group of sentences, for sentences of length

---

[3] http://potassco.sourceforge.net/
[4] http://www.dlvsystem.com/

Table 1. Performance comparison on Section A (newspaper) of Brown corpus using C&C for tagging. Times and timeouts are for the task of enumerating up to 100 parse trees per sentence.

Group: words in sentence	#	1-10	11-15	16-20	21-25	26-30	31-35	36-40	41+
Sentences in group	#	983	684	779	704	526	396	258	281
Words in sentence	avg #	5	13	17	22	27	32	37	48
CCG categories[†]	avg #	14	25	37	48	59	77	87	122
Parse Trees	avg #	6	38	65	81	80	80	79	72
Sentences with parse tree	%	64.2	84.9	84.7	85.9	82.1	81.1	79.1	73.0
CYK+ASP parse time	avg sec	0.6	0.7	1.2	1.9	3.5	6.7	11.9	42.0
CYK+ASP timeouts	#	0	0	0	0	0	0	0	8
Planning+ASP parse time	avg sec	0.8	4.3	19.1	62.1	110.1	135.7	156.5	205.5
Planning+ASP timeouts	#	0	0	4	51	137	157	118	149

[†] Category set with smallest β value that is sufficient for finding a parse tree, $\beta = 0.001$ if no parse tree can be found with tags provided by the C&C supertagger.

21-25 the CYK approach gives an answer within 1.9 seconds while planning requires more than one minute. Moreover, the planning approach suffers from timeouts already with sentences of length 16-21 while the CYK approach has no timeout for any sentence smaller than 41 tokens.

The CYK approach requires a maximum of 1.5GB of memory with an average of 1.1GB over all sentences, where as the planning approach requires up to 5.8GB of memory with an average of 3.8GB.

Parse Effort Profile. Figure 2 gives a diagram of parsing time for all sentences in our benchmark, first grouped by the β value required to find a parse tree, then by the time required to enumerate up to 100 parse trees. E.g., 2596 sentences obtain a parse tree with $\beta = 0.075$ and 3639 out of 4611 sentences obtain a parse tree with $\beta \geq 0.001$. We plot total time required (dashed red) and solver time required (solid blue).

The graph shows that parse time is not distributed evenly among the sentences in the benchmark: a majority of sentences can be parsed in a comparatively short time, in particular for sentences with the most probable tags ($\beta = 0.075$), while there are few sentences in the benchmark that take a disproportionately high amount of time. There is only a weak correlation between difficulty of a sentence and its length or amount of tags (not shown), therefore additionally plotting the amount of tags and/or the length of each sentence in the figure would make it unreadable. Furthermore we see that solving time is negligible compared to grounding time.

Additionally we measured the time for grounding Π_{CYK} and the time for grounding $\Pi_{CYK} \cup \Pi_{Tree}$ (not shown here); these times are nearly the same independent from the length of the input. This shows that the main computational effort is due to Π_{CYK}.

Comparison with C&C. The popular C&C parser is designed as a highly efficient CCG wide coverage parser [11, 13] and it operates on the same C&C tagger output as our parser. This makes it suitable for a comparison: in [13] C&C is reported to parse the whole section 00 of CCGBank (1913 sentences with a similar distribution of sentence lengths as in the Brown corpus) within less than 100 seconds on a slower computer than

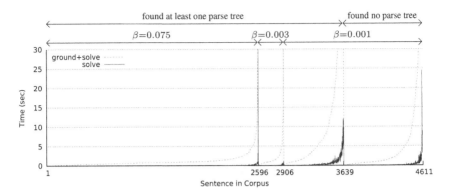

Fig. 2. Total time and solve time for parsing Section A of the Brown corpus, grouped by the tagger β value required to parse each sentence and sorted by parse time (timeout 300 seconds)

ours. Therefore our approach clearly cannot compete with the performance of C&C. However, the aim of this work was not to build the fastest parser, but to build a flexible parser with reasonable performance that can return multiple parse trees and can easily be extended with reasoning capabilities that go beyond what C&C can do.

Stratification and DLV **vs** GRINGO+CLASP. Apart from benchmarking with GRINGO and CLASP we also considered DLV. Our first observation was that DLV does not recognize our encoding to be finitely groundable due to rule (5) and other raising rules which contain a higher level of function symbol nesting in their head than in the body. However due to the *raiseCategory* predicate (which does not depend on the *applicableU* predicate in the head of (5)) the program clearly has a finite instantiation and using DLV with the option -nofinitecheck leads to a finite grounding.

In an early version of the encoding, computing the set of possible categories of a token was performed with a non-stratified rule. With this encoding, DLV performed consistently better than GRINGO for the task of grounding. Replacing this rule by a few stratified rules changed the situation: now DLV produced a slightly smaller grounding in about the same time, but GRINGO became so much faster that it consistently performed better than DLV. (All results in this paper were produced using GRINGO+CLASP and the stratified encoding.) We conclude that efficiency of GRINGO is very sensitive to program structure while for DLV we could not observe this in our benchmarks.

Queries and Magic Sets. As Π_{CYK} is stratified it is possible to use Magic Sets [1] for efficient query evaluation, e.g., the query '$grid(n, 1, \text{"}S\text{"})$?' checks whether a parse tree exists. Such a check is important to see whether a given set of tags is sufficient for finding a parse tree, or whether the β value needs to be reduced to obtain more tags. Unfortunately, using DLV with Magic Set for the above query led to much longer grounding times than using DLV without Magic Set; the reason is not clear to us yet.

Grounding vs Solving. Finally we experimented with putting some of the tree normalization constraints, e.g., (15), already into the Π_{CYK} encoding. This requires to define exceptions to rule applicability, therefore the CYK encoding becomes more complicated (it is still stratified). The result of this experiment is a reduced grounding size and (with GRINGO) a significantly increased grounding time. As the time spent in grounding and solving of Π_{Tree} (including all normalization constraints) is negligible, we reverted to the simpler CYK encoding with larger and faster grounding. We conclude that eliminating solutions in solving can perform significantly better than making a program more complex in order to eliminate those solutions already in grounding, even if the complex program has a smaller instantiation.

5   Related Work

CCG-based systems OPENCCG [31] and TCCG [4,5] (implemented in the LKB toolkit) can provide multiple parse trees for a given sentence. Both use chart parsing algorithms with CCG extensions such as modalities or hierarchies of categories. While OPENCCG is primarily geared towards generating sentences from logical forms, TCCG targets parsing. However, both implementations require lexicons with specialized categories.

The wide-coverage CCG parser C&C [9, 10] relies on machine learning techniques for tagging an input sentence with CCG categories as well as for computing the single most likely parse tree with an efficient chart algorithm. In ASPCCGTK we reuse the CCG supertagger of C&C to obtain CCG categories, we also compare ASPCCGTK to C&C performance.

The Grail parser [26] is based on multi-modal categorial grammar (which is able to represent CCG) and contains a graphical user interface for 'interactive parsing'. Grail uses theorem proving techniques based on the Lambek calculus, this makes it very expressive but slow in some cases; therefore in some cases the user must support the search for a parse tree in the user interface. Compared to our work, Grail is more general and has different aims, e.g., being a tool for learning about Lambek calculus.

Transforming context free grammars (CFGs) in Chomsky Normal Form (CNF) using the CYK algorithm and parsing them using SAT solvers has been studied under the name "GRAMMAR constraint" [20, 28], including recent work based on ASP [14]. Results indicate that SAT and ASP solving can perform well for parsing using the CYK algorithm. Two important differences between these studies and our work are: (i) CFGs use atomic categories and a large set of rules that forms the grammar, while CCG uses structured categories and a small set of rule schemas, hence performance observations might not directly carry over and encodings must be significantly different; moreover (ii) our work is about natural language parsing while GRAMMAR studies experiment with artificial grammars that encode solutions to Shift Scheduling problems.

Parsing CCG with CYK is not polynomial if categories are represented explicitly, however recording only changes of categories can make it polynomial [30]; we here represent categories explicitly and consider a more involved encoding as future work.

6   Conclusion

We have presented an encoding for parsing CCG in ASP which — as opposed to a previous approach, and as opposed to usual ASP methodology — puts the major effort of computation into instantiation of the representation. This increased effort of grounding allows the search for an answer set to be fast, empirical results show that the new approach consistently outperforms the former approach that used planning. Experiments show that our approach provides reasonable performance for using it in practice.

The possibility to trade search effort for grounding effort is due to the CYK algorithm which has been around for a long time and can be realized in ASP in a natural way. This approach of gaining efficiency goes against the declarativity of ASP, because our encoding effectively prescribes a way of grounding that reproduces the data structure generated by CYK. Nevertheless the result is a parsing framework that profits from the declarative nature of ASP because reasoning modules that operate on parse trees (i.e., normalization, semantic disambiguation) can be tightly and modularly integrated with the parser without significant changes to the parser encoding.

Future Work. In the future we want to adapt ASPCCGTK to become compatible with Boxer [6] which is a tool for creating semantic representations for sentences in first order logic. Integration with Boxer opens new possibilities for NLP tasks where multiple readings of a sentence must be considered, e.g., for Recognizing Textual Entailment (RTE) or Semantic Evaluation (SemEval) Challenges.

We have done preliminary work on disambiguation of parse trees using semantic information [23], e.g., from FRAMENET [2], such that the large number of parse trees (our experiments enumerated up to 100 trees per sentence) can be reduced to those trees which are consistent with semantic restrictions (see examples in the introduction). In the future we want to continue work in that direction.

If better efficiency becomes an issue, using techniques from [30] and computing the CYK chart in C++ and enumerating parse trees with constraints in ASP are possibilities.

Finally our CYK encoding could be useful for benchmarking ASP grounders.

Acknowledgments. We thank Yuliya Lierler for fruitful discussions related to the topics of this work. We thank the anonymous reviewers for their constructive comments. Peter Schüller is supported by TUBITAK Research Fellowship 2216.

References

1. Alviano, M., Faber, W., Greco, G., Leone, N.: Magic sets for disjunctive datalog programs. Tech. rep., Università della Calabria, Dipartimento di Matematica (2009)
2. Baker, C.F., Fillmore, C.J., Lowe, J.B.: The Berkeley FrameNet Project. In: 36th Annual Meeting of the Association for Computational Linguistics and 17th International Conference on Computational Linguistics, pp. 86–90 (1998)
3. Baral, C.: Knowledge Representation, Reasoning, and Declarative Problem Solving. Cambridge University Press (2003)
4. Beavers, J.: Documentation: A CCG implementation for the LKB. Tech. rep., Stanford University, Center for the Study of Language and Information (2003)
5. Beavers, J.: Type-inheritance combinatory categorial grammar. In: International Conference on Computational Linguistics, COLING 2004 (2004)
6. Bos, J.: Wide-coverage semantic analysis with boxer. In: Semantics in Text Processing, STEP, pp. 277–286. College Publications (2008)
7. Brewka, G., Eiter, T., Truszczynski, M.: Answer set programming at a glance. Commun. ACM 54(12), 92–103 (2011)
8. Calimeri, F., Faber, W., Gebser, M., Ianni, G., Kaminski, R., Krennwallner, T., Leone, N., Ricca, F., Schaub, T.: ASP-Core-2 input language format (2012)

9. Clark, S., Curran, J.R.: Log-linear models for wide-coverage CCG parsing. In: SIGDAT Conference on Empirical Methods in Natural Language Processing, EMNLP 2003 (2003)
10. Clark, S., Curran, J.R.: Parsing the WSJ using CCG and log-linear models. In: 42nd Annual Meeting of the Association for Computational Linguistics, ACL, pp. 104–111 (2004)
11. Clark, S., Curran, J.R.: Wide-coverage efficient statistical parsing with CCG and log-linear models. Computational Linguistics 33(4), 493–552 (2007)
12. Cocke, J., Schwartz, J.T.: Programming Languages and Their Compilers. Courant Institute of Mathematical Sciences, New York (1970)
13. Djordjevic, B., Curran, J.R.: Efficient combinatory categorial grammar parsing. In: Proceedings of the 2006 Australasian Language Technology Workshop, ALTW, pp. 3–10 (2006)
14. Drescher, C., Walsh, T.: Modelling grammar constraints with answer set programming. In: International Conference on Logic Programming, vol. 11, pp. 28–39 (2011)
15. Eisner, J.: Efficient normal-form parsing for combinatory categorial grammar. In: 34th Annual Meeting of the Association for Computational Linguistics, pp. 79–86. ACL (1996)
16. Francis, W.N., Kucera, H.: Brown corpus manual. Letters to the Editor 5(2), 7 (1979)
17. Gebser, M., Kaufmann, B., Neumann, A., Schaub, T.: Conflict-driven answer set solving. In: International Joint Conference on Artificial Intelligence, pp. 386–392 (2007)
18. Gelfond, M., Lifschitz, V.: The stable model semantics for logic programming. In: Proc. International Logic Programming Conference and Symposium, ICLP, pp. 1070–1080 (1988)
19. Kasami, T.: An efficient recognition and syntax analysis algorithm for context-free languages. Tech. Rep. AFCRL-65-758, Air Force Cambridge Research Laboratory (1965)
20. Katsirelos, G., Narodytska, N., Walsh, T.: Reformulating global grammar constraints. In: van Hoeve, W.-J., Hooker, J.N. (eds.) CPAIOR 2009. LNCS, vol. 5547, pp. 132–147. Springer, Heidelberg (2009)
21. Lange, M., Leiß, H.: To CNF or not to CNF? An efficient yet presentable version of the CYK algorithm. In: Informatica Didactica 8. Universität Potsdam (2009)
22. Lierler, Y., Schüller, P.: Parsing Combinatory Categorial Grammar via planning in Answer Set Programming. In: Erdem, E., Lee, J., Lierler, Y., Pearce, D. (eds.) Correct Reasoning. LNCS, vol. 7265, pp. 436–453. Springer, Heidelberg (2012)
23. Lierler, Y., Schüller, P.: Towards a tight integration of syntactic parsing with semantic disambiguation by means of declarative programming. In: Erk, K., Koller, A. (eds.) 10th International Conference on Computational Semantics (2013)
24. Lifschitz, V.: Answer set programming and plan generation. Artif. Intel. 138, 39–54 (2002)
25. Marek, V., Truszczyński, M.: Stable models and an alternative logic programming paradigm. In: The Logic Programming Paradigm: A 25-Year Perspective, pp. 375–398. Springer (1999)
26. Moot, R.: Proof Nets for Linguistic Analysis. Ph.D. thesis, Utrecht Institute of Linguistics OTS (2002)
27. Niemelä, I.: Logic programs with stable model semantics as a constraint programming paradigm. Annals of Mathematics and Artificial Intelligence 25, 241–273 (1999)
28. Quimper, C.-G., Walsh, T.: Decomposing global grammar constraints. In: Bessière, C. (ed.) CP 2007. LNCS, vol. 4741, pp. 590–604. Springer, Heidelberg (2007)
29. Steedman, M.: The syntactic process. MIT Press, London (2000)
30. Vijay-Shanker, K., Weir, D.J.: Polynomial time parsing of combinatory categorial grammars. In: 28th Annual Meeting of the Association for Computational Linguistics, pp. 1–8 (1990)
31. White, M., Baldridge, J.: Adapting chart realization to CCG. In: European Workshop on Natural Language Generation, EWNLG 2003 (2003)
32. Wittenburg, K.: Predictive combinators: a method for efficient processing of combinatory categorial grammars. In: 25th Annual Meeting of the Association for Computational Linguistics, ACL, pp. 73–80 (1987)
33. Younger, D.H.: Recognition and parsing of context-free languages in time n^3. Information and Control 10(2), 189–208 (1967)

Early Recovery in Logic Program Updates[*]

Martin Slota[1], Martin Baláž[2], and João Leite[1]

[1] CENTRIA & Departamento de Informática, Universidade Nova de Lisboa
[2] Faculty of Mathematics, Physics and Informatics, Comenius University

Abstract. We pinpoint the limitations of existing approaches to the treatment of *strong* and *default negation* in answer-set program updates and formulate the *early recovery principle* that plausibly constrains their interaction.

Keywords: answer-set programming, updates, strong negation, default negation.

1 Introduction

The increasingly common use of rule-based knowledge representation languages in highly dynamic and information-rich contexts, such as the Semantic Web [1], requires standardised support for updates of knowledge represented by rules. Answer-set programming [2, 3] forms the natural basis for investigation of rule updates, and various approaches to answer-set program updates have been explored [4–16].

The most straightforward kind of conflict arising between an original rule and its update occurs when the original conclusion logically contradicts the newer one. Though the technical realisation and final result may differ significantly, depending on the particular rule update semantics, this kind of conflict is resolved by letting the newer rule prevail over the older one. Actually, under most semantics, this is also the *only* type of conflict that is subject to automatic resolution [4–6, 9, 10, 13, 14].

From this perspective, allowing for both *strong* and *default negation* to appear in heads of rules is essential for an expressive and universal rule update framework [7]. While strong negation is the natural candidate here, used to express that an atom *becomes explicitly false*, default negation allows for more fine-grained control: the atom only *ceases to be true*, but its truth value may not be known after the update. The latter also makes it possible to move between any pair of epistemic states by means of updates, as illustrated in the following example:

Example 1 (Railway crossing [7]). Suppose that we use the following logic program to choose an action at a railway crossing:

$$\text{cross} \leftarrow \neg\text{train}. \qquad \text{wait} \leftarrow \text{train}. \qquad \text{listen} \leftarrow {\sim}\text{train}, {\sim}\neg\text{train}.$$

[*] M. Slota and J. Leite were supported by Fundação para a Ciência e a Tecnologia under project "ERRO – Efficient Reasoning with Rules and Ontologies" (PTDC/EIA-CCO/121823/2010). The collaboration between the co-authors resulted from the Slovak–Portuguese bilateral project "ReDIK – Reasoning with Dynamic Inconsistent Knowledge", supported by the APVV agency under SK-PT0-0028-10 and by Fundação para a Ciência e a Tecnologia (FCT/2487/3/6/2011/S).

P. Cabalar and T.C. Son (Eds.): LPNMR 2013, LNAI 8148, pp. 512–517, 2013.

The intuitive meaning of these rules is as follows: one should cross if there is evidence that no train is approaching; wait if there is evidence that a train is approaching; listen if there is no such evidence.

Consider a situation where a train is approaching, represented by the fact (train.). After this train has passed by, we want to update our knowledge to an epistemic state where we lack evidence with regard to the approach of a train. If this was accomplished by updating with the fact (¬train.), we would cross the tracks at the subsequent state, risking being killed by another train that was approaching. Therefore, we need to express an update stating that all past evidence for an atom is to be removed, which can be accomplished by allowing default negation in heads of rules. In this scenario, the intended update can be expressed by the fact ($\sim$train.).

With regard to the support of negation in rule heads, existing rule update semantics fall into two categories: those that only allow for strong negation, and those that primarily consider default negation. As illustrated above, the former are unsatisfactory as they render many belief states unreachable by updates. As for the latter, they optionally provide support for strong negation by means of a syntactic transformation.

Two such transformations are known from the literature, both of them based on the principle of coherence: if an atom p is true, its strong negation $\neg p$ cannot be true simultaneously, so $\sim\neg p$ must be true, and also vice versa, if $\neg p$ is true, then so is $\sim p$. The first transformation, introduced in [17], encodes this principle directly by adding, to both the original program and its update, the rules $(\sim\neg p \leftarrow p.)$ and $(\sim p \leftarrow \neg p.)$ for every atom p. This way, every conflict between an atom p and its strong negation $\neg p$ directly translates into two conflicts between the objective literals p, $\neg p$ and their default negations. However, the added rules lead to undesired side effects that stand in direct opposition with basic principles underlying updates. Specifically, despite the fact that the empty program does not encode any change in the modelled world, the stable models assigned to a program may change after an update by the empty program.

This undesired behaviour is addressed in an alternative transformation from [7] that encodes the coherence principle more carefully. Nevertheless, this transformation also leads to undesired consequences, as demonstrated in the following example:

Example 2 (Faulty sensor). Suppose that we collect data from sensors and multiple sensors are used to supply information about the critical fluent p. In case of a malfunction of one of the sensors, we may end up with the inconsistent logic program $\{p., \neg p.\}$. At this point, no stable model of the program exists and action needs to be taken to find out what is wrong. If a problem is found in the sensor that supplied the first fact $(p.)$, after the sensor is repaired, this information needs to be reset by updating the program with the fact $(\sim p.)$. Following the universal pattern in rule updates, where recovery from conflicting states is always possible, we would expect that some stable model be assigned to the updated program. However, the transformational semantics for strong negation defined in [7] still does not provide any stable model – we remain without a valid epistemic state despite the fact that all conflicts have been solved.

In this short paper we discuss the issues with combining strong and default negation in the context of rule updates. Namely, after presenting the necessary preliminaries, we formulate a generic desirable principle that is violated by the existing approaches.

2 Preliminaries

We assume that a countable set of propositional atoms $\mathcal{A}$ is given and fixed. An *objective literal* is an atom $p \in \mathcal{A}$ or its strong negation $\neg p$. We denote the set of all objective literals by $\mathcal{L}$. A *default literal* is an objective literal preceded by $\sim$ denoting default negation. A *literal* is either an objective or a default literal. We denote the set of all literals by $\mathcal{L}^*$. As a convention, double negation is absorbed, so that $\neg\neg p$ denotes the atom p and $\sim\sim l$ denotes the objective literal l. Given a set of literals S, we introduce the following notation: $S^+ = \{ l \in \mathcal{L} \mid l \in S \}$, $S^- = \{ l \in \mathcal{L} \mid \sim l \in S \}$, $\sim S = \{ \sim L \mid L \in S \}$.

A *rule* is a pair $\pi = (\mathsf{H}_\pi, \mathsf{B}_\pi)$ where H_π is a literal, referred to as the *head of* π, and B_π is a finite set of literals, referred to as the *body of* π. Usually we write π as $(\mathsf{H}_\pi \leftarrow \mathsf{B}_\pi^+, \sim\mathsf{B}_\pi^-.)$. A *fact* is a rule whose body is empty. A *program* is a set of rules.

An *interpretation* is a consistent subset of the set of objective literals, i.e., a subset of $\mathcal{L}$ does not contain both p an $\neg p$ for any atom p. The satisfaction of an objective literal l, default literal $\sim l$, set of literals S, rule π and program P in an interpretation J is defined in the usual way: $J \models l$ iff $l \in J$; $J \models \sim l$ iff $l \notin J$; $J \models S$ iff $J \models L$ for all $L \in S$; $J \models \pi$ iff $J \models \mathsf{B}_\pi$ implies $J \models \mathsf{H}_\pi$; $J \models P$ iff $J \models \pi$ for all $\pi \in P$. Also, J is a *model of* P if $J \models P$, and P is *consistent* if it has a model. Furthermore, the set $[\![P]\!]_{\mathsf{SM}}$ of *stable models of* P consists of all interpretations J such that $J^* = \mathsf{least}(P \cup \mathsf{def}(J))^1$ where $\mathsf{def}(J) = \{ \sim l. \mid l \in \mathcal{L} \setminus J \}$, $J^* = J \cup \sim(\mathcal{L} \setminus J)$ and $\mathsf{least}(\cdot)$ denotes the least model of the argument program in which all literals are treated as propositional atoms.

Turning our attention to rule updates, a *rule update semantics* assigns stable models to a sequence of programs where each component represents an update of the preceding ones. Formally, a *dynamic logic program* (DLP) is a finite sequence of programs and by $\mathsf{all}(\boldsymbol{P})$ we denote the multiset of all rules in the components of $\boldsymbol{P}$. A rule update semantics S assigns a *set of S-models*, denoted by $[\![\boldsymbol{P}]\!]_{\mathsf{S}}$, to $\boldsymbol{P}$. We concentrate on semantics based on the causal rejection principle [4–7,9,10,13] which states that a rule is *rejected* if it is in a direct conflict with a more recent rule. The fundamental type of conflict between rules π and σ occurs when they have complementary heads, i.e. $\mathsf{H}_\pi = \sim\mathsf{H}_\sigma$. We define the most mature of these semantics, the *refined dynamic stable models* [9,10]. Let $\boldsymbol{P} = \langle P_i \rangle_{i<n}$ be a DLP without strong negation. Given an interpretation J, the multisets of *rejected rules* $\mathsf{rej}(\boldsymbol{P},J)$ and of *default assumptions* $\mathsf{def}(\boldsymbol{P},J)$ are defined by:

$$\mathsf{rej}(\boldsymbol{P},J) = \big\{ \pi \in P_i \mid i < n \wedge \exists j \geq i \, \exists \sigma \in P_j : \mathsf{H}_\pi = \sim\mathsf{H}_\sigma \wedge J \models \mathsf{B}_\sigma \big\} \ ,$$
$$\mathsf{def}(\boldsymbol{P},J) = \big\{ \sim l \mid l \in \mathcal{L} \wedge \neg(\exists \pi \in \mathsf{all}(\boldsymbol{P}) : \mathsf{H}_\pi = l \wedge J \models \mathsf{B}_\pi) \big\} \ .$$

The set $[\![\boldsymbol{P}]\!]_{\mathsf{RD}}$ of *RD-models of* $\boldsymbol{P}$ consists of all interpretations J such that

$$J^* = \mathsf{least}\left([\mathsf{all}(\boldsymbol{P}) \setminus \mathsf{rej}(\boldsymbol{P},J)] \cup \mathsf{def}(\boldsymbol{P},J) \right) \ .$$

Support for strong negation can be added to this semantics by performing a syntactic transformation that translates conflicts between opposite objective literals l and $\neg l$ into conflicts between objective literals and their default negations. Two such transformations have been suggested based on the principle of coherence [7,17]. For any program P and DLP $\boldsymbol{P} = \langle P_i \rangle_{i<n}$ they are defined as follows: $\boldsymbol{P}^\dagger = \langle P_i^\dagger \rangle_{i<n}$, $\boldsymbol{P}^\ddagger = \langle P_i^\ddagger \rangle_{i<n}$,

$$P^\dagger = P \cup \big\{ \sim\neg l \leftarrow l. \mid l \in \mathcal{L} \big\} \ , \quad P^\ddagger = P \cup \big\{ \sim\neg\mathsf{H}_\pi \leftarrow \mathsf{B}_\pi. \mid \pi \in P \wedge \mathsf{H}_\pi \in \mathcal{L} \big\} \ .$$

---

[1] The original definition based on reducts [2,3,18] is equivalent to the one we use here [7].

3   Early Recovery Principle

The problem with existing semantics for strong negation in rule updates is that semantics based on the first transformation ($P^\dagger$) assign too many models to some DLPs, while semantics based on the second transformation ($P^\ddagger$) sometimes do not assign any model to a DLP that should have one. The former is illustrated in the following example:

Example 3. Consider the DLP $P_1 = \langle P, U \rangle$ where $P = \{p., \neg p.\}$ and $U = \emptyset$. Since P has no stable model and U does not encode any change in the represented domain, it should follow that P_1 has no stable model either. However, $[\![P_1^\dagger]\!]_{RD} = \{\{p\}, \{\neg p\}\}$, i.e. two models are assigned to P_1 when using the first transformation to add support for strong negation. To verify this, observe that $P_1^\dagger = \langle P^\dagger, U^\dagger \rangle$ where

$$P^\dagger: \qquad p. \qquad\qquad \neg p. \qquad\qquad\qquad U^\dagger: \qquad {\sim}p \leftarrow \neg p. \qquad {\sim}\neg p \leftarrow p.$$
$$\qquad\quad {\sim}p \leftarrow \neg p. \qquad {\sim}\neg p \leftarrow p.$$

Consider the interpretation $J_1 = \{p\}$. It is not difficult to verify that $\mathsf{rej}(P_1^\dagger, J_1) = \{\neg p., {\sim}\neg p \leftarrow p.\}$ and $\mathsf{def}(P_1^\dagger, J_1) = \emptyset$, so it follows that

$$\mathsf{least}\left(\left[\mathsf{all}(P_1^\dagger) \setminus \mathsf{rej}(P_1^\dagger, J_1)\right] \cup \mathsf{def}(P_1^\dagger, J_1)\right) = \{p, {\sim}\neg p\} = J_1^* \ .$$

In other words, $J_1 \in [\![P_1^\dagger]\!]_{RD}$ and similarly it can be verified that $\{\neg p\} \in [\![P_1^\dagger]\!]_{RD}$.

Thus, the problem with the first transformation is that an update by an empty program, which does not express any change in the represented domain, may affect the original semantics. This behaviour goes against basic and intuitive principles underlying updates, grounded already in the classical belief update postulates [19, 20] and satisfied by virtually all belief update operations [21] as well as by the vast majority of existing rule update semantics, including the original RD-semantics.

This undesired behaviour can be corrected by using the second transformation instead. The more technical reason is that it does not add any rules to a program in the sequence unless that program already contains some original rules. However, its use leads to another problem: sometimes *no model* is assigned when in fact one is expected.

Example 4. Consider again Example 2, formalised as the DLP $P_2 = \langle P, V \rangle$ where $P = \{p., \neg p.\}$ and $V = \{{\sim}p.\}$. It is reasonable to expect that since V resolves the conflict present in P, a stable model should be assigned to P_2. However, $[\![P_2^\ddagger]\!]_{RD} = \emptyset$. To verify this, observe that $P_2^\ddagger = \langle P^\ddagger, V^\ddagger \rangle$ where $P^\ddagger = \{p., \neg p., {\sim}p., {\sim}\neg p.\}$ and $V^\ddagger = \{{\sim}p.\}$. Given an interpretation J, we conclude that $\mathsf{rej}(P_2^\ddagger, J) = P^\ddagger$ and $\mathsf{def}(P_2^\ddagger, J) = \emptyset$, so J cannot belong to $[\![P_2^\ddagger]\!]_{RD}$ since

$$\mathsf{least}\left(\left[\mathsf{all}(P_2^\ddagger) \setminus \mathsf{rej}(P_2^\ddagger, J)\right] \cup \mathsf{def}(P_2^\ddagger, J)\right) = \{{\sim}p\} \neq J^* \ .$$

Based on this example, in the following we formulate a generic *early recovery principle* that formally identifies conditions under which *some* stable model should be assigned to a DLP. For the sake of simplicity, we concentrate on DLPs of length 2 which are composed of facts.

We begin by defining, for every objective literal l, the sets of literals $\overline{l}$ and $\overline{\sim l}$ as follows: $\overline{l} = \{\sim l, \neg l\}$ and $\overline{\sim l} = \{l\}$. Intuitively, for every literal L, $\overline{L}$ denotes the set of literals that are in conflict with L. Furthermore, given two sets of facts P and U, we say that U *solves all conflicts in* P if for each pair of rules $\pi, \sigma \in P$ such that $H_\sigma \in \overline{H_\pi}$ there is a fact $\rho \in U$ such that either $H_\rho \in \overline{H_\pi}$ or $H_\rho \in \overline{H_\sigma}$.

Considering a rule update semantics S, the new principle simply requires that when U solves all conflicts in P, S will assign *some model* to $\langle P, U \rangle$. Formally:

Early recovery principle: If P is a set of facts and U is a consistent set of facts that solves all conflicts in P, then $[\![\langle P, U \rangle]\!]_S \neq \emptyset$.

We conjecture that rule update semantics should generally satisfy the above principle. In contrast with the usual behaviour of belief update operators, the nature of existing rule update semantics ensures that recovery from conflict is always possible, and this principle simply formalises the sufficient conditions for such recovery.

The introduced principle can guide the future addition of full support for both kinds of negations in other approaches to rule updates, such as those proposed in [8, 11, 14, 16]. Stronger versions of the principle that apply to DLPs of arbitrary length and with programs other than just sets of facts are also conceivable.

Furthermore, these considerations are also interesting in the context of updates of hybrid knowledge bases [22, 23] and for the development of well-founded rule and hybrid update semantics [24, 25]. An interesting path for future work is also the study of updates of expressive extensions of logic programs [26] in which different negations besides the classical one can be considered.

References

1. Berners-Lee, T., Hendler, J., Lassila, O.: The semantic web. Scientific American 284(5), 28–37 (2001)
2. Gelfond, M., Lifschitz, V.: The stable model semantics for logic programming. In: Kowalski, R.A., Bowen, K.A. (eds.) Proceedings of the 5th International Conference and Symposium on Logic Programming, ICLP/SLP 1988, pp. 1070–1080. MIT Press (1988)
3. Gelfond, M., Lifschitz, V.: Classical negation in logic programs and disjunctive databases. New Generation Computing 9(3-4), 365–385 (1991)
4. Leite, J.A., Pereira, L.M.: Generalizing updates: From models to programs. In: Dix, J., Moniz Pereira, L., Przymusinski, T.C. (eds.) LPKR 1997. LNCS (LNAI), vol. 1471, pp. 224–246. Springer, Heidelberg (1998)
5. Alferes, J.J., Leite, J.A., Pereira, L.M., Przymusinska, H., Przymusinski, T.C.: Dynamic updates of non-monotonic knowledge bases. The Journal of Logic Programming 45(1-3), 43–70 (2000)
6. Eiter, T., Fink, M., Sabbatini, G., Tompits, H.: On properties of update sequences based on causal rejection. Theory and Practice of Logic Programming (TPLP) 2(6), 721–777 (2002)
7. Leite, J.A.: Evolving Knowledge Bases. Frontiers of Artificial Intelligence and Applications, vol. 81, xviii + 307 p. Hardcover. IOS Press (2003)
8. Sakama, C., Inoue, K.: An abductive framework for computing knowledge base updates. Theory and Practice of Logic Programming (TPLP) 3(6), 671–713 (2003)
9. Alferes, J.J., Banti, F., Brogi, A., Leite, J.A.: The refined extension principle for semantics of dynamic logic programming. Studia Logica 79(1), 7–32 (2005)

10. Banti, F., Alferes, J.J., Brogi, A., Hitzler, P.: The well supported semantics for multidimensional dynamic logic programs. In: Baral, C., Greco, G., Leone, N., Terracina, G. (eds.) LPNMR 2005. LNCS (LNAI), vol. 3662, pp. 356–368. Springer, Heidelberg (2005)

11. Zhang, Y.: Logic program-based updates. ACM Transactions on Computational Logic 7(3), 421–472 (2006)

12. Šefránek, J.: Irrelevant updates and nonmonotonic assumptions. In: Fisher, M., van der Hoek, W., Konev, B., Lisitsa, A. (eds.) JELIA 2006. LNCS (LNAI), vol. 4160, pp. 426–438. Springer, Heidelberg (2006)

13. Osorio, M., Cuevas, V.: Updates in answer set programming: An approach based on basic structural properties. Theory and Practice of Logic Programming 7(4), 451–479 (2007)

14. Delgrande, J.P., Schaub, T., Tompits, H.: A preference-based framework for updating logic programs. In: Baral, C., Brewka, G., Schlipf, J. (eds.) LPNMR 2007. LNCS (LNAI), vol. 4483, pp. 71–83. Springer, Heidelberg (2007)

15. Šefránek, J.: Static and dynamic semantics: Preliminary report. In: Mexican International Conference on Artificial Intelligence, pp. 36–42 (2011)

16. Krümpelmann, P.: Dependency semantics for sequences of extended logic programs. Logic Journal of the IGPL 20(5), 943–966 (2012)

17. Alferes, J.J., Pereira, L.M.: Update-programs can update programs. In: Dix, J., Przymusinski, T.C., Moniz Pereira, L. (eds.) NMELP 1996. LNCS, vol. 1216, pp. 110–131. Springer, Heidelberg (1997)

18. Inoue, K., Sakama, C.: Negation as failure in the head. Journal of Logic Programming 35(1), 39–78 (1998)

19. Keller, A.M., Winslett, M.: On the use of an extended relational model to handle changing incomplete information. IEEE Transactions on Software Engineering 11(7), 620–633 (1985)

20. Katsuno, H., Mendelzon, A.O.: On the difference between updating a knowledge base and revising it. In: Allen, J.F., Fikes, R., Sandewall, E. (eds.) Proceedings of the 2nd International Conference on Principles of Knowledge Representation and Reasoning, KR 1991, pp. 387–394. Morgan Kaufmann Publishers (1991)

21. Herzig, A., Rifi, O.: Propositional belief base update and minimal change. Artificial Intelligence 115(1), 107–138 (1999)

22. Slota, M., Leite, J.: Towards Closed World Reasoning in Dynamic Open Worlds. Theory and Practice of Logic Programming 10(4-6), 547–564 (2010)

23. Slota, M., Leite, J., Swift, T.: Splitting and updating hybrid knowledge bases. Theory and Practice of Logic Programming 11(4-5), 801–819 (2011)

24. Banti, F., Alferes, J.J., Brogi, A.: Well founded semantics for logic program updates. In: Lemaître, C., Reyes, C.A., González, J.A. (eds.) IBERAMIA 2004. LNCS (LNAI), vol. 3315, pp. 397–407. Springer, Heidelberg (2004)

25. Knorr, M., Alferes, J.J., Hitzler, P.: Local closed world reasoning with description logics under the well-founded semantics. Artificial Intelligence 175(9-10), 1528–1554 (2011)

26. Gonçalves, R., Alferes, J.J.: Parametrized logic programming. In: Janhunen, T., Niemelä, I. (eds.) JELIA 2010. LNCS, vol. 6341, pp. 182–194. Springer, Heidelberg (2010)

Preference Handling
for Belief-Based Rational Decisions*

Samy Sá** and João Alcântara**

Universidade Federal do Ceará
MDCC, Campus do Pici, Bl 910
Fortaleza, Brazil
`samy@ufc.br, jnando@lia.ufc.br`

Abstract. We introduce an approach to preferences suitable for agents
that base decisions on their beliefs. In our work, agents' preferences are
perceived as a consequence of their beliefs, but at the same time are
used to feed the knowledge base with beliefs about preferences. As a result, agents can reason with preferences to hypothesize, explain decisions,
and review preferences in face of new information. Finally, we integrate
utility-based to reasoning-based criteria of decision making.

Keywords: Reasoning with Preferences, Preference Handling, Decision
Making.

1   Introduction

Autonomous agents are frequently required to make decisions and expected to do
so according to their beliefs, goals, and preferences, however, beliefs are rarely
connected to the preferences of the agent. Dietrich and List argue in [2] that
logical reasoning and the economic concept of rationality are almost entirely
disconnected in the literature: while logical accounts of reasoning rarely gets to
deal with rational decisions in the economic sense, social choice is never wor-
ried about the origin of agents' preferences. But if preferences are disconnected
from beliefs, how can an agent explain their decisions? How could we model the
influence of new information in an agent's preferences?

  In order to reason about options available in a decision, an agent needs a
complimentary theory of what the best options are. Since rationality involves
trying to maximize gains (utility), the successful integration of rationality and
beliefs requires quantifying the utility of some of those beliefs. Further, reasoning
with preferences under uncertainty is a feature rarely considered in the literature (as mentioned in [3]), even though it plays a key role in rational choice and
game theory [9]. Our approach also considers agent theories with multiple models, therefore accounting for decisions under uncertainty while making it easy to

---

* This is a reviewed and extended version of [10] focused on single agent decisions.
** This research is supported by CNPq (Universal 2012 - Proc. n 473110/2012-1),
  CAPES (PROCAD 2009), CNPq/CAPES (Casadinho/PROCAD 2011).

P. Cabalar and T.C. Son (Eds.): LPNMR 2013, LNAI 8148, pp. 518–523, 2013.

perform reasoning tasks such as abduction and belief revision as means to evaluate or update an agent's preferences. In this paper, we show how comparing available options by weighing their relevant features can successfully connect rationality and qualitative preferences in a way to relate rational decision criteria [9] and an inherently qualitative criteria for reasoning-based decisions [5].

2   Preferences as Utility + Beliefs

In this section, we introduce our notions of preference profiles and quality thresholds. Agent preferences are a consequence of their beliefs as we attribute weights to unary predicates meaning features relevant to judge options (alike to *weighted propositional formulas* [8]) to build an unary utility function. Quality thresholds are used to classify options as good, poor or neutral in the eyes of the agent.

Definition 1. *(preference profile) Let Pred be the set of all unary predicates $P(x)$ used to express possible features of objects in the language of an agent. A* preference profile *is a triple $Pr = \langle Ut, Up, Lw \rangle$, involving a utility function $Ut : Pred \rightarrow \mathbb{R}$, and upper and lower utility thresholds $Up, Lw \in \mathbb{R}$, $Up \geq Lw$.*

An agent can have as many preference profiles as desired for each kind of decision the agent may get involved (as proposed in [2]). Each profile (with utilities and thresholds) can be obtained, for instance, from design (just as with the knowledge base) or by modeling user preferences. In each case, a preference profile gives us two perspectives of preferences: quantitative and qualitative.

Given an agent theory and a preference profile $Pr = \langle Ut, Up, Lw \rangle$, ranking possible outcomes is straightforward: Let $Alt = \{o_1, \ldots, o_n\}$ be the set of available options, each option o_i, $1 \leq i \leq n$, has attributed utility

$$Ut^S(o_i) = \sum_{P(x),\ P(o_i)\ \in\ S} Ut(P(x)),$$

where S is a model of the agent's knowledge base. In the context of a particular model S, o_i is a *good option* if $Ut^S(o_i) \geq Up$, a *poor option* if $Ut^S(o_i) < Lw$, and neutral (neither good or poor) otherwise. In that sense, if $Up = Lw$, there are only good and poor options in the eyes of the agent.

The concept of preference profile in Definition 1 induces, for each model S of the agent theory, a *total preorder*[1] over the set of available options

$$\succcurlyeq^S = \{(o_i, o_j) \mid Ut^S(o_i) \geq Ut^S(o_j)\}.$$

The proposition $o_i \succcurlyeq^S o_j$ is usually read as "o_i is *weakly preferred* to o_j [6] in S". If the agent is *indifferent* about two options o_i, o_j, i.e., if both $o_i \succcurlyeq^S o_j$ and $o_j \succcurlyeq^S o_i$ hold, we write $o_i \sim^S o_j$.

From now on, we will account for agents as a pair $Ag = (KB, Prefs)$ involving a program KB and a set $Prefs$ of preference profiles. Also, whenever we mention an utility function Ut or thresholds Up, Lw, the assumption that $Pr = \langle Ut, Up, Lw \rangle$ is the selected preference profile in a decision is left implicit.

---

[1] A *total preorder* is a relation that is transitive, reflexive and in which any two elements are related. Total preorders are also called *weak orders*.

3   Beliefs about Preferences

We now show how to build a general theory of preferences to reason about available options. We use a preference profile (Sec. 2) to devise a set of special rules (a module) about the quality of available options. The rules encoding preferences are not to interfere with the models of the original program: It is only tolerated to add conclusions about the quality of options. New predicates $G(x), N(x)$ are introduced to KB, standing for *good* $(G(x))$ and *neutral* $(N(x))$ options, while options that are not good $(\neg G(x))$ are *poor*. Such predicates are the only ones allowed in the head of rules encoding preferences.

Let R^* be the set of all predicates relevant to the decision, i.e., any $P(x)$ with $Ut(P(x)) \neq 0$ in Pr. To build the set KB_{Pr} of preference rules, first compute

$$^R Ut = \sum_{P(x) \in R} Ut(P(x)),$$

for each $R \subseteq R^*$. Then, for each such R, KB_{Pr} should have a rule r in the general form $\bigwedge(\{P(x)|P(x) \in R\} \cup \{\neg P(x)|P(x) \in R^* \setminus R\}) \rightarrow conc(r)$, where $conc(r) = G(x)$ if $^R Ut \geq Up$, $conc(r) = \neg G(x)$ if $^R Ut < Lw$, and $conc(r) = N(x)$ otherwise.

If the agent $Ag = (KB, Prefs)$ considers a particular preference profile Pr, its reasoning will be modeled by inference in $KB' = KB \cup KB_{Pr}$. Observe that KB' is consistent if KB is consistent, since the models are the same with added conclusions on the quality of options. Also, the number of rules is exponential in $|R^*|$, but they only need to be calculated once (we can save it in a file, for instance) and $|R^*|$ is not likely to be large.

By employing rules as above, we have preferences integrated with beliefs in two ways. First, beliefs promote the utility of each option, so preferences are based on them. Second, the calculated utilities are used to complement the knowledge base with predicated formulas regarding beliefs about them.

4   On Multiple Preference Profiles

In [2], Dietrich and List propose that the preferences of an agent are conditioned by a set of reasons M and each subset of M induces a different preference ordering. Two axioms are proposed to govern the relationship between an agent's set of beliefs and their preferences across variations of their motivational state:

Axiom 1. *([2]) For any two options $o_i, o_j \in O$ and any $M \subseteq Pred$, if $\{P(x) \in M| P(o_i) \text{ holds}\} = \{P(x) \in M \mid P(o_j) \text{ holds}\}$, then $x \sim y$.*

Axiom 2. *([2]) For any $o_i, o_j \in O$ and any $M, M' \subseteq Pred$ such that $M' \supseteq M$, if $\{P(x) \in M' \setminus M| P(o_i) \text{ or } P(o_j) \text{ hold}\} = \emptyset$, then $x \succcurlyeq_M y \Leftrightarrow x \succcurlyeq_{M'} y$.*

Roughly speaking, Axiom 1 states that two options with the exact same characteristics should be equally preferred. Axiom 2 treats the case where new attributes become relevant to the decision. In that case, the preferences over any two options that do not satisfy the extra criteria should remain unchanged.

Theorem 1. *Each preference relation $\succ^S$ (Sec. 2) satisfies Axioms 1 and 2.*

Proof. Consider any consistent answer set S of the agent's knowledge base KB. If two options satisfy the exact same predicates in S, the utility attributed to both options will be the same, so Axiom 1 is satisfied. Now suppose a preference profile Pr is updated in a way the predicate $P(x)$ becomes relevant, i.e., in the updated profile Pr', $Ut'(P(x)) \neq 0$, while originally, $Ut(P(x)) = 0$. If neither options o_i, o_j satisfy $P(x)$ in S, i..e, if $P(o_i), P(o_j) \notin S$, then their utility remains unchanged. Therefore, Axiom 2 is satisfied.

As a consequence, our approach to model preferences of an agent will adequately relate their set of beliefs to their preferences. Variations of the agent's motivational state are modeled as different profiles for the same decisions. Preferences are integrated to the knowledge base as in Section 3.

5   Making Decisions

When there is no uncertainty, the decision of a single agent should be straightforward: Just maximize utility. Since options in our approach can satisfy positive and negative features, maximizing utility can be perceived as comparing and weighting such features, as in the principle of *Bivariate monotonicity* [5].

Definition 2. *(bivariate monotonicity) A decision criteria satisfies bivariate monotonicity if, for each answer set S of KB, whenever $o_i^{+s} \supseteq o_j^{+s}$ and $o_i^{-s} \subseteq o_j^{-s}$ available, it prefers o_i at least as much as o_j.*

Bivariate monotonicity states that whenever an option has all advantages of another (possibly more), but no disadvantages the second does not (possibly less), the first option is possibly better. Of course, if the set of pros and cons is exactly the same, one option is just as good as the other.

In the following, consider o_i and o_j are two options, KB denotes a knowledge base, and the preference profile $Pr = \langle Ut, Up, Lw \rangle$ is selected. Let o_i^{+s} denote the set of positive features satisfied by o_i in S, i.e., those $P(x)$ such that $P(o_i) \in S$ and $Ut(P(x)) > 0$, and o_i^{-s} denote the set of negative features satisfied by o_i in S, i.e., those $P(x)$ such that $P(o_i) \in S$ and $Ut(P(x)) < 0$.

Theorem 2. *When there is no uncertainty, i.e., there is a single plausible scenario, maximizing utility satisfies bivariate monotonicity.*

Proof. In any specific scenario S, if $o_i^+ \supseteq o_j^+$ while $o_i^- \subseteq o_j^-$, we will have $Ut^S(o_i) \geq Ut^S(o_j)$, so maximizing utility satisfies bivariate monotonicity.

6   Decisions under Uncertainty

Our notion of preference profile induces the construction of a payoff matrix in case of multiple plausible scenarios, so we employ decision criteria from game theory [9], namely the *maximin* criteria.

Definition 3. *(maximin criteria) In a decision problem with uncertainty, let AS denote the set of possible scenarios (models of a theory) according to the beliefs of an agent. An option o_i is at least as preferred as the option o_j by the maximin criteria, i.e. $o_i \succcurlyeq_{MAXIMIN} o_j$ iff*

$$MIN(\{Ut^S(o_i) \mid S \in AS\}) \geq MIN(\{Ut^S(o_j) \mid S \in AS\}).$$

The best options are those o_i such that $o_i \succcurlyeq_{MAXIMIN} o_j$ for all o_j, $j \neq i$.

Theorem 3. *The maximin criteria for decision making under uncertainty satisfies bivariate monotonicity (Definition 2) in the case of multiple scenarios.*

Proof. Let o_i, o_j be any two available options, and suppose $o_i^{+s} \supseteq o_j^{+s}$ and $o_i^{-s} \subseteq o_j^{-s}$ hold in every $S \in AS$ (if any different, the options cannot be directly compared). Let the quality of any options with the least quality be $MIN(\{Ut^S(o_j) \mid S \in AS\}) = m$. Let $M \in AS$ be any answer set such that $Ut^M(o_j) = m$. By Theorem 2, $Ut_M(o_i) \geq Ut_M(o_j)$, so the least utility attributed to o_i is $MIN(\{Ut^S(o_i) \mid S \in AS\}) \geq m$ and $o_i \succcurlyeq_{MAXIMIN} o_j$. As a consequence, $o_i \succcurlyeq_{MAXIMIN} o_j$, i.e., o_i is at least as preferred as o_j.

7   Related Work

Dietrich and List, in [2], observe that logical reasoning and the economic concept of rationality are almost entirely disconnected in the literature and propose preference orderings based on alternative logical contexts as being different psychological states of the agent. We connect logics and utility-based decisions in this paper by attaching utilities to predicates, so the best outcomes satisfy the most interesting combinations of predicates. We then connect rational behavior from game theory to bipolar preferences due to Dubois et al. [5]. On the other hand, Dietrich and List are not concerned with reasoning about preferences in [2], but restrict their analysis to how beliefs can influence decisions.

Lafage and Lang propose in [8] an approach to group decisions based on weighted logics in which formulas represent constraints and weights are attributed to each formula to quantify importance. This is related to the way we build utility functions and preferences, however they allow formulas with connectives, while we restrict utility evaluation to ground formulas with unary predicates. Brafman proposes in [1] a relational language of preferences where rules may include utility values in the conclusions. The author argues the approach is flexible, as their value functions can handle a dynamic universe of objects. We consider our approach is also very flexible, as an agent can quickly evaluate a new option, since all the weight is in the features relevant to the agent.

Our proposal takes advantage of multiple models of a theory to reason over uncertainty as decision are utility-based and built over unary predicates. This kind of approach is surprisingly not common, as also stated by Doyle in [4]. A notable exception is [7], where Labreuche deals with the issue of explaining a decision made in a multi-attribute decision model where attributes are weighted in an attempt to circumvent the difficulties of properly explaining utility-based decisions. Anyway, our work has an alike motivation: arguments are better related to logic-based formalisms, so we attach weights to attributes (predicates) to calculate preferences and translate them back into a logical language. For an account of approaches and applications of preferences in Artificial Intelligence, see the more recent survey from Domshlak et al. [3].

8   Conclusions and Future Work

We have introduced an approach to integrate beliefs and rationality in a way utility-based criteria for decisions is shown to satisfy criteria from reasoning-based decision making. This is achieved by attributing utility to unary predicates used to encode relevant qualities of a decision, together with the notion of *qualitative thresholds*, both parts of our preference profiles. By doing so, agent preferences are simultaneously perceived in different perspectives: a utility-based cardinal order, a regular ordinal order and a classification in good/poor/neutral options. All three perspectives are integrated as we devise rules that encode the preferences of the agent and append them to its knowledge base for reasoning purposes. The result is a formalism capable of modeling reasoning about preferences where beliefs are the base for rational decisions. Next, we intend to explore how this model can be employed in group decisions by deliberation and voting.

References

1. Brafman, R.I.: Relational preference rules for control. Artif. Intell. 175(7-8), 1180–1193 (2011)
2. Dietrich, F., List, C.: A reason-based theory of rational choice*. Noûs 47(1), 104–134 (2013)
3. Domshlak, C., Hüllermeier, E., Kaci, S., Prade, H.: Preferences in AI: An overview. Artif. Intell. 175(7-8), 1037–1052 (2011)
4. Doyle, J.: Prospects for preferences. Computational Intelligence 20(2), 111–136 (2004)
5. Dubois, D., Fargier, H., Bonnefon, J.-F.: On the qualitative comparison of decisions having positive and negative features. J. Artif. Intell. Res. (JAIR) 32, 385–417 (2008)
6. Fishburn, P.C.: Preference structures and their numerical representations. Theor. Comput. Sci. 217(2), 359–383 (1999)
7. Labreuche, C.: A general framework for explaining the results of a multi-attribute preference model. Artif. Intell. 175(7-8), 1410–1448 (2011)
8. Lafage, C., Lang, J.: Logical representation of preferences for group decision making. In: Cohn, A.G., Giunchiglia, F., Selman, B. (eds.) KR, pp. 457–468. Morgan Kaufmann (2000)
9. Osborne, M., Rubinstein, A.: A Course in Game Theory. MIT Press (1994)
10. Sá, S., Alcântara, J.: Preferences with qualitative thresholds and methods for individual and collective decisions (extended abstract). In: Ito, onker, Gini, Shehory (eds.) AAMAS. IFAAMAS (2013)

Logic-Based Techniques
for Data Cleaning: An Application
to the Italian National Healthcare System[*]

Giorgio Terracina[1], Alessandra Martello[2], and Nicola Leone[1]

[1] Università della Calabria
[2] DLVSystem s.r.l.
{terracina,leone}@mat.unical.it, martello@dlvsystem.com

Abstract. In this paper we present a technique based on logic programming for data cleaning, and its application to a real use case from the Italian Healthcare System. The use case is part of a more complex project developing a business intelligence suite for the analysis of distributed archives of tumor-based diseases.

1 Introduction

Data anomalies can be roughly classified in syntactic (attribute-level), semantic (tuple-level), and coverage anomalies (e.g., missing values). Data cleaning tasks may also be classified as single-source or multi-source, depending on the number of sources they are able to deal with. Proposed approaches span over several methods, such as *parsing* [5], *integrity constraint enforcement* [4], *duplicate elimination* [2], and *statistical methods* [3]. Only few data cleaning approaches are based on *computational logic*. In fact, data cleaning activities inherently involve procedural tasks and mathematical computations, and require access to possibly distributed databases; this does not fit well with many logic-based frameworks.

In this paper we address the issue of data cleaning from a logic-based perspective. In particular, we resort to recent extensions for logic programming [1] and to evaluation engines supporting them, like DLV^{DB} [6], allowing to both access possibly distributed databases, and include inherently procedural subtasks or mathematical computations directly into logic programs specification. The proposed approach is logic-based and addresses multi-source data cleaning for syntactic and semantic anomaly detection, where a reference dictionary is used for anomaly detection and their potential correction.

The proposed approach should be considered complementary to the existing ones, and capable to provide simplified and flexible specifications of the logic of the data cleaning task. The approach has been implemented as a plugin of Pentaho Data Integration (Kettle)[1] and applied to the AIRTUM real use case.

---

[*] This work has been partially funded by the Calabria Region under the project "Piano Interaziendale di Innovazione del Contratto di Investimento Industria, Artigianato e Servizi" N. 1220000408.

[1] http://kettle.pentaho.com/

P. Cabalar and T.C. Son (Eds.): LPNMR 2013, LNAI 8148, pp. 524–529, 2013.

2   A Logic-Based Approach for Data Cleaning

The main goal of our approach is the definition of an automatic procedure for generating logic programs able to identify and, whenever possible, correct errors within the data. Another objective is to provide an easily configurable solution, also adaptable to several domain-specific applications.

In order to perform record validation and comparison our approach assumes to exploit additional knowledge bases (also called *dictionaries*); moreover, since exact matches are usually not sufficient to detect inter-source matchings and errors in the data, and since we want to provide also a semi-automatic correction strategy, we resort to logic programming language extensions allowing function calls within logic programs [1,6].

Data involved in the cleaning process can be categorized in several ways, such as *correct, not correct but with values within the reference domain, not correct with values outside the reference domain, certainly repairable* (i.e. data is wrong, but it is possible to identify exactly one correction), *possibly repairable* (i.e. data is wrong but there is ambiguity on the proper correction), and *wrong and not repairable*. Then, analyzed data are classified and streamed out separately.

Starting from some basic input specifications, the approach builds the logic program for automatic error detection and semi-automatic error correction.

In order for the approach to work properly, the user should specify:

(1) the table $t(\overline{K_t}, \overline{J_t})$ to be validated;
(2) the reference dictionary $d(\overline{K_d}, \overline{J_d})$;
(3) join conditions between t and d (possibly expressed through a matching function f_M);
(4) the comparison function f_R, and the corresponding threshold th_R, to be used for suggesting corrections;
(5) the reference to the DBMS functions corresponding to f_M and f_R.
(6) ODBC access paths for DBMS data sources.

The notation $t(\overline{K_t}, \overline{J_t})$ stands for table t having the list of key attributes $\overline{K_t}$; attributes $\overline{J_t}$ are those involved in some join. We will refer to the i-th variable in $\overline{J_t}$ as $\overline{J_t}[i]$. Each table may have some other attributes not involved in the cleaning process. Without loss of generality we do not explicitly mention them.

Join conditions specify the mappings between attributes in t and attributes in d, possibly transformed by some matching function. As an example, an attribute in t representing a date, may be required to be transformed in some standard format first, in order to be matched with a dictionary of dates, and the corresponding join condition may be expressed as:

$t(\dots myDate \dots), d(\dots refDate \dots), \#standardize(myDate, refDate)$

The algorithm we have designed for the automatic generation of logic programs for data cleaning is shown in Procedure RuleBuilder. The first step (Row 3) identifies the set of tuples *tuple_ok* without errors. Observe that, if an exact matching is sufficient, $\#f_M$ can be safely downgraded to the identity operator. Row 5 identifies tuple-level errors (named *tuple_not_ok*).These are then detailed at attribute-level by considering two kinds of attribute errors: (*i*) *in_dic*, i.e. the

---

Procedure RuleBuilder(Tables t, d, Functions $\#f_M, \#f_R$)

  Input : Set of specifications
  Output: Set R of rules
1 **begin**
2 $R := \emptyset$;
3 $R := R \cup$
   $tuple\_ok(\overline{K_t}, \overline{J_t}) \leftarrow t(\overline{K_t}, \overline{J_t}, \_), d(\overline{K_d}, \overline{J_d}, \_), \#f_M(\overline{J_t}[1], \overline{J_d}[1]), \dots \#f_M(\overline{J_t}[m], \overline{J_d}[m]);$
4 % *error detection phase;*
5 $R := R \cup \ tuple\_not\_ok(\overline{K_t}, \overline{J_t}) \leftarrow t(\overline{K_t}, \overline{J_t}, \_), not \ tuple\_ok(\overline{K_t}, \overline{J_t});$
6 **foreach** $1 \leq i \leq |\overline{J_t}|$ **do**
7   $R := R \cup \ in\_dic\_J_i(\overline{K_t}, \overline{J_t}[i]) \leftarrow tuple\_not\_ok(\overline{K_t}, \overline{J_t}), d(\overline{K_d}, \overline{J_d}, \_), \#f_M(\overline{J_t}[i], \overline{J_d}[i]);$
8 $R := R \cup \ inconst\_in\_dic(\overline{K_t}, \overline{J_t}) \leftarrow in\_dic\_J_1(\overline{K_t}, \overline{J_t}[1]), \dots, in\_dic\_J_m(\overline{K_t}, \overline{J_t}[m]);$
9 $R := R \cup \ inconst\_out\_dic(\overline{K_t}, \overline{J_t}) \leftarrow tuple\_not\_ok(\overline{K_t}, \overline{J_t}), not \ inconst\_in\_dic(\overline{K_t}, \overline{J_t});$
10 **foreach** $1 \leq i \leq |\overline{J_t}|$ **do**
11   $R := R \cup \ out\_dic\_J_i(\overline{K_t}, \overline{J_t}[i]) \leftarrow inconst\_out\_dic(\overline{K_t}, \overline{J_t}), not \ in\_dic\_J_i(\overline{K_t}, \overline{J_t}[i]);$
12 % *error correction phase ;*
13 **foreach** $1 \leq i \leq |\overline{J_t}|$ **do**
14   $R := R \cup \ repair\_tab\_attr\_J_i(\overline{K_t}, \overline{J_t}, \overline{K_t}, \overline{J_t}[1], \dots \overline{J_d}[i], \dots, \overline{J_t}[m], R) \leftarrow$
15    $tuple\_not\_ok(\overline{K_t}, \overline{J_t}), d(\overline{K_d}, \overline{J_d}, \_), \#f_M(\overline{J_t}[1], \overline{J_d}[1]), \dots$
16    $\#f_M(\overline{J_t}[i-1], \overline{J_d}[i-1]), \#f_M(\overline{J_t}[i+1], \overline{J_d}[i+1]) \dots \#f_M(\overline{J_t}[m], \overline{J_d}[m]),$
17    $\#f_R(\overline{J_t}[i], \overline{J_d}[i], R), R < th_R;$

---

attribute value is wrong but it falls within the dictionary domain, (ii) *out_dic*, i.e. the attribute value is wrong and outside the dictionary domain. Rows 6-7 and 10-11 classify each attribute of a wrong tuple accordingly. Based on this attribute classification, the approach classifies wrong tuples as either having all attribute values within the dictionary (*inconst_in_dic* in Row 8) or with at least one attribute value falling outside the dictionary (*inconst_out_dic* in Row 9).

The next step, which is carried out in Rows 13–17, identifies possible automatic attribute corrections for wrong tuples. Specifically, given the i-th join attribute of t, a *repair_tab_attr_J_i* table is built which contains the original (wrong) tuple $\langle \overline{K_t}, \overline{J_t} \rangle$ and the proposed modification as $\langle \overline{K_t}, \overline{J_t}[1], \dots \overline{J_d}[i], \dots, \overline{J_t}[m] \rangle$, where $\overline{J_d}[i]$ is a value from the dictionary. A measure of confidence for the proposed modification is given by the value R in *repair_tab_attr_J_i*, whose computation is specified next.

A tuple is considered repairable if exactly one attribute is wrong w.r.t. the dictionary. Observe that there may be several tuples in the dictionary satisfying the join condition imposed for d and t in Rows 14–17. Then, the potential corrections are those dictionary values $\overline{J_d}[i]$ sufficiently similar to the wrong $\overline{J_t}[i]$; function $\#f_R$ has precisely the role of measuring the distance R between the two[2]. If R is below the threshold th_R, $\overline{J_d}[i]$ is proposed as a correction, along with R. The value R could also be used as a quality measure to rank the proposed correction in subsequent steps of the computation.

---

[2] In Procedure RuleBuilder, we assumed $\#f_R$ to provide a distance measure R (this could be, e.g., Hamming or Levenshtein for strings, or a simple difference for integers); a switch to similarity measures is straightforward.

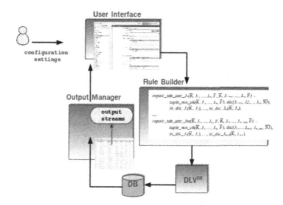

Fig. 1. DLVCleaner Plugin Architecture

The approach has been implemented as a Pentaho Kettle plugin, we called DLVCleaner, whose architecture is shown in Figure 1.

3 Application to the AIRTUM Use Case

In this section we describe the AIRTUM Tumor Registry use case we exploited as an application of our approach. This use case is part of a project funded by the Calabria Region which involves several IT companies and institutions from Calabria. The AIRTUM tumor registry used in the project considers information pertaining to several local healthcare centers (called ASL) from Calabria. Data are collected from many different sources, including public hospitals, ASL, regional sanitary agencies, family doctors, etc. Collected information include the kind of diagnosed cancer, personal data of the patient, current clinical conditions, past and current treatments, disease evolution, etc. All such information are extremely important to analyze causes and evolutions of cancer diseases, in order to study proper treatments, prevention policies, and to schedule sanitary budgets. Overall, the project considered more than 200 tables as sources of data.

Almost all of this information should be inter-linked by the identity of the patient. However, such an information is often imprecise in local sources, since in many cases data are loaded manually. Even patient identification codes, like Tax Codes or Social Security Numbers are often absent or wrong. As a consequence, one of the crucial tasks in this project was the proper identification of each mentioned patient through a subset of its attributes. However, even different official municipality registries used different schemas and standards to represent data; this required a careful analysis of official data too.

The approach followed in the project consisted in first gathering a set of basic information of each patient mentioned among the sources into a local registry of individuals, called *stg_airtum_registry*, where stg stands for staging area. Then, this registry has been cleansed using official data as dictionaries.

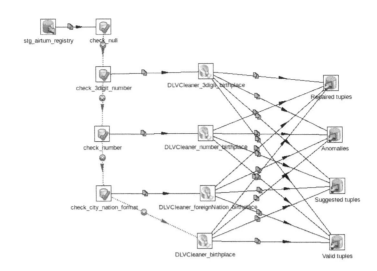

Fig. 2. Example of a Kettle workflow using the DLVCleaner plugin in AIRTUM

As it should be clear from this brief presentation, the overall project is quite complex and involves several integration and cleaning tasks. We next present the general philosophy followed in the project and the adoption of our approach for cleaning tasks, by showing the cleaning steps for the *birthplace* of a patient.

The dictionary table exploited in this case includes several attributes, such as *city name, ISTAT code, region, province, nation* etc. From a careful analysis of involved sources, it has been possible to point out that the *birthplace* is expressed in a very heterogeneous way. In fact, sometimes it is specified by a three-digit number, which often corresponds to an official code for foreign nationality (used in place of city name). Sometimes it is expressed by a number corresponding to the Italian ISTAT classification of cities. In other cases, it is described by a text pair *(city - nation)* containing either Italian or foreign cities with the corresponding nation. Finally, sometimes the birthplace is not specified at all.

Figure 2 shows how the DLVCleaner plugin has been used to address the cleaning of the *birthplace* attribute in *stg_airtum_registry* table. From the previous analysis, it emerged that it is profitable to carry out specific cleaning activities depending on the format of each birthplace instance. In fact, the workflow shown in Figure 2 first discards tuples from *stg_airtum_registry* having NULL birthplaces (these are handled in a different sub-task not shown in Figure 2). Then it partitions remaining *stg_airtum_registry* tuples singling out (*i*) the three-digit birthplaces, (*ii*) the numbered birthplaces, (*iii*) the foreign cities, and (*iv*) Italian cities (all the rest). Each of these data flows is sent to a specifically configured DLVCleaner instance which, based on the stream classifications introduced in previous sections, outputs results onto one of four tables, namely *valid tuples, corrected tuples, suggested tuples,* and *anomalies*. The four DLVCleaner plugin instances have been suitably configured for the specific birthplace format.

As an example, in the *DLVCleaner_3digit_birthplace* transformation the birthplace is mapped onto the *nationality* attribute of the reference dictionary, whereas in *DLVCleaner_number_birthplace* the birthplace is mapped onto the *ISTAT code* dictionary attribute. Analogously, in *DLVCleaner_birthplace* the pair *(city - nation)* is handled by a matching function that first tokenizes the string, singling out the city name, and then matches it to the *city name* dictionary attribute. In order to detect potential corrections, the most proper comparison function is applied, depending on data format; as an example, for the three-digit birthplaces, we used the Hamming distance whereas for city names we used the Levenshtein distance.

Setting up the workflow shown in Figure 2 takes only few minutes and it is possible to follow a try-and-error approach. In the described use case, the input table was composed of 1.000.000 tuples collecting records from 155 municipalities, whereas the dictionary stored about 15.000 tuples. From the application of the transformation shown in Figure 2 we obtained that almost 50% of input tuples were wrong. 72% of wrong tuples have been automatically corrected, whereas 24% had multiple corrections. Only 2% of input tuples have been detected as wrong and not repairable.

Clearly, the cleaning step for birthplaces shown above is only one small step in a more complete workflow dealing with the overall database.

4   Conclusion

In this paper we illustrated the application of a logic-based approach to data cleaning in the real world AIRTUM use case. We have shown that extensions of classical logic programming, such as external function calls and database interaction, may enable logic programming to be effectively exploited in practice. As for future work, we plan to further extend the approach to accommodate also cleaning strategies not necessarily based on dictionaries.

References

1. Calimeri, F., Cozza, S., Ianni, G., Leone, N.: Enhancing ASP by functions: Decidable classes and implementation techniques. In: AAAI (2010)
2. Hernandez, M., Stolfo, S.: The merge/purge problem for large databases. In: Proceedings of the ACM SIGMOD Conference (1995)
3. Maletic, J., Marcus, A.: Data cleansing: Beyond integrity analysis. In: Proceedings of the Conference on Information Quality (October 2000)
4. Mayol, E., Teniente, E.: A survey of current methods for integrity constraint maintenance and view updating. In: Kouloumdjian, J., Roddick, J., Chen, P.P., Embley, D.W., Liddle, S.W. (eds.) ER Workshops 1999. LNCS, vol. 1727, pp. 62–73. Springer, Heidelberg (1999)
5. Raman, V., Hellerstein, J.: Potter's wheel: An interactive framework for data transformation and cleaning. In: Proc. of VLDB 2001, Roma, Italy (2001)
6. Terracina, G., De Francesco, E., Panetta, C., Leone, N.: Enhancing a DLP system for advanced database applications. In: Calvanese, D., Lausen, G. (eds.) RR 2008. LNCS, vol. 5341, pp. 119–134. Springer, Heidelberg (2008)

Justifications for Logic Programming

Carlos Viegas Damásio[1], Anastasia Analyti[2], and Grigoris Antoniou[3]

[1] CENTRIA, Departamento de Informática Faculdade de Ciências e Tecnologia
Universidade Nova de Lisboa, 2829-516 Caparica, Portugal
cd@fct.unl.pt
[2] Institute of Computer Science, FORTH-ICS, Crete, Greece
analyti@ics.forth.gr
[3] Department of Informatics, University of Huddersfield, Huddersfield, UK, and
Institute of Computer Science, FORTH-ICS, Crete, Greece
G.Antoniou@hud.ac.uk

Abstract. Understanding why and how a given answer to a query is generated from a deductive or relational database is fundamental to obtain justifications, assess trust, and detect dependencies on contradictions. Propagating provenance information is a major technique that evolved in the database literature to address the problem, using annotated relations with values from a semiring. The case of positive programs/relational algebra is well-understood but handling negation (or set difference in relational algebra) has not been addressed in its full generality or has deficiencies. The approach defined in this work provides full provenance information for logic programs under the least model, well-founded semantics and answer set semantics, and is related to the major existing notions of justifications for all these logic programming semantics.

1 Introduction

An essential problem that users of logic programming systems face is the understanding of why a given query is true or false in a model of a program, under a given particular semantics. This problem has received attention for quite a long time, and has been addressed for the case of definite programs under least model semantics in [20,15], stratified negation in [20,19], for the case of well-founded semantics in [16,19,17], and for the answer set semantics in [17]. Most of the approaches resort to the non-deterministic construction of complex structures, usually graph based in order to obtain justifications for programs [19,15,17], or provide algorithmic approaches [20,2]. In this paper it is presented a fully declarative and logical approach to the problem, by constructing provenance formulae from which justifications can be extracted. Even though the problem of finding justifications is related to the debugging of logic programming theories [16,2,1,5], the exact relationship will be deferred to a subsequent work.

In the current work is defined a declarative logical approach able to extract provenance information for logic programs. Using values of the Lindenbaum-Tarski algebra as annotation tags for atoms, we are able to specify

---

[1] Partially supported by FCT Project ERRO PTDC/EIACCO/121823/2010.

P. Cabalar and T.C. Son (Eds.): LPNMR 2013, LNAI 8148, pp. 530–542, 2013.

why-provenance both for definite and normal logic programs under well-founded semantics, and relate it to abduction and calculation of prime implicants. The approach is subsequently generalised to the case of answer set programming. Moreover, why-provenance for the case of well-founded semantics is novel, extending the results known for the case of relational algebra, since the case of bag semantics with positive recursion has been addressed in [10], but to the best of our knowledge no work in the literature considers the case of recursion over negation under set (or bag) semantics. We will use these provenance formulae to obtain justifications for literals true in a given model, extending the approaches of evidence graphs [15] and offline justifications [17].

The background is introduced in the next section. Section 3 specifies the why-provenance approach for the case of definite programs, covering the case of positive relational algebra, introducing also why-provenance for the case of an atom not belonging to the model. The technique is generalized afterwards to the well-founded semantics in Section 4. Why-provenance for the answer set semantics can be obtained from the provenance formulas for the well-founded model. We discuss our results and summarize the main conclusions in the last section. It is assumed that the reader is acquainted with the major semantics for logic programs: least model [21], well-founded [7] and answer set semantics [8].

2   Preliminaries and Background

Normal logic programs are sets of rules. A rule r has the following syntax: $A_1 :\!- A_2, \ldots, A_m, {\sim}A_{m+1}, \ldots, {\sim}A_n (n \geq m \geq 0)$, where each A_i is a logical atom without occurrence of function symbols. As usual, define $Head(r) = A_1$, $Body^+(r) = \{A_2, \ldots, A_m\}$, $Body^-(r) = \{A_{m+1}, \ldots, A_n\}$, and $Body(r) = \{A_2, \ldots, A_m, {\sim}A_{m+1}, \ldots, {\sim}A_n\}$. A program is definite (or positive) if there are no occurrences of weakly negated atoms. Without loss of generality, it is assumed that programs are ground (no variables in the rules). Given a set of literals J let ${\sim}J = \{a \mid {\sim}a \in J\} \cup \{{\sim}a \mid a \in J \wedge \forall_b\, a \neq {\sim}b\}$. The Herbrand Base $\mathbb{H}_P$ of a program P is formed by the set of atoms occurring in it. A two-valued interpretation is a subset of $\mathbb{H}_P$ specifying the true atoms, and a partial interpretation is a subset of $\mathbb{H}_P \cup {\sim}\mathbb{H}_P$. A two-valued interpretation I corresponds to the partial interpretation $I \cup {\sim}(\mathbb{H}_P \setminus I)$. The least model $least(P)$ of a definite program P is the least fixpoint of operator $T_P(I) = \{Head(r) \mid r \in P \wedge Body(r) \subseteq I\}$. The answer sets of normal logic program P are the fixpoints of $\Gamma(I) = least(P^I)$, where $P^I = \{Head(r) \leftarrow Body^+(r) \mid r \in P, Body^-(r) \cap I = \emptyset\}$. The well-founded model WFM_P of P is $T \cup {\sim}F$ where T and F are interpretations such that $T \cap F = \emptyset$, T is the least fixpoint $T = \Gamma(\Gamma(T))$ and $F = \mathbb{H}_P \setminus \Gamma(T)$.

Example 1. Consider the following logic program

$$a :\!- c, {\sim}b. \qquad b :\!- {\sim}a. \qquad d :\!- {\sim}c, {\sim}d. \qquad c :\!- {\sim}e. \qquad e :\!- f. \qquad f :\!- e.$$

This program has two answer sets: $M_1 = \{a, c\}$ and $M_2 = \{b, c\}$ (absent atoms are false). Its well-founded model is $WFM = \{c, {\sim}d, {\sim}e, {\sim}f\}$ (absent literals are undefined). A program may not have answer sets: by adding the fact e to the previous program, c becomes false, but then d can only be true iff it is false.

Determining why a given atom belongs to the model of a program is not a trivial task, due to the mutual dependencies that can occur in programs. This is fundamental to users to be able to debug or to check their programs. In [15] is defined the notion of evidence for tabled definite logic programs, extending the previous work in [19], where it is assumed a left-to-right execution order of goals. This approach has been generalised in [17] to the case of normal logic programs by introducing offline-justifications.

Definition 1 (Offline explanation graph). *Let P be a program, J a partial interpretation, U a set of atoms, and $b^{\pm}$ an element of $\mathbb{H}_P^+ \cup \mathbb{H}_P^-$, where $\mathbb{H}_P^+ = \{a^+ \mid a \in \mathbb{H}_P\}$ and $\mathbb{H}_P^- = \{a^- \mid a \in \mathbb{H}_P\}$. An offline explanation graph G of b with respect to J and U is a labeled, directed graph $G = (N, E)$, where $N \subseteq \mathbb{H}_P^+ \cup \mathbb{H}_P^- \cup \{assume, \top, \bot\}$ and $E \subseteq N \times N \times \{+, -\}$, satisfying:*

1. *Element $b^{\pm}$ belongs to N and every node of G is reachable from $b^{\pm}$;*
2. *The only sinks in the graph are: $assume$, $\top$, and $\bot$;*
3. *For every $h^+ \in N$ then $h \in J$, and there is $h :- b_1, \ldots, b_m, \sim c_1, \ldots, \sim c_n$ in P such that $\{b_1, \ldots, b_m\} \subseteq J$ and $\{\sim c_1, \ldots, \sim c_n\} \subseteq J \cup \sim U$.*
   Additionally, there is an arc $(h^+, b_i^+, +)$ for each $1 \leq i \leq m$, and an arc $(h^+, c_j^-, -)$ for each $1 \leq j \leq n$; if the rule has an empty body (it is a fact), then h^+ is $(h^+, \top, +) \in E$. No more arcs have source h^+.
4. *If $h^- \in N$ and $h \in U$ then $(h^-, assume, -)$ is the only arc with source h^-;*
5. *For every $h^- \in N$ such that $h \notin U$ then $\sim h \in J$, and the successors of h^- is the least set of arcs such that for every rule $h :- b_1, \ldots, b_m, \sim c_1, \ldots, \sim c_n$:*
   - *$\exists_{1 \leq i \leq m}$ such that $(h^-, b_i^-, +)$ and $\sim b_i \in J$ or $b_i \in U$, or*
   - *$\exists_{1 \leq j \leq n}$ such that $(h^-, c_j^+, -)$ and $c_j \in J$.*
   If there are no rules for h then the only arc with source h^- is $(h^-, \bot, +)$;
6. *There are no cycles in the subgraph involving nodes in $\mathbb{H}_P^+$ and containing arcs labeled only with $+$;*

The labels of the arcs in an offline explanation graph indicate if the dependency is positive or negative. There are special nodes used to denote true facts ($\top$), and atoms without rules ($\bot$). The set of assumptions U captures the literals which are assumed false, and there is a special sink node for those ($assume$). In order to make an atom true, a rule with a true body must exist (true positive atoms and false weakly negated literals). We have a dual reason for a literal being false: all rules must have a falsified positive atom b_i (thus b_i^- must belong to the graph) or a falsified negated literal c_j (thus c_j^+ must belong to G).

An offline justification of an answer set M of a program P is an offline explanation graph for M using the set of assumptions U containing the false atoms in

Fig. 1. Offline justification graphs for the program of Example 1

M that are undefined in the well-founded model WFM_P. Moreover, this graph does not have negative cycles. It is also shown in [17] that every non-undefined atom in WFM_P can be justified without any assumptions.

Example 2. The offline justification for a in the answer set $\{a, c\}$ of Example 1 can be found in the left hand side of Figure 1. So, a is true because c is true and b is assumed false. Atom c is true because e is false due to a positive mutual dependency between e and f. The offline justification for b in the answer set $\{b, c\}$ is simpler, and rests solely in the assumption that a is false. Intuitively, an offline justification graph represents a conjunction of literals true in the model supporting the conclusion plus dependency information.

Relational algebra has expressive power equivalent to acyclic datalog programs, and the translation of relational algebra queries into datalog is immediate. There is an extensive work on provenance for the case of relational algebra, that is summarised here to motivate the use of the proposed algebraic structure to represent why-provenance for logic programs, and simultaneously relate to this area of research. A general data model for annotated relations has been introduced in [10], for positive relational algebra (i.e., excluding the difference operator). These annotations can be used to check derivability of a tuple, lineage, and provenance, and perform query evaluation over incomplete/probabilistic databases. The main concept is the notion of $\mathcal{K}$-relation where tuples are annotated with values (tags) of a commutative semiring $\mathcal{K}$, while positive relational algebra operators semantics are extended and captured by corresponding compositional operations over $\mathcal{K}$. A commutative semiring is an algebraic structure $\mathcal{K} = (\mathbb{K}, \oplus, \otimes, 0, 1)$ where $(\mathbb{K}, \oplus, 0)$ is a commutative monoid ($\oplus$ is associative and commutative, to capture union of relations) with identity element 0, $(\mathbb{K}, \otimes, 1)$ is a commutative monoid with identity element 1, to encode natural join, and thus operation $\otimes$ distributes over $\oplus$, and 0 is an annihilating element of $\otimes$. The obtained algebra on $\mathcal{K}$-relations is expressive enough to capture different kinds of annotations with set or bag semantics, and it is shown that the semiring of polynomials with integer coefficients is the most general semiring.

Example 3. A photo sharing site allows users to register, and users can upload photos of users. Of course, all registered users are users. Users are represented in a relation $u(\text{Name})$, registered users in $r(\text{Name})$, guest users in $g(\text{Name})$, and $p(\text{U1}, \text{U2})$ stores the information that user U1 has uploaded a photo of user U2. Ann, Bob and David are registered users, Ann uploaded a photo of David, and Bob uploaded a photo of himself. Consider $\mathcal{K}$-relations $r = \{a : t_1, b : t_2, d : t_3\}$, and $p = \{(a, d) : p_1, (b, b) : p_2\}$, where t_1, t_2, t_3, p_1 and p_2 are tuple identifiers. Let $u = r \cup g$ be a view and query $\Pi \left[\rho_{x \leftarrow \text{Name}}(u) \bowtie \rho_{x \leftarrow \text{U1}, y \leftarrow \text{U2}}(p) \bowtie \rho_{y \leftarrow \text{Name}}(r) \right]$ to check if there is an user that uploaded a photo of a registered user. In the most general semiring of polynomials with integer coefficients, the provenance for the query under bag semantics is $p_1 \times t_1 \times t_3 + p_2 \times t_2{}^2$, showing that a join of the tuples identified by p_1, t_1 and t_3, or the join of the tuple annotated of p_2 with t_2 (two times) are the ways to construct a solution to the query. Why-provenance returns the annotation $\{\{p_1, t_1, t_3\}, \{p_2, t_2\}\}$, or equivalently, the boolean formula $(p_1 \wedge t_1 \wedge t_3) \vee (p_2 \wedge t_2)$ showing which tuples have been used for obtaining each answer under set semantics; using lineage semiring one gets simply $\{p_1, p_2, t_1, t_2, t_3\}$, i.e. the tuples supporting the query. This situation can be encoded in datalog by the set of facts (extensional part) $\{r(a), r(b), r(d), p(a,d), p(b,b)\}$ and intensional rules: $c :- u(X), p(X,Y), r(Y).$; $u(X) :- r(X).$; $u(X) :- g(X)$. Because we have facts, $r(a), r(d), p(a,d)$, or $r(b)$ and $p(b,b)$ we can conclude that c holds. This corresponds exactly to what is obtained with the why-provenance semiring of Boolean formulas (see [10]).

To be able to capture relational difference, i.e. negation, the authors in [6] assume the $\mathcal{K}$ semiring is naturally ordered (i.e. binary relation $x \preceq y$ is a partial order, where $x \preceq y$ iff there exists a $z \in \mathbb{K}$ such that $x \oplus z = y$), and require additionally that for every pair x and y there is a least z such that $x \preceq y \oplus z$, defining in this way $x \ominus y$ to be such smallest z. A $\mathcal{K}$ semiring with such monus $\ominus$ operator is designated by m-semiring. An important m-semiring is obtained from the above Boolean formulas semiring, being powerful enough to represent why-provenance for relational algebra under set semantics. The shortcomings of the database approach is that it can only handle negation over acyclic programs, and it is not able to indicate which rules have been used to derive an answer. In Example 3, c holds because the ground program contains rule $u(b) : -r(b)$ and the above mentioned facts (or similarly, for the cases of a and c).

3   Provenance for Definite Logic Programming

Provenance in Logic Programming is captured by adapting the Boolean formulas m-semiring for tackling full why-provenance information, i.e. both negative and positive. This is achieved by annotating every literal in the language by an identifier, as well as every rule with an identifier r_i, where i is the rule number in some ordering of the program. Our setting is restricted to the case of finite logic programs, and is inspired in the debugging transformation presented in [16].

Definition 2. *Given a logic program P over the Herbrand Base $\mathbb{H}_P$, let B_P be the free Boolean algebra generated by the propositional variables $\mathbb{H}_P \cup not(\mathbb{H}_P) \cup \{r_i | 1 \leq i \leq | P |\}$, i.e. the Lindenbaum-Tarski algebra of the propositional language $\mathbb{H}_P \cup not(\mathbb{H}_P) \cup \{r_i | 1 \leq i \leq | P |\}$. The elements of B_P are the equivalence classes of propositional formulas under logical equivalence. Meet ($\wedge$), join ($\vee$), and complementation ($'$) are defined by: $[\phi] \vee [\psi] = [\phi \vee \psi]$, $[\phi] \wedge [\psi] = [\phi \wedge \psi]$, $[\phi]' = [\neg\phi]$. The bottom element is $0 = [\phi \wedge \neg\phi] = [\mathbf{f}]$, and the top element $1 = [\phi \vee \neg\phi] = [\mathbf{t}]$. The partial ordering of B_P is entailment: $[\phi] \preceq [\psi]$ iff $\phi \models \psi$. The $\mathcal{K}_{WhyNot}$ provenance m-semiring is obtained from B_P by letting $[\phi] \oplus [\psi] = [\phi] \vee [\psi] = [\phi \vee \psi]$, $[\phi] \otimes [\psi] = [\phi] \wedge [\psi] = [\phi \wedge \psi]$ and $[\phi] - [\psi] = [\phi \wedge \neg\psi]$.*

Notice that in the above definition propositional variables of the form $\mathbf{at}$ and $\mathbf{not(at)}$ are introduced for every atom at in the Herbrand Base. There is no relationship between $\mathbf{at}$ and $\mathbf{not(at)}$ since they are different propositional symbols; thus $[\neg\mathbf{not(at)}]$ and $[\mathbf{at}]$ are distinct elements of $\mathcal{K}_{WhyNot}$. An identifier is also introduced for each rule in the program. A provenance formula is simply an arbitrary Boolean formula over the atoms, default negation of atoms and rule identifiers. We could use as annotation of literals any set of identifiers just taking care that identifiers are in one-to-one correspondence with $\mathbb{H}_P \cup not(\mathbb{H}_P)$.

In order to extract provenance information for definite logic programs, we introduce why-not provenance programs:

Definition 3. *A why-not provenance program is a finite set of why-not provenance rules of the form $A \Leftarrow [J] \otimes B_1 \otimes \ldots \otimes B_m$ where $A, B_1, \ldots, B_m$ ($m \geq 0$) are ground atoms, and $[J]$ is an element of $\mathcal{K}_{WhyNot}$.*

Why-provenance programs have rules which are monotonic, and thus the standard results of multivalued logic programming apply [4], namely the existence of a least model that can be obtained by iterating a modified T_P operator starting from the interpretation that maps every atom to 0.

Definition 4. *Consider a logic program P and a why-not provenance program $\mathfrak{P}$ over $\mathbb{H}_P$. An interpretation I for why-not provenance program $\mathfrak{P}$ is a mapping $I : \mathbb{H}_P \to B_P$. The set of all interpretations is a lattice with pointwise ordering: given any interpretations I_1 and I_2 we say that $I_1 \preceq I_2$ iff for every $A \in \mathbb{H}_P$ it is the case that $I_1(A) \preceq I_2(A)$, i.e. $I_1(A) \models I_2(A)$.*

An interpretation I satisfies a rule $A \Leftarrow [J] \otimes B_1 \otimes \ldots \otimes B_m$ of why-not program $\mathfrak{P}$ iff $I(A) \succeq [J] \otimes I(B_1) \otimes \ldots \otimes I(B_m)$ iff $J \wedge I(B_1) \wedge \ldots \wedge I(B_m) \models I(A)$. Intepretation I is a model of $\mathfrak{P}$ iff I satisfies all the rules of $\mathfrak{P}$.

Lemma 1. *Consider a why-not provenance program $\mathfrak{P}$. Program $\mathfrak{P}$ has a least model $M_\mathfrak{P}$ which can be obtained by iterating through the following operator starting from the least interpretation I_0, which maps every atom to 0:*

$$T_\mathfrak{P}(I)(A) = \bigoplus \{J \wedge I(B_1) \wedge \ldots \wedge I(B_m) \mid A \Leftarrow [J] \otimes B_1 \otimes \ldots \otimes B_m \in \mathfrak{P}\}$$
$$= \left[\bigvee_{A \Leftarrow [J] \otimes B_1 \otimes \ldots \otimes B_m \in \mathfrak{P}} J \wedge I(B_1) \wedge \ldots \wedge I(B_m) \right]$$

Determining why-not provenance for logic programs rests in the following techniques. First, every non-factual rule is annotated with the formula $[r_i]$. Second, for every atom A if there is a fact for it in the program, a rule $A \Leftarrow [A]$ will be introduced in the why-provenance program, otherwise the rule $A \Leftarrow [\neg not(A)]$ is added. We could have used a rule $A \Leftarrow [r_i]$ for annotating facts, but the outcome intepretation would be less readable.

Definition 5. *Let P be a definite logic program and $\mathfrak{P}(P)$ the why-not provenance program constructed as follows:*

- *For the ith rule $A :- B_1, \ldots, B_m$ that is not a fact in P (i.e., $m \geq 1$) add the why-not provenance rule $A \Leftarrow [r_i] \otimes B_1 \otimes \ldots \otimes B_m$ to $\mathfrak{P}(P)$;*
- *For every atom A in $\mathbb{H}_P$ if there is a fact A in P then add $A \Leftarrow [A]$ to $\mathfrak{P}(P)$, otherwise add $A \Leftarrow [\neg not(A)]$ to $\mathfrak{P}(P)$.*

The why-not provenance information $Why(A)$ for an atom A is given by $Why_P(A) = M_{\mathfrak{P}(P)}(A)$, and for literal $\sim A$ is given by $Why_P(\sim A) = [\neg M_{\mathfrak{P}(P)}(A)]$.

The rationale is: if there is a fact A and possibly rules $r_i, \ldots, r_j$ for it, the why-provenance formula for A has the shape $[(\mathbf{r_i} \wedge Why_i) \vee \ldots \vee (\mathbf{r_j} \wedge Why_j) \vee \mathbf{A}]$; otherwise, if there is no fact for A it has the form $[(\mathbf{r_i} \wedge Why_i) \vee \ldots \vee (\mathbf{r_j} \wedge Why_j) \vee \neg\mathbf{not(A)}]$. In the former, one justification for A being true is $[\mathbf{A}]$ meaning that there is a fact for A, other justifications are obtained by using a given rule r_k and justifying why the body is true. The latter case is better understood if we look at the justification for $\sim A$ having why-provenance formula $[\neg(\mathbf{r_i} \wedge Why_i) \wedge \ldots \wedge \neg(\mathbf{r_j} \wedge Why_j) \wedge \mathbf{not(A)}]$, expressing that all bodies must be falsified and that $[\mathbf{not(A)}]$ holds (there is no fact for A).

Definition 6. *Let P be a logic program, and C be a conjunction of literals of $\mathcal{K}_{WhyNot}$. Define the following sets of facts and rules of program P:*

$KeepFacts(C) = \{A. \mid \mathbf{A} \in C\}$ $RemoveFacts(C) = \{A. \mid \neg\mathbf{A} \in C\}$

$MissingFacts(C) = \{A. \mid \neg\mathbf{not(A)} \in C\}$ $NoFacts(C) = \{A. \mid \mathbf{not(A)} \in C\}$

$KeepRules(C) = \{A :- Body \mid \mathbf{r_i} \in C \text{ and } A :- Body \text{ is the } i^{th} \text{ rule of } P\}$

$RemoveRules(C) = \{A :- Body \mid \neg\mathbf{r_i} \in C \text{ and } A :- Body \text{ is the } i^{th} \text{ rule of } P\}$

The first major result relates why-not provenance information with changes to the original program:

Theorem 1. *Let P be a definite logic program, A an arbitrary atom, G a set of facts not in P, F a subset of facts of P and R a subset of rules of P. Then:*

- *Atom A belongs to the least model of $P \backslash (F \cup R) \cup G$ iff there is a conjunction $C = \mathbf{A_1} \wedge \ldots \wedge \mathbf{A_m} \wedge \mathbf{r_{i_1}} \wedge \ldots \wedge \mathbf{r_{i_k}} \wedge \neg\mathbf{notQ_1} \wedge \ldots \wedge \neg\mathbf{notQ_n}$ such that $C \models Why_P(A)$, $MissingFacts(C) \subseteq G$, $KeepFacts(C) \cap F = \emptyset$ and $KeepRules(C) \cap R = \emptyset$. A conjunction of this form is said to be a truth-support for A.*
- *Atom A does not belong to the least model of $P \backslash (F \cup R) \cup G$ iff there is a conjunction $C = \neg\mathbf{A_1} \wedge \ldots \wedge \neg\mathbf{A_m} \wedge \neg\mathbf{r_{i_1}} \wedge \ldots \wedge \neg\mathbf{r_{i_k}} \wedge \mathbf{notQ_1} \wedge \ldots \wedge \mathbf{notQ_n}$ such that $C \models Why_P(\sim A)$, $RemoveFacts(C) \subseteq F$, $RemoveRules(C) \subseteq R$ and $NoFacts(C) \cap G = \emptyset$. Conjunction C is said to be a falsity-support for A.*

By letting $F = R = G = \{\}$ no changes to the program are permitted, and thus justifications for the literals true in the least model of P are:

Corollary 1. *Let P be a definite logic program and M its least model. Then:*

- *An atom A belongs to M (A is true) iff there is a conjunction of propositional variables $C = \mathbf{A_1} \wedge \ldots \wedge \mathbf{A_m} \wedge \mathbf{r_{i_1}} \wedge \ldots \wedge \mathbf{r_{i_k}}$ such that $C \models Why_P(A)$;*
- *An atom A does not belong to M (A is false) iff there is a conjunction of propositional variables $C = \mathbf{notQ_1} \wedge \ldots \wedge \mathbf{notQ_n}$ such that $C \models Why_P(\sim A)$.*

The results in the above corollary are very interesting, meaning that the minimal justification for literals in the least model of a logic program are the prime implicants[1] containing no negations of the corresponding why-not provenance formula. The fundamental use of prime implicates/implicants as justifications goes back to Assumption-Truth Maintenance Systems [18], and has been recently used to capture the notion of causality in databases [14].

---

[1] A prime implicant of F is a minimal conjunction of literals C such that $C \models F$.

Example 4. Consider the logic program P, where rules are numbered:

(1) $a :- b.$ $a.$ (2) $b :- a.$ (3) $c :- b, d.$ (4) $c :- e, f.$ $d.$ (5) $e :- f.$ $f.$

All atoms hold in the least model of P. Why-not provenance program $\mathfrak{P}(P)$ is:

$$a \Leftarrow [\mathbf{r_1}] \otimes b. \quad b \Leftarrow [\mathbf{r_2}] \otimes a. \quad c \Leftarrow [\mathbf{r_3}] \otimes b \otimes d. \quad d \Leftarrow [\mathbf{d}]. \quad e \Leftarrow [\mathbf{r_5}] \otimes f$$
$$a \Leftarrow [\mathbf{a}]. \quad b \Leftarrow [\neg \mathbf{not(b)}]. \quad c \Leftarrow [\mathbf{r_4}] \otimes e \otimes f. \qquad\qquad e \Leftarrow [\neg \mathbf{not(e)}].$$
$$c \Leftarrow [\neg \mathbf{not(c)}]. \qquad\qquad f \Leftarrow [\mathbf{f}].$$

$$Why(a) = [(\mathbf{r_1} \wedge \neg \mathbf{not(b)}) \vee \mathbf{a}] \qquad Why(\sim a) = [\neg((\mathbf{r_1} \wedge \neg \mathbf{not(b)}) \vee \mathbf{a})]$$
$$= [(\neg \mathbf{a} \wedge \neg \mathbf{r_1}) \vee (\neg \mathbf{a} \wedge \mathbf{not(b)})]$$
$$Why(b) = [(\mathbf{r_2} \wedge \mathbf{a}) \vee \neg \mathbf{not(b)}] \qquad Why(\sim b) = [\neg((\mathbf{r_2} \wedge \mathbf{a}) \vee \neg \mathbf{not(b)})]$$
$$= [(\neg \mathbf{r_2} \wedge \mathbf{not(b)}) \vee (\neg \mathbf{a} \wedge \mathbf{not(b)})]$$
$$Why(d) = [\mathbf{d}] \qquad\qquad Why(\sim d) = [\neg \mathbf{d}]$$
$$Why(e) = [(\mathbf{r_5} \wedge \mathbf{f}) \vee \neg \mathbf{not(e)}] \qquad Why(\sim e) = [(\neg \mathbf{f} \wedge \mathbf{not(e)}) \vee (\neg \mathbf{r_5} \wedge \mathbf{not(e)})]$$
$$Why(f) = [\mathbf{f}] \qquad\qquad Why(\sim f) = [\neg \mathbf{f}]$$

$$Why(c) = [(\mathbf{r_3} \wedge ((\mathbf{r_2} \wedge \mathbf{a}) \vee \neg \mathbf{not(b)}) \wedge \mathbf{d}) \vee (\mathbf{r_4} \wedge ((\mathbf{r_5} \wedge \mathbf{f}) \vee \neg \mathbf{not(e)}) \wedge \mathbf{f}) \vee \neg \mathbf{not(c)}]$$
$$= [(\mathbf{r_2} \wedge \mathbf{r_3} \wedge \mathbf{a} \wedge \mathbf{d}) \vee (\mathbf{r_3} \wedge \mathbf{d} \wedge \neg \mathbf{not(b)}) \vee (\mathbf{r_4} \wedge \mathbf{r_5} \wedge \mathbf{f}) \vee (\mathbf{r_4} \wedge \mathbf{f} \wedge \neg \mathbf{not(e)}) \vee \neg \mathbf{not(c)}]$$

We can conclude that d and f are true, because they are stated as facts, and thus to make them false is required their removal. Regarding e, it can be concluded that e is true because of rule 5 and that the fact f is true; anyway, we can make it true by adding the fact e. In order to make e false, it cannot be added as a fact, and fact f or rule r_5 should be removed (or both). Atom a is true because there is a fact for it, or if the rule r_1 is kept and a fact for b is added (a may be removed). To make a false we always have to remove the fact for a, and additionally remove rule r_1 or not introduce a fact for b. The situation for c is rather complex, but it holds because rules r_2 and r_3 are present, as well as facts a and d, or because rules r_4 and r_5 are in the program as well as the fact for f. There are many ways of making c false, for instance by removing facts d and f.

Example 5. Consider the following definite logic program where all atoms are false in the least model: $\{(1)\ a :- b.\ \ (2)\ b :- a, c.\ \ (3)\ b :- d.\}$. It can be checked that $\mathbf{not(a)} \wedge \mathbf{not(b)} \wedge \mathbf{not(d)} \models Why(\sim a)$, $\mathbf{not(a)} \wedge \mathbf{not(b)} \wedge \mathbf{not(d)} \models Why(\sim b)$ and $\mathbf{not(b)} \wedge \mathbf{not(c)} \wedge \mathbf{not(d)} \models Why(\sim b)$. If we do not allow changes to the program, this provenance information is interpreted as follows: a is false because we do not have a fact for a, b and d; while b is false because we do not have a fact for b, d and a or c. Thus, one needs to add a, b, or d to make a true, while it is required to add b, d, or a and c to turn b true. This can be confirmed from $Why(a) = [(\mathbf{r_1} \wedge \mathbf{r_3} \wedge \neg \mathbf{not(d)}) \vee (\mathbf{r_1} \wedge \neg \mathbf{not(b)} \vee \neg \mathbf{not(a)})]$ and $Why(b) = [(\mathbf{r_2} \wedge \neg \mathbf{not(a)} \wedge \neg \mathbf{not(c)}) \vee (\mathbf{r_3} \wedge \neg \mathbf{not(d)} \vee \neg \mathbf{not(b)})]$.

The relationship to evidence graphs (see [15]) is stated in the next theorem:

Theorem 2. *Consider a definite logic program P. For every evidence graph for A (resp. $\sim A$) it is possible to construct a truth-support (resp. falsity-support) formula C such that $C \models Why_P(A)$ (resp. $C \models Why_P(\sim A)$). For every, prime implicant truth-support $C = A_1 \wedge \ldots \wedge A_m \wedge r_{i_1} \wedge \ldots \wedge r_{i_k}$ of $Why_P(A)$ it is possible to construct an evidence graph for A.*

Our approach allows us to capture all the evidence according to [15], but this work lacks some of our justifications mostly for the case of false atoms. This is expected since evidence graphs are constructed for the negative case just by looking at the first false atom in the body of rules, ignoring possibly other false atoms. The comparison to offline justifications is deferred to the next since our results will cover both definite and normal logic programs, generalizing Th. 2.

4   Provenance for Well-Founded Semantics

By mimicking the iteration of Γ^2 operator, we can obtain the provenance information for logic programs under well-founded semantics by defining a corresponding Gelfond-Lifschitz like operator. The correctness and extra motivation for this approach can be found in [3].

Definition 7. *Let P be a logic program and I a why-not provenance interpretation. Construct provenance program $\frac{\mathfrak{P}}{I}$ as follows:*

- *For the ith rule $A :- B_1, \ldots, B_m, \sim C_1, \ldots, \sim C_n$ $(m + n \geq 1)$ in P add provenance rule $A \Leftarrow [r_i \wedge \neg I(C_1) \wedge \ldots \neg I(C_n)] \otimes B_1 \otimes \ldots \otimes B_m$ to $\frac{\mathfrak{P}}{I}$;*
- *For every atom A in $\mathbb{H}_P$ if there is a fact A in P then add $A \Leftarrow [A]$ to $\frac{\mathfrak{P}}{I}$, otherwise add $A \Leftarrow [\neg \mathtt{not}(A)]$ to $\frac{\mathfrak{P}}{I}$.*

Operator $\mathfrak{G}_P(I) = M_{\frac{\mathfrak{P}}{I}}$ returns the least model of why-not program $\frac{\mathfrak{P}}{I}$.

It is clear that the operator $\mathfrak{G}_P$ is anti-monotonic, and therefore $\mathfrak{G}_P^2$ is monotonic having a least model $\mathfrak{T}_P$, corresponding to provenance information for what is true in the well-founded model, while $\mathfrak{T}\mathfrak{U}_P = \mathfrak{G}_P(\mathfrak{T}_P)$ contains the why-not provenance of what is true or undefined in the well-founded model of P.

Definition 8. *Let P be a normal logic program, and $\mathfrak{T}_P$ the least model of $\mathfrak{G}_P^2$, and $\mathfrak{T}\mathfrak{U}_P = \mathfrak{G}_P(\mathfrak{T}_P)$, and A an atom. The why-not provenance information under the well-founded semantics is defined as follows: $Why_P(A) = [\mathfrak{T}_P(A)]$; $Why_P(\sim A) = [\neg \mathfrak{T}\mathfrak{U}_P(A)]$; and $Why_P(\mathtt{undef}\ A) = [\neg \mathfrak{T}_P(A) \wedge \mathfrak{T}\mathfrak{U}_P(A)]$.*

As usual, for the case of stratified programs (no cycles through negations) we obtain a model on which for every atom A we have $Why_P(A) = [\mathfrak{T}_P(A)] = [\mathfrak{T}\mathfrak{U}_P(A)] = [\neg Why_P(\sim A)]$, and thus $Why_P(\mathtt{undef}\ A) = 0$. The why-not provenance for undefined literals is obtained from the why-not provenance for truth or undefinedness minus the why-not provenance of truth.

Theorem 3. *Let P be a normal logic program, G a set of facts not in P, F a subset of facts of P, and R a subset of rules of P. A literal L belongs to the WFM of $(P \setminus (F \cup R)) \cup G$ iff there is a conjunction of literals $C \models Why_P(L)$, such that $RemoveFacts(C) \subseteq F$, $KeepFacts(C) \cap F = \emptyset$, $RemoveRules(C) \subseteq R$, $KeepRules(C) \cap R = \emptyset$, $MissingFacts(C) \subseteq G$, and $NoFacts(C) \cap G = \emptyset$.*

The above result, generalizing Th. 2, is a fundamental new contribution to the literature of provenance in logic programming, and in particular for datalog. Since any relational algebra query can be translated into an acyclic logic program, thus it is possible to extract for the first time complete provenance information for a more expressive extension of full relational algebra.

Example 6. Consider the logic program $P = \{(1)\ a :- \sim a, b.\quad (2)\ b :- \sim c.\}$. In the well-founded model of P we have a undefined, b true, and c false. Thus:

$$\mathfrak{T}_P(a) = [\neg \mathtt{not(a)}] \qquad \mathfrak{T}\mathfrak{U}_P(a) = [(\mathbf{r_1} \wedge \mathbf{r_2} \wedge \mathtt{not(c)}) \vee (\mathbf{r_1} \wedge \neg\mathtt{not(b)} \vee \neg\mathtt{not(a)})]$$
$$\mathfrak{T}_P(b) = [(\mathbf{r_2} \wedge \mathtt{not(c)}) \vee \neg\mathtt{not(b)}] \qquad \mathfrak{T}\mathfrak{U}_P(b) = [(\mathbf{r_2} \wedge \mathtt{not(c)}) \vee \neg\mathtt{not(b)}]$$
$$\mathfrak{T}_P(c) = [\neg\mathtt{not(c)}] \qquad \mathfrak{T}\mathfrak{U}_P(c) = [\neg\mathtt{not(c)}]$$

The why-not provenance information for negated literals is:

$$Why_P(\sim a) = [(\mathtt{not(a)} \wedge \mathtt{not(b)} \wedge \neg\mathtt{not(c)}) \vee (\neg\mathbf{r_2} \wedge \mathtt{not(a)} \wedge \mathtt{not(b)}) \vee (\neg\mathbf{r_1} \wedge \mathtt{not(a)})]$$
$$Why_P(\sim b) = [(\mathtt{not(b)} \wedge \neg\mathtt{not(c)}) \vee (\neg\mathbf{r_2} \wedge \mathtt{not(b)})]$$
$$Why_P(\sim c) = [\mathtt{not(c)}]$$

Thus, $Why_P(\mathtt{undef}\ a) = [(\mathbf{r_1} \wedge \mathbf{r_2} \wedge \mathtt{not(a)} \wedge \mathtt{not(c)}) \vee (\mathbf{r_1} \wedge \mathtt{not(a)} \wedge \neg\mathtt{not(b)})]$, $Why_P(\mathtt{undef}\ c) = 0$ and $Why_P(\mathtt{undef}\ b) = 0$. The interpretation of these results is now clear. Atom a is undefined since there is no positive prime implicant both for $Why_P(a)$ and $Why_P(\sim a)$; the justification for a being undefined is that there is no fact for a and for c and both rules are in the program, or if we keep only rule r_1 then a fact for b must be added, as can be extracted from $Why_P(\mathtt{undef}\ a)$. Moreover, in order to make a true we need to add fact a, while to make it false one solution is to make c true or remove the rule for b in order to make b false, and not adding facts for a and b; alternatively, we remove the rule for a and do not add a fact for it. There is no way of making b and c undefined.

Theorem 4. *Let $G = (N, E)$ be an offline justification for a literal L true in the well-founded model M of a program P, with respect to M and empty set of assumptions. Let $C = \bigwedge_{(\mathtt{h}+,\top,+)\in E} \mathtt{h} \wedge \bigwedge_{\mathtt{h}-\in N} \mathtt{not(h)} \wedge \bigwedge_{\mathtt{h}+\in N} \mathbf{r_{i_h}}$, where r_{i_h} is the identifier of a rule for h satisfied by G, then $C \models Why_P(L)$.*

The converse direction does not hold, since we have more justifications for a literal being true. An important particular example is a program containing rules $a :- \sim b$ and $a :- b$. We have $Why_P(A) = [(\mathbf{r_1} \wedge \neg\mathtt{not(b)}) \vee (\mathbf{r_2} \wedge \mathtt{not(b)})]$, but $\mathbf{r_1} \wedge \mathbf{r_2} \models Why_P(A)$, thus if both rules are kept, a holds independently of any changes to the program. This cannot be obtained from offline-justifications.

5 Provenance for Answer Set Semantics

The extension of our approach to answer set programming is now straightforward due to the previous results.

Definition 9. *Let P be a logic program, and L a literal. The answer set why-not provenance for L is $AnsWhy_P(L) = Why_P(L) \wedge \bigwedge_{A \in H_P} \neg Why_P(\mathtt{undef}\,A)$.*

We combine the provenance of the literal obtained under well-founded semantics, and impose that all literals cannot be undefined. Note that this contrasts with traditional approaches where the justification is *local* in the sense that only a subset of the dependency graph is included in the justification. This is required since a literal can occur in the body of a constraint, or the program may be inconsistent because of odd-loops, and is formally supported by the results in [11] showing that there is no modular transformation of answer set semantics into propositional theories (this requirement is illustrated in a subsequent example).

Theorem 5. *Let P be a program, M an answer set of P, and L a literal true in M. Then there is a conjunction $C \models AnsWhy_P(L)$ that does not contain any negative literals (obtained as in Th. 4) for literals true in the WFM_P. For every atom $\sim A \in M$ that is undefined in WFM_P, C includes $\mathtt{not(A)} \wedge \neg \mathtt{r_{i_1}} \wedge \ldots \wedge \neg \mathtt{r_{i_k}}$ where $\{r_{i_1}, \ldots, r_{i_k}\}$ is the set of identifiers of all rules for A.*

The above theorem follows from the result in [17] stating that there is an offline justification with respect to M and the set of assumptions containing the literals false in M that are undefined in the well-founded model of P. Moreover, this justification does not have cycles. Therefore, by representing these assumptions as a conjunction of literals, we can then construct such a C. In order to assume a literal false we cannot add a fact for it $(\mathtt{not(A)})$, and must remove all existing rules for it in the program $(\neg \mathtt{r_{i_1}} \wedge \ldots \wedge \neg \mathtt{r_{i_k}})$.

Example 7. Consider the simplified version of the program in Example 1.

$$(1)\ a :- c, \sim b. \qquad (2)\ b :- \sim a. \qquad (3)\ d :- \sim c, \sim d. \qquad c.$$

It has two stable models $\{a, c\}$ and $\{b, c\}$, and its why-not provenance is:

$$
\begin{aligned}
AnsWhy(a) &= [(\mathtt{c} \wedge \neg\mathtt{not(a)}) \vee (\mathtt{r_1} \wedge \neg\mathtt{r_2} \wedge \mathtt{c} \wedge \mathtt{not(b)}) \vee (\neg\mathtt{not(a)} \wedge \neg\mathtt{not(d)})] \\
AnsWhy(\sim a) &= [\mathtt{not(a)} \wedge (\neg\mathtt{r_1} \vee (\neg\mathtt{c} \wedge \neg\mathtt{not(d)}) \vee (\mathtt{c} \wedge \neg\mathtt{not(b)}) \vee (\neg\mathtt{not(b)} \wedge \neg\mathtt{not(d)}))] \\
AnsWhy(b) &= [\mathtt{r_2} \wedge \mathtt{not(a)} \wedge (\neg\mathtt{r_1} \vee \neg\mathtt{c} \vee \neg\mathtt{not(d)}) \wedge \neg(\mathtt{r_1} \wedge \neg\mathtt{c} \wedge \mathtt{not(d)})] \\
AnsWhy(\sim b) &= [\mathtt{not(b)} \wedge (\neg\mathtt{r_2} \vee \neg\mathtt{not(a)}) \wedge \neg(\mathtt{r_1} \wedge \neg\mathtt{c} \wedge \mathtt{not(d)})] \\
AnsWhy(c) &= [(\mathtt{c} \wedge \neg(\mathtt{r_1} \wedge \mathtt{r_2} \wedge \mathtt{not(a)} \wedge \mathtt{not(b)})] \\
AnsWhy(\sim c) &= [(\neg\mathtt{c} \wedge \neg\mathtt{not(d)}) \vee (\neg\mathtt{c} \wedge \neg\mathtt{r_1})] \\
AnsWhy(d) &= [(\neg\mathtt{c} \wedge \neg\mathtt{not(d)}) \vee (\neg\mathtt{not(d)} \wedge \neg(\mathtt{r_1} \wedge \mathtt{r_2} \wedge \mathtt{not(a)} \wedge \mathtt{not(b)}))] \\
AnsWhy(\sim d) &= [(((\mathtt{c} \wedge \mathtt{not(d)}) \vee (\neg\mathtt{r_1} \wedge \mathtt{not(d)})) \wedge \neg(\mathtt{r_1} \wedge \mathtt{r_2} \wedge \mathtt{not(a)} \wedge \mathtt{not(b)})]
\end{aligned}
$$

The results are expected and intuitive. Regarding c, it holds because we have a fact for it. The formula $\neg(\mathtt{r_1} \wedge \mathtt{r_2} \wedge \mathtt{not(a)} \wedge \mathtt{not(b)})$ guarantees that a and b are not undefined: we have to remove at least one of the first two rules, or to add the fact a or b. More interestingly is $AnsWhy(\sim c)$: it expresses that to make c false, it is necessary of course to remove the fact for c, and remove the rule for d or introduce a fact for d, otherwise an odd-loop would appear. By making c false, a becomes false and b true, and thus it is no longer required to guarantee that they are not undefined. The justifications for d and $\sim d$ are similar. The truth of a in the first model is justified by the fact c, and by assuming b to be false as captured by the conjunction $\mathtt{r_1} \wedge \neg\mathtt{r_2} \wedge \mathtt{c} \wedge \mathtt{not(b)}$; if c is kept then a can be added to make it true. If c is removed, then a can be made true by adding it, as well as the fact for d in order to avoid inconsistency. Regarding the truth of b in the second answer set, it is required to assume a false as encoded by $\texttt{not(a)} \wedge \neg r_1$ in the conjunction $r_2 \wedge \texttt{not(a)} \wedge \neg r_1$, and corresponds exactly to the offline justification for b in Fig. 1.

6 Discussion and Conclusions

The applications of why-provenance for logic programs are manifold. The first, and our main motivation, is to understand how a literal is derived from the program. This extends to the case of queries with arbitrary relational algebra operators. Other important application, is the study of why-provenance in the Semantic Web: our approach can detect monotonic and non-monotonic dependencies in SPARQL queries over arbitrary graphs.

It is defined for the first-time a complete provenance model for the major semantics of logic programs. However, there are some important questions remaining to be addressed. In particular, the size of the formulas generated is not guaranteed to be polynomial on the size of the program. This is not a surprise since for some programs it may be required an exponential number of propositional formulas to capture the answer set semantics [12]. However, there are also known polynomial encodings by introducing extra symbols in the translation [13,11], and for the case of programs without positive loops we can construct a polynomial formula and thus obtaining a polynomial representation for full relational algebra. Additionally, we need only a linear number of iterations of the immediate consequences operator to reach the fixpoint in the case of definite programs, and a polynomial time number of iterations for the case of the well-founded semantics. Our approach contrasts with evidence graphs [15] and offline-justifications [17] by being more informative. In fact, our provenance formulas are a declarative representation of all such justifications, which can be exponential in number since the number of prime implicants is exponential in the size of the formula, and the problem of generating them is co-NP-hard. Additionally, the complexity of conjunctive query answering over K-relations most of the times is NP-complete (see [9]).

Why-not provenance formulas resort to a program transformation previously defined for declarative debugging of logic programs [16]. The debugging of Answer Set Programs has been addressed in the literature by several authors, and the most effective approaches resort to meta-transformations to address the several forms of anomalies that can be found in programs [1,5]. In fact, our approach is capable of providing the corrections (adding or removing facts, and removing rules) in order to obtain what is desired. The approaches presented in [1,5] are more fine grained, and are designed to detect errors in programs. Nevertheless, we conjecture the approaches to be related and this is left for future work. For instance, justifications for violation of integrity constraints can be obtained by putting the head $false$ in all ICs, and determine why-not justification for $false$. In general, a direction to explore, consists of translating why-provenance into ASP, apply the debugging transformations and extract the corresponding propositional theories. The generalization to the first-order case is an open research issue, but first-order abduction or constructive negation may be necessary.

References

1. Brain, M., Gebser, M., Pührer, J., Schaub, T., Tompits, H., Woltran, S.: Debugging ASP programs by means of ASP. In: Baral, C., Brewka, G., Schlipf, J. (eds.) LPNMR 2007. LNCS (LNAI), vol. 4483, pp. 31–43. Springer, Heidelberg (2007)
2. Brain, M., Vos, M.D.: Debugging logic programs under the answer set semantics. In: Proc. of ASP 2005 Workshop. CEUR Workshop Proceedings, vol. 142 (2005)
3. Damásio, C.V., Pereira, L.M.: Antitonic logic programs. In: Eiter, T., Faber, W., Truszczyński, M. (eds.) LPNMR 2001. LNCS (LNAI), vol. 2173, pp. 379–392. Springer, Heidelberg (2001)
4. Damásio, C.V., Pereira, L.M.: Monotonic and residuated logic programs. In: Benferhat, S., Besnard, P. (eds.) ECSQARU 2001. LNCS (LNAI), vol. 2143, pp. 748–759. Springer, Heidelberg (2001)
5. Gebser, M., Pührer, J., Schaub, T., Tompits, H.: A meta-programming technique for debugging answer-set programs. In: AAAI 2008, pp. 448–453. AAAI Press (2008)
6. Geerts, F., Poggi, A.: On database query languages for K-relations. J. Applied Logic 8(2), 173–185 (2010)
7. Gelder, A.V., Ross, K.A., Schlipf, J.S.: The Well-Founded Semantics for General Logic Programs. Journal of the ACM 38(3), 620–650 (1991)
8. Gelfond, M., Lifschitz, V.: The Stable Model Semantics for Logic Programming. In: Proc. of ICLP 1988, pp. 1070–1080. MIT Press (1988)
9. Green, T.J.: Containment of conjunctive queries on annotated relations. In: Proc. of Database Theory - ICDT 2009, vol. 361, pp. 296–309 (2009)
10. Green, T.J., Karvounarakis, G., Tannen, V.: Provenance semirings. In: Proc. of PODS 2007, pp. 31–40. ACM, New York (2007)
11. Janhunen, T.: Some (in)translatability results for normal logic programs and propositional theories. Journal of Applied Non-Classical Logics 16(1-2), 35–86 (2006)
12. Lifschitz, V., Razborov, A.: Why are there so many loop formulas? ACM Trans. Comput. Logic 7(2), 261–268 (2006)
13. Lin, F., Zhao, J.: On tight logic programs and yet another translation from normal logic programs to propositional logic. In: Proc. of IJCAI 2003, pp. 853–858. Morgan Kaufmann Publishers Inc. (2003)
14. Meliou, A., Gatterbauer, W., Halpern, J.Y., Koch, C., Moore, K.F., Suciu, D.: Causality in databases. IEEE Data Eng. Bull. 33(3), 59–67 (2010)
15. Pemmasani, G., Guo, H.-F., Dong, Y., Ramakrishnan, C.R., Ramakrishnan, I.V.: Online justification for tabled logic programs. In: Kameyama, Y., Stuckey, P.J. (eds.) FLOPS 2004. LNCS, vol. 2998, pp. 24–38. Springer, Heidelberg (2004)
16. Pereira, L.M., Damásio, C.V., Alferes, J.J.: Diagnosis and debugging as contradiction removal. In: Proc. of LPNMR 1993, pp. 316–330 (1993)
17. Pontelli, E., Son, T.C., El-Khatib, O.: Justifications for logic programs under answer set semantics. TPLP 9(1), 1–56 (2009)
18. Reiter, R., de Kleer, J.: Foundations of assumption-based truth maintenance systems: Preliminary report. In: Proc. of AAAI 1987, pp. 183–189 (1987)
19. Roychoudhury, A., Ramakrishnan, C.R., Ramakrishnan, I.V.: Justifying proofs using memo tables. In: Proc. of PPDP, pp. 178–189 (2000)
20. Specht, G.: Generating explanation trees even for negations in deductive database systems. In: Proc. of LPE, pp. 8–13 (1993)
21. van Emden, M.H., Kowalski, R.A.: The semantics of predicate logic as a programming language. J. ACM 23(4), 733–742 (1976)

Belief Change in Nonmonotonic Multi-Context Systems

Yisong Wang[1], Zhiqiang Zhuang[2], and Kewen Wang[2]

[1] Department of Computer Science, Guizhou University, Guiyang, 550025, China
[2] School of Information and Communication Technology, Griffith University, Australia

Abstract. Brewka and Eiter's nonmonotonic multi-context system is an elegant knowledge representation framework to model heterogeneous and nonmonotonic multiple contexts. Belief change is a central problem in knowledge representation and reasoning. In this paper we follow the classical AGM approach to investigate belief change in multi-context systems. Specifically, we formulate semantically the AGM postulates of belief expansion, revision and contraction for multi-context systems. We show that the change operations can be characterized in terms of minimal change by ordering equilibria of multi-context systems. Two distance based revision operators are obtained and related to the classical Satoh and Dalal revision operators (via loop formulas).

1 Introduction

Knowledge Representation and Reasoning (KR) is a long-standing and traditional research area, which plays a crucial role in artificial intelligence and computer science. A mass of logical theories have been proposed for KR, including monotonic and nonmonotonic ones [18]. Among the latter, logic programming based on answer set semantics (ASP) is distinguished due to its elegant theoretical foundation and efficient implementations for various applications [16,6].

Over the last decade, there has been increasing interest in KR systems comprising multiple knowledge bases. It leads to the development of Multi-Context Systems (MCS), which builds up on several theories (the *contexts*) that are interlinked with *bridge rules* so that it allows to incorporate knowledge into a context according to knowledge in other contexts [28,3]. The abstract and general MCS framework of Brewka and Eiter [3] is of special interest since it allows for heterogeneous and nonmonotonic MCS in two aspects. On the one hand, every context may have different and nonmonotonic logics; on the other hand, bridge rules may use default negation. It raises many attractive applications in interesting scenarios [4].

As knowledge is continually evolving and subject to change, change of logical theories and knowledge bases is a central issue in KR. Three major operations on belief change are *expansion, revision* and *contraction* [25]. Among the work on belief change, the most influential approach is AGM model, where a set of postulates is proposed for rational belief change operations [1]. Those postulates for propositional knowledge bases are semantically characterized in terms of minimal change [19].

In nonmonotonic logics and ASP in particular, the problem of knowledge base change appears to be intrinsically more difficult than in a monotonic setting. For answer set programming, there has been substantial efforts in developing belief change,

P. Cabalar and T.C. Son (Eds.): LPNMR 2013, LNAI 8148, pp. 543–555, 2013.

several approaches have been proposed, and some of them are under the title of *update* [29,2,13,26,10]. As multi-context systems may refer to multiple individual (nonmonotonic) contexts, belief change in MCS is highly possible and intricate because of nonmonotonic bridge rules, thus a challenge. Let us consider the following scenario.

Example 1. [A running example] John went to hospital for a high fever lasting more than a week. Doctors Smith and Alice were assigned to diagnose John's high fever in two rounds. In the first round of diagnosis, while Smith believes that there is no bird flu, Alice thinks that it is due to either bird flue or pneumonia, and she believes that pneumonia means no swine flu. They both agree that if one concludes with bird/swine flu then the other does.

In the second round of diagnosis, while Alice has the same knowledge as in the first round, Smith believes that John got either bird flu or swine flu. And the agreement between Smith and Alice are same as in first round. In terms of the second round of diagnosis, how should we update/change the beliefs about John's high fever in the first round of diagnosis?

In this paper we follow the classical AGM approach to address the above belief change problem in multi-context systems. We formulate the postulates of belief change (expansion, revision and contraction) for multi-context systems in a semantic manner. By orderings over equilibria of multi-context systems, we characterize the postulates of belief change operators in terms of minimal change. According to the generalized Satoh and Dalal distance [27,7] among equilibria, two specific revision operators for multi-context systems are obtained. Since the equilibria of a multi-context system can be captured by the models of a corresponding propositional theory, i.e. loop formulas [8], we establish a connection between the revisions for multi-context systems and propositional theories.

The rest of the paper is organized as follows. In the next section, we briefly recall the basic notions of AGM model for belief change and multi-context systems. In Section 3, we formulate the belief change postulates for expansion, revision and contraction. Their properties are investigated. Related work and future work are discussed in Section 4. Due to space constraints, we only give a stench of the proofs.

2   Preliminaries

In the section we briefly recall basic notions of belief change [1,19] and multi-context systems [3,4]. For a comprehensive overview of AGM model, see [14].

2.1   Classical Belief Revision

The three major belief changes are expansion, revision and contraction. While expansion means addition of a belief without checking inconsistency, contraction means removal of a belief, revision stands for addition of a belief while maintaining consistency.

The most recognized belief change theory is AGM model [1]. Katsuno and Mendelzon semantically formulated AGM belief revision postulates as an operator $\dotplus : L \times L \to L$ satisfying the following conditions [19]:

(R1) $\alpha \dotplus \beta \vdash \beta$.

(R2) If $\alpha \wedge \beta$ is satisfiable then $\alpha \dotplus \beta \equiv \alpha \wedge \beta$.

(R3) If β is satisfiable then $\alpha \dotplus \beta$ is also satisfiable.

(R4) If $\alpha_1 \equiv \alpha_2$ and $\beta_1 \equiv \beta_2$ then $\alpha_1 \dotplus \beta_1 \equiv \alpha_2 \dotplus \beta_2$.

(R5) $(\alpha \dotplus \beta) \wedge \gamma \vdash \alpha \dotplus (\beta \wedge \gamma)$.

(R6) If $(\alpha \dotplus \beta) \wedge \gamma$ is satisfiable then $\alpha \dotplus (\beta \wedge \gamma) \vdash (\alpha \dotplus \beta) \wedge \gamma$.

Here we assume that L is a propositional language over a finite signature and we fix a way of representing any deductively closed set K of formulas by a propositional formula ψ such that $K = \{\phi | \psi \vdash \phi\}$. The intuition behind the first four postulates can be easily read out, e.g., the new knowledge β is kept in the updated knowledge base by (R1). The last two postulates express that the revision by a conjunction is the same as revision by one conjunct conjoined with the other conjunct if the result is satisfiable.

In proportional logic, the belief expansion is quite natural and trivial which is defined as $\alpha + \beta =_{def} Cn(\alpha \cup \beta)$, i.e. $\{\phi | \alpha \wedge \beta \vdash \phi\}$. The revision and contraction ($\dotminus$) can be defined by each other via identities and expansion as follows:

- Levi identity: $\alpha \dotplus \beta =_{def} (\alpha \dotminus \neg \beta) + \beta$,
- Harper identity: $\alpha \dotminus \beta =_{def} \alpha \cap (\alpha \dotplus \neg \beta)$.

Recall that formulas are identified as deductively closed sets (called *knowledge sets*).

2.2   Nonmonotonic Multi-Context Systems

We recall the basic notations of nonmonotonic multi-context systems [3]. An *(abstract) logic L* is a tuple $L = (\mathbf{KB}_L, \mathbf{BS}_L, \mathbf{ACC}_L)$, where

- $\mathbf{KB}_L$ is a set of *well-formed knowledge bases*, each being a set (of formulas),
- $\mathbf{BS}_L$ is a set of *possible belief sets*, each being a set (of formulas),
- $\mathbf{ACC}_L : \mathbf{KB}_L \to 2^{\mathbf{BS}_L}$ assigns to each $kb \in \mathbf{KB}_L$ a set of acceptable belief sets.

For instance, if L is ASP then $\mathbf{KB}_L$ is a set of ASP programs, $\mathbf{BS}_L$ is a collection of sets of atoms and, $\mathbf{ACC}_L$ assigns every ASP programs a set of answer sets. For our belief change purpose, we assume that (i) the signature of L is finite, (ii) each belief set of L can be represented as a set of atoms and, (iii) for every collection $\mathbb{M}_L$ of belief sets of L there exists a knowledge base kb such that $\mathbf{ACC}_L(kb) = \mathbb{M}_L$. The ASP logic [15] and propositional logic satisfy the conditions (ii) and (iii). Note that if L is a propositional logic over finite signature then every formula α of L can be identified with its deductive closure, i.e. $\{\phi | \alpha \vdash \phi\}$. In the following we identify a formula (and a single element set of formulas) α with $Cn(\alpha)$ when it is clear from context.

Given a sequence $\mathcal{L}$ of abstract logics $(L_1, \ldots, L_n)$, an *indexed atom* of $\mathcal{L}$ is an expression of the form $(i : p)$ where $1 \leq i \leq n$ and p is an element of some belief sets of L_i. An L_i-*bridge rule* r over $\mathcal{L}$ is an expression of the form

$$p \leftarrow (c_1 : p_1), \ldots, (c_m : p_m), not\,(c_{m+1} : p_{m+1}), \ldots, not\,(c_k : p_k) \qquad (1)$$

where p is a formula of $L_i = (\mathbf{KB}_i, \mathbf{BS}_i, \mathbf{ACC}_i)$ such that $\{p\} \cup kb_i \in \mathbf{KB}_i$ for any $kb_i \in \mathbf{KB}_i$, and each $(c_i : p_i)$ is an indexed atom of $\mathcal{L}$. We denote $Head(r) = p$, $Pos(r) = \{(c_i : p_i) | 1 \leq i \leq m\}$ and $Neg(r) = \{(c_j : p_j) | m + 1 \leq j \leq k\}$.

Definition 1. *A* multi-context system *(MCS in short)* $\alpha = (C_1, \ldots, C_n)$ *consists of contexts* $C_i = (L_i, kb_i, br_i)$, *where* $L_i = (\mathbf{KB}_i, \mathbf{BS}_i, \mathbf{ACC}_i)$ *is an abstract logic,* $kb_i \in \mathbf{KB}_i$ *is a knowledge base of* L_i, *and* br_i *is a set of* L_i-*bridge rules over* $\mathcal{L} = (L_1, \ldots, L_n)$ *and* $1 \le i \le n$.

We usually abbreviate a context (L, kb, br) as (kb, br) if the underlying logic L is clear from the context, unless explicitly stated otherwise. A *belief state* (of $\mathcal{L}$) is a sequence $S = (S_1, \ldots, S_n)$ such that S_i is an element of $\mathbf{BS}_i$ for every i $(1 \le i \le n)$. By $\mathbb{BS}_\mathcal{L}$ we denote the *belief state space* of $\mathcal{L}$, i.e., $\mathbb{BS}_\mathcal{L} = \{S | S \text{ is a belief state of } \mathcal{L}\}$. For a collection $\mathbb{M} \subseteq \mathbb{BS}_\mathcal{L}$ of belief states, we denote $\overline{\mathbb{M}}$ the *complement* of $\mathbb{M}$, i.e., $\overline{\mathbb{M}} = \mathbb{BS}_\mathcal{L} \setminus \mathbb{M}$. The belief state S *satisfies* an indexed atom $(c : p)$, written $S \models (c : p)$, if $p \in S_c$. A bridge rule r of the form (1) is *applicable* w.r.t. S if S satisfies $(c_i : p_i)$ for every i $(1 \le i \le m)$ and it does not satisfy $(c_j : p_j)$ for every j $(m + 1 \le j \le k)$. By $app(br, S)$ we denote the set of bridge rules in br that are applicable w.r.t. S.

Definition 2. *A belief state* $S = (S_1, \ldots, S_n)$ *of* $\mathcal{L}$ *is an* equilibrium *of* α *iff* $S_i \in \mathbf{ACC}_i(kb_i \cup \{Head(r) | r \in app(br_i, S)\})$ *for every* i $(1 \le i \le n)$.

In what follows we denote $EQ(\alpha)$ the set of equilibria of the multi-context system α. If $EQ(\alpha) \ne \emptyset$ then it is *consistent*, otherwise it is *inconsistent*. Since we assume that every logic L_i in a multi-context system has the ability to express an arbitrary given collection of belief sets by a knowledge base of L_i, for any collection $\mathbb{S}$ of belief states of $\mathcal{L}$, there exists a multi-context system α such that $EQ(\alpha) = \mathbb{S}$. For convenience, we denote the multi-context system by *form*($\mathbb{S}$).

Example 2. [Continued from Example 1] We can use two multi-context systems α and β to model the first and second rounds of diagnosis respectively as follows. Here, the underlying logic of each context is propositional logic. The multi-context system $\alpha = (C_1, C_2)$ where C_1 is for Smith, while C_2 for Alice, $C_i = (L_i, kb_i, br_i)$ for $i = 1, 2$,

- $kb_1 = \{\neg bird\_flu\}$,
- $br_1 = \{bird\_flu \leftarrow (2 : bird\_flue); \quad swine\_flu \leftarrow (2 : swine\_flu)\}$,
- $kb_2 = \{(bird\_flu \lor pneumonia) \land (pneumonia \leftrightarrow \neg swine\_flu)\}$,
- $br_2 = \{bird\_flu \leftarrow (1 : bird\_flu); \quad swine\_flu \leftarrow (1 : swine\_flu)\}$.

Please note here that the signature of L_1 does not contain *pneumonia*. It is not difficult to check that α has a unique equilibrium $(\emptyset, \{pneumonia\})$.

The multi-context system $\beta = (C_1', C_2')$, where C_1' is for Smith, while C_2' for Alice, $C_i' = (L_i, kb_i', br_i')$ for $i = 1, 2$, and

- $kb_1' = \{bird\_flu \lor swine\_flu\}$, $br_1' = br_1$,
- $kb_2' = kb_2$, $br_2' = br_2$.

One can verify that β has two equilibria (S_1, S_2) and (S_1', S_2') where

- $S_1 = \{bird\_flu\}$, $S_2 = \{bird\_flu, pneumonia\}$,
- $S_1' = \{bird\_flu, swine\_flu\}$, $S_2' = \{bird\_flu, swine\_flu\}$.

Now the problem in Example 1 becomes that how should we change the multi-context system α using β? Though $kb_1 \land kb_1'$ is obviously consistent, it is a wishful thinking to change only kb_1 with $kb_1 \dotplus kb_1'$ in the multi-context system α, because both α and β have some acceptable belief states (equilibria) but such a revised result has no acceptable belief state. This violates the traditional intuition of belief revision.

3 Belief Change for Nonmonotonic Multi-Context Systems

In the following we consider the three major belief changes, expansion, revision and contraction in multi-context systems, starting from the expansion. We assume that the underlying logical language of multi-context systems is $\mathcal{L} = (L_1, \ldots, L_n)$ and belief change operators are mappings of the form $\mathcal{L} \times \mathcal{L} \to \mathcal{L}$.

3.1 Expansion

Recall that the intuition of expansion is to add belief without checking inconsistency. Semantically speaking, for the expansion of α by β, written $\alpha + \beta$, the acceptable belief states of the expansion result are exactly those that are acceptable belief states of both α and β. This motivates the postulates for expansion in multi-context systems:

(mcs-E1) $EQ(\alpha + \beta) \subseteq EQ(\alpha) \cap EQ(\beta)$;
(mcs-E2) If $EQ(\alpha) \subseteq EQ(\beta)$ then $EQ(\alpha + \beta) = EQ(\alpha)$;
(mcs-E3) If $EQ(\alpha') \subseteq EQ(\alpha)$ then $EQ(\alpha' + \beta) \subseteq EQ(\alpha + \beta)$.

Recall that $EQ(\alpha)$ is the set of the equilibria of α.

Theorem 1. *The expansion operator $+$ satisfies (mcs-E1) – (mcs-E3) iff $EQ(\alpha + \beta) = EQ(\alpha) \cap EQ(\beta)$ for any multi-context systems α and β.*

Proof Sketch: We prove "only if". By (mcs-E1) we obtain $EQ(\alpha + \beta) \subseteq EQ(\alpha) \cap EQ(\beta)$. Let γ be an MCS satisfying $EQ(\gamma) = EQ(\alpha) \cap EQ(\beta)$. By (mcs-E2) we have $EQ(\gamma + \beta) = EQ(\gamma)$. Note that $EQ(\gamma + \beta) \subseteq EQ(\alpha + \beta)$ holds by (mcs-E3). It follows that $EQ(\alpha) \cap EQ(\beta) = EQ(\gamma + \beta) \subseteq EQ(\alpha + \beta)$.   ∎

It shows that the expansion of two multi-context systems are semantically unique. For the multi-context systems α and β in Example 2, they have no common equilibria. It is not appropriate to model the belief change by the expansion, in the sense that the expansion result is inconsistent.

Recall that, in propositional logic, the expansion of a formula α by a formula β can be obtained by $\alpha \wedge \beta$. As the underlying logics in multi-context systems may be nonmonotonic, such a simple approach does not work for multi-context systems. For instance in ASP logic, let $\alpha = \{p \leftarrow not\, p\}$ and $\beta = \{p\}$, here α and β can be taken as multi-context systems with only one context and having no bridge rules. Then $\alpha \cup \beta = \{p \leftarrow not\, p;\quad p\}$ which has an answer set $\{p\}$ but it is not an answer set of α, which has no answer set. It illustrates that an expansion for nonmonotonic logics, and thus multi-context systems, is fundamentally different from that of propositional logic.

However, if the expansion can be relaxed as $EQ(\alpha) \cap EQ(\beta) \subseteq EQ(\alpha + \beta)$, i.e., the postulate (mcs-E1) is discarded, then we can have a simple expansion like the one for propositional logic. We will make this point clear in the following. A logic L is *quasi-cumulative* if for any two knowledge bases kb and kb' of L, and a belief set S of L such that $S \in \mathbf{ACC}_L(kb) \cap \mathbf{ACC}_L(kb')$ then $S \in \mathbf{ACC}_L(kb \cup kb')$. It is clear that propositional logic, first-order logic and description logic are all semi-cumulative. We can show that ASP logic define in [15] is quasi-cumulative.

Proposition 1. *Let Π_1 and Π_2 be two propositional theories and S an answer set of Π_1 and Π_2. Then S is an answer set of $\Pi_1 \cup \Pi_2$.*

Let $\alpha = (C_1, \ldots, C_n)$ and $\beta = (C'_1, \ldots, C'_n)$ be two multi-context systems over the logic languages $\mathcal{L} = (L_1, \ldots, L_n)$, where the underlying logics L_is of the contexts are quasi-cumulative. The *semi-expansion* of α by β, written $\alpha \mp \beta$, is the multi-context system $(C_1 + C'_1, \ldots, C_n + C'_n)$ where $C_i + C'_i$ is the context whose knowledge base (resp. bridge rule) is the union of the knowledge bases (resp. bridge rules) from the contexts C_i and C'_i for each i $(1 \leq i \leq n)$.

Theorem 2. *Let α and β be two multi-context systems where the underlying logics of the contexts are quasi-cumulative. If $\mathcal{S}$ is an equilibrium of α and β then it is an equilibrium of $\alpha \mp \beta$.*

Proof Sketch: Let $\mathcal{S} = (S_1, \ldots, S_n)$. It suffices to prove $S_i \in \mathbf{ACC}_{L_i}(kb_i \cup kb'_i \cup \{Head(r)|r \in app(br_i, \mathcal{S}) \cup app(br'_i, \mathcal{S})\})$ for every i $(1 \leq i \leq n)$. Recall that $S_i \in \mathbf{ACC}_{L_i}(kb_i \cup \{Head(r)|r \in app(br_i, \mathcal{S})\})$ and $S_i \in \mathbf{ACC}_{L_i}(kb'_i \cup \{Head(r)|r \in app(br'_i, \mathcal{S})\})$. By the quasi-cumulative property of the underlying logic L_i, we have $S_i \in \mathbf{ACC}_{L_i}(kb_i \cup \{Head(r)|r \in app(br_i, \mathcal{S})\} \cup kb'_i \cup \{Head(r)|r \in app(br'_i, \mathcal{S})\}$, i.e. $S_i \in \mathbf{ACC}_{L_i}(kb_i \cup kb'_i \cup \{Head(r)|r \in app(br_i \cup br'_i, \mathcal{S})\})$. ∎

3.2 Revision

Note that, in the revision of a knowledge base α (a set of formulas) by a formula β, the intent is that the resulted knowledge base contains β and is consistent (unless β is not), while keeping whatever information of α can be "reasonably" retained [1]. Semantically speaking, the revision of α by β is a knowledge base α' whose intended models are just those of β that are "closest" to those of α [19]. This motivates the following revision postulates for multi-context systems:

(mcs-R1) $EQ(\alpha \dotplus \beta) \subseteq EQ(\beta)$;
(mcs-R2) If $EQ(\alpha) \cap EQ(\beta) \neq \emptyset$ then $EQ(\alpha \dotplus \beta) = EQ(\alpha) \cap EQ(\beta)$;
(mcs-R3) If $EQ(\beta) \neq \emptyset$ then $EQ(\alpha \dotplus \beta) \neq \emptyset$;
(mcs-R4) If $EQ(\alpha_1) = EQ(\alpha_2)$ and $EQ(\beta_1) = EQ(\beta_2)$
  then $EQ(\alpha_1 \dotplus \beta_1) = EQ(\alpha_2 \dotplus \beta_2)$;
(mcs-R5) $EQ(\alpha \dotplus \beta_1) \cap EQ(\beta_2) \subseteq EQ(\alpha \dotplus \beta')$;
(mcs-R6) If $EQ(\alpha \dotplus \beta_1) \cap EQ(\beta_2) \neq \emptyset$ then $EQ(\alpha \dotplus \beta') \subseteq EQ(\alpha \dotplus \beta_1) \cap EQ(\beta_2)$

where β' is a multi-context system such that $EQ(\beta') = EQ(\beta_1) \cap EQ(\beta_2)$. The intended meaning of these revision postulates for multi-context systems is similar to that of propositional logic in Section 2.1.

Orders between Belief States. Let L be an abstract logic. A *pre-order* $\preceq_L$ over the possible belief sets[1] of L is a reflexive and transitive binary relation on $\mathbf{BS}_L$. And we define $\prec_L$ as $S \prec_L S'$ iff $S \preceq_L S'$ and $S' \npreceq_L S$. A pre-order $\preceq_L$ is *total* if for every two belief sets S and S' of L, either $S \preceq_L S'$ or $S' \preceq_L S$. A function that assigns a knowledge base kb of L a pre-order $\preceq_{kb}$ is *faithful* if the following conditions hold:

(1) If $S, S' \in \mathbf{ACC}_L(kb)$ then $S \preceq_{kb} S'$ does not hold.

---
[1] Recall that the notion of belief sets corresponds to "interpretations" in propositional logic.

(2) If $S \in \mathbf{ACC}_L(kb)$ and $S' \notin \mathbf{ACC}_L(kb)$ then $S \preceq_{kb} S'$.

(3) If $\mathbf{ACC}_L(kb) = \mathbf{ACC}_L(kb')$ then $\preceq_{kb}$ is same to $\preceq_{kb'}$.

Let $\mathbb{M}_L$ be a collection of belief sets of L. A belief set $S \in \mathbb{M}_L$ is *minimal w.r.t.* $\preceq_{kb}$ if there is no $S' \in \mathbb{M}_L$ such that $S' \prec_{kb} S$. We denote

$$Min(\mathbb{M}_L, \preceq_{kb}) = \{S|S \text{ is minimal in } \mathbb{M}_L \text{ w.r.t. } \preceq_{kb}\}.$$

Let $\alpha = (C_1, \ldots, C_n)$ be an MCS over langues $\mathcal{L} = (L_1, \ldots, L_n)$ with $C_i = (L_i, kb_i, br_i)$ $(1 \le i \le n)$, $\mathcal{S} = (S_1, \ldots, S_n)$ and $\mathcal{S}' = (S'_1, \ldots, S'_n)$ two belief states of α. A *pre-order* $\preceq_{\mathcal{L}}$ over the space of belief states of $\mathcal{L}$ is a reflexive and transitive binary relation such that $\mathcal{S} \preceq_{\mathcal{L}} \mathcal{S}'$ if $S_i \preceq_{L_i} S'_i$ for every i $(1 \le i \le n)$. Similarly, $\prec_{\mathcal{L}}$ is defined as $\mathcal{S} \prec_{\mathcal{L}} \mathcal{S}'$ if $\mathcal{S} \preceq_{\mathcal{L}} \mathcal{S}'$ and $\mathcal{S}' \npreceq_{\mathcal{L}} \mathcal{S}$. A pre-order $\preceq_{\mathcal{L}}$ is *total* if either $\mathcal{S} \preceq_{\mathcal{L}} \mathcal{S}'$ or $\mathcal{S}' \preceq_{\mathcal{L}} \mathcal{S}$ for every two of belief states $\mathcal{S}$ and $\mathcal{S}'$ of $\mathcal{L}$. A function that assigns α a pre-order $\preceq_\alpha$ is *faithful* whenever the assigned pre-order $\preceq_{kb_i}$ is faithful for every i $(1 \le i \le n)$.

Let $\mathbb{S}$ be a collection of belief states of $\mathcal{L}$. A belief state $\mathcal{S} \in \mathbb{S}$ is *minimal w.r.t.* $\preceq_\alpha$ if there is no $\mathcal{S}' \in \mathbb{S}$ such that $\mathcal{S}' \prec_\alpha \mathcal{S}$. We define

$$Min(\mathbb{S}, \preceq_\alpha) = \{\mathcal{S}|\mathcal{S} \text{ is minimal in } \mathbb{S} \text{ w.r.t. } \preceq_\alpha\}.$$

Theorem 3. *A revision operator $\dot{+}$ satisfies the conditions (mcs-R1)–(mcs-R6) iff there exists a faithful assignment that maps each MCS α of $\mathcal{L}$ to a total pre-order $\preceq_\alpha$ such that $EQ(\alpha \dot{+} \beta) = Min(EQ(\beta), \preceq_\alpha)$.*

Proof Sketch: The overall proof is similar to that of Theorem 3.3 of [19]. We outline the proof of "only if". Firstly we define $\preceq_\alpha$ over the space of belief states as $\mathcal{S} \preceq_\alpha \mathcal{S}'$ if either $\mathcal{S} \in EQ(\alpha)$ or $\mathcal{S} \in EQ(\alpha \dot{+} \beta')$ where β' is an MCS such that $EQ(\beta') = \{\mathcal{S}, \mathcal{S}'\}$. We can show that $\preceq_\alpha$ is a total pre-order and the assignment that maps α to $\preceq_\alpha$ is faithful. Finally we can prove $EQ(\alpha \dot{+} \beta) \subseteq Min(EQ(\beta), \preceq_\alpha)$ and $Min(EQ(\beta), \preceq_\alpha) \subseteq EQ(\alpha \dot{+} \beta)$. ∎

This theorem shows that our belief revision operator for multi-context systems obeys the principle of minimal change in the sense that the acceptable belief states (equilibria) of the revision result $\alpha \dot{+} \beta$ are exactly the ones of β that are "closet" to those of α. There are many approaches of "closeness" in propositional logic, we consider two distance-based operators for multi-context systems in the following.

Two Distance Based Revision Operators. Given two belief states $\mathcal{S} = (S_1, \ldots, S_n)$ and $\mathcal{S}' = (S'_1, \ldots, S'_n)$ of $\mathcal{L}$, we define the following notations,

- $|\mathcal{S}| = \sum_{i=1}^{n} |S_i|$, where $|S|$ denotes the cardinality of the set S,
- $\mathcal{S} \subseteq \mathcal{S}'$ if $S_i \subseteq S'_i$ for every i $(1 \le i \le n)$;
- $\mathcal{S} \subset \mathcal{S}'$ if $\mathcal{S} \subseteq \mathcal{S}'$ and $S_i \subset S'_i$ for some i $(1 \le i \le n)$;
- $\mathcal{S} \ominus \mathcal{S}' = (S_1 \ominus S'_1, \ldots, S_n \ominus S'_n)$, where $\ominus$ is the symmetric difference operator between sets, i.e. $X \ominus Y = (X \setminus Y) \cup (Y \setminus X)$.

Let α and β be two multi-context systems. We define

$$\ominus^{min}(\alpha, \beta) = min_\subseteq(\{\mathcal{S} \ominus \mathcal{S}' : \mathcal{S} \in EQ(\alpha) \& \mathcal{S}' \in EQ(\beta)\}),$$

$$|\ominus|^{min}(\alpha, \beta) = min_\le(\{|\mathcal{S} \ominus \mathcal{S}'| : \mathcal{S} \in EQ(\alpha) \& \mathcal{S}' \in EQ(\beta)\})$$

where $min_{\subseteq}(\mathbb{X})$ denotes the set of minimal (under set inclusion) elements in collection $\mathbb{X}$ of sets, and $min_{\leq}(X)$ denotes the least number in the set X of numbers.

Definition 3. *The revision operators $\dotplus_s$ and $\dotplus_d$ for multi-context systems are defined respectively as follows:*

$$EQ(\alpha \dotplus_s \beta) = \{\mathcal{S} \in EQ(\beta) | \exists \mathcal{S}' \in EQ(\alpha) \ s.t. \ \mathcal{S} \ominus \mathcal{S}' \in \ominus^{min}(\alpha, \beta)\},$$
$$EQ(\alpha \dotplus_d \beta) = \{\mathcal{S} \in EQ(\beta) | \exists \mathcal{S}' \in EQ(\alpha) \ s.t. \ |\mathcal{S} \ominus \mathcal{S}'| = |\ominus|^{min}(\alpha, \beta)\}.$$

Intuitively, the revision result $\alpha \dotplus_\star \beta$ has the equilibria of β that are "closest" to some equilibrium of α, where $\star \in \{s, d\}$. One can see that the operator $\dotplus_s$ (resp. $\dotplus_d$) is identical to Sato revision operator $\circ_S$ (resp. Dalal revision operator $\circ_D$) in [19] when there is only one context with underlying proportional logic and without bridge rules. Therefore, the revision operators $\dotplus_s$ and $\dotplus_d$ are generalizations of propositional knowledge base revision operator $\circ_S$ and $\circ_D$ respectively.

Example 3 (Continued from Example 2). According to the operators $\dotplus_s$ and $\dotplus_d$, one can check that the unique equilibrium of $\alpha \dotplus_\star \beta$ is (S_1, S_2), where $S_1 = \{bird\_flu\}$, $S_2 = \{bird\_flu, pneumonia\}$ and $\star \in \{s, d\}$. A multi-context system corresponding to $\alpha \dotplus_\star \beta$ can be (C_1'', C_2'') with $C_i'' = (kb_i'', br_i'')$ for $i = 1, 2$ where

- $kb_1'' = \{bird\_flu \wedge \neg swine\_flu\}$, note that $pneumonia$ is not a symbol of L_1,
- $kb_2'' = \{bird\_flu \wedge pneumonia \wedge \neg swine\_flu\}$,
- $br_1'' = br_2'' = \emptyset$, or alternatively $br_1'' = br_1$ and $br_2'' = br_2$.

Theorem 4. *Suppose that α is consistent. The revision operators $\dotplus_s$ satisfies (mcs-R1) – (mcs-R5); and the revision operator $\dotplus_d$ satisfies (mcs-R1) – (mcs-R6).*

Proof Sketch: We prove that $\dotplus_d$ satisfies (mcs-R6). By $EQ(\beta') = EQ(\beta_1) \cap EQ(\beta_2)$, we have $|\ominus|^{min}(\alpha, \beta_1) \leq |\ominus|^{min}(\alpha, \beta')$. Recall that $EQ(\alpha \dotplus_d \beta_1) \cap EQ(\beta_2) \neq \emptyset$. It implies that there exists $\mathcal{S} \in EQ(\beta_2) \cap EQ(\beta_1)$ and $|\mathcal{S} \ominus \mathcal{S}'| = |\ominus|^{min}(\alpha, \beta_1)$ for some $\mathcal{S}' \in EQ(\alpha)$, from which it follows that $|\ominus|^{min}(\alpha, \beta') \leq |\ominus|^{min}(\alpha, \beta_1)$. Thus $|\ominus|^{min}(\alpha, \beta') = |\ominus|^{min}(\alpha, \beta_1)$. Let $\mathcal{S} \in EQ(\alpha \dotplus_d \beta')$. We have $\mathcal{S} \in EQ(\beta_1) \cap EQ(\beta_2)$ and $\exists \mathcal{S}' \in EQ(\alpha)$ s.t. $|\mathcal{S} \ominus \mathcal{S}'| = |\ominus|^{min}(\alpha, \beta')$. It implies $\mathcal{S} \in EQ(\beta_1)$ and $\mathcal{S}' \in EQ(\alpha)$ s.t. $|\mathcal{S} \ominus \mathcal{S}'| = |\ominus|^{min}(\alpha, \beta_1)$. Thus $\mathcal{S} \in EQ(\alpha \dotplus_d \beta_1)$. ∎

The operator $\dotplus_s$ may falsify (mcs-R6) due to the fact that the corresponding revision operator $\circ_S$ in propositional logic may falsify the corresponding postulate (R6) (cf. Example 4.1 of [19]).

Relating to Classical Revision via Loop Formulas. Recall that every context in a multi-context system has its own language, which has its own signature too. In this sense, the signatures for contexts are pairwise disjoint. In the following we assume that the signatures for the langues in $\mathcal{L}$ share no common symbols, unless explicitly stated otherwise. In the following readers are assumed being familiar with answer set programming [16] and the basic notions of loops and loop formulas of logic programs [22].

We note that if $\mathcal{L}$ consists of ASP logics, then every MCS can be translated into a propositional theory via loop formulas [8]. Briefly, given a multi-context system

$\alpha = (C_1, \ldots, C_n)$ with $C_i = (L_i, kb_i, br_i)$ where $L_i s$ are ASP logics, by identifying C_i with the logic program $\Pi_i = kb_i \cup \ell(br_i)$ where $\ell(br_i)$ is obtained from br_i by replacing every indexed atom of the form $(n : p)$ with the atom p,

- the *loops* of C_i are the loops of Π_i, [2]
- the *loop formula* $\lambda(X, C_i)$ for loop X of C_i is the same as the loop formula $LF(X, \Pi_i)$ for the loop X of Π_i with the exception that it allows for circular supports[3] from $\ell(br_i)$,
- the *loop completion* $\pi(C_i)$ of C_i is the following theory

$$\{\lambda(X, C_i)|X \text{ is a loop of } C_i\} \cup \kappa(\Pi_i).$$

where $\kappa(\Pi_i)$ is obtained from Π_i by replacing every rule as a formula[4],
- the *loop completion* $\pi(\alpha)$ of α is $\pi(C_1) \cup \cdots \cup \pi(C_n)$.

It is proved that the equilibria of α correspond one-to-one to the models of $\pi(\alpha)$ (cf. Theorem 5 of [8]). To illustrate the notion of loop formulas for multi-context systems, let us consider the following example.

Example 4. Let $\alpha = (C_1, C_2)$ and $\beta = (C_1', C_2')$ be two multi-context systems with $C_i = (L_i, kb_i, br_i)$ and $C_i' = (L_i, kb_i', br_i')$ for $i = 1, 2$ where

- both L_1 and L_2 are ASP logics,
- $kb_1 = \{p\}$, $kb_2 = \{p'\}$, $kb_1' = \{\leftarrow p\}$ and $kb_2' = \{\leftarrow p'\}$,
- $br_1 = \{p \leftarrow not\,(2 : p')\}$, $br_2 = \{p' \leftarrow (1 : p)\}$, $br_1' = br_2' = \emptyset$.

One can check that the unique equilibrium of α is $\mathcal{S} = (\{p\}, \{p'\})$ and the unique equilibrium of β is $\mathcal{S}' = (\emptyset, \emptyset)$. Note that the loop of C_1 is $X = \{p\}$, which is the unique loop of C_1', and $X' = \{p'\}$ is the unique loop of C_2 and C_2'. Now we have:

- $\pi(C_1) \equiv p \wedge (\neg p' \supset p) \wedge (p \supset \top \vee \neg p')$, the last conjunct is $\lambda(X, \alpha)$,
- $\pi(C_2) \equiv p' \wedge (p \supset p') \wedge (p' \supset \top \vee p)$, the last conjunct is $\lambda(X', \beta)$,
- $\pi(\alpha) \equiv \pi(C_1) \wedge \pi(C_2) \equiv p \wedge p'$,
- $\pi(C_1') \equiv \neg p \wedge (p \supset \bot)$, the last conjunct is $\lambda(X, C_1')$,
- $\pi(C_2') \equiv \neg p' \wedge (p' \supset \bot)$, the last conjunct is $\lambda(X', C_2')$,
- $\pi(\beta) \equiv \neg p \wedge \neg p'$.

One can verify that, over the signature $\{p, p'\}$, the unique model of $\pi(\alpha)$ is $\{p, p'\}$, and the unique model of $\pi(\beta)$ is $\emptyset$.

We will show that belief revision in multi-context systems with ASP logics can be achieved by revision in propositional logic via loop formulas.

Lemma 1. *Let $\mathcal{S} = (S_1, \ldots, S_n)$ and $\mathcal{S}' = (S_1', \ldots, S_n')$ be two belief states of $\mathcal{L}$.*

(i) $\mathcal{S} \subseteq \mathcal{S}'$ iff $(\bigcup_{1 \leq i \leq n} S_i) \subseteq (\bigcup_{1 \leq i \leq n} S_i')$.

---

[2] Here, every single element set is a loop as defined in [20].
[3] E.g., the rule "$a \leftarrow a, b, not\,c$" is a circular supporting rule for a.
[4] E.g., the rule "$a \leftarrow b, c, not\,d$" is replaced by $\neg d \wedge b \wedge c \supset a$.

(ii) $|\mathcal{S}| \leq |\mathcal{S}'|$ *iff* $|(\bigcup_{1 \leq i \leq n} S_i)| \leq |(\bigcup_{1 \leq i \leq n} S_i')|$.

The following theorem shows that, a belief revision for multi-context systems with loop definable logics can be achieved via a belief revision in propositional logic.

Theorem 5. *Let α and β be two multi-context systems where the underlying logics are ASP. Then the equilibria of $\alpha \dotplus_s \beta$ (resp. $\alpha \dotplus_d \beta$) correspond one-to-one to the models of $\pi(\alpha) \circ_S \pi(\beta)$ (resp. $\pi(\alpha) \circ_D \pi(\beta)$).*

Proof Sketch: We prove the case $\dotplus_s$. Let $S = (S_1, \ldots, S_n)$ be a belief state of $\mathcal{L}$. We define $\pi(\mathcal{S}) = \bigcup_{1 \leq i \leq n} S_i$. Clearly, π is a one-to-one mapping from $\mathcal{S}$ to $\pi(\mathcal{S})$. We have that $\mathcal{S} \in EQ(\alpha \dotplus_s \beta)$
iff $\mathcal{S} \in EQ(\beta)$ and $\exists \mathcal{S}' \in EQ(\alpha)$ s.t $\mathcal{S} \ominus \mathcal{S}' \in \ominus^{min}(\alpha, \beta)$
iff $\pi(\mathcal{S}) \in Mod(\pi(\beta))$ and $\exists \pi(\mathcal{S}') \in Mod(\pi(\alpha))$ s.t $\pi(\mathcal{S}) \ominus \pi(\mathcal{S}') \in \ominus^{min}(\pi(\alpha),$
$\pi(\beta))$ by (i) of Lemma 1, the fact $\pi(\mathcal{S}) \ominus \pi(\mathcal{S}') = \pi(\mathcal{S} \ominus \mathcal{S}')$ and Theorem 5 of [8]
iff $\pi(\mathcal{S}) \in Mod(\pi(\alpha) \circ_s \pi(\beta))$. ∎

Example 5. [Continued from Example 4] In terms of the revision operators $\dotplus_s$ and $\dotplus_d$, we evidently have that the unique equilibrium of $\alpha \dotplus_s \beta$ and $\alpha \dotplus_d \beta$ is $\mathcal{S}'$. Thus we can take β as one of its revision result. One can check that $\pi(\alpha) \circ_\star \pi(\beta) = (p \wedge p') \circ_\star (\neg p \wedge \neg p')$, the unique model of $\pi(\alpha) \circ_\star \pi(\beta)$ is the model of $\pi(\beta)$ for $\star \in \{S, D\}$.

Partial Order Revisions. To characterize revision operators for multi-context systems that satisfy the postulates (mcs-R1) – (mcs-R6) in terms of pre-orders among belief states, it requires that every two belief states must be comparable, i.e. the pre-orders are total. Similar to the belief revision for propositional logic, it needs some partial orders among interpretations instead of total ones sometimes. We follow Katsuno and Mendelzon's approach to relax the totality conditions on orders over belief states. This motivates the following two postulates as a replacement of (mcs-R6):

(mcs-R6a) If $EQ(\alpha \dotplus \beta_1) \subseteq EQ(\beta_2)$ and $EQ(\alpha \dotplus \beta_2) \subseteq EQ(\beta_1)$ then $EQ(\alpha \dotplus \beta_1) = EQ(\alpha \dotplus \beta_2)$.
(mcs-R6b) $EQ(\alpha \dotplus \beta_1) \cap EQ(\alpha \dotplus \beta_2) \subseteq EQ(\alpha \dotplus \beta)$ where β is an MCS such that $EQ(\beta) = EQ(\beta_1) \cup EQ(\beta_2)$.

There are two alternatives to the postulate (mcs-R6a):

(mcs-R6w)   If $EQ(\alpha \dotplus \beta_1) \subseteq EQ(\beta_2)$, then $EQ(\alpha \dotplus \beta) \subseteq EQ(\alpha \dotplus \beta_1) \cap EQ(\beta_2)$ where β is an MCS satisfying $EQ(\beta) = EQ(\beta_1) \cap EQ(\beta_2)$.
(mcs-Rt)   If $EQ(\alpha \dotplus \beta) = EQ(\beta_1)$ and $EQ(\alpha \dotplus \gamma) = EQ(\beta_2)$ then $EQ(\alpha \dotplus \zeta) = EQ(\beta_1)$ where β, γ, ζ are MCSs satisfying $EQ(\beta) = EQ(\beta_1) \cup EQ(\beta_2)$, $EQ(\gamma) = EQ(\beta_2) \cup EQ(\beta_3)$ and $EQ(\zeta) = EQ(\beta_1) \cup EQ(\beta_3)$.

Lemma 2. *Assume that a revision operator $\dotplus$ for multi-context systems satisfies (mcs-R1) – (mcs-R5). Then the following conditions are equivalent:*

(i) The revision operator $\dotplus$ satisfies (mcs-R6a).
(ii) The revision operator $\dotplus$ satisfies (mcs-R6w).

(iii) The revision operator $\dotplus$ satisfies (mcs-Rt).

Theorem 6. *A revision operator $\dotplus$ satisfies conditions (mcs-R1)–(mcs-R5) and (mcs-R6a)–(mcs-R6b) iff there exists a faithful assignment that maps each multi-context system α to a partial pre-order $\preceq_\alpha$ such that $EQ(\alpha \dotplus \beta) = Min(EQ(\beta), \preceq_\alpha)$.*

Proof Sketch: It can be proved as that of Theorem 5.2 of [19]. ∎

3.3 Contraction

In propositional logic belief contraction is definable via Harper identity. It has independently desirable properties. In the following we adapt these properties for nonmonotonic multi-context systems and relate it with the revision and expansion for MCSs.

Suppose $\dotdiv$ is a contraction operator for multi-context systems. The contraction postulates for multi-context systems are formulated below from a model theoretical view:

(mcs-C1) $EQ(\alpha) \subseteq EQ(\alpha \dotdiv \beta)$;
(mcs-C2) If $EQ(\alpha) \not\subseteq EQ(\beta)$ then $EQ(\alpha \dotdiv \beta) = EQ(\alpha)$;
(mcs-C3) If β is not a tautology then $EQ(\alpha \dotdiv \beta) \not\subseteq EQ(\beta)$;
(mcs-C4) If $EQ(\beta) = EQ(\gamma)$ then $EQ(\alpha \dotdiv \beta) = EQ(\alpha \dotdiv \gamma)$;
(mcs-C5) $EQ(\alpha \dotdiv \beta) \cap EQ(\beta) \subseteq EQ(\alpha)$;
(mcs-C6) $EQ(\alpha \dotdiv \beta) \subseteq EQ(\alpha \dotdiv \beta_1) \cup EQ(\alpha \dotdiv \beta_2)$;
(mcs-C7) $EQ(\alpha \dotdiv \beta_1) \subseteq EQ(\alpha \dotdiv \beta)$ whenever $EQ(\alpha \dotdiv \beta) \not\subseteq EQ(\beta_1)$

where β is a multi-context system with $EQ(\beta) = EQ(\beta_1) \cap EQ(\beta_2)$ in (mcs-C6) and (mcs-C7). In terms of Harper identity, a belief contraction $\dotdiv$ can be defined via a belief revision $\dotplus$ as follows:

$$EQ(\alpha \dotdiv \beta) = EQ(\alpha) \cup EQ(\alpha \dotplus \beta') \tag{2}$$

where β' is a multi-context system satisfying $EQ(\beta') = \overline{EQ(\beta)}$.

Proposition 2. *The belief contraction defined by equation (2) satisfies the postulates (mcs-C1) – (mcs-C7).*

Based on the contraction and the above expansion, the revision operator can be alternatively defined by Levi identity as follows:

$$EQ(\alpha \dotplus' \beta) = EQ((\alpha \dotdiv \beta') \dotplus \beta) \tag{3}$$

where β' is a multi-context system satisfying $EQ(\beta') = \overline{EQ(\beta)}$.

Proposition 3. *The revision operator $\dotplus'$ is identical to $\dotplus$ for multi-context systems.*

4 Discussion and Conclusion

Related Work. Belief revision has been investigated in multiple (heterogeneous) information sources, multi-agent systems in particular [23,11,12]. On the one hand, there

is no AGM style postulates for those approaches; on the other hand, both knowledge
bases and bridge rules of agents are monotonic.

Note that each knowledge base in propositional logic can be seen as an MCS. One
can easily see that the belief revision operator $\dotplus$ for multi-context systems is a generalization to the one for propositional logic. As every answer set program can be taken
as an MCS, the proposed belief change theory for MCS is applicable for ASP as well.
This approach is substantially different from the existing ones, however. Many of them
are under the title *update* [29,2,13,26,9,24,17], in which a update of a sequence of logic
programs is focused and founded on the notion of *causal rejection* – some rules in a
logic program will be rejected since they *conflict* with some others in a higher priority
logic program. Thus these approaches fall outside the AGM belief revision paradigm
and are syntactic in nature. Delgrande *et al.* proposed a model-theoretical belief revision model for ASP [10], in which they concentrated on "SE-models" that capture the
strong equivalence[5] of logic programs [21]. When applying our belief change theory of
MCS to ASP, it considers answer sets instead of SE-models.

Conclusion and Future Work. Following AGM model we proposed a model-
theoretical belief expansion, revision and contraction theories for nonmonotonic multi-
context systems under equilibria semantics. Two distance-based revision operators were
presented. There are two other important semantics – grounded equilibria and well-
founded model [3]. We believe that it is possible to establish similar results under the
two semantics. Some interesting and challenging issues remain. For instance, in the
case that every underlying logics of MCS has a belief revision operator, such as managed multi-context systems [5], is it possible to obtain a rational revision operator for
MCS by *combining* these individual revision operators?

Acknowledgement. This work was partially supported by Australian Research Council under grants DP1093652 and DP110101042. Yisong Wang was also partially supported by NSFC under grants 60963009 and 61262029 and Stadholder Fund of Guizhou
Province under grant (2012)62.

References

1. Alchourrón, C.E., Gärdenfors, P., Makinson, D.: On the logic of theory change: Partial meet
   contraction and revision functions. JSL 50(2), 510–530 (1985)
2. Alferes, J.J., Leite, J.A., Pereira, L.M., Przymusinska, H., Przymusinski, T.C.: Dynamic updates of non-monotonic knowledge bases. JLP 45(1), 43–70 (2000)
3. Brewka, G., Eiter, T.: Equilibria in heterogeneous nonmonotonic multi-context systems. In:
   AAAI 2007, Vancouver, Canada, pp. 385–390. AAAI Press (2007)
4. Brewka, G., Eiter, T., Fink, M.: Nonmonotonic multi-context systems: A flexible approach
   for integrating heterogeneous knowledge sources. In: Balduccini, M., Son, T.C. (eds.) Logic
   Programming, Knowledge Representation, and Nonmonotonic Reasoning. LNCS, vol. 6565,
   pp. 233–258. Springer, Heidelberg (2011)
5. Brewka, G., Eiter, T., Fink, M., Weinzierl, A.: Managed multi-context systems. In: Walsh, T.
   (ed.) IJCAI 2011, Barcelona, Spain, pp. 786–791. IJCAI/AAAI (2011)

---

[5] Two logic programs Π_1 and Π_2 are *strongly equivalent* if $\Pi_1 \cup \Pi$ and $\Pi_2 \cup \Pi$ have the same
answer sets for any logic program Π.

6. Brewka, G., Eiter, T., Truszczynski, M.: Answer set programming at a glance. Communications of the ACM 54(12), 92–103 (2011)
7. Dalal, M.: Investigations into a theory of knowledge base revision. In: AAAI 1988, St. Paul, MN, pp. 475–479. AAAI Press/The MIT Press (1988)
8. Dao-Tran, M., Eiter, T., Fink, M., Krennwallner, T.: Distributed nonmonotonic multi-context systems. In: KR 2010, Toronto, pp. 60–70. AAAI Press (2010)
9. Delgrande, J.P., Schaub, T., Tompits, H.: A preference-based framework for updating logic programs. In: Baral, C., Brewka, G., Schlipf, J. (eds.) LPNMR 2007. LNCS (LNAI), vol. 4483, pp. 71–83. Springer, Heidelberg (2007)
10. Delgrande, J.P., Schaub, T., Tompits, H., Woltran, S.: A model-theoretic approach to belief change in answer set programming. TOCL 14(2), A:1–A:42 (2012)
11. Dragoni, A.F., Giorgini, P.: Toward a revision for multi-context systems. In: Workshop on Logics for Agent-Based Systems, LABS 2002, Tolose, France (2002)
12. Dragoni, A.F., Giorgini, P.: Distributed belief revision. In: AAMAS 2003, vol. 6(2), pp. 115–143 (2003)
13. Eiter, T., Fink, M., Sabbatini, G., Tompits, H.: On properties of update sequences based on causal rejection. TPLP 2, 711–767 (2002)
14. Fermé, E.L., Hansson, S.O.: AGM 25 years - twenty-five years of research in belief change. JPL 40(2), 295–331 (2011)
15. Ferraris, P.: Logic programs with propositional connectives and aggregates. TOCL 12(4), 25:1–25:40 (2011)
16. Gelfond, M., Lifschitz, V.: The stable model semantics for logic programming. In: ICLP 1988, Seattle, Washington, pp. 1070–1080. MIT Press (1988)
17. Guadarrama, J.C.A.: On Updates of Epistemic States Belief Change under Incomplete Information. PhD thesis, Clausthal University of Technology (2010)
18. Van Harmelen, F., Lifschitz, V., Porter, B. (eds.): Handbook of Knowledge Representation. Elsevier (2008)
19. Katsuno, H., Mendelzon, A.O.: Propositional knowledge base revision and minimal change. Artificial Intelligence 52(3), 263–294 (1992)
20. Lee, J., Lifschitz, V.: Loop formulas for disjunctive logic programs. In: Palamidessi, C. (ed.) ICLP 2003. LNCS, vol. 2916, pp. 451–465. Springer, Heidelberg (2003)
21. Lifschitz, V., Pearce, D., Valverde, A.: Strongly equivalent logic programs. TOCL 2(4), 526–541 (2001)
22. Lin, F., Zhao, Y.: ASSAT: computing answer sets of a logic program by SAT solvers. Artificial Intelligence 157(1-2), 115–137 (2004)
23. Liu, W., Williams, M.-A.: A framework for multi-agent belief revision. Studia Logica 67(2), 291–312 (2001)
24. Osorio, M., Cuevas, V.: Updates in answer set programming: An approach based on basic structural properties. TPLP 7(4), 451–479 (2007)
25. Peppas, P.: Belief Revision. In: Handbook of Knowledge Representation, ch. 8. Foundations in Artificial Intelligence, pp. 317–360. Elsevier (2008)
26. Sakama, C., Inoue, K.: An abductive framework for computing knowledge base updates. TPLP 3(6), 671–713 (2003)
27. Satoh, K.: Nonmonotonic reasoning by minimal belief revision. In: FGCS 1988, Tokyo, Japan, pp. 455–462 (1988)
28. Serafini, L., Bouquet, P.: Comparing formal theories of context in AI. Artificial Intelligence 155(1-2), 41–67 (2004)
29. Zhang, Y., Foo, N.Y.: Towards generalized rule-based updates. In: IJCAI (1), pp. 82–88. Morgan Kaufmann (1997)

On Optimal Solutions of Answer Set Optimization Problems

Ying Zhu and Miroslaw Truszczynski

Department of Computer Science, University of Kentucky, Lexington, KY 40506, USA

Abstract. In 2003, Brewka, Niemelä and Truszczynski introduced answer-set optimization problems. They consist of two components: a logic program and a set of preference rules. Answer sets of the program represent possible outcomes; preferences determine a preorder on them. Of interest are answer sets of the program that are optimal with respect to the preferences. In this work, we consider computational problems concerning optimal answer sets. We implement and study several methods for the problems of computing an optimal answer set; computing another one, once the first one is found; and computing an optimal answer set that is similar to (respectively, dissimilar from) a given interpretation. For the problems of the existence of similar and dissimilar optimal answer set we establish their computational complexity.

1 Introduction

Preferences play an important role in AI applications that involve decision making [7,8]. In many practical problems, hard constraints still leave many feasible solutions and a mechanism for selecting those with some desirable properties is needed. A typical approach consists of eliciting from the user *preferences* (sometimes referred to as *soft constraints*) on the space of solution candidates and returning to the user only those solutions that "score" high on the user's preference criteria, in most cases, actually only those that are optimal. To provide a formal basis to that approach, researchers proposed several preference representation formalisms [8]. However, in most cases, preference theories still have multiple optimal solutions and the final selection has to be performed by user.

To help the user make that selection, one needs computational support for the key preference reasoning tasks. They include computing an optimal solution, and computing an alternative optimal solution once the first optimal solution was found. In some cases, the user knows a combination of desirable properties and would like to pose that as a query to which the system would respond with optimal solutions that come close (are similar). On the flip side, the user may know a combination of undesirable properties and would like to see optimal solutions that are unlike that undesirable one (are dissimilar).

The problem of computing similar/dissimilar solutions was identified as important in the setting of problems defined by hard constraints, and was studied in the context of answer sets of programs by Eiter et al. [5]. We consider it in a more general setting of preference formalisms. Specifically, our work concerns

P. Cabalar and T.C. Son (Eds.): LPNMR 2013, LNAI 8148, pp. 556–568, 2013.

the problems mentioned above in the preference formalism of *answer set optimization* (ASO) [2]. We are interested in the computational complexity of these problems, and in effective computing methods. In particular, we study applications of answer set programming [10,11] and answer-set programming tools to solve them.

The main contributions of our paper are as follows. (1) We show that for ASO programs, the problems of the existence of optimal answer sets that are similar to (respectively, dissimilar from) a given interpretation are Σ_2^p-complete. (2) We propose several methods, exploiting answer set solvers, to support optimization tasks in ASO. Some of these methods are fully declarative, that is, represent optimization problems in terms of single disjunctive logic programs. However, in one of our methods, we execute the optimization task in an iterative way. (3) We give several methods of generating random instances of ASO programs for testing and present experimental results. They show that for non-ranked preferences, encodings in terms of disjunctive programs are competitive, that is, disjunctive answer set solvers work well. On the other hand, if ASO programs have ranked preferences, the solvers are no longer effective on the encodings we used. The results suggest a class of challenging benchmarks for ASP solvers developed for disjunctive logic programs.

2   Preliminaries

An answer set optimization (ASO) program P [2] consists of two parts: a *generator* P_{gen} and a *selector* P_{pref}. The generator is a propositional non-disjunctive answer set program[1] or a propositional theory. It is used to represent hard constraints. The complexity results we present below do not depend on the exact form of the generator as the complexity of model generation task is the same for the two types of generators we allow. However, in the experimentation part of the work, for the sake of concreteness we consider as generators only propositional CNF theories.

The selector is a collection of preference rules of the form:

$$C_1 > \cdots > C_k \leftarrow a_1, \ldots, a_n, \neg\, b_1, \ldots, \neg\, b_m \tag{1}$$

where a_is and b_is are literals and C_is are boolean combinations of atoms. The selector represents preferences or soft constraints of the problem. Informally, the rule above reads: if an answer set contains $a_1, \ldots, a_n$ and does not contain any of the literals $b_1, \ldots, b_m$, then C_1 is preferred over C_2, C_2 is preferred over C_3, etc.

The formal semantics of rule (1) is based on the notion of the *satisfaction degree*. Let r be a rule of the form (1). If an interpretation S does not satisfy the body of r or if it does not satisfy any of the options in the head of r, then r is *irrelevant* to S and the satisfaction degree of S on r, $v_S(r)$, is set to 1. Otherwise,

---

[1] Allowing disjunctive programs as generators is possible but that affects the complexity results as well as computational methods used.

we define the satisfaction degree of S on r by setting $v_S(r) = min\{i : S \models C_i\}$. Now, given two answer sets S_1 and S_2, we say that S_1 is preferred over S_2, written as $S_1 \succeq S_2$, if $v_{S_1}(r) \leq v_{S_2}(r)$, for every rule $r \in P_{pref}$. We say that S_1 is strictly preferred to S_2, written as $S_1 \succ S_2$, if $S_1 \succeq S_2$ and for some rule $r \in P_{pref}$, $v_{S_1}(r) < v_{S_2}(r)$. An answer set (a model, if the generator is a propositional theory) S is *optimal* if there is no answer set (model) S' of the generator such that $S' \succ S$.

To illustrate the ASO formalism, we consider a simple example. Let us assume P_{gen} is any theory generating 4 answer sets:

$$S_1 = \{soup, beef\}, \ S_2 = \{salad, beef\},$$
$$S_3 = \{soup, fish\}, \ S_4 = \{salad, fish\}.$$

For example, we can take for P_{gen} an answer set program:

$$1\{soup, salad\}1$$
$$1\{beef, fish\}1$$

or a propositional theory:

$$(soup \lor salad) \land (beef \lor fish) \land \neg(soup \land salad) \land \neg(beef \land fish).$$

Assuming P_{pref} is:

$$soup > salad$$
$$beef > fish,$$

the satisfaction vectors for the four answer sets are $V_1 = (1,1)$, $V_2 = (2,1)$, $V_3 = (1,2)$, $V_4 = (2,2)$, respectively. Thus, S_1 is the optimal answer set, S_4 is the worst answer set, and S_2 and S_3 are incomparable.

The formalism we just presented is rather weak in that it is based on the Pareto Principle. Consequently, it renders many answer sets of P_{gen} optimal. To strengthen it one may consider ranked preferences, that is, preferences that differ in importance. A *ranked* ASO program is a tuple (P_{gen}, P_{pref}), where P_{gen} is as before and P_{pref} is a collection of *ranked* preference rules, that is, rules of the form

$$C_1 > \cdots > C_k \overset{j}{\leftarrow} a_1, \ldots, a_n, \neg \, b_1, \ldots, \neg \, b_m \tag{2}$$

where the notation is as above, the only difference being the presence of a positive integer j indicating the rank (with 1 being the highest rank possible.) For a rule r, we write $rank(r)$ to denote its rank.

The satisfaction degree is defined as before. Given two answer sets S_1 and S_2, we say that S_1 is preferred over S_2, written as $S_1 \succeq S_2$ if $v_{S_1}(r) = v_{S_2}(r)$, for every rule $r \in P_{pref}$, or if there is a rule r_0 such that

1. $v_{S_1}(r_0) < v_{S_2}(r_0)$

2. $v_{S_1}(r) \leq v_{S_2}(r)$, for every rule r such that $rank(r) = rank(r_0)$
3. $v_{S_1}(r) = v_{S_2}(r)$, for every rule r such that $rank(r) < rank(r_0)$.

Moreover, S_1 is strictly preferred over S_2, $S_1 \succ S_2$, if $S_1 \succeq S_2$ holds due to the existence of the rule r_0 above. As before, an answer set (or model) S is *optimal* if there is no answer set (model) S' such that $S' \succ S$.

3   Problems and Complexity

In the paper we consider the propositional case only and assume all programs are finite. The two fundamental optimization problems for any preference framework concern finding an optimal solution and an alternative one. We state them below for ASO programs.

OPTIMAL SOLUTION   Given an ASO program P, decide whether an optimal answer set (model) S for P exists.

ANOTHER OPTIMAL SOLUTION   Given an ASO program P, and one optimal answer set (model) S, decide whether a different optimal answer set (model) S' ($S' \neq S$) for P exists.

Clearly, if the generator of an ASO program P has answer sets (models, respectively), P has an optimal answer set (model). Thus the complexity of deciding whether an optimal solution exists is the same as the complexity of deciding whether the generator P_{gen} is satisfiable (relative the semantics of choice). That latter problem is NP-complete for both generator formalisms (we recall that generators are restricted to be non-disjunctive programs) and therefore, so is the former.

For the problem ANOTHER OPTIMAL SOLUTION, if we can find an answer set of P_{gen}, which is not worse than the given optimal one S and different from it, then there exists another optimal answer set for P. The complexity of deciding whether there is an answer set of P_{gen} that is not worse than S is clearly in NP. The hardness can be proved by reduction from the answer set existence problem. The same arguments apply to the case of propositional theories as generators. Thus, we have the following result.

Theorem 1. *The* ANOTHER OPTIMAL SOLUTION *problem is NP-complete.*

We also consider the problems of finding a similar/dissimilar optimal solution to/from a given one. In each problem we assume that the distance is determined by some measure of distance between interpretations, say Δ. Moreover, we assume that the input consists of an ASO program P, an interpretation S, and a nonnegative integer k.

k-SIMILAR OPTIMAL SOLUTION   Decide whether there is an optimal answer set (model) S' such that $\Delta(S, S') \leq k$.

k-DISSIMILAR OPTIMAL SOLUTION   Decide whether there is an optimal answer set (model) S' such that $\Delta(S, S') \geq k$.

The complexity of these problems depends on the complexity of deciding whether $\Delta(S, S') \leq k$. Assuming that problem is in P, the problems to find a similar/dissimilar optimal answer set are Σ_2^p-complete. The membership can be showed by guessing an interpretation S', verifying whether S' is an optimal answer set (that problem is coNP-complete [2]), and checking whether $\Delta(S, S') \leq k$ (in polynomial time). For the hardness, we construct a reduction from the problem to decide whether there is an optimal answer set including a given literal, which is Σ_2^p-complete [2]. The details depend on the definition of the distance function. The theorems we present below assume that the Hamming distance HD is used to measure how far from each other are the solutions.

Theorem 2. *Given an ASO program P, an interpretation S, and a nonnegative integer k, deciding whether there is an optimal answer set (model) S' such that $HD(S, S') \leq k$ is Σ_2^p-complete.*

Proof: (Membership) The problem is in Σ_2^p because we can guess an interpretation S', verify in polynomial time that it is an answer set, then use a coNP-oracle to verify that S' is an optimal answer set [2] and, finally, check whether $HD(S, S') \leq k$ in polynomial time.

(Hardness) Given an ASO program P and a literal l, it is Σ_2^p-hard to decide whether there is an optimal answer set M, such that $l \in M$ [2]. We construct an ASO program P' and an interpretation S so that there is an optimal answer set S' for P' with $HD(S, S') \leq k$ if and only if there is an optimal answer set M for P and $l \in M$.

To this end, we first introduce $k+1$ fresh atoms $a_1, a_2, \ldots, a_{k+1}$. We construct the program P' and an interpretation S as follows. We define

$$P'_{gen} = P_{gen} \cup \{a_1 \leftarrow not\ l, a_2 \leftarrow not\ l, \ldots, a_{k+1} \leftarrow not\ l\}$$

and set k to the number of atoms in P, $P'_{pref} = P_{pref}$ and $S = \emptyset$.

($\Leftarrow$) We show that if there is an optimal answer set M such that $l \in M$, then there is an optimal answer set S' with $HD(S, S') \leq k$.

Let M be an optimal answer set for P and $l \in M$. Let $S' = M$. Clearly, S' is an optimal answer set of P' and $a_i \notin S'$ for $i = 1, \ldots, k+1$. Since $|S'| = |M| \leq k$ and $|S| = 0$, we have $HD(S, S') \leq k$.

($\Rightarrow$) Next, we show that if there is an optimal answer set S', and $HD(S, S') \leq k$, then there is an optimal answer set M, and $l \in M$.

Let S' be an optimal answer set of P' and $HD(S, S') \leq k$. If some a_i in S', then all a_i in S'. Thus $|S'| \geq k+1$ and $HD(S, S') \geq k+1$, which is contradicts to our assumption. It follows that $a_i \notin S'$, for every i, and that $l \in S'$. Let $M = S'$. Clearly, M is an optimal answer set of P and $l \in M$.

The next theorem concerns the problem of finding a dissimilar optimal answer set. The proof is similar and we omit it.

Theorem 3. *Given an ASO program P, an interpretation S, and a nonnegative integer k, deciding whether there is an optimal answer set (model) S' such that $HD(S, S') \geq k$ is Σ_2^p-complete.*

4   Computational Methods

In this paper, we study three methods to solve the problems we discussed above. We describe them under the assumption that the generator is an answer set program but the discussion extends literally to the case when we use propositional theories for generators. The first method uses an iterative way to find optimal answer sets in ASO programming. The second method modifies the ASP solver *clasp* [6]. The third method encodes the problems as disjunctive logic programs. We used *dlv* [9] and *claspD* [3] to process them.

4.1   Iterative Method

This method is based on an answer set program, a "tester", to decide whether an answer set of a generator program is optimal with respect to the preference relation determined by the selector. The tester program is designed to have no answer sets when the input answer set is optimal; otherwise it returns a strictly better answer set. We designed the tester program following the method described by Brewka et al. [2].

To find an optimal answer set, the iterative method starts with any answer set to the generator (if none exists, it terminates with failure) and runs the tester program on it. If a better answer set is found, the tester program is run on it. This iterative improvement process terminates as the space of answer sets is finite. When it does, the method returns the last answer set generated.

To find another optimal answer set, we first find an answer set which is not worse than and different from a given optimal answer set, say M. Such an answer set will be incomparable to M, or will be different from M but have the same satisfaction vector as M. These constraints can be modeled as rules. We add them to the generator program. Answer sets of the program that results are precisely the answer sets of the original generator program that are not worse than and different from M. Starting the iterative process described above from any of them (if there are none, M is the only optimal answer set) results in an optimal answer set different from M.

To solve the problem finding a similar or dissimilar optimal answer set, the iterative method computes optimal answer sets one by one, following the method described above, until it finds an optimal answer set which satisfies the distance limitation or until no more optimal answer sets can be found. As the method proceeds, all optimal answer sets found are stored and more and more constraints representing computed optimal answer sets are added to the original generator. Thus, the method may become infeasible for problems with the large number of optimal answer sets (however, that does not show in the range of problems we experimented with).

4.2   Modified *clasp*

The iterative method needs to record many optimal answer sets when solving the problems to find a similar/dissimilar optimal answer set. The second method modifies the ASP solver *clasp* to solve the problems without the need to store intermediate results. To this end, we modify *clasp* so that when each time it finds an answer set, it checks its optimality using the tester program described above. When an optimal answer set is found, the program terminates.

To find another optimal answer set, we call the modified *clasp* on the problem with the generator modified to reflect the constraints for an answer set not to be identical to or worse than the given optimal one (described above). To find a similar (dissimilar) optimal answer set to a given interpretation, we add distance constraints to the generator program and run the modified *clasp* on the program that results.

4.3   The Disjunctive Logic Program Encoding

Both the iterative way and the modified *clasp* need multiple calls to an ASP solver to solve our preference optimization problems. Moreover, the iterative method organizes the calls in a procedural (imperative) manner rather than declaratively. Our third method compiles the entire computation into a single disjunctive logic program so that a single call to a program such as *dlv* or *claspD* could solve it (the complexity result guarantee the existence of such representations). It does not need to store any intermediate results.

To this end, we first describe the problem in terms of a quantified boolean formula (QBF) and then apply a version of the Eiter-Gottlob translation [4] to produce the program. We provide the details for the problem of computing a single optimal answer set. The other problems can be handled similarly, as they essentially only differ in additional constraints one needs to impose on the generator.

It is convenient to assume now that the generator is a propositional formula (the case of a logic program is not much different). Thus, let G be a propositional formula representing the generator. We write $G(X)$ to stress that it is built of propositional atoms in some set $X = \{x_1, \ldots, x_m\}$. By $G(Y)$ we will denote the same formula but with y_i replacing x_i, $1 \leq i \leq m$. Finally, let $F(X, Y)$ be a propositional formula representing the statement that the interpretation Y is strictly better than X (one can construct it from the selector part of the ASO program). The property that an interpretation over X is an optimal answer set can be stated as follows: an interpretation X is an answer set and for every other interpretation Y, Y is not an answer set or Y is not strictly better than X. It can be expressed formally by the QBF

$$\Phi = \exists X \forall Y (G(X) \wedge (\neg G(Y) \vee \neg F(X, Y))).$$

Applying the Eiter-Gttlob translation [4] to Φ (it should be adjusted to the structure of the QBF Φ, part of which is subject only to the existential quantifier) results in a disjunctive logic program whose answer sets correspond to the answer sets of the given optimization problem.

If we modify $G(X)$ (but not $G(Y)$) with constraints that X be not identical to and not worse than a given interpretation, or that X be at distance at least (at most) k from a given interpretation, we obtain disjunctive logic programs encoding the remaining problems of interest. We omit the details due to the page limit.

5   Experiments and Analysis

All the methods we developed were implemented in C/C++. All experiments were conducted on an Intel processor clocked at 2.40GHz with 2GB memory.

We experimented with the three computational methods described in Section 4. For the iterative method we used *clasp-2.1.1* as the ASP solver. With the third method we used two disjunctive ASP solvers: *claspD-1.1.2* and *dlv-2012-12-17*. In the discussion below we simply write *claspD* and *dlv*, respectively to denote the appropriate version of the method described in Section 4.3. Similarly, we write *iterative* and *mclasp* (modified clasp) for the other two methods.

In experiments we used ASO problems in which generators are represented by 3-CNF formulas, because there are well understood random models to generate them. To construct them, we randomly generated 3-CNF formulas with n atoms and $3n$ and $4n$ clauses respectively. The numbers of clauses are below the threshold ratio of $4.25n$ to ensure that generators have models (the space of feasible solutions is not empty). This is important as we are interested in studying optimization problems of selecting optimal solutions and not the satisfiability problem. We note that the formulas in the first class have more models than those in the second class. Thus, the two classes are qualitatively different.

We generated preference rules for the selectors through the following three mechanisms:

1. randomly generating $3n$ preference rules without rank (n being as before the number of atoms)
2. randomly generating $3n$ rules with two ranks, half of the rules having rank 1 and half of the rules having rank 2
3. extracting preference rules from two *lexicographic preference trees* (LP-trees, for short) [1].

In the first two cases, we randomly choose a variable, say x, and form the head to be of the form $x > \neg x$ or $\neg x > x$, choosing between them with equal probability. Each rule has no condition or has a condition of at most two literals. With the probability 0.5 it has no condition, having a one literal condition or two literal condition are equally likely. Literals for the conditions are generated uniformly at random.

LP trees are concise representations of strict total orders [1]. Preferences encoded by LP trees can be represented as ranked preference rules, with as many ranks as there are atoms in the language. For the third method, for each instance we randomly generate two LP trees (we subject them to some restrictions to make their sizes linear in the number of atoms: each tree may have at most one node where it splits into two branches, and each node's preferences depend on values in at most one ancestor node). To form a selector, we represent preferences encoded by the two trees as ranked preference rules.

We considered four computational problems: OPTIMAL SOLUTION, ANOTHER OPTIMAL SOLUTION, 3-SIMILAR OPTIMAL SOLUTION, and 8-DISSIMILAR OPTIMAL SOLUTION. For the problems to find a similar/dissimilar optimal answer set, we used the Hamming distance to measure the distance between two answer sets.

We prepared data sets for unranked and ranked cases. For unranked case, n ranged with step 10 from 20 to 120 and for the ranked one, from 20 to 50. For each n, 20 instances were generated and the total computation time was calculated. We only present the experimental results for the three data sets with generators consisting of $4n$ three-literal clauses. The relative behavior of all the methods is the same for $3n$-clause generators as for $4n$-clause generators. However, the problems based on $3n$-clause generators are harder as the space of feasible solutions is larger.

The first data set consists of problems with $4n$ clauses in generators and with $3n$ random preference rules all of rank 1. Table 1 shows the computation time for OPTIMAL SOLUTION and ANOTHER OPTIMAL SOLUTION problems. The iterative method scales up the best and is a clear winner. The reasons seem to be that its iterative improvement process terminates with no backtracks and, since there are many optimal solutions for this data set, few iterations are required. For the same reasons, the method is similarly effective on the ANOTHER OPTIMAL SOLUTION problem. The *mclasp* method quickly becomes impractical. It simply searches through the space of all answer sets seeking an optimal one, and does not take advantage of the information gathered earlier. We see how that may be a problem when we look at times for the problem ANOTHER OPTIMAL SOLUTION, where *mclasp* has a smaller search space to consider (only of those answer sets that are different and not worse from the given one). Indeed for that problem, *mclasp* does slightly better. The third method that compiles the problems into single disjunctive programs, behaves better and, especially when implemented with *claspD*, scales up reasonably well. It is slower on the ANOTHER OPTIMAL SOLUTION problem than on the OPTIMAL SOLUTION one. That may be caused by the fact that the ANOTHER OPTIMAL SOLUTION problem requires additional constraints and programs the two methods have to work with get larger.

Table 1. Results of experiments on data set 1 for the first two problems

	OPTIMAL SOLUTION				ANOTHER OPTIMAL SOLUTION			
nVar	Iterative	mclasp	claspD	dlv	Iterative	mclasp	claspD	dlv
20	4.14	7.64	1.97	1.46	2.86	5.84	1.80	1.61
30	4.86	8.82	2.52	1.91	4.59	12.64	2.46	2.74
40	6.41	21.03	3.19	2.24	4.81	11.45	3.46	2.48
50	6.79	32.95	3.61	2.92	5.35	13.19	4.11	3.55
60	11.19	500+	9.48	8.63	9.74	277.73	9.56	8.75
70	12.77	500+	12.82	7.27	10.67	500+	12.84	7.30
80	13.99	500+	11.54	9.92	12.99	500+	11.50	11.88
90	11.29	500+	15.33	302.25	9.24	500+	16.57	301.01
100	11.17	500+	13.86	129.83	11.03	500+	14.07	500+
110	14.16	500+	19.24	179.35	11.08	500+	21.31	500+
120	14.83	500+	48.64	500+	15.11	500+	51.04	500+

Table 2. Results of experiments on data set 1 for the last two problems

	3-SIMILAR OPTIMAL SOLUTION				8-DISSIMILAR OPTIMAL SOLUTION			
nVar	Iterative	mclasp	claspD	dlv	Iterative	mclasp	claspD	dlv
20	11.80	6.37	2.16	1.97	11.34	5.56	2.09	1.83
30	97.93	7.13	2.80	2.84	12.70	10.25	3.23	2.46
40	159.74	14.35	3.22	3.50	15.98	38.89	3.04	5.89
50	500+	25.83	4.19	4.40	24.84	33.92	4.02	4.31
60	500+	43.75	6.72	6.49	96.35	162.91	8.77	14.53
70	500+	93.52	7.48	6.80	16.75	500+	11.62	19.70
80	500+	122.10	8.26	10.99	66.63	500+	12.22	16.63
90	500+	500+	6.15	8.22	12.33	500+	11.85	330.39
100	500+	500+	7.62	7.33	22.09	500+	12.60	154.13
110	500+	500+	8.09	9.20	73.29	500+	72.71	34.31
120	500+	500+	8.90	12.74	17.08	500+	58.61	280.67

Next, we tested all approaches on the two remaining problems: 3-SIMILAR OP-
TIMAL SOLUTION, and 8-DISSIMILAR OPTIMAL SOLUTION. The results are shown
in Table 2. In all approaches we modeled the distance constraints using aggregates. On both problems, the *mclasp* method does not perform well but slightly
better than on the first two problems. The difference seems to be caused by the
fact that the search space now is slightly smaller (*mclasp* checks the optimality
as it searches for answer sets already subject to the distance condition). The
iterative method does not perform well on the 3-SIMILAR OPTIMAL SOLUTION
problem (it times out when $n \geq 50$) but is the fastest on the 8-DISSIMILAR
OPTIMAL SOLUTION problem. To find a similar/dissimilar optimal answer set,
the iterative method keeps finding optimal answer sets one by one, remembering
those found along the way until it finds an answer set that satisfies the distance
constraint. For instances with a large number of variables, the constraint that
the distance is ≤ 3 is substantially stricter than that the distance is ≥ 8. As its
"improvement" search is essentially uninformed, the iterative method performs
poorly on the former problem and very well on the latter one. The third method
shows an orthogonal behavior. It works very well on the 3-SIMILAR OPTIMAL
SOLUTION problem but, while still acceptable, much worse on the 8-DISSIMILAR
OPTIMAL SOLUTION problem. It may be that while a tightly constraint search
space is a problem for the iterative method, which in some way is blind to
the distance constraint until the very end, it is in fact beneficial for the third
method.

The data set 2 is the same as the data set 1, except that preference rules
are now assigned ranks. Half of the rules have rank 1 and the other half have
rank 2. The results of experiments on that data set 2 are shown in Tables 3
and 4. It is evident that except for the iterative method, other methods become
much slower in comparison with their performance on the data set 1. For the
data set 2, because of the ranks, the number of optimal answer sets is much
smaller than for the corresponding instances in the data set 1. Thus, with the
same search space of answer sets, it is much more difficult for *mclasp* to find an
optimal one. In contrast, the change in the number of optimal answer sets does
not have a major influence on the performance of the iterative method. For the
two implementations of the third method, in the presence of ranks the theory

Table 3. Results of experiments on data set 2 for the first two problems

nVar	OPTIMAL SOLUTION				ANOTHER OPTIMAL SOLUTION			
	Iterative	*mclasp*	*claspD*	*dlv*	*Iterative*	*mclasp*	*claspD*	*dlv*
20	3.74	7.21	15.78	15.87	2.64	5.10	20.57	14.78
30	4.81	21.29	33.61	168.63	4.43	24.01	31.47	179.82
40	6.51	82.28	63.33	61.57	4.68	31.62	63.70	54.61
50	7.70	322.17	93.22	200.35	5.62	72.48	106.79	281.17

Table 4. Results of experiments on data set 2 for the last two problems

nVar	3-SIMILAR OPTIMAL SOLUTION				8-DISSIMILAR OPTIMAL SOLUTION			
	Iterative	*mclasp*	*claspD*	*dlv*	*Iterative*	*mclasp*	*claspD*	*dlv*
20	7.49	25.80	21.04	24.43	11.10	8.23	19.48	16.36
30	26.59	35.08	33.02	70.41	13.19	19.13	35.59	104.65
40	169.50	100.94	50.53	139.73	6.64	73.46	60.84	326.03
50	450.06	124.11	76.20	152.77	12.23	211.83	107.33	225.17

modeling the "not worse" constraint gets more complicated. That results in a larger program and poorer performance of *claspD* and *dlv*. We observe similar results for the problems to find a similar/dissimilar optimal answer set. The only interesting thing to note here is that the iterative method performs better than in the case of the data set 1. This is due to the smaller number of optimal answer sets for the corresponding instances. Consequently, there are fewer of them to search through.

Table 5 shows the results for data set 3, in which the selector of each instance consists of ranked preference rules extracted from two randomly generated LP-trees. Because one LP-tree has only one optimal answer set (it defines a total strict order), most of the instances in this data set have very few optimal answer sets. The experiments show that all methods perform worse here than on the corresponding problems for the other two data sets. Because of many levels of ranks, the performance of *claspD* and *dlv* deteriorates rapidly and in most cases even for relatively small values of n they time out. We do not report these results here. The other two methods perform better, with the iterative method performing well on all these problems. In fact, due to a small number of optimal answer sets, it runs faster for the 3-SIMILAR OPTIMAL SOLUTION, and 8-DISSIMILAR OPTIMAL SOLUTION problems than it did on instances from the other two data sets.

Table 5. Results of experiments on data set 3

nVar	OPTIMAL SOLUTION		ANOTHER OPTIMAL SOLUTION		3-SIMILAR OPTIMAL SOLUTION		8-DISSIMILAR OPTIMAL SOLUTION	
	Iterative	*mclasp*	*Iterative*	*mclasp*	*Iterative*	*mclasp*	*Iterative*	*mclasp*
20	6.18	26.35	3.66	8.01	5.21	20.62	5.22	15.50
30	9.61	304.27	6.78	33.28	8.45	46.17	7.83	330.08
40	13.39	500+	7.02	30.95	7.80	93.98	7.63	500+
50	27.70	500+	14.92	500+	16.62	310.89	13.32	500+

6   Discussion and Conclusions

We studied three kinds of computational problems related to reasoning with preferences in the ASO formalism: to find an optimal answer set, to find an alternative optimal answer set, and to find a similar/dissimilar optimal answer set. We extended results known previously by showing that the problems of deciding the existence of a similar (dissimilar, respectively) optimal answer set are Σ_2^p-complete. In this way, all problems considered are located within the second level of the polynomial hierarchy. Thus, they can be modeled as disjunctive logic programs under the answer-set semantics and disjunctive ASP solvers can be used to solve them. This observation formed the basis of one of the approaches we developed and experimented with in the paper, employing the solvers *claspD* and *dlv*. We also developed a more naive method of using ASP in which an ASP solver searches for answer sets and tests them for optimality as it finds them (the *mclasp* approach). Finally, we proposed a mix of imperative and declarative approaches (the iterative method), which consists of an imperative algorithm that uses ASP solvers for solving some basic optimization tasks that are at the first level of the polynomial hierarchy.

According to the experimental results, with unranked preferences, the iterative method works the best for problems to find optimal answer sets. It is due to the fact that it implements a simple iterative improvement process that is bound to terminate. Since no additional constraints are required to be satisfied, any final result is as good as any other. However, when additional requirements are imposed (like particular distance from another optimal outcome), the method does not perform well. Instead, the method compiling the entire reasoning task into a disjunctive logic program (and then using *claspD* or *dlv*) is most promising overall for the problems to find similar/dissimilar optimal answer sets. That points to the potential of declarative approaches, as they win on that class of instances with the iterative method, that implements in a procedural way a reasonable heuristics of zooming in on an optimal answer set. However, even if there are preferences of two different ranks the picture starts to change and for instances with very many ranks changes drastically. The method based on modeling problems as disjunctive logic programs lags much behind the iterative one, the reason being much larger sizes of programs needed to model comparisons of interpretations when multiple ranks are present.

Lastly, we note that for the types of preference rules we considered, the iterative method works well for the problems of finding an optimal answer set and an alternative one, once the first one was found. For problems to find similar/dissimilar answer sets its performance may be negatively affected when the search space of answer sets is large and the distance constraint is relatively tight.

On the one hand, our work demonstrates the effectiveness of ASP tools in addressing preference optimization problems. On the other hand, it brings up classes of challenging benchmarks based on problems that are Σ_2^p-complete (ranked ASO problems) that can stimulate further research on solver enhancements.

In the future work, we will study possible improvements to our third method based on disjunctive logic program encodings. We will also consider additional optimization problems (for instance, finding a set of p diverse optimal answer sets). Finally, as part of the present work we experimented also with QBF encodings and QBF solvers. The performance of that approach was not satisfactory mostly due to the fact that the solvers we considered do not provide support for aggregates. We intend to study QBF encodings with aggregates as well as extensions to QBF solvers capable to support aggregates.

Acknowledgments. This work was supported by the NSF grant IIS-0913459.

References

1. Booth, R., Chevaleyre, Y., Lang, J., Mengin, J., Sombattheera, C.: Learning conditionally lexicographic preference relations. In: ECAI, pp. 269–274 (2010)
2. Brewka, G., Niemelä, I., Truszczynski, M.: Answer set optimization. In: IJCAI, pp. 867–872 (2003)
3. Drescher, C., Gebser, M., Grote, T., Kaufmann, B., König, A., Ostrowski, M., Schaub, T.: Conflict-driven disjunctive answer set solving. In: KR 2008, pp. 422–432. AAAI Press (2008)
4. Eiter, T., Gottlob, G.: On the computational cost of disjunctive logic programming: propositional case. Annals of Mathematics and Artificial Intelligence 15(3-4), 289–323 (1995)
5. Eiter, T., Erdem, E., Erdoğan, H., Fink, M.: Finding similar or diverse solutions in answer set programming. In: Hill, P.M., Warren, D.S. (eds.) ICLP 2009. LNCS, vol. 5649, pp. 342–356. Springer, Heidelberg (2009)
6. Gebser, M., Kaufmann, B., Neumann, A., Schaub, T.: Conflict-driven answer set solving. In: Twentieth International Joint Conference on Artificial Intelligence, IJCAI 2007, pp. 386–392. MIT Press (2007)
7. Goldsmith, J., Junker, U.: Special Issue on Preferences. AI Magazine 29(4) (2008)
8. Kaci, S.: Working with Preferences: Less Is More. Cognitive Technologies. Springer (2011)
9. Leone, N., Pfeifer, G., Faber, W., Eiter, T., Gottlob, G., Perri, S., Scarcello, F.: The dlv system for knowledge representation and reasoning. ACM Transactions on Computational Logic 7(3), 499–562 (2006)
10. Marek, V., Truszczynski, M.: Stable models and an alternative logic programming paradigm. In: Apt, K., Marek, V., Truszczynski, M., Warren, D. (eds.) The Logic Programming Paradigm: a 25-Year Perspective, pp. 375–398. Springer, Berlin (1999)
11. Niemelä, I.: Logic programming with stable model semantics as a constraint programming paradigm. Annals of Mathematics and Artificial Intelligence 25(3-4), 241–273 (1999)

Author Index